# PETROLEUM ENGINEERING HANDBOOK FOR THE PRACTICING ENGINEER

PennWell Books

PENNWELL PUBLISHING COMPANY
TULSA, OKLAHOMA

# PETROLEUM ENGINEERING HANDBOOK FOR THE PRACTICING ENGINEER

VOLUME I

M.A. MIAN

PennWell Publishing Company
1421 South Sheridan/P.O. Box 1260
Tulsa, Oklahoma 74101

Library of Congress cataloging in publication data

Mian, M. A. (Mohammed A.)
Petroleum engineering handbook for the practicing engineer/M.A. Mian.
p. cm.
Includes bibliographical references and index.
ISBN 0-87814-370-X
1. Petroleum engineering—Handbooks, manuals, etc. I. Title.
TN870.M58 1991
665.5—dc20 91–26283
CIP



Printed in the United States of America

1 2 3 4 5 95 94 93 92 91

## DEDICATION

I wish to dedicate this book to the teachers, engineers, and friends from whom I have learned. Some of whom I have developed high regards for because of their capabilities are:

My parents, M. Y. Mian and J. Y. Mian, who have always been a major source of inspiration and who made me capable of what I am now.

My uncle Brig. H. U. Mian, whose personality and capabilities I will always admire.

Dr. Daniel M. Bass, Jr., who taught me reservoir engineering.

Dr. Douglas W. Hilchie, who taught me log analysis.

Dr. Charles A. Kohlhass, who taught me well test analysis.

Kyle R. Miller, President of Miller Oil Co., whom I worked for.

H. F. (Kep) Keplinger, Chairman of Keplinger & Associates, Inc., in whose company I gained my first experience.

Mr. Nasser K. Jaidah, Technical Manager (QGPC Offshore Operations).

Mr. Abdul Aziz Al-Dulaimi, Manager, Exploration & Production (QGPC Headquarters).

Dr. Khaled A. Hamam, Head of Production Geology (QGPC Offshore Operations).

and last, but not least

My wife and four children whose patience, understanding, and encouragement were essential to completion of this book. They contributed to this effort in ways that I probably will never know or understand.

CONTENTS

It is not claimed that there are no good books on the market to be used by petroleum engineers. However, each one of these books deals with a specific subject and therefore serves a specific objective. The major objective of this book is to provide the practicing petroleum engineer a very comprehensive *practical* reference manual which could be used in day-to-day operations.

Some years ago when I was taking the Professional Engineering Examination, I noticed that each engineer brought dozens of books to the examination. With all these books on hand, the frustration of not finding the appropriate information in due time was always there. This scene of frustration tempted me to prepare this manual so that a majority of the information needed in such situations and in day-to-day operations can be accessed easily rather than going back and forth between a dozen books. I hope this book will eliminate such frustrations, and an engineer will be able to find appropriate information easily when needed.

The book can be used by engineers working towards graduate degrees, by those preparing for competitive examinations, by those needing a practical reference manual on the job, and by professors who need a high-level undergraduate text. Above all, it will come in handy in routine use by a practicing engineer whether he/she is in the office, in the field, in a court room as an expert witness, or on a business trip utilizing his/her time on the airplane without paying excess baggage charges for carrying dozens of books.



Since a lot of important and practical information had to be added, I had to compromise between the amount of information presented and the conceptual details, keeping in mind the handiness of the completed book. Therefore each chapter covers a brief explanation of the principles involved, followed by appropriate equations and solved examples covering different aspects of the subjects. Mathematical derivations of the equations used and various tools' descriptions are avoided. Some 250 solved examples are provided to give hands-on application of the various equations and concepts.

I have tried to present the most widely used, generally accepted engineering principles and practical techniques, equations, working charts/type curves, and empirical correlations available to the petroleum engineers. I have put together topics, equations, working charts/type curves, and empirical correlations which otherwise require searching several references. The book is self-contained; no other chart books/type curves, etc., are required.

I can state with confidence that answers to at least 80% of the day-to-day problems regarding reservoir engineering, openhole well-log interpretation, transient test analysis, secondary oil recovery calculations, drilling technology, production, and economics will be provided by this book. The remaining 20% are considered specialized problems which will not be available in any book and most likely will not be part of the day-to-day operations.

The book is essentially a reference manual and assumes that the engineer using it is already reasonably familiar with the subjects discussed. If you are having a problem understanding the concepts from the brief discussion provided, it is recommended that a text on the specific subject be consulted. The end of chapter references will provide you with appropriate references on the subject.

Much of the material presented in this book is a reorganization and restatement of concepts presented in the publications of the Society of Petroleum Engineers of AIME, the American Gas

Association, American Petroleum Institute, Schlumberger, Atas Wireline Services, the Gas Processors Suppliers Association, and the concepts I have learned from the many people I have studied under and worked with, as well as those countless authors who have put their ideas in writing. I gratefully acknowledge their contributions. It is these organizations and authors who have written this book: I have merely put it together.

In writing a book it is always necessary to ask permission from numerous organizations and individuals for use of various material. I have tried throughout to acknowledge credit for this aid. I have tried also to be scrupulously honest in the numerous references cited, and I hope no one has been overlooked. If oversights exist, I hope they are few and excusable.

In developing this comprehensive work, I have kept in mind the scope of the text and length of topic discussion. I sincerely welcome all comments concerning improvements, corrections, or expansion of any topic.



**M. A. Mian, P.E.**

# CHAPTER ONE

# ENGINEERING ECONOMICS



In this chapter we will discuss the time value of money concepts, different types of interest arrangements, converting cash flows (cash inflows or outflows) from one period to another, methods used to select the best economic service producing investments, methods used to select the best economic revenue producing investments (mutually exclusive and non-mutually exclusive), the development of after–income tax cash flows, expected value concept, and other useful statistical techniques.

All of the concepts in this chapter are presented in a hands-on manner. You may be already reasonably familiar with some of the concepts; in which case, you may skip those and proceed to the ones that you are not familiar with. You will notice that most of the methods presented, if properly used, will give the same results. The reason for including so many methods is to complete the presentation. In some cases it may be advantageous to use one method over the other. Therefore, it is necessary that you be familiar with all of the methods presented. For further reference the reader is referred to an excellent book by Franklin J. Stermole.[1] In Chapter 4, one more section is presented on economics which deals exclusively with oil and gas investments.

## TIME VALUE OF MONEY CONCEPTS

If you were asked, would you rather have $100 today or $100 one year from today? your obvious answer would most likely be, I would rather have $100 today. This reply involves the consideration of time value of money. The difference in the worth between the two amounts is interest.

You know in the back of your mind that if you receive $100 today, you can invest it at 6% interest per year and have $106 ($100 × 1.06 = $106) one year from today. On the other hand, what if you were asked, would you rather have \$100 today or \$106 one year from today? You will probably be indifferent. In the same way, if you were asked, would you rather have \$100 today or \$130 one year from today? you will most likely settle for \$130. This is because you know that you cannot find an investment that will yield 30% return if \$100 is invested today. This is why it is important to consider time value of money when considering investments.

## TIME VALUE OF MONEY

Interest is the cost of using someone's money, as rent is the cost of using someone's house, appliances, tools, or vehicle. This interest becomes cost to the borrower and revenue to the lender. For example, if you borrow \$1,000 from a bank at an effective interest rate of 12% per year for a period of one year, at the end of one year you must pay back the bank \$1,120 of which \$1,000 is the *borrowed principal* and \$120 is the *accrued interest.* The \$120 is the cost to the borrower for using \$1,000 for one year, and it is a revenue to the lender for letting you use their money. In time value of money consideration, the interest rate, the principal amount of money, and the period money is used for must be specified.



In investment decision making, the term *interest* will be used interchangeably as the minimum rate of return, cost of capital, and opportunity cost of capital. To corporations, this rate will be the weighted average of the costs of money generated from different sources, such as borrowing money from a bank, issuing stocks and bonds, retained earnings, etc. The term *opportunity cost of capital* is used when the opportunity for investment of a certain amount of capital exists and must be bypassed by investing the capital elsewhere.

Most investments are long-term investments: you invest money (capital) now and receive benefits (revenues or services) in future years. The benefits, depending on the type of investment, will be received in a certain pattern over the future life of the investment.

To compare different investments or to determine the yield on each investment, we have to convert capital outlays and the net revenues (revenues less expenses) paid or received at different times to a common time period and then compared. The conversion is done by either removing interest from the future value dollars thus converting them to the present value dollars, or by adding interest to the present value dollars and converting them to the future value dollars. The removal of interest is called *discounting,* and the addition of interest is called *compounding.*

In the next few sections, we will learn how to convert cash outflows or inflows at different points in time to a common point in time. These conversion techniques will serve as the basis for investment evaluations.

## SIMPLE AND COMPOUND INTEREST

Simple interest is the interest paid or received on the original principal regardless of the number of time periods that have passed or the amount of interest that has been paid or accrued in the past. Interest rates are reported as annual percentage. The simple interest is determined using the following equation:

$$\text{Interest} = \text{Principal} \times \text{Rate} \times \text{Time} \tag{1.1}$$

The rate is a fraction (% interest/100), and the time is either a fraction of a year or multiple of years. For simple interest computations, year is assumed to be 12 months with each month being 30 days. Compound interest is the interest that accrues on both the principal and the past unpaid balance. The accumulated interest depends on the number of compounding periods per year. In this type of interest, the interest applied to the principal at the end of one period is added to the existing principal. The total is then treated as principal again for the second interest period, and so on. The formula used to compute the compound interest is as follows:

$$\text{Interest} = [(1 + i)^n - 1] \times \text{Principal} \tag{1.2}$$

where $i$ is the interest rate per period (fraction) and $n$ is the number of periods. Interest rate is normally reported as a nominal interest per year compounded a certain number of times per year. For example, if the interest rate is reported as 12% per year compounded monthly, this means that the interest will be paid at the rate of $i = 1\%$ per month for 12 periods. In the same way, interest is compounded daily, quarterly, and semiannually. In any of the above cases, the total yield per year will be greater than nominal annual interest rate (i.e., 12% in this case). 

This leads us to define *effective interest* rate. Effective interest rate is the rate when applied to a sum of money once per year will give the same amount of interest as will be given by a nominal interest rate compounded $n$ periods per year. Effective and nominal interest rates are equal if interest is compounded annually. The two interests are related to each other by the following equation:

$$i_e = \left(1 + \frac{i_n}{n}\right)^n - 1 \tag{1.3}$$

where $i_e$ is the effective interest rate (fraction), $i_n$ is the nominal interest rate (fraction) and $n$ is the number of compounding periods per year ($n = 2$ for semiannual, $n = 4$ for quarterly, and $n = 12$ for monthly compounding). In working with time value of money concepts, it is useful to draw a *cash flow diagram*. A cash flow diagram is a diagram showing the cash outflows and inflows at each point in time throughout the life of the investment. A typical cash flow diagram for a five-year investment is shown in Figure 1–1. In this figure, PV (present value) is the time zero at which capital outlay takes

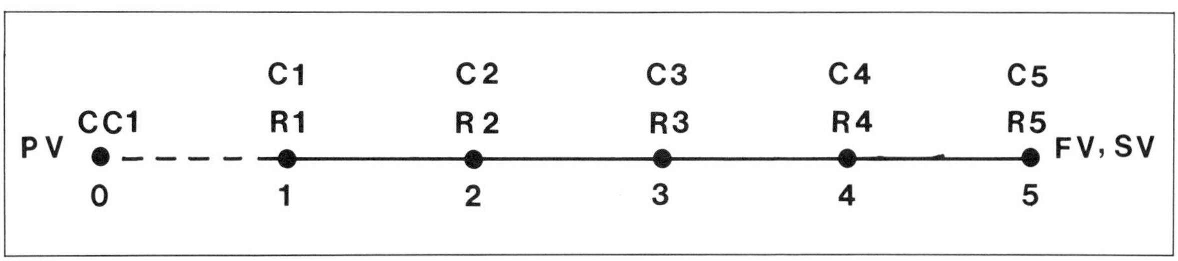


**Figure 1–1.** A typical cash flow diagram for a 5-year investment.

place, the numbers one through five represent the end of each year (year), *R1* through *R5* represent the cash inflows, CC*1* represent the capital cost, and C*1* through C5 represent the annual costs. The FV shows the future value or salvage value. Salvage value, also referred to as residual value, is the estimated proceeds from the sale or disposal of an asset at the end of its estimated service life. The cash flow diagram is very helpful in visualizing how and when each event is taking place. You may define costs by C and revenues by R.

In working with time value of money concepts, interest tables are more frequently utilized. The interest tables provide different factors at different interest rates and periods. A comprehensive set of such interest tables is computer generated and presented in Appendix A. You have to be careful in using such tables because there are various types of tables available. The basic difference is in the assumption of the *flow of funds. The tables presented here assume that cash outflows and inflows take place at the end of the period with discrete periodic interest compounding.* We will discuss the use of these tables in the next few sections.



Another concept in handling the time value of money that finds considerable use in the industry is the assumption that all values of cost and income after time zero occur at the middle of periods instead of at the end of period. This treats all profits and costs after time zero as being incurred half a period sooner than with the end of period convention. If midperiod assumption was used, the values obtained by using the end of period assumption can be easily converted to values obtained by multiplying the former value by a factor $(1 + i)^{0.5}$.

**Example 1–1.** If a sum of $2,000 is invested in a bank at a simple interest rate of 12%, determine (a) the interest if the money is kept in the bank for one year, (b) the interest if the money is kept for only six months, and (c) the interest if the money is kept in the bank for 90 days.

Solution:

(a) Interest = $2,000 × 0.12 × 1 = $240.00
(b) Interest = $2,000 × 0.12 × 6/12 = $120.00
(c) Interest = $2,000 × 0.12 × (90/360) = $60.00

**Example 1–2.** A sum of $2,000 is invested in a savings account yielding an annual interest rate of 12%. Compute the interest amount for one year if (a) the interest is compounded monthly, (b) the interest is compounded quarterly, (c) the interest is compounded annually, and (d) the interest is compounded semiannually.

Solution:
(a)

$$i = \frac{12}{12} = 1\% \quad \text{for monthly compounding}$$

$$n = 1 \text{ year} \times 12 = 12 \quad \text{compounding periods per year}$$

From Equation (1.2),

$$\begin{aligned}\text{Interest} &= [(1 + 0.01)^{12} - 1] \times \$2{,}000 \\ &= \$253.65\end{aligned}$$

Alternatively, we may convert the nominal interest rate of 12% into effective interest rate and compute the accrued interest. Using Equation (1.3), compute the effective interest rate as follows:

$$\begin{aligned}i_e &= \left(1 + \frac{0.12}{12}\right)^{12} - 1 \\ &= (1.01)^{12} - 1 \\ &= 0.1268 \quad \text{or} \quad 12.68\%\end{aligned}$$



Applying this effective interest rate to the sum of $2,000 will give us the same value as calculated above.

$$\begin{aligned}\text{Interest} &= [(1 + i_e)^{1.0} - 1] \times \$2{,}000 \\ &= [(1 + 0.1268) - 1] \times \$2{,}000 \\ &= \$253.60 \quad \text{(within rounding error)}\end{aligned}$$

(b) If interest is compounded quarterly,

$$\begin{aligned}i &= \frac{0.12}{4} = 0.03 \quad \text{per period} \\ n &= \frac{12}{3} = 4 \quad \text{periods} \\ \text{Interest} &= [(1 + 0.03)^4 - 1] \times \$2{,}000 \\ &= \$251.02\end{aligned}$$

(c) If interest is compounded annually,

$$\begin{aligned}i_e &= i_n = 12\% \\ n &= 1 \\ \text{Interest} &= [(1 + 0.12)^1 - 1] \times \$2{,}000 \\ &= (1.12 - 1) \times \$2{,}000 \\ &= \$240.00\end{aligned}$$

(d) If interest is compounded semiannually, let us solve this part without using Equation (1.2). Interest (simple interest) at the end of the six-month period would be

$$\text{Interest}_1 = \$2{,}000 \times 0.06$$
$$= \$120.00$$

This interest is added to the original principal and treated as principal amount of $2,120 for the second six-month period.

$$\text{Interest}_2 = \$2{,}120 \times 0.06$$
$$= \$127.20$$

The total accrued interest at the end of the year would be

$$\text{Interest}_{total} = \text{Interest}_1 + \text{Interest}_2$$
$$= \$120.00 + \$127.20$$
$$= \$247.20$$



Using Equation (1.2), we must get the same interest.

$$\text{Interest}_{total} = [(1 + 0.06)^2 - 1] \times \$2{,}000$$
$$= (1.1236 - 1) \times \$2{,}000$$
$$= \$247.20$$

## PRESENT VALUE OF FUTURE SUM

This concept deals with the removing of interest from the future sum and bringing it to a common point in time called *present value* or *present worth*. The process is more commonly referred to as *discounting* the future sum. To convert the future sum of money to the present value, the following equation is used:

$$\text{PV} = \text{FV} \times \frac{1}{(1 + i)^n} \quad \textbf{(1.4)}$$

In Equation (1.4), the factor $1/(1 + i)^n$ is represented by an interest factor *P/F* in the interest tables in Appendix A.

The easy way to remember which interest factor to use in which situation is to put the abbreviation for the value to be *determined* on the left hand side and the abbreviation for the value *given* on the right hand side of the equal sign, respectively. The interest factor would be designated by a factor as shown below.

$$\overbrace{\text{PV} = \text{FV}\left(\underbrace{\frac{P}{F_{i,n}}}\right)}$$

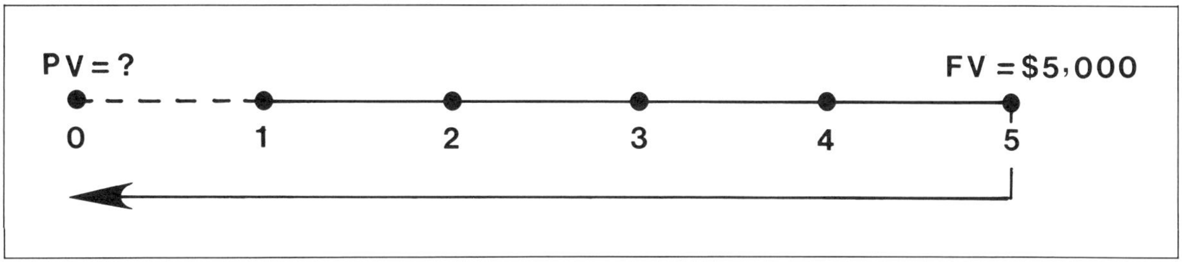


**Figure 1–2.** Cash flow diagram for investment in Example 1–3.

The numerator designates what we are calculating, and the denominator designates what we are calculating it from. The subscripts $i$ and $n$ designate the interest rate per period and the number of compounding periods, respectively.

To find the appropriate interest factor, go to the interest table for the interest $i$, go down the appropriate interest factor column (*P/F* in this case) to the intersection of row $n$ and factor column, read the factor, and multiply the given quantity by the factor to compute the required quantity.



**Example 1–3.** Compute the present value of a sum of $5,000 to be received in one lump sum five years from today. Assume the interest rate is 12% compounded (a) annually and (b) semiannually. The cash flow diagram for the problem is shown in Figure 1–2.

Solution:

(a)

$$\begin{aligned} PV &= FV\,(P/F_{i,n}) \\ &= \$5{,}000\,(P/F_{0.12,5}) \\ &= \$5{,}000 \times 0.5674 \\ &= \$2{,}837 \end{aligned}$$

(b) In this part, the interest is compounded semiannually; therefore, $i = 6\%$ ($12\%/2 = 6\%$) and $n = 10$ periods (5 years $\times$ 2 periods per year $= 10$).

$$\begin{aligned} PV &= FV\,(P/F_{6,10}) \\ &= \$5{,}000 \times 0.5584 \\ &= \$2{,}792 \end{aligned}$$

Note how the interest and the periods are handled for different compounding periods per year and how it affects the present value of the same sum with the same investment life.

## PRESENT VALUE OF ANNUAL SUMS

In this type of computation, the values at the end of each period are given, all of which have to be converted to their equivalent value in present value dollars. It is also referred to as the present value of annuity. The following equation is used to calculate the present value of an annuity or annual sums:

$$PV = AV \times \frac{(1 + i)^n - 1}{i(1 + i)^n} \quad \textbf{(1.5)}$$

In the interest tables, the factor used to convert annuities to present value is represented by the factor *P/A*. In order to apply Equation (1.5) or the interest factor *P/A*, the following conditions have to be met:

1. The regular periodic payments or receipts must be of equal amount. They may increase or decrease in a definite arithmetic or geometric fashion, in which case we have to first convert them into their equivalent annual values and then convert these equivalent annual values to present value.
2. The time intervals between the payments must be of equal length. These intervals may be monthly periods, quarterly periods, semiannual periods, or yearly periods.
3. The interest rate through each period throughout the total investment life must be the same.
4. The interest compounding must be at the end of each period.

**Example 1–4.** Investment in an oil well will yield \$10,000 per year for each of the next five years. If the desired minimum rate of return of the investor is 15% per year, what would be the present value of the revenues received over the five-year period? The cash flow diagram for the investment is presented in Figure 1–3.

Solution:

$$\begin{aligned} PV &= AV\,(P/A_{15,5}) \\ &= \$10{,}000 \times 3.3522 \\ &= \$33{,}522 \end{aligned}$$

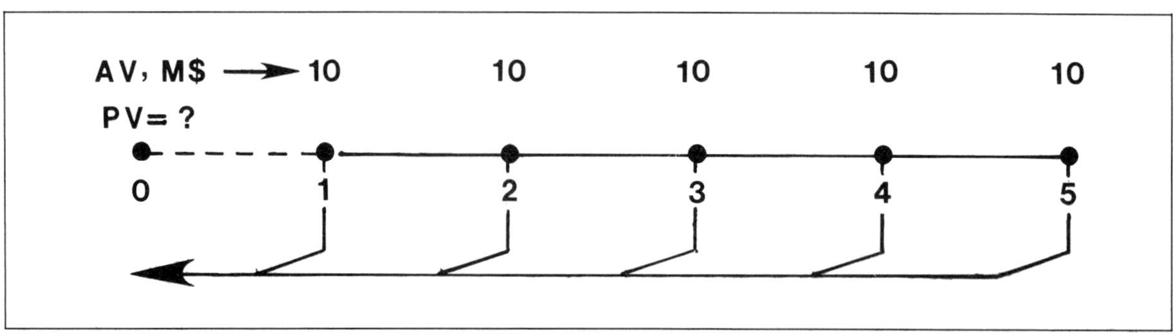


**Figure 1–3.** Cash flow diagram showing annuity.

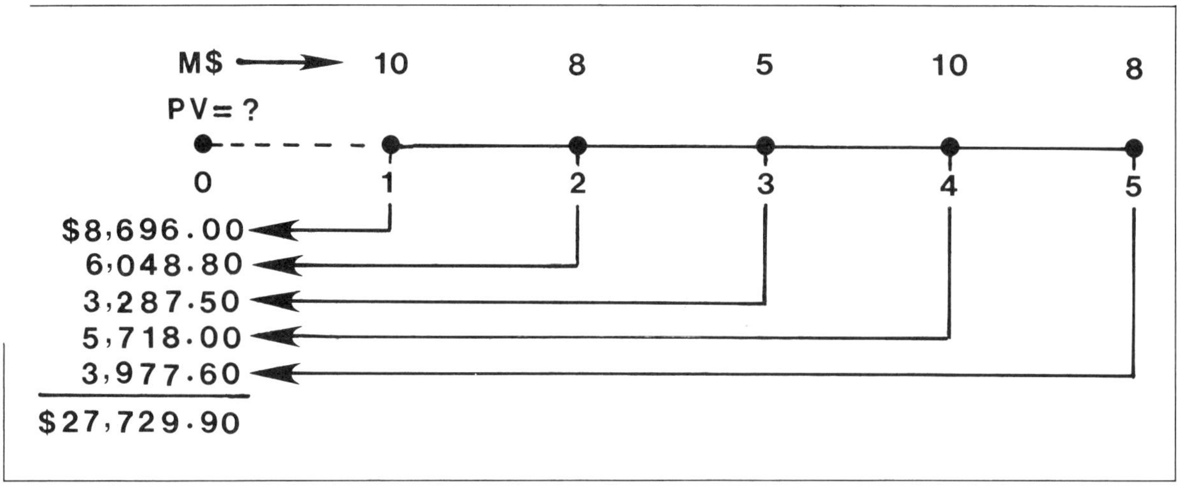


**Figure 1–4.** Cash flow diagram for the investment in Example 1–5.



If the revenues were not equal in value, and if they were not increasing or decreasing in a specific arithmetic or geometric fashion, then the present value will be calculated for each year individually and added together to give the total present value. In such case the unequal annual sums are not termed as annuity. The following example will clarify this procedure.

**Example 1–5.** Let us assume that the revenues in Example 1–4 are not $10,000 per year; instead the revenues are received as shown in Figure 1–4. Compute the present value assuming the same minimum rate of return is applicable.

Solution: In this case, each end of year revenue will be treated as a future sum and the present value computed as

$$\begin{aligned} PV_1 &= FV_1\,(P/F_{15,1}) \\ &= \$10{,}000 \times 0.8696 \\ &= \$8{,}696 \\ PV_2 &= FV_2\,(P/F_{15,2}) \\ &= \$8{,}000 \times 0.7561 \\ &= \$6{,}048.80 \\ PV_3 &= FV_3\,(P/F_{15,3}) \\ &= \$5{,}000 \times 0.6575 \\ &= \$3{,}287.5 \end{aligned}$$

In the same way, the other two revenues are converted to PV and all five present values are then added together. The PV for this investment is shown in Figure 1–4.

## FUTURE VALUE OF PRESENT SUM

In one of the previous sections we learned how to convert future sum into its equivalent present value or discounting future sum. In this section we will deal with the conversion of present value into its equivalent future value or compounding the present sum. The following equation is used to convert present sum to its future value. The interest factor in the interest tables is represented by the factor *F/P*.

$$FV = PV\,(1 + i)^n \tag{1.6}$$

**Example 1–6.** If a sum of $10,000 is invested today at an interest rate of 12% compounded quarterly, what would be the future value of this sum at the end of five years? Figure 1–5 shows the cash flow diagram for the investment.

Solution:

$$\begin{aligned} i_n &= 12\% \\ i &= 12/4 = 3\% \\ n &= 5 \quad \text{years} \times 4 \quad \text{periods per year} = 20 \quad \text{periods} \\ FV &= PV\,(F/P_{3,20}) \\ &= \$10{,}000 \times 1.8061 \\ &= \$18{,}061 \end{aligned}$$

## FUTURE VALUE OF ANNUAL SUMS

The future value of annual sums or annuity is similar to the present value of annuity calculations as discussed in the previous sections. The same conditions have to be met as in the case of calculating present value of annual sums. The following equation is used to compute the future value of annuity:

$$FV = AV\left[\frac{(1 + i)^n - 1}{i}\right] \tag{1.7}$$

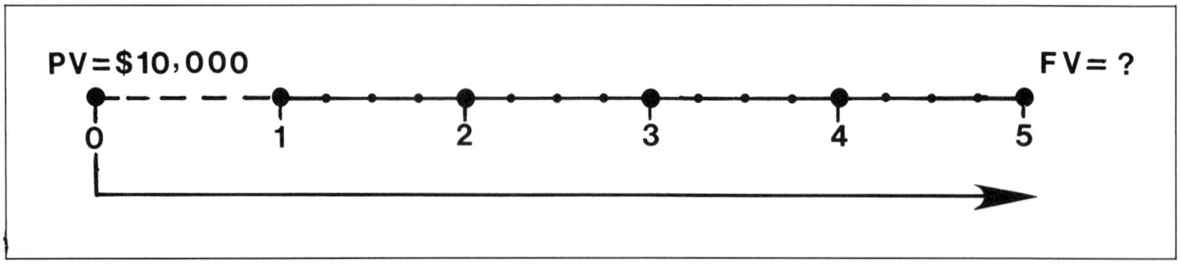


**Figure 1–5.** Cash flow diagram for investment in Example 1–6.

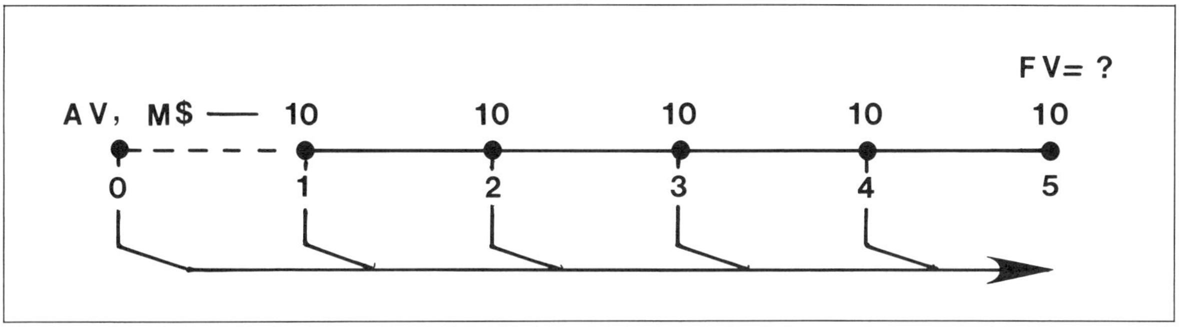


**Figure 1–6.** Cash flow diagram for the investment in Example 1–7.

The interest factor is represented by the factor *F/A* in the interest tables. The factor is also referred to as the sinking fund factor.



**Example 1–7.** Consider the investment of Example 1–4, compute the future value of the $10,000 annuity at the minimum *rate of return* of 15%. The cash flow diagram for the investment is presented in Figure 1–6.

Solution:

$$\begin{aligned} FV &= AV\,(F/A_{15,5}) \\ &= \$10{,}000 \times 6.7424 \\ &= \$67{,}424 \end{aligned}$$

## ANNUAL EQUIVALENT OF FUTURE SUMS

In this type, the future sum will be given and its equivalent annual values have to be determined or the future value to be converted into its equivalent annuity. The following equation is used to convert future sum into its equivalent annuity:

$$AV = FV\left[\frac{i}{(1+i)^n - 1}\right] \qquad \textbf{(1.8)}$$

The interest factor used is represented by the factor *A/F* in the interest tables. Use the value $67,424 computed in Example 1–7 and convert it into its equivalent annuity (i.e., $67,424 × 0.14832 = $10,000).

## ANNUAL EQUIVALENT OF PRESENT SUMS

This type is similar to the computation of annual equivalents of a given future value. Our house mortgage loans, auto loans, and other installment loans represent this kind of problem. In this type, a certain present value is converted into its equivalent annuity or equal end of period payments or receipts. The following equation is used to convert present value into its equivalent annuity:

$$AV = PV\left[\frac{i(1+i)^n}{(1+i)^n - 1}\right] \tag{1.9}$$

The interest factor used to convert present value into its equivalent annuity is represented by the factor $A/P$ in the interest tables.

**Example 1–8.** You bought a car for $10,000, the car was financed by a bank at an interest rate of 12% compounded monthly. The loan has to be paid off in 24 equal monthly installments. Compute (a) your monthly payment and (b) the interest and principal paid in the first three months.

Solution:

(a)

$$\begin{aligned} i &= 12\%/12 = 1\% \\ n &= 24 \quad \text{periods (months)} \\ AV &= PV\ (A/P_{1,24}) \\ &= \$10{,}000 \times 0.04707 \\ &= \$470.70 \quad \text{per month} \end{aligned}$$

(b)

| (1) Beginning of Month | (2) Outstanding Balance | (3) Interest | (4) Principal | (5) Balance End of Month |
|---|---|---|---|---|
| 1 | $10,000.00 | $100.00 | $370.70 | $9,629.30 |
| 2 | 9,629.30 | 96.29 | 374.41 | 9,254.89 |
| 3 | 9,254.89 | 92.55 | 378.15 | 8,876.74 |

Column (3) = Column (2) × Interest/month
Column (4) = Monthly payment − Column (3)
Column (5) = Column (2) − Column (4)

## PRESENT VALUE OF DEFERRED ANNUITY

In this type of calculation, the annual payments or receipts are deferred until some time in the future, and then the deferred annuity is converted into its equivalent present value. In this type, two interest factors ($P/A$ and $P/F$) are used simultaneously.

The annuity is first converted into its present value at the beginning of the period in which the annuity starts. This present value is then treated as a future value and thus converted into its present value. The following example will clarify the conversion calculations.

**Example 1–9.** An investment will generate zero revenue for the first three years. In the fourth through eighth years, the investment will generate $10,000 per year. Compute the present value of this annuity if the desired minimum rate of return from the investment is 12% per year. The cash flow diagram for the investment is shown in Figure 1–7.

Solution:

$$\begin{aligned} PV_3 &= AV\,(P/A_{12,5}) \\ &= \$10{,}000 \times 3.6048 \\ &= \$36{,}048 \end{aligned}$$

Now, let $PV_3 = FV$.

$$\begin{aligned} PV &= FV\,(P/F_{12,3}) \\ &= \$36{,}048 \times 0.7118 \\ &= \$25{,}659 \end{aligned}$$



This problem can also be solved by computing the future value of the annuity and then converting this future value to its equivalent present value by utilizing interest factors *F/A* and *P/F*, respectively. The present value using this procedure would be

$$\begin{aligned} PV &= AV\,(F/A_{12,5})(P/F_{12,8}) \\ &= \$10{,}000(6.3528)(0.4039) \\ &= \$25{,}659 \end{aligned}$$

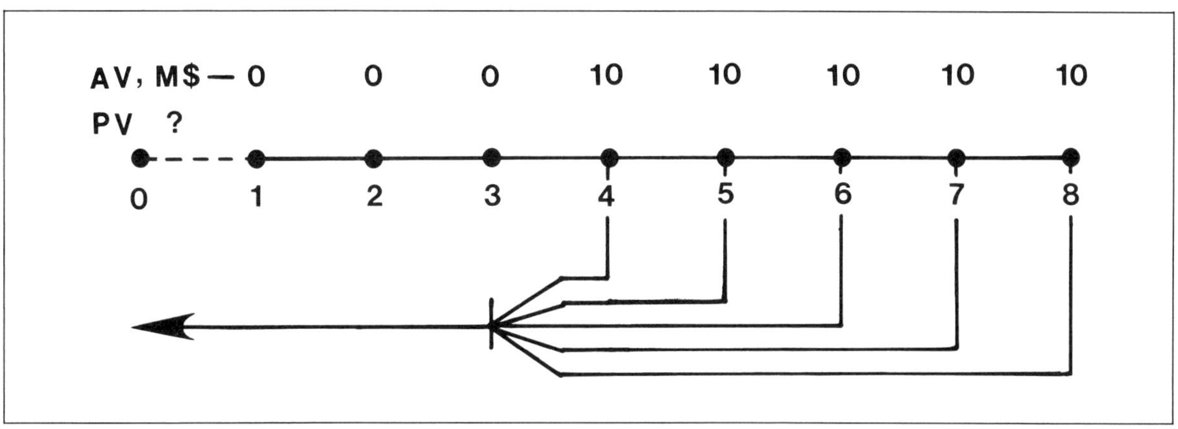


**Figure 1–7.** Cash flow diagram for the investment in Example 1–9.

This value is the same (and it should be the same) as the one obtained above. In solving these type of problems, you have to be very careful in choosing the interest factor and the value of $n$.

## ARITHMETIC GRADIENT SERIES

If the periodic payments or receipts are increasing or decreasing in a specific arithmetic fashion, we can convert these cash flows into ordinary annuity (equal periodic payments or receipts) and use the $P/A$ or $F/A$ factors to convert this ordinary annuity into present or future value, respectively.

The following equation will be used to convert arithmetic gradient series (positive or negative) into its equivalent ordinary annuity:

$$AV_G = AV \pm \left\{ \frac{1}{i} - \frac{n}{i} \left[ \frac{i}{(1+i)^n - 1} \right] \right\} \quad \textbf{(1.10)}$$



Once again, interest factors are available in the interest tables to facilitate the computations. The factor is represented by the A/G factor in the interest tables. The following example will clarify the use of arithmetic gradient series:

**Example 1–10.** A pumping unit is installed on an oil well. The estimated maintenance cost of the unit in the first year is \$2,000. Starting the second year, the maintenance cost will increase by \$300 per year for the next four years. Compute (a) the ordinary annuity and (b) the present value of costs. The interest rate is 15%.

Solution: The problem is diagrammatically presented in Figure 1–8. Note that the gradient in this problem is a positive (increasing) gradient of \$300 per year.

(a)

$$\begin{aligned} AV &= \$2{,}000 \\ G &= +\$300 \\ AV_G &= AV + G(A/G_{15,5}) \\ &= \$2{,}000 + \$300(1.7228) \\ &= \$2{,}516.84 \quad \text{or} \quad \$2{,}517 \end{aligned}$$

The new cash flow diagram for the ordinary annuity is shown in Figure 1–9.

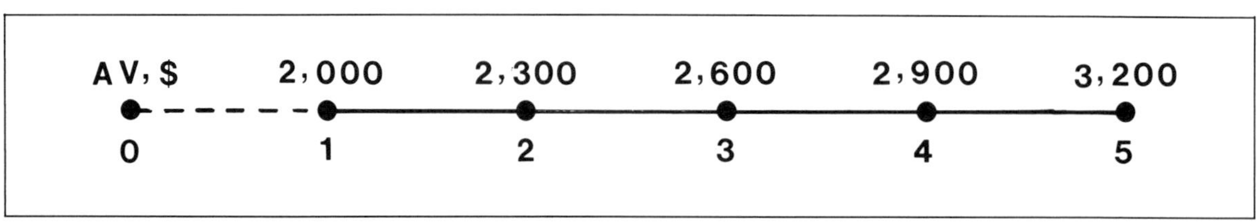


**Figure 1–8.** Cash flow diagram showing arithmetic gradient series.

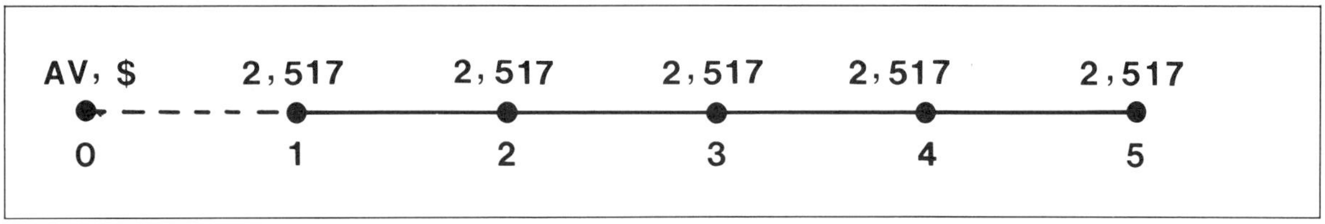


**Figure 1–9.** Cash flow diagram showing ordinary annuity from gradient series.

(b)

$$\begin{aligned} PV &= AV_G(P/A_{15,5}) \\ &= \$2{,}517(3.3522) \\ &= \$8{,}437.49 \end{aligned}$$

## CAPITALIZED COST CALCULATION



Capitalized cost is the present value of the costs of an investment with infinite (perpetual, permanent) life. Capitalized cost is determined by dividing the annual cost by the discount rate or minimum rate of return. Investment projects such as railroads, dams, and recreation parks are normally evaluated using capitalized cost concept.

**Example 1–11.** A dam is being constructed and expected to have a permanent life. The interest rate is 5% and the initial investment is $200,000 plus additional costs as (1) an additional one-time cost of $300,000 in 15 years, (2) an annual cost of $40,000 per year for the first 15 years and $60,000 per year thereafter, and (3) the installation of a new generator every 30 years at a cost of $30,000 each installation. What is the capitalized cost of the foregoing?

Solution:

| | |
|---|---|
| Initial investment | $ 200,000 |
| PV of $300,000 in 15 years: | |
| $300,000 ($P/F_{5,15}$) = $300,000(0.4810) | 144,300 |
| PV of $40,000 for 15 years: | |
| $40,000 ($P/A_{5,15}$) = $40,000(10.3796) | 415,184 |
| Capitalized cost of $60,000 per year ($60,000/0.05) | |
| PV of capitalized cost ($1,200,000 × 0.4810) | 577,200 |
| AV of $30,000 installation every 30 years: | |
| $30,000 ($A/F_{5,30}$) = $30,000(0.01505) = $451.50 | |
| Capitalized cost of $30,000 = $451.5/0.05 | 9,030 |
| Total capitalized cost | $1,345,714 |

This means that if we invest $1,345,714 at 5% interest now, it will pay for the dam and the associated future costs forever.

## CONTINUOUS COMPOUNDING AND DISCOUNTING

In the previous sections, we discussed compounding and discounting with the interest being compounded a certain fixed number of times per year. In this section we will deal with the interest being compounded infinite number of times per year.

Continuous compounding and discounting is further divided into two subcategories as (a) when money is received or paid at the end of a period and (b) when money is received or paid continuously over a period (continuous flow of funds). Let us represent the continuous interest rate by $r$. The following equations will be used to do the various conversions discussed in the previous sections. The continuous interest rate may be converted into an effective interest rate compounded annually using the relationship $i_e = e^r - 1$.

$$FV = PV(e^{rn}) \tag{1.11}$$

$$PV = FV\left(\frac{1}{e^{rn}}\right) \tag{1.12}$$

$$FV = AV\left(\frac{e^{rn} - 1}{e^r - 1}\right) \tag{1.13}$$

$$AV = FV\left(\frac{e^r - 1}{e^{rn} - 1}\right) \tag{1.14}$$

$$PV = AV\left[\frac{e^{rn} - 1}{(e^r - 1)e^{rn}}\right] \tag{1.15}$$

$$AV = PV\left[\frac{(e^r - 1)e^{rn}}{e^{rn} - 1}\right] \tag{1.16}$$

**Example 1–12.** Compute the future value five years from now of a sum of $3,000 invested today at an interest rate of 12% (a) compounded annually, (b) compounded continuously, and (c) compounded daily.

Solution:

(a) Interest compounded annually, $i_e = i_n = i$ and $n = 5$.

$$FV = PV\,(F/P_{12,5}) = \$3{,}000(1.7623) = \$5{,}286.90$$

(b)

$$FV = PV\,(e^{rn}) = \$3{,}000(e^{0.12\times 5}) = \$3{,}000(1.8221) = \$5{,}466.30$$

OR

$$\begin{aligned} i_e &= e^r - 1 \\ &= e^{0.12} - 1 \\ &= 0.127497 \quad \text{or} \quad 12.7479\% \\ \text{FV} &= \text{PV}\,(1 + 0.127479)^5 \\ &= \$3{,}000(1.8221) \\ &= \$5{,}466 \end{aligned}$$

(c) For daily compounding, calculate $i$ as

$$\begin{aligned} i &= 0.12/365 = 0.0003288 \quad \text{and} \quad n = 5 \times 365 = 1{,}825 \\ \text{FV} &= \text{PV}\,(1 + i)^n \\ &= \$3{,}000(1 + 0.0003288)^{1825} \\ &= \$3{,}000(1.8221) \\ &= \$5{,}466 \end{aligned}$$



*You can see from parts (b) and (c) that the continuous compounding with the end of period payments or receipts is similar to end of period compounding with interest compounded daily. This, however, is not true at higher interest rates.*

In continuous compounding and discounting, since interest is compounded and discounted continuously, it is reasonable to assume that the funds will also flow in and out of a treasury continuously. The subscript $c$ indicates continuous flow of funds in the following equations:

$$\text{FV} = \text{AV}_c\left[\frac{(e^{rn} - 1)}{r}\right] \quad \textbf{(1.17)}$$

$$\text{AV}_c = \text{FV}\left[\frac{r}{(e^{rn} - 1}\right] \quad \textbf{(1.18)}$$

$$\text{AV}_c = \text{PV}\left[\frac{re^{rn}}{(e^{rn} - 1)}\right] \quad \textbf{(1.19)}$$

$$\text{PV} = \text{AV}_c\left[\frac{(e^{rn} - 1)}{re^{rn}}\right] \quad \textbf{(1.20)}$$

$$\text{PV} = \text{FV}\left[\frac{(e^{r} - 1)}{re^{rn}}\right] \quad \textbf{(1.21)}$$

## DETERMINING INTEREST RATES

This section deals with computing the unknown interest rate that has resulted in two values at two different points in time. For example, you borrow \$200 from a bank; after two years you pay back the bank \$264.50. In this case, you know the PV (\$200) and the FV (\$264.50), but the rate of interest that the bank has charged you is not known.

In order to determine the unknown interest rate, we have to first compute the appropriate interest factor, (i.e., if PV and FV are given, *P/F* or *F/P* factor has to be computed). Once the factor is determined, then go to the interest tables and locate the factor at the given $n$. If the factor was found in the interest table for $i = 10\%$, that would be the unknown interest rate. If you end up with a factor that is between two interest rates, the required interest rate is then determined by interpolation between the two rates.

**Example 1–13.** Mr. Lucky borrowed \$5,000 from a bank. The bank told him that he has to pay back the loan in 24 equal end-of-month payments with each payment of \$235.35 per month. Compute the interest rate at which the amortization schedule was developed.

Solution:

$$PV = \$5{,}000$$
$$AV = \$235.35$$
$$n = 24 \quad \text{periods}$$
$$i = ?$$
$$PV = AV\,(P/A_{i,n})$$

OR

$$\begin{aligned}(P/A_{i,24}) &= PV/AV\\ &= \$5{,}000/\$235.35\\ &= 21.2450\end{aligned}$$

The *P/A* factor is 21.2450, starting with the interest table for $i = 1\%$, we go to *P/A* and $n = 24$ and look for the value of 21.2450. In this case the factor at *P/A* and $n = 24$ in the $i = 1\%$ table is 21.2434, which is approximately equal to the above factor of 21.2450. Therefore the bank has used 12% interest (nominal) rate per year compounded monthly or 1% per month.

## NPV, NAV, AND NFV COMPUTATIONS

So far we have been converting either cash inflows or cash outflows from one point in time to another. In this section, we will deal with both of them simultaneously. Net present value (NPV) also referred to as present worth or present value cash surplus is obtained by subtracting the present value of all cash outflows from the present value of all cash inflows. The NPV is accepted widely in the industry as a reliable criterion of profitability.

When NPV of an investment at a certain discount rate is positive, we know that the investment will pay for the cost of financing the investment (the rate of return on the investment will be at least equal to the discount rate), *and the investment will generate revenue the present worth of which will be equal to the positive value.* The following example will clarify the concept.

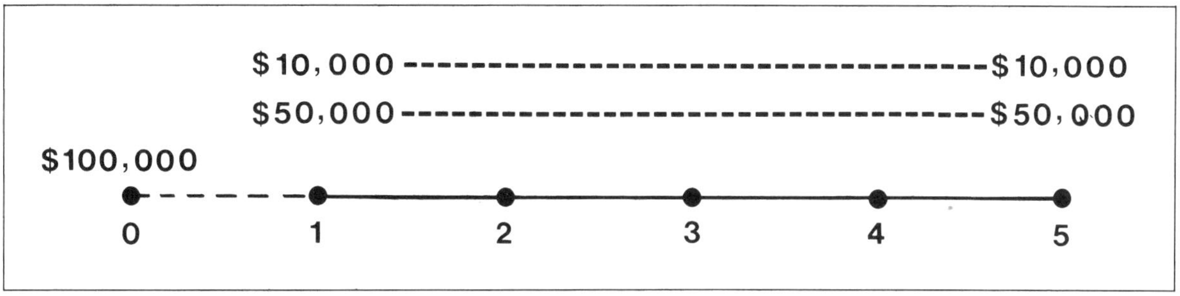


**Figure 1–10.** Cash flow diagram showing the investment in Example 1–14.

**Example 1–14.** An investment requires a capital outlay of $100,000 at time zero. The investment will generate $50,000 per year for each of the next five years. The annual expenses required to keep the business running will be $10,000 per year. If the desired minimum rate of return is 15%, compute (a) NPV, (b) NAV, and (c) NFV. Figure 1–10 shows the cash flow diagram for the investment. 

Solution:

(a) To simplify the computations, we will subtract the annual costs from the annual revenues, i.e., the net income per year would be $40,000 ($50,000 − $10,000 = $40,000). *Note that the capital outlay is already at the present value.*

$$\begin{aligned}\text{NPV} &= \$40{,}000\ (P/A_{15,5}) - \$100{,}000\\ &= \$40{,}000(3.3522) - \$100{,}000\\ &= \$134{,}088 - \$100{,}000\\ &= \$34{,}088\end{aligned}$$

(b) To calculate the NAV, the capital cost is first converted into its equivalent annual costs, and then these equivalent annual costs plus the annual costs of $10,000 are subtracted from the annual net revenues of $40,000.

$$\begin{aligned}\text{NAV} &= \$40{,}000 - \$100{,}000(A/P_{15,5})\\ &= \$40{,}000 - \$100{,}000(0.29832)\\ &= \$40{,}000 - \$29{,}832\\ &= \$10{,}168\end{aligned}$$

(c) To calculate the NFV, both the capital cost and the annual income have to be converted into the future.

$$\begin{aligned}\text{NFV} &= \$40{,}000(F/A_{15,5}) - \$100{,}000(F/P_{15,5})\\ &= \$40{,}000(6.7424) - \$100{,}000(2.0114)\\ &= \$269{,}696 - \$201{,}140\\ &= \$68{,}556\end{aligned}$$

## RATE OF RETURN (ROR) CALCULATIONS

Rate of return is defined as that interest or discount rate at which the net present value is exactly equal to zero, or the present value of cash inflows is equal to the present value of cash outflows. The ROR is given different names including the internal yield, internal rate of return (IRR), discounted cash flow rate of return (DCFROR), marginal efficiency of capital, and the investors method.

The rate of return can be determined in one of two ways (a) by interpolation and (b) by graphs. To determine the ROR by interpolation, we need to find two interest rates that will give us a positive and a negative NPV. The ROR will be between the two interest rates, which is determined by linear interpolation using the following equation:

$$\text{ROR} = \frac{\text{NPV}_\text{A}}{\text{NPV}_\text{A} - \text{NPV}_\text{B}}(i_\text{B} - i_\text{A}) + i_\text{A} \qquad \textbf{(1.22)}$$

where $\text{NPV}_\text{A}$ is the positive net present value, $\text{NPV}_\text{B}$ is the negative net present value, $i_\text{A}$ is the interest rate associated with $\text{NPV}_\text{A}$, and $i_\text{B}$ is the interest rate associated with $\text{NPV}_\text{B}$.

*Note that the relationship between interest rate* i *and NPV, NAV, and NFV is not always linear; therefore, linear interpolation is not recommended over more than five or at the most 10 percentage points.*

To determine the ROR graphically, three to five NPVs are calculated at different interest rates. These NPVs are then plotted versus their respective interest rates. A line is drawn through these points, the intersection of this line and the line of NPV = 0 will give the ROR.

ROR basically is a good idea because a rate criterion is easy to visualize and is independent of the cost of money; however, the criterion has serious limitations:

1. The approach is not very good for ranking projects because ROR is not reliable for comparative purposes if it is much different from the cost of money.
2. The method also sometimes gives multiple solutions. This is especially troublesome in acceleration projects and in cases where investment occurs after a project is under way.
3. Moreover, ROR in itself tells nothing about other desirable or undesirable aspects of a project.

**Example 1–15.** Consider the investment of Example 1–14. Determine the ROR of the investment using (a) linear interpolation and (b) graphs.

Solution:

(a) Compute NPV at $i$ = 25%

$$\begin{aligned}\text{NPV}_{25\%} &= \$40{,}000(P/A_{25,5}) - \$100{,}000\\ &= \$40{,}000(2.6893) - \$100{,}000\\ &= \$7{,}572\end{aligned}$$

Compute NPV at $i = 30\%$

$$\begin{aligned}\text{NPV}_{30\%} &= \$40{,}000(P/A_{30,5}) - \$100{,}000\\ &= \$40{,}000(2.4356) - \$100{,}000\\ &= -\$2{,}576\end{aligned}$$

By linear interpolation, the ROR is

$$\begin{aligned}\text{ROR} &= \frac{\text{NPV}_{25}}{\text{NPV}_{25} - \text{NPV}_{30}}(i_{30} - I_{25}) + i_{25}\\ &= \frac{\$7{,}572}{\$7{,}572 + \$2{,}576}(0.30 - 0.25) + 0.25\\ &= 0.7462(0.05) + 0.25\\ &= 0.2873 \quad \text{or} \quad 28.73\%\end{aligned}$$



(b) The graphical determination of ROR is shown in Figure 1–11. The NPVs at 15%, 25%, and 30% are plotted.

## COMPARISON OF SERVICE PRODUCING INVESTMENTS

In this section we will learn how to select the most economic service producing investment alternative. The different alternatives considered will be of equal life and will provide the same service. Five methods are provided that can be used to select the most economic service producing investment. Each one of the methods when used properly should provide the same selection. The five methods are:

1. PV of costs comparison
2. AV of costs comparison
3. FV of costs comparison
4. ROR analysis of incremental benefits
5. Service life breakeven analysis

### PV, AV, AND FV OF COSTS ANALYSIS

In this type, we will compute the PV of all costs, or the AV of all costs, or the FV of all costs in the same way as discussed in the previous sections. The investment alternative with the lowest PV of costs, AV of costs, or the lowest FV of costs will be selected as the most economic choice. The following example will clarify the computation.

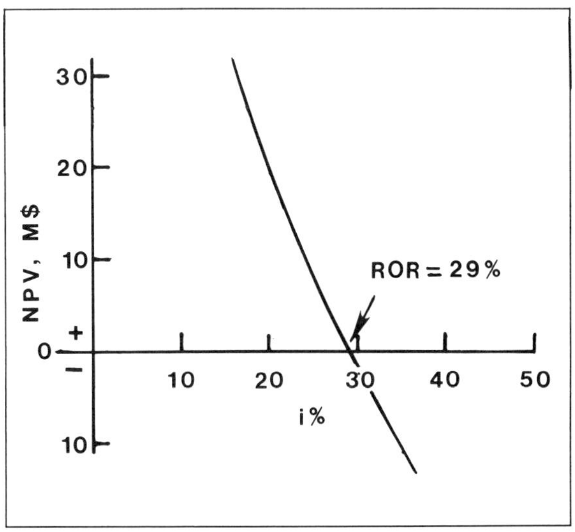


**Figure 1–11.** Graphical determination of ROR.

**Example 1–16.**[2] A choice is to be made between a diesel engine and a secondhand gasoline engine for a service that is to last seven years. Minimum attractive rate of return is 8%. Select the most economical alternative using (a) PV of costs, (b) AV of costs, and (c) FV of costs. The two alternatives are diagrammatically presented in Figure 1–12.

| | A<br>Diesel | B<br>Gasoline |
|---|---|---|
| First cost | \$1,500 | \$500 |
| Salvage value after 7 years | 500 | 0 |
| Annual fuel cost | 250 | 600 |
| Annual repair cost | 50 | 100 |

Solution:

(a)

$$
\begin{aligned}
PV_A &= \$1{,}500 + \$300(P/A_{8,7}) - \$500(P/A_{8,7}) \\
&= \$1{,}500 + \$300(5.2064) - \$500(0.5835) \\
&= \$2{,}770.17 \\
PV_B &= \$500 + \$700(P/A_{8,7}) - 0(P/F_{8,7}) \\
&= \$500 + \$700(5.2064) \\
&= \$4{,}144.48
\end{aligned}
$$

Select Alternative A, i.e., diesel engine.

(b)

$$
\begin{aligned}
AV_A &= \$300 + \$1{,}500(A/P_{8,7}) - \$500(A/F_{8,7}) \\
&= \$300 + \$1{,}500(0.19207) - \$500(0.11207) \\
&= \$532.07
\end{aligned}
$$

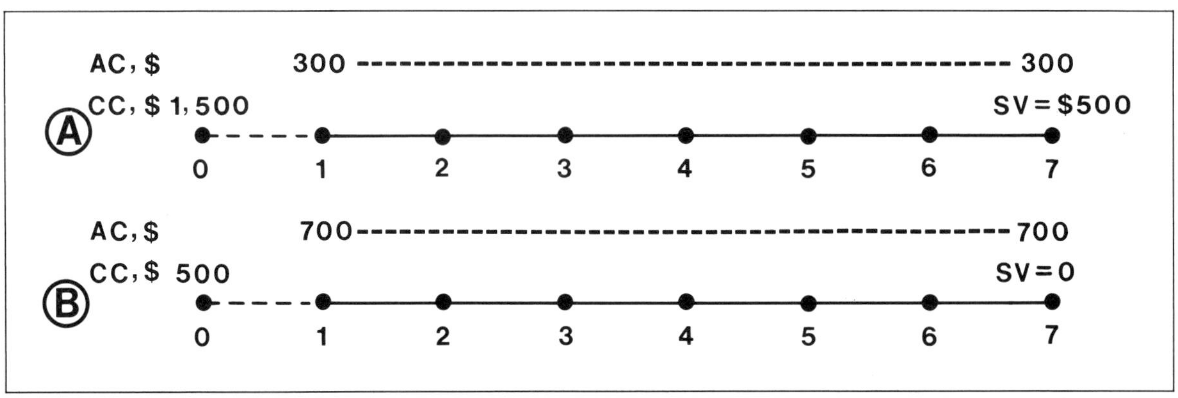


**Figure 1–12.** Cash flow diagram for investments in Example 1–16.

$$AV_B = \$700 + \$500(A/P_{8,7}) - 0(A/F_{8,7})$$
$$= \$700 + \$500(0.19207)$$
$$= \$796.04$$

Select Alternative A because of lower annual cost.

$$FV_A = \$1,500(F/P_{8,7}) + \$300(F/A_{8,7}) - \$500$$
$$= \$1,500(1.7138) + \$300(8.9228) - \$500$$
$$= \$4,747.54$$
$$FV_B = \$500(1.7138) + \$700(8.9228) - 0$$
$$= \$7,102.80$$

Select Alternative A because of lower future cost. We have seen in this example that any one of the above methods will give the same selection. So an appropriate alternative will be selected using any one of the above methods. 

## RATE OF RETURN ANALYSIS

The rate of return computation was discussed in the section on Rate of Return (ROR) Calculation. By applying this concept to the selection of service producing investments, you will notice that the ROR will be infinite. This is because we are considering only costs (negative cash) and not any revenues (positive cash) to offset the costs.

To avoid the problem of obtaining infinite ROR, we perform incremental ROR analysis. The costs of one alternative are subtracted from the costs of the other alternative, the difference of the two costs will give us savings (positive and negative). The ROR analysis are performed on these incremental costs. If the incremental ROR is greater than the minimum acceptable ROR, select the high initial cost alternative, and vice versa. If there are more than two alternatives, the survivor is then compared with another alternative. This is repeated until all the alternatives are exhausted. Remember that the negative incremental costs are treated as positive savings, and vice versa.

**Example 1–17.** Using incremental ROR analysis on the investments presented in Example 1–16, which alternative will be the economic choice? Is the selection consistent with the selection made in Example 1–16? The incremental cash flow diagram for the investment is shown in Figure 1–13.

Solution: Try $i = 8\%$

$$PV_{(A-B)} = \$1,000 + \$400(P/A_{8,7}) - \$500(P/F_{8,7})$$
$$= -\$1,000 + \$400(5.2064) - \$500(0.5835)$$
$$= \$790.81$$

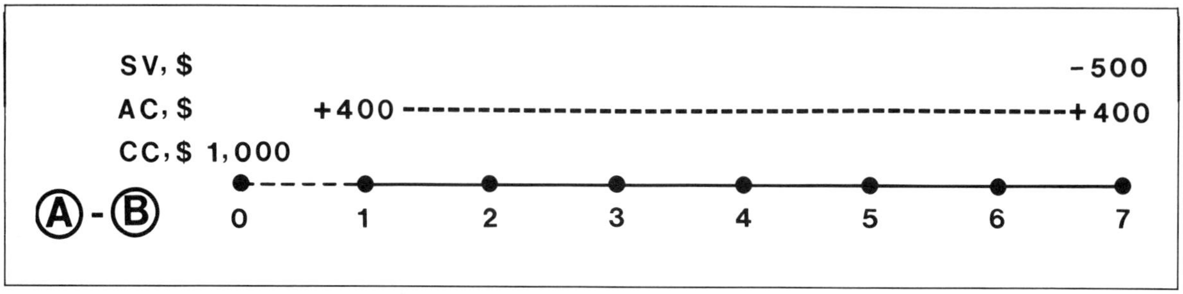


**Figure 1–13.** Cash flow diagram showing incremental costs of investments in Example 1–16.

Since the present value of the incremental costs is positive at $i = 8\%$ (minimum rate of return), we know that the incremental ROR will be greater than 8%; therefore, select alternative A. The selection is consistent with the selection made previously. The incremental ROR is 32.07%.

## SERVICE LIFE BREAKEVEN ANALYSIS

The use of this method will give the same selection as will be obtained by using the methods previously discussed. In this method, (1) select one alternative and calculate the AV of its costs; (2) for the second alternative, compute $n$ that will give the AV equal to the AV calculated in (1); and (3) if $n$ is less than the life of the first alternative, select the alternative in (2). If $n$ in (2) is greater than the life in (1), then select (1). Let us look at the following example.

**Example 1–18.** Use service life breakeven analysis to select the less expensive investment alternative given in Example 1–16.

Solution:

$$AV_A = \$532.07 \quad \text{(from Example 1–16b)}$$
$$n = 7 \quad \text{years}$$

Compute $n$ for Alternative $B$ which will make $AV_A = AV_B$. For $n = 10$,

$$\begin{aligned} AV_B &= \$500(A/P_{8,10}) + \$700(P/A_{8,7})(A/P_{8,10}) \\ &= \$500(0.14903) + \$700(5.2064)(0.14903) \\ &= \$74.52 + \$543.14 \\ &= \$617.66 \end{aligned}$$

For $n = 15$,

$$\begin{aligned} AV_B &= \$500(A/P_{8,15}) + \$700(P/A_{8,7})(A/P_{8,15}) \\ &= \$500(0.11683) + \$700(5.2064)(0.11683) \\ &= \$484.20 \end{aligned}$$

By linear interpolation,

$$n = \frac{532.07 - 484.20}{617.66 - 532.07}(15 - 10) + 10$$
$$= \frac{47.87}{85.59}(5) + 10 = 12.796 \quad \text{or} \quad 13 \text{ years}$$

The service life of Alternative *B* (13 years) is greater than the life of Alternative A for the same annual costs; therefore, select Alternative A. This selection is consistent with the selection made using the other methods.

## UNEQUAL LIFE SERVICE PRODUCING ALTERNATIVES

So far, we dealt with the selection of service producing investment alternatives with equal service life. The same techniques are used to select the most economic unequal life service producing investment alternatives, provided the lives of the alternatives are made equal for comparison purposes.

The common service life is obtained by first carefully defining the time span for which the service will be required, for instance, 10 years. If we have two alternatives, one with a 10-year life and the other with a 5-year life, to make the lives common, we assume that the 5-year alternative will be either replaced by another similar 5-year alternative and then compared, or a major repair is done at the end of Year 5 to extend its life to 5 more years.

The situation can be the other way around, i.e., the service is required for only 5 years. In this case, the extra 5 years of the 10-year life alternative is ignored and some salvage value for it at the end of Year 5 is considered. The alternative may be either sold at the end of Year 5 or it will be used elsewhere. In either case, it will have an alternative value.

If capitalized cost analysis is used for comparison, the life of the permanent life (perpetual) alternative is used as a common life and the above assumptions are not required.

## ANALYSIS OF INCOME PRODUCING INVESTMENTS

In the preceding sections, we learned how to convert the value of money at one point in time to another and then applied these concepts to the investment analysis of service producing investments. In this section, we will deal with the analysis of income producing investment alternatives. The following methods will be used for the analysis of income producing investment analysis:

1. Net present value (NPV)
2. Rate of return (ROR)
3. Present value ratio (PVR)
4. Benefit cost ration (BCR)

The net present value and the rate of return computations are discussed in the preceding sections. In this section, we will apply these concepts to the selection of the most profitable revenue producing investments.

Present value ratio (PVR) is the ratio of the net present value to the present value of investment. Present value ratio provides a measure of profitability per dollar invested. There is another analogous technique referred to as present value index (PVI). PVI is the ratio of the PV of net cash inflows to the present worth of investment. The PVR is equal to PVI + 1. Some books have defined the two in exactly the opposite way; however, both methods will provide the same selection or ranking of investments.

Benefit cost ratio is the technique most frequently used for the selection of public projects such as roads, bridges, airports, and recreation areas. There are many variations of this technique, but the most common version is presented by the following equation:

$$BCR = \frac{\text{Benefits to Public}}{\text{Cost to Public}} \tag{1.23}$$

OR

$$BCR = \frac{\text{PV of Benefits}}{\text{PV of Investments}} \tag{1.23a}$$

You will notice that Equation (1.23a) is similar to the definition of present value index (PVI).

## SINGLE INDEPENDENT INVESTMENT ANALYSIS

This is the type where only one investment opportunity is available and a decision is made either to select or not to select it, i.e., accept-reject decision.

In this case the first step is to carefully select the minimum rate of return or cost of capital. Then using this minimum rate of return, compute the four parameters mentioned above. Each of these methods, if properly used, should give the same selection. The decision rules are summarized below.

| Accept if | Reject if |
|---|---|
| $NPV > 0$ | $NPV < 0$ |
| $ROR > i_{min}$ | $ROR < i_{min}$ |
| $PVR > 0$ | $PVR < 1$ |
| $PVI > 1$ | $PVI < 1$ |
| $BCR > 1$ | $BCR < 1$ |

## MUTUALLY EXCLUSIVE INVESTMENT ANALYSIS

In this type of investments, several opportunities will exist out of which only one investment will be selected. The selection of one alternative will abandon the other alternatives. The following rules will apply. The first step would always be to apply the accept-reject criteria. The accepted projects are then further analyzed. Once again, any of the methods discussed above if properly used will give the same accept-reject selection of the available alternatives.

1. Select the alternative with the highest NPV. The NAV and the NFV may be used in the same way accept for unequal life investments, in which case the life of the longest life alternative should be used as a common life. The disadvantages in using only the NPV criteria is that it does not provide the magnitude of the return per dollar invested.
2. Perform incremental ROR analysis on two alternatives at a time. If the incremental ROR is greater than the minimum acceptable ROR, select the investment with the higher initial cost, or vice versa. The survivor is then compared with other available alternatives, and so on. The use of ROR by itself will not give consistent selection in each case.
3. The PVR, PVI, and BCR criteria are also applied on incremental basis. If the incremental PVR is greater than zero, then select the capital intensive alternative, or vice versa, and so on.

## NONMUTUALLY EXCLUSIVE INVESTMENT ANALYSIS

In this type of investments, several investment alternatives will be available and more than one may be selected, depending on the availability of capital. It is used in capital budgeting or capital rationing situations, the projects have to be selected in such a way as to maximize the return within the available investment dollars. The first step is the same as in the previous section, i.e., apply accept-reject criteria. The following rules will apply: 

1. Select the projects that will yield the highest cumulative NPV within the budget constraint.
2. ROR analysis cannot be applied to these type of investments.
3. Projects are selected in the order of decreasing PVR, PVI, and BCR. This is the easiest way of ranking nonmutually exclusive investments. When there are several alternatives, the NPV criteria becomes cumbersome because several different combinations will be tried to maximize the cumulative NPV.

Remember that in the selection of nonmutually exclusive investments, all of the available cash may not be evenly invested in the available investments. The remaining cash will be invested elsewhere at the minimum rate of return.

**Example 1–19.** A company has an opportunity to invest money. Two alternatives are considered with the investments and expected cash flows as given below. If the company's cost of capital is 15%, which alternative should be selected? The company has enough funds to invest in both alternatives. Base your decision on (a) NPV analysis, (b) ROR analysis, (c) PVR analysis, and (d) BCR analysis. The cash flow diagrams for the two investments are shown in Figure 1–14.

| | Investment A | Investment B |
|---|---|---|
| Initial investment | $500,000 | $800,000 |
| Annual revenues | 100,000 | 150,000 |
| Annual expenses | 10,000 | 13,000 |
| Investment life (years) | 20 | 20 |
| Salvage value | 50,000 | 75,000 |

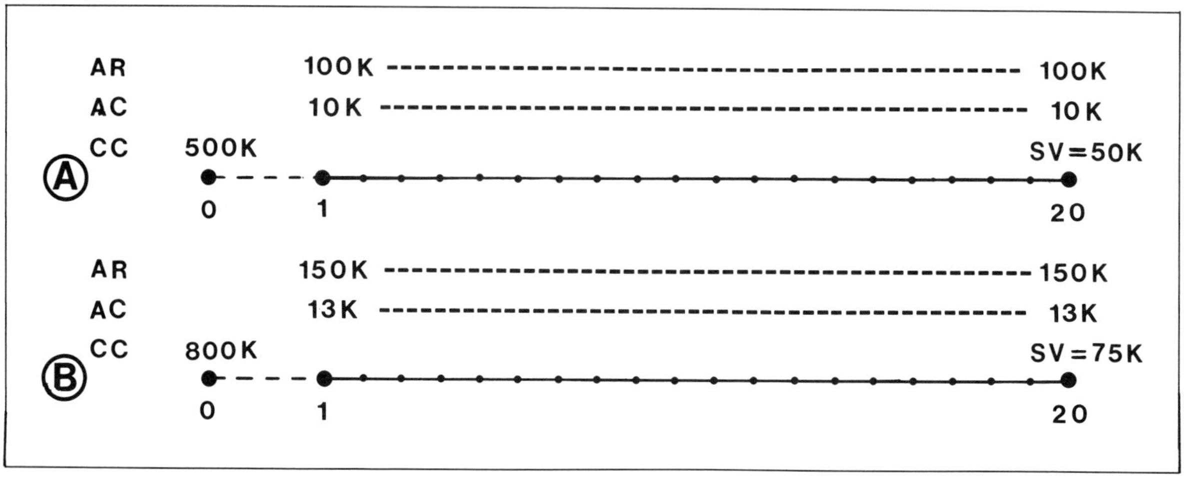


**Figure 1–14.** Cash flow diagram for investments in Example 1–19.

Solution:

(a)

$$
\begin{aligned}
NPV_A &= -\$500{,}000 + \$90{,}000(P/A_{15,20}) + \$50{,}000(P/F_{15,20}) \\
&= -\$500{,}000 + \$90{,}000(6.2593) + \$50{,}000(0.0611) \\
&= \$66{,}392 \\
NPV_B &= -\$800{,}000 + \$137{,}000(6.2593) + \$75{,}000(0.0611) \\
&= \$62{,}107
\end{aligned}
$$

Both investments have positive NPV; therefore, both are accepted.

(b) Using ROR analysis, the ROR is determined by trial and error. Consider the NPV of Part (a) as the first trial.

For $i = 20\%$

$$
\begin{aligned}
NPV_A &= -\$500{,}000 + \$90{,}000(4.8696) + \$50{,}000(0.0261) \\
&= -\$60{,}431 \\
ROR_A &= \frac{\$66{,}392}{\$66{,}392 + \$60{,}431}(20 - 15) + 15 \\
&= 17.62\% \\
NPV_B &= -\$800{,}000 + \$137{,}000(4.8696) + \$75{,}000(0.0261) \\
&= -\$130{,}907 \\
ROR_B &= \frac{\$62{,}107}{\$62{,}107 + \$130{,}907}(20 - 15) + 15 \\
&= 16.61\%
\end{aligned}
$$

Both RORs are greater than the cost of capital of the company of 15%; therefore, the investments are accepted.

(c)

$$PVR_A = \frac{NPV_A}{\text{Investment}} = \frac{\$66,392}{\$500,000} = 0.13$$

$$PVR_B = \frac{NPV_B}{\text{Investment}} = \frac{\$62,107}{\$800,000} = 0.08$$

Both $PVR_A$ and $PVR_B$ are greater than zero; therefore, the investments are accepted. 

(d)

$$BCR_A = \frac{\text{NPV} + \text{Investment}}{\text{Investment}} = \frac{\$66,392 + \$500,000}{\$500,000} = 1.13$$

$$BCR_B = \frac{\$62,107 + \$800,000}{\$800,000} = 1.08$$

Again, BCRs of both investments are greater than 1; therefore, accept the investments.

**Example 1–20.** Rework Example 1–19 keeping in mind that the company has to invest in only one of the two investments. Which investment alternative should be selected? Base your decision on (a) NPV analysis, (b) appropriate ROR analysis, and (c) appropriate PVR analysis. Remember, the alternatives are mutually exclusive.

Solution:

(a)

$$NPV_A = \$66,392 \quad \text{(from Example 1–19a)}$$
$$NPV_B = \$62,107 \quad \text{(from Example 1–19a)}$$

$NPV_A$ is greater than $NPV_B$; therefore, select Investment A.

(b) Use incremental ROR analysis. Cash flow diagram for the incremental investments is shown in Figure 1–15.

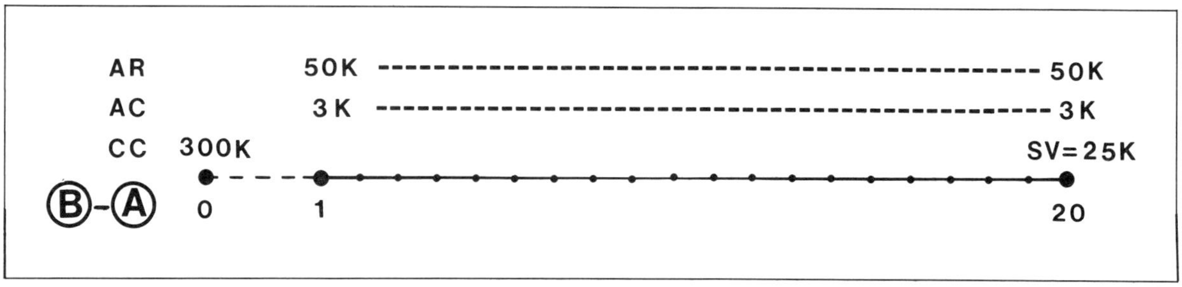


**Figure 1–15.** Cash flow diagrams for incremental investments of Example 1–20.

For $i = 15\%$



$$\begin{aligned} NPV_{(B-A)} &= -\$300{,}000 + \$47{,}000(P/A_{15,20}) + \$25{,}000(P/F_{15,20}) \\ &= -\$300{,}000 + \$47{,}000(6.2593) + \$25{,}000(0.0611) \\ &= -\$4{,}285 \end{aligned}$$

Since the $NPV_{(B-A)}$ is negative, we know that the $ROR_{(B-A)}$ will be less than 15%, therefore, select the less capital intensive investment, i.e., Investment A.

(c)

$$\begin{aligned} PVR_{(B-A)} &= \frac{NPV_{(B-A)}}{\text{Investment}_{(B-A)}} \\ &= \frac{-\$4{,}285}{\$300{,}000} \\ &= -0.014 \end{aligned}$$

$PVR_{(B-A)}$ is less than zero; therefore, select the less capital-intensive investment, Alternative A. The BCR if applied on incremental basis will give the same selection.

*In this example, we note that even the nonincremental RORs, PVRs, and BCRs give the same selections as the NPV or the incremental analysis. This, however, will not always be the case. One way to find out if incremental analysis are required is to draw a NPV profile.*

Net present value profile is a graphic presentation of the NPVs at different discount rates for an investment. If several investment alternatives are under consideration, the NPV of each alternative at three to five discount rates is determined and then all of these NPVs plotted versus their respective rate (NPV vs. $i$) on the same graph paper as shown in Figure 1–16. The NPV profile is a very useful graphic illustration and has the following advantages:

1. The profile shows the sensitivity of each investment to the timing and discount rate. In Figure 1–16b, Investment $D$ is more sensitive to discounting than Investment C. In Figure 1–16a, both investments have the same behavior.
2. The profile tells us if we will get a unique result by using any of the regular (nonincremental)

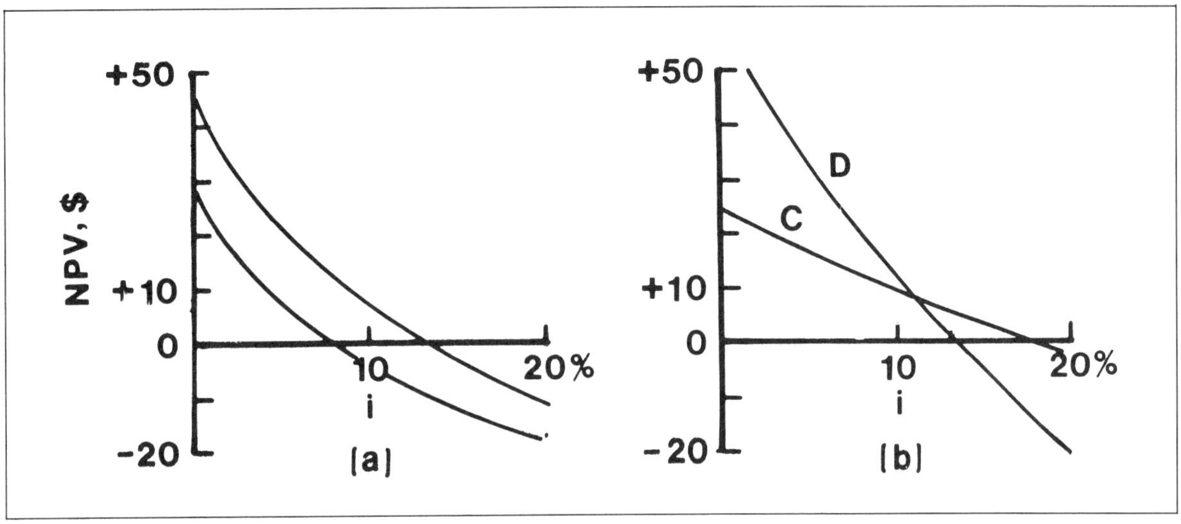


**Figure 1–16.** Net present value (NPV) profile.

investment analysis techniques, if we have two investments that will give the NPV profile such as shown in Figure 1–16a, then using any of the regular investment analysis techniques will provide consistent selections. When the profiles intersect each other such as shown in Figure 1–16b, then incremental analysis are required.

3. The NPV profile shows the range of minimum acceptable rate of return in which each project will be preferred, so if you do not have a fixed minimum guideline, the profile will provide you with a range. In Figure 1–16b, if the cost of capital is less than 11% (the intersection of two profiles), then Project *D* is preferred. If the cost of capital is between 11% and 18%, then Project C is preferred.

**Example 1–21.** Consider the two investments of Example 1–19 and a third investment C (with an initial investment of $1,000,000 which will generate revenues of $200,000 per year with annual expenses of $35,000). The salvage value of Investment C will be $100,000 after 20 years. If the company has only $1,500,000 to invest, which of the three investments will maximize their return on investment?

Solution:

$$
\begin{aligned}
NPV_C &= -\$1{,}000{,}000 + \$165{,}000(P/A_{15,20}) + \$100{,}000(P/F_{15,20}) \\
&= -\$1{,}000{,}000 + \$165{,}000(6.2593) + \$100{,}000(0.0611) \\
&= \$38{,}895 \\
NPV_A &= \$66{,}392 \quad \text{(from Example 1–19a)} \\
NPV_B &= \$62{,}107 \quad \text{(from Example 1–19a)}
\end{aligned}
$$

Investments A and B will maximize cumulative NPV for initial investment of $1,300,000. The remaining $200,000 will be invested elsewhere at 15%. Let us apply PVR to the same investment and see if we get the same result.

$$PVR_A = 0.133 \quad \text{(from Example 1–19c)}$$

$$PVR_B = 0.078 \quad \text{(from Example 1–19c)}$$

$$PVR_C = \frac{\$38{,}895}{\$1{,}000{,}000} = 0.039$$

The selection is made in descending order of PVR; therefore, select Investments A and B. The selection is consistent with the NPV method above.



## DEPRECIATION

Property, plant, and equipment are categories of assets held for use in trade or business or are held for the production of income. If a business property has a useful life of more than one year, a limited amount of the cost may be deducted as an expense for income tax purposes. The remaining cost of these properties is recovered through depreciation each year by reducing the cash outflow through reduction in income tax. In other words, depreciation is a process of allocating, in a systematic and rational manner, the total expense attributed to the acquisition of the asset to each period benefited by the asset.

Depreciable property may be *tangible* or *intangible.* Tangible property is any property that can be seen or touched. Intangible property is property, such as a copyright or franchise, that is not tangible. Depreciable property may be real or personal. Personal property is property such as machinery or equipment that is not real estate. Real property is land and generally anything that is erect on, growing on, or attached to land. However, land itself is never depreciable.

Property is depreciable if it meets the following requirements. In general, if a property does not meet all these conditions, then it is not depreciable.

1. It must be used in business or held for the production of income.
2. It must have a determinable life and that life must be longer than one year.
3. It must be something that wears out, decays, gets used up, becomes obsolete, or loses value from natural causes.

There are several methods available for depreciating assets. In 1981, a new depreciation method was introduced referred to as the Accelerated Cost Recovery System (ACRS). The pre-1981 methods may still be used for financial purposes, but ACRS must be used for income tax reporting purposes. ACRS method cannot be used for property placed in service before 1981. The tax reform act of 1986 introduced another method referred to as the Modified Accelerated Cost Recovery System (MACRS); it may be elected for property placed in service after July 31, 1986.

In most cases, four factors are involved in the computation of the periodic charge for a depreciable asset: (1) cost of asset or depreciable base, (2) service life of asset, (3) residual (salvage) value, and (4) method of cost allocation. The following methods are discussed in this section:

1. Straight line depreciation
2. Declining balance depreciation
3. Sum of the years' digits depreciation
4. Unit of production depreciation
5. ACRS and MACRS methods

The cost of certain qualifying property may be deducted as an expense in the year the property is placed in service. The total cost that you may elect to deduct in 1986 was $5,000. For property placed in service after December 31, 1986, the total cost that you elect to deduct increased from $5,000 to $10,000. For more detailed discussion of depreciation, the reader is referred to *Publication 534* published by the Internal Revenue Service (IRS).[3]



## STRAIGHT LINE DEPRECIATION

In this method of depreciation, the cost is equally distributed over the useful life of the asset. The following equation is used:

$$D_n = \frac{C - SV}{n} \tag{1.24}$$

where $D_n$ is the depreciation in $n$th year, C is the total cost of the depreciable asset, SV is the salvage value, and $n$ is the useful life of the asset.

**Example 1–22.** Compute the straight line depreciation for an asset costing $12,000 with a salvage value of $2,000 at the end of its useful life of five years.

Solution:

$$D_n = \frac{\$12{,}000 - \$2{,}000}{5}$$
$$= \$2{,}000 \quad \text{per year}$$

## DECLINING BALANCE DEPRECIATION

This method is also referred to as the accelerated depreciation method. The method is appropriate when it can be reasonably estimated that the benefits derived from an asset will decline with time. A fixed percentage is applied to the book value (total value of asset less accumulated depreciation) of the asset each year.

The declining balance depreciation is further divided into different percentages, such as 200% declining balance also referred to as the *double declining balance,* 175% declining balance, 150% declining balance, and 125% declining balance. The fixed percentage is calculated by dividing the above mentioned percentages by the life of the asset. The following example will clarify the computations.

*You do not subtract salvage value when you figure your yearly depreciations under the declining balance method. However, you may not depreciate the property below its reasonable salvage value.*

**Example 1–23.** An asset was purchased for $12,000 on January 1, 1985. It has an estimated useful service life of five years with $2,000 salvage value at the end of the useful life. Compute the double declining balance depreciation for the five years.

Solution:

$$\text{Fixed Percentage} = \frac{200}{5} = 40\%$$

| Year | Book Value of Asset at Beginning of Year | Rate | Depreciation Charge | Book Value at End of Year |
|---|---|---|---|---|
| 1985 | $12,000 | 0.40 | $4,800 | $7,200 |
| 1986 | 7,200 | 0.40 | 2,880 | 4,320 |
| 1987 | 4,320 | 0.40 | 1,728 | 2,592 |
| 1988 | 2,592 | 0.40 | 592 | 2,000 |
| 1989 | 2,000 | | | |

## SUM OF YEARS' DIGITS DEPRECIATION

This method produces a declining depreciation charge each year by applying a declining charge to the total cost of asset (depreciation base). The declining charge is determined each year by dividing the remaining life of asset by sum of years' digits. For an asset with a five-year life, the sum of years' digits is 5 + 4 + 3 + 2 + 1 = 15, and the declining charge for the first year is 5/15, and so on. The following equation is used to calculate the sum of years' digits.

$$\text{SYD} = \frac{n(n + 1)}{2} \tag{1.25}$$

where SYD is the sum of years' digits, and $n$ is the useful life of the asset.

**Example 1–24.** Compute the sum of years' digits depreciation for the asset in Example 1–23.

Solution:

$$SYD = \frac{5(5 + 1)}{2} = 15$$

| Year | Depreciation Base | Fraction | Depreciation Charge | Book Value at Year End |
|---|---|---|---|---|
| 1985 | $10,000 | 5/15 | $3,333 | $8,667 |
| 1986 | 10,000 | 4/15 | 2,667 | 6,000 |
| 1987 | 10,000 | 3/15 | 2,000 | 4,000 |
| 1988 | 10,000 | 2/15 | 1,333 | 2,667 |
| 1989 | 10,000 | 1/15 | 667 | 2,000 |



## UNITS OF PRODUCTION DEPRECIATION

This method is used when it is determined that the life of the asset is dependent on how much the asset is used or it will produce, rather than the passage of time. The activity may be measured in hours, days, months, years, the number of items produced, or other units. The following equation is used to establish depreciation rate. The depreciation charge for each year is calculated by multiplying the activity per period by the depreciation rate.

$$\text{Depreciation Rate} = \frac{\text{Cost} - \text{Salvage Value}}{\text{Total Lifetime Activity Level}} \quad \textbf{(1.26)}$$

**Example 1–25.** A drilling rig is purchased for $1,300,000. It is estimated that the rig will be able to drill a total of 10,000,000 ft. After drilling that many feet, the rig can be sold for $150,000. In the first year of service, the rig drills 155,000 ft, and in the second year it drills 100,000 ft. What is the depreciation for the two years?

Solution:

$$\begin{aligned}
\text{Depreciation Rate} &= \frac{\$1{,}300{,}000 - \$150{,}000}{10{,}000{,}000} \\
&= \$0.115/\text{ft} \\
\text{Depreciation for Year 1} &= 155{,}000 \times \$0.115 \\
&= \$17{,}825 \\
\text{Depreciation for Year 2} &= 100{,}000 \times \$0.115 \\
&= \$11{,}500
\end{aligned}$$

## ACCELERATED COST RECOVERY SYSTEM (ACRS)

Historically, depreciation has been calculated using the methods discussed in the preceding few sections. According to the ACRS system, the depreciable assets are divided into six property classes. Each class has been assigned a preset life by IRS ranging from a 3-year property to 19-year property. ACRS applies equally to both new and used recovery property. However, you cannot use ACRS for property placed in service before 1981. ACRS also cannot be used for *intangible* depreciable property or for tangible property that is acquired in certain types of transactions. For more information refer to *Publication* 534.

1. **3-year property.** This class includes personal property with a short useful life such as automobiles, tractor units for use over the road, light duty trucks (weight under 13,000 lbs), and equipment used in research and development (R&D).
2. **5-year property.** This class includes personal property that is not 3-year property or 10-year property. It includes most equipment such as pumping units, compressors, separators, petroleum storage facilities, and other office furniture and fixtures.
3. **10-year property.** This class includes certain real property such as certain public utility property and theme park structures. Manufactured homes, including mobile homes, are also designated as 10-year property.
4. **15-year real property.** This class includes all real property placed in service before March 16, 1984, such as buildings other than any designated as 10-year property. This class also includes low-income housing and certain public utility property.
5. **18-year real property.** This class includes all real property placed in service after March 15, 1984, and before May 9, 1985, such as buildings other than those designated as 10-year property.
6. **19-year real property.** This class includes all real property placed in service after May 8, 1985, such as buildings other than any property designated as 10-year property.

The deduction under ACRS is figured by multiplying your *unadjusted basis* in the recovery property by a certain percentage. Do not subtract the expected salvage value of the property from its basis when figuring your deduction under ACRS. This percentage varies from year to year during the recovery period. For 3-, 5-, and 10-year property, the full first year percentage applies no matter in which month the property is placed in service. The percentages for 3-, 5-, and 10-year recovery property are:

**3-Year Property:**

| | |
|---|---|
| 1st year | 25% |
| 2nd year | 38% |
| 3rd year | 37% |

**5-Year Property:**

| | |
|---|---|
| 1st year | 15% |
| 2nd year | 22% |
| 3rd through 5th year | 21% |

**10-Year Property:**

| | |
|---|---|
| 1st year | 8% |
| 2nd year | 14% |
| 3rd year | 12% |
| 4th through 6th year | 10% |
| 7th through 10th year | 9% |

## MODIFIED ACCELERATED COST RECOVERY SYSTEM (MACRS)

The Tax Reform Act of 1986 provided a modified accelerated cost recovery system (MACRS). This method of depreciation generally applies to properties placed in service after December 31, 1986. However, you may make a property-by-property election in service after July 31, 1986.

Under MACRS, the new classes of property are 3-, 5-, 7-, 10-, 15- and 20-year property. In addition, most real property is classified as residential rental or nonresidential real property. The different classes of property are described below. For all classes, salvage value is treated as zero. Prior to this, most oilfield equipment was placed in the 5-year property class. Under the new classification, this same property is classified as 7-year property.



1. **3-Year Property:** This class includes property with a class life of four years or less, such as tractor units for use over the road and, as designated, any race horse that is over two years old when placed in service and any other horse that is 12 years old when placed in service.
2. **5-Year property:** This class includes property with a class life of more than four years but less than 10 years, such as heavy general-purpose trucks (actual unloaded weight of 13,000 lbs or more), computers and peripheral equipment, and office machinery (typewriters, calculators, etc.), and, as designated, any automobile and light general-purpose truck (actual unloaded weight of under 13,000 lbs).
3. **7-Year property:** This class includes property with a class life of 10 years or more but less than 16 years, such as office furniture and fixtures (desks, files, etc.) and, as designated, any single-purpose agricultural or horticultural structure. Any property which does not have a class life and which has not been designated by law as being in any other class is also included in this class.
4. **10-Year property:** This class includes property with a class life of 16 years or more but less than 20 years. It includes water transportation equipment.
5. **15-Year property:** This class includes property with a class life of 20 years or more but less than 25 years. It includes, as designated, any municipal waste water treatment plant.
6. **20-Year property:** This class includes property with a class life of 25 years or more, such as farm buildings and, as designated, any municipal sewers.

The *class life* of an asset is its guideline period which is set by IRS. For the 3-, 5-, 7-, or 10-year property, use double (200%) declining balance method over the respective life and a *half-year* convention (defined later). For property in the 15- or 20-year class, use the 150% declining balance method over 15 or 20 years and a half-year convention. In addition, depreciation is switched to *straight line* depreciation in the year for which this method when applied to the adjusted basis at the beginning of such year will yield a larger deduction. Straight line depreciation must be used for nonresidential real property and residential rental property.

Under MACRS, the half-year convention treats all property placed in service or disposed of during a tax year as placed in service or disposed of in the midpoint of that tax year.

## AFTER-TAX CASH FLOW ANALYSIS

Taxes are disbursements that must be included in the economic evaluations. Of course, the introduction of taxes in economic evaluation increases the complexity of the study; however, performing economic evaluation without accounting for tax effects may be misleading. In some cases incorporating taxes in economic evaluation may reverse the before tax decisions.



In the previous section we discussed various depreciation methods; in this section we will present several more topics which form the basis for after-tax calculations. A simple after–federal income tax model incorporating the various deductions is presented. After-tax cash flow is prepared and analyzed for discounted cash flow rate of return.

### AMORTIZATION AND DEPLETION

A part of a certain business capital expenditures may be amortized (deducted) each year. Amortization generally allows a write-off of costs that are nor ordinarily deductible. On the other hand, if you have an exhaustible natural resource or timber, you are allowed a depletion deduction.

**Amortization.** All costs incurred to get a business started are treated as capital expenditures and are part of your basis in the business. If you elect to amortize, these costs are deducted in equal amounts over a period of 60 months or more.

Business startup costs and organizational expenses for a corporation or partnership are capital expenditures. To be amortizable the costs must be paid or incurred before the day on which the active trade or business begins. However, startup costs do not include deductible interests, taxes, and research and experimental costs.

Startup costs include cost of items such as survey of potential market; analysis of available facilities, labor supply, etc.; initial advertisements; salaries and wages for training employees and their instructors; and salaries and fees for executives, consultants, and other professional services.

Organizational costs include the cost of temporary directors; organizational meetings; incorporation fee; accounting services; and legal services for drafting the charter, bylaws, terms of the original stock certification, and minutes of organizational meeting. However, costs of issuing and selling stock or securities, such as commissions, professional fees, and printing costs, are not amortizable.

The amortization deduction each year is figured out by dividing the total of startup and organizational costs by the number of months in the amortization period, the resulting amount can be deducted each month.

**Depletion.** Depletion permits the owner of an economic interest in mineral deposits, oil well, gas wells, geothermal deposits, or standing timber to deduct the cost over the economic life of the property. There are two ways of figuring depletion: (1) *Percentage Depletion*, and (2) *Cost Depletion*.

Cost depletion is figured out by dividing the adjusted basis of the mineral property by the total number of recoverable units in the deposit, and multiplying the resulting rate per unit by:

1. The number of units sold and for which you received payment during your tax year, if you use the cash method of accounting, or
2. The number of units sold, if you use the accrual method of accounting.

In equation form, cost depletion can be represented by: 

$$\text{Cost Depletion} = \text{AB}\left(\frac{P}{\text{RR} + P}\right) \quad \textbf{(1.27)}$$

where AB is the adjusted basis for the taxable year, $P$ is the number of units sold during the year, and RR is the number of units of remaining reserves at end of the taxable year. Items included in the depletable basis for cost depletion are geologic and geophysical (G&G) costs, lease bonus paid by the lessee, and capitalized intangible costs not represented by physical property.

Percentage depletion is a certain percentage, specified for each mineral, of your gross income from the property during the tax year. However, the deduction under this method cannot be more than 50% of your taxable income from the property, figured out without the deduction for depletion.

In figuring the percentage depletion deduction and the 50% limit for percentage depletion, the gross income from the property should be reduced by the allocable part of any bonus that is paid for a mineral lease or an oil and gas lease on the property. When figuring percentage depletion, rents and royalties (which are depletable income to the payee) paid or incurred for the property must be excluded from gross income from the property.

If you qualify for percentage depletion and it is less than the cost depletion for any year, you must use cost depletion for that year. To figure your depletion allowance for mineral properties, you should figure it under both depletion methods for each property, and claim a deduction for the greater of the two amounts.

The use of percentage depletion for oil and gas is not allowed, except for certain domestic oil and gas production. In general the percentage depletion rate is 15% of the gross income from the *qualifying* production from oil and gas property. Gross income from the property does not include lease bonuses, advance royalties, or other amounts payable without regard to production from the property received or occurred after *August 16, 1986*. Independent producers and royalty owners (small producers) qualify for the 15% depletion rate for a maximum daily average production of 1,000 barrels of oil equivalent (6,000 $ft^3$ of natural gas are considered equivalent to a barrel of oil). This is the small producer exemption. The deduction for depletion may not be more than 65% of taxable income before depletion. Some of the other conditions and percentages as applied to oil and gas percentage depletion are as follows:

1. Income from natural gas sold under a fixed contract from a well located in the United States or its possessions qualified for percentage depletion at 22%.
2. Independent producers and small royalty owners qualify for 15% rate.
3. Gross income from oil and gas property (including geothermal steam wells) is the amount received from the sale of oil or gas at the wellhead.
4. Qualified natural gas from geopressured brine is eligible for percentage depletion at the rate of 10%.
5. For partnership oil and gas properties, depletion allowance, whether by cost or percentage, must be figured out separately by each partner and not by the partnership as a whole.



**Example 1–26.** It is estimated that a taxpayer has remaining reserves (net to his interest) at the end of his tax year of 50,000 barrels of oil. His production share in the year was 6,000 barrels sold at $18 per barrel. His net taxable income before depletion from the property was $80,000, and the taxpayer's total taxable income was $450,000. Compute the appropriate depletion charge for the year. The adjustable basis of the capitalized leasehold costs are $45,000.

Solution:

$$\text{Cost Depletion} = AB\left(\frac{P}{RR + P}\right)$$
$$= \$45{,}000\left(\frac{6{,}000}{50{,}000 + 6{,}000}\right)$$
$$= \$4{,}821$$

Percentage depletion computation and limitations check:

(a) $$\text{Percentage Depletion} = 0.15(6{,}000 \times 18) = \$16{,}200$$

(b) $$\text{50\% Net Income Limitation} = \$40{,}000$$

(c) $$\text{65\% Taxable Income Limitation} = \$292{,}500$$

Therefore, allowable percentage depletion is $16,200 since percentage depletion is greater than cost depletion of $4,821. Note that the adjustable basis (AB) for next year's cost depletion calculations will be reduced by $16,200 (i.e., AB for next year will be $28,800).

## INVESTMENT TAX CREDIT (ITC)

The investment tax credit (ITC) is generally eliminated for assets placed in service after *December 13, 1985*. Prior to the tax reform act of 1986, the ITC was 10% for a five-year property. If the full 10% ITC rate was used, the basis of the property must be reduced for depreciation calculations. A reduced ITC rate of 8% could then be taken in lieu of the depreciation basis adjustment.

## INTANGIBLE DRILLING COSTS (IDCs)

The pre-1986 law allows tax payers to expense domestic intangible drilling costs. Under this law, the amount that the integrated producer can expense is 80% of its IDCs with the difference being amortized over 36 months beginning in the month costs are incurred. The independent producer may elect to expense IDCs in the year incurred. The Tax Reform Act of 1986 changes the rules as summarized below.



1. The independent producer may elect to expense IDCs in the year incurred for domestic wells. For foreign IDCs, the cost may be recovered over a 10-year period using straight line amortization or may be added to its depletable basis.
2. The integrated producer may elect to expense 70% of IDCs in the year incurred and amortize the remaining 30% over 60 months for domestic wells. The foreign IDCs must be capitalized over a 10-year period using straight line amortization or may be added to its depletable basis.

*Dry hole costs* may be expensed in the year the costs are incurred regardless of the way in which IDCs are treated. This is applicable to both the integrated and independent producers. The breakdown of various expenditures involved in drilling a well and their respective tax treatment is shown in Table 1–1.[4]

## MINIMUM TAX AND TAX PREFERENCE ITEMS

The tax laws give special treatment to some kinds of income and allow special deductions for some kinds of expenses. So that taxpayers (including corporations) who benefit from these laws will pay at least a minimum amount of tax. *Alternative Minimum Taxable Income* (AMTI) is a regular taxable income less certain deductions plus items of *tax preference.*

The alternative minimum tax is figured out on benefits received in the form of deductions, lower tax rates, and exclusion from tax. These benefits are known as tax preference or tax preference items because they result from the preferential treatment given in the tax law. However, generally only the amount of the benefit received minus the benefit that would otherwise be allowed is counted as the tax preference item.

Generally, every taxable corporation is subject to the minimum tax. This tax is in addition to any other tax owed by the corporation and must be added to the corporation's *regular income tax.* The minimum tax is figured by multiplying the minimum taxable income (reduced by an exemption amount of $40,000) by 20% tax rate. The exemption amount is reduced by (but not below zero) by

**TABLE 1–1.[4] Breakdown of various expenditures involved in drilling a well and their respective tax treatment**

| | Tax Treatment | |
|---|---|---|
| | **Dry Hole** | **Completion** |
| Administrative Costs | E | IDC |
| Location—Stake/Permit/Damages | E | IDC |
| Rig Tariff | E | IDC |
| Rig Transportation | E | IDC |
| Company Labor—Drilling | E | IDC |
| Cementing—Surface Casing | E | IDC |
| Cementing—Production Casing | | IDC |
| Bits and Reamers | E | IDC |
| Logging | E | IDC |
| Coring Services | E | IDC |
| Mudlogger | E | IDC |
| Engineering while Drilling | E | IDC |
| Mud, Water, and Fuel | E | IDC |
| Drillstem Tests | E | IDC |
| Casing Crew—Surface Casing | E | IDC |
| Casing Crew—Production Casing | | IDC |
| Completion Unit | | IDC |
| Completion Costs | | IDC |
| Swabbing Costs | | IDC |
| Company Labor—Completion | | IDC |
| Perforating | | IDC |
| Surface Casing | | Tangible |
| Product Casing | | Tangible |
| Tubing | | Tangible |
| Wellhead Equipment | | Tangible |
| Float Equipment, Centralizers | | Tangible |
| Packers | | Tangible |
| Line Pipe | | Tangible |
| Tanks and Fittings | | Tangible |
| Separators and Fittings | | Tangible |
| Pumping Unit | | Tangible |

25% of the amount by which alternative minimum taxable income exceeds $150,000. The taxpayer is liable for the *larger of the regular tax liability or the minimum tax liability*. When the AMT is larger than the regular tax, the difference is referred to as the *net minimum tax*. It is the additional tax due because of the minimum tax calculations.

The following tax preference items will apply under the new corporate minimum tax law and will be figured under this new law:

1. Excess depreciation is treated as a tax preference item. It is the amount the depreciation taken exceeds the amount of depreciation that would have been taken if 150% declining balance was used while using the alternative depreciation life.
2. Percentage depletion to the extent that it exceeds the depletable basis is an item of tax preference.
3. Excess intangible drilling costs is also considered as a tax preference item. The 1986 Tax Reform Act states that to the extent the excess intangible drilling costs exceed 65% of the net income from all oil and gas properties is an item of tax preference. Excess IDCs are the difference between the amount taken and the amount which would have been taken if the IDCs were capitalized and amortized over 120 months using straight line amortization (or cost depletion).



When additional tax is paid as a result of the AMT calculations, this amount can be used to offset the amount that the regular tax exceeds the AMT in future years. This new feature is referred to as the *minimum tax credit.*

## AFTER-TAX CASH FLOW

In this section we will discuss how to generate after–federal income tax cash flows. The emphasis is on project evaluation as opposed to determination of a corporation's tax liability. The preceding sections briefly summarized the topics which form the basis for tax liability calculations and the after-tax cash flow. This section is intended to be brief and is not an exhaustive analysis of the more complex tax laws. Specific tax questions should be addressed to a tax professional. A framework for the development of federal income tax model is illustrated in Figures 1–17 through 1–20.

The graduated corporate tax rates under the 1986 Tax Reform Act as applied to U.S. corporations are summarized below.

| Taxable Income | Tax Rate (%) |
|---|---|
| Not over than $50,000 | 15 |
| Over $50,000 but less than $75,000 | 25 |
| Over $75,000 | 34 |

For corporate taxable income in excess of $100,000 there is an additional 5% tax but this rate cannot exceed $11,750.

Once the tax liability is determined, the noncash deductions, such as depreciation, depletion, and capitalized costs, are added back to the net after tax cash flow. This cash flow is then used to calculate the net present value and discounted cash flow rate of return (DCFROR). The following simplified example will illustrate some of the features of after–federal income tax investment analysis.

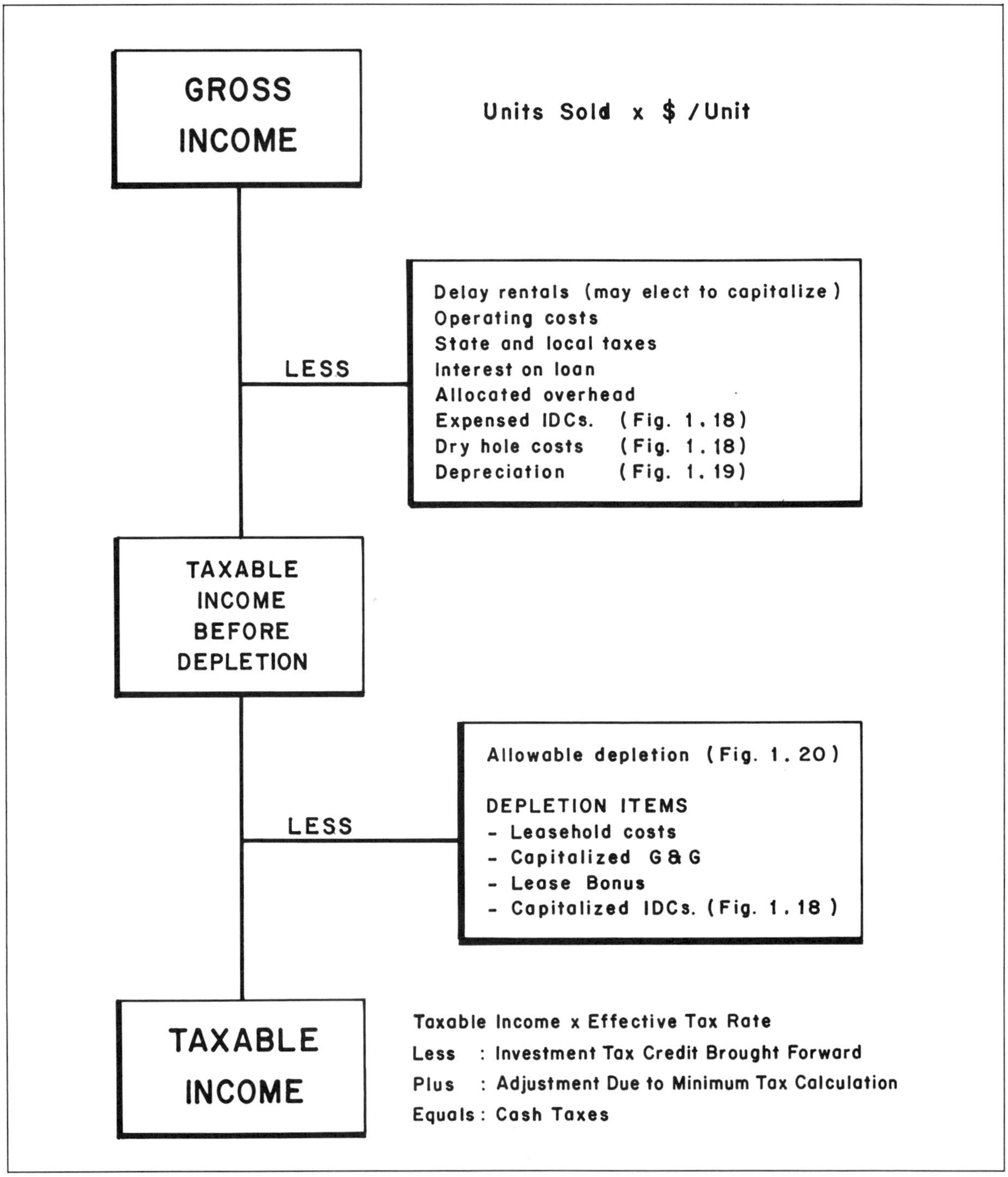


**Figure 1–17.** Federal income tax model for oil and gas transactions (after Thompson and Wright).[4]

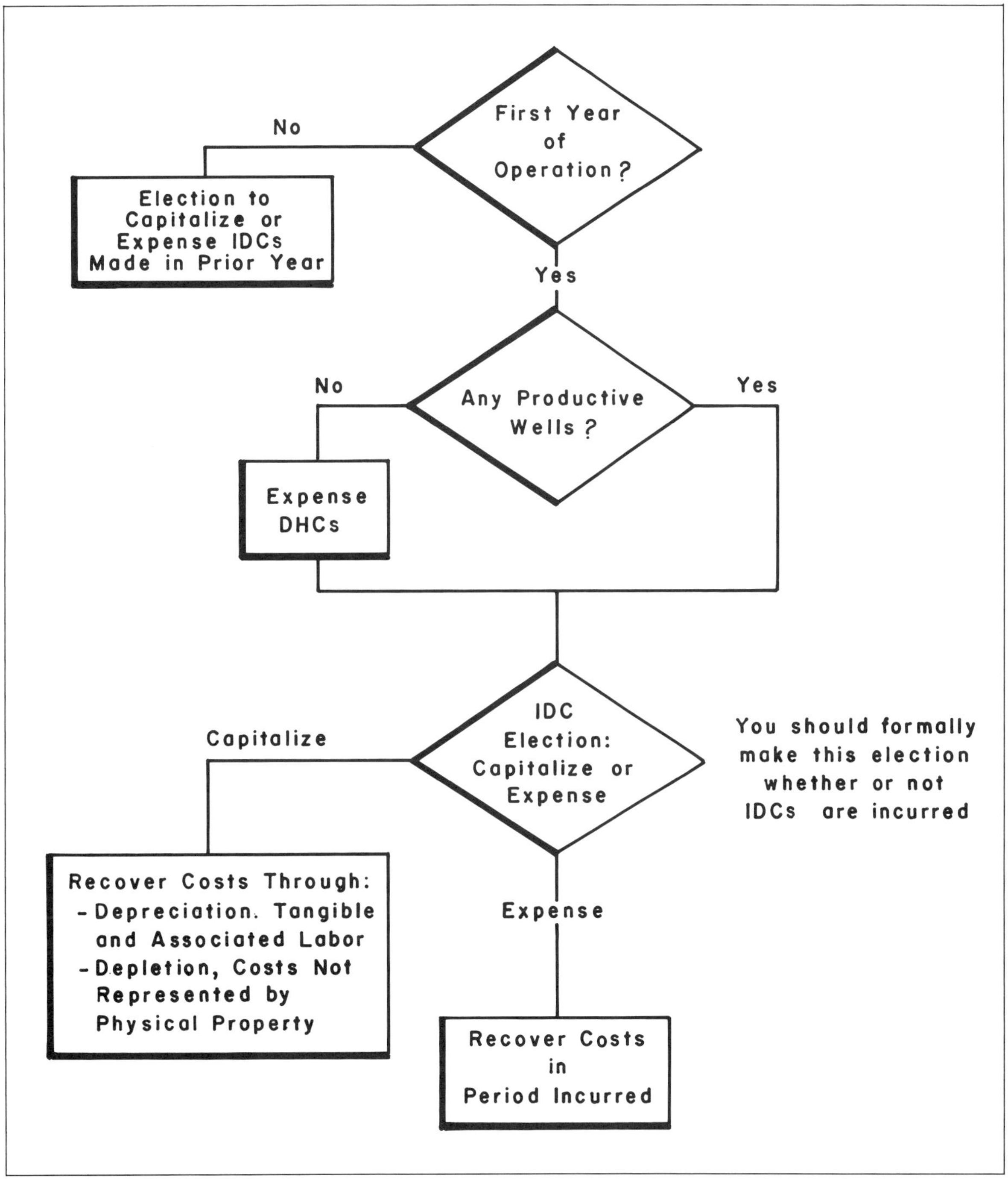


**Figure 1–18.** Recovery of drilling and completion costs (after Thompson and Wright).[4]

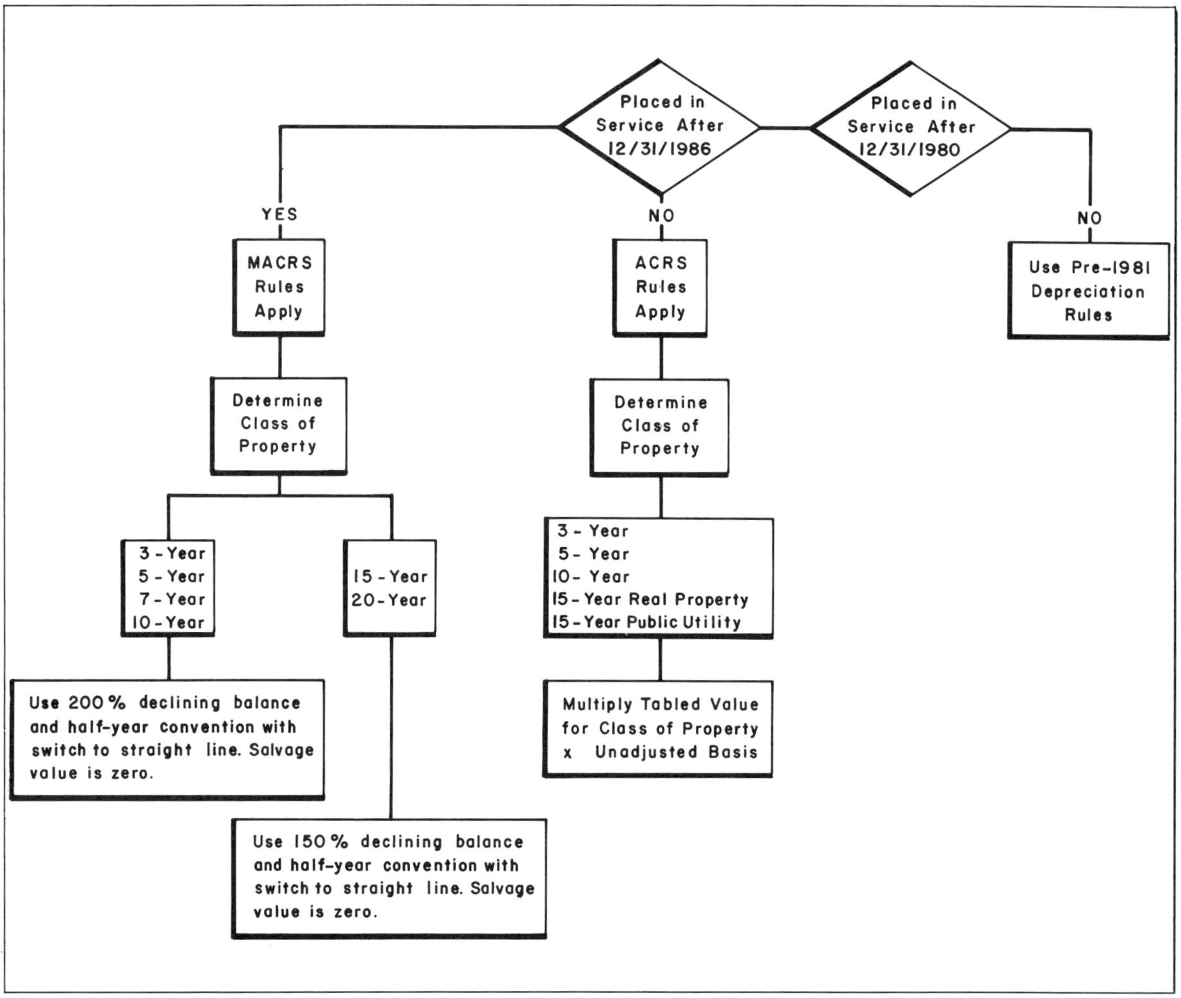


**Figure 1–19.** Recovery of tangible costs (after Thompson and Wright).[4]

**Example 1–27.** An independent oil company acquired mineral rights for \$500,000 in 1988. The company drilled and completed a well on the lease for a total of \$3,000,000 out of which \$1,000,000 were depreciable completion costs. The drilling costs of \$2,000,000 will be treated as operating cost for tax purposes and deducted against income in 1989 because the drilling costs were incurred right after the end of 1988. Initial well tests show an estimated daily production of 750 barrels per day, declining at a rate of 50 barrels per day in each succeeding year. Lease operating costs are projected at \$400,000 per year, escalating by \$50,000 per year in each succeeding year. The crude oil will be sold at \$10 per barrel. Recoverable oil reserves are estimated to be 1,500,000 barrels. Assume that the company has 100% WI and 87.5% NRI in the lease, salvage value is zero, the income tax rate is 30%, and there is no production tax. Disregard minimum tax calculations and use double declining balance depreciation. Calculate the after tax cash flow for the first five years and calculate the DCFROR for the project using the generated cash flows.

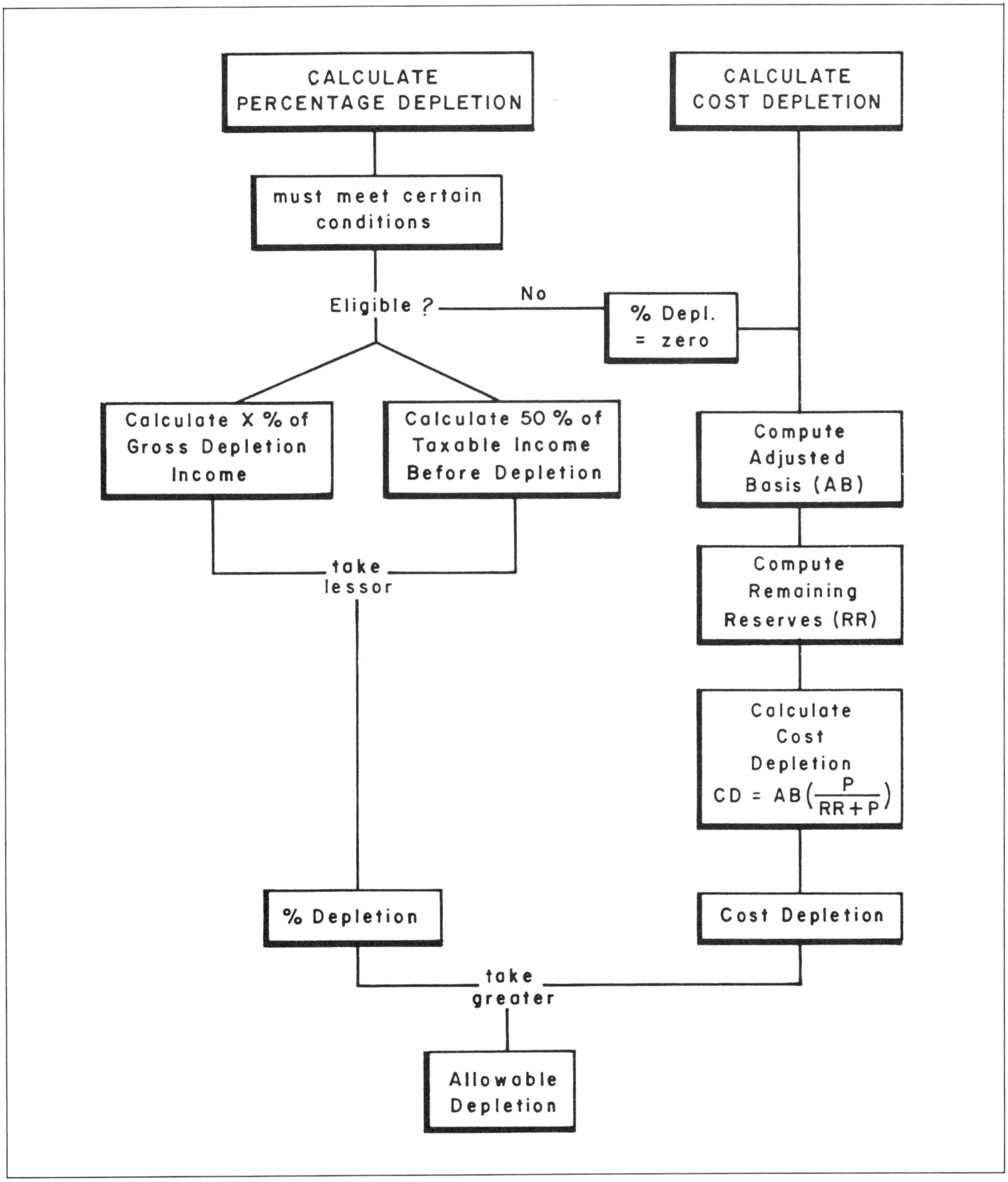


**Figure 1–20.** Depletion allowance (after Thompson and Wright).[4]

Solution: First calculate cost depletion for the five years as

$$CD_{1989} = 500{,}000\left[\frac{273{,}750}{(1{,}500{,}000 - 273{,}750) + 273{,}750}\right] = \$91{,}250$$

$$CD_{1990} = (500{,}000 - 91{,}250)\left[\frac{255{,}500}{(1{,}500{,}000 - 529{,}250) + 255{,}500}\right] = \$85{,}167$$

$$CD_{1991} = (500{,}000 - 176{,}417)\left[\frac{237{,}250}{(1{,}500{,}000 - 766{,}500) + 237{,}250}\right] = \$79{,}083$$

Similarly for 1992 and 1993,

$$CD_{1992} = \$73{,}000$$

$$CD_{1993} = \$66{,}917$$

$$\begin{aligned} NPV_{@20\%} &= -3{,}500{,}000 + (1{,}995{,}313)(0.8338) + (1{,}075{,}385)(0.6944) + (1{,}239{,}773)(0.5787) \\ &\quad + (1{,}068{,}526)(0.4823) + (904{,}191)(0.4019) \\ &= \$505{,}643 \end{aligned}$$

$$NPV_{@30\%} = -\$146{,}966$$

By interpolation, DCFROR = 27.75%

| | 1988 | 1989 | 1990 | 1991 | 1992 | 1993 |
|---|---|---|---|---|---|---|
| Revenue | 0 | 2,737,500 | 2,555,000 | 2,372,500 | 2,190,000 | 2,007,500 |
| Royalty | 0 | −342,188 | −319,375 | −296,563 | −273,750 | −250,938 |
| Net Revenue | 0 | 2,395,313 | 2,235,625 | 2,075,937 | 1,916,250 | 1,756,562 |
| Operating Costs | 0 | −400,000 | −450,000 | −500,000 | −550,000 | −600,000 |
| Depreciation | 0 | −400,000 | −240,000 | −144,000 | −86,400 | −51,840 |
| IDCs | 0 | −2,000,000 | 0 | 0 | 0 | 0 |
| Taxable Before Depl. | 0 | −404,687 | 1,545,625 | 1,431,937 | 1,279,850 | 1,104,722 |
| % Depletion (15%) | 0 | 359,297 | −335,344 | −311,391 | −287,438 | −263,484 |
| 50% Limit | 0 | 0 | 790,772 | 715,083 | 631,064 | 552,361 |
| Cost Depletion | 0 | −91,250 | 85,167 | 79,083 | 73,000 | 66,917 |
| Loss Forward | 0 | 0 | −495,937 | 0 | 0 | 0 |
| Taxable Income | 0 | −495,937 | 714,344 | 1,120,546 | 992,412 | 841,238 |
| Tax @ 30% | 0 | 0 | 214,303 | 336,164 | 297,724 | 252,371 |
| Net Profit | 0 | −495,937 | 500,041 | 784,382 | 694,688 | 588,867 |
| + Depreciation | 0 | 400,000 | 240,000 | 144,000 | 86,400 | 51,840 |
| + Depletion | 0 | 91,250 | 335,344 | 311,391 | 287,438 | 263,484 |
| + IDCs | 0 | 2,000,000 | 0 | 0 | 0 | 0 |
| + Leasehold Costs | 500,000 | 0 | 0 | 0 | 0 | 0 |
| IDCs + Completion | 3,000,000 | 0 | 0 | 0 | 0 | 0 |
| Cash Flow | −3,500,000 | 1,995,313 | 1,075,385 | 1,239,773 | 1,068,526 | 904,191 |

## EXPECTED VALUE CONCEPT

Most decision analysis provides a rational methodology for decision making in the face of uncertainty. It enables one to choose among alternatives in an optimal fashion taking into account the worth of acquiring experimental data to reduce the uncertainty.

The *expected value* can be interpreted as a long-run average outcome of a project resulting from various expected outcomes when all outcomes are weighted on the basis of their respective probabilities. In other words, expected value is a tool which accounts for risk and uncertainty into economic evaluations. The words *risk* and *uncertainty* are considered to be synonymous. Expected value can be expressed in any quantity, such as monetary value, barrels, feet of pay, porosity, and acres, and at the same time account for time value of money.

All the decision-making tools presented in the previous sections can be used on the basis of expected value or expected cost and the same accept/reject criteria applied. In this section we will discuss the definitions of probability, expected net present value concept, expected net present value profile, expected present value ratio, and the use of decision trees in expected value decision making. For further reference, the reader is referred to an excellent work by Paul D. Newendorp, published by PennWell Books.[5]



Two definitions are important to the understanding of the expected value concept:

1. *Expected value of an event.* The product obtained by multiplying the probability of occurrence of the outcome and the conditional value (or worth) that is received if the outcome occurs.
2. *Expected value of a decision alternative.* The algebraic sum of the expected values of each possible outcome that could occur if the decision alternative is accepted.

Expected value concept is an extremely useful decision-making tool; however, the usefulness is dependent upon accurately assigning probability to each outcome involved.

## DEFINITIONS OF PROBABILITY

Probability is merely a way of expressing the likelihood of an event. Qualitatively, probability can be modified with an adjective expressing the chance of occurrence, such as remote, poor, weak, good, or strong. In quantitative terms, probability is expressed as a numerical value, either a fraction or decimal, between 0 and 1.0. Probability of 0 implies that an event can not occur and a probability of 1.0 means that it is certain to occur.

This section shows how to combine probabilities by using certain rules. A list of all possible outcomes of an experiment is a universal set called the *sample space,* which we will call *S*. The entries in *S* are called *elements.* Any subset of *S* is called an *event.* If the occurrence of one event is made dependent on the occurrence of a second event, the events are called *conditional events.* An event is usually our main interest, but it is always necessary to know the sample space to work with different events from this space.

For the purpose of determining probabilities, it is helpful to classify events as *simple* or *compound.* A simple event cannot be further subdivided into component events, and a compound event always has two or more component simple events. For example, the event of "spade" in a deck of cards is a simple event but the "ace of spades" in the same deck could be considered as a compound event. A compound event specifies *without replacement.*

For simple events, probabilities may be derived *logically*, observed *empirically*, or assigned *subjectively*.

1. Some processes are so predictable and so rigidly defined that the probabilities of their outcomes can be deduced logically. By definition, the logical probability of an event is the ratio of the number of ways that event can occur to the total number of ways all outcomes to an experiment can be generated.
2. In some situations it is just not apparent what the probabilities are. In such cases, it is appropriate to refer to observed, or empirical, data. By definition, the empirical probability of an event is the ratio of the observed frequency of that event to the total number of observations.
3. When a process is not predictable and observed data are not available, a decision maker must assign probabilities subjectively. This assignment may be based on qualitative factors, experience in similar situations, and even intuition.

For compound events with which we shall be concerned, there are two subcategories. In the first, *every* specified simple event must occur to satisfy the compound event. In the second category, the occurrence of *any one or any combination* of simple events satisfies the compound event. Probabilities of compound events of the first kind are determined by the *Laws of Multiplication* while those of the second kind follow the *Laws of Addition*. The two laws are discussed in the following paragraphs while summary of the mathematical relationships involved are given in Table 1–2.

The probability that two events will both occur is called a *joint probability* and is denoted by P(AB). Probability of a single event is denoted by P(A), the probability of event A and/or B occurring is denoted by P(A + B), and the conditional probability of A occurring given that B has occurred is denoted by P(A/B).

The *General Law of Mutliplication* states that the joint probability of two events is the product of the probability of the first event and the probability of the second event, *given that the first event*

**TABLE 1–2. Summary of probability formulas**

| | P(A + B) | | P(AB) |
|---|---|---|---|
| Events | Mutually Exclusive | Nonmutually Exclusive | Either Mutually Exclusive or Nonmutually Exclusive |
| Independent | P(A) + P(B) | P(A) + P(B) − P(A) · P(B) | P(A) · P(B) |
| Nonindependent | P(A) + P(B) | P(A) + P(B) − P(A/B) · P(B) | P(A/B) · P(B) |

*has occurred*. The Laws of Multiplication typically apply when both (or all) simple events must occur to satisfy the compound event. Mathematically,

$$P(AB) = P(B/A) \times P(A) \quad \textbf{(1.28)}$$

The General Law of Multiplication always works for joint probabilities. However, if the compound event specifies *with replacement*, a simple form of multiplication law applies in which case the probability of A and B is the same whether B occurs or not. In this case A and B are called independent events. This gives rise to the *Special Law of Multiplication*, which states that the joint probability of two independent events is equal to the product of the probabilities of the simple events. Mathematically,

$$P(AB) = P(B) \times P(A) \quad \textbf{(1.29)}$$



The *General Law of Addition* states that the probability of either of two events is equal to the sum of the probabilities of each of the events *minus their joint probability*. Laws of Addition apply when *either* (or any) simple event satisfies the compound event. Mathematically,

$$P(A + B) = P(A) + P(B) - P(AB) \quad \textbf{(1.30)}$$

If the events are mutually exclusive, the probability of either of two mutually exclusive events is equal to the sum of the probabilities of each. This is the *Special Law of Addition*.

$$P(A + B) = P(A) + P(B) \quad \textbf{(1.31)}$$

One last probability rule that needs to be presented here is called the *Bayes' Formula*. This formula includes the conditional probabilities of both B/A and A/B as

$$P(A/B) = \frac{P(A) \times P(B/A)}{P(B)} \quad \textbf{(1.32)}$$

**Example 1–28.** Past experience indicates a 40% chance of finding recoverable precious metals (M) and a 30% chance of finding them in combination with copper ore (C + M). If there is 80% chance of finding copper ore (C) deposits in a particular area, what is the probability of finding (a) copper or precious metals, (b) neither copper nor precious metals, and (c) copper only?

Solution: (a) Copper or metals is found by using Equation (1.30) as

$$\begin{aligned} P(C + M) &= P(C) + P(M) - P(CM) \\ &= (0.80) + (0.40) - (0.30) \\ &= 0.70 \end{aligned}$$

(b) Neither copper nor precious metals

$$= 1 - P(C + M)$$
$$= 1 - 0.90 = 0.10$$

(c) The event C (copper only) means that no metals are found. By subtraction we have

$$\begin{aligned} P(C) &= P(C) - P(C + M) \\ &= 0.80 - 0.30 \\ &= 0.50 \end{aligned}$$

## EXPECTED NET PRESENT VALUE E(NPV)

Expected net present value E(NPV) is the product obtained by multiplying the probability of occurrence of the outcome and the *conditional net present value* that is received if the outcome occurs. The word *conditional* means that the value will be received only if that particular outcome occurs.

In decision making the E(NPV) is used in the same way as explained in the section on Analysis of Income Producing Investments, and the expected present value of costs E(PV) is used in the same way as explained in the section on Comparison of Service Producing Investments. The following example will clarify the use of expected value analysis.

**Example 1–29.**[5] An exploration staff of an oil company has been evaluating a proposed drilling prospect in Eastern Colorado. The proposed well has to be drilled on a 640-ac gas unit of which only a 320-ac block is currently leased by the company. The drilling would be subject to leasing the remaining 320 ac. The completed-well costs and the dry-hole costs are estimated to be \$250,000 and \$150,000, respectively. By careful analysis of risk and uncertainty, the possible outcomes with their associated NPV and probability of occurrence are estimated as shown in Table 1–3. Assuming that the 320 ac are acquired, three mutually exclusive decision alternatives are identified as (1) drill the well with a 50% working interest, (2) farm out the acreage and retain a one-eighth of seven-eighths

**TABLE 1–3. Data for proposed drilling prospect in Example 1–29**

| Possible Outcomes | Probability of Occurrence | Conditional NPV at 10% | | |
|---|---|---|---|---|
| | | (1) | (2) | (3) |
| Dry Hole | 0.40 | \$150,000 | 0 | 0 |
| 2 BSCF Reserves | 0.30 | 60,000 | 14,000 | 12,000 |
| 3 BSCF Reserves | 0.20 | 90,000 | 17,000 | 38,000 |
| 4 BSCF Reserves | 0.10 | 120,000 | 21,000 | 55,000 |

ORRI on 320 net ac, or (3) participate with a backin privilege of acquiring 50% working interest after the participating parties recover 150% of their investment. If the 320 ac are acquired, which of the three decision alternatives will maximize the E(NPV), and how much (if any) should be paid for the 320-ac leasehold rights?

Solution:

$$
\begin{aligned}
E(NPV)_{Drill} &= 0.4(-150{,}000)(0.5) + (0.3)(60{,}000) + (0.2)(90{,}000) \\
&\quad + (0.1)(120{,}000) \\
&= \$18{,}000 \\
E(NPV)_{Farm\text{-}out} &= (0.3)(14{,}000) + (0.2)(17{,}000) + (0.1)(21{,}000) \\
&= \$9{,}700 \\
E(NPV)_{Backin} &= (0.3)(12{,}000) + (0.2)(38{,}000) + (0.1)(55{,}000) \\
&= \$16{,}700
\end{aligned}
$$



Since $E(NPV)_{Drill}$ is the highest, it is the preferred choice. The answer to the second question is that as much as $18,000 could be paid for the lease bonus and still expect a rate of return equal to the discount rate of 10%.

## EXPECTED PRESENT VALUE RATIO E(PVR)

Expected present value ratio E(PVR) is the ratio of the expected net present value to the expected present value of investments. As previously mentioned, E(PVR) provides a measure of expected profitability per dollar invested.

In the accept/reject decisions, a project is accepted if E(PVR) is greater than zero and rejected if it is less than zero. In dealing with mutually exclusive investment analysis, E(PVR) is applied on incremental basis, i.e., if the incremental E(PVR) is greater than or equal to zero, then the most capital intensive alternative is selected, and vice versa. In dealing with nonmutually exclusive investment analysis, projects are selected in order of decreasing E(PVR) until the total investment is within the budget constraint.

## EXPECTED NET PRESENT VALUE PROFILE

Expected net present value profile is a graph of E(NPV) as a function of the probability of finding hydrocarbons. This very useful correlation shows the effect of different probabilities on E(NPV). In majority of cases, an analyst may not be able to pinpoint the exact magnitude of probability associated with each outcome; however, by specifying a general range of probabilities for each possible outcome, a decision could be made using this profile.

The correlation is also a useful tool to portray to management the sensitivity of various E(NPV)s of the probability of finding hydrocarbons. The following example shows the steps required to construct the E(NPV) profile and interpret the results thus obtained.

**Example 1–30.**[5] An oil and gas company's geologist is evaluating a drilling prospect in a field having complex stratigraphic and lithologic variations within the field. Currently 15 wells have been drilled in the field ranging in estimated gas reserves from 2 to 5 BSCF per well. From the reservoir control developed by these wells and the heterogeneous nature of the accumulation, the geologist estimates the probability of drilling another producer is only 30%. The company's engineering staff developed the data presented in Table 1–4 for two strategies. The management wants to know (1) what is the best decision choice based on maximizing E(NPV), and (2) what minimum probability of finding gas is required to even justify the risk of drilling?

Solution:

(a) **Step 1.** Calculate probabilities for each possible outcome as

$$(P)_{\textit{Dry-hole}} = (1 - 0.30) = 0.70$$
$$(P)_{\text{2-BSCF}} = (5/15)(0.3) = 0.10$$
$$(P)_{\text{3-BSCF}} = (4/15)(0.3) = 0.08$$
$$(P)_{\text{4-BSCF}} = (3/15)(0.3) = 0.06$$
$$(P)_{\text{5-BSCF}} = (3/15)(0.3) = 0.06$$

**Step 2.** Calculate E(NPV) for each decision alternative as

$$\begin{aligned} E(NPV)_{\textit{Drill}} &= (0.7)(-150{,}000) + (0.1)(75{,}000) + (0.08)(120{,}000) \\ &\quad + (0.06)(150{,}000) + (0.06)(210{,}000) \\ &= -\$66{,}300 \\ E(NPV)_{\textit{Farm-out}} &= (0.1)(18{,}000) + (0.08)(29{,}000) + (0.06)(36{,}000) + (0.06)(50{,}000) \\ &= \$9{,}280 \end{aligned}$$

Therefore, farming out is the preferred choice.

**TABLE 1–4. Tabulation of gross per well gas reserves and associated economics for Example 1–30**

| Recoverable Gas, BSCF | Wells Having Given Reserves | NPV of Decision Alternatives: Drill | NPV of Decision Alternatives: Farm Out |
|---|---|---|---|
| 2 | 5 | $ 75,000 | $18,000 |
| 3 | 4 | 120,000 | 29,000 |
| 4 | 3 | 150,000 | 36,000 |
| 5 | 3 | 210,000 | 50,000 |
| | Dry Hole Cost = $150,000 | | |

(b) To construct the E(NPV) profile, calculate the E(NPV) for each strategy using, say, $(P)_{Dry\text{-}hole}$ = 0.50. Based on this, probabilities for each possible outcome have to be recalculated as

**Step 1.**

$$
\begin{aligned}
(P)_{Dry\text{-}hole} &= (1 - 0.05) = 0.50 \\
(P)_{2\text{-}BSCF} &= (5/15)(0.5) = 0.167 \\
(P)_{3\text{-}BSCF} &= (4/15)(0.5) = 0.133 \\
(P)_{4\text{-}BSCF} &= (3/15)(0.5) = 0.10 \\
(P)_{5\text{-}BSCF} &= (3/15)(0.5) = 0.10
\end{aligned}
$$

**Step 2.** Recalculate E(NPV) for each decision alternative as

$$
\begin{aligned}
E(NPV)_{Drill} &= (0.5)(-150{,}000) + (0.167)(75{,}000) + (0.1)(120{,}000) \\
&\quad + (0.1)(150{,}000) + (0.1)(210{,}000) \\
&= -\$10{,}515 \\
E(NPV)_{Farm\text{-}out} &= (0.167)(18{,}000) + (0.133)(29{,}000) + (0.1)(36{,}000) \\
&\quad + (0.1)(50{,}000) \\
&= \$15{,}463
\end{aligned}
$$

**Step 3.** Using the two E(NPV)s for each strategy, prepare a graph as shown in Figure 1–21.

In Figure 1–21, the intersection of the two straight lines (Point A) shows that the minimum probability of finding gas required to justify the risk of drilling is 0.61, i.e., $(P)_{Dry\text{-}hole}$ = 0.39. For probabilities of finding gas greater than 0.61, the drill strategy is preferred, and for any probability of finding gas less than 0.61, the farmout strategy will be the preferred choice.

The minimum probability of finding hydrocarbons required to justify drilling, i.e., the point of Intersection A in Figure 1–21, can also be mathematically calculated by equating the *conditional probabilities* of the two strategies.

$$
\begin{aligned}
E(NPV)_{Drill} &= P(G)[(5/15)(75{,}000) + (4/15)(120{,}000) + (3/15)(150{,}000) \\
&\quad + (3/15)(210{,}000)] + [1 - P(G)](-150{,}000) \\
E(NPV)_{Farm\text{-}out} &= P(G)[(5/15)(18{,}000) + (4/15)(29{,}000) + (3/15)(36{,}000) \\
&\quad + (3/15)(50{,}000)]
\end{aligned}
$$

By equating $E(NPV)_{Drill}$ and $E(NPV)_{Farm\text{-}out}$ we solve for $P(G)$ as

$$
\begin{aligned}
248{,}078\, P(G) &= 150{,}000 \\
P(G) &= 0.605 \approx 0.61
\end{aligned}
$$

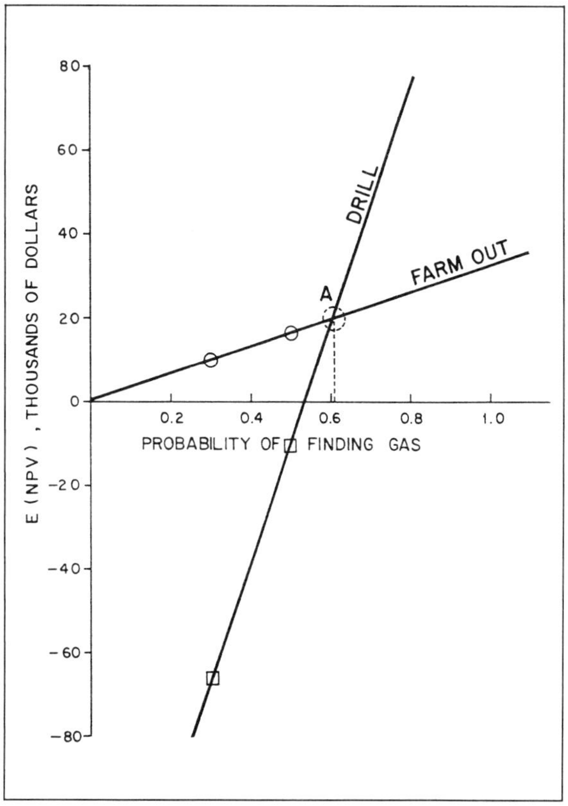


**Figure 1–21.** Expected net present value profile for data in Example 1–30.

## DECISION TREE ANALYSIS

An alternative method to the analysis of expected value calculations presented in this section is through the use of *decision tree.* A decision tree is a diagrammatic presentation expressing in chronological order the alternative actions that are available to the decision maker and the choices determined by chance.

Decision trees permit a ready analysis of what might be an otherwise confusing and tortuous sequence of decision alternatives. Decision trees consist of forks (nodes) and branches. The forks are subdivided into two types as decision forks and chance forks, the two are typically represented by *square* and *circle,* respectively. Branches are straight lines that emanate from decision forks or chance forks. The ends of a decision tree are called *terminal points;* i.e., there are no further contingencies, decisions, or chance events beyond this point. A typical, rather simple decision tree is shown in Figure 1–22.

When a decision maker encounters a decision fork he must choose one of the alternative branches to travel on. When a chance fork is encountered, there is no control over which branch to travel on. Instead, the travel path is determined by chance events whose probabilities are those associated with the branches that emanate from the chance fork.

Decision tree analysis is one of the few things that is analyzed by starting at the end and working backward. To explain decision tree analysis, let us solve the problem presented in Example 1–30, using the decision tree and the following steps:

**Step 1.** Construct the decision tree showing the events with the chance nodes and decision nodes.

**Step 2.** Associate probabilities of occurrence to all the branches emanating from chance nodes. The sum of the probabilities around a given chance node must add to 1.0. Note that no probabilities are shown on the branches emanating from a decision tree.

**Step 3.** Specify the conditional values (NPVs in this case) received at the terminal points.
The constructed decision tree is shown in Figure 1–23. Now we start with solving this decision tree. The E(NPV) at the chance node 1 is called the *conditional* E(NPV) *given* that gas is found for the drill strategy.

**Step 4.** Compute E(NPV) at chance node 1 as

$$
\begin{aligned}
E(NPV)_1 &= (0.333)(75{,}000) \\
&\quad + (0.267)(120{,}000) \\
&\quad + (0.20)(150{,}000) \\
&\quad + (0.2)(210{,}000) \\
&= \$129{,}015
\end{aligned}
$$

**Step 5.** In the same way compute $E(NPV)_3$ for the chance node 3:

$$
\begin{aligned}
E(NPV)_3 &= (0.333)(18{,}000) \\
&\quad + (0.267)(29{,}000) \\
&\quad + (0.2)(36{,}000) \\
&\quad + (0.2)(50{,}000) \\
&= \$30{,}937
\end{aligned}
$$

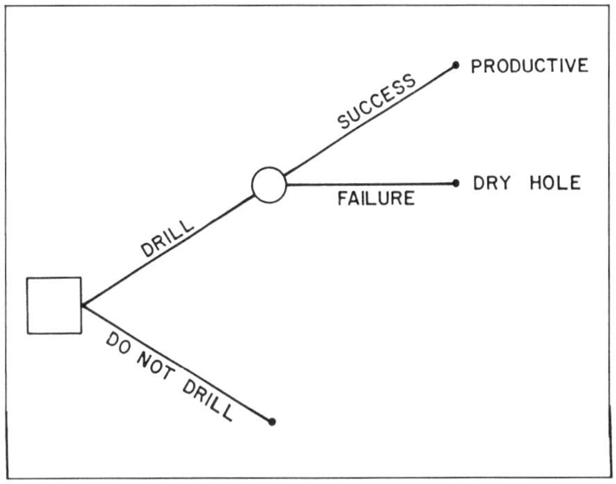


**Figure 1–22.** A typical decision tree.



The $E(NPV)_3$ at chance node 3 is also called the *conditional* $E(NPV)_3$, given that gas is found for the farmout strategy.

**Step 6.** Now we start with the second roll back at chance nodes 2 and 4. Compute the $E(NPV)_2$ at chance node 2:

$$
\begin{aligned}
E(NPV)_2 &= (0.7)(-70{,}000) + (0.30)(129{,}015) \\
&= -\$10{,}296
\end{aligned}
$$

**Step 7.** Compute the $E(NPV)_4$ at chance node 4:

$$
\begin{aligned}
E(NPV)_4 &= (0.7)(0) + (0.3)(30{,}937) \\
&= \$9{,}281
\end{aligned}
$$

**Step 8.** This is the decision-making step since we have rolled back to the decision node. Since $E(NPV)_2$ at chance node 2 is negative (i.e., suboptimal), we discard the drill strategy option and select the farmout option.

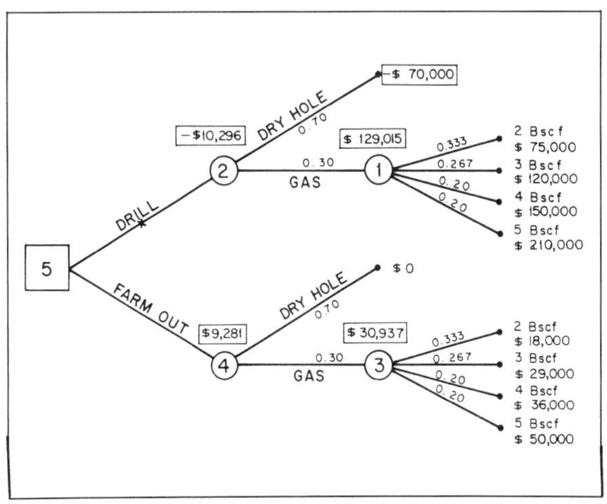


**Figure 1–23.** Completed decision tree for the situation in Example 1–30 (after Paul D. Newendorp).[5]

## EQUATIONS AND MATHEMATICAL TECHNIQUES

This section summarizes some of the important statistical concepts that are widely used in day-to-day engineering analysis. The topics discussed are regression analysis (obtaining a best-fit line through data); some common characteristics of univariate data sets, such as mean, variance, and standard deviation; log normal distribution—the most commonly encountered probability distribution in petroleum engineering problems; and a technique which determines the unknown constants in Y or X for producing Y versus X as a straight line.

## REGRESSION ANALYSIS

In this section we will deal with acquiring a prediction equation based upon one or more independent variables. We shall show how to fit a *simple linear* model to a set of data, a process called a *regression analysis*. The application of simple linear regression to the following correlations will be presented:

1. Straight line linear regression, $y = a + bx$
2. Exponential curve, $y = ae^{bx}$
3. Logarithmic curve, $y = a + b \ln x$
4. Power curve, $y = ax^b$

A least-squares method of regression analysis can be used either to the original equations (straight line and logarithmic curve) or to the transformed equations (exponential and power curves). This statistical procedure for finding the "best fitting" straight line for a set of points is, in many respects, a formalization of the procedure mostly applied by petroleum engineers when they fit a line by eye (eyeballing).

The principle of least squares is commonly used in regression analysis, which states, *choose as the best fitting line the line that minimizes the sum of squares of the deviations of the observed values of* Y *from those predicted.* Expressed mathematically, we wish to choose values for $a$ and $b$ in the correlations above that will minimize

$$\text{SSE} = \sum_{i=1}^{n} (y_i - \hat{y}_i)^2 \tag{1.33}$$

where SSE represents the sum of squares of deviations, $y_i$ and $\hat{y}_i$ represent the observed and predicted values of the variable $y$, respectively. The graph of the regression function is commonly referred to as the *regression line* or *curve*. The parameters $a$ and $b$ are called *regression parameters*, where $a$ is intercept and $b$ is the slope of the regression line. These two parameters are obtained by applying the following equations to the sample data to be regressed:

$$b = \frac{\sum_{i=1}^{n} (x_i - \bar{x})(y_i - \bar{y})}{\sum_{i=1}^{n} (x_i - \bar{x})^2} \tag{1.34}$$

$$a = \bar{y} - b\bar{x} \tag{1.35}$$

We now present two more measures that are widely used to describe the degree of relationship between the variables X and Y as reflected in the simple linear regression model. These measures are called *coefficient of simple determination* ($r^2$) and *coefficient of simple correlation* ($r$). Mathematically,

$$r^2 = \frac{a \sum_{i=1}^{n} y_i + b \sum_{i=1}^{n} x_i y_i - \frac{1}{n}(\sum_{i=1}^{n} y_i)^2}{\sum_{i=1}^{n} (y_i)^2 - \frac{1}{n}(\sum_{i=1}^{n} y_i)^2} \quad \textbf{(1.36)}$$

and

$$r = \pm \sqrt{r^2} \quad \textbf{(1.37)}$$

The $r^2$ values range from 0 to 1 ($0 \leq r^2 \leq 1$) where $r^2 = 1$ shows perfect correlation and $r^2 = 0$ shows no correlation in the observations. The $r$ values range from $-1 \leq r \leq 1$, and the sign attached to $r$ simply shows whether the slope of the regression line is positive or negative. The closer the observed value of $r$ is to 1, the greater is the degree of linear statistical relation in the sample observations.



In Chapter 4, we will see the application of regression analysis to gas material balance problem where we regress $P/Z$ versus $G_p$. Other applications include regressing rate-time data and straight line portion of Horner plot.

## MEAN, VARIANCE, AND STANDARD DEVIATION

In this section we will discuss properties of collected data and show how to compute and interpret commonly used measures for central tendency (mean) and dispersion around the mean (variance and standard deviation). All the properties discussed are frequently used to describe the outcome of experiments or tests.

The *average, mean,* or *expected value* are all terms used to describe a very common measure of *central tendency* of data. Central tendency is a definite indication that the data seems to *cluster* around some central value. The average of a variable $X$, symbol $\overline{X}$, and a sample of size $n$ is computed by summing the values and dividing by $n$:

$$\overline{X} = \frac{\sum_{i=1}^{n} X_i}{n} \quad \textbf{(1.38)}$$

When a set of items is divided into two or more subsets, the overall mean is a weighted average of the means of the subsets:

$$\overline{X} = \frac{f_1\overline{X}_1 + f_2\overline{X}_2 + \cdots f_k\overline{X}_k}{n} \quad \textbf{(1.39)}$$

where $k$ is the number of subsets, $\overline{X}_i$ is the mean of the $i$th subset, $f_i$ is the number of items in the $i$th subset, and $n = \sum_{i=1}^{k} f_i$ is the total number of items. When data are presented only in terms of a frequency distribution, an approximation of the mean $\overline{X}$ can be obtained as follows:

$$\overline{X} \approx \frac{\sum_{i=1}^{k} f_i M_i}{n} \tag{1.40}$$

where $k$ is the number of classes in the frequency distribution, $M_i$ is the midpoint of the $i$th class, $f_i$ is the frequency of the $i$th class, and $n = \sum_{i=1}^{k} f_i$. When there is wide variation in data points, the simple arithmetic average discussed above will give optimistic mean values. In such cases *geometric mean* and *harmonic mean* may provide appropriate measures of central tendency.

Geometric Mean:

$$G = N\sqrt{X_1.X_2.X_3 \cdots X_4} \tag{1.41}$$



Harmonic Mean:

$$H = \frac{N}{\sum_{i=1}^{N}\left(\frac{1}{x_i}\right)} \tag{1.42}$$

The most commonly used measure of dispersion in statistical analysis is called the variance. It is a measure that takes into account all the values in a set of items and provides quantified information about the variability in a data set. The variance, symbol $s^2$, of a set of values $X_1, X_2, \cdots, X_n$ is defined as:

$$s^2 = \frac{\sum_{i=1}^{k} X_i^2}{n-1} - \frac{n}{n-1}\overline{X}^2 \tag{1.43}$$

When the data are presented in a frequency distribution, only an approximation to the variance can be obtained as:

$$s^2 \approx \frac{\sum_{i=1}^{k} f_i (M_i - \overline{X})^2}{n-1} \tag{1.44}$$

Another measure of dispersion is called *standard deviation* around a mean. The positive square root of the variance is called standard deviation and is denoted by $s = \sqrt{s^2}$.

With variance and standard deviation, the greater the dispersion of the values around their mean, the larger the variance and standard deviation. If there is no dispersion (i.e., if all values are equal), the standard deviation and variance are zero. A large standard deviation is an indication of a wide spread of data with a possibility of some extreme values. If the standard deviation is small, the data points will be more closely clustered around the center of the data.

## NORMAL PROBABILITY DISTRIBUTION

Normal probability distribution is a typical bell-shaped curve symmetric about the mean as shown in Figure 1–24a. The distribution is defined by the two single value parameters, mean and standard deviation. The first standard deviation on either side of the mean includes 34.13% of the elements in the distribution, and the probability of the random variable falling *within* one standard deviation is therefore 68.26%. Similarly the probability of being within two standard deviations of the mean is 95.44%. Only 26% of the elements are more than three standard deviations from the mean. The total area under the curve is unity with an area of 0.5 on each side of the mean.

Cumulative probability or area under the curve ranging from $-\infty$ up to the particular value of $z$ is given in Table 1–5. Probabilities for any *normal random variable* (data must be normally distributed) can be obtained from Table 1–5 for the standard normal variable $z$ defined as

$$z = \frac{X - \overline{X}}{s} \tag{1.45}$$



The following example will clarify the use of Table 1–5 and the normal probability distribution.

**Example 1–31.** A certain data set of core porosities showed an average porosity of 15% with a standard deviation of 4%. What is the probability that a porosity will be within the range of 12% and 22%? What is the probability that a porosity will be less than 12%?

Solution: First we must compute $z$ for each $X$ as

$$z_{12\%} = \frac{12 - 15}{4} = -0.750$$

$$z_{22\%} = \frac{22 - 15}{4} = 1.750$$

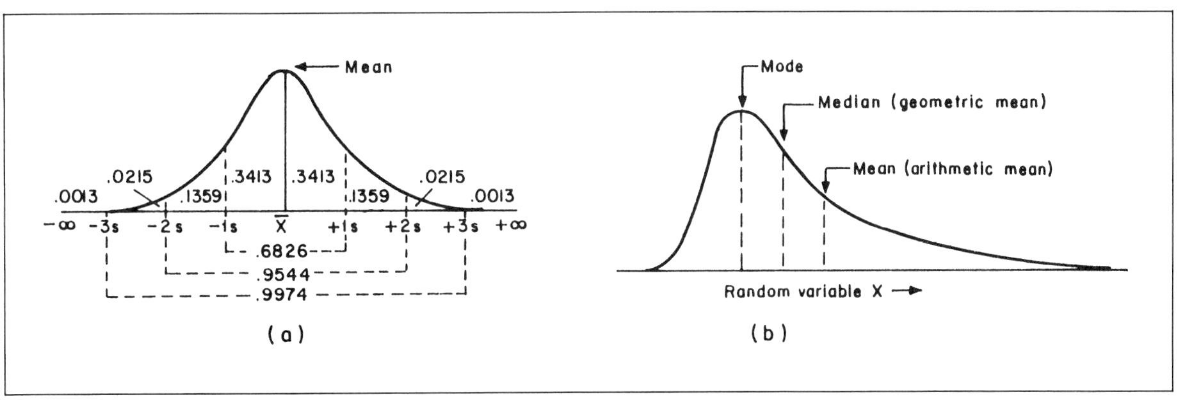


**Figure 1–24.** Examples of probability distribution.

**TABLE 1–5. Areas under the normal distribution less than or equal to the value of $X$ used to compute $z$**

| $z$ | .00 | .01 | .02 | .03 | .04 | .05 | .06 | .07 | .08 | .09 |
|---|---|---|---|---|---|---|---|---|---|---|
| −3.5 | .0002 | .0002 | .0002 | .0002 | .0002 | .0002 | .0002 | .0002 | .0002 | .0002 |
| −3.4 | .0003 | .0003 | .0003 | .0003 | .0003 | .0003 | .0003 | .0003 | .0003 | .0002 |
| −3.3 | .0005 | .0005 | .0005 | .0004 | .0004 | .0004 | .0004 | .0004 | .0004 | .0003 |
| −3.2 | .0007 | .0007 | .0006 | .0006 | .0006 | .0006 | .0006 | .0005 | .0005 | .0005 |
| −3.1 | .0010 | .0009 | .0009 | .0009 | .0008 | .0008 | .0008 | .0008 | .0007 | .0007 |
| −3.0 | .0013 | .0013 | .0013 | .0012 | .0012 | .0011 | .0011 | .0011 | .0010 | .0010 |
| −2.9 | .0019 | .0018 | .0018 | .0017 | .0016 | .0016 | .0015 | .0015 | .0014 | .0014 |
| −2.8 | .0026 | .0025 | .0024 | .0023 | .0023 | .0022 | .0021 | .0021 | .0020 | .0019 |
| −2.7 | .0035 | .0034 | .0033 | .0032 | .0031 | .0030 | .0029 | .0028 | .0027 | .0026 |
| −2.6 | .0047 | .0045 | .0044 | .0043 | .0041 | .0040 | .0039 | .0038 | .0037 | .0036 |
| −2.5 | .0062 | .0060 | .0059 | .0057 | .0055 | .0054 | .0052 | .0051 | .0049 | .0048 |
| −2.4 | .0082 | .0080 | .0078 | .0075 | .0073 | .0071 | .0069 | .0068 | .0066 | .0064 |
| −2.3 | .0107 | .0104 | .0102 | .0099 | .0096 | .0094 | .0091 | .0089 | .0087 | .0084 |
| −2.2 | .0139 | .0136 | .0132 | .0129 | .0125 | .0122 | .0119 | .0116 | .0113 | .0110 |
| −2.1 | .0179 | .0174 | .0170 | .0166 | .0162 | .0158 | .0154 | .0150 | .0146 | .0143 |
| −2.0 | .0228 | .0222 | .0217 | .0212 | .0207 | .0202 | .0197 | .0192 | .0188 | .0183 |
| −1.9 | .0287 | .0281 | .0274 | .0268 | .0262 | .0256 | .0250 | .0244 | .0239 | .0233 |
| −1.8 | .0359 | .0351 | .0344 | .0336 | .0329 | .0322 | .0314 | .0307 | .0301 | .0294 |
| −1.7 | .0446 | .0436 | .0427 | .0418 | .0409 | .0401 | .0392 | .0384 | .0375 | .0367 |
| −1.6 | .0548 | .0537 | .0526 | .0516 | .0505 | .0495 | .0485 | .0475 | .0465 | .0455 |
| −1.5 | .0668 | .0655 | .0643 | .0630 | .0618 | .0606 | .0594 | .0582 | .0571 | .0559 |
| −1.4 | .0808 | .0793 | .0778 | .0764 | .0749 | .0735 | .0721 | .0708 | .0694 | .0681 |
| −1.3 | .0968 | .0951 | .0934 | .0918 | .0901 | .0885 | .0869 | .0853 | .0838 | .0823 |
| −1.2 | .1151 | .1131 | .1112 | .1093 | .1075 | .1056 | .1038 | .1020 | .1003 | .0985 |
| −1.1 | .1357 | .1335 | .1314 | .1292 | .1271 | .1251 | .1230 | .1210 | .1190 | .1170 |
| −1.0 | .1587 | .1562 | .1539 | .1515 | .1492 | .1469 | .1446 | .1423 | .1401 | .1379 |
| −0.9 | .1841 | .1814 | .1788 | .1762 | .1736 | .1711 | .1685 | .1660 | .1635 | .1611 |
| −0.8 | .2119 | .2090 | .2061 | .2033 | .2005 | .1977 | .1949 | .1922 | .1894 | .1867 |
| −0.7 | .2420 | .2389 | .2358 | .2327 | .2296 | .2266 | .2236 | .2206 | .2177 | .2148 |
| −0.6 | .2743 | .2709 | .2676 | .2643 | .2611 | .2578 | .2546 | .2514 | .2483 | .2451 |
| −0.5 | .3085 | .3050 | .3015 | .2981 | .2946 | .2912 | .2877 | .2843 | .2810 | .2776 |
| −0.4 | .3446 | .3409 | .3372 | .3336 | .3300 | .3264 | .3228 | .3192 | .3156 | .3121 |
| −0.3 | .3821 | .3783 | .3745 | .3707 | .3669 | .3632 | .3594 | .3557 | .3520 | .3483 |
| −0.2 | .4207 | .4168 | .4129 | .4090 | .4052 | .4013 | .3974 | .3936 | .3897 | .3859 |
| −0.1 | .4602 | .4562 | .4522 | .4483 | .4443 | .4404 | .4364 | .4325 | .4286 | .4247 |
| −0.0 | .5000 | .4960 | .4920 | .4880 | .4840 | .4801 | .4761 | .4721 | .4681 | .4641 |

**TABLE 1–5. (continued)**

| z | .00 | .01 | .02 | .03 | .04 | .05 | .06 | .07 | .08 | .09 |
|---|---|---|---|---|---|---|---|---|---|---|
| +0.0 | .5000 | .5040 | .5080 | .5120 | .5160 | .5199 | .5239 | .5279 | .5319 | .5359 |
| +0.1 | .5398 | .5438 | .5478 | .5517 | .5557 | .5596 | .5636 | .5675 | .5714 | .5753 |
| +0.2 | .5793 | .5832 | .5871 | .5910 | .5948 | .5987 | .6026 | .6064 | .6103 | .6141 |
| +0.3 | .6179 | .6217 | .6255 | .6293 | .6331 | .6368 | .6406 | .6443 | .6480 | .6517 |
| +0.4 | .6554 | .6591 | .6628 | .6664 | .6700 | .6736 | .6772 | .6808 | .6844 | .6879 |
| +0.5 | .6915 | .6950 | .6985 | .7019 | .7054 | .7088 | .7123 | .7157 | .7190 | .7224 |
| +0.6 | .7257 | .7291 | .7324 | .7357 | .7389 | .7422 | .7454 | .7486 | .7517 | .7549 |
| +0.7 | .7580 | .7611 | .7642 | .7673 | .7704 | .7734 | .7764 | .7794 | .7823 | .7852 |
| +0.8 | .7881 | .7910 | .7939 | .7967 | .7995 | .8023 | .8051 | .8078 | .8106 | .8133 |
| +0.9 | .8159 | .8186 | .8212 | .8238 | .8264 | .8289 | .8315 | .8340 | .8365 | .8389 |
| +1.0 | .8413 | .8438 | .8461 | .8485 | .8508 | .8531 | .8554 | .8577 | .8599 | .8621 |
| +1.1 | .8643 | .8665 | .8686 | .8708 | .8729 | .8749 | .8770 | .8790 | .8810 | .8830 |
| +1.2 | .8849 | .8869 | .8888 | .8907 | .8925 | .8944 | .8962 | .8980 | .8997 | .9015 |
| +1.3 | .9032 | .9049 | .9066 | .9082 | .9099 | .9115 | .9131 | .9147 | .9162 | .9177 |
| +1.4 | .9192 | .9207 | .9222 | .9236 | .9251 | .9265 | .9279 | .9292 | .9306 | .9319 |
| +1.5 | .9332 | .9345 | .9357 | .9370 | .9382 | .9394 | .9406 | .9418 | .9429 | .9441 |
| +1.6 | .9452 | .9463 | .9474 | .9484 | .9495 | .9505 | .9515 | .9525 | .9535 | .9545 |
| +1.7 | .9554 | .9564 | .9573 | .9582 | .9591 | .9599 | .9608 | .9616 | .9625 | .9633 |
| +1.8 | .9641 | .9649 | .9656 | .9664 | .9671 | .9678 | .9686 | .9693 | .9699 | .9706 |
| +1.9 | .9713 | .9719 | .9726 | .9732 | .9738 | .9744 | .9750 | .9756 | .9761 | .9767 |
| +2.0 | .9772 | .9778 | .9783 | .9788 | .9793 | .9798 | .9803 | .9808 | .9812 | .9817 |
| +2.1 | .9821 | .9826 | .9830 | .9834 | .9838 | .9842 | .9846 | .9850 | .9854 | .9857 |
| +2.2 | .9861 | .9864 | .9868 | .9871 | .9875 | .9878 | .9881 | .9884 | .9887 | .9890 |
| +2.3 | .9893 | .9896 | .9898 | .9901 | .9904 | .9906 | .9909 | .9911 | .9913 | .9916 |
| +2.4 | .9918 | .9920 | .9922 | .9925 | .9927 | .9929 | .9931 | .9932 | .9934 | .9936 |
| +2.5 | .9938 | .9940 | .9941 | .9943 | .9945 | .9946 | .9948 | .9949 | .9951 | .9952 |
| +2.6 | .9953 | .9955 | .9956 | .9957 | .9959 | .9960 | .9961 | .9962 | .9963 | .9964 |
| +2.7 | .9965 | .9966 | .9967 | .9968 | .9969 | .9970 | .9971 | .9972 | .9973 | .9974 |
| +2.8 | .9974 | .9975 | .9976 | .9977 | .9977 | .9978 | .9979 | .9979 | .9980 | .9981 |
| +2.9 | .9981 | .9982 | .9982 | .9983 | .9984 | .9984 | .9985 | .9985 | .9986 | .9986 |
| +3.0 | .9987 | .9987 | .9987 | .9988 | .9988 | .9989 | .9989 | .9989 | .9990 | .9990 |
| +3.1 | .9990 | .9991 | .9991 | .9991 | .9992 | .9992 | .9992 | .9992 | .9993 | .9993 |
| +3.2 | .9993 | .9993 | .9994 | .9994 | .9994 | .9994 | .9994 | .9995 | .9995 | .9995 |
| +3.3 | .9995 | .9995 | .9995 | .9996 | .9996 | .9996 | .9996 | .9996 | .9996 | .9997 |
| +3.4 | .9997 | .9997 | .9997 | .9997 | .9997 | .9997 | .9997 | .9997 | .9997 | .9998 |
| +3.5 | .9998 | .9998 | .9998 | .9998 | .9998 | .9998 | .9998 | .9998 | .9998 | .9998 |

Reprinted with permission from *Engineering Statistics and Quality Control*, by Irving W. Burr, McGraw-Hill, 1953

Next we find the cumulative area to the left of each X value from Table 1–5 as

$$a_{12\%} \;@\; z_{12\%} = 0.2266$$
$$a_{22\%} \;@\; z_{22\%} = 0.9599$$

Since we are interested in the area between $a_{12\%}$ and $a_{22\%}$, we compute this area by simply substracting area of $a_{12\%}$ from the area of $a_{22\%}$ as

$$\begin{aligned} a_{22\%} - a_{12\%} &= 0.9599 - 0.2266 \\ &= 0.7333 \quad \approx 0.73 \end{aligned}$$



So the probability of finding porosity values between 12% and 22% is 73%. The probability of finding porosity less than 12% is 22.66%, and the probability of finding porosity greater than 22% is only 4% (1 − 0.9599 = 0.0401).

The $\overline{X}$ and $s$ parameters are computed using the equations presented in the previous section, or a graphical method may be used. The graphical method involves either (1) plotting cumulative frequency percent versus random variable X on a special probability-linear graph paper or (2) plotting cumulative frequency percent versus random variable on coordinate graph paper. The cumulative frequency distribution plotted in (1) should give a straight line and in (2) it will be an S-shaped curve. To illustrate the use of one of the graphs, let us consider the porosity data given in Table 1–6.

The first step in preparing the normal probability plot is to group the data as shown in Table 1–7. The computed data is then plotted on probability-linear graph paper as shown in Figure 1–25.

**TABLE 1–6. Porosity data from core analysis[5]**

| Depth | ϕ | Depth | ϕ | Depth | ϕ |
|---|---|---|---|---|---|
| 4,812.5 | 24.0 | 4,825.5 | 15.1 | 4,838.5 | 20.0 |
| 4,813.5 | 23.3 | 4,826.5 | 14.0 | 4,839.5 | 16.8 |
| 4,814.5 | 16.1 | 4,827.5 | 15.6 | 4,840.5 | 13.3 |
| 4,815.5 | 17.2 | 4,828.5 | 15.5 | 4,841.5 | 18.0 |
| 4,816.5 | 15.3 | 4,829.5 | 19.4 | 4,842.5 | 16.1 |
| 4,817.5 | 15.9 | 4,830.5 | 17.5 | 4,843.5 | 15.1 |
| 4,818.5 | 18.6 | 4,831.5 | 16.4 | 4,844.5 | 18.0 |
| 4,819.5 | 16.2 | 4,832.5 | 17.2 | 4,845.5 | 15.7 |
| 4,820.5 | 20.0 | 4,833.5 | 15.5 | 4,847.5 | 17.8 |
| 4,821.5 | 16.9 | 4,834.5 | 20.2 | 4,848.5 | 9.2 |
| 4,822.5 | 18.1 | 4,835.5 | 18.3 | 4,849.5 | 20.5 |
| 4,823.5 | 20.3 | 4,836.5 | 19.6 | 4,850.5 | 8.4 |
| 4,824.5 | 15.3 | 4,837.5 | 17.7 | 4,851.5 | 1.1 |

**TABLE 1–7. Cumulative frequency computations for the data in Table 1–6**

| Porosity Interval (%) | Frequency | Cumulative Frequency | Cumulative Frequency (%) |
|---|---|---|---|
| 7.0—under 10.0 | 2 | 2 | 5.3 |
| 10.0—under 13.0 | 0 | 2 | 5.3 |
| 13.0—under 16.0 | 11 | 13 | 34.2 |
| 16.0—under 19.0 | 16 | 29 | 76.3 |
| 19.0—under 22.0 | 7 | 36 | 94.7 |
| 22.0—under 25.0 | 2 | 38 | 100.0 |

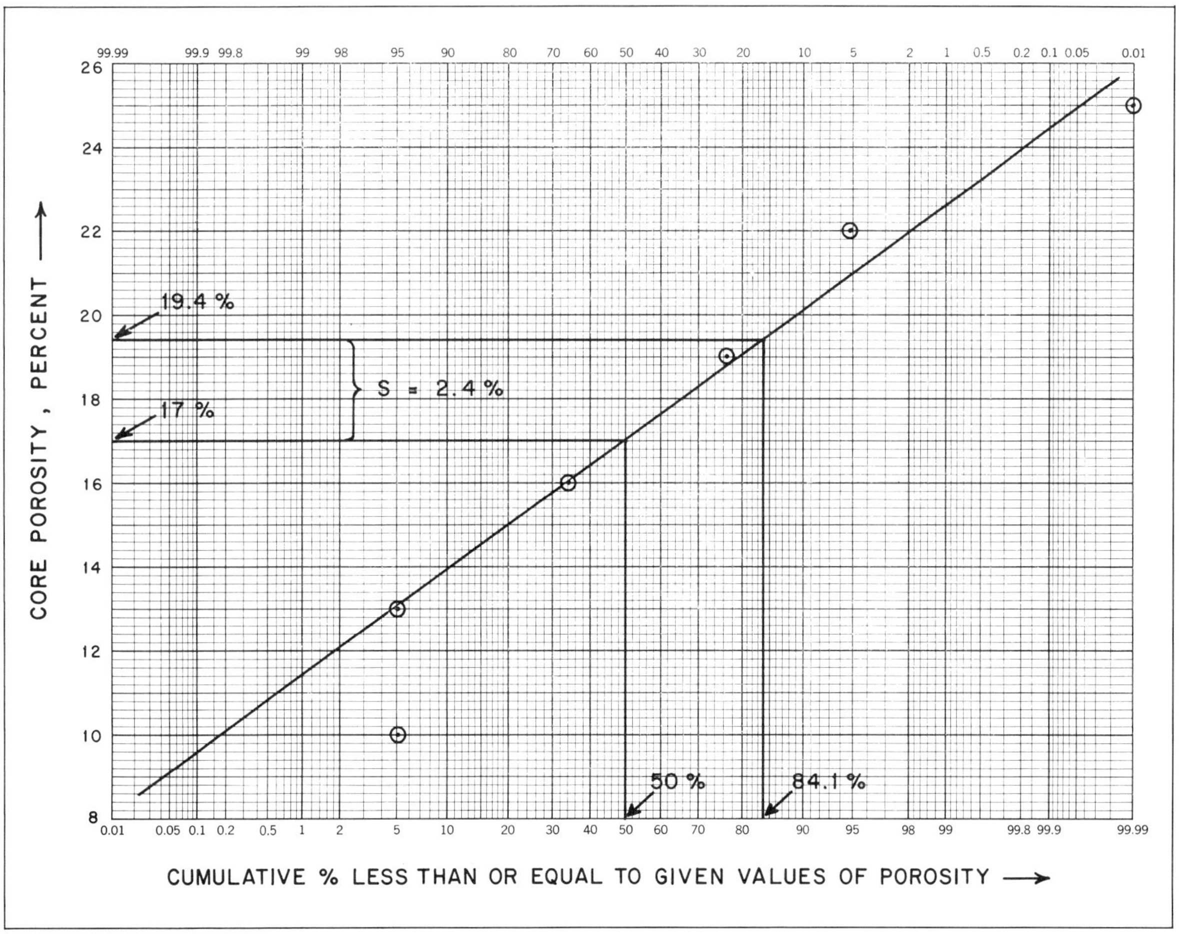


**Figure 1–25.** Cumulative frequency graph of the porosity data in Table 1–7.

A best-fit line is then drawn through the data points as shown. The mean $\overline{X}$ then is read directly from the graph at 50 cumulative frequency percent, and the standard deviation is the difference between the porosity value read at 84.1 cumulative frequency percent and the mean.

The mean and standard deviation for the same data could also be calculated by using Equations (1.40) and (1.44), respectively. To do this, prepare the data of Table 1–7 as shown in Table 1–8.

Using Equation (1.44),

$$s = \sqrt{\frac{387.44}{38 - 1}} \approx 3.24$$

By using Equations (1.38) and (1.43), the porosity data given in Table 1–6 calculates a mean of 17.06% and a standard deviation of 3.05%. By using the frequency distribution, we obtain a mean of 17.03% and an approximate standard deviation of 3.24%. *From the graph, however, the standard deviation is not even close to 3.0%.* This is obvious from the wide scatter of data points in Figure 1–25, which means that the data is not perfectly normally distributed.

## LOG-NORMAL PROBABILITY DISTRIBUTION

Log-normal distribution is the transformed version of the normal distribution, i.e., if the *logarithm* of a variable is *normally distributed,* it will show log-normal distribution. The distribution appears similar to the normal distribution except that it is skewed in one direction as shown in Figure 1–24b. The log-normal distribution usually characterizes a wide variation in the data. In the oil industry, permeability, reservoir thickness, field size, ultimate recovery, etc., are believed to be log-normally distributed.

The log-normally distributed data can be analyzed using Equations (1.40) and (1.44) and

**TABLE 1–8. Computation of mean and standard deviation of porosity data in Table 1–7**

| Frequency $f_i$ | Class Mid-point, $M_i$ | $f_iM_i$ | Deviation $(M_i - \overline{X})$ | $(M_i - \overline{X})^2$ | $f_i(M_i - \overline{X})^2$ |
|---|---|---|---|---|---|
| 2 | 8.5 | 17 | −8.53 | 72.76 | 145.52 |
| 0 | 11.5 | 0 | −5.53 | 30.58 | 0 |
| 11 | 14.5 | 159.5 | −2.53 | 6.40 | 70.40 |
| 16 | 17.5 | 280 | 0.47 | 0.22 | 3.52 |
| 7 | 20.5 | 143.5 | 3.47 | 12.04 | 84.28 |
| 2 | 23.5 | 47.0 | 6.47 | 41.86 | 83.72 |
| 38 | | 647.0 | | | 387.44 |

$\overline{X}$ = 647.0/38 = 17.03

transforming $y = \ln x$. The distribution of the transformed variable $y$ will be normally distributed with $x$. The data could be interpreted graphically by either (1) preparing cumulative frequency graph on a *special log-probability* paper or (2) plotting *percentage of variable greater than or less than* versus the variable on log-probability graph paper. This author has experienced that plotting *percent of sample greater than* (linear) versus variable (log scale) on a regular semilogarithmic graph paper in most cases will provide satisfactory results.

Statistical properties of interest can be calculated from the graphs by using the following relationships. The mean read from the log-normal graph at 50% is the geometric mean (or median), the same as would be calculated by using Equation (1.41).

$$\overline{X} = e^{\alpha + 0.5\beta^2} \quad \textbf{(1.46)}$$

$$s^2 = \overline{X}^2(e^{\beta^2} - 1) \quad \textbf{(1.47)}$$

$$\text{mode of } X = \frac{\overline{X}}{e^{1.5\beta^2}} \quad \textbf{(1.48)}$$



where $\alpha$ is the natural logarithm of the value read at 50% probability line, and $\beta$ is the standard deviation of the logarithmic distribution. The standard deviation $\beta$ can be calculated by reading two variable values ($P_1$ and $P_2$) at any two selected percentiles and then using the following equation:

$$\beta = \frac{\ln(P_1) - \ln(P_2)}{z(a_2) - z(a_1)} \quad \textbf{(1.49)}$$

where $z(a_2)$ and $z(a_1)$ are read from the percentiles of the standard normal distribution at the respective percentiles from Table 1–9.

**Example 1–32.** Using the concept of log-normal probability distribution, compute the variable parameters for the estimated ultimate recovery of fields in the Los Angeles Basin given in Table 1–10. Use the log-probability graph paper.

Solution: The data of Table 1–10 are plotted on a log-probability graph paper as shown in Figure 1–26.

**TABLE 1–9. Percentiles of the standard normal distribution[7]**

| a: | 0.10 | 0.05 | 0.025 | 0.02 | 0.01 | 0.005 | 0.001 |
|---|---|---|---|---|---|---|---|
| z(a) | −1.282 | −1.645 | −1.960 | −2.054 | −2.326 | −2.576 | −3.090 |
| a: | 0.90 | 0.95 | 0.975 | 0.98 | 0.99 | 0.995 | 0.999 |
| z(a) | 1.282 | 1.645 | 1.960 | 2.054 | 2.326 | 2.576 | 3.090 |

**TABLE 1–10. Estimated ultimate recovery for fields in the Los Angeles Basin**

| Case | Ultimate Recovery (MB) | Percent Greater Than | Case | Ultimate Recovery (MB) | Percent Greater Than |
|---|---|---|---|---|---|
| 1 | 2,600,813 | 3 | 8 | 241,466 | 54 |
| 2 | 970,197 | 10 | 9 | 194,826 | 61 |
| 3 | 896,821 | 17 | 10 | 189,909 | 68 |
| 4 | 611,928 | 25 | 11 | 187,607 | 75 |
| 5 | 367,800 | 32 | 12 | 183,217 | 83 |
| 6 | 316,803 | 39 | 13 | 111,647 | 90 |
| 7 | 268,497 | 46 | 14 | 102,470 | 97 |

Source: Calculated from data in Oil & Gas Journal, Jan. 25, 1971, p. 124.

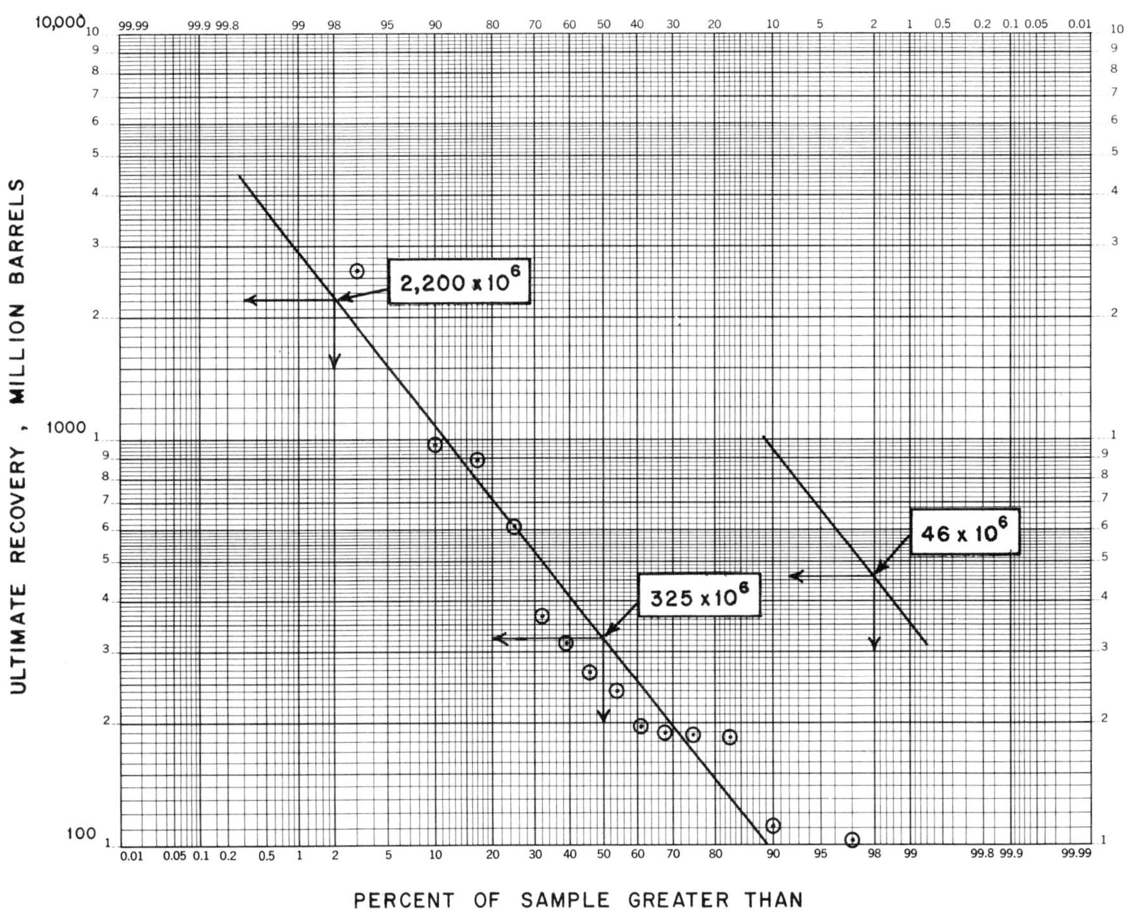


**Figure 1–26.** Ultimate recovery versus percent sample greater than plot for the data in Table 1–10.

After drawing a best-fit line through the data points, values at 2, 50, and 98 percentiles are read directly from the curve as 2,200 × $10^6$, 325 × $10^6$, and 46 × $10^6$, respectively.

$$\alpha = \ln(325 \times 10^6)$$
$$= 19.60$$

Referring to Table 1–9, standard deviations at 2 and 98 percentiles are −2.054 and +2.054, respectively.

$$\beta = \frac{\ln(2200 \times 10^6) - \ln(46 \times 10^6)}{2.054 - (-2.054)}$$
$$= \frac{21.512 - 17.644}{4.108}$$
$$= 0.942$$
$$\beta^2 = 0.887$$
$$\overline{X} = e^{(19.6 + 0.5 \times 0.887)}$$
$$= 507 \quad \text{million barrels}$$
$$s = [(507 \times 10^6)^2(e^{0.887} - 1)]^{.5}$$
$$= 606 \quad \text{million barrels}$$



By using the equations presented in the previous section, different results are obtained. The difference is mainly due to the wide spread of data points. Thus a poor correlation is obtained.

## SOLVING PROBLEMS REQUIRING TRIAL-AND-ERROR CALCULATIONS

It is a common practice in many petroleum engineering problems to plot a certain function A versus another function B such that a straight line results. However, in some cases there is an unknown factor in either A or B which has to be determined in order to get a straight line. The method most commonly used is to determine the unknown factor by trial and error, which at times could be very tedious.

The method presented here provides a simple technique for a faster and reliable computation of the unknown factor. Three data points and the following relationship are required for using this technique.

$$f(c) = \frac{y_2(c) - y_1(c)}{x_2 - x_1} - \frac{y_3(c) - y_1(c)}{x_3 - x_1} = 0 \qquad \textbf{(1.50)}$$

This equation is used when $y$ is comprised of some unknown factor, say $c$, which makes it a straight line with $x$. Similar expression can be used for $x$ being a function of $c$. Any three data points out of a set of $n$ data points can be chosen and used in Equation (1.50) to calculate the unknown. However, in order to span the complete data set, it is recommended to use the two end points and the median or central point from the given set. The technique will provide a satisfactory unknown variable if the chosen data points are correct.

Some of the specific petroleum engineering problems in which the technique can be useful are

1. The after-flow analysis of Russel for pressure buildup analysis, which requires a plot of $\Delta p/[1 - (1/C\Delta t)]$ versus $\log \Delta t$
2. Determination of non-Darcy flow coefficient F in multirate gas well test analysis, which requires a plot of $[\psi(p_i) - \psi(p_{wfn}) - FQ_n^2]/Q_n$ versus $\Sigma_{j=1}^{n} \Delta Q_j/Q_n \log(t_n - t_{j-1})$
3. Finding the water influx $W_e$ in waterdrive gas reservoir material balance calculations, which requires a plot of $G_p$ versus $(W_eB_g)/1 - B_g/B_{gi}$
4. Finding the gas cap size $m$ in gas cap drive reservoir material balance calculations, which requires a plot of F versus $E_o + mE_g$
5. Determination of the initial pressure $p_i$ in multigas well tests analysis, which requires plotting $[\psi(p_i) - \psi(p_{wfn}]/Q_n$ versus $Q_n$

An example is provided here to clarify the use of the technique. The other cases provided above will provide more examples of the use of this technique in their respective sections in Chapters 4 and 6. A computer program written in BASIC is provided in Table 1–12 which can be used to solve for the unknown in either $x$ or $y$ very rapidly.

**Example 1–33.** A four-point test is conducted on a gas well. The rate, times, and observed pressure during the flow test are listed in Table 1–11. Analyze this multirate test to determine the non-Darcy flow coefficient F. The initial reservoir pressure $p_i = 4{,}290$ psia, and $\psi(p_i) = 1{,}068 \times 10^6$ psia²/cp.

**TABLE 1–11. Four-point test data for Example 1–33[10]**

| Rate (Q) MSCF/d | Cum. Flow Time, hrs | $p_{wf}$ psia | $\psi(p_{wf})$ $\psi^2$/cp | $\Sigma \frac{Q_j}{Q_n} \log(t_n - t_{j-1})$ |
|---|---|---|---|---|
| $10 \times 10^3$ | 1 | 4,160 | $1{,}023.35 \times 10^6$ | 0 |
| $20 \times 10^3$ | 2 | 3,997 | $967.00 \times 10^6$ | 0.1505 |
| $30 \times 10^3$ | 3 | 3,802 | $899.59 \times 10^6$ | 0.2594 |
| $40 \times 10^3$ | 4 | 3,577 | $821.81 \times 10^6$ | 0.3450 |

## TABLE 1–12. Computer program used to solve problems requiring trial-and-error calculations[9]

```
1000 CLEAR: KEY OFF: Z=2.3026
1010 W$="BASIC PROGRAMME SIMPLIFIES TRIAL AND ERROR CALCULATIONS"
1020 W1$="MAIN MENU"
1030 W2$="1. Russell's Afterflow Analysis"
1040 W3$="2. Hyperbolic Decline Exponent b"
1050 W4$="3. Average Pressure from Buildup Test"
1060 W5$="4. Matrix Properties from Log Analysis"
1070 W6$="5. Determine Non-Darcy Flow Coefficient F"
1080 W7$="6. Determine Gascap Size m"
1090 W8$="7. Plot of [m(Pi)—m(Pwfn)]/Qn Vs. Qn"
1100 TB=LEN(W$):TB1=LEN(W1$):TB2=LEN(W2$):TB3=LEN(W3$):
 TB8=LEN(W8$)
1110 TB4=LEN(W4$):TB5=LEN(W5$):TB6=LEN(W6$):TB7= LEN(W7$)
1120 TB=(80-TB)/2: TB1=(80-TB1)/2: TB2=(80-TB2)/2
1130 TB3=(80-TB3)/2:TB4=(80-TB4)/2:TB5=(80-TB5)/2
1140 TB6=(80-TB6)/2:TB7=(80-TB7)/2:TB8=(80-TB8)/2
1150 CLS: SEL=0: PRINT TAB(TB);W$: PRINT: PRINT
1160 DEF SEG=0
1170 PRINT TAB(TB1);W1$
1180 LOCATE 7,20: PRINT W2$: LOCATE 9,20: PRINT W3$
1190 LOCATE 11,20: PRINT W4$: LOCATE 13,20: PRINT W5$
1200 LOCATE 15,20: PRINT W6$: LOCATE 17,20: PRINT W7$
1210 LOCATE 19,20: PRINT W8$: LOCATE 21,20: PRINT "8. End of Session"
1220 LOCATE 23,25: PRINT "Make Your Selection ===>"
1230 LOCATE 23,50: PRINT " ";: LOCATE 23,50
1240 I$=INKEY$: IF I$="" THEN 1240 ELSE PRINT I$;
1250 IF MID$(I$,1,1)="1" THEN SEL=1
1260 IF MID$(I$,1,1)="2" THEN SEL=2
1270 IF MID$(I$,1,1)="3" THEN SEL=3
1280 IF MID$(I$,1,1)="4" THEN SEL=4
1290 IF MID$(I$,1,1)="5" THEN SEL=5
1300 IF MID$(I$,1,1)="6" THEN SEL=6
1310 IF MID$(I$,1,1)="7" THEN SEL=7
1320 IF MID$(I$,1,1)="8" THEN SEL=8
1330 IF SEL<1 OR SEL>8 THEN BEEP: GOTO 1360
1340 ON SEL GOSUB 1860,2590,1420,1600,2050,2260,2420,2910
1350 GOTO 1150
1360 FOR I=1 TO 60
1370 LOCATE 23,25: PRINT "INVALID SELECTION, PLEASE TRY AGAIN!"
1380 NEXT I
1390 LOCATE 23,25: PRINT "                                        "
1400 GOTO 1220
1410 '
1420 '***** AVERAGE PRESSURE FROM BUILDUP TEST *****"
1430 CLS: PRINT:PRINT: PRINT TAB(TB4);W4$
1440 LOCATE 8,21: PRINT "POINT    ELAPSED TIME    Pws (psi)"
1450 GOSUB 2790
1460 DT1=Z(1):DT2=Z(2):DT3=Z(3):PWS1=Z1(1):PWS2=Z1(2):PWS3=Z1(3)
1470 PB=PWS3+1
1480 FPB=(LOG(PB-PWS2)/Z-LOG(PB-PWS1)/Z)/(DT2-DT1)
1490 FPB=FPB-((LOG(PB-PWS3)/Z-LOG(PB-PWS1)/Z)/(DT3-DT1))
1500 IF ABS(FPB) <=.0001 THEN 1550
1510 FFPB=(1/(PB-PWS2)-1/(PB-PWS1))/(DT2-DT1)
1520 FFPB=FFPB-((1/(PB-PWS3)-1/(PB-PWS1))/(DT3-DT1))
1530 PB=PB-(FPB/FFPB)
1540 GOTO 1480
1550 LOCATE 16,21: PRINT "AVERAGE RESERVOIR PRESSURE IS ";PB;"psi"
1560 LOCATE 22,21:INPUT"RUN THE SAME PROGRAM AGAIN (Y/N) ===> ",R$
1570 IF LEFT$(R$,1)="y" OR LEFT$(R$,1)="Y" THEN 1420
1580 RETURN
1590 '
1600 '******** LOG ANALYSIS ********
1610 CLS:PRINT:PRINT:PRINT TAB(TB5);W5$
1620 LOCATE 8,21: PRINT "POINT    LOG READING    RESISTIVITY"
1630 GOSUB 2790
1640 DT1=Z(3):DT2=Z(2):DT3=Z(1):R1=Z1(3):R2=Z1(2):R3=Z1(1)
1650 LOCATE 14,21: PRINT "POROSITY LOG TYPE: 1. SONIC"
1660 LOCATE 15,40: PRINT "2. DENSITY":LOCATE 16,40:PRINT"3. NLL"
1670 LOCATE 18,21: INPUT "INPUT TYPE OF LOG ===> ",TY
1680 LOCATE 19,21: INPUT"GUESS ESTIMATED MATRIX VALUE ===> ",TM
1690 FTM=(LOG (DT2-TM)/Z-LOG(DT1-TM)/Z)/(LOG(R2)/Z-LOG(R1)/Z)
1700 FTM=FTM-(LOG(DT3-TM)/Z-LOG(DT1-TM)/Z)/(LOG(R3)/Z-LOG(R1)/
 Z))
1710 IF ABS(FTM) <=.0001 THEN 1760
1720 FFTM=(1/(DT2-TM)-1/(DT1-TM))/(LOG(R2)/Z-LOG(R1)/Z)
1730 FFTM=FFTM-((1/(DT3-TM)-1/(DT1-TM))/(LOG(R3)/Z-LOG(R1)/Z))
1740 TM=TM+(FTM/FFTM)
1750 GOTO 1690
1760 ON TY GOTO 1770,1790,1810
1770 LOCATE 21,21: PRINT"MATRIX TRAVEL TIME IS "; TM;"MICROSEC/FT"
1780 GOTO 1820
1790 LOCATE 21,21: PRINT"MATRIX DENSITY IS "; TM;"GRAMS/CC"
1800 GOTO 1820
1810 LOCATE 21,21: PRINT "MATRIX CAPTURE CROSS SECTION IS "; TM
1820 LOCATE 23,21: INPUT "RUN THE SAME PROGRAM AGAIN (Y/N)
 ===> ",R$
1830 IF LEFT$(R$,1)="y" OR LEFT$(R$,1)="Y" THEN 1600
1840 RETURN
1850 '
1860 ******** AFTERFLOW ANALYSIS ********
1870 CLS:PRIN:PRINT: PRINT TAB(TB2);W2$
1880 LOCATE 8,21: PRINT "POINT    ELAPSED TIME    Pws - Pwf (psi)"
1890 GUSUB 2790
1900 DT1=Z(1):DT2=Z(2):DT3=Z(3):DP1=Z1(1):DP2=Z1(2):DP3=Z1(3):C=.5
1910 FC=(DP2/(1-1/(DT2*C))-DP1/(1-1/(DT1*C)))/(LOG(DT2)/Z-LOG(DT1)/
 Z)
1920 FC=FC-((DP3/(1-1/(DT3*C))-DP1/(1-1/(DT1*C)))/(LOG(DT3)/
 Z-LOG(DT1)/Z))
1930 IF ABS(FC) <= .001 THEN 2000
1940 X=DP2/((DT2*C)+1/(DT2*C)-2): X1=DP3/((DT3*C)+1/(DT3*C)-2)
1950 X2=DP1/((DT1*C)+1/(DT1*C)-2)
1960 FFC=(X-X2)/(LOG(DT2)/Z-LOG(DT1)/Z)
1970 FFC=FFC-((X1-X2)/(LOG(DT3)/Z-LOG(DT1)/Z))
1980 C=C+(FC/FFC)
1990 GOTO 1910
2000 LOCATE 16,21: PRINT "THE VALUE OF CONSTANT C IS ";C;"UNITS"
2010 LOCATE 22,21: INPUT "RUN THE SAME PROGRAM AGAIN (Y/N)
 ===> ",R$
2020 IF LEFT$(R$,1)="y" OR LEFT$(R$,1)="Y" THEN 1860
2030 RETURN
2040 '
2050 '******** TO DETERMINE NON-DARCY FLOW COEFFICIENT F ********
2060 CLS: PRINT:PRINT: PRINT TAB(TB6);W6$
2070 LOCATE 8,12: PRINT "POINT m(Pi) - m(Pwf) Qn (Mscf/d) [sum Qj/Qn ";
2080 PRINT "Log(tn - tj-1)]"
2090 FOR I=1 TO 3
2100 LOCATE (I+9),13:PRINT I:LOCATE (I+9),24:INPUT " ",Z(I):
2110 LOCATE (I+9),41:INPUT " ",Z1(I):LOCATE (I+9),60:INPUT " ",Z2(I)
2120 NEXT I
2130 A1=Z(1):A2=Z(2):A3=Z(3):Q1=Z1(1):Q2=Z1(2):Q3=Z1(3): F=0
2140 X1=Z2(1):X2=Z2(2):X3=Z2(3)
2150 FFF=((Q1-Q2)/(X2-X1))-((Q1-Q3)/(X3-X1))
2160 FF=((A2-F*Q2^2)/Q2-(A1-F*Q1^2)/Q1)/(X2-X1)
2170 FF=FF-((A3-F*Q3^2)/Q3-(A1-F*Q1^2)/Q1)/(X3-X1)
2180 IF ABS(FF) <= .0001 THEN 2210
2190 F=F-(FF/FFF)
2200 GOTO 2160
2210 LOCATE 16,21: PRINT "THE NON-DARCY FLOW COEFFICIENT F IS ";F
2220 LOCATE 22,21: INPUT "RUN THE SAME PROGRAM AGAIN (Y/N)
 ===> ",R$
2230 IF LEFT$(R$,1)="y" OR LEFT$(R$,1)="Y" THEN 2260
2240 RETURN
2250 '
2260 ' ******** TO DETERMINE GASCAP SIZE m ********
2270 CLS: PRINT:PRINT: PRINT TAB(TB7);W7$
2280 LOCATE 8,21: PRINT "POINT F Eo Eg"
2290 M=O:GOSUB 2790
2300 F1=0:F2=Z(2):F3=Z(3):E01=0:E02=Z1(2):E03=Z1(3):EG1=0:
 EG2=Z2(2):EG3=Z2(3)
2310 X=(E02+M*EG2)-(E01+M*EG1)
2320 X1=(E03+M*EG3)-(E01+M*EG1)
2330 FM=(F2-F1)/X-(F3-F1)/X1
2340 FFM=((F1-F2)*(EG2-EG1))/X^2-((F1-F3)*(EG3-EG1))/X1^2
2350 IF ABS(FM) <=.001 THEN 2370
2360 M=M-(FM/FFM): GOTO 2310
2370 LOCATE 16,21: PRINT "THE GASCAP SIZE m IS ";M
2380 LOCATE 22,21: INPUT "RUN THE SAME PROGRAM AGAIN (Y/N)
 ===> ",R$
2390 IF LEFT$(R$,1)="y" OR LEFT$(R$,1)="Y" THEN 2260
2400 RETURN
2410 '
2420 ' ***** TO DETERMINE Pi, PLOT OF M(Pi)-M(Pwfn)/Qn VS. Qn *****
2430 CLS: PRINT:PRINT: PRINT TAB(TB8);W8$
2440 LOCATE 8,21: PRINT "POINT m(Pwfn) Qn(Mscf/d)"
2450 GOSUB 2790
2460 MPWF1=Z(1):MPWF2=Z(2):MPWF3=Z(3):Q1=Z1(1):Q2=Z1(2):Q3=Z1(3)
2470 LOCATE 15,21: INPUT "INPUT INITIAL GUESS OF m(Pi) ===> ",MPI
2480 FFMPI=((1/Q2-1/Q1)/(Q2-Q1))-((1/Q3-1/Q1)/(Q3-Q1))
2490 FMPI=((MPI-MPWF2)/Q2-(MPI-MPWF1)/Q1)/(Q2-Q1)
2500 FMPI=FMPI-(((MPI-MPWF3)/Q3-(MPI-MPWF1)/Q1)/(Q3-Q1))
2510 IF ABS(FMPI) <= .0001 THEN 2530
2520 MPI=MPI-(FMPI/FFMPI): GOTO 2490
2530 LOCATE 18,21: PRINT "THE CALCULATED m(Pi)  IS ";
2540 PRINT USING "#.###.###^^^^";MPI
2550 LOCATE 23,21: INPUT "RUN THE SAME PROGRAM AGAIN (Y/N)
 ===> ",R$
2560 IF LEFT$(R$,1)="y" OR LEFT$(R$,1)="Y" THEN 2420
2570 RETURN
2580 '
2590 ' ***** TO DETERMINE THE HYPERBOLIC EXPONENT B *****
2600 CLS:PRINT:PRINT:PRINT TAB(TB3);W3$
2610 LOCATE 8,21: PRINT "POINT TIME RATE"
2620 GOSUB 2790
2630 A=0:T1=Z(3):T2=Z(2):T3=Z(1):Q1=Z1(3):Q2=Z1(2):Q3=Z1(1)
2640 FA1=(LOG(Q2)/Z-LOG(Q1)/Z)/(LOG(T2+A)/Z-LOG(T1+A)/Z)
2650 FA2=(LOG(Q3)/Z-LOG(Q1)/Z)/(LOG(T3+A)/Z-LOG(T1+A)/Z)
2660 FA=FA1-FA2
2670 FFA1=((LOG(Q1)/Z-LOG(Q2)/Z)*(1/(T2+A)-1/(T1+A)))
```

## TABLE 1–12. (continued)

```
2680  FFA1=FFA1/((LOG(T2+A)/Z-LOG(T1+A)/Z)^2)
2690  FFA2=((LOG(Q1)/Z-LOG(Q3)/Z)*(1/(T3+A)-1/(T1+A)))
2700  FFA2=FFA2/((LOG(T3+A)/Z-LOG(T1+A)/Z)^2)
2710  FFA=FFA1-FFA2
2720  IF ABS(FA) <=.001 THEN 2740
2730  A=A-(FA/FFA): GOTO 2640
2740  LOCATE 16,21: PRINT "THE CURVE IS SHIFTED BY A FACTOR OF ";A
2750  LOCATE 17,21: PRINT "THE HYPERBOLIC EXPONENT b IS ";-1/FA1
2760  LOCATE 22,21: INPUT "RUN THE SAME PROGRAM AGAIN (Y/N) ===> ",R$
2770  IF LEFT$(R$,1)="y" OR LEFT$(R$,1)="Y" THEN 2590
2780  RETURN
2790  FOR I=1 TO 3
2800  IF SEL=6 THEN 2840
2810  LOCATE (I+9),22: PRINT I: LOCATE (I+9),35: INPUT " ",Z(I)
2820  LOCATE (I+9),53: INPUT " ",Z1(I)
2830  GOTO 2890
2840  IF I>1 THEN 2870
2850  LOCATE (I+9),22: PRINT I: LOCATE (I+9),34: PRINT F1
2860  LOCATE (I+9),46: PRINT E01: LOCATE (I+9),59: PRINT EG1:GOTO 2890
2870  LOCATE (I+9),22: PRINT I: LOCATE (I+9),31: INPUT " ",Z(I)
2880  LOCATE (I+9),43: INPUT " ",Z1(I): LOCATE (I+9),56: INPUT " ",Z2(I)
2890  NEXT I
2900  RETURN
2910  CLS:END
```

Solution: In this example, we have

$$Y = \frac{\psi(p_i) - \psi(p_{wf}) - FQ_n^2}{Q_n}$$



and

$$X = \sum \frac{Q_j}{Q_n} \log (t_n - t_{j-1})$$

Choose three points at flow times 1, 2, and 4 hours. Substituting in Equation (1.50) we get:

$$\begin{aligned} f(F) = & \left\{\left[\frac{(1{,}068.29 - 967.00)10^6 - F(20 \times 10^3)^2}{20 \times 10^3}\right.\right. \\ & \left.\left. - \frac{(1{,}068.29 - 1{,}023.35)10^6 - F(10 \times 10^3)^2}{10 \times 10^3}\right] \div (0.1505 - 0)\right\} \\ & - \left\{\left[\frac{(1{,}068.29 - 821.81)10^6 - F(40 \times 10^3)^2}{40 \times 10^3}\right.\right. \\ & \left.\left. - \frac{(1{,}068.29 - 1{,}023.35)10^6 - F(10 \times 10^3)^2}{10 \times 10^3}\right] \div (0.3450 - 0)\right\} = 0 \end{aligned}$$

$$2.051 \times 10^4 F = 1.041 \times 10^3$$

$$F = 0.051$$

## REFERENCES


1. Franklin J. Stermole, *Economic Evaluation and Investment Decision Methods*, Investment Evaluations Corporation, Golden, Colorado, 1974.
2. *Fundamentals of Engineering Sample Examinations*, The National Council of Engineering Examiners, Clemson, South Carolina, 1983.
3. *Tax Guide for Small Business, Publication* 534, Department of the Treasury, Internal Revenue Service, Rev. Nov., 1986.
4. Robert S. Thompson and John D. Wright, *Oil Property Evaluation*, Thompson-Wright Associates, Golden, Colorado, 1984.

[5] Paul D. Newendorp, *Decision Analysis for Petroleum Exploration*, PennWell Publishing Company, Tulsa, Oklahoma, 1975.

[6] Irving W. Baker, *Engineering Statistics and Quality Control*, McGraw-Hill, New York, 1953.

[7] John Neter, et al., *Applied Statistics*, Allyn and Bacon, Inc., Boston, Massachusetts, Second Edition, 1982.

[8] Arthur W. McCray, *Petroleum Evaluations and Economic Decisions*, Prentice-Hall, Inc., New Jersey, 1975.

[9] M. A. Mian, "Program Quickly Solves Trial-and-Error Problems," *World Oil*, Sept. 1990, pp. 67–75.

[10] L. P. Dake, *Fundamentals of Reservoir Engineering*, Elsevier Scientific Publishing Company, The Netherlands, 1978, pp. 275–278.

CHAPTER TWO

# BASIC ROCK AND FLUID PROPERTIES



This chapter presents concepts and applications of rock and fluid properties usually required for solving reservoir engineering and transient well test analysis problems. The engineering equations and correlations presented in this chapter represent technical papers well known to the petroleum engineers.

In reservoir engineering and transient test analysis problems, the basic information required pertains to the physical properties of reservoir rocks and fluids and to the manner in which these properties change with the change in pressure and temperature.

In most of this chapter, two approaches are adopted for computing or finding the various properties: (a) equation approach and (b) chart or table approach. For the properties that require iterative solutions of the equations and/or the equations require extensive mathematical manipulations, the equations are ignored and only charts or tables are presented. For the properties where the equations require simple mathematical manipulations, both the equations and the charts are presented. You may use either the equations or the charts.

Each property computation and its use are illustrated by a solved example. Detailed equations for each property are presented in Appendix B which can be used in computer programming, etc.

## GAS PROPERTIES

In this section we will deal only with the gas PVT properties correlations such as pseudocritical pressure and temperature, gas deviation factor ($z$), gas formation volume factor ($B_g$), gas viscosity ($\mu_g$), and gas isothermal compressibility ($c_g$). In most cases the required data will be gas gravity or molecular weight, reservoir temperature and pressure, and standard temperature and pressure. A section is presented which shows how to calculate the above properties if gas composition is available. If gas composition is available, it is strongly recommended to compute gas properties from the composition.

Most gas laws, such as Charles' Law and Boyles' Law, are applicable to ideal gases at low pressures. Natural gases in the reservoir are at very high temperatures and reservoir pressures and their behavior deviates from that of the ideal gas. To use the pressure-volume-temperature (PVT properties) relationships of ideal gas laws, some correction factors have to be incorporated into the equations of state (equations correlating pressure-volume-temperature).

## PSEUDOCRITICAL PROPERTIES

Each component of a gas mixture has its own critical temperature ($T_c$) and critical pressure ($p_c$). When the individual critical property is multiplied by the mole fraction (fraction of the whole gas mixture) of the respective component, we get pseudocritical temperature ($T_{pc}$) and pseudocritical pressure ($p_{pc}$). In this section, $T_{pc}$ and $p_{pc}$ will be estimated from empirical correlations developed by Brown et al. using equations or charts.[1] If gas composition is available, more accurate properties will be calculated using the composition.

For condensate fluids:

$$T_{pc} = 187 + 330(\gamma_g) - 71.5(\gamma_g)^2 \quad \textbf{(2.1)}$$

$$p_{pc} = 706 - 51.7(\gamma_g) - 11.1(\gamma_g)^2 \quad \textbf{(2.2)}$$

For miscellaneous gases:

$$T_{pc} = 168 + 325(\gamma_g) - 12.5(\gamma_g)^2 \quad \textbf{(2.3)}$$

$$p_{pc} = 677 + 15.0(\gamma_g) - 37.5(\gamma_g)^2 \quad \textbf{(2.4)}$$

where $\gamma_g$ is the specific gravity of whole gas mixture (air = 1), $T_{pc}$ is the pseudocritical temperature for reservoir gas in °R, and $p_{pc}$ is the pseudocritical pressure for reservoir gas in pounds per square inch absolute. Specific gravity is a ratio of the molecular weight of gas to the molecular weight of air.

Equations (2.1) and (2.2) are used for calculations when gas is in equilibrium with crude oil or condensate in the reservoir. Equations (2.3) and (2.4) are used for dry surface gases.

The above equations were developed for gases with no contaminants, such as $H_2S$, $CO_2$, and $N_2$. If any or all of these contaminants are present, the values obtained above must be corrected using Wichert-Aziz correction.[2]

In Equations (2.1) through (2.4), the $\gamma_g$ is replaced by $\gamma_{hc}$ (gravity of the hydrocarbon portion of gas), where $\gamma_{hc}$ is given by

$$\gamma_{hc} = \frac{\gamma_g - 0.9672 y_{N_2} - 1.5195 y_{CO_2} - 1.1765 y_{H_2S}}{A} \quad \textbf{(2.5)}$$

and the pseudocritical temperature and pressure for the whole gas mixture are

$$T_{pc1} = A(T_{pc}) + 227.3y_{N_2} + 547.6y_{CO_2} + 672.4y_{H_2S} \quad (2.6)$$

$$p_{pc1} = A(p_{pc}) + 493.0y_{N_2} + 1071y_{CO_2} + 1306y_{H_2S} \quad (2.7)$$

where $y_{N_2}$, $y_{CO_2}$, and $y_{H_2S}$ are the mole fractions of nitrogen, carbon dioxide, and hydrogen sulfide in the mixture, respectively. The $T_{pc1}$ and $p_{pc1}$ are the pseudocritical properties of the whole mixture. The term A is given by $A = (1 - y_{N_2} - y_{CO_2} - y_{H_2S})$.

The pseudocritical properties $T_{pc1}$ and $p_{pc1}$ are then corrected for acid gases ($CO_2$ and $H_2S$) by means of the Wichert-Aziz correction factor. The Wichert-Aziz correction factor is given by the following equation:

$$CWA = 120(y_{CO_2} + y_{H_2S})^{0.90} - (y_{CO_2} + y_{H_2S})^{1.6} + 15(y_{H_2S})^{0.5} - (y_{H_2S})^4 \quad (2.8)$$



The corrected pseudocritical temperature and pressure are:

$$T^*_{pc} = T_{pc1} - CWA \quad (2.9)$$

$$p^*_{pc} = \frac{p_{pc1} \times T^*_{pc}}{T_{pc1} + y_{H_2S}(1 - y_{H_2S})CWA} \quad (2.10)$$

The pseudocritical temperature and pressure may also be determined graphically using gas gravity and Figure 2–1. Figure 2–1 represents the Equations (2.1) through (2.4). If there are contaminants present, the pseudocritical properties are corrected by determining the Wichert-Aziz temperature correction factor from Figure 2–2 and using Equations (2.9) and (2.10).

## PSEUDOREDUCED PROPERTIES

The second step in most gas PVT properties calculations is to calculate the pseudoreduced pressure ($p_{pr}$) and pseudoreduced temperature ($T_{pr}$). These pseudoreduced properties are related with the pseudocritical properties by the following equations:

$$T_{pr} = \frac{T_R + 460}{T_{pc}} \quad (2.11)$$

$$p_{pr} = \frac{p_R}{p_{pc}} \quad (2.12)$$

where $T_{pr}$ is the pseudoreduced temperature in °R and $p_{pr}$ is the pseudoreduced pressure in psia, $T_R$ is the reservoir temperature in °F, and $p_R$ is the reservoir pressure in psia.

If $T_{pr}$ and $p_{pr}$ are the same for two gases, their physical properties will be equivalent (same state) even if their absolute pressure and temperature are not the same. These properties are widely

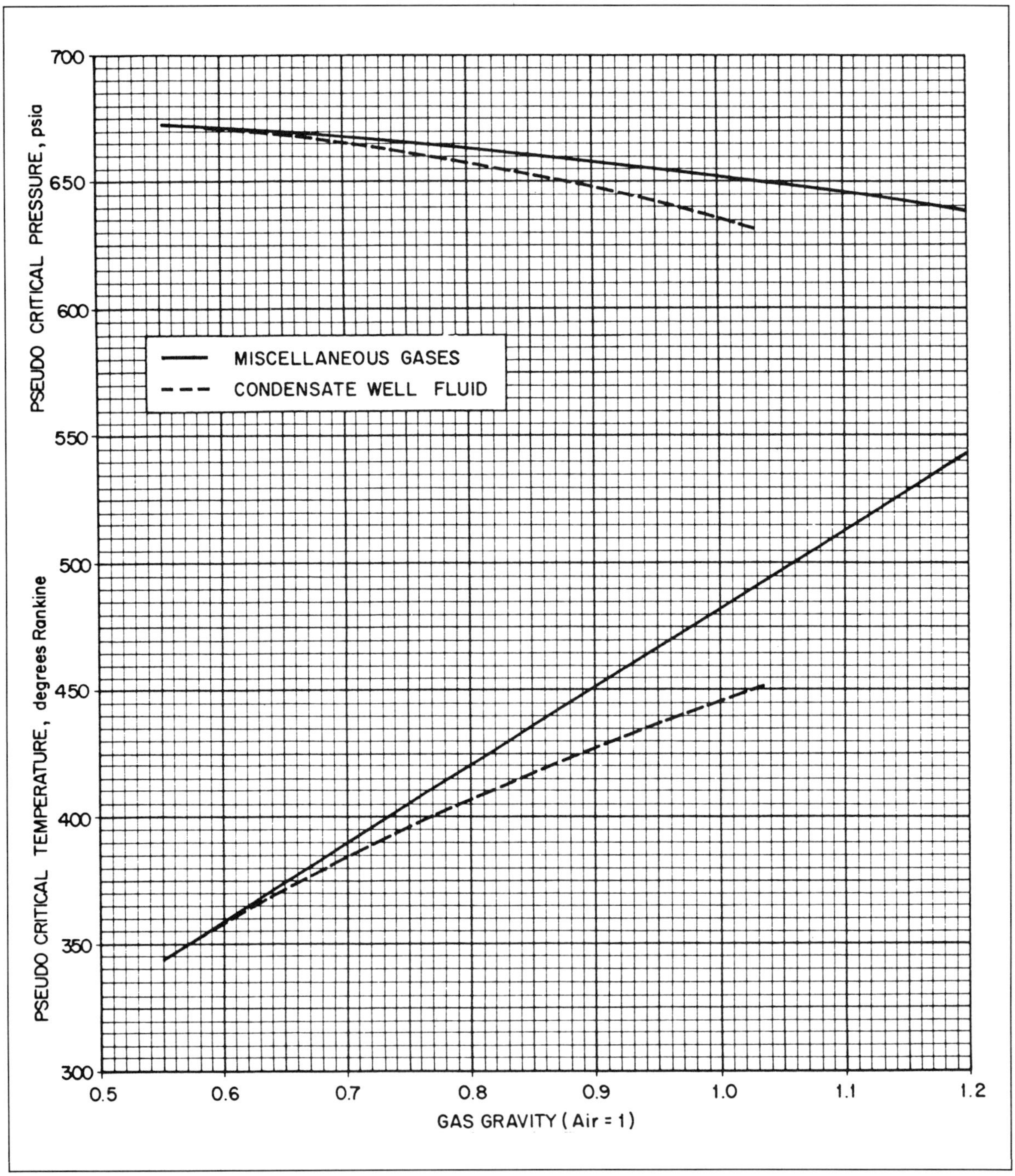


**Figure 2–1.** Pseudocritical properties of natural gases (after Brown et. al.).[1]

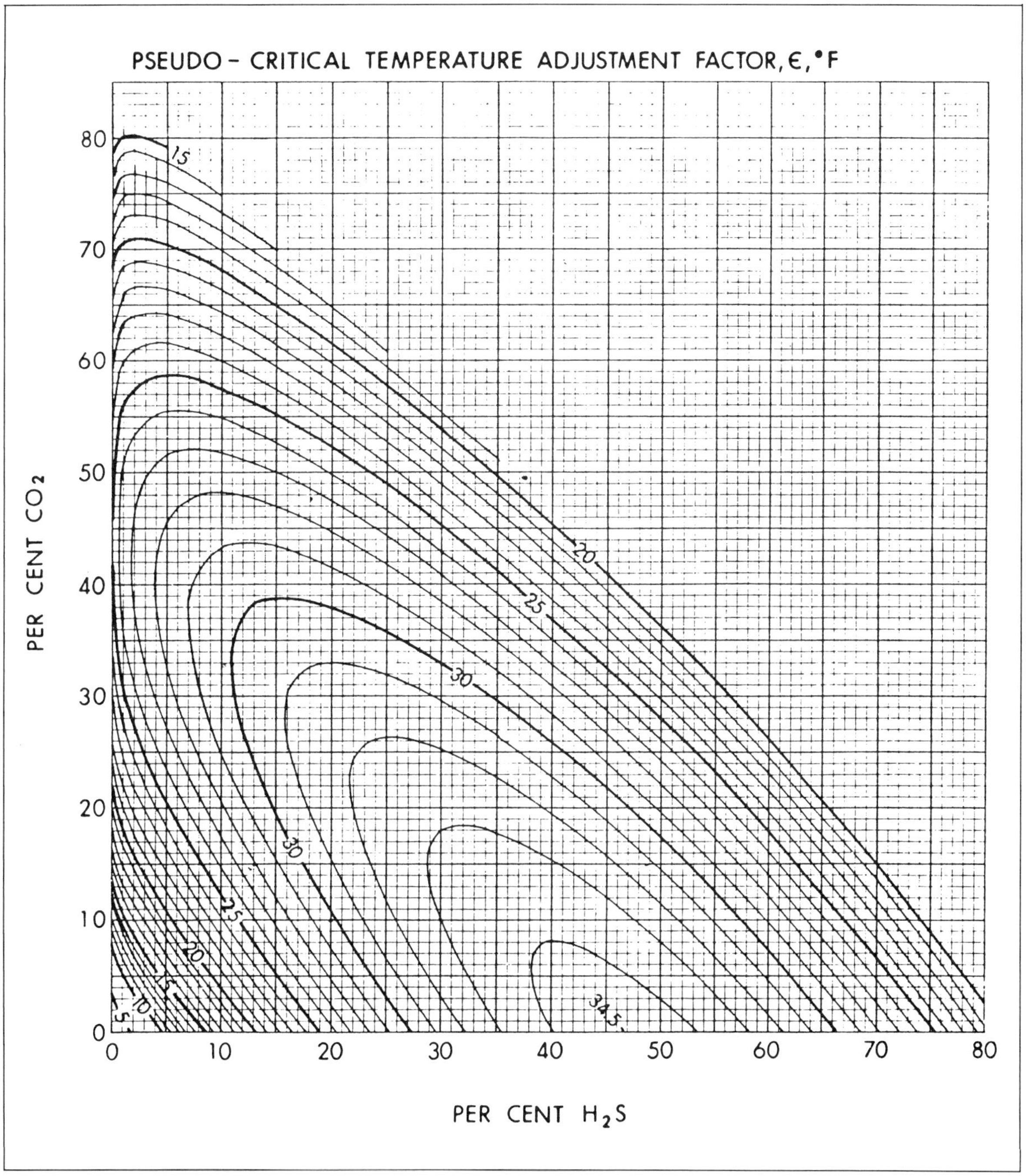


**Figure 2–2.** Correction factor chart for sour gases (courtesy of GPSA).[3]

used in gas properties determination. If contaminants are present, then $T^*_{pc}$ and $p^*_{pc}$ are used in calculating pseudoreduced properties.

**Example 2–1.** Compute pseudocritical temperature and pressure for a dry gas of 0.75 gravity (air = 1), using (a) appropriate equations and (b) charts.

Solution:

(a) Use Equations (2.3) and (2.4) to compute critical properties.

$$\begin{aligned} T_{pc} &= 168.0 + 325(0.75) - 12.5(0.75)^2 \\ &= 168.0 + 243.75 - 7.031 \\ &= 404.719°\text{R} \\ p_{pc} &= 677.0 + 15.0(0.75) - 37.5(0.75)^2 \\ &= 677.0 + 11.25 - 21.094 \\ &= 667.156 \quad \text{psia} \end{aligned}$$

Since there are no contaminant gases present, $T_{pc} = T^*_{pc}$ and $p_{pc} = p^*_{pc}$.

(b) Use Figure 2–1 to determine critical properties. Enter Figure 2–1 with gas gravity of 0.75 from $x$-axis, go to $T_c$ line for miscellaneous gases and read $T_{pc}$ on the $y$-axis of 405°R. Go on 0.75 gravity line to $p_c$ for miscellaneous gases and read $p_{pc}$ on the $y$-axis of 667 psia.

**Example 2–2.** Rework Example 2–1 for the same gravity gas assuming that the gas contains contaminant gases. The contaminants reported are $H_2S$ = 5%, $N_2$ = 2%, and $CO_2$ = 3%. Determine pseudocritical pressure and temperature by using (a) equations and (b) charts. Also compute pseudoreduced pressure and temperature if the reservoir temperature and pressure are 250°F and 7,500 psia, respectively.

Solution:

$$\begin{aligned} A &= 1 - y_{N_2} - y_{CO_2} - y_{H_2S} \\ &= 1 - 0.02 - 0.03 - 0.05 \\ &= 0.90 \\ \gamma_{hc} &= \frac{0.75 - 0.9672(0.02) - 1.5195(0.03) - 1.1765(0.05)}{0.90} \\ &= \frac{0.75 - 0.0193 - 0.0456 - 0.0588}{0.90} \\ &= 0.6959 \end{aligned}$$

For miscellaneous gases:

$$\begin{aligned}
T_{pc} &= 168.0 + 325(0.6959) - 12.5(0.6959)^2 \\
&= 388.114°R \\
p_{pc} &= 677.0 + 15(0.6959) - 37.5(0.6959)^2 \\
&= 669.278 \quad \text{psia} \\
T_{pc1} &= 0.9(388.114) + 227.3(0.02) + 547.6(0.03) + 672.4(0.05) \\
&= 403.897°R \\
p_{pc1} &= 0.9(669.278) + 493(0.02) + 1{,}071(0.03) + 1{,}306(0.05) \\
&= 709.640 \quad \text{psia}
\end{aligned}$$

Compute Wichert-Aziz correction using Equation (2.8).



$$\begin{aligned}
\text{CWA} &= 120(0.03 + 0.05)^{0.9} - (0.03 + 0.05)^{1.6} + 15(0.05^{0.5} - 0.05^4) \\
&= 120(0.1030 - 0.0176) + 15(0.2236) \\
&= 10.2492 + 3.3540 \\
&= 13.6032°F
\end{aligned}$$

Compute the corrected $T^*_{pc}$ and $p^*_{pc}$ using Equations (2.9) and (2.10):

$$\begin{aligned}
T^*_{pc} &= 403.897 - 13.6032 \\
&= 390.294°R \\
p^*_{pc} &= \frac{709.64 \times 390.294}{403.897 + 0.05(1 - 0.05)13.6032} \\
&= \frac{276{,}968.234}{404.5432} \\
&= 684.645 \quad \text{psia}
\end{aligned}$$

(b) In this case, use the following steps:

**Step 1:** Using Figure 2–1 and $\gamma_g = 0.75$, determine $T_{pc}$ and $p_{pc}$.

$$\begin{aligned}
T_{pc} &= 405°R \\
p_{pc} &= 665 \quad \text{psia}
\end{aligned}$$

**Step 2:** Compute A.

$$A = 0.90 \quad \text{from Part (a)}$$

**Step 3:** Compute $p_{pc1}$ using Equation (2.7):

$$p_{pc1} = 0.9(665) + 493(0.02) + 1{,}071(0.03) + 1{,}306(0.05)$$
$$= 705.79 \quad \text{psia}$$

**Step 4:** Read Wichert-Aziz correction from Figure 2–2.

$$CWA = 13.6$$

**Step 5:** Compute corrected $T^*_{pc}$ and $p^*_{pc}$ using Equations (2.9) and (2.10).

$$T^*_{pc} = 405 - 13.6$$
$$= 391.4°R$$
$$p^*_{pc} = \frac{705.79 \times 391.4}{405 + 0.05(1 - 0.05)13.6}$$
$$= \frac{276{,}246.206}{405.646}$$
$$= 681 \quad \text{psia}$$



You will note that the difference between the values in Parts (a) and (b) is very small, the pseudocritical temperature is within 0.283%, and the pseudocritical pressure is within 0.535%. This difference may be slightly more if higher content of nitrogen is present. To compute pseudoreduced pressure and temperature, use the corrected pseudocritical pressure and temperature from Part (a).

$$T_{pr} = \frac{250 + 460}{390.294}$$
$$= 1.819$$
$$p_{pr} = \frac{7{,}500}{684.645}$$
$$= 10.955$$

## GAS DEVIATION FACTOR

It was mentioned earlier that the behavior of natural gases deviate from the behavior of ideal gases at the elevated reservoir temperatures and pressures. In order to facilitate computations of PVT properties of natural gases using ideal gas laws, a gas deviation factor, symbol $z$, is used. A natural gas with a deviation factor of "1" will behave in the same way as an ideal gas would.

The deviation factor in most cases is determined from empirical correlations developed by Standing and Katz as a function of pseudoreduced pressure and temperature.[4] The Standing and Katz $z$-factor chart and the extended Standing and Katz $z$-factor chart developed by Yarborough are presented in Figures 2–3 and 2–3a, respectively.[5] These charts will be utilized in finding the gas deviation factor.

Equations have been developed to fit these correlations, but they are nonlinear, and therefore

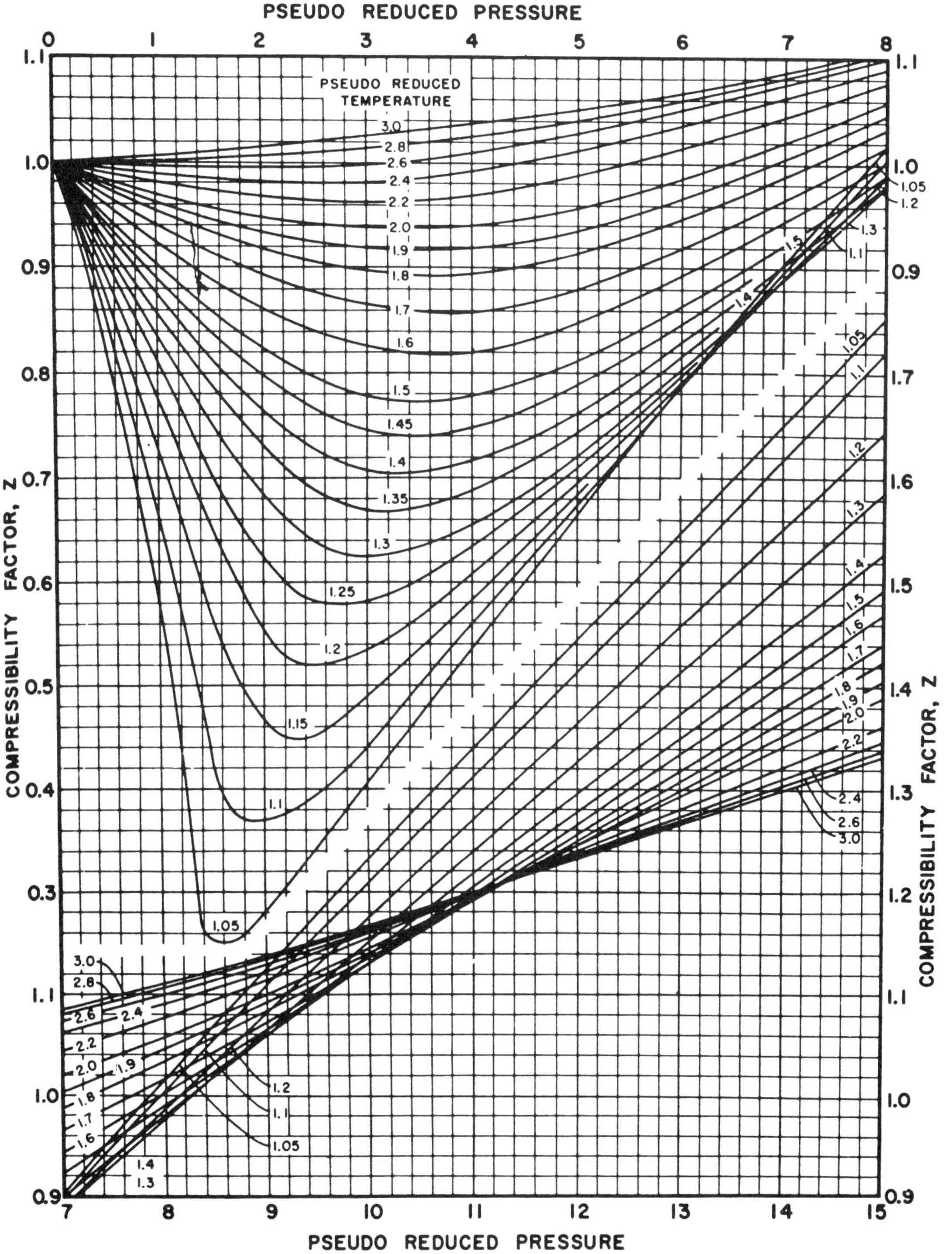




**Figure 2–3.** Gas deviation factors for natural gases (after Standing and Katz, Trans. AIME).[4]

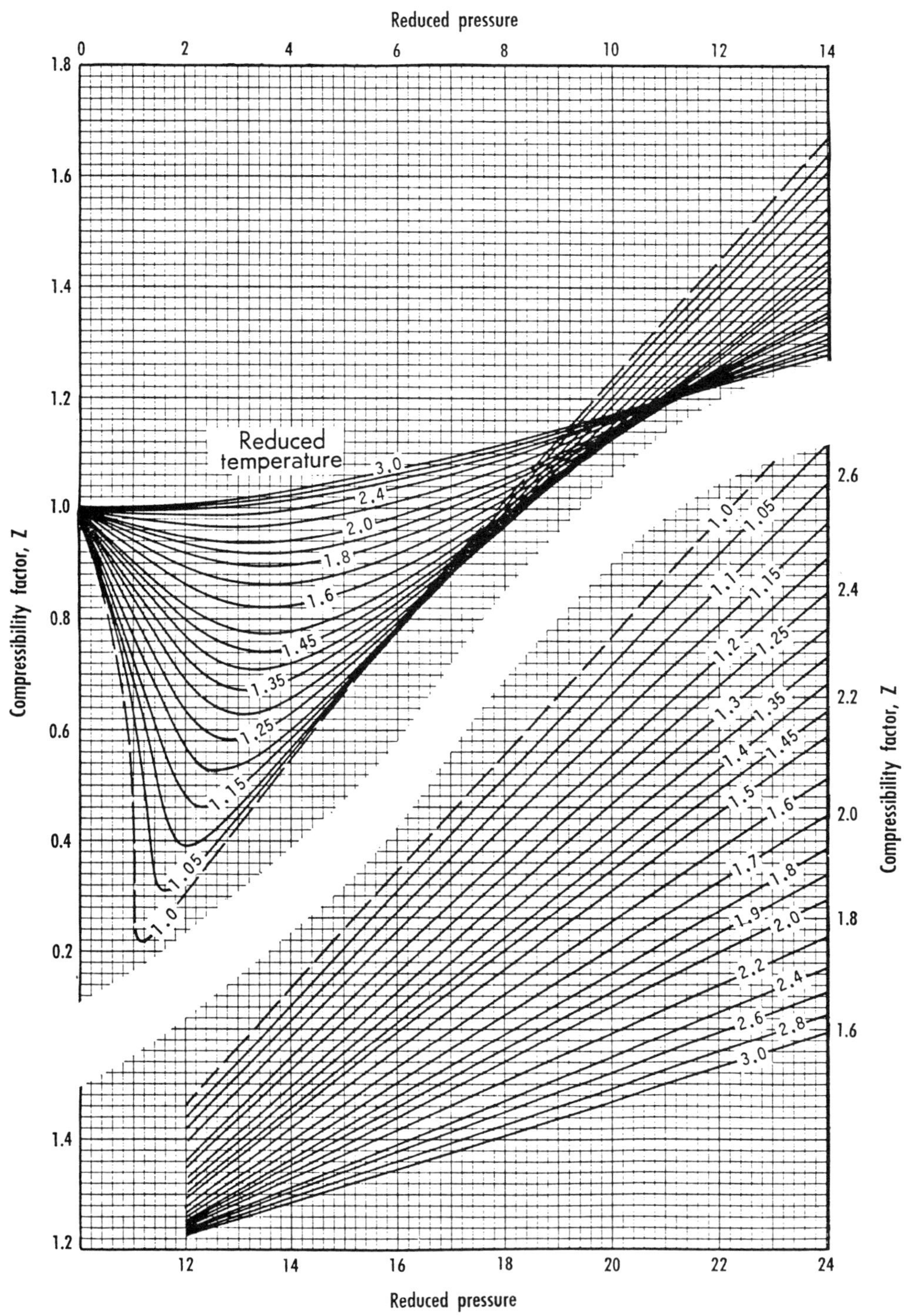


**Figure 2–3a.** Extended Standing and Katz $z$-factor chart (after Yarborough).[5]

iterative solutions are required. Solving the equations by hand will be very time consuming. These equations are provided in Appendix B.

**Example 2–3.** Compute the gas deviation factor $z$ for a 0.7 gravity dry gas at reservoir pressure and temperature of 2,500 psia and 180°F, respectively.

Solution: The following steps will be used.

**Step 1:** Determine pseudocritical pressure and temperature using equations or Figure 2–1.

$$T_{pc} = 389°\text{R} \quad \text{(from Fig. 2–1)}$$
$$p_{pc} = 669 \quad \text{psia} \quad \text{(from Fig. 2–1)}$$

**Step 2:** Compute pseudoreduced pressure and temperature using Equations (2.11) and (2.12). 

$$T_{pr} = \frac{180 + 460}{389} = 1.65$$

$$p_{pr} = \frac{2{,}500}{669} = 3.74$$

**Step 3:** Read the gas deviation factor $z$ from Figure 2–3 or Figure 2–3a as

$$z = 0.84$$

## GAS FORMATION VOLUME FACTOR

The gas formation volume factor, symbol $B_g$, is used to convert the reservoir gas volume (at reservoir temperature and pressure) to gas volume at standard conditions. The standard conditions are at a temperature of 60°F and atmospheric pressure. The factor is expressed in standard cubic feet (SCF) per reservoir cubic feet or reservoir barrels and vice versa. The atmospheric pressure varies (most common is 14.7 psi), depending upon location. The standard pressures used in various locations in the United States are presented in Table 2–1. The following equations are used to compute $B_g$:

$$B_g = \frac{T_{sc} p_R}{p_{sc} z T_R} \tag{2.13}$$

where $B_g$ is the gas formation volume factor in SCF/ft$^3$, $p_{sc}$ is the standard pressure (atmospheric) in psia, and $T_{sc}$ is the standard temperature in °R. If both oil and gas are present in the reservoir, the

**TABLE 2–1. Standard pressures by location**

| Location | $p_{sc}$, (psia) | Location | $p_{sc}$, (psia) |
|---|---|---|---|
| Arkansas | 14.65 | New Mexico | 15.025 |
| California | 14.73 | Oklahoma | 14.65 |
| Colorado | 15.025 | Texas | 14.65 |
| Illinois | 14.65 | Utah | 15.025 |
| Kansas | 14.65 | West Virginia | 14.85 |
| Louisiana | 15.025 | Wyoming | 15.025 |
| Michigan | 14.73 | U.S. Federal Lands | 14.73 |
| Mississippi | 15.025 | Canada | 14.696 |



$B_g$ can be expressed in bbl/SCF or SCF/bbl as

$$B_g = \frac{5.615 T_{sc} p_R}{p_{sc} z T_R}, \qquad \text{SCF/bbl} \tag{2.14}$$

**Example 2–4.** Compute the gas formation volume factor $B_g$ in SCF/ft$^3$ for the gas in Example 2–3. The standard temperature and pressure are 60°F and 14.65 psia, respectively.

Solution: Given: $p_R = 2{,}500$ psia $\quad T_R = 180°\text{F} \quad z = 0.84$

$$B_g = \frac{(60 + 460) \times 2{,}500}{14.65 \times 0.84 \times (180 + 460)}$$
$$= 165.062 \quad \text{SCF/ft}^3$$

## GAS VISCOSITY

Viscosity of natural gases, symbol $\mu_g$, at reservoir pressure and temperature can be estimated from correlations developed by Carr et al. and presented in Figures 2–4 and 2–5.[6] The correlations require gas gravity or molecular weight, pseudoreduced pressure and temperature, and reservoir pressure and temperature. If the gas contains any contaminant gases, the viscosity read from Figure 2–4 must be corrected using the correction factors from the insets. The correction factors may be calculated using Equations (2.15a) through (2.15c). The gas viscosity is reported in centipoise (cp). The following steps are required to obtain gas viscosity:

**Step 1.** Using gas gravity or molecular weight ($MW = \gamma_g \times 28.97$) and Figure 2–4, determine gas viscosity ($\mu_1$) at atmospheric pressure and reservoir temperature.

**Step 2.** Correct ($\mu_1$) for contaminant gas, if any, using insets in Figure 2–4 or Equations (2.15a) through (2.15c). Corrected $\mu_1 = \mu_1$ (uncorrected) $+ N_2$ correction $+ H_2S$ correction $+ CO_2$ correction.

**Step 3.** Read viscosity ratio ($\mu/\mu_1$) from Figure 2–5 and convert gas viscosity at atmospheric pressure (from Step 1 or 2) to reservoir pressure using the following equation.

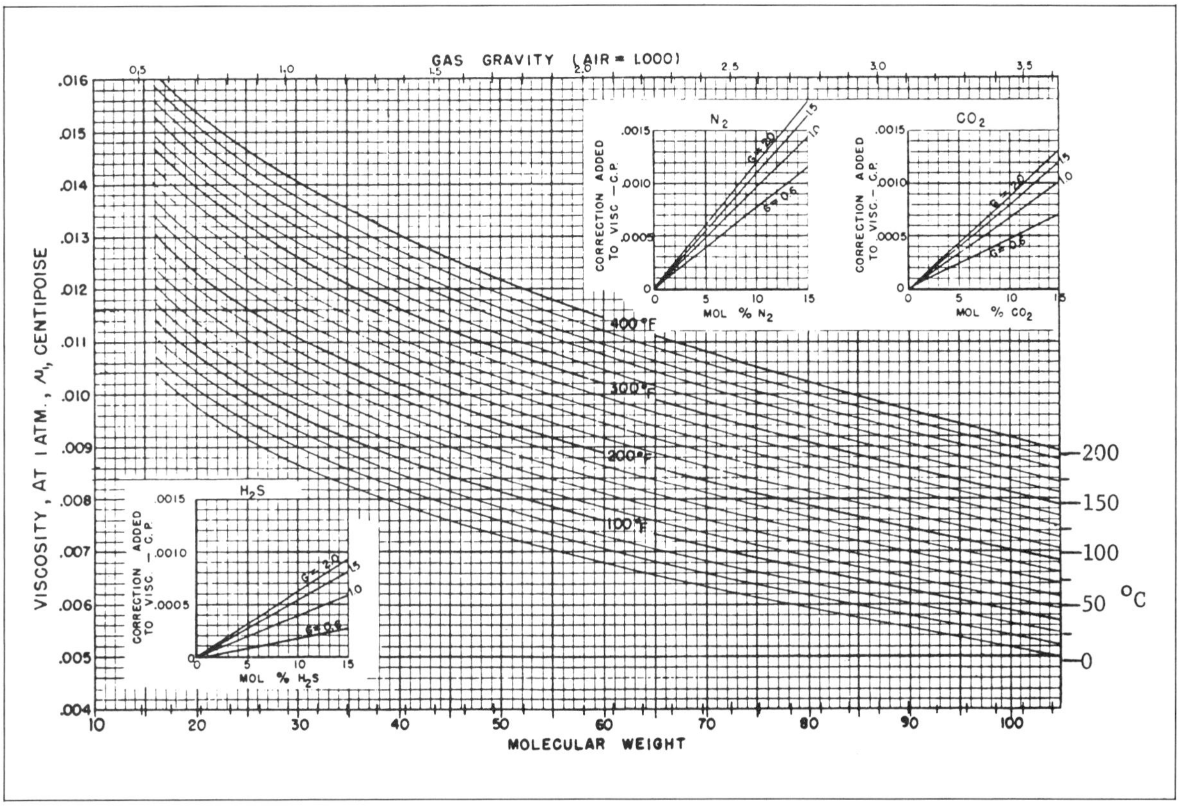


**Figure 2–4.** Viscosity of natural gases at one atmosphere (© SPE, Trans. AIME, 1954).[6]

$$\mu_g = (\mu/\mu_1) \times \mu_1 \tag{2.15}$$

where $\mu_g$ is the gas viscosity at reservoir conditions in centipoise and $\mu_1$ is the gas viscosity at one atmosphere in centipoise. Equations used to compute correction factors are:

$$N_2 \text{ correction} = y_{N_2}\, 8.48 \times 10^{-3} \log(\gamma_g) + 9.59 \times 10^{-3} \tag{2.15a}$$

$$CO_2 \text{ correction} = y_{CO_2}\, 9.08 \times 10^{-3} \log(\gamma_g) + 6.24 \times 10^{-3} \tag{2.15b}$$

$$H_2S \text{ correction} = y_{H_2S}\, 8.49 \times 10^{-3} \log(\gamma_g) + 3.73 \times 10^{-3} \tag{2.15c}$$

**Example 2–5.** Estimate gas viscosity $\mu_g$ for the gas in Example 2–3.

Solution:

$$T_{pr} = 1.65$$
$$p_{pr} = 3.74$$
$$\gamma_g = 0.7$$
$$p_R = 2{,}500 \quad \text{psia}$$
$$T_R = 180°F$$

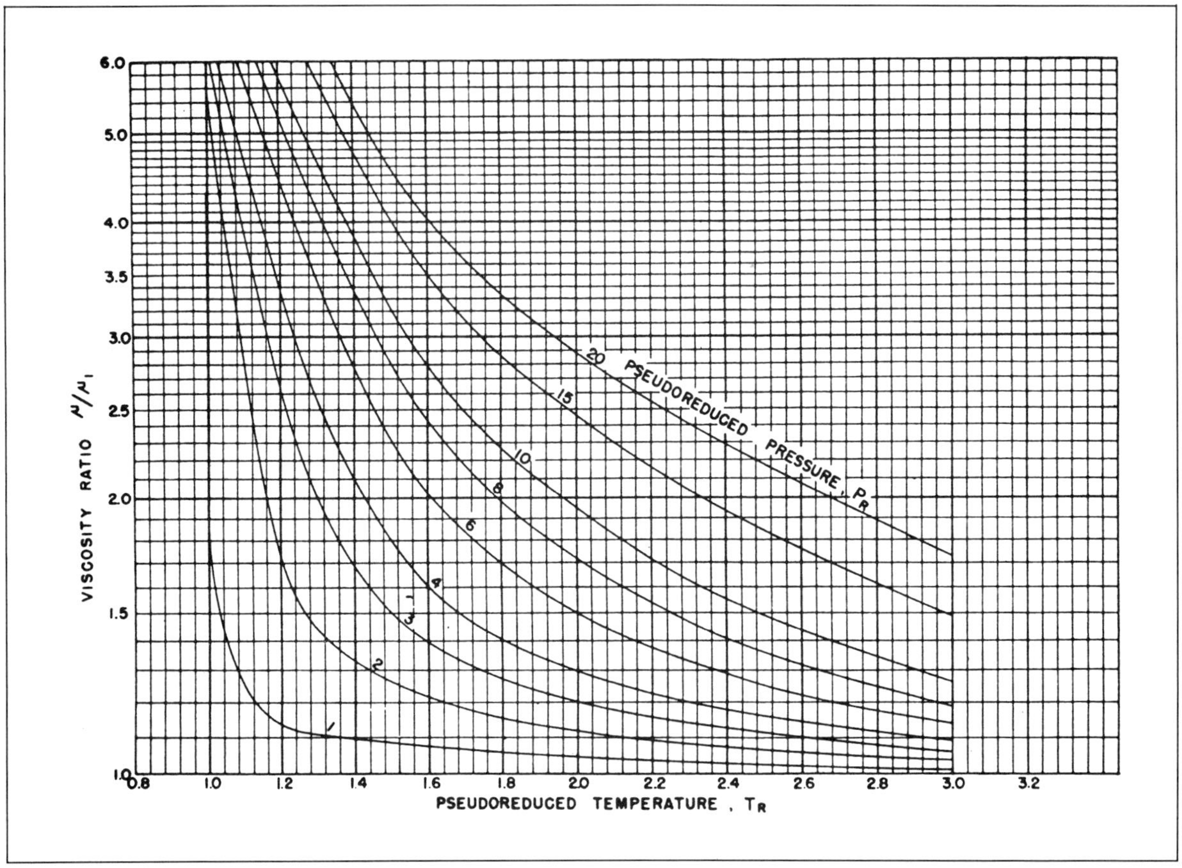


**Figure 2–5.** Effect of pressure and temperature on gas viscosity (© SPE, Trans. AIME, 1954).[6]

**Step 1.** Read $\mu_1$ from Figure 2–4, using $\gamma_g = 0.70$ and $T_R = 180$ °F.

$$\mu_1 = 0.012 \quad \text{cp}$$

**Step 2.** No contaminants are present, so proceed to Step 3.

**Step 3.** Read the ratio $(\mu/\mu_1)$ from Figure 2–5, using $T_{pr}$ and $p_{pr}$. Compute $\mu_g$ using Equation (2.15) as

$$\begin{aligned} \mu/\mu_1 &= 1.48 \quad \text{(from Fig. 2–5)} \\ \mu_g &= 1.48 \times 0.012 \\ &= 0.01776 \quad \text{cp} \end{aligned}$$

## GAS ISOTHERMAL COMPRESSIBILITY

Gas isothermal compressibility, symbol $c_g$, is extensively used in transient well test analysis and other reservoir engineering problems. Correlations developed by Trube as presented in Figure

2–6 are utilized.[7] The required parameters are $T_R$, $p_R$, $T_{pc}$, $p_{pc}$, $T_{pr}$, and $p_{pr}$. Figure 2–6 is used to determine pseudoreduced compressibility $c_{pr}$ as a function of $T_{pr}$ and $p_{pr}$. The $c_g$ is then estimated using the following equation:

$$c_g = \frac{c_{pr}}{p_{pc}} \tag{2.16}$$

where $c_g$ is the isothermal gas compressibility in 1/psia and $c_{pr}$ is the pseudoreduced gas compressibility in 1/psia.

**Example 2–6.** Determine the gas isothermal compressibility for a 0.65 gravity dry gas. The reservoir pressure and temperature are 1,800 psia and 170°F, respectively.



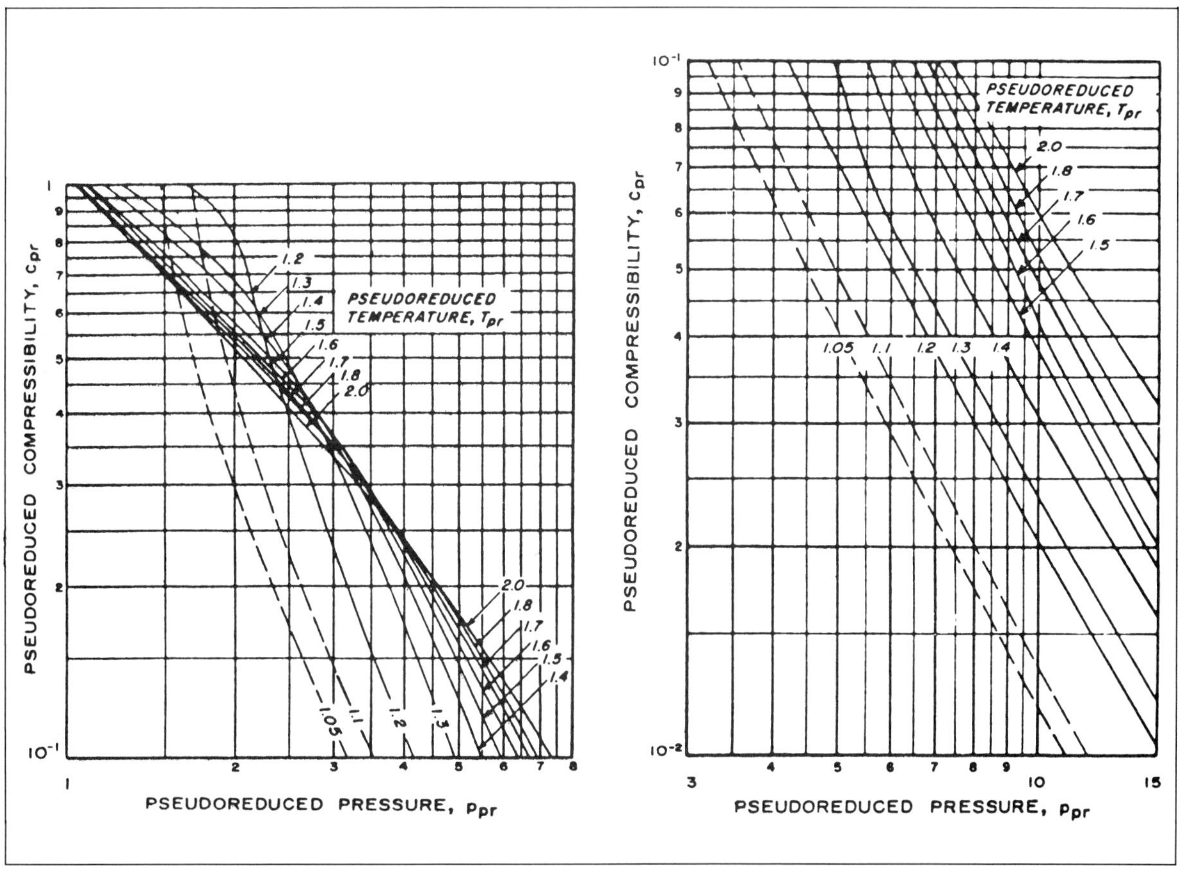


**Figure 2–6.** Correlation of pseudoreduced compressibility for natural gases (© SPE, Trans. AIME, 1957).[7]

Solution:

$$T_{pc} = 374°\text{R} \quad \text{(from Fig. 2–1)}$$
$$p_{pc} = 671 \quad \text{psia} \quad \text{(from Fig. 2–1)}$$
$$T_{pr} = \frac{170 + 460}{374} = 1.68$$
$$p_{pr} = \frac{1{,}800}{671} = 2.68$$

Determine pseudoreduced compressibility from Figure 2–6, and using Equation (2.16), compute $c_g$.

$$c_{pr} = 0.41 \quad \text{(from Fig. 2–6)}$$



$$c_g = \frac{c_{pr}}{p_{pc}} = 0.41/671 = 6.11 \times 10^{-4} \quad 1/\text{psia}$$

## GAS PROPERTIES FROM GAS COMPOSITION

If gas composition is available, the gas gravity and the pseudocritical properties must be determined from the composition, instead of the empirical correlations discussed in the previous sections. In addition, several other properties, such as net and gross heating values, may be determined from the composition.

Natural gas consists of several other gaseous components, such as nitrogen, carbon dioxide, hydrogen sulfide, methane, ethane, propane, isobutane, n-butane, isopentane, n-pentane, n-hexane, n-heptane, n-octane, n-nonane, n-decane, oxygen, hydrogen, helium, and water vapor. Each of these constituents has its own physical properties. Each constituent will be in a certain percentage in the whole gas mixture; therefore, it will constitute its properties in that proportion to the gas mixture. The physical properties of various constituents are given in Table 2–2. The following equations are used to compute the physical properties of a gas using gas composition:

$$\gamma_g = \sum_{i=1}^{n} \gamma_{gi}(y_i) \tag{2.17}$$

$$T_{pc} = \sum_{i=1}^{n} T_{ci}(y_i) \tag{2.18}$$

$$p_{pc} = \sum_{i=1}^{n} p_{ci}(y_i) \tag{2.19}$$

$$\text{NHV} = \sum_{i=1}^{n} \text{NHV}_i(y_i) \tag{2.20}$$

$$\text{GHVD} = \sum_{i=1}^{n} \text{GHVD}_i(y_i) \tag{2.21}$$

**TABLE 2–2. Properties of different natural gas constituents**

| Constituent, *i* | $\gamma_{gi}$ | $T_{ci}$ | $p_{ci}$ | $NHV_i$ | $GHV_i$ |
|---|---|---|---|---|---|
| Nitrogen, $N_2$ | 0.9672 | 227.30 | 493.00 | 0 | 0 |
| Carbon dioxide, $CO_2$ | 1.5195 | 547.60 | 1,071.0 | 0 | 0 |
| Hydrogen sulfide, $H_2S$ | 1.1765 | 672.40 | 1,306.0 | 588.0 | 637.0 |
| Methane, $C_1$ | 0.5539 | 343.04 | 667.8 | 909.1 | 1,009.7 |
| Ethane, $C_2$ | 1.0382 | 549.76 | 707.8 | 1,617.8 | 1,768.8 |
| Propane, $C_3$ | 1.5225 | 665.68 | 616.3 | 2,316.1 | 2,517.4 |
| Isobutane, $i$-$C_4$ | 2.0068 | 734.65 | 529.1 | 3,001.1 | 3,252.7 |
| N-Butane, $n$-$C_4$ | 2.0068 | 765.32 | 550.7 | 3,010.4 | 3,262.1 |
| Isopentane, $i$-$C_5$ | 2.4911 | 828.77 | 490.4 | 3,698.3 | 4,000.3 |
| N-Pentane, $n$-$C_5$ | 2.4911 | 845.40 | 486.6 | 3,707.5 | 4,009.5 |
| N-Hexane, $n$-$C_6$ | 2.9753 | 913.40 | 436.9 | 4,403.7 | 4,756.1 |
| N-Heptane, $n$-$C_7$ | 3.4596 | 972.50 | 396.8 | 5,100.2 | 5,502.9 |
| N-Octane, $n$-$C_8$ | 3.9439 | 1,023.89 | 360.6 | 5,796.7 | 6,249.7 |
| N-Nonane, $n$-$C_9$ | 4.4282 | 1,070.35 | 332.0 | 6,493.3 | 6,996.6 |
| N-Decane, $n$-$C_{10}$ | 4.9125 | 1,111.80 | 304.0 | 7,186.6 | 7,742.3 |
| Oxygen, $O_2$ | 1.1048 | 278.60 | 736.9 | 0 | 0 |
| Hydrogen, $H_2$ | 0.0696 | 59.90 | 188.1 | 274 | 324 |
| Helium, He | 0.1380 | 9.50 | 33.2 | 0 | 0 |
| Water Vapor, $H_2O$ | 0.6220 | 1,165.30 | 3,208.0 | 0 | 0 |



$$GHVW = 0.9826(GHVD) \tag{2.22}$$

where $y_i$ is the mole fraction of the $i$th component in the mixture, $\gamma_{gi}$ is the gravity of the $i$th component, $T_{ci}$ is the critical temperature of the $i$th component, $p_{ci}$ is the critical pressure of the $i$th component, $NHV_i$ is the net heating value of the $i$th component in BTU/SCF, and $GHVD_i$ is the gross heating value (dry gas) of the $i$th component in BTU/SCF. GHVW is the gross heating value (wet gas) of the total mixture in BTU/SCF, and NHV is the net heating value of the gas mixture in BTU/SCF.

The pseudocritical pressure and temperature thus obtained must be corrected for sour gases using the Wichert-Aziz correction. The sum of the mole fractions of the total gas mixture should be equal to 1.0.

**Example 2–7.** A sour natural gas mixture consists of 15% carbon dioxide, 13% hydrogen sulfide, 4% nitrogen, 62% methane, and 6% ethane. Compute gas gravity, pseudocritical pressure and temperature (corrected for sour gases), net heating value, and gross heating value of the gas mixture.

Solution: The best way to solve this type of problem is to tabulate all the properties as shown below.

| Component | (1) $y_i$ | (2) $T_{ci}$ | (3) $p_{ci}$ | (4) $T_{pc}$ | (5) $p_{pc}$ | (6) $\gamma_{gi}$ | (7) $\gamma_g$ |
|---|---|---|---|---|---|---|---|
| $CO_2$ | 0.15 | 547.60 | 1,071.0 | 82.14 | 160.65 | 1.5195 | 0.228 |
| $H_2S$ | 0.13 | 672.40 | 1,306.0 | 87.41 | 169.78 | 1.1765 | 0.153 |
| $N_2$ | 0.04 | 227.30 | 493.0 | 9.09 | 19.72 | 0.9672 | 0.039 |
| $C_1$ | 0.62 | 343.04 | 667.8 | 212.68 | 414.04 | 0.5539 | 0.343 |
| $C_2$ | 0.06 | 549.76 | 707.8 | 32.99 | 42.47 | 1.0382 | 0.062 |
| Total | 1.00 | | | 424.31 | 806.66 | | 0.825 |

Column (1) = mole fraction of individual gas component
Column (2) = critical temperature of individual components
Column (3) = critical pressure of individual components
Column (4) = Column (2) × Column (1)
Column (5) = Column (3) × Column (1)
Column (6) = gas gravity of individual component
Column (7) = Column (6) × Column (1)

$$\gamma_g = 0.825$$
$$T_{pc} = 424.31°R$$
$$p_{pc} = 806.66 \quad \text{psia}$$

Wichert-Aziz correction from Figure 2–2 is equal to 27.85°F.

$$T^*_{pc} = 424.31 - 27.85 = 396.46°R$$

$$p^*_{pc} = \frac{806.66 \times 396.46}{424.31 + 0.2(1 - 0.13)27.85} = 745.203 \quad \text{psia}$$

Net and gross heating values are determined in the same way.

$$\text{NHV} = (0 \times 0.15) + (588 \times 0.13) + (0 \times 0.04) + (909.1 \times 0.62) + (1{,}617.8 \times 0.06) = 737.15 \quad \text{BTU/SCF}$$

$$\text{GHVD} = (0 \times 0.15) + (637 \times 0.13) + (0 \times 0.04) + (1{,}009.7 \times 0.62) + (1{,}768.8 \times 0.06) = 814.95 \quad \text{BTU/SCF}$$

$$\text{GHVW} = 0.9826 \times 814.95 = 800.77 \quad \text{BTU/SCF}$$

## DEW POINT PRESSURE FROM GAS COMPOSITION

Nemeth and Kennedy developed an empirical correlation for the determination of the dew point pressure as a function of gas composition and temperature.[8] The correlation can be used when more complete and expensive laboratory measurements are unavailable.

$$p_d = \exp\{A[0.2\%N_2 + \%CO_2 + \%H_2S + 0.4\%C_1 + \%C_2 + 2(\%C_3 + \%iC_4 + \%nC_4) + \%iC_5 + \%nC_5 + \%nC_6] + B\rho_{C_{7+}} + C\left(\frac{\%C_1}{\%C_{7+} + 0.2}\right) + D(T_R + 460) + EL + FL^2 + GL^3 + HM + IM^2 + JM^3 + K\}$$

where

$$\rho_{C_{7+}} = \frac{[0.6882(\%nC_7) + 0.7068(\%nC_8) + 0.7217(\%nC_9) + 0.7342(\%nC_{10})]}{\%C_{7+}}$$

$$MW_{C_{7+}} = \frac{100.2(\%nC_7) + 114.2(\%nC_8) + 128.3(\%nC_9) + 142.3(\%nC_{10})}{\%C_{7+}}$$

$$\%C_{7+} = \%nC_7 + \%nC_8 + \%nC_9 + \%nC_{10}$$

$$M = \frac{MW_{C_{7+}}}{\rho_{C_{7+}} + 0.0001}$$



and

| | | |
|---|---|---|
| $A = -2.062305 \times 10^{-2}$ | $E = 3.2673714 \times 10^{-2}$ | $I = 6.2476497 \times 10^{-4}$ |
| $B = 6.6259728$ | $F = -3.6453277 \times 10^{-3}$ | $J = -1.0716866 \times 10^{-6}$ |
| $C = -4.4670559 \times 10^{-3}$ | $G = 7.4299951 \times 10^{-5}$ | $K = 10.746622$ |
| $D = 1.0448346 \times 10^{-4}$ | $H = -0.11381195$ | $L = (C_{7+})(MW_{C_{7+}})$ |

## PSEUDOPRESSURE, $\psi(p_R)$, CALCULATIONS

The concept of pseudopressure or real gas potential, symbol $\psi(p)$, is used in gas well testing (pressure buildup), particularly in very low-permeability gas reservoirs when they are produced at high flow rates. The concept was originally defined by Al-Hussainy and Ramey.[9]

In this section we will learn how to compute the pseudopressure at different reservoir pressures. The required data are gas gravity, pseudocritical pressure and temperature, gas deviation factor, viscosity, reservoir pressure and temperature, and pseudoreduced pressure and temperature.

The calculations are time consuming, in practice computer programs are available or developed to handle the computations. The pseudopressure is defined by the following equation:

$$\psi(p) = \int_{p_b}^{p} \frac{2p}{\mu_g z}\, dp \tag{2.23}$$

where $\psi(p_R)$ is the pseudopressure or real gas potential in psia/cp, $p_b$ is an arbitrary base pressure, and $p_R$ is the reservoir pressure in psia.

**Example 2–8.** Convert the reservoir pressure of 500, 1,000, 1,500, and 2,000 psia into their respective pseudopressures. The gas gravity is 0.80 (air = 1) and reservoir temperature is 273°F. Use base pressure of 15 psia.

Solution: Develop a table as shown below. The various gas properties are determined using the methods discussed in the previous sections.

$$\gamma_g = 0.80 \quad (\text{air} = 1) \qquad T_{pr} = 1.75$$
$$T_{pc} = 420°\text{R} \qquad \mu_g = 0.01416 \quad \text{cp}$$
$$p_{pc} = 665 \quad \text{psia}$$

| Pressure psia | $p_{pr}$ | $z$ | $\mu/\mu_1$ | $\mu_g$ (cp) | $2p/\mu_g z$ | $\psi(p)$ (psia/cp) |
|---|---|---|---|---|---|---|
| 15 | 0.023 | 0.999 | 1.0 | 0.01416 | 2.121E3 | 0 |
| 500 | 0.752 | 0.957 | 1.03 | 0.01459 | 71.620E3 | 17.88E6 |
| 1,000 | 1.504 | 0.920 | 1.086 | 0.01538 | 141.347E3 | 71.12E6 |
| 1,500 | 2.256 | 0.894 | 1.163 | 0.01647 | 203.746E3 | 157.39E6 |
| 2,000 | 3.008 | 0.880 | 1.259 | 0.01783 | 254.933E3 | 272.06E6 |



$$\psi(p)_{15} = 0$$

$$\psi(p)_{500} = \int_{p_b=15}^{p=500} \frac{2.121 + 71.620}{2}\, dp + \psi(p)_{15}$$
$$= 36.87 \times 10^3(500 - 15) + 0$$
$$= 17.88 \times 10^6$$

$$\psi(p)_{1000} = \int_{500}^{1000} \frac{71.620 + 141.347}{2}(1000 - 500) + \psi(p)_{500}$$
$$= (106.484 \times 10^3)(500) + 17.88 \times 10^6$$
$$= 71.12 \times 10^6 \quad \text{and so on.}$$

## GAS DENSITY AND WELLSTREAM GRAVITY

Gas density, $\rho_g$, at standard conditions can be calculated using the equation of state as follows:

$$(\rho_g)_{sc} = \frac{M_g p_{sc}}{z_{sc} R(T_{sc} + 460)} \quad \text{lb/ft}^3 \tag{2.24}$$

where $z_{sc} = 1$ and $R = 10.732$. The molecular weight of gas, $M_g$, can be calculated as the product of molecular weight of air ($M_{air} = 28.97$), and the gas gravity, $\gamma_g$ (air = 1). Alternatively, it is useful to know that the density of air, $\rho_{air}$, at standard conditions is 0.0763 lb/ft$^3$. Thus if the gas gravity is known, gas density at standard conditions is then given by

$$(\rho_g)_{sc} = 0.0763\gamma_g \tag{2.25}$$

Once the gas density at standard conditions is known, the gas density at reservoir conditions can be directly calculated using Equation (2.24) as

$$(\rho_g)_R = \frac{35.35(\rho_g)_{sc}p_R}{z(T_R + 460)} \tag{2.26}$$

**Example 2–9.** Calculate the specific gravity and density of a reservoir gas with molecular weight of 19.91. The reservoir temperature and pressure are 180°F and 2,000 psia, respectively. The standard temperature and pressure are 60°F and 14.7 psi, respectively. Gas deviation factor, $z$, at reservoir condition is 0.852.
Solution:

$$\begin{aligned}\rho_g &= \frac{M_g}{M_{air}}\\ &= \frac{19.91}{28.97}\\ &= 0.687\end{aligned}$$



Using Equation (2.24),

$$\begin{aligned}(\rho_g)_{sc} &= \frac{(19.91)(14.7)}{(1)(10.732)(60 + 460)}\\ &= 0.0524 \quad \text{lb/ft}^3\end{aligned}$$

Using Equation (2.26),

$$\begin{aligned}(\rho_g)_R &= \frac{(35.35)(0.0524)(2{,}000)}{(0.852)(180 + 460)}\\ &= 6.794 \quad \text{lb/ft}^3.\end{aligned}$$

In dealing with gas-condensate reservoirs, both retrograde and nonretrograde, the knowledge of average gas specific gravity (air = 1) and the wellstream gravity of the total well fluid is required. Rzasa and Katz developed a chart showing the relationship between the ratio of well fluid gravity (as a vapor) to trap gas gravity and the barrels of surface condensate per million cubic feet of surface gas, such as shown in Figure 2–7.[10] Mathematically, the ratio of well fluid gravity to trap gas gravity is given by the following equation:

$$\frac{\text{Well Fluid Gravity}}{\text{Trap Gas Gravity}} = \frac{M_o}{28.97\gamma_g}\left(\frac{76.4R_g\gamma_g + 350\gamma_o}{2.64R_gM_o + 350\gamma_o}\right) \tag{2.27}$$

where $R_g$ is the surface GOR in MSCF/bbl of condensate, $\gamma_o$ is the surface gravity of condensate (tank oil), $\gamma_g$ is the specific gravity of dry gas produced from separator and stock tank, and $M_o$ is the molecular weight of condensate (tank oil).

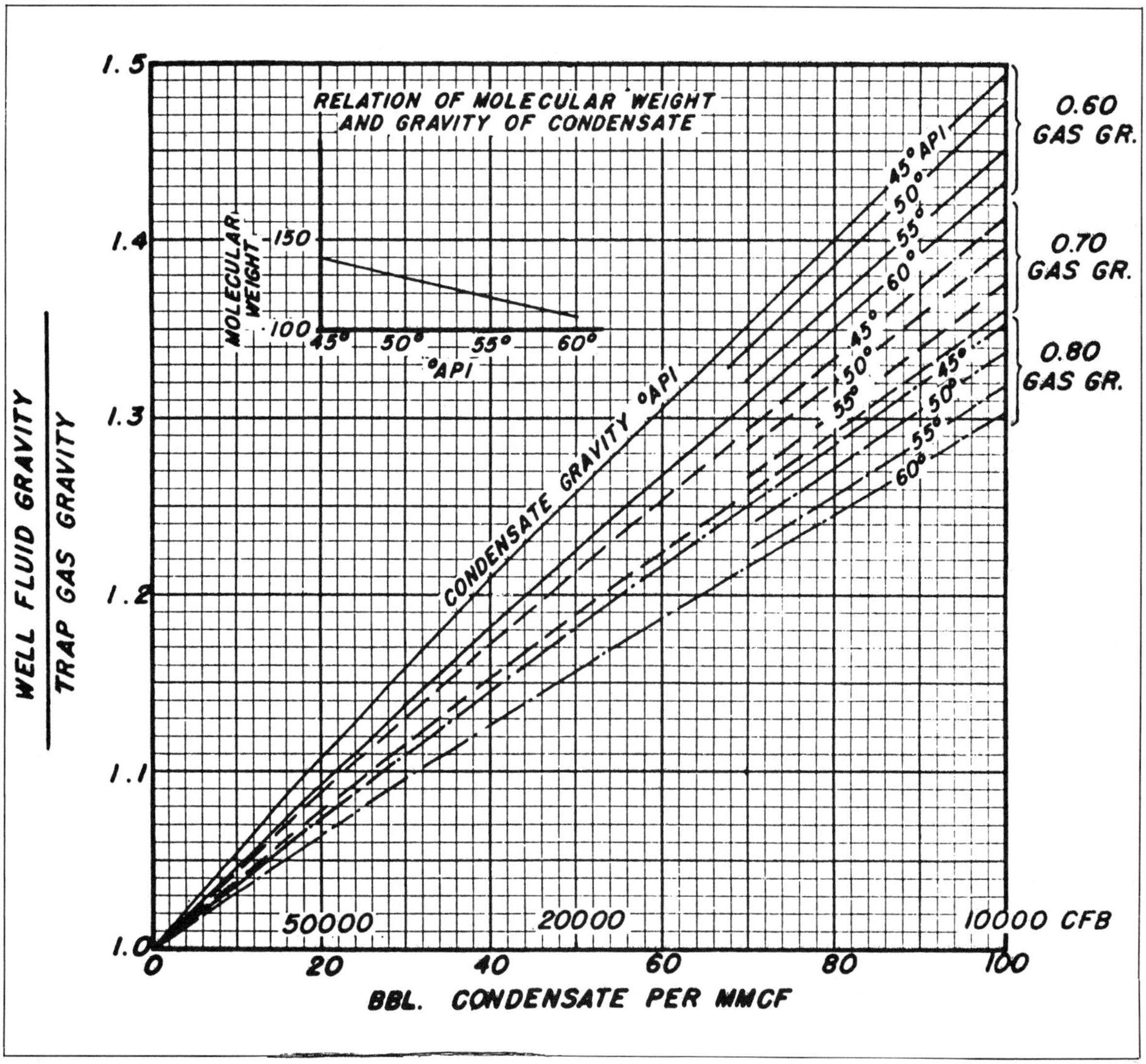


**Figure 2–7.** Effect of condensate volume on the ratio of surface gas gravity to well fluid gravity (© SPE, 1977).[11]

Hence the specific gravity of well fluid is given by

$$\gamma_w = \frac{R_g(\gamma_g)_{ave} + 4{,}584\gamma_o}{R_g + 132{,}800\left(\dfrac{\gamma_o}{M_o}\right)} \tag{2.28}$$

If the molecular weight of the condensate (tank oil) is unknown, it may be calculated using the following equation:

$$M_o = \frac{44.29\gamma_o}{1.03 - \gamma_o} = \frac{6{,}084}{°API - 5.9} \tag{2.29}$$

The average gas gravity is calculated by recombining the produced gas and oil in the correct ratio as

$$(\gamma_g)_{ave} = \frac{q_{SG}\gamma_{SG} + q_{TG}\gamma_{TG}}{q_{SG} + q_{TG}} \quad \textbf{(2.30)}$$

where the subscripts SG and TG refer to the daily separator gas and the daily tank gas production, respectively.

**Example 2–10.** Calculate the average gas gravity and the wellstream gravity from the following field data:

Daily tank oil, $q_o$ = 350 bbls
Produced tank oil gravity = 50°API
Daily separator gas, $q_{SG}$ = 3,200 MSCF
Produced separator gas gravity, $\gamma_{SG}$ = 0.70 (air = 1)
Daily tank gas, $q_{TG}$ = 300 MSCF
Produced tank gas gravity, $\gamma_{TG}$ = 1.18 (air = 1)

Solution: Using Equation (2.30),

$$(\gamma_g)_{ave} = \frac{(3{,}200)(0.70) + (300)(1.18)}{3{,}200 + 300}$$
$$= 0.741$$
$$\gamma_o = \frac{141.5}{°API + 131.5}$$
$$= \frac{141.5}{50 + 131.5} = 0.780$$

Using Equation (2.29),

$$M_o = \frac{6{,}084}{50 - 5.9}$$
$$= 137.959 \text{ lbm/lb-mole}$$

The GOR is calculated as

$$R_g = \frac{q_{SG} + q_{TG}}{q_O}$$
$$= \frac{3{,}200 + 300}{350}$$
$$= 10 \text{ MSCF/bbl} \quad \text{OR} \quad 10{,}000 \text{ SCF/bbl}$$

Then by using Equation (2.28), we get

$$\gamma_w = \frac{(10{,}000)(0.741) + (4{,}584)(0.780)}{(10{,}000) + 132{,}800\left(\dfrac{0.780}{137.959}\right)}$$

$$= \frac{10{,}985.520}{10{,}750.832}$$

$$= 1.022$$

## OIL PROPERTIES

This section deals with the computation of oil PVT properties from empirical correlations. The properties discussed in this section include oil gravity (°API), oil viscosity ($\mu_o$), oil formation volume factor ($B_0$), bubblepoint pressure ($p_b$), and two-phase formation volume factor ($B_t$). The properties will be determined for oil at bubblepoint pressure, below bubblepoint pressure, and above bubblepoint pressure. The latest developments in the oil PVT correlations, including viscosity calculations of heavy oil, bitumen, and petroleum fractions, are also presented.

### SPECIFIC GRAVITY OF OIL

Crude oils are classified according to their physical properties, such as oil specific gravity, viscosity, surface tension, density, boiling point, and freezing point, etc. The most commonly used properties for classification are viscosity and oil gravity.

The crude oil gravity is reported in degrees API (°API), ranging from 8°API to 58°API. The higher the API gravity, the lighter the crude, and vice versa. The API gravity and crude oil specific gravity (water = 1) are related by the following equations:

$$\gamma_o = \frac{141.5}{131.5 + °\text{API}} \tag{2.31}$$

and

$$°\text{API} = \frac{141.5}{\gamma_o} - 131.5 \tag{2.32}$$

### BUBBLEPOINT PRESSURE

Bubblepoint as defined by Standing is the state at which an infinitesimal quantity of gas is in equilibrium with a large quantity of fluid.[11] Bubblepoint pressure, symbol $p_b$, is the fluid pressure in a system at its bubblepoint. Bubblepoint pressure is used synonymously with saturation pressure.

When the pressure is above the bubblepoint pressure (undersaturated), the fluid is capable of holding additional gases or liquids at the existing pressure and temperature. When oil and gas are produced and the reservoir pressure drops below the bubblepoint pressure, gas is liberated from oil. The following equation or Figure 2–8 is used to estimate the bubblepoint pressure:

$$p_b = 18.2\left\{\left[\frac{R_s}{\gamma_g}\right]^{0.83} \times 10^{[0.00091(T_R) - 0.0125(°\text{API})]} - 1.4\right\} \quad \textbf{(2.33)}$$

where $R_s$ is the produced gas/oil ratio in SCF/STB and $T_R$ is the reservoir temperature in °F. If the separator temperature and pressure are known, the gas gravity should be corrected for separator conditions, using the following equation:

$$\gamma_{g-corrected} = \gamma_g\left[1 + \frac{5.912}{10^5} \times °\text{API} \times T_{sp} \times \left(\log\frac{P_{sp}}{114.7}\right)\right] \quad \textbf{(2.33a)}$$



where $\gamma_{g-corrected}$ is the corrected gas gravity for separator temperature ($T_{sp}$) in °F and separator pressure ($p_{sp}$) in psia.

**Example 2–11.** Determine the bubblepoint pressure of a 38°API crude oil at reservoir temperature of 200°F. The gas/oil ratio is 400 SCF/bbl and the gas gravity is 0.65.

Solution: Using Equation (2.33),

$$\begin{aligned} p_b &= 18.2\left\{\left[\frac{400}{0.65}\right]^{0.83} \times 10^{(0.00091 \times 200 - 0.0125 \times 38)} - 1.4\right\} \\ &= 18.2[(206.53)(10^{-0.293}) - 1.4] \\ &= 18.20(103.792) \\ &= 1{,}889 \quad \text{psia} \quad (\text{from Fig. 2–8, } p_b = 1{,}900 \text{ psia}) \end{aligned}$$

## OIL VISCOSITY

Three figures are provided to help in estimating oil viscosity. Figure 2–9 or Equation (2.34) is used to estimate dead oil (gas-free) viscosity, symbol $\mu_{od}$, at reservoir temperature and °API. Figure 2–10 is used to correct the $\mu_{od}$ to account for the amount of dissolved gas at reservoir pressure (gas-saturated oil viscosity, $\mu_{ob}$). If the reservoir pressure is above the bubblepoint pressure ($p_R > p_b$), then Figure 2–11 and Equation (2.35) are used to account for the increase in viscosity above the $p_b$. The crude oil viscosities are reported in centipoise (cp).

$$\mu_{od} = \left[0.32 + \frac{1.8(10^7)}{°\text{API}^{4.53}}\right]\left(\frac{360}{T_R + 200}\right)^a \quad \textbf{(2.34)}$$

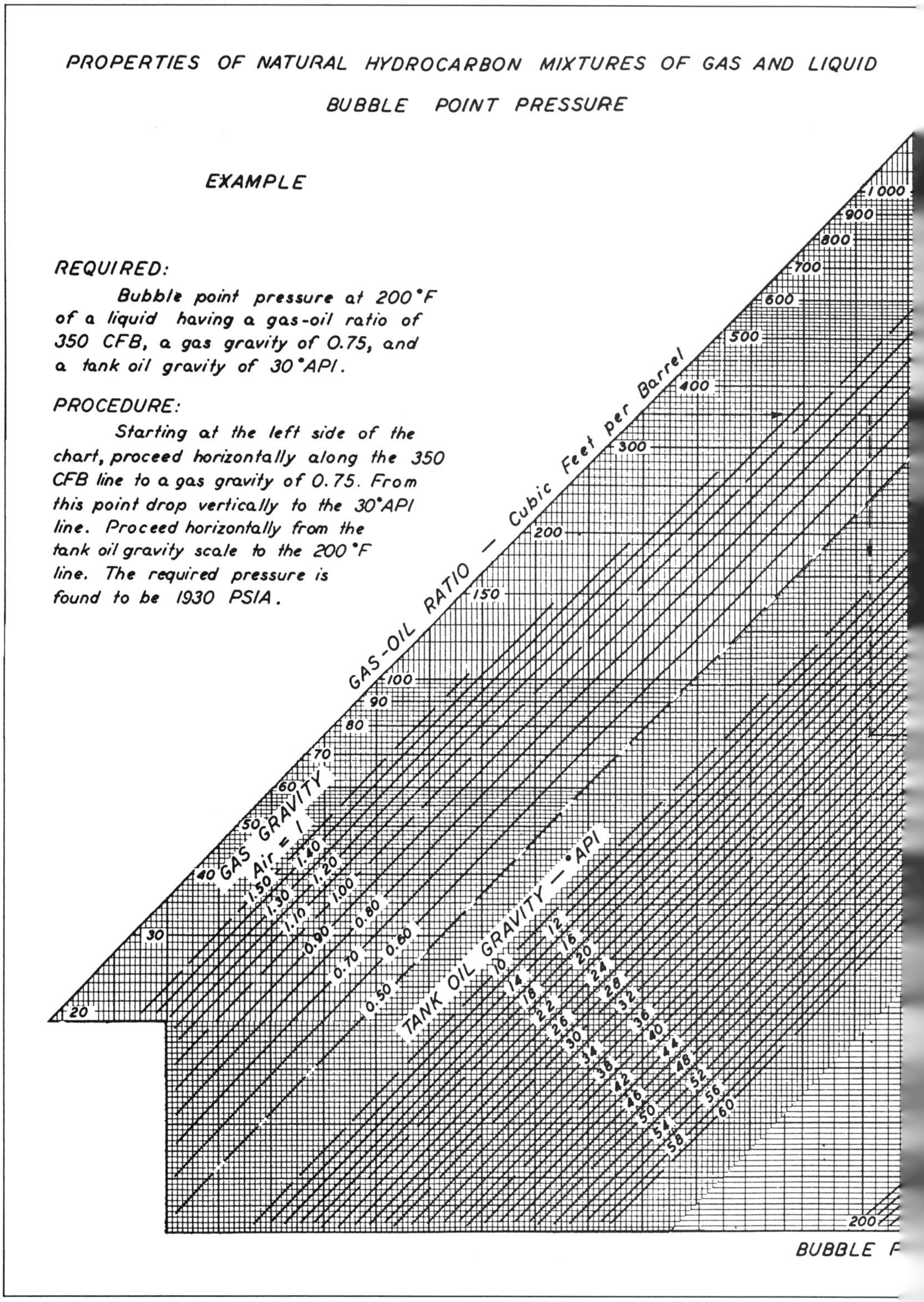


**Figure 2–8.** Bubblepoint pressure or dissolved gas/oil ratio chart.

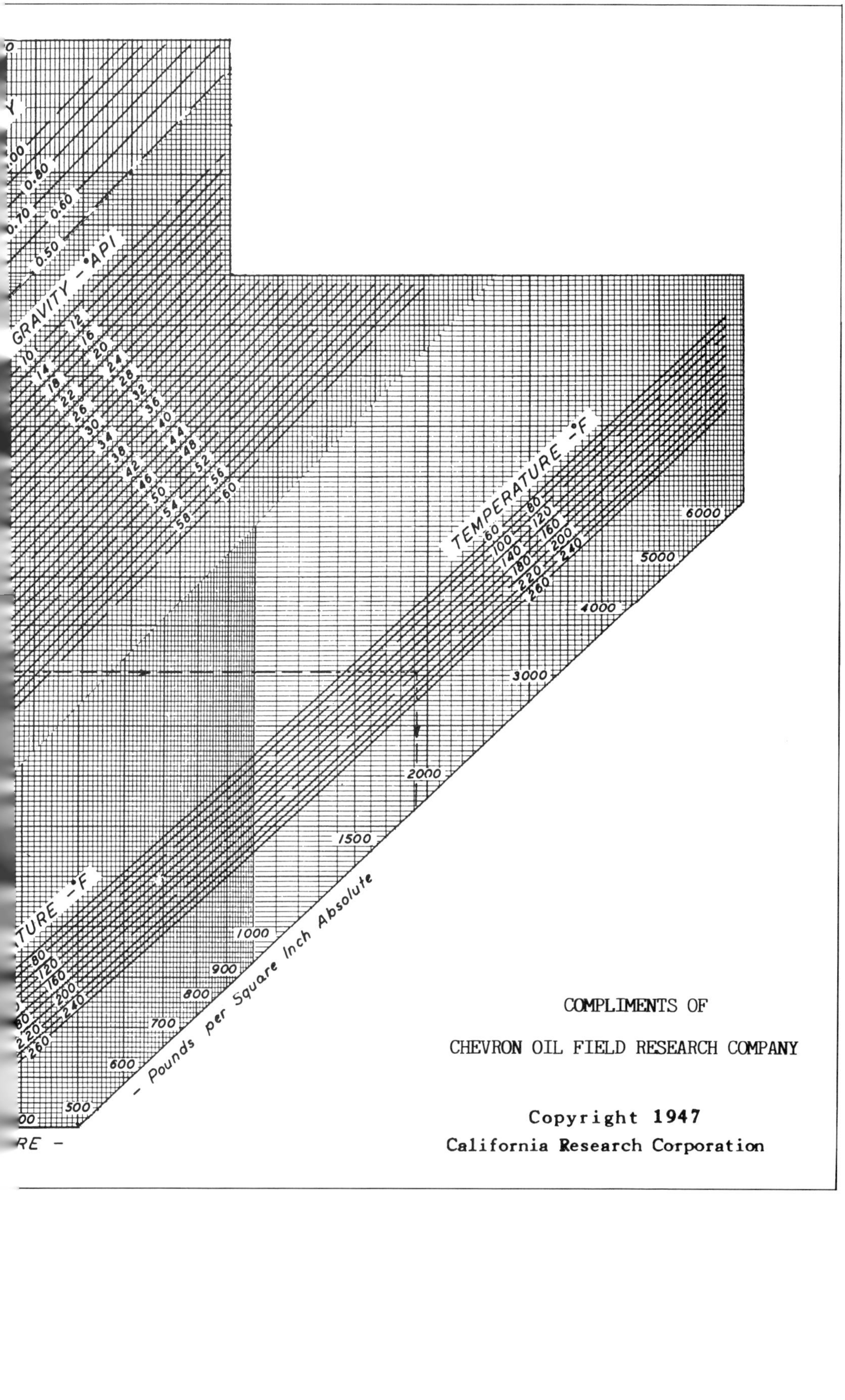
GRAVITY – °API
TEMPERATURE – °F
– Pounds per Square Inch Absolute
RE –
COMPLIMENTS OF
CHEVRON OIL FIELD RESEARCH COMPANY
Copyright 1947
California Research Corporation

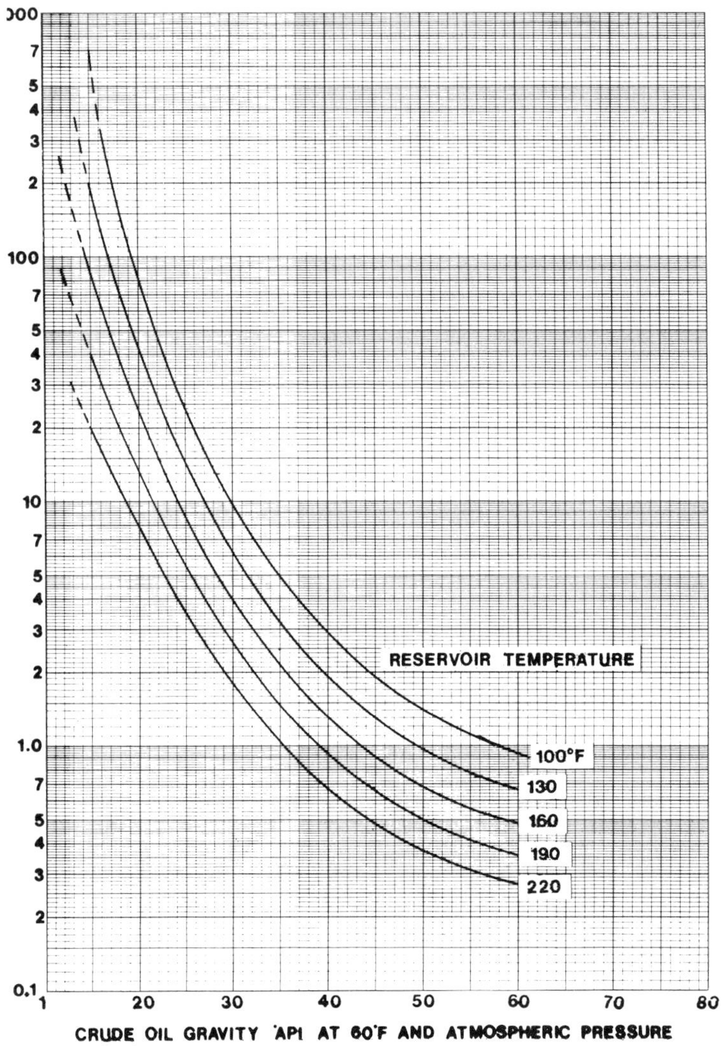


**Figure 2–9.** Dead oil viscosity at reservoir temperature and pressure (after Beal).[12]

where

$$a = 10^{(0.43 + 8.33/°API)} \tag{2.34a}$$

$$\mu_o = 1 + (\Delta\mu)\left(\frac{p_R - p_b}{1{,}000}\right) \tag{2.35}$$

where $\Delta\mu$ is estimated from Figure 2–11.

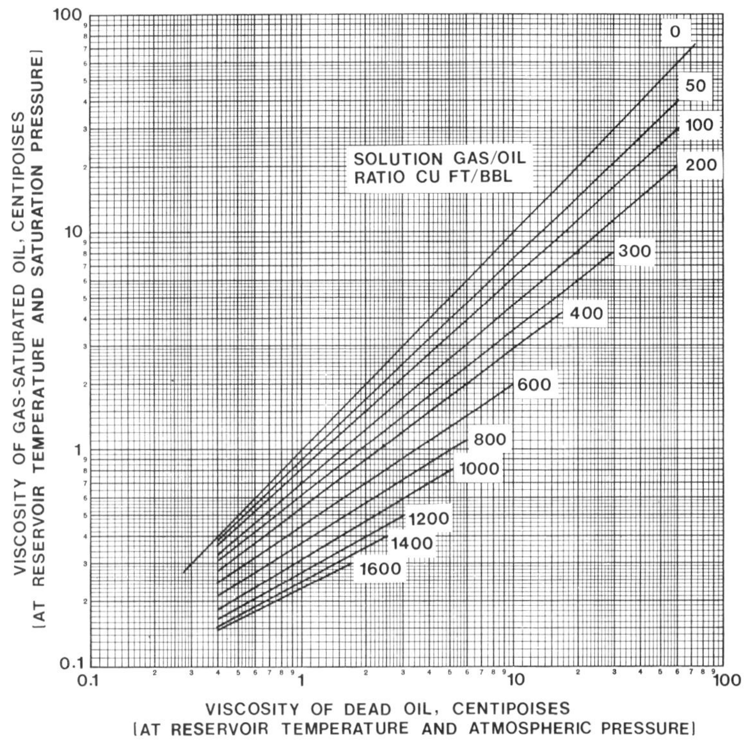


**Figure 2–10.** Viscosity of gas saturated crude oil at reservoir temperature and pressure (after Chew and Conally).[13]

**Example 2–12.** Estimate the viscosity of an undersaturated ($p_R < p_b$) crude oil at a reservoir pressure of 4,000 psia and a bubblepoint pressure of 2,000 psia. Crude oil gravity is 35°API, reservoir temperature is 150°F, and solution gas/oil ratio is 400 SCF/STB.

Solution:

**Step 1.** Estimate $\mu_{od}$ from Figure 2–9 or Equation (2.34).

$$
\begin{aligned}
a &= 10^{(0.43+8.33/35)} = 10^{0.668} \\
&= 4.656 \\
\mu_{od} &= \left[0.32 + \frac{1.8(10^7)}{35^{4.35}}\right]\left(\frac{360}{150+200}\right)^{4.656} \\
&= (2.142)(1.0286)^{4.656} \\
&= 2.443 \quad \text{cp} \quad (\text{from Fig. 2–8, } \mu_{od} = 2.5 \text{ cp})
\end{aligned}
$$

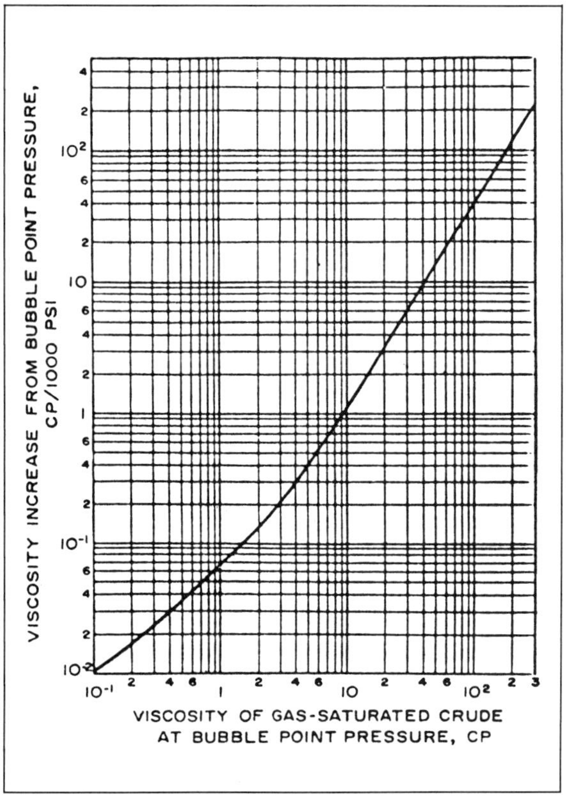


**Figure 2–11.** Change in oil viscosity above bubblepoint pressure (after Beal).[12]

**Step 2.** Correct the dead oil viscosity for dissolved gas using Figure 2–10.

$$\mu_{ob} = 1.05 \quad \text{cp}$$

**Step 3.** Using Figure 2–11 and Equation (2.35), estimate the viscosity ($\mu_o$) of undersaturated crude oil at a reservoir pressure of 4,000 psia.

$$\Delta\mu = 0.067 \quad \text{(from Fig. 2–11)}$$

$$\mu_o = 1 + (0.067)\left(\frac{4{,}000 - 2{,}000}{1{,}000}\right)$$

$$= 1.134 \quad \text{cp}$$

## OIL FORMATION VOLUME FACTOR

Oil formation volume factor, symbol $B_o$, is defined as the ratio of the liquid volume at reservoir conditions to the liquid volume at stock-tank (standard) conditions. This factor is used to convert reservoir barrels to stock-tank barrels. Figure 2–12 or Equation (2.36) developed by Standing is used to estimate $B_o$ at or below the bubblepoint pressure.

$$B_{ob} = 0.972 + \frac{1.47}{10^4}\left[(R_s)\sqrt{\frac{\gamma_g}{\gamma_o}} + 1.25(T_R)\right]^{1.175} \qquad \textbf{(2.36)}$$

where $B_{ob}$ is the formation volume factor at the bubblepoint pressure ($p_b$) in RB/bbl and $T_R$ is reservoir temperature in °F. If the reservoir pressure is above the bubblepoint pressure, then Equation (2.37) is used to compute $B_o$. This equation requires oil isothermal compressibility to be discussed in the next section.

$$B_o = B_{ob}e^{c_o(p_b - p_R)} \qquad \textbf{(2.37)}$$

where $B_o$ is the oil formation volume factor at $p_R > p_b$ in RB/bbl and $c_o$ is the oil isothermal compressibility in 1/psia.

**Example 2–13.** Estimate oil formation volume factor of a saturated crude oil from a reservoir with solution gas/oil ratio of 350 SCF/bbl, gas gravity of 0.75 (air = 1), tank oil gravity of 30°API, and reservoir temperature of 200°F.

Solution: Use Equation (2.36).

$$\gamma_o = \frac{141.5}{131.5 + 30}$$
$$= 0.876$$

$$B_{ob} = 0.972 + \frac{1.47}{10^4}\left[(350)\sqrt{\frac{0.750}{0.876}} + 1.25(200)\right]^{1.175}$$
$$= 0.972 + \frac{1.47}{10^4}(323.852 + 250)^{1.175}$$
$$= 1.23 \quad \text{RB/bbl} \quad (\text{from Fig. 2–12, } B_{ob} = 1.22 \text{ RB/bbl})$$

## OIL ISOTHERMAL COMPRESSIBILITY



Oil isothermal compressibility, symbol $c_o$, at or above the bubblepoint pressure is estimated from the following equation:

$$c_o = \frac{-1{,}433 + 5(R_s) + 17.2(T_R) - 1{,}189(\gamma_g) + 12.61(\gamma_o)}{10^5(p_R)} \tag{2.38}$$

where $c_o$ is the oil isothermal compressibility at $p_R \geq p_b$, in 1/psia, $T_R$ is in °F, $R_s$ is in SCF/bbl, and $\gamma_g$ is the gas gravity corrected for separator conditions.

To estimate the oil isothermal compressibility bellow the bubblepoint pressure ($p_R < p_b$), the required data will be $T_R$, $p_R$, solution GOR ($R_s$), $\gamma_o$, $\gamma_g$, $B_o$, and $B_g$ in bbl/SCF. The following equation will be used:

$$c_o = \frac{1}{B_o}\frac{dR_s}{dp_R}\left(B_g - \frac{dB_o}{dR_s}\right) \tag{2.39}$$

where

$$\frac{dR_s}{dp_R} = \frac{R_s}{(0.83p_R + 21.75)} \tag{2.39a}$$

and $dB_o/dR_s$ is estimated from Figure 2–13.

**Example 2–14.** Estimate $c_o$ for an undersaturated crude oil of 30°API gravity at a reservoir temperature of 200°F and pressure of 4,500 psia, with a GOR of 500 SCF/bbl, and gas gravity of 0.75.

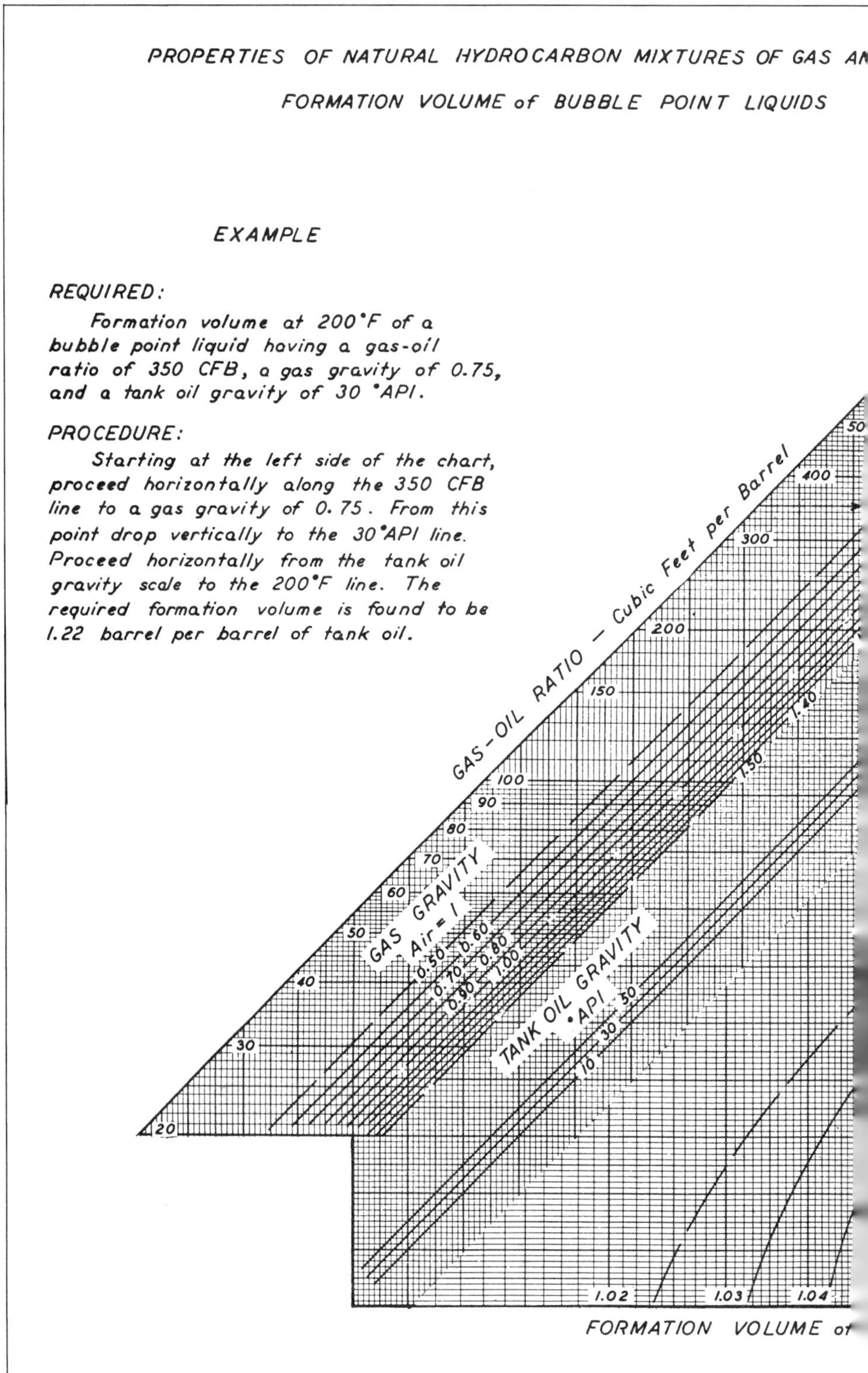


**Figure 2–12.** Oil formation volume factor at the bubblepoint pressure.

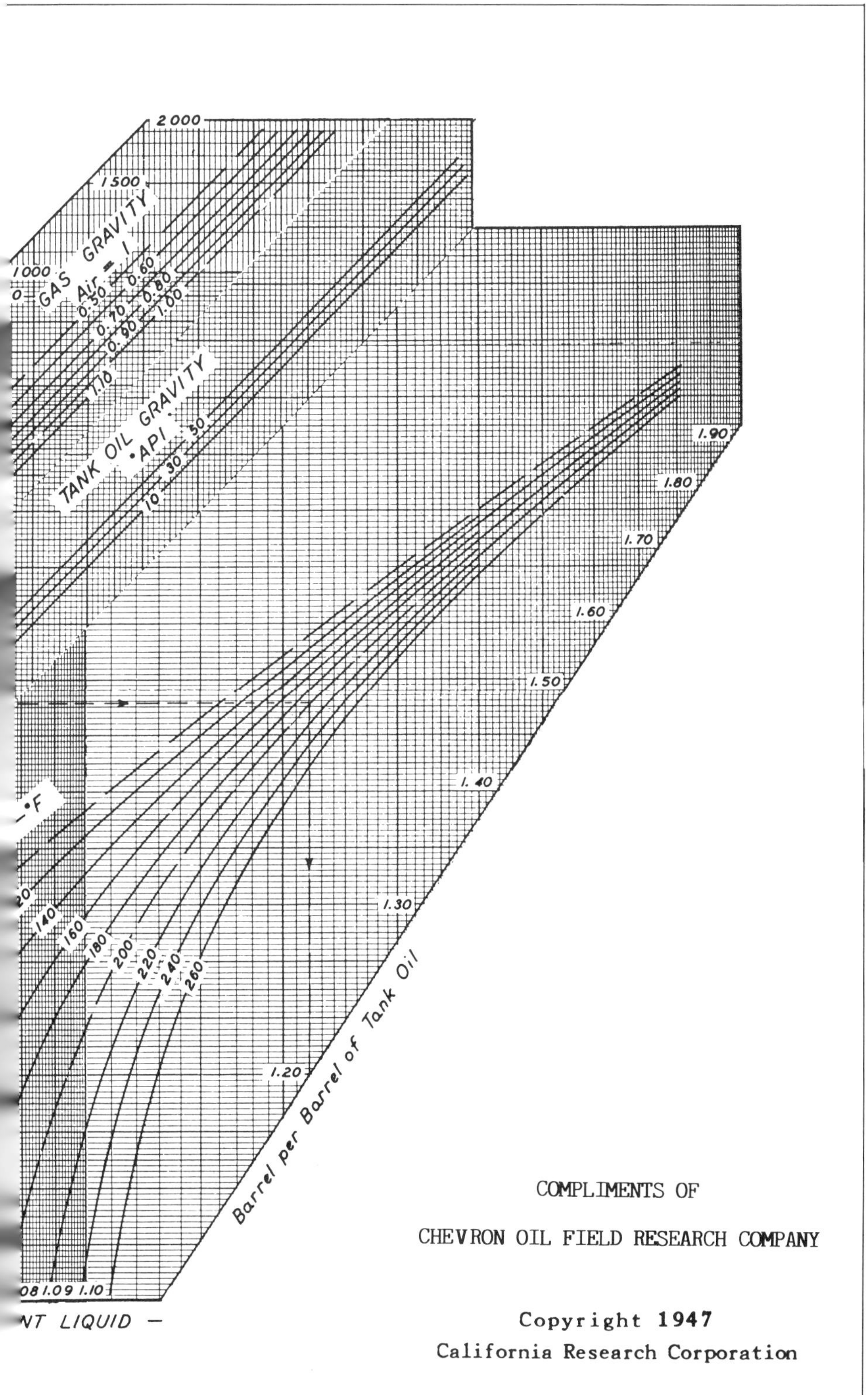
2 000
1 500
1000
GAS GRAVITY
Air = 1
0.50
0.60
0.70
0.80
0.90
1.00
1.10
TANK OIL GRAVITY
°API
10
30
50
1.90
1.80
1.70
1.60
1.50
1.40
1.30
1.20
140
160
180
200
220
240
260
1.09
1.10
Barrel per Barrel of Tank Oil
COMPLIMENTS OF
CHEVRON OIL FIELD RESEARCH COMPANY
Copyright 1947
California Research Corporation

Solution: Use Equation (2.38) as

$$\gamma_o = \frac{141.5}{131.5 + 30} = 0.876$$

$$c_o = \frac{-1{,}433 + 5(500) + 17.2(200) - 1{,}180(0.75) + 12.61(0.876)}{10^5 \times 4{,}500}$$

$$= \frac{-1{,}433 + 2{,}500 + 3{,}440 - 885 + 11.046}{10^5 \times 4{,}500}$$

$$= 8.0734 \times 10^{-6} \quad 1/\text{psia}$$

**Example 2–15.** Estimate $c_o$ below the bubblepoint pressure for the data in Example 2–14. The reservoir pressure has dropped to 1,500 psia ($p_b$ = 2,724 psia).

Solution:



**Step 1.** First estimate gas PVT properties.

$$T_{pc} = 404.72°\text{R} \quad \text{(from Fig. 2–1)}$$

$$p_{pc} = 667.16 \quad \text{psia} \quad \text{(from Fig. 2–1)}$$

$$T_{pr} = 1.63$$

$$p_{pr} = 2.248$$

$$z = 0.86 \quad \text{(from Fig. 2–3)}$$

$$B_g = \frac{0.00504(T_R + 460)z}{p_R}$$

$$= \frac{0.00504(200 + 460)0.86}{1{,}500}$$

$$= 0.0019 \quad \text{bbl/SCF}$$

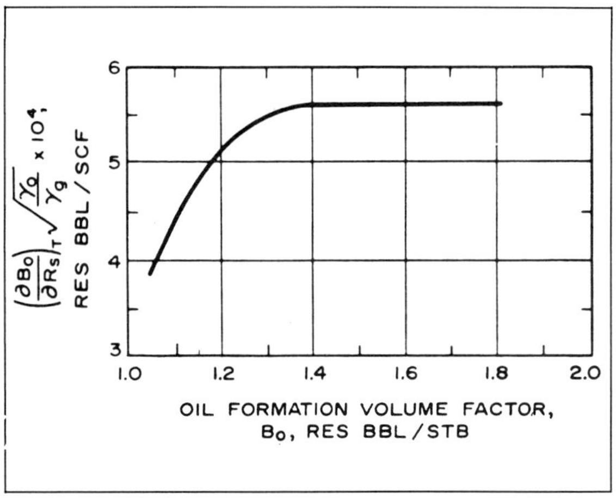


**Figure 2–13.** Change in oil formation volume factor with gas in solution versus $B_O$ (after Ramey).[14]

**Step 2.** Estimate $R_s$ at 1,500 psia.

$$R_s = 260 \quad \text{SCF/bbl} \quad \text{(from Fig. 2–7)}$$

**Step 3.** Estimate $B_o$ from Figure 2–12.

$$B_o = 1.182 \quad \text{RB/bbl}$$

**Step 4.** Compute $dR_s/dp_R$ using Equation (2.39a).

$$dR_s/dp_R = \frac{260}{(0.83 \times 1{,}500 + 21.75)}$$

$$= 0.2052 \quad \text{SCF/bbl-psia}$$

**Step 5.** Estimate change of formation volume factor with gas in solution, using Figure 2–13.

$$\frac{dB_o}{dR_s}\frac{\gamma_o}{\gamma_g} \times 10^4 = 5.0$$

$$\frac{dB_o}{dR_s} = 5 \times 10^{-4}\frac{0.876}{0.750}$$
$$= 5.4037 \times 10^{-4} \quad \text{bbl/SCF}$$

**Step 6.** Now compute $c_o$ from Equation (2.39).

$$c_o = \frac{1}{1.182} \times 0.2052(0.0019 - 5.4037 \times 10^{-4})$$
$$= 0.1736 \times 0.00136$$
$$= 2.36 \times 10^{-4} \quad 1/\text{psia}$$

## OIL DENSITY DETERMINATION

The knowledge of oil density at various reservoir pressures and a reservoir temperature is required for most reservoir engineering calculations. An equation for oil density is

$$\rho_o = \frac{350\gamma_o + 0.0764\gamma_g R_s}{5.615B_o}$$

where 350 = density of water at standard conditions (lb/STB), $\rho_o$ = oil density (lb/ft$^3$), $R_s$ = solution or dissolved gas (SCF/STB), and 0.0764 = density of air at standard conditions (lb/SCF).

For the case when $p_R > p_b$, the $\rho_o$ is given by

$$\rho_o = \rho_{ob} \exp\left[c_o(p_R - p_b)\right]$$

## DISSOLVED GAS/OIL RATIO

To estimate oil viscosity and formation volume factor at reservoir pressure below the bubblepoint pressure ($p_R < p_b$), the free gas must be considered. The gas/oil ratio below the bubblepoint pressure, symbol $R_{sb}$, is estimated from Figure 2–7 or the following equations. The required data are bubblepoint pressure, reservoir pressure, reservoir temperature, gas gravity (corrected for separator conditions), and stock-tank oil gravity in °API. *If the reservoir pressure is above the bubblepoint pressure, the observed gas/oil ratio is equal to* $R_s$.

If oil gravity is less than or equal to 30°API,

$$R_{sb} = 0.0362(\gamma_g)(p_R^{1.0937}) \exp\left(25.7240\frac{°\text{API}}{T_R + 460}\right) \tag{2.40}$$

If oil gravity is greater than 30°API,

$$R_{sb} = 0.0178(\gamma_g)(p_R^{1.1870}) \exp\left(23.9310 \frac{°\text{API}}{T_R + 460}\right) \quad \textbf{(2.40a)}$$

**Example 2–16.** The initial gas/oil ratio of a 30°API crude oil is 350 SCF/bbl. The reservoir temperature and pressure are 200°F and 1,200 psia, respectively. The specific gravity of the dissolved gas corrected for separator conditions is 0.75. The bubblepoint pressure of the reservoir is 1,900 psia. Estimate the gas/oil ratio at reservoir pressure of (a) 1,200 psia and (b) 2,100 psia.

Solution: (a) $p_R = 1{,}200$ psia is below the bubblepoint pressure; therefore

$$\begin{aligned} R_{sb} &= 0.0362(0.75)\,(1{,}200^{1.0937}) \exp\left(25.724 \frac{30}{200 + 460}\right) \\ &= 0.0272(2{,}331.85)e^{1.1693} \\ &= 204.21 \quad \text{SCF/bbl} \end{aligned}$$

(b) Since $p_R = 2{,}100$ psia is above the bubblepoint pressure of 1,900 psia, $R_s = R_{si} = 350$ SCF/bbl.

## TWO-PHASE FORMATION VOLUME FACTOR

The two-phase formation volume factor, symbol $B_t$, is a combination of the individual gas and oil formation volume factors. The $B_t$ is required in transient-well test analysis when the flow rate consists of both oil and gas, and in material balance calculations.

Figure 2–14 is used to estimate $B_t$ when reservoir pressure is above the bubblepoint pressure. Equation (2.41) will be used to estimate $B_t$ at and below the bubblepoint pressure.

Below bubblepoint:

$$B_{tb} = B_{ob} + (R_{si} - R_{sb})B_g \quad \textbf{(2.41)}$$

where $B_g$ is in RB/SCF, $B_{tb}$ is the two-phase formation volume factor below $p_b$ in RB/bbl. If the reservoir is at the bubblepoint pressure ($p_R = p_b$), $B_{tb}$ will be equal to $B_t$ estimated at the bubblepoint pressure ($p_R = p_b$), i.e., $R_{sB} = R_{si}$.

Instead of using Figure 2–14, the two-phase formation volume factor ($B_t$) above the bubblepoint pressure may be estimated using the following equation:

$$B_t = B_{tb}\,(e^{c_o(p_b - p_R)}) \quad \textbf{(2.41a)}$$

## VISCOSITY OF HEAVY OIL, BITUMEN, AND PETROLEUM FRACTIONS

This section presents a generalized correlation for predicting viscosities of binary mixtures of heavy oil, bitumen, and petroleum fractions, with emphasis on heavy-oil/solvent systems. The correlations presented here require only the input of densities and viscosities of the crude and solvent that form the mixture.

Calculation of mixture viscosities is often required when solvents are injected into the reservoir for well cleaning, stimulation, fracturing, and as input to reservoir simulators for heavy-oil recovery processes.

Lederer published the following equations for mixture viscosity, $\mu_m$:[15]

$$\ln \mu_m = X_A \ln \mu_A + X_B \ln \mu_B \tag{2.42}$$

$$X_A = \frac{\alpha V_A}{\alpha V_A + V_B} \tag{2.43}$$

and

$$X_B = 1 - X_A \tag{2.44}$$

where $\alpha$ is an empirical constant having a value between zero and unity, $A$ designates the more viscous component (oil) and $B$ the less viscous component (solvent), thus giving the viscosity ratio $\mu_A/\mu_B$. The $V_A$ and $V_B$ represent the volume composition for a given mixture. In application to heavy-oil systems, mixtures with viscosity ratio $(\mu_A/\mu_B)$ of $10^3$ or higher are generally of interest. Successful application of the Lederer equation also have been reported in studies of miscible displacement.

Recently Shu developed the following empirical correlation to obtain the empirical constant $\alpha$ in Equation (2.43):[16]

$$\alpha = \frac{17.04 \Delta\rho_m^{0.5237} \rho_A^{3.2745} \rho_B^{1.6316}}{\ln\left(\dfrac{\mu_A}{\mu_B}\right)} \tag{2.45}$$

The use of Equation (2.45) showed the root-mean-square error of $\alpha$-prediction to be 0.033, i.e., a maximum error of 0.022 ln $(\mu_A/\mu_B)$ for prediction of ln $\mu_m$ by means of Equation (2.42). The error is well within the experimental uncertainty of the data, particularly in the high-viscosity region $(\mu_m > 10^4)$.

Although the results obtained by using Equations (2.42) through (2.45) have shown good comparison with the viscosities obtained using other methods, the following limitations were noted:

1. Caution should be exercised when using the equations with low-viscosity ratio systems.
2. The $\alpha$-correlation was based on data at temperatures from 75° to 82°F. Since the viscosity composition varies with temperature, this variation should be kept in mind. The $\alpha$-correlation was compared with mixture viscosities obtained at 77°, 140°, and 180°F, the predicted viscosities matched the data equally well. This is because the temperature dependence somewhat cancels out when applied to the log-type equation such as Equation (2.42).[16]

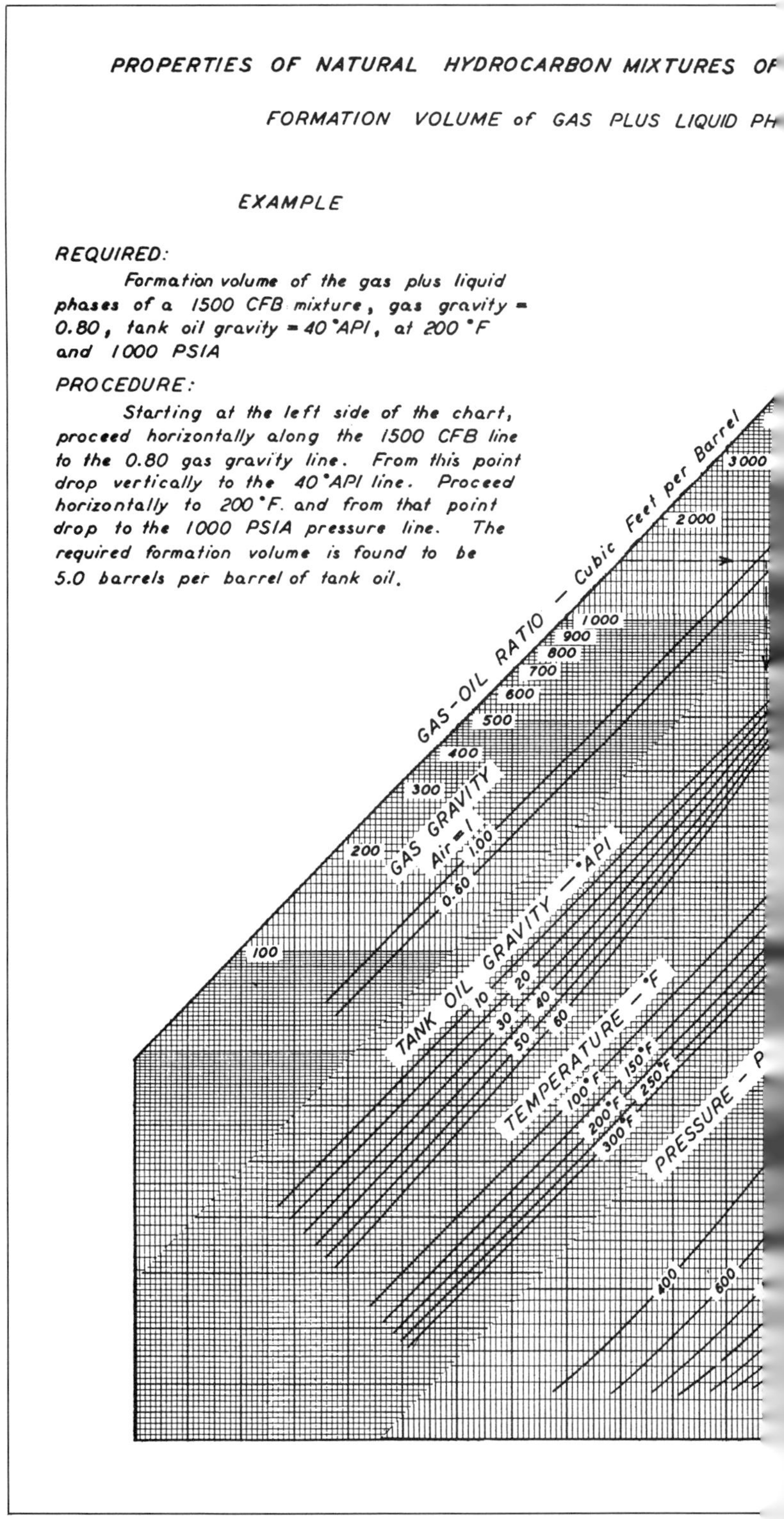


**Figure 2–14.** Formation volume factor of gas plus liquid phas

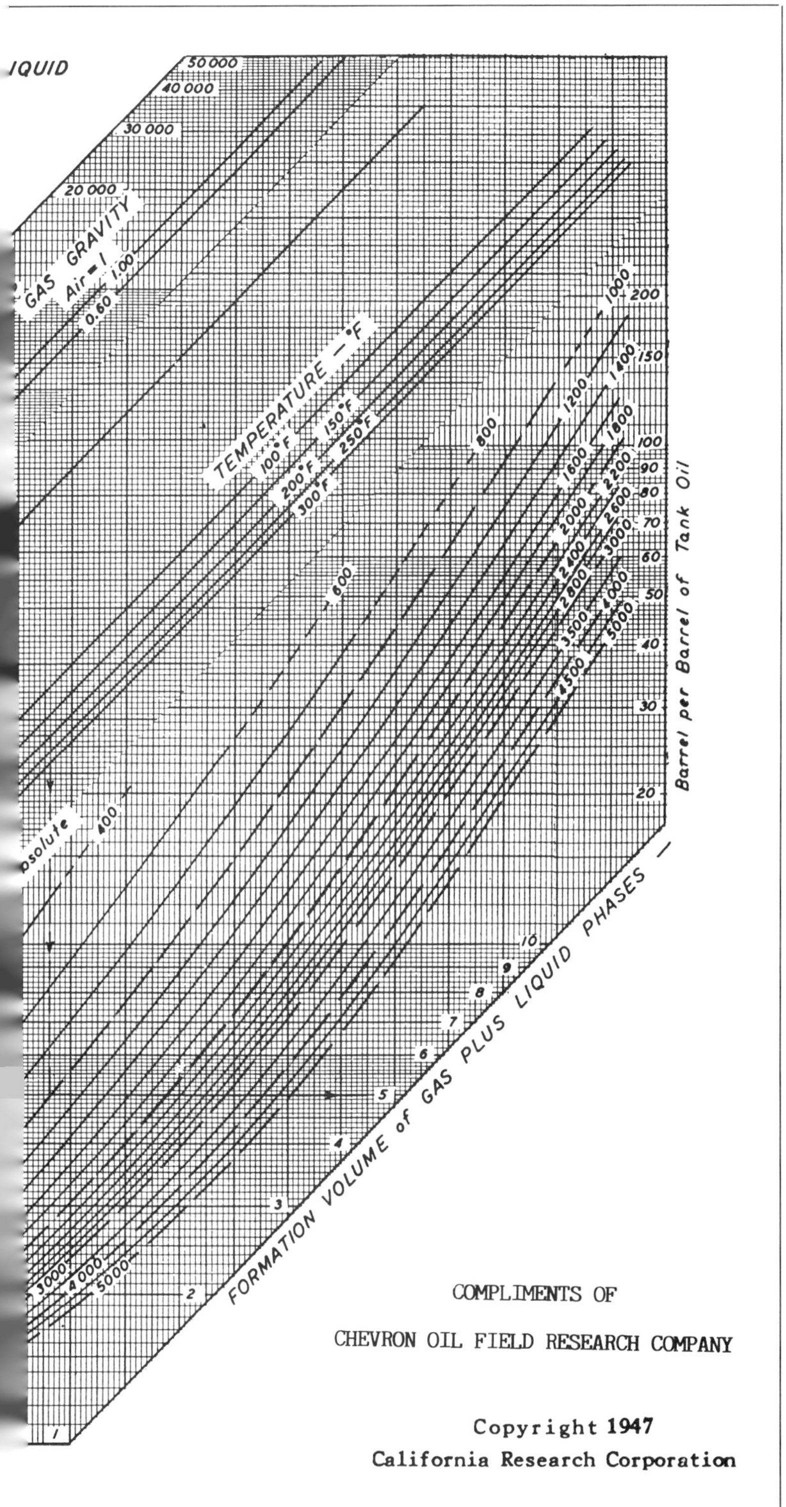
IQUID
50 000
40 000
30 000
20 000
GAS GRAVITY
Air = 1
1.00
0.60
TEMPERATURE — °F
100°F
150°F
200°F
250°F
300°F
400
600
800
1000
1200
1400
1600
1800
2000
2200
2400
2600
2800
3000
3500
4000
4500
5000
200
150
100
90
80
70
60
50
40
30
20
Barrel per Barrel of Tank Oil
1
2
3
4
5
6
7
8
9
10
FORMATION VOLUME of GAS PLUS LIQUID PHASES —
COMPLIMENTS OF
CHEVRON OIL FIELD RESEARCH COMPANY
Copyright 1947
California Research Corporation

3. The correlation does not apply to systems in which excessive asphaltene precipitation occurs upon blending.

## LATEST ADDITIONS TO OIL PVT CORRELATIONS

For PVT correlations, Standing's work is perhaps the most widely used.[11] However, Standing's correlations were developed for California oil and do not account for any corrections for oil type and nonhydrocarbon content. On the other hand, Beal and Chew and Conally's correlations are widely used in oil viscosity calculations as presented in the previous sections.[12,13]

In this section three recent contributions to the oil PVT correlations developed over a wider geographical area are presented. The correlations developed by (1) Glaso, (2) Vasquez and Beggs, and (3) Mohammad Ali Al-Marhoun are presented.[17–19]



**Glaso's Correlation.** Glaso presented empirical equations for estimating $p_b$, $B_{ob}$, and $B_t$ as a function of reservoir temperature, total surface gas gravity, producing GOR, and stock-tank oil gravity. Glaso also presented equations accounting for nonhydrocarbons in surface gases and paraffinicity of oil. These correlations were developed using exclusively North Sea oils. These equations are

For bubblepoint (saturation) pressure:

$$\log p_b = 1.7669 + 1.7447 \log p_b^* - 0.30218(\log p_b^*)^2 \quad \textbf{(2.46)}$$

where

$$p_b^* = \left(\frac{R_s}{\overline{\gamma}_g}\right)^{0.816} \left(\frac{T_R^{0.172}}{°\mathrm{API}^{0.989}}\right) \quad \textbf{(2.46a)}$$

For oil formation volume factor at saturation pressure:

$$\log (B_{ob} - 1) = -6.58511 + 2.91329 \log B_{ob}^* - 0.27683(\log B_{ob}^*)^2 \quad \textbf{(2.47)}$$

where

$$B_{ob}^* = R_s \left(\frac{\overline{\gamma}_g}{\gamma_o}\right)^{0.526} + 0.968T_R \quad \textbf{(2.47a)}$$

For two-phase (total) formation volume factor,

$$\log B_t = 8.0135 \times 10^{-2} + 4.7257 \times 10^{-1}(\log B_t^*) + 1.7351 \times 10^{-1}(\log B_t^*)^2 \quad \textbf{(2.48)}$$

where

$$B_t^* = R_s \left(\frac{T_R^{0.5}}{\overline{\gamma}_g^{0.3}}\right) \gamma_o^{2.9 \times 10^{-0.00027R_s}} p_R^{-1.1089} \quad \textbf{(2.48a)}$$

In all of the above equations oilfield units are used, where $R_s$ is the producing GOR in SCF/bbl, $\overline{\gamma}_g$ is the average specific gravity of the total surface gases, $T_R$ is the reservoir temperature in °F, °API is the stock-tank oil gravity from flash separation, $\gamma_o$ is the specific gravity of stock-tank oil from flash separation, and $B_{ob}$ and $B_t$ are in RB/STB.

In addition, the following correlations were presented for correcting flash API gravity for paraffinicity. The correction procedure requires (1) differential separation viscosity and (2) residual oil gravity. A dead oil viscosity correlation was developed to correlate $\mu_{oD}$, °API, and $T_R$ for North Sea oils with paraffinicity of 11.9 in the temperature range of 50° to 300°F.

$$\mu_{oD} = c(\log {}^\circ\text{API})^d \tag{2.49}$$

where

$$c = 3.141 \times 10^{10}(T_R)^{-3.444}$$

and

$$d = 10.313(\log T_R) - 36.447$$

The dead oil gravity can be calculated directly by rearranging Equation (2.49) as

$$^\circ\text{API}^*_{corr} = \text{antilog}\,[(a\mu_{oD})^b] \tag{2.50}$$

where

$$a = 3.184(10 \times {}^{-11})(T_R)^{3.444}$$

and

$$b = [10.213(\log T_R) - 36.447]^{-1}$$

Using Equation (2.50), the API gravity of stock-tank oil from a reservoir with paraffnicity not characterized by 11.9 can be corrected.[17] The corrected flash stock-tank API gravity can then be calculated using the following equation:

$$^\circ\text{API}_{corr} = \frac{^\circ\text{API}^*_{corr}}{^\circ\text{API}^*} \times {}^\circ\text{API} \tag{2.51}$$

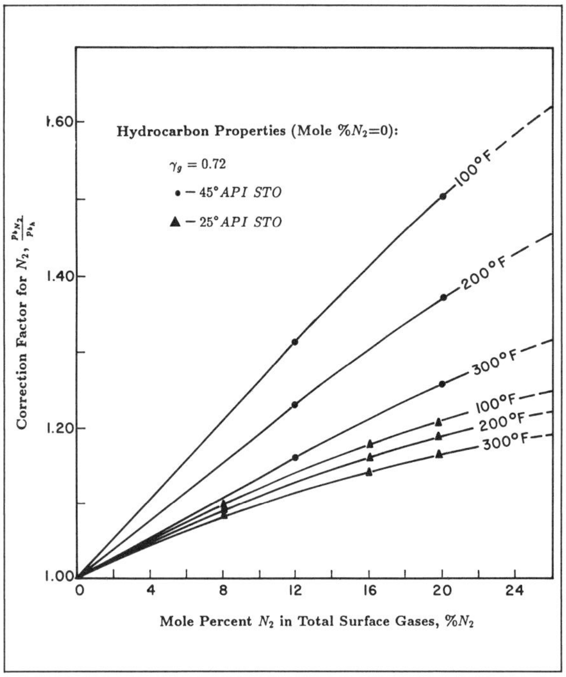


**Figure 2–15.** Effect of $N_2$ on $p_b$ for two API gravity oils at various temperatures (© SPE, JPT, May 1980).[17]

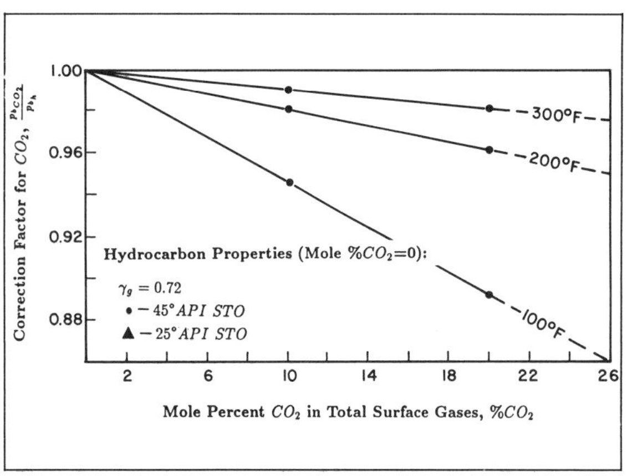


**Figure 2–16.** Effect on $CO_2$ on $p_b$ for two API gravity oils at various temperatures (© SPE, JPT, May 1980).[17]

where °API* is the gravity of residual oil from differential separation, $°API^*_{corr}$ is the corrected gravity of residual oil from Equation (2.50), and °API is the gravity of stock-tank oil from flash separation.

Glaso also noted that the nonhydrocarbon content in surface gas affects the bubblepoint pressure. Hence the $p_b$ estimated by Equation (2.46) should be corrected for nonhydrocarbons ($N_2$, $CO_2$, and $H_2S$), if any are present. Figures 2–15 to 2–17 show the effect of nonhydrocarbons on $p_b$. The following equations represent these figures:

For nitrogen, Figure 2–15:

$$\frac{p_{b_{N_2}}}{p_{b_h}} = 1.0\,[(-2.65 \times 10^{-4}°\text{API} + 5.5 \times 10^{-3})T_R + (0.0931°\text{API} - 0.8295)]\,y_{N_2} + [(1.954 \times 10^{-11}°\text{API}^{4.699})T_R + (0.027°\text{API} - 2.366)]\,(y_{N_2})^2 \quad \textbf{(2.52)}$$

For carbon dioxide, Figure 2–16:

$$\frac{p_{b_{CO_2}}}{p_{b_h}} = 1.0 - 693.8(y_{CO_2})T_R^{-1.553} \quad \textbf{(2.53)}$$

For hydrogen sulfide, Figure 2–17:

$$\frac{p_{b_{H_2S}}}{p_{b_h}} = 1.0 - (0.9035 + 0.0015°\text{API})y_{H_2S} + 0.019(45 - °\text{API})(y_{H_2S})^2 \quad \textbf{(2.54)}$$

where $p_{b_h}$ is the bubblepoint pressure for a sample containing only hydrocarbons. The bubblepoint pressure obtained using Equation (2.46) is then corrected by multiplying it by the correction factors calculated by Equations (2.52) through (2.54).

**Milton Vasquez and H. Dale Beggs Correlations.**[18] They developed oil PVT correlations using 600 laboratory PVT analyses from fields all over the world. Their analysis encompasses very wide ranges of pressure, temperature, and oil properties and included more than 6,000 measurements of $R_s$, $B_o$, and $\mu_o$ at various pressures. The following correlations were developed. These correlations should be used with gas gravity corrected for separator conditions by using Equation (2.33a).

For dissolved gas/oil ratio:

$$R_s = C_1 \gamma_g p_R^{C_2} \exp\left[C_3\left(\frac{°API}{T_R + 460}\right)\right] \quad \textbf{(2.55)}$$

where values for the coefficients are as follows:

| Coefficient | °API ≤ 30 | °API > 30 |
|---|---|---|
| $C_1$ | 0.0362 | 0.0178 |
| $C_2$ | 1.0937 | 1.1870 |
| $C_3$ | 25.7240 | 23.9310 |

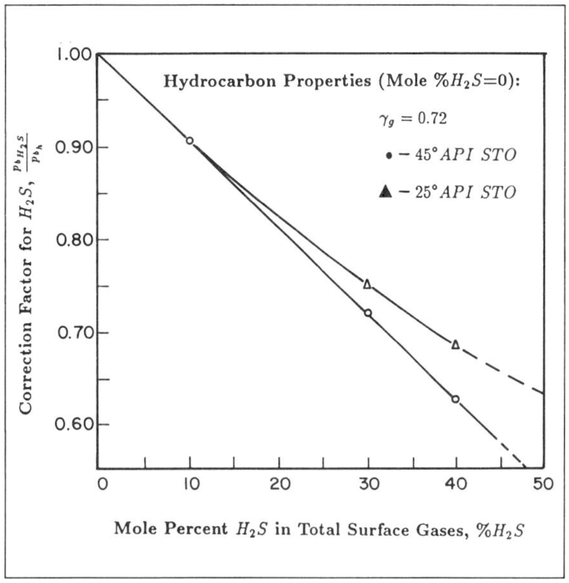


**Figure 2–17.** Effect of $H_2S$ on $p_b$ for two API gravity oils; temperature has no effect (© SPE, JPT, May 1980).[17]

For oil formation volume factor, $p_R \leq p_b$:

$$B_{ob} = 1 + C_1 R_s + C_2(T_R - 60)(°API/\gamma_g) + C_3 R_s (T_R - 60)\left(\frac{°API}{\gamma_g}\right) \quad \textbf{(2.56)}$$

where values for the coefficients are as follows:

| Coefficient | °API ≤ 30 | °API > 30 |
|---|---|---|
| $C_1$ | $4.677 \times 10^{-4}$ | $4.670 \times 10^{-4}$ |
| $C_2$ | $1.751 \times 10^{-5}$ | $1.100 \times 10^{-5}$ |
| $C_3$ | $-1.811 \times 10^{-8}$ | $1.337 \times 10^{-9}$ |

For oil formation volume factor, $p_R \geq p_b$:

$$B_o = B_{ob} \exp\left[C_o(p_b - p_R)\right] \quad \textbf{(2.57)}$$

where

$$C_o = \frac{a_1 + a_2R_s + a_3T_R + a_4\gamma_g + a_5{}^\circ\text{API}}{a_6p_R}$$

$$\begin{aligned} a_1 &= -1{,}433 \\ a_2 &= 5.0 \\ a_3 &= 17.20 \\ a_4 &= -\ 1{,}180 \\ a_5 &= 12.61 \\ a_6 &= 10^5 \end{aligned} \tag{2.58}$$

For undersaturated oil viscosity, $p_R \geq p_b$:



$$\mu_o = \mu_{ob}(p_R/p_b)^m \tag{2.59}$$

where

$$m = C_1 p_R^{C_2} \exp\,(C_3 + C_4 p_R) \tag{259a}$$

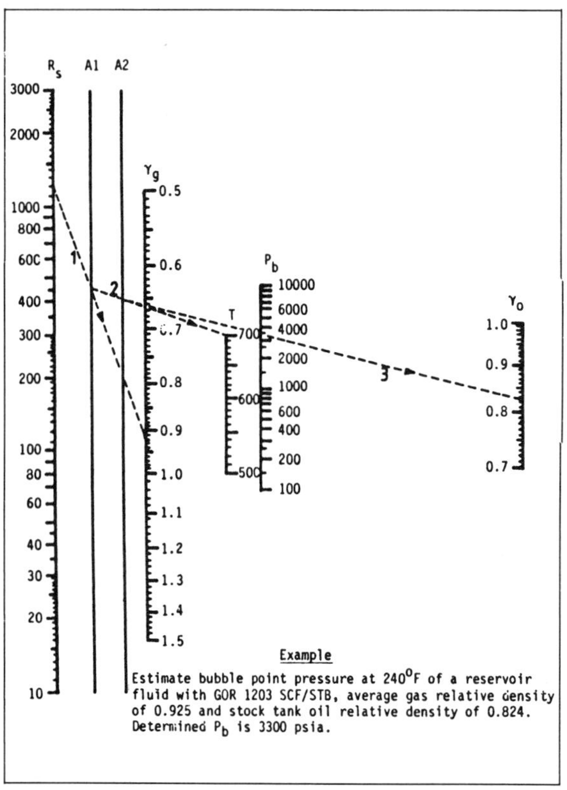


**Figure 2–18.** Correlation chart for bubble-point pressure (© SPE, JPT, May 1988).[19]

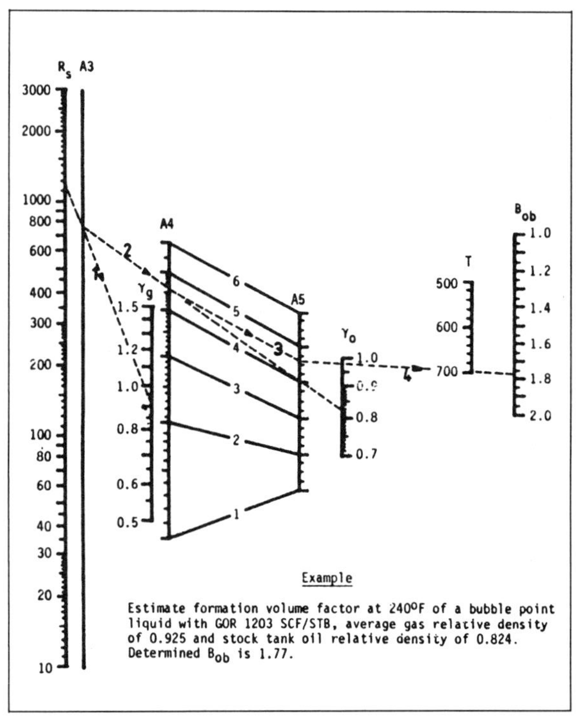


**Figure 2–19.** Correlation chart for bubble-point oil formation volume factor (© SPE, JPT, May 1988).[19]

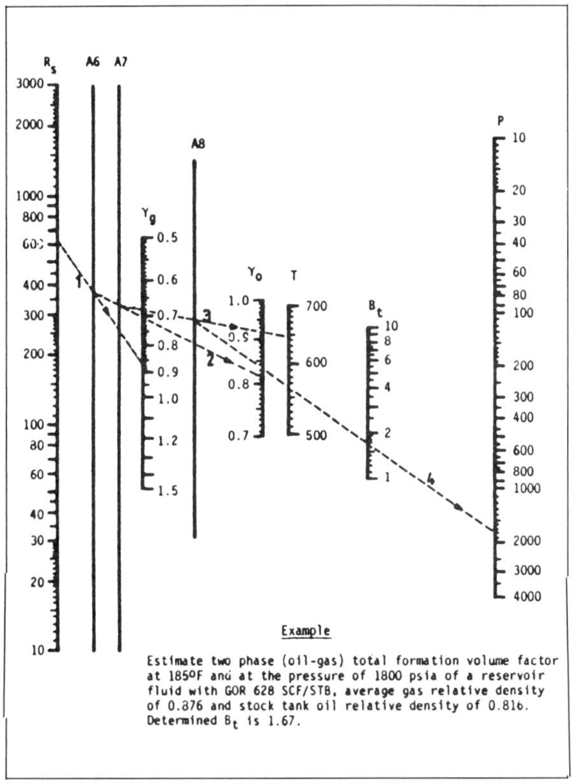

**Figure 2–20.** Correlation chart for two-phase formation volume factor (© SPE, JPT, May 1988).[19]

and

$$C_1 = 2.6$$
$$C_2 = 1.187$$
$$C_3 = -11.513$$
$$C_4 = -8.98 \times 10^{-5}$$

**Muhammad Ali Al-Marhoun's Correlation.**[19] He developed oil PVT correlations using samples exclusively from the Middle East. A total of 69 different fluid samples from 69 Middle East reservoirs were analyzed. His correlations are as follows:

$$p_b = 5.38088 \times 10^{-3} R_s^{0.715082} \gamma_g^{-1.877840} \gamma_o^{3.14370} T_R^{1.326570} \quad \textbf{(2.60)}$$

$$B_{ob} = 0.497069 + 0.862963 \times 10^{-3} T_R + 0.182594 \times 10^{-2} F + 0.318099 \times 10^{-5} F^2 \quad \textbf{(2.61)}$$

where

$$F = R_s^{0.742390} \gamma_g^{0.323294} \gamma_o^{-1.202040}$$

$$B_t = 0.314693 + 0.160253 \times 10^{-4} F_t + 0.188830 \times 10^{-10} F_t^2 \quad \textbf{(2.62)}$$

where

$$F_t = R_s^{0.644516} \gamma_g^{-1.079340} \gamma_o^{0.724874} T_R^{2.00621} p_R^{-0.761916}$$

On the basis of the mathematically developed PVT correlations, correlation charts have been developed for each PVT property as presented in Figures 2–18, 2–19, and 2–20.

## WATER PROPERTIES

This section deals with the properties of reservoir brines. The properties discussed are solubility of gas in water, water isothermal compressibility, water formation volume factor, and water viscosity. In each case the property is estimated for gas-free water, a correction is then applied accounting for gas saturation and salinity.

## SOLUBILITY OF GAS IN WATER

Dodson and Standing's correlations presented in Figures 2–21 and 2–22 are used to estimate solubility of gas in water (gas/water ratio), symbol $R_{sw}$.[20] Figure 2–21 is used to estimate gas/water ratio, symbol $R_{swp}$, in pure water. This $R_{swp}$ is then corrected for salinity using Figure 2–22. The required data are reservoir temperature, reservoir pressure, and salt content of water. Example 2–17 illustrates the procedure for estimating $R_{sw}$.

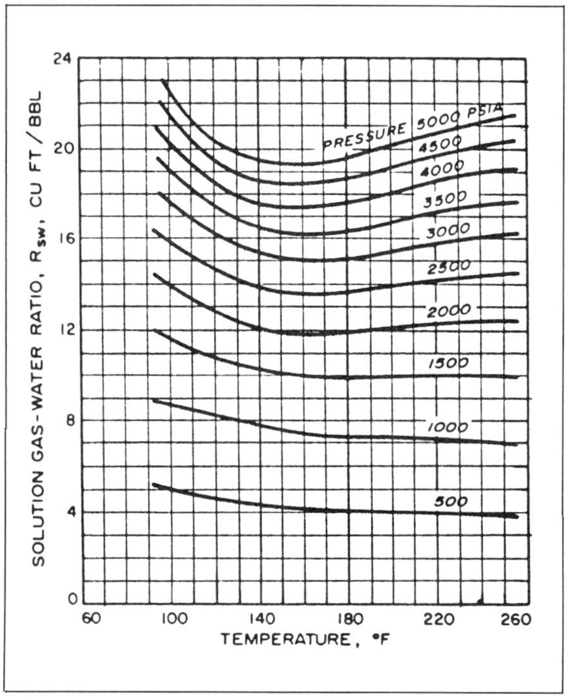


**Figure 2–21.** Solubility of natural gas in pure water (after Dodson and Standing).[20]

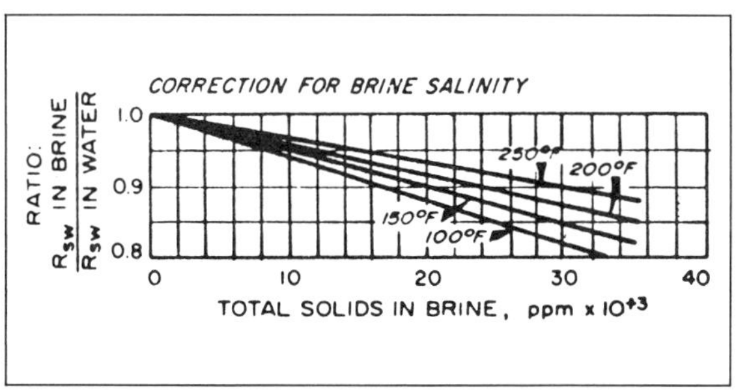


**Figure 2–22.** Correction of natural gas solubility for salt content (after Dodson and Standing).[20]

**Example 2–17.** Estimate the gas/water ratio for reservoir water with 30,000 ppm sodium chloride at a reservoir temperature and pressure of 170°F and 3,000 psia, respectively.

Solution: Estimate $R_{swp}$ from Figure 2–21 as

$$R_{swp} = 15 \quad \text{ft}^3/\text{bbl}$$

Correct $R_{swp}$ for salinity using Figure 2–22:

$$R_{sw}/R_{swp} = 0.86 \quad \text{(from Fig. 2–22)}$$

then

$$\begin{aligned} R_{sw} &= R_{swp} \times \frac{R_{sw}}{R_{swp}} \\ &= 15 \times 0.86 \\ &= 12.9 \quad \text{SCF/bbl} \end{aligned}$$

## WATER FORMATION VOLUME FACTOR

Water formation volume factor, symbol $B_w$, is defined as the ratio of reservoir barrels to stock-tank barrels. The $B_w$ is estimated using the correlations developed by Dodson and Standing as shown in Figure 2–23. In addition, Figure 2–22 is used to account for the necessary corrections. The data required are the same as used in estimating $R_{sw}$. The following steps will be used:

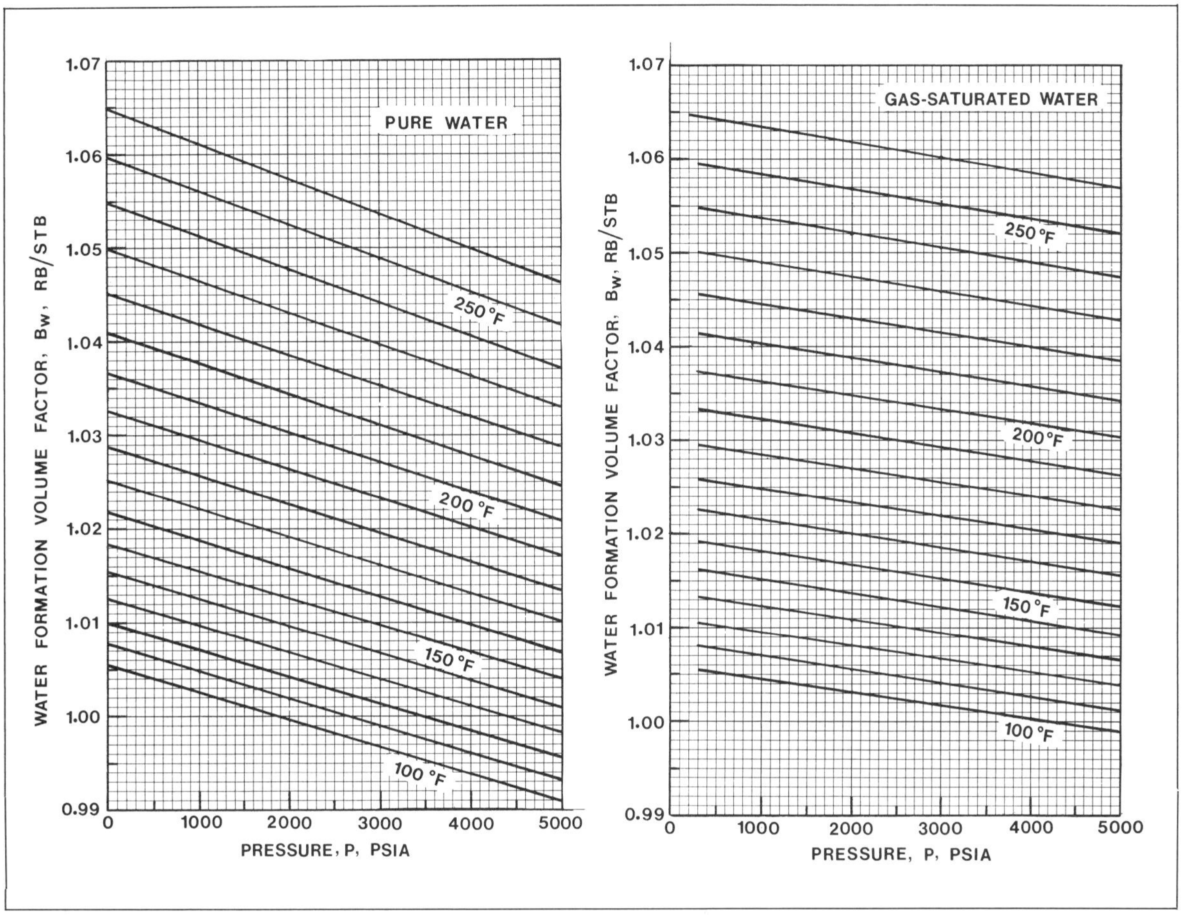


**Figure 2–23.** Formation volume factor for pure and gas saturated water (after Dodson and Standing).[20]

**Step 1.** Estimate $B_{wp}$ of pure water from Figure 2–23a.

**Step 2.** Estimate $B_{wgs}$ of gas-saturated water from Figure 2–23b.

**Step 3.** Estimate the correction of natural gas solubility ($R_{sw}/R_{swp}$) of salt content from Figure 2–22.

**Step 4.** Compute water formation volume factor using the following equation:

$$B_w = B_{wgs} + (B_{wgs} - B_{wp})(R_{sw}/R_{swp}) \tag{2.63}$$

**Example 2–18.** Estimate water formation volume factor for 30,000 ppm reservoir brine at reservoir temperature and pressure of 180°F and 4,500 psia, respectively. Reservoir brine is gas saturated.

Solution:

**Step 1.** Estimate $B_{wp}$ for pure water using Figure 2–23a.

$$B_{wp} = 1.0152 \quad \text{RB/bbl}$$

**Step 2.** Estimate $B_{wgs}$ for gas-saturated water using Figure 2–23b.

$$B_{wgs} = 1.0234 \quad \text{RB/bbl}$$

**Step 3.** Estimate correction for salt content using Figure 2–22.

$$R_{sw}/R_{swp} = 0.86$$

**Step 4.** Compute water formation volume factor $B_w$ using Equation (2.63).

$$\begin{aligned} B_w &= 1.0234 + (1.0234 - 1.0152)(0.86) \\ &= 1.0305 \quad \text{RB/bbl} \end{aligned}$$

## WATER ISOTHERMAL COMPRESSIBILITY

Water isothermal compressibility, symbol $c_w$, may be estimated for undersaturated and saturated reservoirs. Figures 2–24 and 2–25 developed by Dodson and Standing are used to estimate

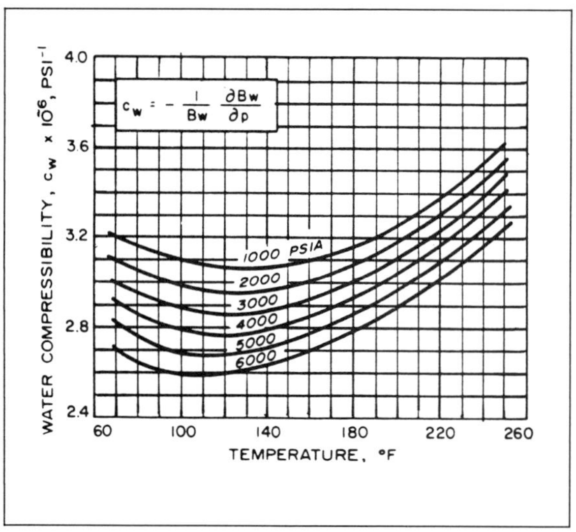


**Figure 2–24.** Compressibility of gas-free water (after Dodson and Standing).[20]

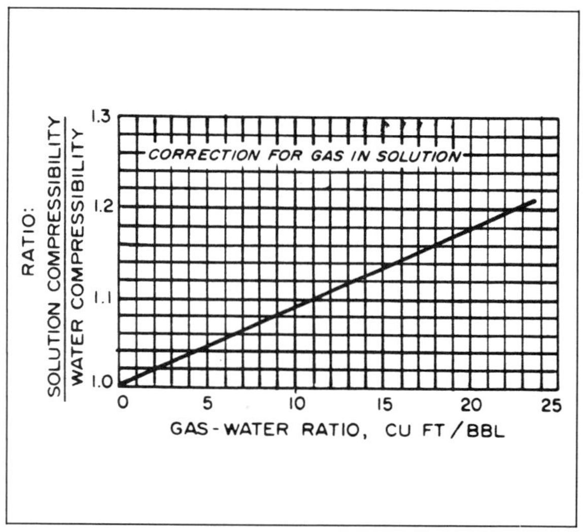


**Figure 2–25.** Effect of dissolved gas on $c_w$ (after Dodson and Standing).[20]

$c_w$ for undersaturated water. To estimate $c_w$ in saturated reservoirs, the following equation is used:

$$c_w = -\frac{1}{B_w}\frac{dB_w}{dp_R} + \frac{B_g}{B_w}\frac{dR_{sw}}{dp_R} \tag{2.64}$$

In the equation above, the term $(-1/B_w dB_w/dp_R)$ is estimated from Figure 2–24 and it is corrected for dissolved gas content (if any) using Figure 2–25. The factor $dR_{sw}/dp_R$ is estimated from Figure 2–26 developed by Ramey and corrected for salt content using Figure 2–22. The $B_w$ is estimated as described in the previous section, and $B_g$ (in RB/SCF) is calculated.

**Example 2–19.** Estimate water isothermal compressibility of the formation water in Example 2–17. The bubblepoint pressure is estimated to be 2,000 psia.

Solution:

**Step 1.** Use Figure 2–24 to estimate $c_{wp}$ for gas-free water as

$$c_{wp} = 3 \times 10^{-6} \quad 1/\text{psia}$$

**Step 2.** Estimate $R_{sw}$ using Figure 2–21 and 2–24.

$$R_{sw} = 12.9 \quad \text{SCF/bbl} \quad \text{(from Example 2–17)}$$

Correct $c_{wp}$ for dissolved gas using Figure 2–25.

$$\frac{c_w}{c_{wp}} = 1.118$$

$$\begin{aligned} c_w &= (c_w)(c_w/c_{wp}) \\ &= (3 \times 10^{-6})(1.118) \\ &= 3.354 \times 10^{-6} \quad 1/\text{psia} \end{aligned}$$

**Example 2–20.** Estimate water isothermal compressibility for formation water in Examples 2–17 and 2–19. The reservoir pressure and temperature are 1,500 psia and 170°F, respectively. The associated gas has $B_g$ of 0.001132 bbl/SCF.

Solution:

**Step 1.** Estimate the term $[(-1/B_w)(dB_w/dp_R)]$ from Figure 2–24:

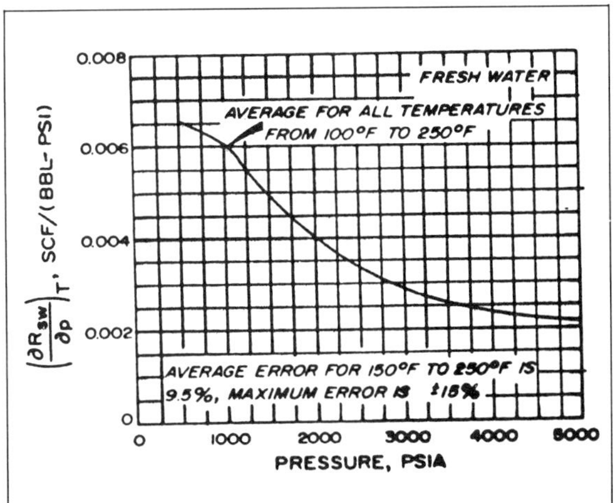


**Figure 2–26.** Change of natural gas in solution in formation water with pressure versus pressure (after Ramey).

$$-\frac{1}{B_w}\frac{dB_w}{dp_R} = 3.354 \times 10^{-6} \quad \text{(from Example 2–19)}$$

**Step 2.** Estimate the term $dR_{sw}/dp_R$ from Figure 2–26 as

$$dR_{sw}/dp_R = 0.0048 \quad \text{SCF/bbl-psia}$$

**Step 3.** Estimate correction factor for salt content (if any) using Figure 2–22.

$$R_{sw}/R_{swp} = 0.86 \quad \text{(from Example 2–17)}$$

Therefore,



$$\begin{aligned} dR_{sw}/dp_R &= 0.0048(0.86) \\ &= 0.0041 \quad \text{SCF/bbl-psia} \end{aligned}$$

**Step 4.** Estimate $B_w$ using Figure 2–23 and Equation (2.63).

$$\begin{aligned} B_w &= 1.022 + (1.027 - 1.022)0.86 \\ &= 1.0263 \quad \text{RB/bbl} \end{aligned}$$

**Step 5.** Finally, compute $c_w$ using Equation (2.64) as

$$\begin{aligned} c_w &= 3.354 \times 10^{-6} + \frac{0.001132 \times 0.0041}{1.0263} \\ &= 7.876 \times 10^{-6} \quad \text{1/psia} \end{aligned}$$

## WATER VISCOSITY

Reservoir water viscosity, symbol $\mu_w$, is estimated using correlations developed by Mathews and Russel, as presented in Figure 2–27.[21] Water viscosity is reported in centipoise (cp). The data required are reservoir pressure and temperature and formation water salinity. Figure 2–27 provides water viscosity ($\mu_w^*$) at reservoir temperature and atmospheric pressure. The inset in the figure is then used to estimate the correction factor $f$ for converting viscosity at atmospheric pressure to viscosity at reservoir pressure, $\mu_w = (\mu_w^* \times f)$.

**Example 2–21.** Estimate water viscosity for a 40,000 ppm formation water at reservoir pressure of 6,000 psia and reservoir temperature of 200°F.

Solution: 40,000 ppm = 4% NaCl.

**Step 1.** Estimate $\mu_w^*$ at $T_R$ = 200°F, NaCl = 4%, and atmospheric pressure using Figure 2–27.

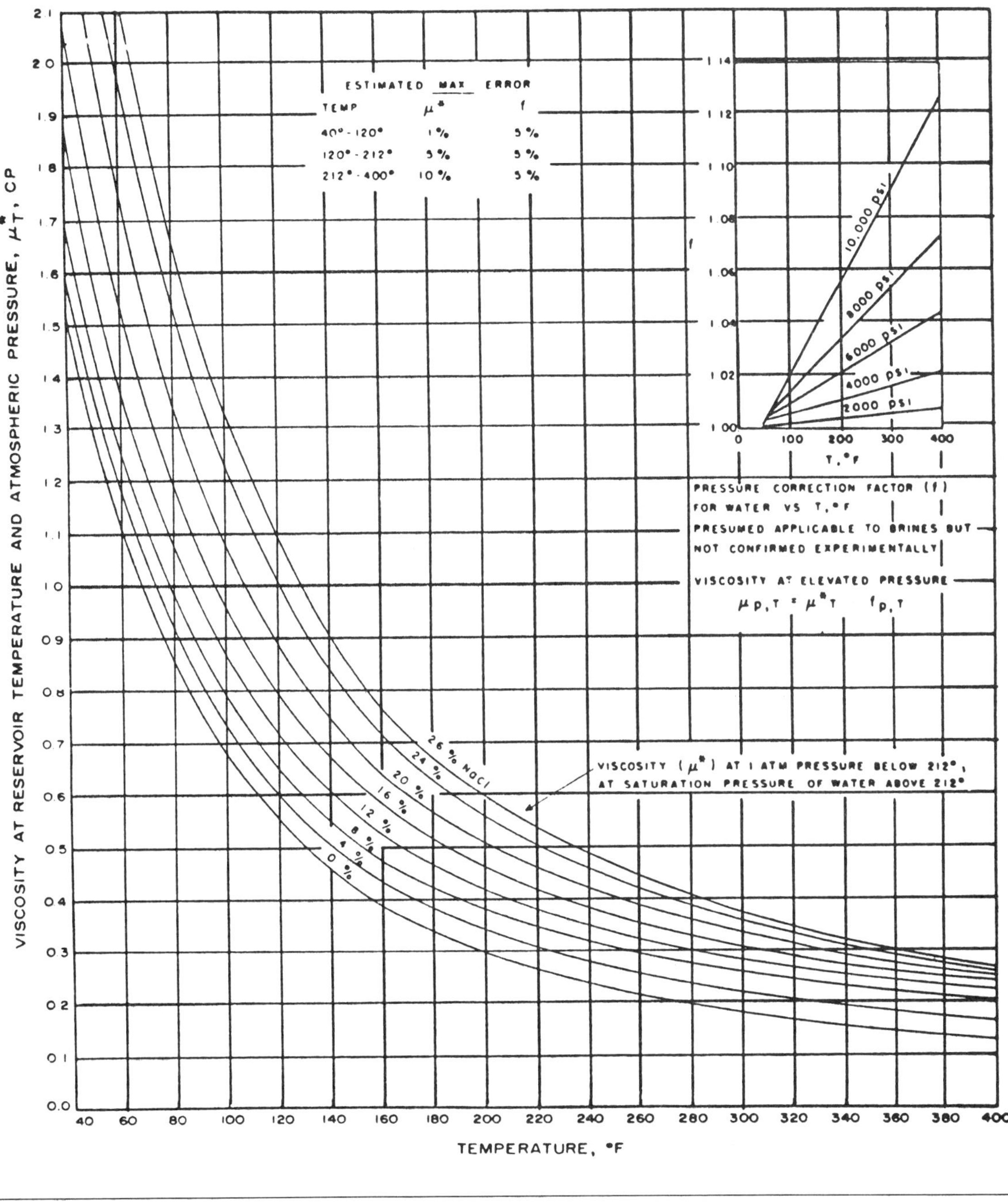


**Figure 2–27.** Water viscosity at various salinities and reservoir temperature (after Mathews and Russel).[21]

**Step 1.** Estimate $\mu_w^*$ at $T_R = 200°F$, NaCl = 4%, and atmospheric pressure using Figure 2–27.

$$\mu_w^* = 0.34 \quad \text{cp}$$

**Step 2.** Estimate correction factor $f$ from the inset in Figure 2–27 as

$$f = 1.02$$

**Step 3.** Compute $c_w$ at reservoir pressure and temperature as

$$\begin{aligned}\mu_w &= 0.34 \times 1.02\\ &= 0.3468 \quad \text{cp}\end{aligned}$$



## ROCK PROPERTIES

Naturally occurring reservoir rocks contain fluid, water, oil, gas, or a combination of these fluids. The petroleum engineer is obviously concerned with the properties of these rocks and the fluids contained within the rocks. In the previous sections fluid properties were discussed; this section deals with reservoir rock properties.

The properties discussed are the formation compressibility, effective isothermal compressibility, hydrostatic and overburden pressure, permeability, relative permeability, and capillary pressure. These properties constitute a set of fundamental parameters by which the rock can be quantitatively described.

### FORMATION COMPRESSIBILITY

Formation compressibility, symbol $c_f$, in most cases is estimated from the correlation developed by Hall presented as Equation (2.65) below:[22]

$$c_f = \frac{1.87}{10^6} \times \phi^{-0.415} \tag{2.65}$$

where $\phi$ is the formation porosity in fraction and $c_f$ is the formation isothermal compressibility in 1/psia.

### TOTAL AND EFFECTIVE ISOTHERMAL COMPRESSIBILITY

The total isothermal coefficient of compressibility, symbol $c_t$, is the weighted sum of the compressibilities of the separate phases. The compressibilities of the individual phases are estimated using the techniques discussed in the previous sections. The following equations are utilized for estimating $c_t$:

Above bubblepoint pressure:

$$c_t = c_O S_O + c_w S_w + c_f \quad \textbf{(2.66)}$$

Below bubblepoint pressure:

$$c_t = c_{Ob} S_O + c_g S_g + c_w S_w + c_f \quad \textbf{(2.66a)}$$

No oil present:

$$c_t = c_g S_g + c_w S_w + c_f \quad \textbf{(2.66b)}$$

where $S_O$ is formation oil saturation (fraction), $S_g$ is formation gas saturation (fraction), and $S_w$ is formation water saturation (fraction).

The effective compressibility, symbol $c_e$, is the effective compressibility of the mobile phase and is estimated by dividing the total compressibility ($c_t$) by the saturation of that phase. The effective compressibility of oil above the bubblepoint pressure is given by Equation (2.66c).

$$c_e = c_O + \frac{S_w}{1 - S_w}(c_w) + \frac{c_f}{1 - S_w} \quad \textbf{(2.66c)}$$

## HYDROSTATIC AND OVERBURDEN PRESSURE

Hydrostatic pressure is the pressure at a certain point in a wellbore exerted by a column of fluids (oil, gas, or water) above that point. Overburden pressure is the pressure at a certain point in a reservoir exerted by the weight of material (weight of minerals plus weight of water) above that point. The hydrostatic and overburden pressures are useful parameters used in most drilling, completion, and reservoir engineering operations.

The hydrostatic pressure is determined using the following equation, and the overburden pressure, in most cases, is assumed as 1 psia per foot of depth.

$$p_h = 0.433(\gamma)(D) \quad \textbf{(2.67)}$$

where $p_h$ is the hydrostatic pressure at depth $D$ in psia, $\gamma$ is the specific gravity of fluid, $D$ is the depth at which pressure is to be determined (feet), and 0.433 is the fresh water gradient (62.4/144) in psia/ft.

To calculate the hydrostatic pressure created by drilling mud or cement slurries, the $\gamma$ in Equation (2.67) is replaced by density of mud or cement in pounds per gallon (ppg) and the constant 0.433 is replaced by 0.052.

Formation pressure, most commonly referred to as reservoir pressure, is the pressure created by the fluids (oil, water, and/or gas) which exists within the pores of a rock formation. The reservoirs are further classified based on their initial reservoir pressure as normal pressured, overpressured, or underpressured. The normal pressure corresponds to the pressure exerted by the column of formation water extending from the formation to the surface of the earth. In the Gulf Coast region, the normal pressure corresponds to a pressure gradient of 0.465 psia/ft of depth (salt water of specific gravity 1.074). Reservoir pressure less than normal pressure is termed as underpressured and that greater than normal pressure is termed as overpressured.

In gas wells, gas gradient (psi/ft) due to gas column is computed using the following equation:

$$\text{Gas Gradient} = \frac{0.01875\gamma_g p_R}{z(T_R + 460)} \tag{2.68}$$



**Example 2–22.** Compute the hydrostatic pressure at a depth of 5,000 ft exerted by 35°API crude oil.

Solution:

$$\gamma_o = \frac{141.5}{131.5 + 35} = 0.8498$$

$$p_h = 0.433 \times 0.8498 \times 5{,}000$$

$$= 1{,}839.82 \approx 1{,}840 \quad \text{psia}$$

**Example 2–23.** A reservoir pressure of 4,500 psia is encountered at a depth of 7,500 feet. If the normal pressure gradient corresponds to water of specific gravity 1.085 (water = 1), is the reservoir overpressured, underpressured, or normal?

Solution:

$$p_h = 0.433 \times 1.085 \times 7{,}500$$

$$= 3{,}523.5 \approx 3{,}524 \quad \text{psia}$$

Since the reservoir pressure is higher than the normal pressure of 3,524 psia, the reservoir is overpressured by 976 psia (4,500 − 3,524 = 976).

**Example 2–24.** A well is full of a 10 ppg drilling mud to 3,500 ft of depth. A blowout preventer (BOP) is installed, and it is assumed that the formation at 3,500 ft will hold 70% of the overburden pressure. Determine (a) how much pressure can be held by the blowout preventer and (b) the pump pressure required for successful fracturing if the well has to be hydraulically fractured at 3,500 ft with 0.8 gravity fracturing fluid.

Solution:

(a)

$$\text{Overburden Pressure} = (1 \text{ psia/ft})(3{,}500) = 3{,}500 \quad \text{psia}$$
$$\text{Breakdown Pressure} = 0.75 \times 3{,}500 = 2{,}450 \quad \text{psia}$$
$$p_h = 0.052 \times 10 \times 3{,}500 = 1{,}820 \quad \text{psia}$$
$$\text{Pressure held by BOP} = \text{BDP} - p_h = 2{,}450 - 1{,}820 = 630 \quad \text{psia}$$

(b) Hydrostatic head of fracturing fluid is

$$p_h = 0.433 \times 0.8 \times 3{,}500 = 1{,}212 \quad \text{psia}$$
$$\text{Required pump pressure} = \text{BDP} - p_h$$
$$= 2{,}450 - 1{,}212 = 1{,}238 \quad \text{psia}$$



## ABSOLUTE AND EFFECTIVE PERMEABILITY

The characteristics of a porous medium to allow the flow of a fluid or fluids through it is referred to as *permeability*. For a single-phase flow, the measured permeability is referred to as the *absolute permeability*. When more than one phase is present, the permeability of a porous media to a fluid in the presence of one or more other fluids is termed as effective permeability.

Absolute permeability is measured in routine core analysis. Measurements are usually performed at room temperature with a pressure drop across a core plug of less than 1 atm. Effective permeability is usually measured from transient test analysis. Effective oil and gas permeabilities are needed in evaluating the steady-state flow behavior of oil and gas wells and in developing relative permeability relationship (to be discussed in the next section).

Routine core permeability (absolute) measurements are based on the assumption of (1) single-phase flow, (2) laminar or viscous flow, (3) steady-state flow, (4) no reaction between flowing fluid and porous medium, (5) velocity of flow is equal to $q/A$, and (6) the core is 100% saturated with the flowing fluid. The following equations based on Darcy's law are used for absolute permeability calculations.

Linear flow, when the saturating fluid is liquid,

$$k = \frac{q\mu L}{A(p_1 - p_2)} \tag{2.69}$$

Linear flow, when the saturating fluid is gas,

$$k = \frac{q_m\mu L}{A(p_1 - p_2)} \tag{2.70}$$

OR

$$k = \frac{2q_a\mu L p_{sc}}{A(p_1^2 - p_2^2)} \tag{2.70a}$$

Radial flow, when the saturating fluid is liquid,

$$k = \frac{q\mu \ln (r_e/r_w)}{2\pi h(p_e - p_w)} \tag{2.71}$$

Radial flow, when the saturating fluid is gas,



$$k = \frac{q_m\mu \ln (r_e/r_w)}{2\pi h(p_e - p_w)} \tag{2.72}$$

OR

$$k = \frac{2q_a\mu \ln (r_e/r_w)p_{sc}}{h(p_e^2 - p_w^2)} \tag{2.72a}$$

where the flow rate $q$ is in cc/sec, absolute permeability ($k$) is in darcies, thickness ($h$) is in centimeters, viscosity ($\mu$) is in centipoise, the pressures are in atmospheres, the length of core ($L$) is in centimeters, the core cross-sectional area (A) is in square centimeters, the radii $r_e$ and $r_w$ are in centimeters, $q_m$ is in cc/sec at $p_m = (p_1 + p_2)/2$ and flowing temperature, and $q_a$ is in cc/sec at 1 atmosphere pressure and flowing temperature.

Experiments have shown that gases do not strictly follow a laminar flow in porous media. A slippage effect occurs along the walls of the pores resulting in higher flow rate and thus higher apparent absolute permeability. This effect is referred to as the Klinkenberg effect.[23] The amount of gas slippage is a function of the mean pressure of gas flow and of the particular rock specimen. The change can be represented by the following equation:

$$k_a = k\left(1 + \frac{b}{p}\right) \tag{2.73}$$

where $k_a$ is the absolute permeability calculated using Equation (2.70) or (2.72) at the mean flowing pressure, $\bar{p}$; $k$ is the permeability of the specimen that would have been obtained if liquid were used; and $b$ is a constant depending on the type of rock, type of gas, and permeability.

To correct for the Klinkenberg effect, absolute permeability is plotted versus the reciprocal of mean pressure ($1/\bar{p}$) on coordinate graph paper. The true absolute permeability is obtained by extrapolating the line to ($1/\bar{p} = 0$) on the y-axis. The following example will clarify absolute permeability calculations, correcting for the Klinkenberg effect.

**Example 2–25.** The following data were obtained during a routine laboratory permeability test. Calculate the absolute permeability of the core.

$\mu = 0.0183$ cp $\quad L = 1.30$ cm

$A = 1.36$ cm$^2$ $\quad p_2 = 1$ atm

Atmospheric pressure = 760 mm Hg

| Run | $q_g$ (cc/sec) | $p_1$ (atm) | $1/\bar{p}$ (1/atm) |
|---|---|---|---|
| 1 | 4.05 | 1.13 | 0.94 |
| 2 | 17.90 | 1.50 | 0.80 |
| 3 | 34.56 | 1.86 | 0.70 |
| 4 | 61.79 | 2.33 | 0.60 |

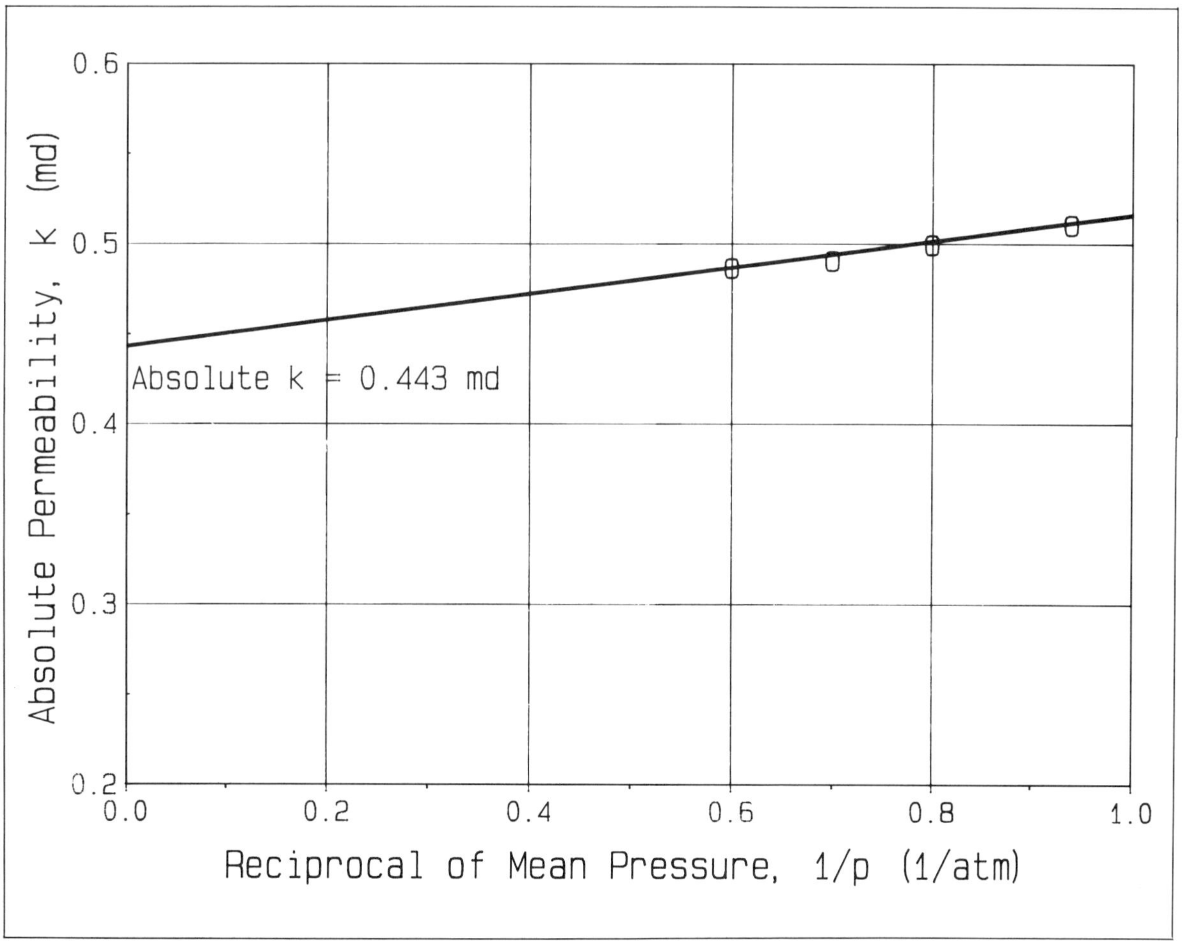


**Figure 2–28.** Illustration of absolute permeability calculation and Klinkenberg effect for the data in Example 2–25.

Solution: Using Equation (2.70a),

$$\begin{aligned} k_1 &= \frac{2(4.05)(0.0183)(1.3)(1)}{(1.36)(1.13^2 - 1^2)} \\ &= 0.512 \quad \text{darcy} \end{aligned}$$

Similarly,

$$\begin{aligned} k_2 &= 0.501 \quad \text{darcy} \\ k_3 &= 0.492 \quad \text{darcy} \\ k_4 &= 0.488 \quad \text{darcy} \end{aligned}$$



These permeabilities are then plotted versus $1/\bar{p}$ as shown in Figure 2–28 and a best-fit line drawn through the points. The true absolute permeability (corrected for slippage) is read at the intercept as 0.443 darcy.

## RELATIVE PERMEABILITY

Relative permeability is defined as the ratio of the ability of a rock to transmit a fluid when partially saturated by that fluid (effective permeability) to its ability to transmit the same fluid when fully saturated by it (absolute permeability).[24] The ratio of a rock's relative permeability to one fluid to its relative permeability to another fluid at the same liquid saturation is the relative permeability ratio for that saturation. The relative permeability ratio equals the effective permeability ratio. Mathematically, relative permeability is given by

$$k_{rw} = \frac{k_w}{k},$$

$$k_{ro} = \frac{k_o}{k},$$

and

$$\frac{k_{rw}}{k_{ro}} = \frac{k_w}{k_o}$$

A typical plot of oil and water relative permeability curves as a function of water saturation for a particular formation is shown in Figure 2–29. It is also helpful to present such plots on a semilog scale to expand the relative permeability characteristics near the endpoint saturations. The main influence on the shapes of the curves depends on the wettability, i.e., which fluid adheres to the rock's surface. The wetting-phase relative permeability is concave upward while the nonwetting-phase relative permeability takes on a S-shape curve. This means that in an oil/water system, the $k_{rw}$ curve is concave upward and the $k_{ro}$ is a S-shape curve, while in gas/oil system, $k_{ro}$ curve is concave upward and $k_{rg}$ is a S-shape curve. In computations concerning the behavior of two-phase flow in porous media, the results may depend strongly on the shape of the relative permeability curves used.

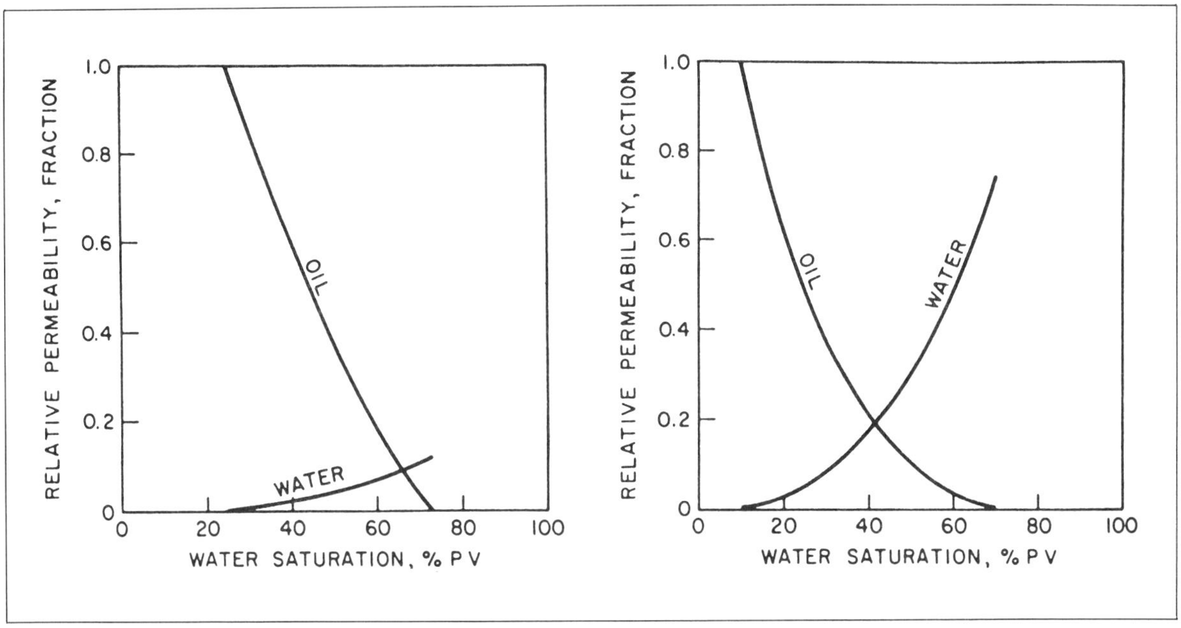


**Figure 2–29.** Water/oil relative permeability curves (a) strongly oil-wet rock, (b) strongly water-wet rock (© SPE, SPE Monograph Series, Vol. 3, 1971).[34]



Relative permeability is determined in the laboratory using steady-state or unsteady-state flow measurements. Steady-state methods for determining permeabilities have the widest application and greatest reliability because the capillary equilibrium prevails, the saturation is measured directly, and the calculation scheme is based on Darcy's law. Unsteady-state techniques present many uncertainties in calculations due to some operational constraints.

Because of the difficulties involved in measurement of relative permeability in the laboratory, empirical models are usually employed to reproduce experimentally determined relative permeability curves, or to approximate them when experimental data are unavailable. Relative permeability may also be determined from the production history of a reservoir and its fluid properties. However, the agreement between laboratory-determined relative permeabilities and those calculated from production data is generally poor. Relative permeability calculations from this method require complete production-history data and provide average values influenced by pressure and saturation gradients, differences in stage of depletion, and saturation variations in stratified reservoirs.

Methods of determining relative permeability from field data along with widely used empirical correlations are presented here. Determination of the $k_g/k_o$ ratio from production and fluid-analysis data involves the use of the following equations:

$$\frac{k_g}{k_o} = (R_i - R_s)\,\frac{\mu_g}{\mu_o}\,\frac{B_g}{B_o} \qquad \textbf{(2.74)}$$

The total liquid saturation corresponding to each respective $k_g/k_o$ is given by the following equation:

$$S_o = S_w + (1 - S_w) \times \frac{N - N_p}{N} \times \frac{B_o}{B_{oi}} \quad \textbf{(2.75)}$$

where $R_i$ is the instantaneous GOR in SCF/STB, $R_s$ is the solution GOR in SCF/STB, $B_o$ is in RB/STB, $B_g$ is in RB/SCF, $\mu_g$ and $\mu_o$ are in cp, N is the original oil in place in STB, $N_p$ is the cumulative oil produced in STB, $S_o$ is the total liquid saturation fraction of pore volume, and $S_w$ is the interstitial water saturation fraction of pore volume.

**Example 2–26.** Given the following production and fluid analysis data on a reservoir, determine the $k_g/k_o$ ratio at a reservoir pressure of 1,625 psia.

| | |
|---|---|
| $N = 116.5$ MMSTB | $N_p = 16.929$ MMSTB @ 1,625 psia |
| $p_b = 3{,}548$ psia | $B_o = 1.298$ RB/STB |
| $S_{iw} = 28.5\%$ | $B_{oi} = 1.450$ RB/STB |
| $R_i = 2{,}800$ SCF/STB | $\mu_o/\mu_g = 58.20$ |
| $R_s = 410$ SCF/STB | $B_g = 0.001795$ RB/SCF |

Solution:

Using Equation (2.74):

$$k_g/k_o = (2{,}800 - 410)\left(0.0172 \times \frac{0.001795}{1.298}\right)$$
$$= 0.05686 \quad \approx 0.057$$

Using Equation (2.75):

$$S_o = 0.285 + (1 - 0.285)\left(\frac{116.5 - 16.929}{116.5}\right)\left(\frac{1.298}{1.450}\right)$$
$$= 0.285 + 0.7150(0.8547)(0.8952)$$
$$= 0.8321 \quad \text{OR} \quad 83.2\% \quad \text{of pore volume}$$

Empirical correlations which can be used to represent experimental data are presented below.

**Wyllie's empirical relative permeability equations.**[25]

1. Oil/gas relative permeabilities (for drainage cycle relative to oil), where $S^* = S_o/(1 - S_{iw})$.

| | $k_{ro}$ | $k_{rg}$ |
|---|---|---|
| Unconsolidated Sand—Well Sorted | $(S^*)^{3.0}$ | $(1 - S^*)^{3.0}$ |
| Unconsolidated Sand—Poorly Sorted | $(S^*)^{3.5}$ | $(1 - S^*)^2(1 - S^{*1.5})$ |
| Cemented Sand, Oolitic Lime, & Vugular Lime | $(S^*)^{4.0}$ | $(1 - S^*)^{2.0}(1 - S^{*2.0})$ |

2. Water/oil relative permeabilities (for drainage cycle relative to water) where $S^* = (S_w - S_{iw})/(1 - S_{iw})$.

| | $k_{ro}$ | $k_{rw}$ |
|---|---|---|
| Unconsolidated Sand—Well Sorted | $(1 - S^*)^{3.0}$ | $(S^*)^{3.0}$ |
| Unconsolidated Sand—Poorly Sorted | $(1 - S^*)^2(1 - S^{*1.5})$ | $(S^*)^{3.5}$ |
| Cemented Sand, Oolitic Lime, & Vugular Lime | $(1 - S^*)^2(1 - S^{*2})$ | $(S^*)^{4.0}$ |

**Mehdi Honarpour et al.'s correlations.**[26] Their correlations were based on extensive data set from all over the world. Their initial findings showed that their correlations better reproduced laboratory relative permeability data than those developed by Wyllie and others in consolidated rocks. These correlations are are follows:



Water/oil system, sandstone and conglomerate (water wet):

$$k_{rw} = 0.035388\left(\frac{S_w - S_{iw}}{1 - S_{iw} - S_{orw}}\right) - 0.010874\left(\frac{S_w - S_{orw}}{1 - S_{iw} - S_{orw}}\right)^{2.90} + 0.56556(S_w)^{3.6}(S_w - S_{iw}) \quad \textbf{(2.76)}$$

Water/oil system, sandstone and conglomerate (oil & intermediate wettability):

$$k_{rw} = 1.5814\left(\frac{S_w - S_{iw}}{1 - S_{iw}}\right)^{1.91} - 0.58617\left(\frac{S_w - S_{orw}}{1 - S_{iw} - S_{orw}}\right)(S_w - S_{iw}) - 1.2484\phi(1 - S_{iw})(S_w - S_{iw}) \quad \textbf{(2.76a)}$$

Water/oil system, sandstone and conglomerate:

$$k_{ro} = 0.76067\left[\frac{\left(\frac{S_o}{1 - S_{iw}}\right) - S_{orw}}{1 - S_{orw}}\right]^{1.8}\left(\frac{S_o - S_{orw}}{1 - S_{iw} - S_{orw}}\right)^{2.0} + 2.6318\phi(1 - S_{orw})(S_o - S_{orw}) \quad \textbf{(2.76b)}$$

Oil/gas system, sandstone and conglomerate:

$$k_{ro} = 0.98372\left(\frac{S_o}{1 - S_{iw}}\right)^4\left(\frac{S_o - S_{org}}{1 - S_{iw} - S_{org}}\right)^2 \quad \textbf{(2.76c)}$$

Oil/gas system, sandstone and conglomerate:

$$k_{rg} = 1.1072\left(\frac{S_g - S_{gc}}{1 - S_{iw}}\right)^2 k_{rg(S_{org})}) + 2.7794 \times \left[\frac{S_{org}(S_g - S_{gc})}{1 - S_{iw}}\right] k_{rg(S_{org})} \quad \textbf{(2.76d)}$$

Water/oil system, limestone and dolomite:

$$k_{rw} = 0.0020525 \times \frac{(S_w - S_{iw})}{\phi^{2.15}} - 0.051371(S_w - S_{iw})\left(\frac{1}{k_a}\right)^{0.43} \quad \textbf{(2.76e)}$$

Water/oil system, limestone and dolomite:



$$k_{rw} = 0.29986\left(\frac{S_w - S_{iw}}{1 - S_{iw}}\right) - 0.32797\left(\frac{S_w - S_{orw}}{1 - S_{iw} - S_{orw}}\right)^2 (S_w - S_{iw}) + 0.413259\left(\frac{S_w - S_{iw}}{1 - S_{iw} - S_{orw}}\right)^4 \quad \textbf{(2.76f)}$$

Water/oil system, limestone and dolomite:

$$k_{ro} = 1.2624\left(\frac{S_o - S_{orw}}{1 - S_{orw}}\right)\left(\frac{S_o - S_{orw}}{1 - S_{iw} - S_{orw}}\right)^2 \quad \textbf{(2.76g)}$$

Oil/gas system, limestone and dolomite:

$$k_{ro} = 0.93752\left(\frac{S_o}{1 - S_{iw}}\right)^4\left(\frac{S_o - S_{org}}{1 - S_{iw} - S_{org}}\right)^2 \quad \textbf{(2.76h)}$$

Oil/gas system, limestone and dolomite:

$$k_{rg} = 1.8655\left[\frac{(S_g - S_{gc})(S_g)}{1 - S_{iw}}\right] k_{rg(S_{org})} + 8.0053\left[\frac{(S_g - S_{gc})(S_{org})^2}{1 - S_{iw}}\right] - 0.025890(S_g - S_{gc})\left(\frac{1 - S_{iw} - S_{org} - S_{gc}}{1 - S_{iw}}\right)^2 \times \left(1 - \frac{1 - S_{iw} - S_{org} - S_{gc}}{1 - S_{iw}}\right)^2\left(\frac{k_a}{\phi}\right)^{0.5} \quad \textbf{(2.76i)}$$

where $k_a$ is air permeability in millidarcies, $k_{o(S_{iw})}$ is oil permeability at irreducible water saturation (md), $k_{rg(Sor)}$ is the gas relative permeability at residual oil saturation (fraction), $S_{gc}$ is the critical gas saturation (fraction), $S_{org}$ is the residual oil saturation to gas (fraction), $S_{orw}$ is the residual oil saturation to water (fraction), and $\phi$ is the porosity (fraction).

Chierici proposed exponential four- and five-parameter equations for gas/oil drainage and water/oil imbibition relative permeability curves.[27] These equations were shown to match the experimentally determined curves, in particular at and near their initial points and endpoints better than many other correlations. These proposed relations are

For gas/oil drainage:

$$k_{ro} = \exp\left(-AR_g^L\right) \quad OR \quad -\ln k_{ro} = AR_g^L \tag{2.77}$$

$$k_{rg} = \exp\left(-BR_g^{-M}\right) \quad OR \quad -\ln k_{rg} = BR_g^{-M} \tag{2.78}$$

where A, B, L, and M are positive numbers, and

$$R_g = \frac{S_g - S_{gc}}{1 - S_{iw} - S_g} \tag{2.79}$$



with the constraint

$$S_g - S_{gc} = 0 \quad \text{for} \quad S_g \geq S_{gc} \tag{2.79a}$$

For water/oil imbibition:

$$K_{ro}^* = \exp\left(-AR_w^L\right) \quad OR \quad -\ln k^*ro = AR_w^L \tag{2.80}$$

$$k_{rw}^* = \exp\left(-BR_w^{-M}\right) \quad OR \quad -\ln K_{rw}^* = BR_w^{-M} \tag{2.81}$$

where A, B, L, and M are positive numbers, and

$$R_w = \frac{S_w - S_{iw}}{1 - S_{or} - S_w} \tag{2.82}$$

$$k_{ro}^* = \frac{k_{ro}}{k_{ro(S_{iw})}} \tag{2.83}$$

$$k_{rw}^* = \frac{k_{rw}}{k_{rw(S_{or})}} \tag{2.83a}$$

The values of the empirical coefficients A, L, and B, M are determined by nonlinear regression on the sets of experimental data points. Where $k_{ro}^*$ and $k_{rw}^*$ are the normalized relative permeabilities to oil and water, respectively, $k_{ro(S_{iw})}$ is the relative permeability to oil at $S_o = 1 - S_{iw}$, and $k_{rw(S_{or})}$ is the relative permeability to water at $S_w = 1 - S_{or}$.

**Example 2–27.** A core is 2 ft long and has a cross-sectional area of 1 ft$^2$. The core is oriented so fluid flows horizontally through it. Pressure difference between the ends is 100 psi. When the core is 100% saturated with $\mu_o = 1.2$ cp viscosity oil, the flow rate is 13.29 ft$^3$/day. Determine (a) absolute permeability, (b) effective permeability to oil, (c) relative permeability to oil, (d) the rate at which water will flow through the core when it is 100% saturated with $\mu_w = 1$ cp viscosity water, (e) the effective permeabilities to oil and water when both oil and water are flowing in the core at rates of 9 ft$^3$/day and 5 ft$^3$/day, respectively, and (f) the relative permeability to oil and water in (e).

Solution: Using Equation (2.69) converted to oilfield terminology as

(a)

$$\begin{aligned} k &= \frac{q_o \mu_o L}{6.323(A)(p_1 - p_2)} \\ &= \frac{(13.29)(1.2)(2)}{6.323(1.0)(100)} \\ &= 0.05 \quad \text{darcy} \end{aligned}$$

(b) Since the core is 100% saturated with oil, $k_o = k = 0.05$ darcy.

(c)

$$\begin{aligned} k_{ro} &= \frac{k_o}{k} \\ &= \frac{0.05}{0.05} = 1.0 \end{aligned}$$

(d)

$$\begin{aligned} q_w &= 6.323 \frac{kA(p_1 - p_2)}{\mu_w L} \\ &= 6.323 \frac{(0.05)(1)(100)}{(1)(2)} \\ &= 15.81 \quad \text{ft}^3/\text{day} \end{aligned}$$

(e) $q_o = 9$ ft$^3$/day and $q_w = 5$ ft$^3$/day.

$$\begin{aligned} k_o &= \frac{(9)(1.2)(2)}{6.323(1.0)(100)} \\ &= 0.034 \quad \text{darcy} \\ k_w &= \frac{(5)(1.0)(2)}{6.323(1.0)(100)} \\ &= 0.016 \quad \text{darcy} \end{aligned}$$

(f)

$$\begin{aligned} k_{ro} &= \frac{k_o}{k} \\ &= \frac{0.034}{0.05} = 0.680 \\ k_{rw} &= \frac{k_w}{k} \\ &= \frac{0.016}{0.05} = 0.320 \end{aligned}$$

**Example 2–28.** A laboratory experiment was run to measure oil and water permeability data on a core sample. The following data were recorded. Plot the relative permeability curves for oil and water. What is the value of irreducible water saturation ($S_{iw}$), what is the value of residual oil saturation ($S_{or}$), and at what $S_w$ does $k_{ro} = k_{rw} \neq 1$?

| Run | $S_w$(%) | Flow Rate, cc/hr Oil | Flow Rate, cc/hr Water | $k_{ro}$ | $k_{rw}$ |
|---|---|---|---|---|---|
| 1 | 0 | 2.00 | 0 | 1.00 | 0 |
| 2 | 10 | 1.98 | 0 | 0.99 | 0 |
| 3 | 20 | 1.94 | 0 | 0.97 | 0 |
| 4 | 24 | 1.90 | 0 | 0.95 | 0 |
| 5 | 30 | 1.78 | 0.06 | 0.89 | 0.01 |
| 6 | 40 | 1.48 | 0.24 | 0.74 | 0.04 |
| 7 | 50 | 1.40 | 0.54 | 0.70 | 0.09 |
| 8 | 60 | 0.38 | 1.02 | 0.19 | 0.17 |
| 9 | 70 | 0.12 | 1.68 | 0.06 | 0.28 |
| 10 | 80 | 0 | 2.64 | 0 | 0.44 |
| 11 | 90 | 0 | 4.02 | 0 | 0.67 |
| 12 | 100 | 0 | 6.00 | 0 | 1.00 |



Solution:

(a) The relative permeability curves are shown in Figure 2–29.

(b) From Figure 2–29, $S_{iw} = 24\%$

(c) From Figure 2–29, $S_{or} = 20\%$

(d) From Figure 2–29, $k_{ro} = k_{rw}$ at $S_w = 61\%$

## CAPILLARY PRESSURE

Capillary pressure for any rock type possessing porosity and permeability can be viewed as the force required to push a fluid through a given pore size. In other words, capillary pressure is defined as the differential pressure between the hydrocarbon and the water (pressure in the nonwetting phase minus the pressure in the wetting phase) at any level in the reservoir, i.e., the water/oil capillary pressure is defined as $p_c = p_o - p_w$ and the gas/water capillary pressure is defined as $p_c = p_g - p_w$. The differential pressure or capillary pressure increases at increasing elevations above the free-water level of the accumulation. The capillary pressure is a function of the fluid properties and of the saturation.

Capillary pressure is determined in the laboratory by injecting mercury at increasing pressures into plugs obtained from core or side wall samples. These pressures when plotted produce a curve of injection pressure versus fluid saturation from which a number of parameters can be extracted to be used in geological mapping and engineering. A typical capillary pressure curve is shown in Figure 2–30. The capillary pressure data is usually plotted on a regular coordinate graph paper; however, for accuracy and interpretation considerations, the semilog format is preferred.

Pore throat sorting, reservoir grade, oil column height (thickness of the hydrocarbon/water transition zone), irreducible water saturation, permeability and displacement pressure are the parameters obtained from capillary pressure curve.

Pore throat sorting (PTS) is a number which measures the sorting of the pore throats within a rock sample and signifies its ability to accept oil saturation. It ranges from 1.0 (perfect sorting) to up to 8 (essentially no sorting). Pore throat sorting indicates the slope of the plateau formed by the capillary pressure curves and is calculated as follows:[28]



$$\text{PTS} = \left(\frac{\text{Third Quartile Pressure}}{\text{First Quartile Pressure}}\right)^{0.5}$$

where the first and third quartile pressures are obtained directly from the capillary pressure curve at 25% and 75% mercury saturation, respectively, and adjusted for irreducible water saturation.

Reservoir grade is a number ranging from 0 to 100 and is the indicator of reservoir quality of a rock. It is simply the percentage of linear area integrated under the capillary pressure curve. While PTS measures the range in pore throat size, reservoir grade provides a relative number to the size of the pore throats.

The displacement pressure, irreducible water saturation, and transition zone can be read directly from the curve as shown in Figure 2–30. The transition zone is defined as the vertical thickness over which the saturations range from 100% water to $S_{wi}$ in the case of water/oil contact and from 100% to $S_{wi}$ in the case of gas/oil contact. Displacement pressure is interpreted to be the pressure at which mercury first begins to imbibe into the rock at reservoir conditions. A high displacement pressure indicates either a strong degree of wetting or small pores, or both.

The shape of the capillary pressure curve depends on the following factors:[29]

1. Changes in pore size and reservoir fluid densities alter the shape of the curve and thus the thickness of the transition zone. When the difference in densities of two fluids increases, the thickness of the transition zone decreases, and vice versa.
2. If all other factors remain unchanged, a low-API gravity oil reservoir with an oil/water contact will have a longer transition zone than a high-API gravity oil reservoir, as shown in Figure 2–31a.
3. The thickness of the transition zone decreases with the increase in radius of the pores. Thus a reservoir rock system with small pore sizes will have a longer transition zone, and vice versa.
4. High-permeability reservoirs will have shorter transition zones than low-permeability reservoirs, as shown in Figure 2–31b.

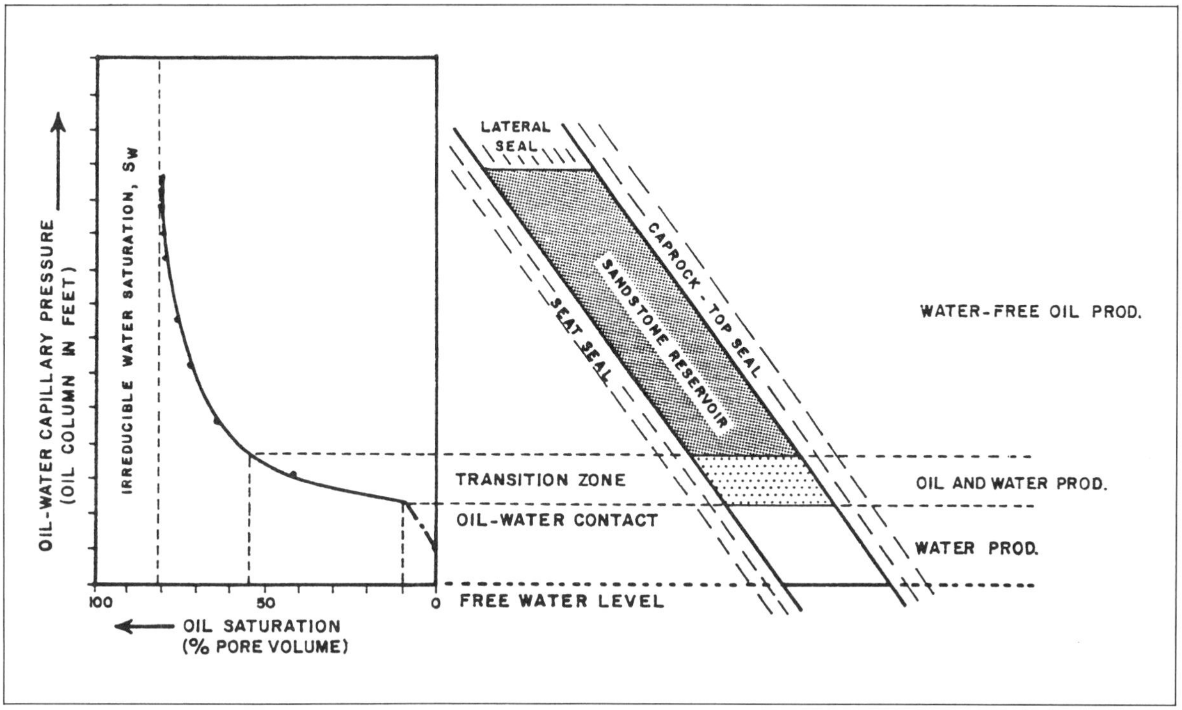


**Figure 2–30.** A typical capillary pressure curve.

A general expression for capillary pressure as a function of interfacial tension and curvature of the interface is shown below.

$$P_c = \sigma\left(\frac{1}{R_1} + \frac{1}{R_2}\right) \qquad \textbf{(2.84)}$$

and

$$P_c = \frac{h}{144}(\rho_1 - \rho_2) \qquad \textbf{(2.85)}$$

where $R_1$ and $R_2$ (in centimeters) are the principal radii of curvature of the interface, $\sigma$ (dynes/cm) is the interfacial tension between the two fluids, $\rho_1$ and $\rho_2$ are the densities of heavier and lighter fluids, respectively, and $h$ is the height above plane of zero capillary pressure between Fluids 1 and 2. Equation (2.85) is in oilfield terms with $h$ in feet, $P_c$ in psi, and densities ($\rho_1$ and $\rho_2$) in pounds per cubic feet (the densities could be replaced by the pressure gradient of the respective fluids). The $P_c$ in Equation (2.84) is in dynes/cm$^2$.

Note that all height measurements are from the free-water level. This level is defined by the zero capillary pressure and not necessarily the level at which only water will produce.

The capillary pressure data are obtained using a gas/water or an oil/water system. In practice these laboratory-derived data are then corrected to reservoir conditions. When considering a specific case wherein the laboratory values are determined with gas and water, the capillary pressure becomes

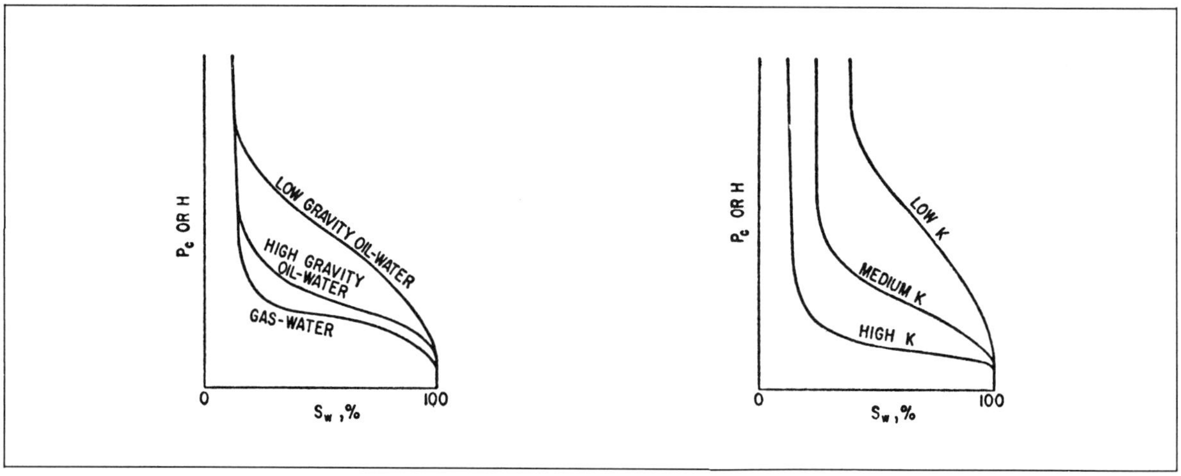




**Figure 2–31.** Variation of $p_c$ with $S_w$ for (a) the same rock with different fluids, and (b) the same fluid but different rock system.[29]

$$P_{cL} = \frac{2\sigma_{wg} \cos \theta_{wg}}{r} \quad \textbf{(2.86)}$$

where $\sigma_{wg}$ is the interfacial tension between gas and water used in laboratory tests, $r$ is the radius of the capillary, and $\theta_{wg}$ is the contact angle of gas and water. If reservoir fluids oil and water were used in the same capillary, the equation becomes

$$P_{cR} = \frac{2\sigma_{wo} \cos \theta_{wo}}{r} \quad \textbf{(2.87)}$$

Combining Equations (2.86) and (2.87) gives

$$P_{cR} = \frac{\sigma_{wo} \cos \theta_{wo}}{\sigma_{wg} \cos \theta_{wg}} (P_{cL}) \quad \textbf{(2.88)}$$

Thus *reservoir* capillary pressure can be calculated from laboratory capillary pressure when the interfacial tensions and contact angle between oil and water in the reservoir and gas and water in the laboratory are known.

Another relationship which neglects contact angle and is based on Equation (2.84) is given below. In case of mercury injection tests, this equation yields better results.

$$P_{CR} = \frac{\sigma_R}{\sigma_L} \times P_{cL} \quad \textbf{(2.88a)}$$

Typical contact angles and interfacial tension values for various fluid combinations in laboratory and reservoir are given in Table 2–3.

Swanson developed correlations between brine and air permeabilities with capillary pressure data.[30] His correlations were developed on samples for which brine permeability under 1,000 psi effective stress and mercury capillary pressure data were available. The correlations developed are presented below in equation form.

For sandstone samples:

$$k_w = 431 \left(\frac{S_b}{P_c}\right)_A^{2.109} \tag{2.89}$$

**TABLE 2–3. Typical Interfacial Tension and Contact Angle Constants**

| System | θ | σ |
|---|---|---|
| **Laboratory** | | |
| Air/water | 0 | 72 |
| Oil/water | 30 | 48 |
| Air/mercury | 140 | 480 |
| Air/oil | 0 | 24 |
| **Reservoir** | | |
| Water/oil | 30 | 30 |
| Water/gas | 0 | 50 |

For carbonate samples:

$$k_w = 290 \left(\frac{S_b}{P_c}\right)_A^{1.901} \tag{2.89a}$$

Overall equation for sandstone and carbonates:

$$k_w = 355 \left(\frac{S_b}{P_c}\right)_A^{2.005} \tag{2.89b}$$

where $(S_b/P_c)_A$ is the correlating parameter taken at Point A of the capillary pressure curve as shown in Figure 2–32, $S_b$ is the mercury saturation in percent of bulk volume, and $P_c$ is the mercury capillary pressure in psi.

Swanson also developed a brine permeability/capillary pressure nomograph as shown in Figure 2–33. The nomograph allows direct estimation of brine permeability of clean sandstones and carbonates directly from a plot of capillary pressure data (plotted on the nomograph). A large part of the stress dependence as well as Klinkenberg corrections for air permeability are accounted for through the use of the nomograph.

Purcell also reported on computation of permeability from mercury capillary pressure data.[31] This equation is given below.

$$k = \left(\frac{\sigma \cos \theta}{2}\right)^2 \phi \sum_{i=1}^{n} \frac{S_i}{(P_c)_i^2} \tag{2.90}$$

OR

$$k = 10.24(\sigma \cos \theta)^2 \phi \lambda \int_{S=0}^{S=1} \frac{dS}{(P_c)^2} \tag{2.90a}$$

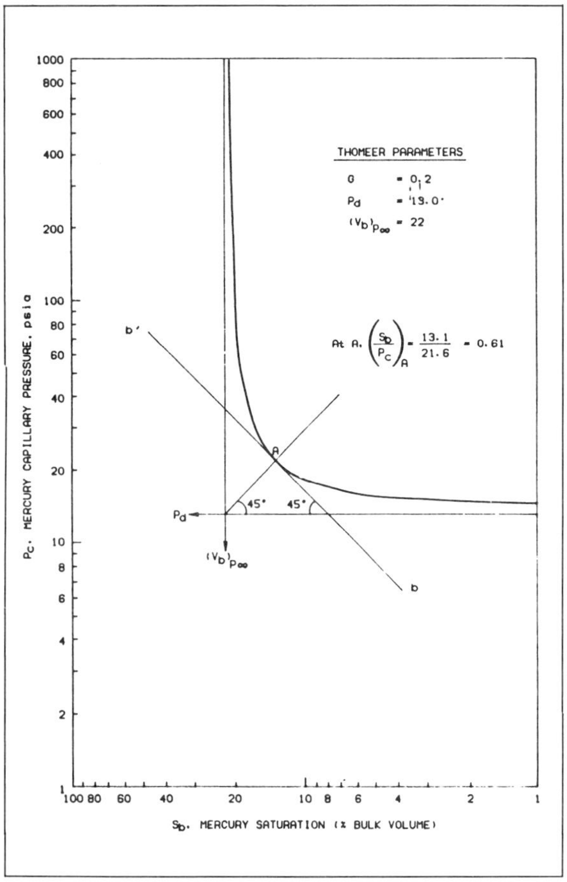


**Figure 2–32.** Correlating parameter determination for Swanson's correlation (© SPE, JPT, Dec. 1981).[30]

OR

$$k = 14{,}260\phi\lambda \int_{S=0}^{S=1} \frac{dS}{(P_c)^2} \qquad \textbf{(2.90b)}$$

where $k$ is in millidarcies, porosity $\phi$ is a fraction of total pore space occupied by liquid injected or forced out of the sample, and $\lambda$ is a lithology factor required to make the calculated and observed permeabilities identical. Equation (2.90) is derived by assuming that the contact angle for mercury was 140° and that the interfacial tension of mercury was 480 dynes/cm.

The value of the integral in Equation (2.90) is the area under the curve $1/(P_c)^2$. The $1/(P_c)^2$ is calculated from the corresponding $P_c$ from the capillary pressure curve at various saturations. The $1/(P_c)^2$ values are then plotted versus percent of pore space occupied by mercury as shown in Figure 2–34.

An approximate linear relationship of the form $S_w = a \log k + C$ usually results if $S_w$ is plotted against the logarithm of permeability for constant values of capillary pressure as shown in Figure 2–35. The series of lines should converge at high permeabilities and low-water saturations as would normally be expected because larger capillaries are associated with high permeabilities. It has been noticed that water saturation at constant capillary pressure is not only a function of permeability but also porosity.[33] Therefore, in some cases the correlation will improve by incorporating the porosity term in the form $S_w = a_1\phi + a_2 \log k + C$.

**Example 2–29.** Typical capillary pressure data on a core sample are given in Table 2–4. Determine the displacement pressure and irreducible water saturation. At what height above oil/water contact would you expect $S_w = 50\%$?

$$\rho_o = 49.92 \text{ lb/ft}^3 \quad \sigma_R = 250 \text{ dynes/cm}$$
$$\rho_w = 65.52 \text{ lb/ft}^3 \quad \sigma_L = 110 \text{ dynes/cm}$$

Solution: The laboratory capillary pressure $P_{cL}$ is first converted to reservoir capillary pressure $P_{cR}$ using Equation (2.88).

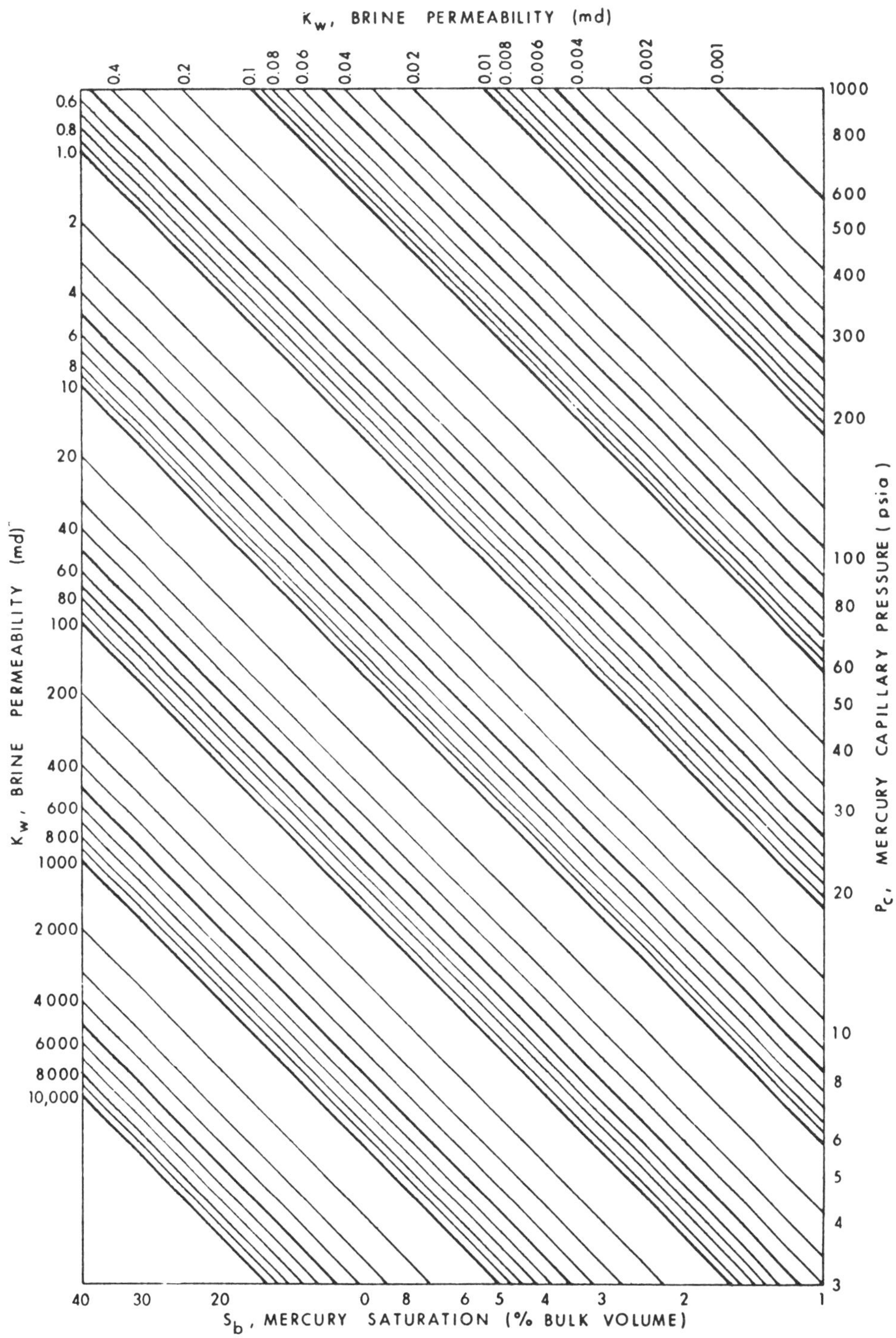


**Figure 2–33.** Stressed (1,000 psi) brine permeability/capillary pressure nomograph (© SPE, JPT, Dec. 1981).[30]

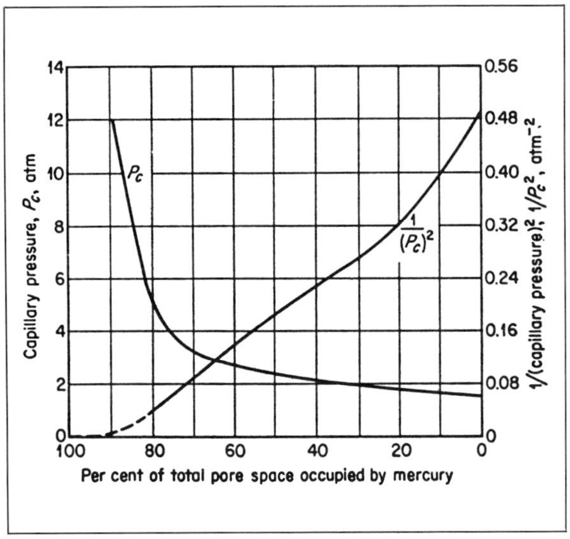


**Figure 2–34.** Graphical presentation of capillary pressure data for calculating permeability.[31]

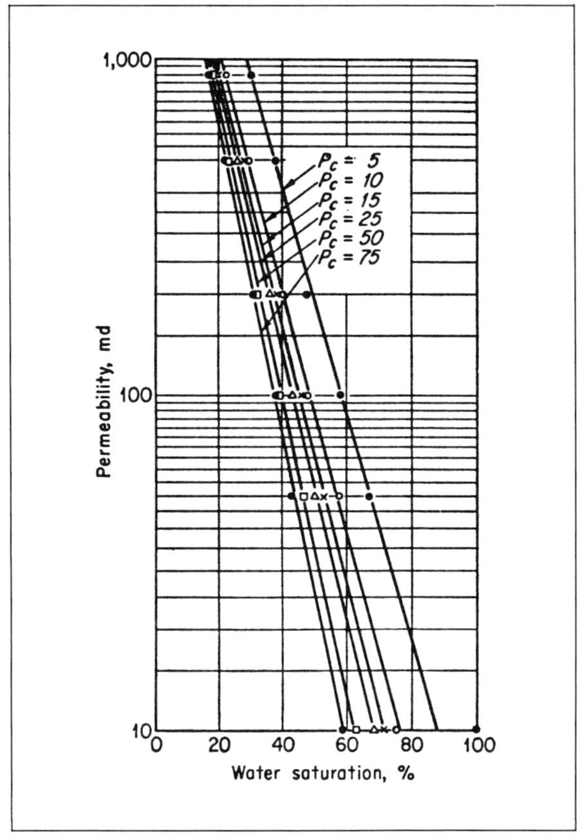


**Figure 2–35.** Correlation of water saturation with permeability for various capillary pressures (from Wright and Wooddy).[32]

$$P_{cR} = P_{cL}\frac{\sigma_R}{\sigma_L}$$

$$= 0.26\,\frac{250}{100}$$

$$= 0.59 \quad \text{psi}$$

**TABLE 2–4. Typical Capillary Pressure Data for Example 2–29**

| $S_w$ (%) | $P_{cL}$ (psi) | $P_{cR}$ (psi) | h (ft) |
|---|---|---|---|
| 100 | 0 | 0 | 0 |
| 95 | 0.26 | 0.59 | 5.45 |
| 80 | 0.63 | 1.43 | 13.20 |
| 65 | 0.75 | 1.70 | 15.69 |
| 50 | 0.90 | 2.05 | 18.92 |
| 30 | 1.13 | 2.57 | 23.72 |
| 25 | 1.38 | 3.14 | 28.98 |
| 21 | 1.58 | 3.59 | 33.14 |
| 20 | 2.50 | 5.68 | 52.43 |
| 19 | 3.00 | 6.82 | 62.95 |

The height above free-water level is calculated using Equation (2.85).

$$h = \frac{144 - P_{cR}}{\rho_w - \rho_o}$$

$$= \frac{144(0.59)}{65.52 - 49.92}$$

$$= 5.45 \quad \text{ft}$$

The $P_{cR}$ is then plotted versus $S_w$ as shown in Figure 2–36.

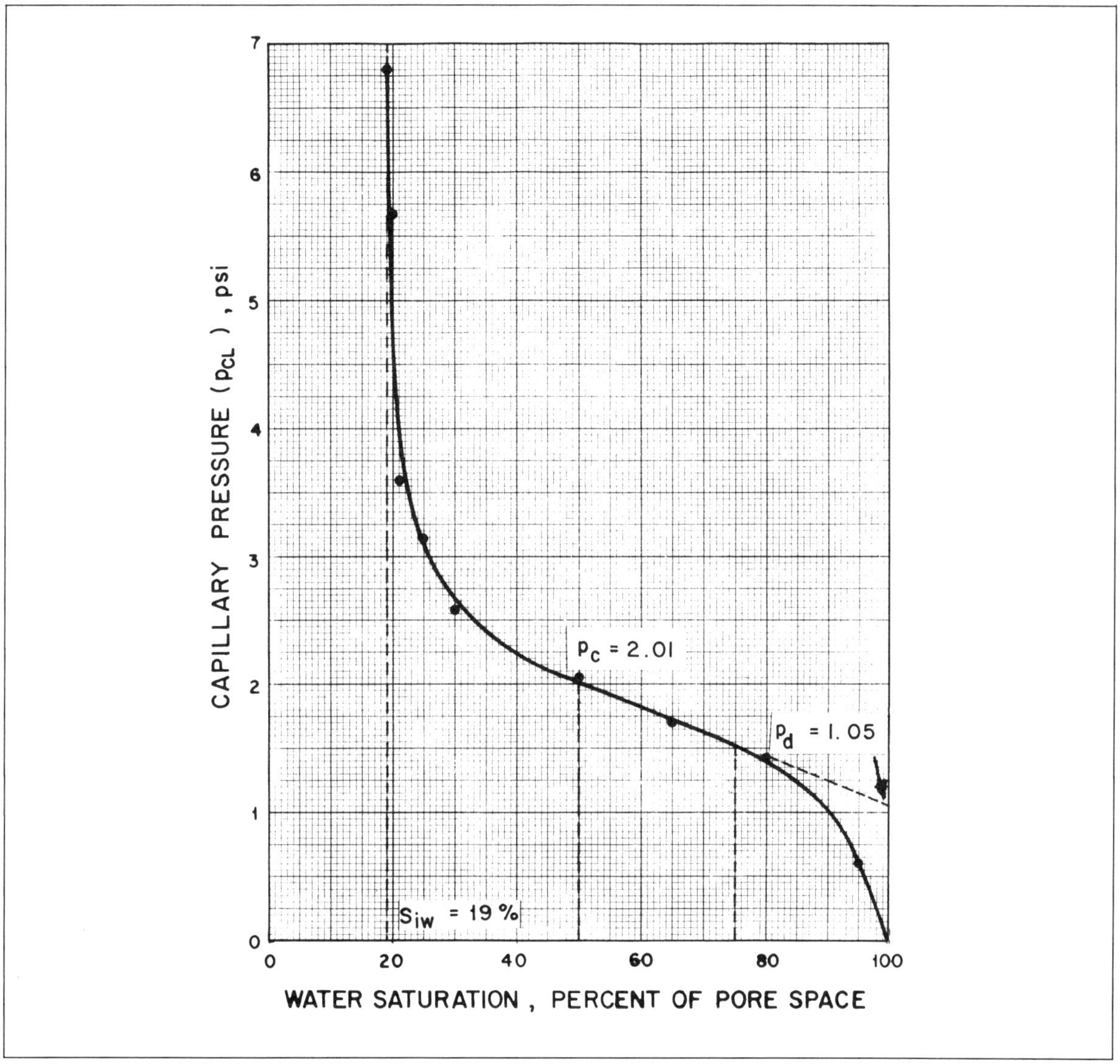


**Figure 2–36.** Typical capillary pressure curve for the data in Example 2–29.

Displacement pressure, $p_D$ = 1.05 psi (from Fig. 2–36)
Irreducible water saturation, $S_{wi}$ = 19% (from Fig. 2–36)

at $S_w$ = 50%, $P_{cR}$ = 2.01 psi:

$$h = \frac{144(2.01)}{65.52 - 49.92}$$
$$= 18.55 \quad \text{ft}$$

At approximately 18 ft above the oil/water contact, the $S_w$ will be 50%. If the $h$ at each $P_{cR}$ is plotted versus the water saturation, the $h$ at any $S_w$ can be read directly from the graph.

**Example 2–30.**[32] For the following capillary pressure data, calculate the water and oil saturation in gas zone. Thickness of oil zone $h_o = 70$ ft.

$$
\begin{array}{llll}
\sigma_{wg} = 72 \text{ dynes/cm} & \rho_o = 53 \text{ lb/ft}^3 \\
\sigma_{og} = 50 \text{ dynes/cm} & \rho_w = 68 \text{ lb/ft}^3 \\
\sigma_{wo} = 25 \text{ dynes/cm} & \rho_g = 7 \text{ lb/ft}^3 \\
P_{cL} = 54 \text{ psi} & h_{fw} = 120 \text{ ft (height above free water level)} \\
P_{cR} = 18 \text{ psi} & S_w = 16\% \text{ (at a height of 70 ft or above)}
\end{array}
$$

Solution:

$$
\begin{aligned}
h_{fo} &= h_{fw} - h_o \\
&= 120 - 70 = 50 \quad \text{ft} \\
(P_{cR})_{og} &= \frac{h_{fo}}{144}(\rho_o - \rho_g) \\
&= \frac{50}{144}(53 - 7) = 15.97 \quad \text{psi} \\
P_{cL} &= P_{cR}\left(\frac{\sigma_{wg}}{\sigma_{og}}\right) \\
&= 15.97\left(\frac{72}{50}\right) = 23 \quad \text{psi}
\end{aligned}
$$

The total wetting saturation $S_t = 18\%$ was read from the capillary pressure curve at $P_{cL} = 23$ psi. Therefore,

$$
\begin{aligned}
S_o &= S_t - S_w \\
&= 18 - 16 = 2\% \\
S_g &= 100 - S_t \\
&= 100 - 18 = 82\%
\end{aligned}
$$

## HYSTERESIS OF CAPILLARY PRESSURE AND RELATIVE PERMEABILITY CURVES

The wettability and direction of saturation change influence the fluid distribution of the wetting or nonwetting phase within the pore spaces. The direction of saturation change is defined by two widely used terms called *drainage* and *imbibition*. This path dependency is called *hysteresis* and is applied to the difference in multiphase rock properties that depends upon the direction of saturation change.

The drainage phenomena occurs when water (wetting phase) is displaced from the core by successive increase in the pressure of the nonwetting phase, thereby reducing the wetting phase saturation. On the contrary imbibition phenomena occurs when wetting phase saturation is increased.

In a strongly water-wet rock, imbibition often occurs spontaneously without additional pressure being applied, or it will occur also by injecting water into the rock to increase water saturation. On the other hand, injecting water into an oil-wet rock will result in a drainage process.

Because of this phenomena, both the capillary pressure curves and the relative permeability curves exhibit hysteresis. Figure 2–37 shows a typical set of drainage and imbibition relative permeability characteristics in terms of the wetting phase saturation.[34] Figure 2–38 shows capillary pressure data for a strongly water-wet rock with hysteresis. Curve 1 is the drainage curve and Curve 2 is the imbibition curve. Figure 2–39 shows capillary pressure data for a rock with intermediate wettability.[34] The curves in Figure 2–39 show (1) drainage, (2) spontaneous imbibition, and (3) forced imbibition. 

In case of the wetting phase relative permeability, the imbibition curve will retrace the drainage curve to the maximum wetting phase saturation (i.e., the saturation corresponding to a zero oil per-

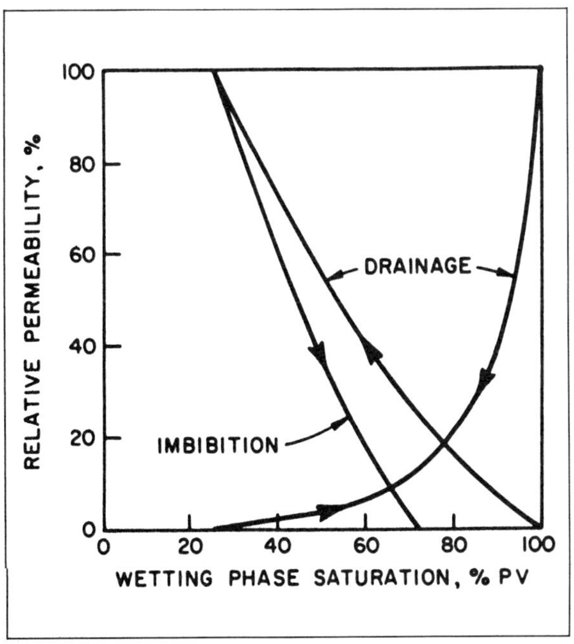


**Figure 2–37.** Drainage and imbibition relative permeability characteristics (© SPE, SPE Monograph Series, Vol. 3, 1971).[34]

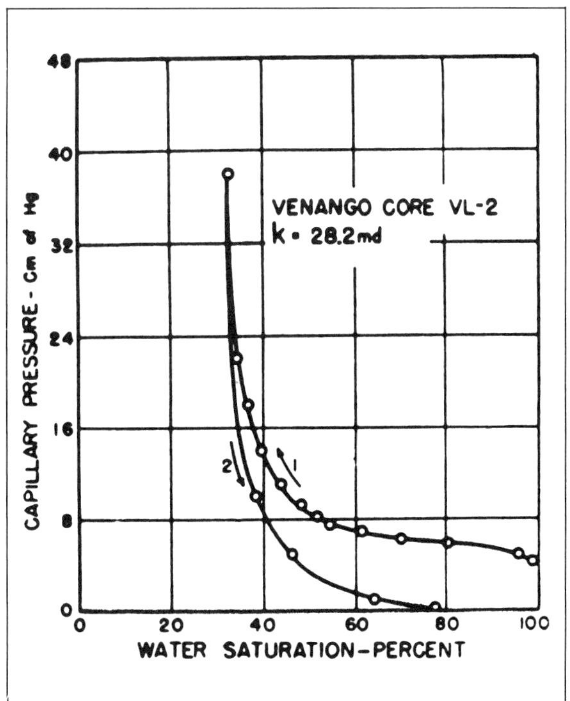


**Figure 2–38.** Capillary pressure characteristics, strongly water-wet rock (© SPE, SPE Monograph Series, Vol. 3, 1971).[34]

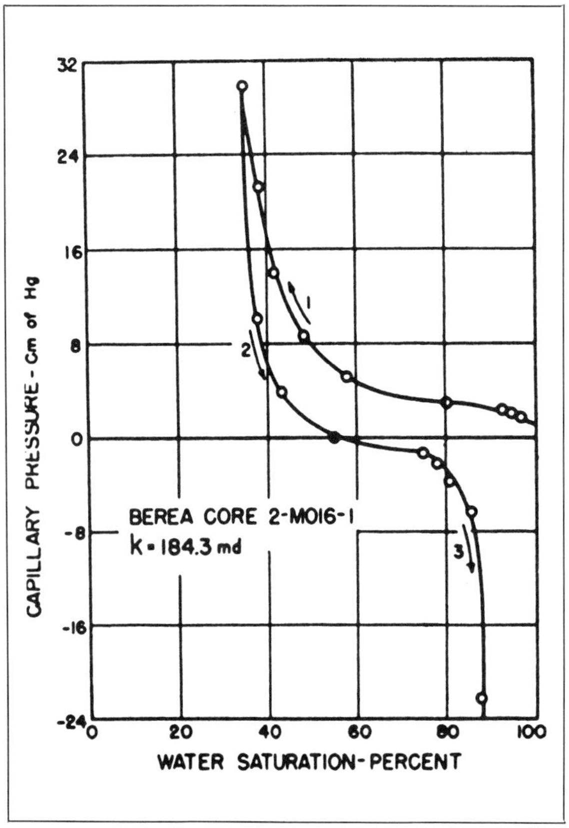


**Figure 2–39.** Capillary pressure characteristics, intermediate wettability (© SPE, SPE Monograph Series, Vol. 3, 1971).[34]

meability). On the other hand, the nonwetting phase imbibition relative permeability curve is lower than the drainage relative permeability curve.

## AVERAGING RESERVOIR PROPERTIES

For most petroleum engineering problems involving porosity, water saturation, permeability, and capillary pressure data, the average of these properties for a producing interval, lease, or entire reservoir is desired because of the heterogeneous nature of most reservoirs.

In this section various methods of averaging porosity, water saturation, permeability, and capillary pressure are illustrated.

### WATER SATURATION AND POROSITY

Porosity and water saturation appear in many engineering equations and form the basis of the volumetric equation. Many of the commonly used procedures for obtaining the average porosity and water saturation of a lease, group of leases, or a reservoir are given below.

Average water saturation:

Arithmetic average

$$S_{w_{avg}} = \frac{\sum_{i=1}^{n} \phi_i S_{wi}}{\sum_{i=1}^{n} n_i} \quad \textbf{(2.91)}$$

Thickness weighted average

$$S_{w_{avg}} = \frac{\sum_{i=1}^{n} \phi_i S_{wi} h_i}{\sum_{i=1}^{n} h_i \phi_i} \quad \textbf{(2.91a)}$$

Area weighted average

$$S_{w_{avg}} = \frac{\sum_{i=1}^{n} \phi_i S_{wi} A_i}{\sum_{i=1}^{n} A_i \phi_i} \quad \textbf{(2.91b)}$$

Area-thickness weighted average

$$S_{w_{avg}} = \frac{\sum_{i=1}^{n} \phi_i S_{wi} A_i h_i}{\sum_{i=1}^{n} A_i h_i \phi_i} \quad \textbf{(2.91c)}$$

Average porosity:

Arithmetic average



$$\phi_{avg} = \frac{\sum_{i=1}^{n} \phi_i}{\sum_{i=1}^{n} n_i} \quad \textbf{(2.92)}$$

Thickness weighted average

$$\phi_{avg} = \frac{\sum_{i=1}^{n} \phi_i h_i}{\sum_{i=1}^{n} h_i} \quad \textbf{(2.92a)}$$

Area weighted average

$$\phi_{avg} = \frac{\sum_{i=1}^{n} \phi_i A_i}{\sum_{i=1}^{n} A_i} \quad \textbf{(2.92b)}$$

Volume weighted average

$$\phi_{avg} = \frac{\sum_{i=1}^{n} \phi_i A_i h_i}{\sum_{i=1}^{n} A_i h_i} \quad \textbf{(2.92c)}$$

In addition, average water saturation and average porosity can be determined using the statistical methods (geometric mean and frequency distribution) described in the previous chapter when warranted. A *cutoff* value of water saturation and porosity is selected so that only samples with water saturation less than and porosities greater than the cutoff value are used as these are considered to be the net contributors in reservoir evaluation.

## PERMEABILITY

To obtain average permeability which will describe the overall performance of a reservoir, the following equations can be used:

Parallel combination of beds (linear or radial flow):

$$k_{avg} = \frac{\sum_{i=1}^{n} k_i h_i}{\sum_{i=1}^{n} h_i} \quad \mathbf{(2.93)}$$

Series combination of beds (linear flow):

$$k_{avg} = \frac{\sum_{i=1}^{n} L_i}{\sum_{i=1}^{n} \frac{L_i}{k_i}} \quad \mathbf{(2.94)}$$



Series combination of beds (radial flow):

$$k_{avg} = \frac{\ln (r_e/r_w)}{\sum_{i=1}^{n} \frac{\ln r_i/r_{i-1}}{k_i}} \quad \mathbf{(2.95)}$$

where $i$ is the number of reservoir layers, $k_i$ is the permeability of individual layer, $h_i$ is the thickness of individual layer in combination of parallel beds, and $L_i$ is the length of each layer in combination of series beds. In series combination of beds and radial flow, $r_e$ is the drainage radius, $r_w$ is the wellbore radius, and $r_i$ refers to the radius of each layer of different permeability from the wellbore.

When there is a wide variation between the permeabilities of each layer (heterogeneous system), the average system permeability is best represented by geometric or harmonic mean as discussed in the previous chapter.

Spherical flow permeabilities may relate horizontal and vertical flow streams so that $k_s = (k_v k_{h1} k_{h2} \ldots k_{hn})^{1/(n+1)}$.

## CAPILLARY PRESSURE

Due to the heterogeneous nature of most reservoirs, once again averaging of the capillary pressure data is also required in order that the reservoir saturation distribution is represented by one average capillary pressure curve. As previously mentioned, fluid saturation-capillary pressure relationships are affected by the permeability of the sample. Therefore, it is necessary to evaluate the various sets of capillary pressure data with respect to the permeability of the core sample from which they are obtained.[33]

One of the classical methods of averaging capillary pressure data is by using the $J$ function. The method provides a good understanding of the relationship between porosity, permeability, and capillary pressure characteristics. The $J$ function correlating term is given by the following equations:

$$J(S_w) = \frac{P_c}{\sigma \cos \theta}\left(\frac{k}{\phi}\right)^{0.5} \tag{2.96}$$

OR

$$J(S_w) = \frac{P_c}{\sigma}\left(\frac{k}{\phi}\right)^{0.5} \tag{2.96a}$$

The normalized capillary pressure curve obtained by plotting $J$ versus wetting phase saturation can be applied to a reservoir with any permeability and porosity containing any combination of wetting and nonwetting fluids characterized by a particular interfacial tension and wetting angle.[36] One simple curve will be obtained representing all of the capillary pressure data for all of the samples measured. A typical $J$ function curve is shown in Figure 2–40.

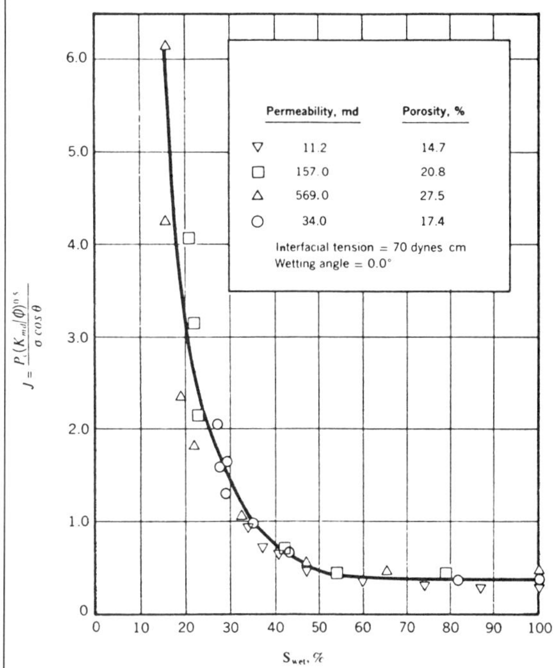


**Figure 2–40.** A typical $J$ function curve (from Slider, courtesy of PennWell Books).[36]

**Example 2–31.**[36] For the capillary pressure data given below, calculate the $J$ function for each sample. Use water/air system with $\sigma$ = 70 dynes/cm and $\theta$ = 1.0.

| (1) Core No. | (2) $P_c$ at $S_w$ = 50% | (3) k, md | (4) $\phi$ |
|---|---|---|---|
| 1 | 3.66 | 11.2 | 0.147 |
| 2 | 2.50 | 34.0 | 0.174 |
| 3 | 1.38 | 157.0 | 0.208 |
| 4 | 0.85 | 569.0 | 0.275 |

Solution: Using Equation (2.96),

$$\begin{aligned} J_1 &= \frac{3.66}{70 \cos(1.0)}\left(\frac{11.2}{0.147}\right)^{0.5} \\ &= 0.052(8.729) \\ &= 0.454 \end{aligned}$$

153

Similarly,

$$J_2 = 0.499$$
$$J_3 = 0.542$$
$$J_4 = 0.552$$

## PERMEABILITY RATIO CURVE

The following equation is used to develop an average permeability ratio curve:

$$\log\left(\frac{k_g}{k_o}\right)_{avg} = \left[\frac{n\Sigma(S_L \log k_g/k_o) - \Sigma S_L \Sigma \log k_g/k_o}{n\Sigma S_L^2 - (\Sigma S_L)^2}\right]\overline{S}_L + \left[\frac{\Sigma S_L^2 \Sigma \log k_g/k_o - \Sigma S_L \Sigma(S_L \log k_g/k_o)}{n\Sigma S_L^2 - (\Sigma S_L)^2}\right] \quad \textbf{(2.97)}$$

where $\overline{S}_L$ is the average saturation corresponding to the average permeability ratio $(k_g/k_o)_{avg}$, $S_L$ is the total liquid saturation, and $k_g/k_o$ is the permeability ratio to be averaged.

Note that Equation (2.97) is the equation of a straight line best fitting the data points. The equation determines the linear portion of the permeability ratio curve. Such an approach is sufficiently accurate since the linear portion of the curve represent the major part of the history of a depletion-type reservoir.

## REFERENCES

[1] G. G. Brown, D. L. Katz, G. B. Oberfell, and R. C. Alden, *Natural Gasoline and Volatile Hydrocarbons,* Natural Gasoline Association of America, Oklahoma, 1948, p. 44.

[2] E. Wiehert, and K. Aziz, "Calculate Z's for Sour Gases," *Hydrocarbon Processing,* May 1972, pp. 51, 119.

[3] *Engineering Data Book,* Gas Processors Suppliers Association, Tulsa, Oklahoma, 9th Edition, 1977.

[4] M. B. Standing, and D. L. Katz, "Density of Natural Gases," *Transactions of AIME,* 1942, Vol. 146, p. 144.

[5] L. Yarborough, and K. R. Hall, "How to Solve Equation of State for Z-Factor," *Oil and Gas Journal,* Feb. 18, 1974, p. 86.

[6] N. L. Carr, R. Kobayashi, and D. B. Burrows, "Viscosity of Hydrocarbon Gases Under Pressure, *Transactions of AIME,* 1954, pp. 264–272.

[7] A. S. Trube, "Compressibility of Natural Gases," *Transactions of AIME,* Vol. 210, 1957, pp. 335–357.

[8] L. K. Nemeth, and H. T. Kennedy, "A Correlation of Dew Point Pressure with Fluid Composition and Temperature," *Journal of Petroleum Technology,* June 1967, p. 99.

[9] R. Al-Hussainy, H. J. Ramey, Jr., and P. B. Crawford, "The Flow of Real Gases Through Porous Media," *Journal of Petroleum Technology,* May 1966, pp. 624–636.

[10] M. J. Rzasa, and D. L. Katz, "Calculation of Static Pressure Gradients in Gas Wells," *Transactions of AIME*, Vol. 160, 1945, p. 100.

[11] M. B. Standing, *Volumetric and Phase Behavior of Oil Field Hydrocarbon Systems*, Society of Petroleum Engineers, Dallas, 1977.

[12] C. Beal, "The Viscosity of Air, Water, Natural Gas, Crude Oil, and Its Associated Gases at Oil Field Temperatures and Pressures," *Transactions of AIME*, 1946, pp. 94–115.

[13] J. N. Chew, and C. A. Connally, "A Viscosity Correlation for Gas-saturated Crude Oils, *Transactions of AIME*, 1959, pp. 23–25.

[14] H. J. Ramey, Jr., "Rapid Method of Estimating Reservoir Compressibility," *Journal of Petroleum Technology*, Vol. 231, April 1964, pp. 447–454.

[15] E. L. Lederer, *Proceedings*, World Petroleum Conference, London, 1933, pp. 526–28.

[16] W. R. Shu, "A Viscosity Correlation for Mixtures of Heavy Oil, Bitumen, and Petroleum Fractions," *Society of Petroleum Engineers Journal*, June 1984, pp. 272–282.

[17] Glaso Oistein, "Generalized Pressure-Volume-Temperature Correlations," *Journal of Petroleum Technology*, May 1980, pp. 785–795.



[18] M. Vasquez, and H. D. Beggs, "Correlations for Fluid Physical Property Predications," *Journal of Petroleum Technology*, June 1980, pp. 968–970.

[19] M. A. Al-Marhoun, "PVT Correlations for Middle East Crude Oils," *Journal of Petroleum Technology*, May 1988, pp. 650–665.

[20] C. R. Dodson, and M. B. Standing, "Pressure-Volume-Temperature and Solubility Relations for Natural Gas-Water Mixtures," *Drilling and Production Practice*, API, 1944, pp. 173–179.

[21] C. S. Mathews and D. G. Russel, *Pressure Buildup and Flow Tests in Wells*, Monograph Series, Society of Petroleum Engineers of AIME, Dallas, Texas, 1967.

[22] H. N. Hall, "Compressibility of Reservoir Rocks," *Transactions of AIME*, Vol. 198, 1953, pp. 309–311.

[23] L. J. Klinkenberg, "The Permeability of Porous Media to Liquids and Gases," *Drilling and Production Practice*, API, 1941, p. 200.

[24] J. C. Calhoun, "Relative Permeability Ratio," *The Oil and Gas Journal Manual on Engineering Fundamentals*, p. 23.

[25] T. C. Frick, *Petroleum Production Handbook*, Society of Petroleum Engineers of AIME, Dallas, Texas, 1962.

[26] L. F. Mehdi Honarpour, L. F. Koedritz, and A. H. Harvey, "Empirical Equations for Estimating Two-Phase Relative Permeability in Consolidated Rock," *Journal of Petroleum Technology*, December 1982, pp. 2,905–2,908.

[27] G. L. Chierici, "Novel Relations for Drainage and Imbibition Relative Permeabilities," *Society of Petroleum Engineers Journal*, June 1984, pp. 275–276.

[28] J. B. Jennings, "Using Capillary Pressure Techniques to Map and Identify an Economic Fairway in the Ordovician Red River on the Western Flank of the Williston Basin," *Fifth International Williston Basin Symposium*, pp. 212–216.

[29] F. W. Cole, *Reservoir Engineering Manual*, Gulf Publishing Company, Houston, Texas, 2nd Edition, 1969, p. 9.

[30] B. F. Swanson, "A Simple Correlation Between Permeabilities and Mercury Capillary Pressures," *Journal of Petroleum Technology*, December 1981, pp. 2,498–2,504.

[31] W. R. Purcell, "Capillary Pressure—Their Measurement Using Mercury and the Calculation of Permeability Therefrom," *Transactions of AIME*, Vol. 186, 1949, pp. 39–48.

[32] H. T. Wright, Jr., and L. D. Wooddy, Jr., "Formation Evaluation of the Borregas and Seeligson Field, Brooks and Jim Wells County, Texas," *Symposium on Formation Evaluation*, AIME, October 1955.

[33] J. W. Amyx, D. M. Bass, Jr., and R. L. Whiting, *Petroleum Reservoir Engineering—Physical Properties,* McGraw-Hill Book Company, New York, 1960.

[34] F. F. Craig, Jr., *The Reservoir Engineering Aspects of Waterflooding,* Monograph Series, Vol. 3, Society of Petroleum Engineers of AIME, Dallas, Texas, 1971.

[35] G. P. Willhite, *Waterflooding,* SPE Textbook Series, Vol. 3, Society of Petroleum Engineers of AIME, Richardson, Texas, 1986.

[36] H. C. Slider, *Practical Petroleum Reservoir Engineering Methods,* PennWell Publishing Company, Tulsa, Oklahoma, 1976.

CHAPTER THREE

# WELL LOG INTERPRETATION



Well logs play an important role in the geological and engineering decisions concerning a particular well and the overall exploration and development process. Well logs represent different formation properties, foot-by-foot, throughout the depth of the well (from top to bottom). The information obtained from well logs helps in developing structural maps, isopach maps, and isovolume maps; identifies the different lithologies through the wellbore; and identifies the thickness and depth of possible productive (permeable) zones. The logs are also used to correlate the productive formations in one well with that of the others in order to get an idea about the continuity of these productive formations.

The logs are used to determine formations' porosity, thickness, and water saturation, which in turn are used in calculating hydrocarbon pore volume. In short, well logs will provide quantitative and qualitative information that enables us to interpret the reservoir characteristics and content. The important step in log interpretation is to understand the limitations of each type of log used: what it is used for, what affects its response, and what corrections are required to arrive at a valid information. The rest of the interpretation is just number crunching in appropriate equations.

In this chapter we will discuss various logs and how they are analyzed. We will avoid the discussion of the different tools used in logging and the mathematical derivations. A majority of the charts used in interpretation are presented so that this chapter is self-contained and no outside chart books are required. In practice the log analyst will find that in every case and every log, all the text book corrections may not cause significant change to his interpretation and can be safely ignored. The information in the chapter may not benefit those who do not have previous knowledge of log analysis; however, it will provide excellent reference for the others. For those without previous knowledge of log analysis, the book by Dr. D. W. Hilchie is strongly recommended.[1] Other handy references are given at the end of this chapter.[2–8]

## BASIC RELATIONSHIPS OF WELL LOGS INTERPRETATION

The primary parameters determined from well log measurements are porosity, the fraction of pore space filled with hydrocarbons, and formation's depth and thickness. Electrical, nuclear, and acoustic or sonic devices are used to determine these properties. During the drilling of a well, the drilling mud alters some of the rock properties surrounding the wellbore and consequently will affect the log measurements. In this section we will discuss the borehole configuration and its contribution to log analysis, formation temperature calculations, and the importance of log heading in log interpretation.



### BOREHOLE ENVIRONMENT

A schematic of the borehole environment is shown in Figure 3–1. When a well is drilled, drilling mud is used to help remove cuttings from the wellbore, lubricate and cool the bit, and maintain an excess of borehole pressure over formation pressure to avoid blow out. The borehole pressure is controlled by controlling mud weight.

When a porous and permeable zone is encountered, some drilling fluid is forced into the formation because of higher wellbore pressure, pushing any hydrocarbons and/or water (formation fluids) away from the proximity of the wellbore. This phenomena is called *invasion*, and the formation invaded by the mud filtrate is called the *invaded zone*. When filtering into the porous and permeable formation, the drilling fluid will leave some of the contained mud and drill cuttings on the wall of the hole. The area from which formation fluids are pushed away is referred to as the *flushed zone*. The mud and drill cuttings on the wall of the wellbore are referred to as *mudcake*. After the flushed zone, we get into a *transition zone* where formation fluids and mud filtrate are mixed. This zone occurs between the uninvaded zone and the flushed zone. The uninvaded zone is defined as the area beyond the invaded zone where the formation's fluids are uncontaminated by mud filtrate. This uninvaded (virgin) zone is of interest to the log analyst, engineers, and geologists.

The depth of mud filtrate invasion into the invaded zone is referred to as the *diameter of invasion*. The borehole diameter is commonly described as the outside diameter of the drill bit. This practice is not correct: the diameter of the borehole may be larger or smaller because of formations' washouts or mudcake buildup. The borehole diameter at the zone of interest must be taken from the caliper log (to be discussed).

The well logs are also affected by the characteristics of the drilling mud used. This factor needs to be considered before selecting which logs to run in a well which will provide the most useful information about the formations throughout the wellbore.

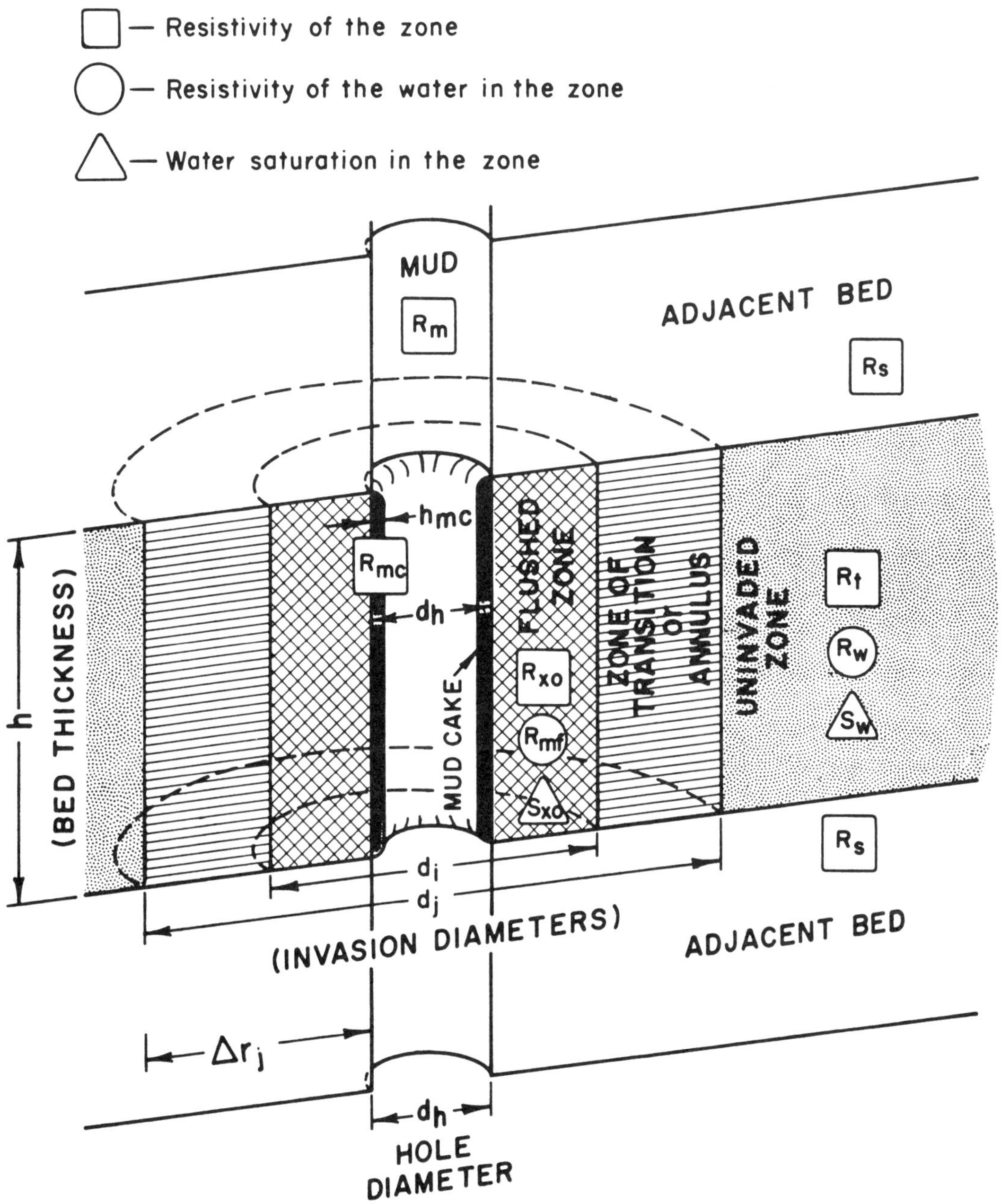


**Figure 3–1.** The borehole environment and symbols used in log interpretation (courtesy of Schlumberger).

## INVASION AND RESISTIVITY PROFILES

Of all the rock parameters measured by today's logging tools, resistivity is of particular importance. Electrical resistivity of a substance is its ability to impede the flow of electrical current through the substance. The unit used in logging is *ohm-meter²/meter* (*ohm-m*). The dry matrix rock will have infinitely high resistivity. Resistivity is a basic measurement of a reservoir's fluid saturation and is a function of porosity, type of fluid (hydrocarbons, salt or fresh water), and type of rock.

Resistivity profiles illustrate the horizontal distribution of the invaded and uninvaded zones and their corresponding relative resistivities. A log analyst should always keep these profiles in mind because by using them, an analyst can quickly scan a log and look for potential zones of interest such as hydrocarbon-bearing zones.

There are three commonly encountered invasion profiles: (1) step profile, (2) transition profile, and (3) annulus profile. These three profiles are illustrated in Figure 3–2. An annulus profile can commonly occur in zones which bear hydrocarbons; however, it rapidly disappears and will be detected only by an induction log run soon after a well is drilled.

Resistivity profiles for both water-bearing and hydrocarbon-bearing zones are illustrated in Figures 3–3 and 3–4, respectively. Always try to visualize these three illustrations when analyzing well logs.

## FORMATION TEMPERATURE

The knowledge of formation temperature is also important in log analysis since the resistivity values vary with temperature and the temperature varies with depth. In log analysis it is always assumed that temperature increases linearly with depth. Using the assumption of linear geothermal gradient and given bottomhole temperature (BHT), total depth of the well (TD), and surface temperature, temperature at any depth can be calculated using the following equation or the chart given in Figure 3–5 (for mean surface temperatures of 30°, 40°, 50°, 60°, and 70°F).

$$T_f = \left(\frac{T_{bh} - T_{ms}}{TD}\right) \times D_f + T_{ms} \qquad \textbf{(3.1)}$$

In Equation (3.1) the total depth and the temperature at this depth are usually taken from the log heading. The mean surface temperature for the area where the well is located should be known. If the mean surface temperature is different than those provided in Figure 3–5, the $T_f$ can be found by linear interpolation between $T_f$ at two different mean surface temperatures.

**Example 3–1.** The bottomhole temperature at a depth of 8,000 ft is 160°F. If the mean surface temperature is 80°F, what will be the formation temperature at 5,000 ft?

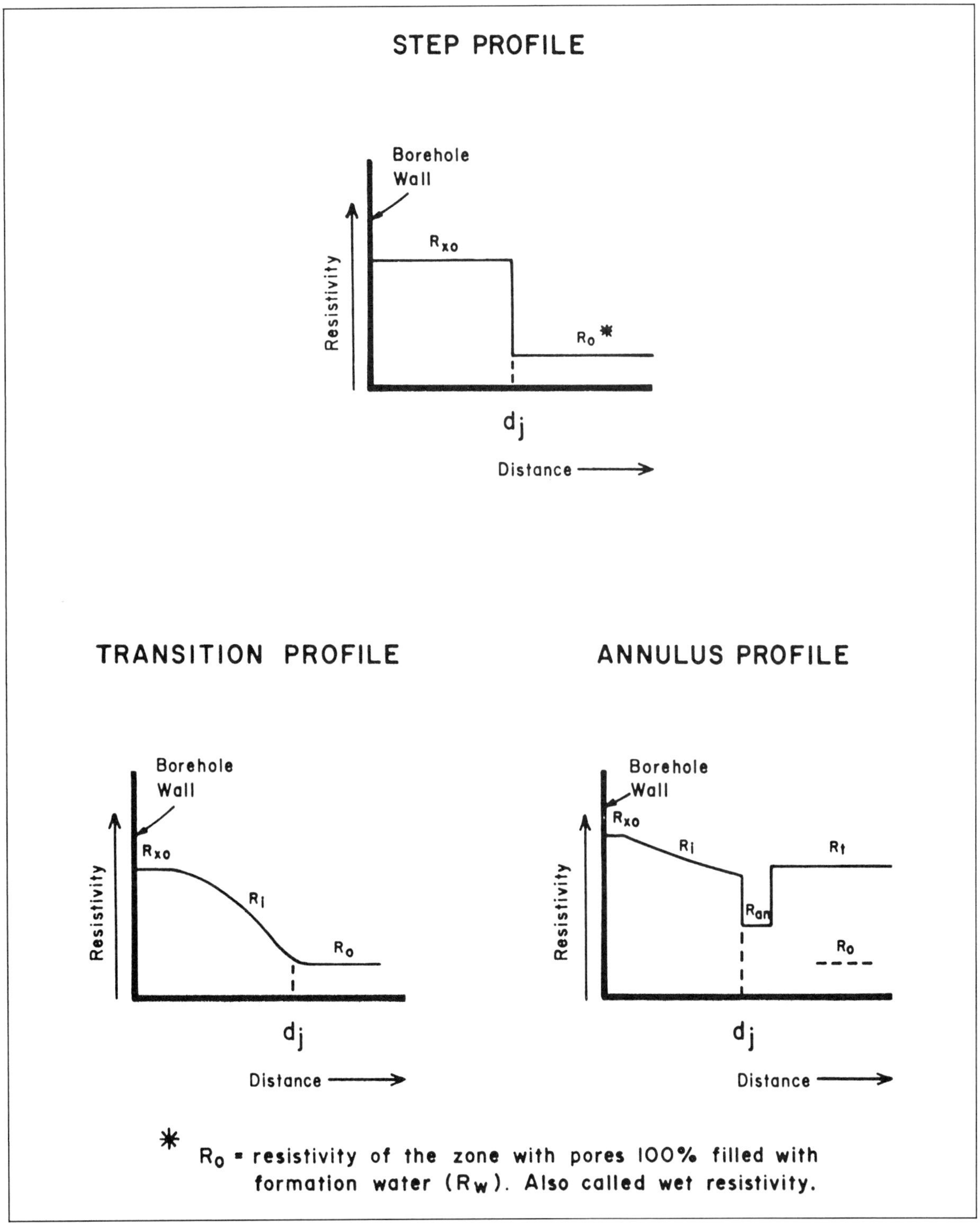


**Figure 3–2.** Typical invasion profiles in the vicinity of the wellbore (Asquith, 1982, reprinted by permission from AAPG).[2]

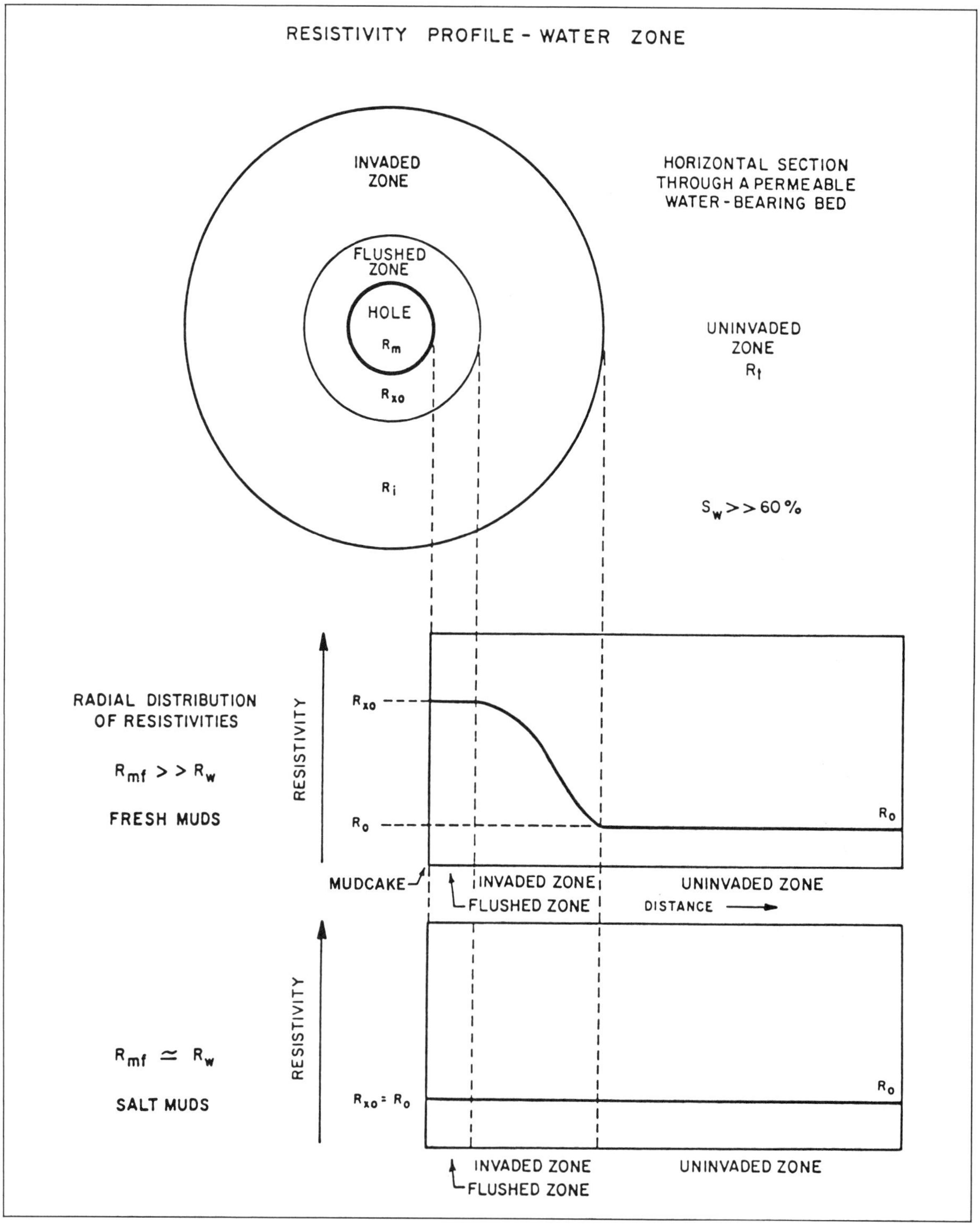


**Figure 3–3.** Resistivity profiles in water-bearing zones (Asquith, 1982, reprinted by permission from AAPG).[2]

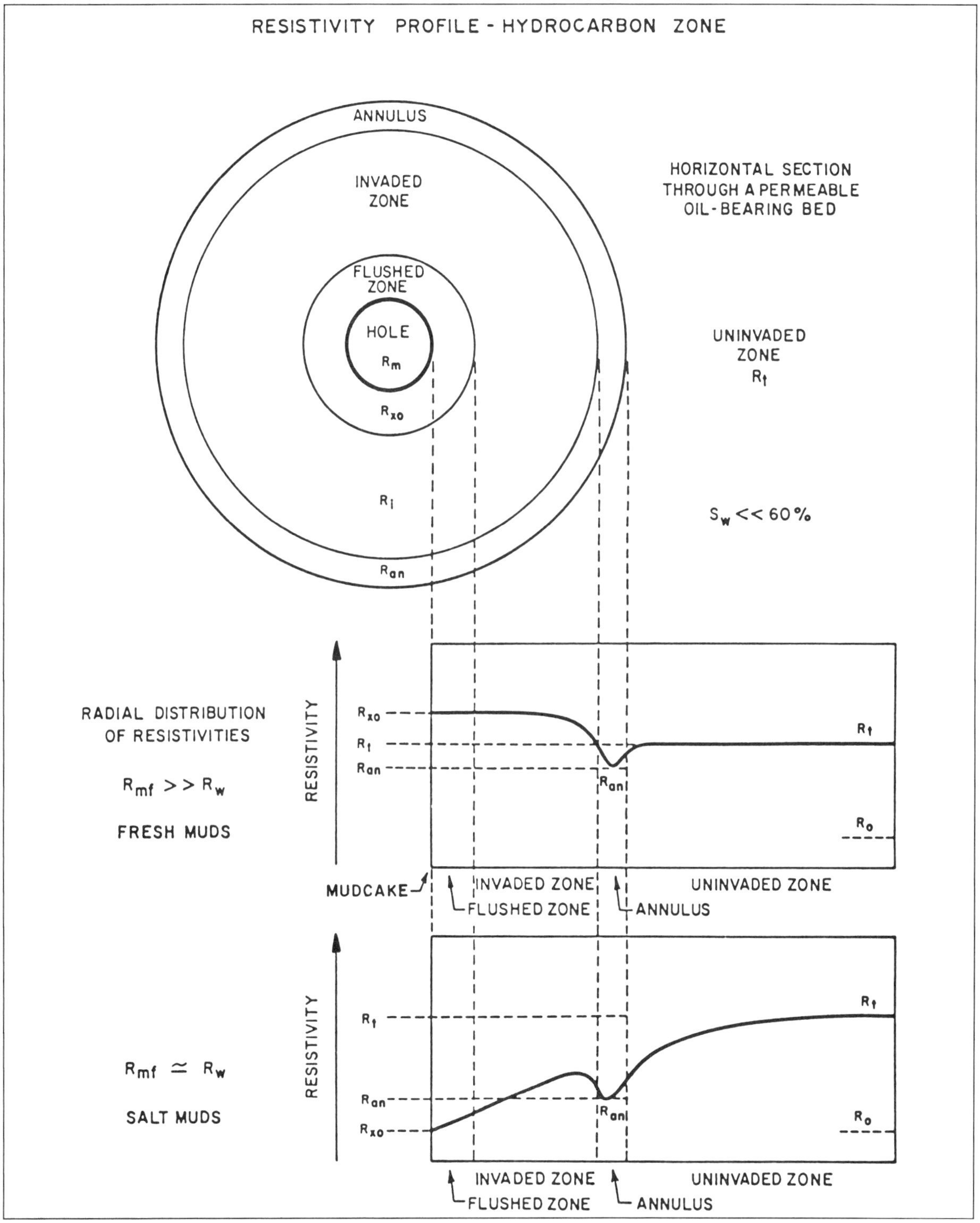


**Figure 3–4.** Resistivity profiles in hydrocarbon-bearing zones (Asquith, 1982, reprinted by permission from AAPG).[2]

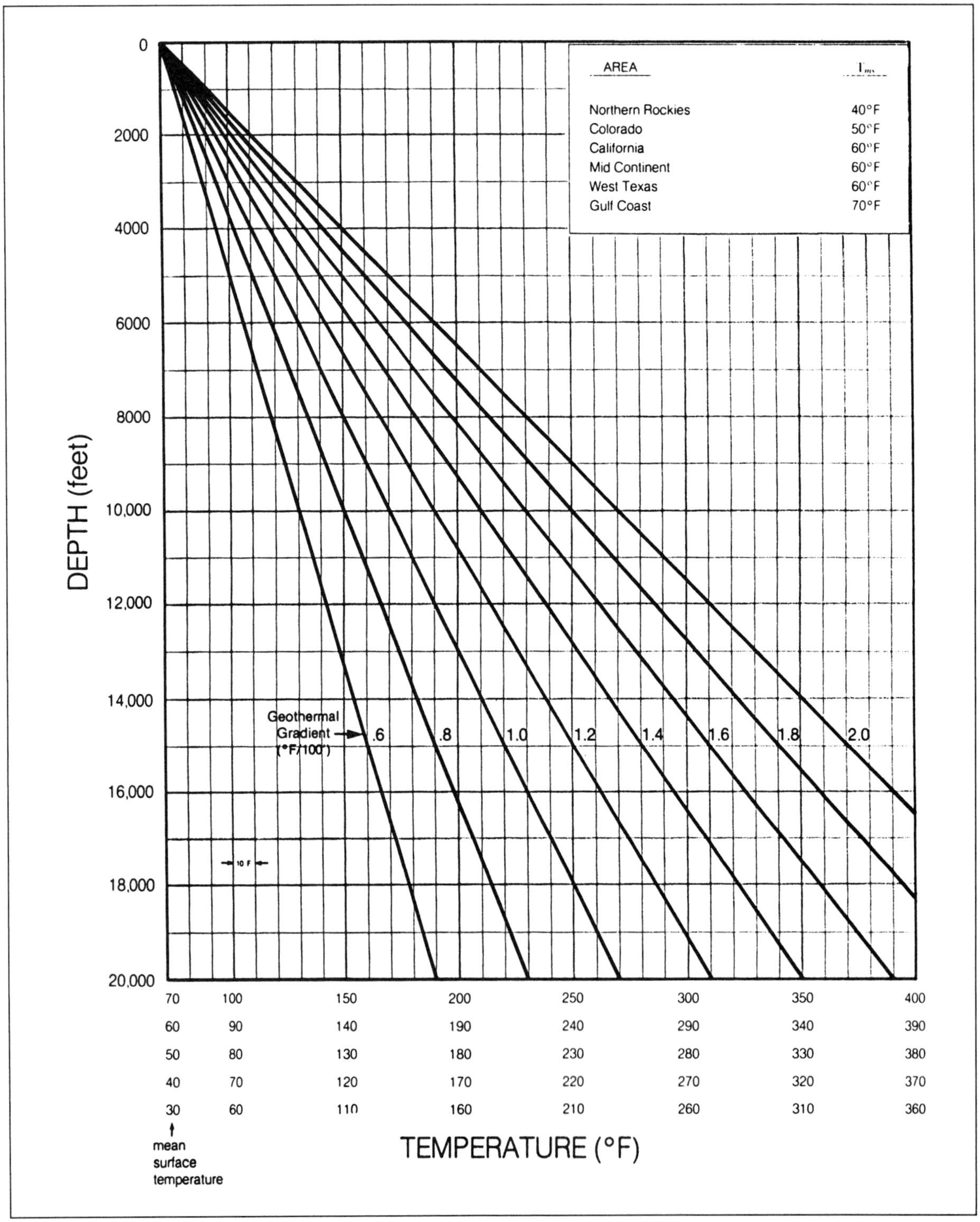


**Figure 3–5.** Chart for estimating formation temperature ($T_f$) with depth (courtesy of Welex).[9]

Solution:

$$T_f = \left(\frac{T_{bh} - T_{ms}}{TD}\right) \times D_f + T_{ms}$$
$$= \left(\frac{160 - 80}{8{,}000}\right) \times 5{,}000 + 80$$
$$= 130^\circ\mathrm{F}$$

Once the formation temperature is determined, the resistivities of the different fluids ($R_m$, $R_{mf}$, or $R_w$) can be corrected to any formation temperature. The following approximate equation or the chart in Figure 3–6 can be used to achieve this objective.

$$R_{Tf} = R_{temp} \times \left(\frac{T_{temp} + 6.77}{T_f + 6.77}\right) \tag{3.2}$$

The chart in Figure 3–6 can also be used to obtain resistivity if sodium chloride concentration and temperature are known. Alternatively, if resistivity is known, sodium chloride concentration can be obtained.

**Example 3–2.** If drilling mud resistivity at 75°F is $R_m = 0.35$ ohm-m, what would be the resistivity of this mud at 190°F? What would be the sodium chloride concentration?

Solution:

$$R_f = R_{temp} \times \left(\frac{T_{temp} + 6.77}{T_f + 6.77}\right)$$
$$R_{m@190} = R_{m@75} \times \left(\frac{75 + 6.77}{190 + 6.77}\right)$$
$$= 0.35 \times \left(\frac{81.8}{196.8}\right)$$
$$= 0.145 \quad \text{ohm-m}$$

Use the following steps to correct the resistivity to formation temperature using Figure 3–6:

1. Locate the resistivity value $R_m = 0.35$ ohm-m on the scale at the bottom of the chart.
2. Follow the vertical line up to 75°F temperature line, Point A.
3. Follow the diagonal line down to 190°F temperature line, Point B.
4. From Point B, drop vertically down to the bottom of the chart, and find the resistivity $R_m = 0.145$ ohm-m.

To find the sodium chloride concentration, follow the diagonal line upward from Point B to the NaCl concentration of 17,000 ppm.

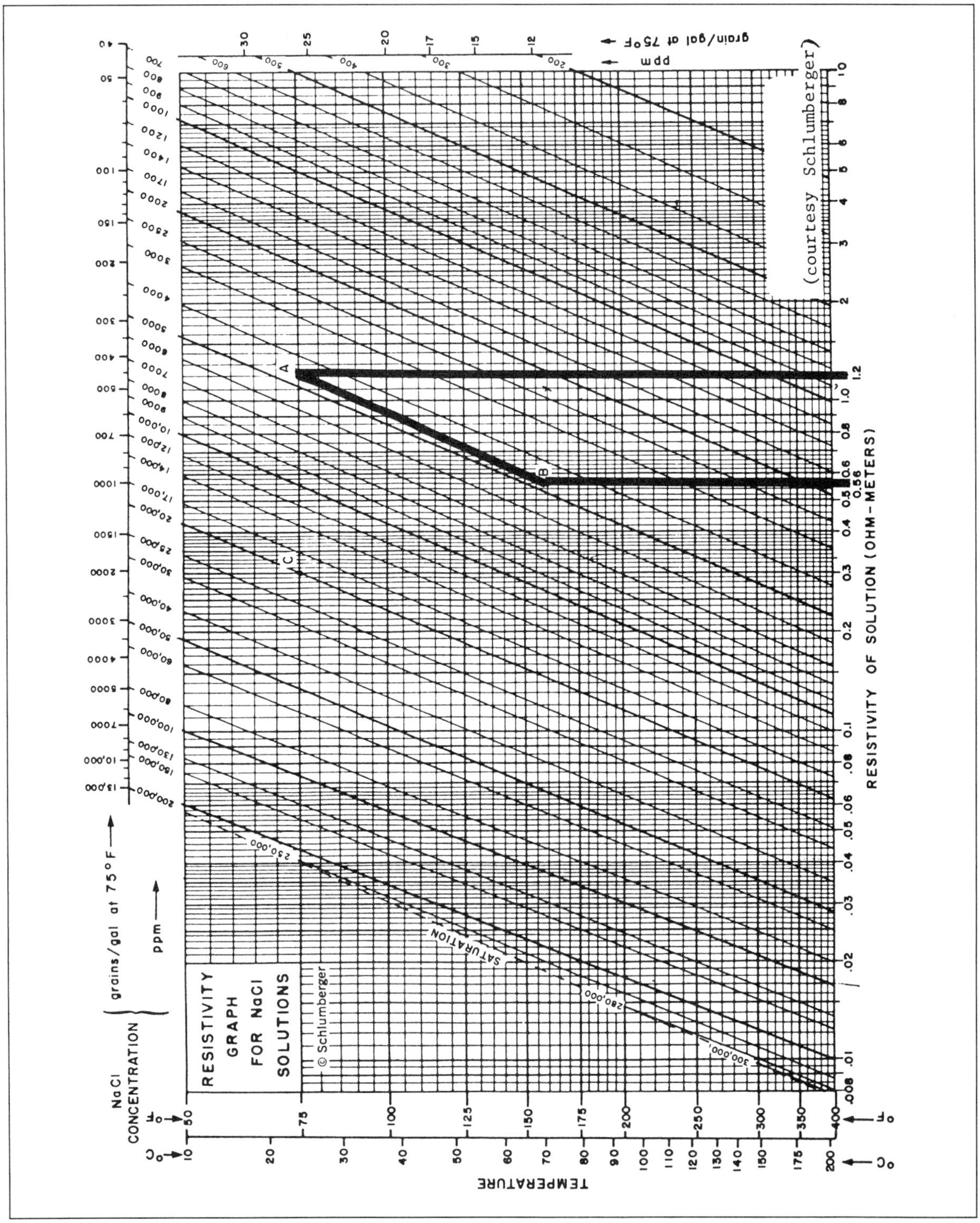


**Figure 3–6.** Chart to correct resistivity for temperature difference (courtesy of Schlumberger).[4]

## THE LOG HEADING

The log heading provides some basic information regarding the well's identity; depth logged; bottomhole temperature; the type of mud in the hole; and resistivities of mud, mud filtrate, and mudcake. The typical log heading is shown in Figure 3–7. Information on the header about the resistivity values for drilling mud ($R_m$) and mud filtrate ($R_{mf}$) are especially useful in log interpretation.

It should be kept in mind that the temperature recorded by well logs may not be the same as the reservoir temperature. Usually this temperature is lower than the actual reservoir temperature. This is because of the cooling effect of the drilling mud in the hole. The DST will usually provide more accurate temperature of the reservoir.



## THE SPONTANEOUS POTENTIAL (SP) LOG

The SP log is used for identifying permeable beds and their thickness and locating their boundaries. The log is also used to compute formation water resistivity $R_w$ and to give a quantitative indication of bed shaliness.

The SP curve records the electrical potential (voltage) produced by the interaction of formation water, drilling mud, and ion-selective rocks such as shales. The SP curve will be featureless (no deflection or very little deflection), even in permeable formations when the resistivity of the mud filtrate and the formation water is about the same. In such cases the SP log is not of much use.

Opposite shales, the SP curve response is relatively constant and is referred to as the *shale baseline*, from which the SP deflections are measured. The shale baseline is arbitrarily set by the logging engineer. A typical SP log is shown in Figure 3–8. The permeable bed boundaries and thickness are measured at the inflection points of the SP curve as shown in Figure 3–8.

## PROPERTIES OF THE SP CURVE

The shape of the SP curve and the amplitude of its deflection opposite a permeable bed depend on several factors. The main features of SP curves may be broadly described as follows:

1. A conductive fluid in the borehole, permeable bed, and shale bed are required for the SP curve to develop. The SP log cannot be used in nonconductive (oil-based) drilling muds.
2. Permeable zones are indicated by a positive (to the right, $R_{mf} < R_w$) or negative (to the left, $R_{mf} > R_w$) deflection of the SP curve from the shale baseline.
3. The maximum SP that a thick, shale-free, porous and permeable formation will generate is called *static spontaneous potential* (SSP), also referred to as the clean sand line.
4. The bed thickness affects the SP deflection. In thin formations (i.e., less than 10 ft thick), the log will read the SP value less than SSP. The chart in Figure 3–9 is used to incorporate thin bed corrections. As a rule of thumb, when the SP curve is pointed in shape, thin-bed correction is required.

FORM 925322 REV. (8-80)

Dresser Atlas

DRESSER

Compensated Densilog®
Compensated Neutron
Gamma Ray

| | |
|---|---|
| FILE NO. | COMPANY KYLE R. MILLER |
| API NO. | WELL NO. 1 MORRIS |
| | FIELD WILDCAT |
| | COUNTY KIMBALL STATE NEBRASKA |
| | LOCATION: C-NE-NW<br>SEC 17 TWP 13N RGE 56W |
| | Other Services<br>DIFL<br>PROLOG |

| | | | |
|---|---|---|---|
| Permanent Datum | G.L. | Elev. 5014 | Elev. KB 5026 |
| Log Measured from | K.B. | 12 Ft. Above Permanent Datum | DF |
| Drilling Measured from | K.B. | | GL 5014 |

| | | | | |
|---|---|---|---|---|
| Date | 7-10-85 | | | |
| Run No. | ONE | | | |
| Service Order | 97589 | | | |
| Depth—Driller | 6820 | | | |
| Depth—Logger | 6788 | | | |
| Bottom Logged Interval | 6786 | | | |
| Top Logged Interval | 5776 | | | |
| Casing—Driller | 8-5/8 @ 502 | @ | @ | @ |
| Casing—Logger | NOT LOGGED | | | |
| Bit Size | 7-7/8 | | | |
| Type Fluid in Hole | CHEM GEL | | | |
| Density and Viscosity | 9.4 \| 67 | \| | \| | \| |
| pH and Fluid Loss | 8.5 \| 5.8 cc | \| cc | \| cc | \| cc |
| Source of Sample | FLOWLINE | | | |
| Rm @ Meas. Temp. | 4.63 @ 92 °F | @ °F | @ °F | @ °F |
| Rmf @ Meas. Temp. | 3.47 @ 92 °F | @ °F | @ °F | @ °F |
| Rmc @ Meas. Temp. | 5.79 @ 92 °F | @ °F | @ °F | @ °F |
| Source of Rmf and Rmc | CALC. \| CALC. | \| | \| | \| |
| Rm @ BHT | 2.63 @ 162 °F | @ °F | @ °F | @ °F |
| Time Since Circ. | 4.5 HOURS | | | |
| Max. Rec. Temp. Deg. F. | 162 °F | °F | °F | °F |
| Equip. No. and Location | 6270 \| F.M. | \| | \| | \| |
| Recorded By | J.L. PATTERSON | | | |
| Witnessed By | K.MILLER/J.DYER | | | |

**Figure 3–7.** A typical log heading (courtesy of Miller Oil Co., Ft. Collins, Colorado).

5. Highly resistive formations except the shales and the permeable zones will affect the shape and magnitude of the SP deflection.
6. Borehole diameter and invasion do affect SP deflection; however, Hilchie indicates that the effects are very small and, in general, can be ignored.[1]
7. The presence of shales in permeable formations will reduce the SP deflection to SP value called pseudostatic spontaneous potential (PSP). In water-bearing formations the amount of SP reduction is proportional to the volume of shale in the formation. In hydrocarbon-bearing formations the amount of SP reduction is greater (less deflection) than the volume of shale and is referred to as *hydrocarbon suppression*.

## $R_w$ FROM THE SP CURVE

The resistivity of formation water ($R_w$), also referred to as connate water or interstitial water (uncontaminated by drilling mud), is an important interpretation variable in log analysis. The use of the SP curve for $R_w$ determination requires a clean, nonshaly, and only water-saturated formation.

The following steps lead to the determination of $R_w$ from SP log:

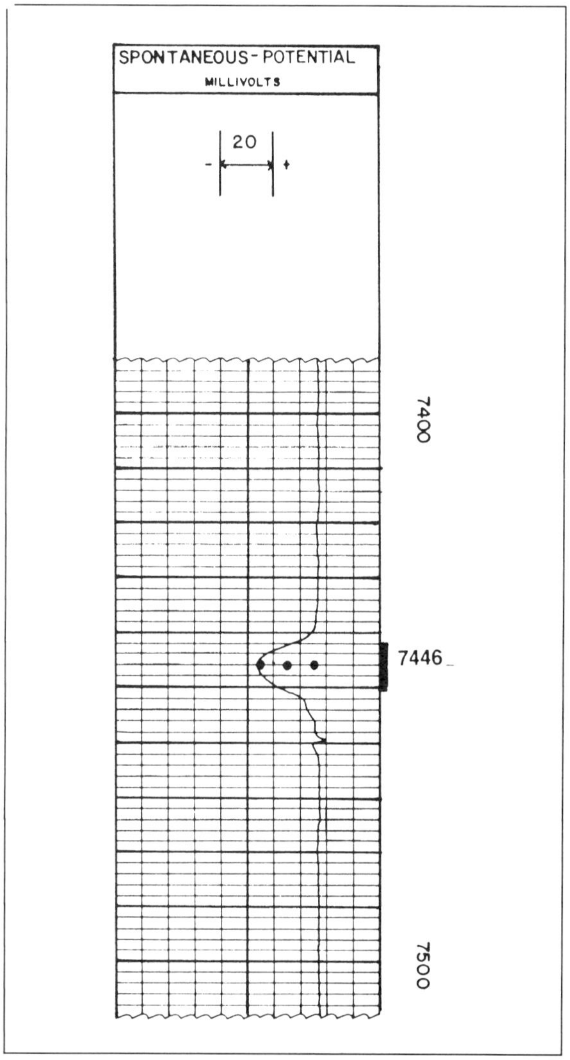


**Figure 3–8.** A typical Spontaneous Potential (SP) log (Asquith, 1982, reprinted by permission from AAPG).[2]

1. Estimate the shale baseline on the SP next to the formation to be used for $R_w$ calculations.
2. Read SSP deflection from shale baseline in millivolts (mv) if the zone is thick and clean, otherwise read SP. If SSP is read then go to Step 4.
3. Using formation thickness; SP; resistivity from short normal, spherically focused log or laterolog; and Figure 3–9, correct SP value for SSP.
4. Obtain $R_{mf}$ and bottomhole temperature from log heading. Calculate formation temperature $T_f$ and adjust $R_{mf}$ to formation temperature. Use Equation (3.1) or Figure 3–5 and Equation (3.2) or Figure 3–6.

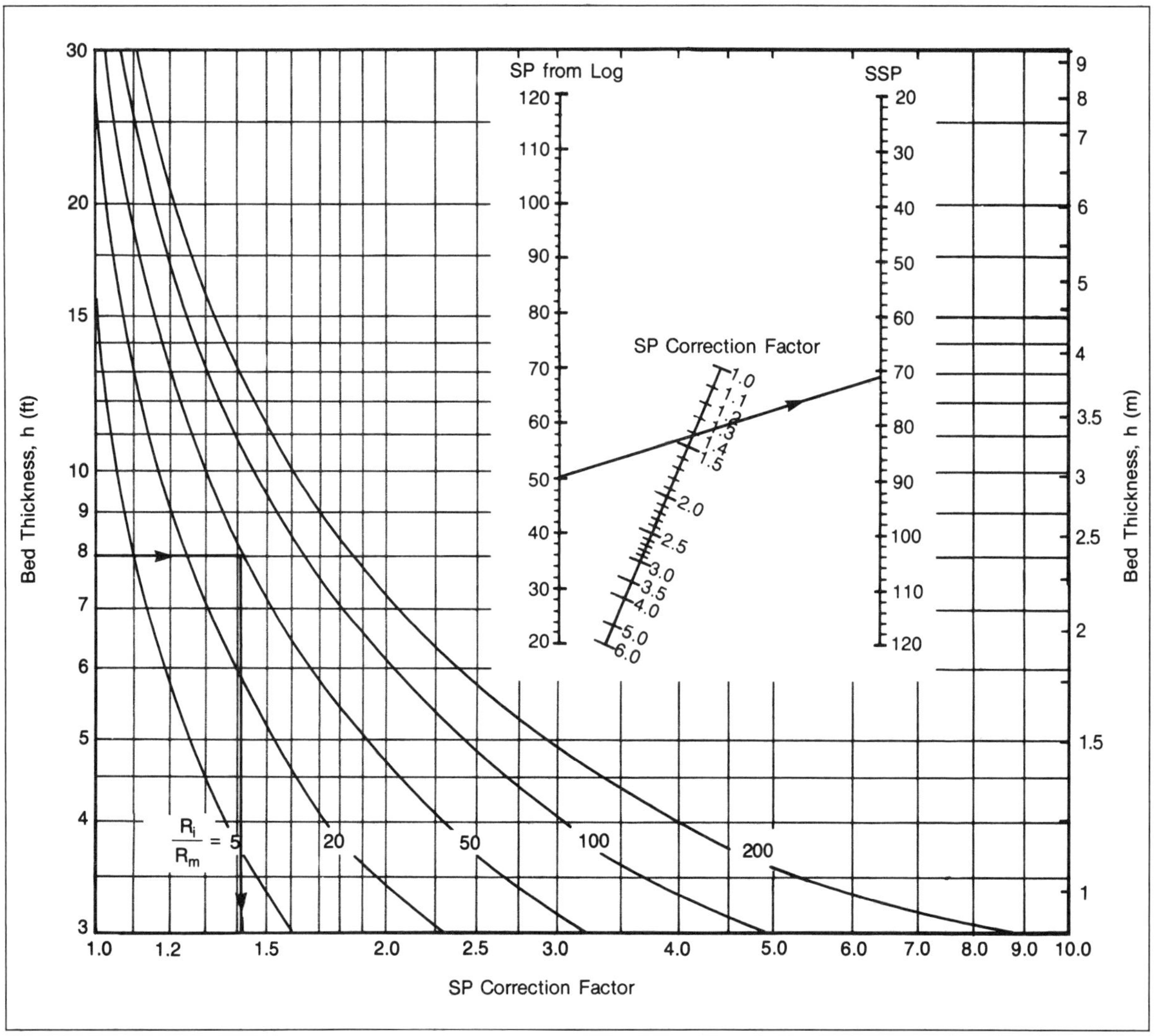


**Figure 3–9.** Chart to correct SP deflection for thin-bed effects (courtesy of Atlas Wireline Services).[6]

5. If $R_{mf} \geq 0.10$, $R_{mf} = R_{mfeq}$. If $R_{mf} < 0.10$, correct $R_{mf}$ to $R_{mfeq}$ using Figure 3–10.
6. Use the following equation to calculate the equivalent water resistivity ($R_{weq}$):

$$\log R_{weq} = \left( \frac{SSP + K \log R_{mfeq}}{K} \right) \tag{3.3}$$

$$K = 60 + (0.133 \times T_f) \tag{3.4}$$

7. If $R_{weq} \geq 0.10$, then $R_{weq} = R_w$, otherwise correct $R_{weq}$ to $R_w$ using Figure 3–10.

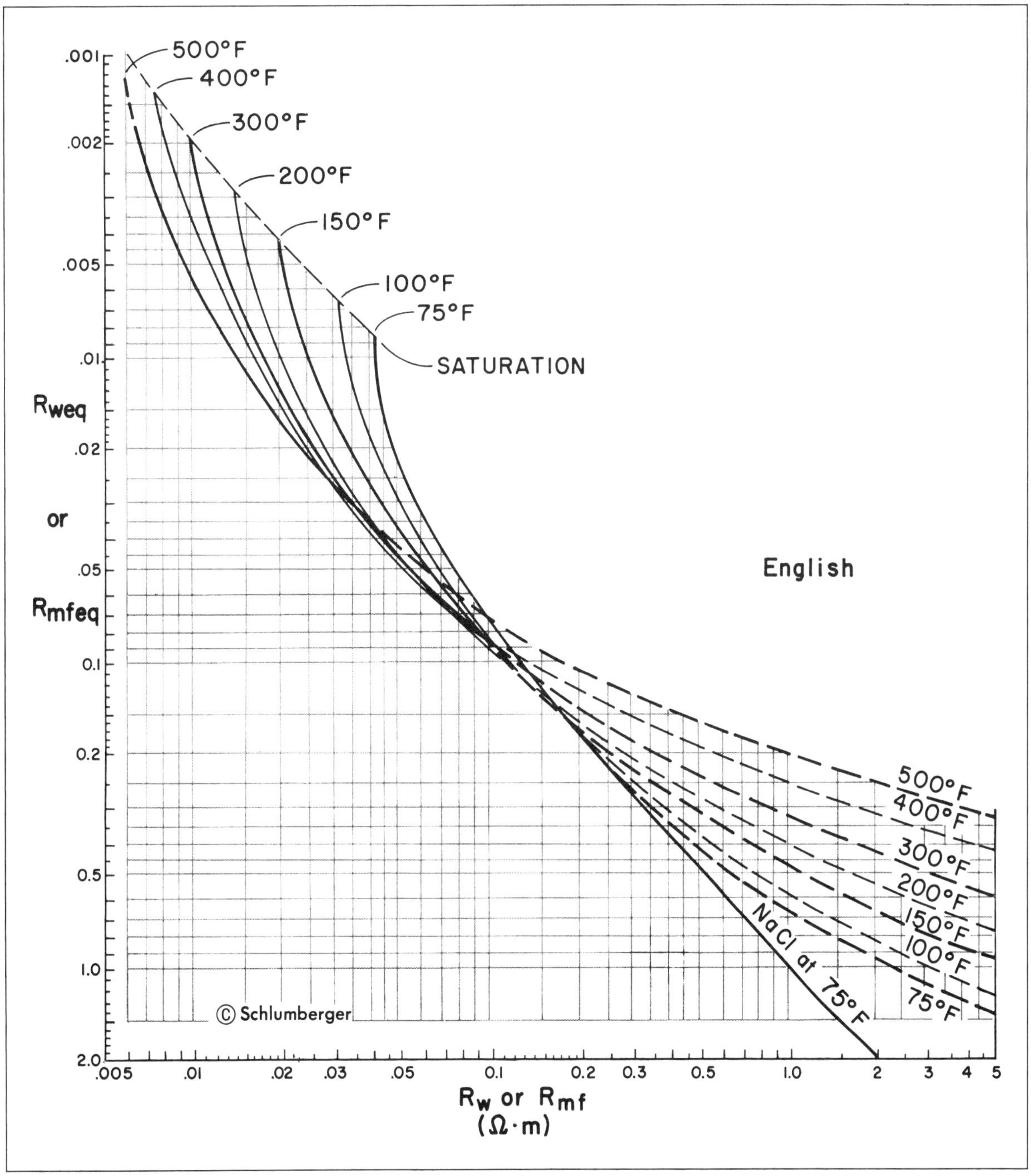


**Figure 3–10.** $R_w$ and $R_{weq}$ (also $R_{mf}$ and $R_{mfeq}$) relationship (courtesy of Schlumberger).

Always check the $R_w$ determined from the SP log against other sources if available. Some other sources of $R_w$ determination are discussed in the following pages. When the measured values of $R_{mf}$ and $R_{mc}$ are not available, the following relationships may be used to estimate these parameters:

$$R_{mc} = 0.69\, R_{mf} \left(\frac{R_m}{R_{mf}}\right)^{2.65}$$

$$R_{mf} = C(R_{mc})^{1.07}$$

where C is dependent on mud weight as shown below.

| Mud Weight (ppg) | C |
|---|---|
| 10 | 0.847 |
| 11 | 0.708 |
| 12 | 0.584 |
| 13 | 0.488 |
| 14 | 0.412 |
| 16 | 0.380 |
| 18 | 0.350 |

**Example 3–3.** You have just drilled a well to a total depth (TD) of 9,500 ft. The following information was obtained from the SP and resistivity log. The zone of interest is from 8,500 to 8,516 ft.

$T_{bh} = 170°F$     $R_t\ @T_f = 65$ ohm-m

$T_{ms} = 80°F$     $R_{mf}\ @70°F = 2.4$ ohm-m

$R_m\ @T_{bh} = 1.2$ ohm-m     $SSP = -70$ mv

Determine the formation temperature and the $R_w$ at the formation temperature.

Solution:

$$T_f = \left(\frac{T_{bh} - T_{ms}}{TD}\right) \times D_f + T_{ms}$$

$$= \left(\frac{170 - 80}{9{,}500}\right) \times 8{,}508 + 80$$

$$= 161°F$$

$R_{mf}\ @161°F = 1.10$ ohm-m $= R_{mfeq}$ (from Fig. 3–6),

$$
\begin{aligned}
K &= 60 + 0.133 \times T_f \\
&= 60 + 0.133 \times 161 \\
&= 81.41 \\
\log R_{weq} &= \left( \frac{SSP + K \log R_{mfeq}}{K} \right) \\
&= \left[ \frac{-70 + (81.41) \log (1.10)}{(81.41)} \right] \\
&= -0.818 \\
R_{weq} &= 0.152
\end{aligned}
$$



Since $R_{weq} > 0.10$, $R_w = R_{weq} = 0.152$ ohm-m

**Example 3–4.**[5] Calculate $R_w$ from the SP and resistivity log shown in Figure 3–11. Other pertinent data from the log heading are as follows:

Formation: Morrow Sand

$$
\begin{aligned}
R_{mf} \ @70°F &= 2.0 \ \text{ohm-m} & TD &= 10{,}500 \ \text{ft} \\
R_m \ @70°F &= 2.5 \ \text{ohm-m} & SP &= -70 \ \text{mv} \ \text{(Fig. 3–11)} \\
T_{ms} &= 60°F & R_i &= 65 \ \text{ohm-m} \ \text{(Fig. 3–11)} \\
T_{bh} &= 164°F & h &= 24 \ \text{ft}
\end{aligned}
$$

Solution:

$$
\begin{aligned}
T_f &= \left( \frac{T_{bh} - T_{ms}}{TD} \right) \times D_f + T_{ms} \\
&= \left( \frac{164 - 60}{10{,}500} \right) \times 8{,}126 + 60 = 140°F
\end{aligned}
$$

Correct $R_m$ and $R_{mf}$ to formation temperature using Figure 3–6.

$$
\begin{aligned}
R_{mf} \ @140°F &= 1.0 \ \text{ohm-m} \\
R_m \ @140°F &= 1.3 \ \text{ohm-m}
\end{aligned}
$$

Correct SP to SSP using Figure 3–9.

$$R_i/R_m = 65/1.3 = 50$$

correction factor from Figure 3–9 = 1.07

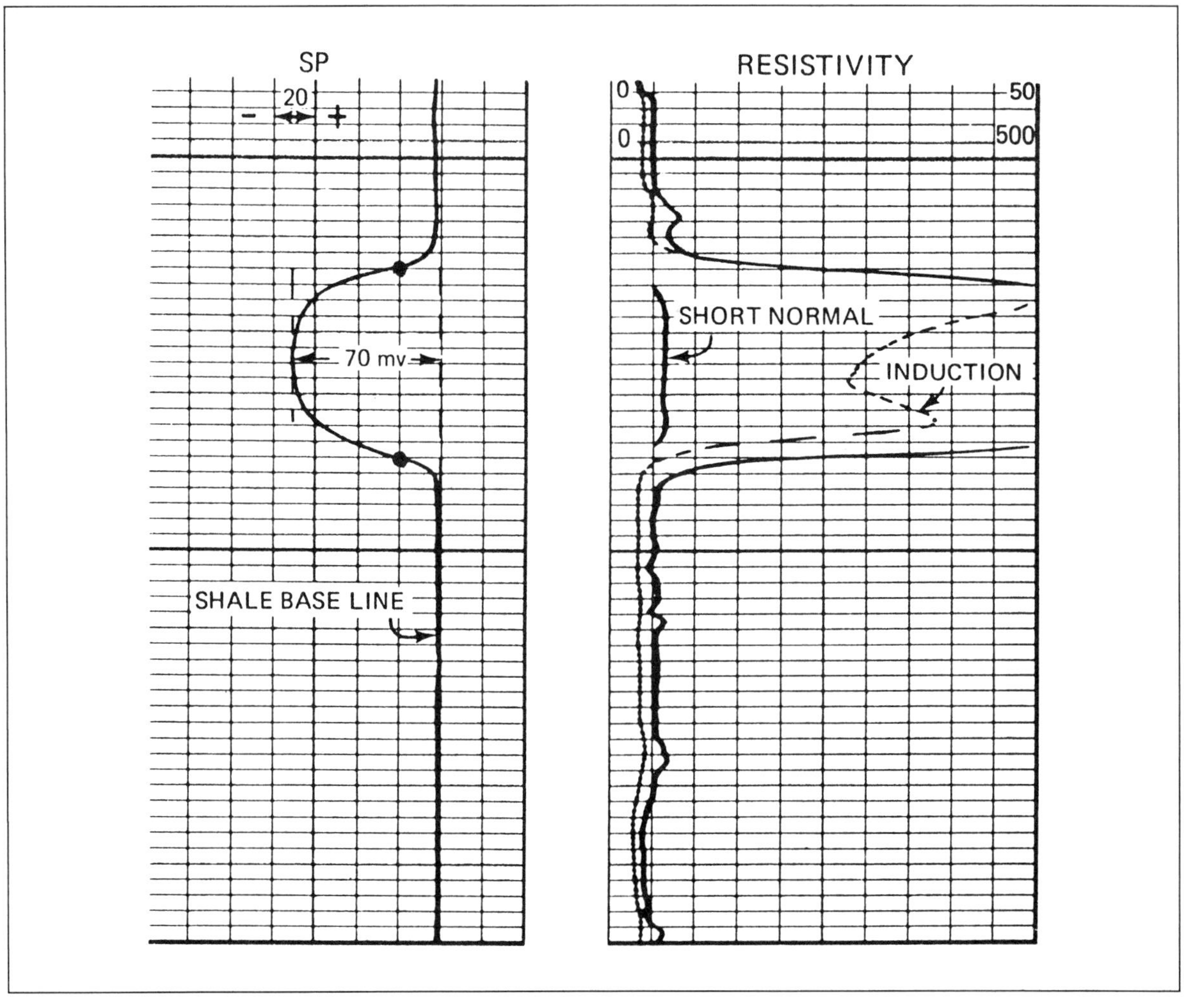


**Figure 3–11.** SP and resistivity log for Example 3–4 (courtesy of Atlas Wireline Services).

$$SSP = 1.07 \times SP$$
$$= 1.07 \times 70 = -75 \quad \text{mv}$$
$$K = 60 + 0.133 \times T_f$$
$$= 60 + 0.133 \times 140 = 78.62$$
$$\log R_{weq} = \left[\frac{-75 + (78.62)\log(1.0)}{78.62}\right]$$
$$R_{weq} = 0.111$$

Since $R_{weq} > 0.10$, therefore, $R_w = R_{weq} = 0.111$ ohm-m.

## VOLUME OF SHALE FROM SP LOG

The SP log can be used to calculate the volume of shale in a permeable zone using the following equation:

$$V_{sh}(\%) = 1.0 - \frac{\text{PSP}}{\text{SSP}} \tag{3.5}$$



The volume of shale determined in the formation can then be used in the shaly formation log interpretation that we will discuss in the following pages.

## GAMMA RAY AND CALIPER LOG

Gamma ray (GR) logs are used to (1) identify lithologies, (2) correlate between formations, and (3) calculate volume of shale. The gamma ray log is a recording of naturally occurring gamma rays in the formations adjacent to the borehole. The log measures the radioactivity of the formation. When it sees a radioactive formation, it will show very high GR value and will deflect to the right on the log. When a clean formation is encountered, the log will deflect to the left (minimum GR). Since radioactive elements tend to concentrate in shales and clays of sedimentary rocks, for this reason this log is used to identify these zones and estimate shale content of the formation.

The magnitude of gamma ray deflection is reported in gamma ray API units. Note that even clean formations containing potassium feldspars, micas, glauconite, or uranium-rich waters will show GR response similar to that of shaly formation. This confusion can be avoided by running Spectrolog, which will identify the radioactive materials. The gamma ray log is effective in any borehole environment. In wells drilled with saturated salt mud, oil-based mud, or air, the SP log fails to provide accurate information, so it is replaced by GR log. Also a clean low-permeability carbonate formation will be more appropriately identified by GR log than by SP log. Figure 3–12 shows a good comparison between SP and gamma ray log.

To determine the volume of shale from a gamma ray log, we first calculate gamma ray index ($I_{GR}$) using the following equation:

$$I_{GR} = \frac{GR_{log} - GR_{min}}{GR_{max} - GR_{min}} \tag{3.6}$$

Once the $I_{GR}$ is calculated, the volume of shale is then determined from Figure 3–13 or the following equations.

For older and consolidated rocks:

$$V_{sh} = 0.33\,[2^{(2 \times I_{GR})} - 1.0] \tag{3.7}$$

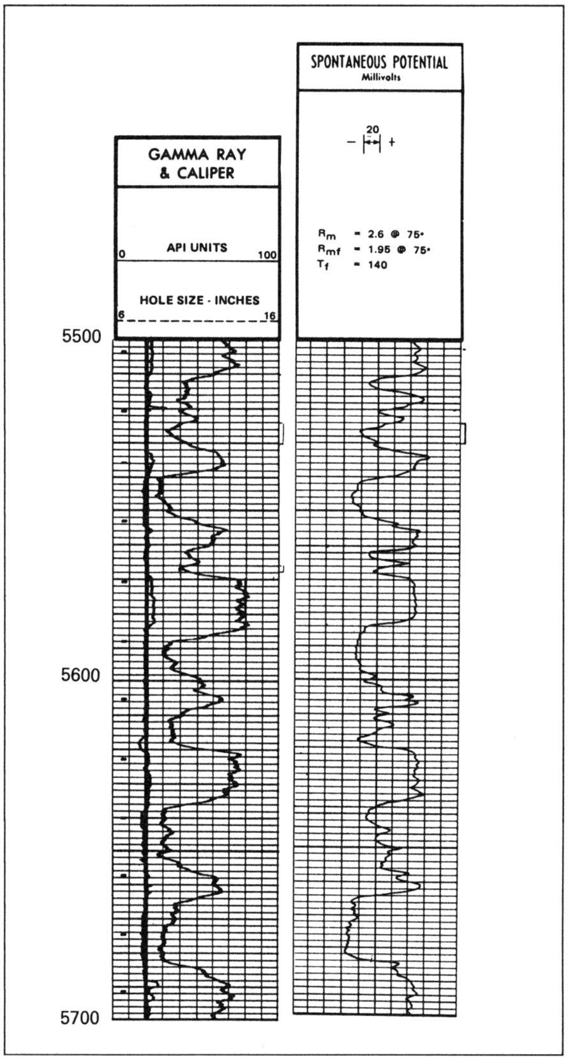


**Figure 3–12.** Comparison of gamma ray and SP log (courtesy of Atlas Wireline Services).[5]

For tertiary and consolidated rocks:

$$V_{sh} = 0.083\ [2^{(3.7 \times I_{GR})} - 1.0] \qquad \textbf{(3.8)}$$

**Example 3–5.** Determine the volume of shale for the zones 5,524–5,530 ft (Zone 1) and 5,566–5,568 ft (Zone 2) in the gamma ray log presented in Figure 3–12. Assume older rocks.

Solution: From the log,

$$GR_{max} = 80 \quad 5{,}573–5{,}583 \text{ ft}$$

$$GR_{min} = 28 \quad 5{,}540-5{,}546 \text{ ft}$$

$$GR_{zone1} = 32 \quad GR_{zone2} = 40$$

$$(I_{GR})_{zone1} = \frac{32 - 28}{80 - 28} = 0.08$$

$$(I_{GR})_{zone2} = \frac{40 - 28}{80 - 28} = 0.23$$

Using Equation (3.7), the volume of shale would be:

$$V_{sh(zone1)} = 0.33\ (2^{(2 \times 0.08)} - 1.0) \times 100$$

$$= 3.87\%$$

$$V_{sh(zone2)} = 12.39\%$$

The caliper log is used to measure the diameter of the wellbore in inches from the top of the hole to the bottom. The size of the hole is required for correcting some of the logs for hole size effects, and it also helps in determining hole volume in order to estimate the amount of cement required for cementing the casing. The log will show if there is any swelling or washout, etc., in the hole. The caliper log is run in conjunction with the SP log or the GR log.

## RESISTIVITY LOGS

As mentioned earlier, formation (uncontaminated) resistivity ($R_t$) is one of the most important measurements used in evaluating a reservoir. In this section we will discuss several different resistivity logs, each of which has a different depth of investigation to analyze zones invaded with mud filtrate. On the other hand, resistivity of the drilling mud in the wellbore influences the selection of a particular resistivity log.

Two types of resistivity-measuring devices are generally used. Induction logging tools are used

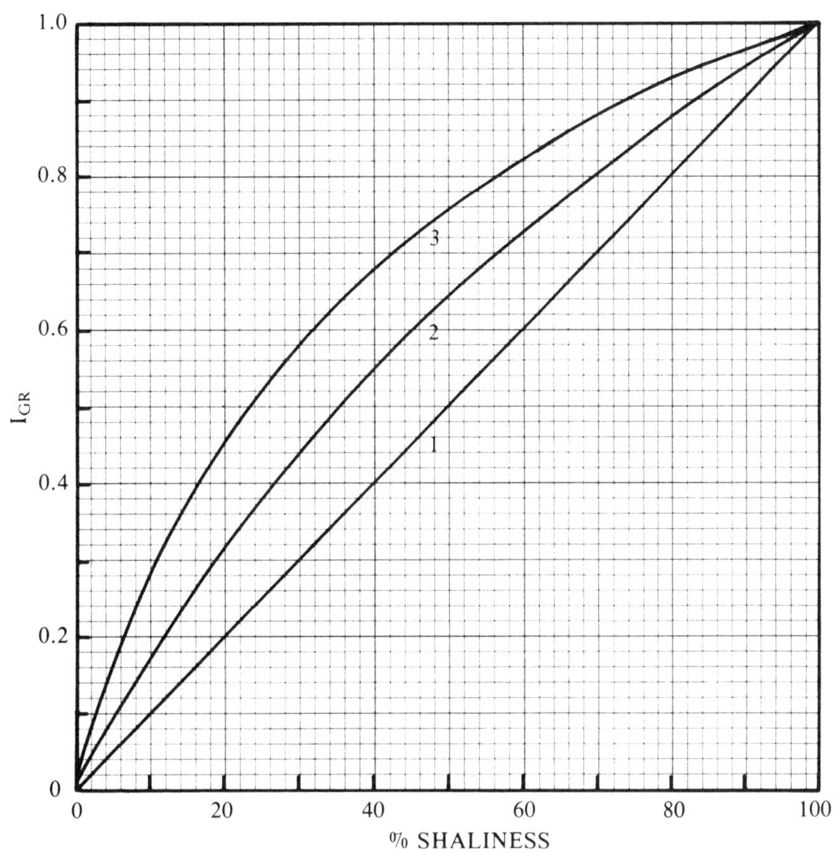


**Figure 3–13.** Chart for correcting the gamma ray index ($I_{GR}$) to the volume of shale ($V_{sh}$) (courtesy of Atlas Wireline Services).

in *fresh mud* ($R_{mf} > 3R_w$) systems, and electrode tools are used in *salt mud* ($R_{mf} \approx R_w$) systems. These two logs used by themselves will record true resistivity of the uninvaded formation. However, this resistivity is influenced by borehole size, formation thickness, and invasion.

To correct the true resistivity ($R_t$) above for the influencing variables, two other resistivity tools are run simultaneously with the induction or laterolog. These additional resistivity tools will provide the resistivities of the invaded zone ($R_i$) and the flushed zone ($R_{xo}$). All these three resistivities are then used to correct the $R_t$ to true resistivity.

The different types of resistivity logs and their depths of investigation are provided in Table 3–1.

## INDUCTION ELECTRIC LOG

The induction electric log is generally composed of four curves: (1) short normal, (2) induction, (3) spontaneous potential (SP), and (4) gamma ray log. All these four curves are generated simultaneously. A typical induction electric log is shown in Figure 3–14.

**TABLE 3–1. Classification of resistivity logs**

| Flushed Zone ($R_{xo}$) | Invaded Zone ($R_i$) | Uninvaded Zone ($R_t$) |
|---|---|---|
| Microlog | Short Normal | Long Normal |
| Microlaterolog | Laterolog-8 (LL8) | Laterolog |
| Proximity Log | Spherically Focused Log (SFL) | Deep Induction Log |
| Microspherically Focused | Medium Induction | Deep Laterolog (LLD) |
| Log (MSFL) | Shallow Laterolog (LLS) | Laterolog-3 (LL3) |
| | | Laterolog-7 (LL7) |
| | | Induction Log (6FF40) |

The vertical resolution of the induction log is about 5 to 6 ft in *resistive beds* (beds with higher resistivity than the adjacent beds) and a little over 2 ft in *conductive beds* (those with lower resistivity than the adjacent beds). This point is worth noting that the induction log will not pick thin resistive beds but it will pick thin conductive beds. The induction tool, a conductivity-sensitive device, will provide accurate values in low-to-medium resistivity formations.

The induction log measures electrical conductivity, resistivity is calculated from conductivity by using the following equation. The conductivity curve is necessary to more accurately determine the $R_t$ value of low-resistivity formations.

$$R_t = \frac{1{,}000}{C_t} \tag{3.9}$$

The short normal tool measures resistivity ($R_i$) of the invaded zone. Invasion is detected by comparing the $R_i$ with the resistivity ($R_t$). To obtain a more accurate value of $R_i$ from the short normal curve, an *amplified short normal* curve is sometimes displayed along with the regular short normal curve.

The induction log must be corrected for (1) borehole effects, (2) bed thickness effects, and (3) invasion effects. The corrections must be made in the stated order. Figure 3–15 is used to correct 6FF40, ID, IM, and 6FF28 for various standoffs. The standoff value is normally specified in the log heading. Figures 3–16 and 3–17 provide bed thickness correction for the deep and medium induction logs, respectively. To correct the $R_{ILD}$ (6FF40 and 6FF28) in thin conductive beds, Figure 3–18 is used. To use Figure 3–16 for 6FF28, the bed thickness depicted by 6FF40 must first be multiplied by the ratio 40/28 = 1.43.

The short normal induction resistivity may be corrected for borehole effects using Figure 3–19. The single induction log resistivity cannot be corrected for invasion. For invasion correction, three resistivity logs are required (i.e., $R_t$, $R_m$, and $R_{xo}$).

**Example 3–6.** A 14-ft thick pay has an apparent resistivity $R_a = R_{ILD} = 15$ ohm-m and the adjacent bed resistivity $\dot{R}_s = 2$ ohm-m.[4] What is the corrected $R_t$?

Solution: From Figure 3–16, $R_t = 18$ ohm-m

**Example 3–7.** For a thin conductive bed, resistivity of the adjacent bed ($R_s$) is 5 ohm-m, the apparent resistivity from the log is $R_a = R_{ILD} =$ 2 ohm-m, and the bed thickness is 2.5 feet. What is the corrected resistivity $R_t$?

Solution:

$$R_{ILD}/R_s = 2/5 = 0.40$$

From Figure 3–18, $R_{ILDcor}/R_{ILD} = 0.81$

$$R_{ILDcor} = 0.81 \times R_{ILD}$$
$$= 0.81 \times 2 = 1.62 \text{ ohm-m}$$

OR $\quad R_t/R_s = 0.33$

Therefore, $\quad R_t = 0.33 \times 5 = 1.65$ ohm-m

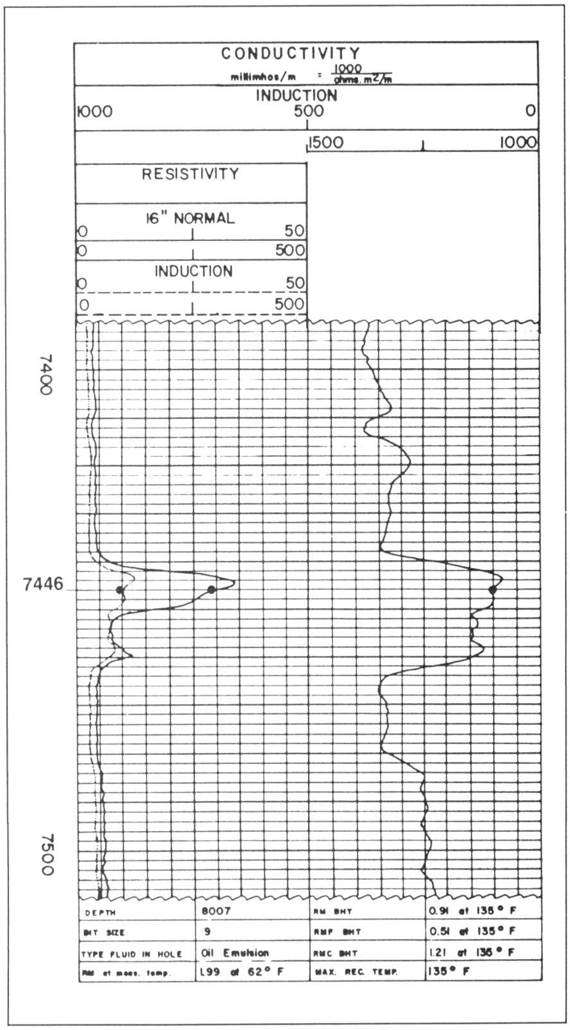


**Figure 3–14.** A typical induction electric log (Asquith, 1982, reprinted by permission from AAPG).[2]

## DUAL INDUCTION FOCUSED LOG

The dual induction focused log is designed to provide resistivity measurements such that the true resistivity could be corrected for invasion effects. The resistivity curves consist of a deep induction curve, a medium induction curve, and a shallow guard-type focused log. The curves are recorded on a four-cycle logarithmic scale, ranging from 0.20 to 2,000 ohm-m.

Visual observation of the separation between the three resistivity curves due to invasion will provide valuable information. The separation between the curves is the log of the ratio of the two curves. The amount of separation between the curves depends on the diameter of invasion and the ratio of $R_{mf}/R_m$. Note the following qualitative observations:

1. In a porous, permeable, moderately invaded, and water-bearing formation, the ratio $R_{FL}/R_{ILD}$ will be close to the ratio $R_{mf}/R_w$.
2. The ratio decreases as the hydrocarbon content increases.
3. The ratio between the $R_{ILM}/R_{ILD}$ is an indication of invasion.

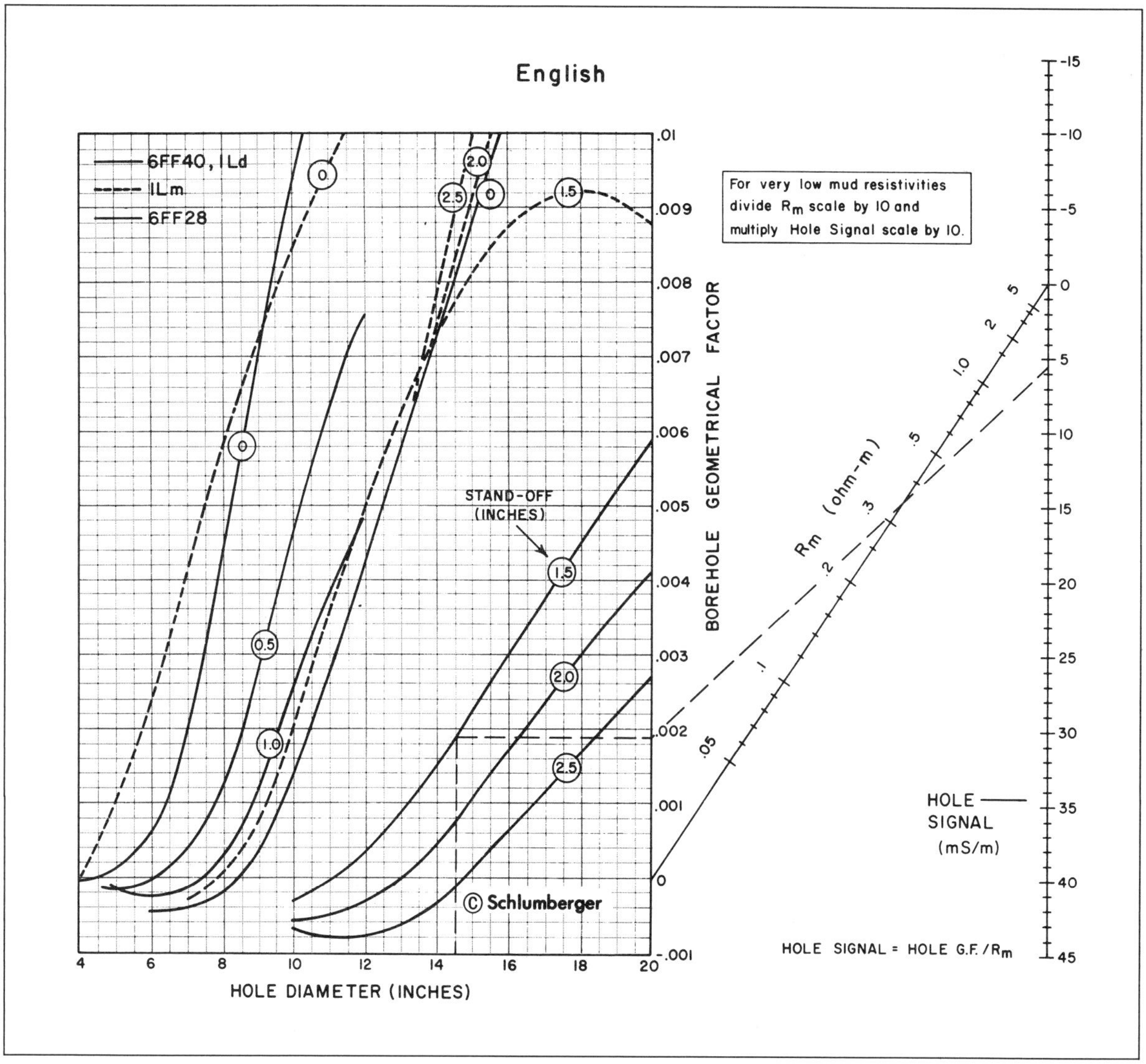


**Figure 3–15.** Induction log borehole correction (courtesy of Schlumberger).[4]

4. When $R_{mf} > R_w$, increasing invasion beyond a moderate depth will cause $R_{ILM}/R_{ILD}$ to increase and $R_{FL}/R_{ILD}$ to decrease.
5. If a medium induction curve reads a lower resistivity than the deep curve ($R_{ILM} < R_{ILD}$), either a severe washout or the possibility of an annulus is indicated.
6. When a change in resistivity accompanied by a decrease in separation between the $R_{FL}$ and $R_{ILD}$ is observed, increase in hydrocarbon or extremely deep invasion is indicated.
7. When $R_{mf} > R_w$ and only the focused curve changes position to read a lower resistivity than the deep induction curve ($R_{FL} < R_{ILD}$), vertical fracture may be indicated.

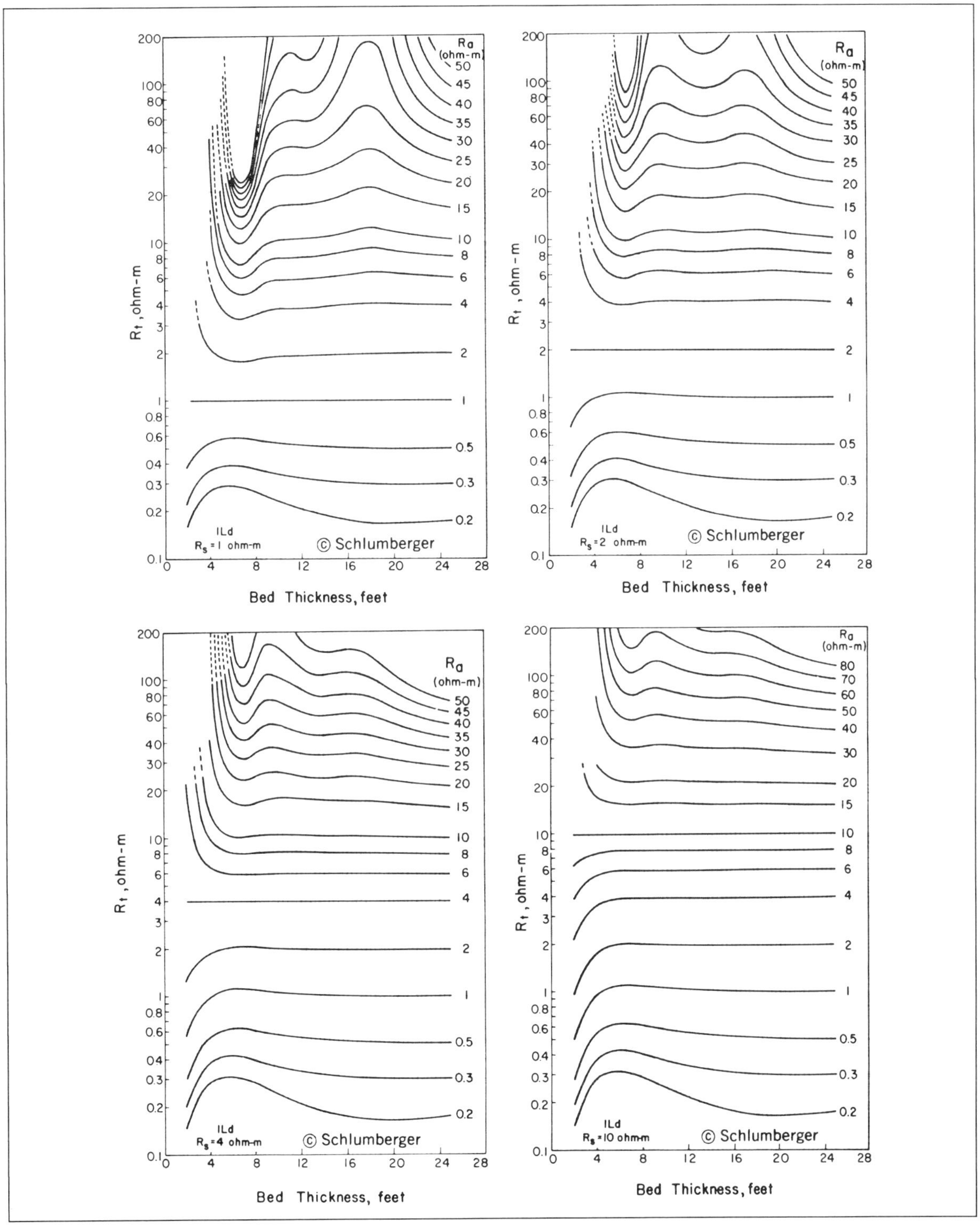


**Figure 3–16.** Induction log bed thickness correction—6FF40 or ILD and 6FF28 (courtesy of Schlumberger).[4]

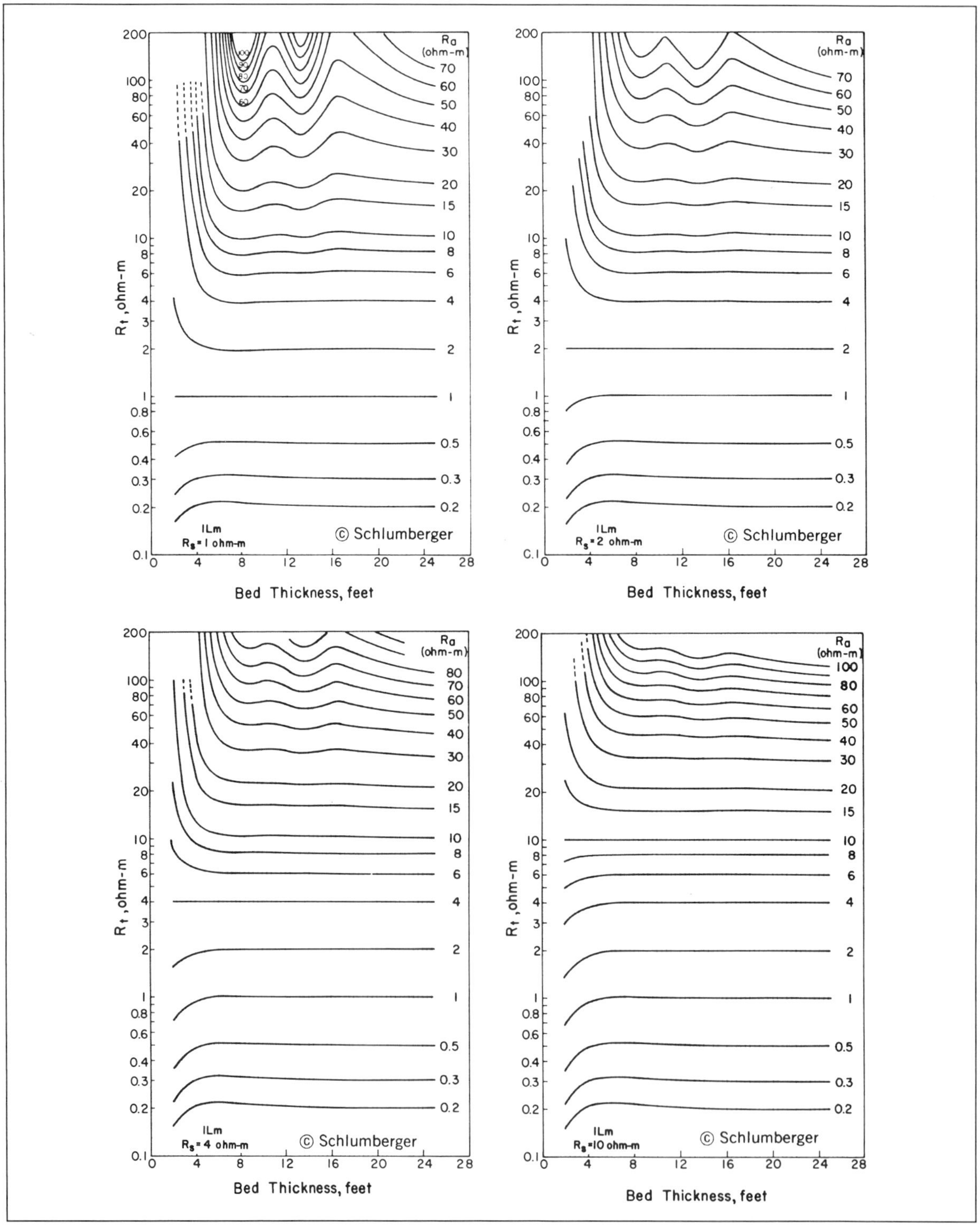


**Figure 3–17.** Induction log bed thickness correction for ILM (courtesy of Schlumberger).[4]

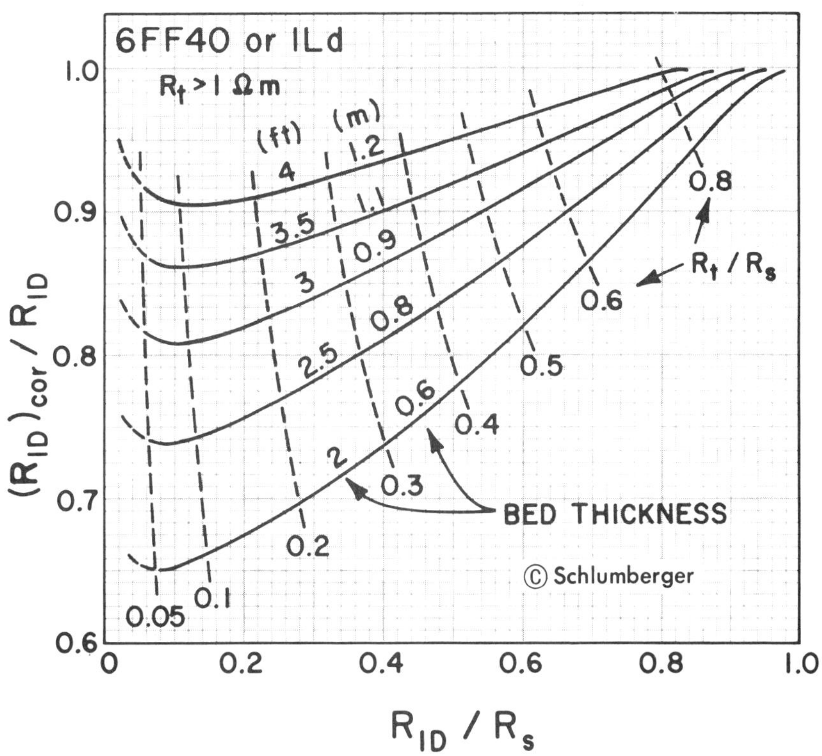


**Figure 3–18.** Correction of thin conductive beds (courtesy of Schlumberger).[4]

In general:

1. When $R_{mf} > R_w$ in a water-bearing zone, the deep induction curve will show lowest resistivity, the focused curve the highest resistivity with the medium induction value between the two ($R_{FL} > R_{ILM} > R_{ILD}$). The curve positions can be reversed in presence of movable hydrocarbons where $R_{FL} < R_{ILD}$. Refer to Figure 3–20.
2. When $R_{mf} \approx R_w$, the curve will read approximately the same in water-bearing formations.
3. When $R_{mf} < R_w$, the position of the curves will reverse (i.e., $R_{FL} < R_{ILM} < R_{ILD}$).
4. The presence of hydrocarbons in either case will cause the shallow and medium resistivity curves to read lower than the deep investigation curve.

The dual induction log resistivities must be corrected for (1) borehole effects, (2) bed thickness effects, and (3) invasion effects. The corrections must be made in the stated order. The induction curves are corrected, where necessary, for *borehole effects* and bed thickness effects using Figures 3–15 through 3–18. The accompanying flushed zone resistivity (Laterolog-8 or SFL) should be corrected for borehole effects using Figure 3–21.

After the borehole and bed thickness effects' corrections are taken care of, the deep resistivity

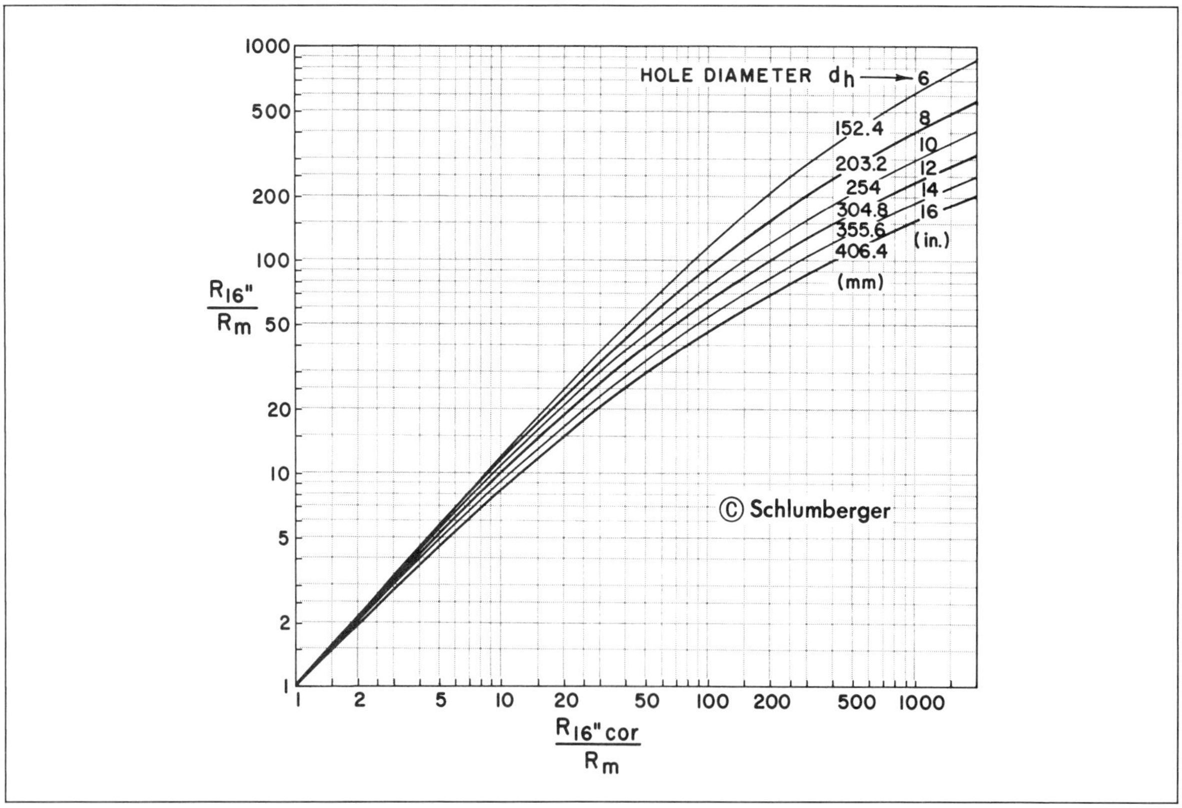


**Figure 3–19.** Borehole correction for 16-inch normal (courtesy of Schlumberger).[4]

($R_{ILD}$) should be corrected, where necessary, for *invasion effects.* The invasion correction charts, also referred to as the *Tornado* or *Butterfly* charts, are given in Figures 3–22 through 3–27. These charts are used to define the depth of invasion $d_i$, the $R_{xo}/R_t$ ratio, and true resistivity $R_t$. All charts assume step-contact profile of invasion.

**Example 3–8.** A well is logged with a combination of dual induction and Laterolog-8 (LL8) logs. The resistivity log readings in the zone of interest (corrected for borehole and bed thickness effect) are:

$$R_{LL8} = 25 \quad \text{ohm-m}$$
$$R_{ILM} = 6.50 \quad \text{ohm-m}$$
$$R_{ILD} = 5.0 \quad \text{ohm-m}$$
$$R_m = 0.40 \quad \text{ohm-m}$$

What is the resistivity $R_t$, diameter of invasion $d_i$, and $R_{xo}$ for the zone of interest?

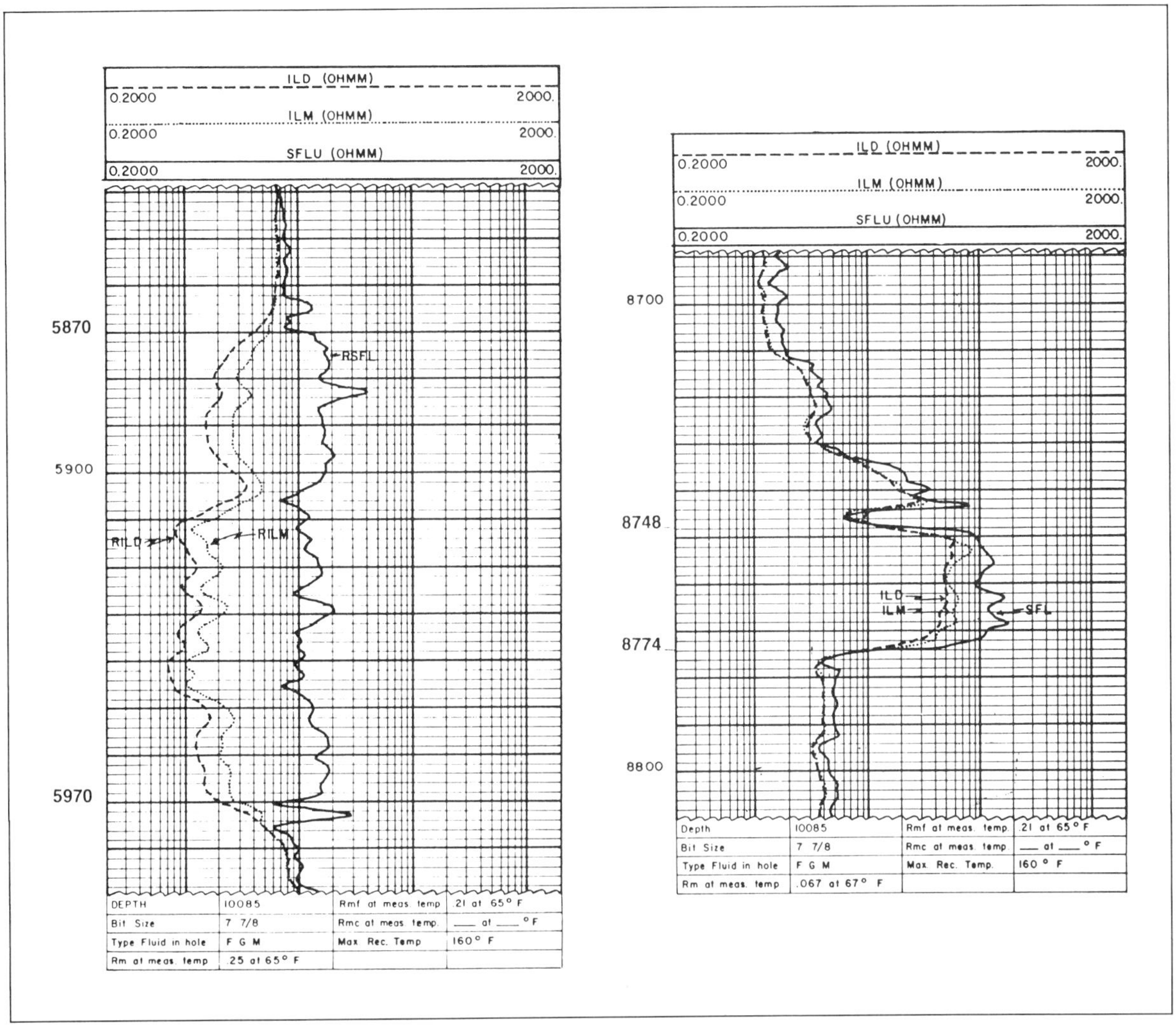


**Figure 3–20.** Dual induction focused logs in fresh water-based mud ($R_{mf} > R_w$) (a) through water-bearing zone and (b) through hydrocarbon-bearing zone (Asquith, 1982, reprinted by permission from AAPG).[2]

Solution: Use Figure 3–23 for $R_{xo}/R_{ILM} = 100$

$$R_{LL8}/R_{ILD} = 25.0/5.0$$
$$= 5.0$$
$$R_{ILM}/R_{ILD} = 6.50/5.0$$
$$= 1.30$$

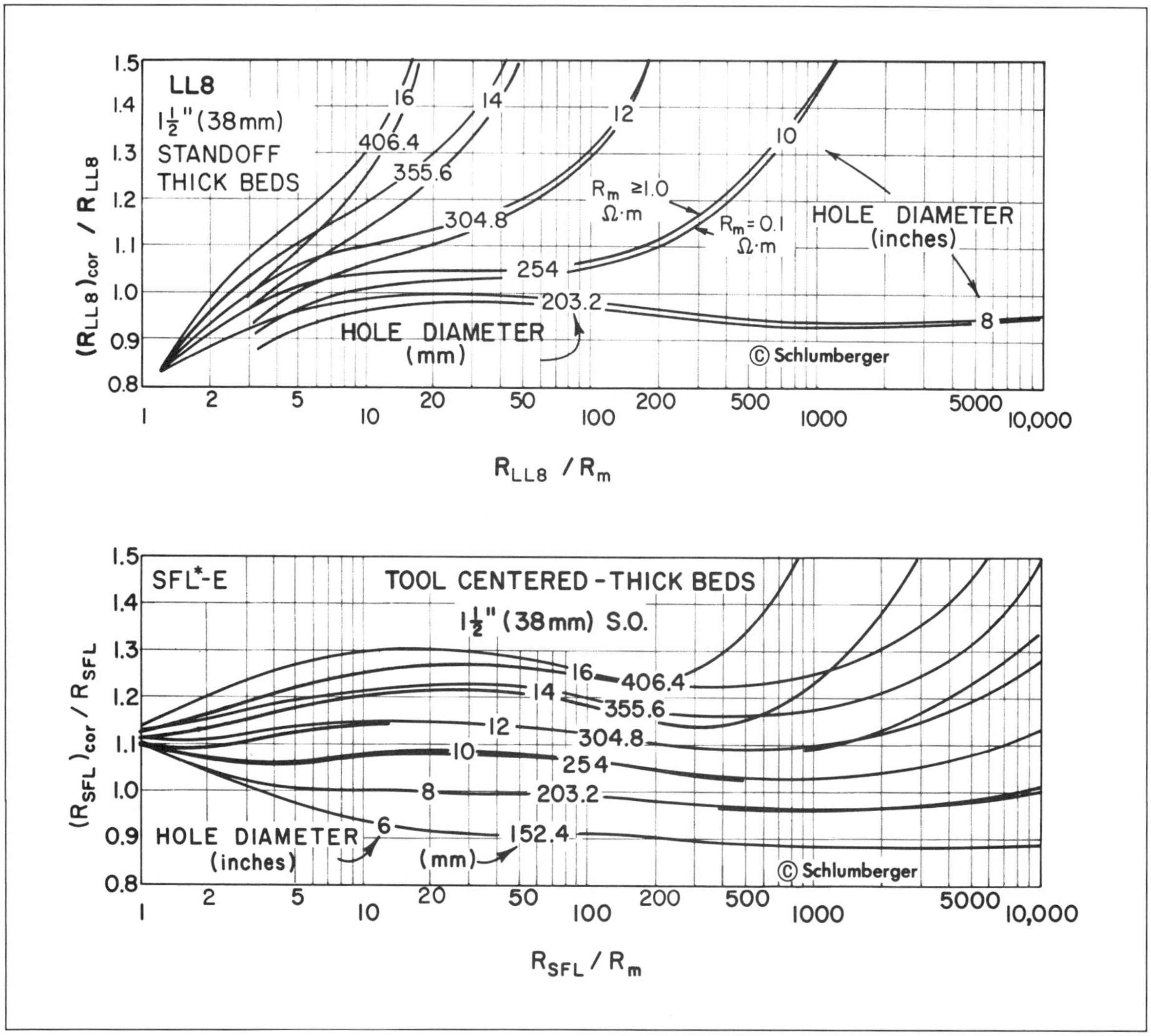


**Figure 3–21.** Borehole correction charts for Laterolog-8 (LL8) and Spherically Focused Log (SFL) (courtesy of Schlumberger).[4]

From Figure 3–23,

$$R_t/R_{ILD} = 0.95$$

$$d_i \approx 46 \quad \text{in.}$$

$$R_{xo}/R_t = 7$$

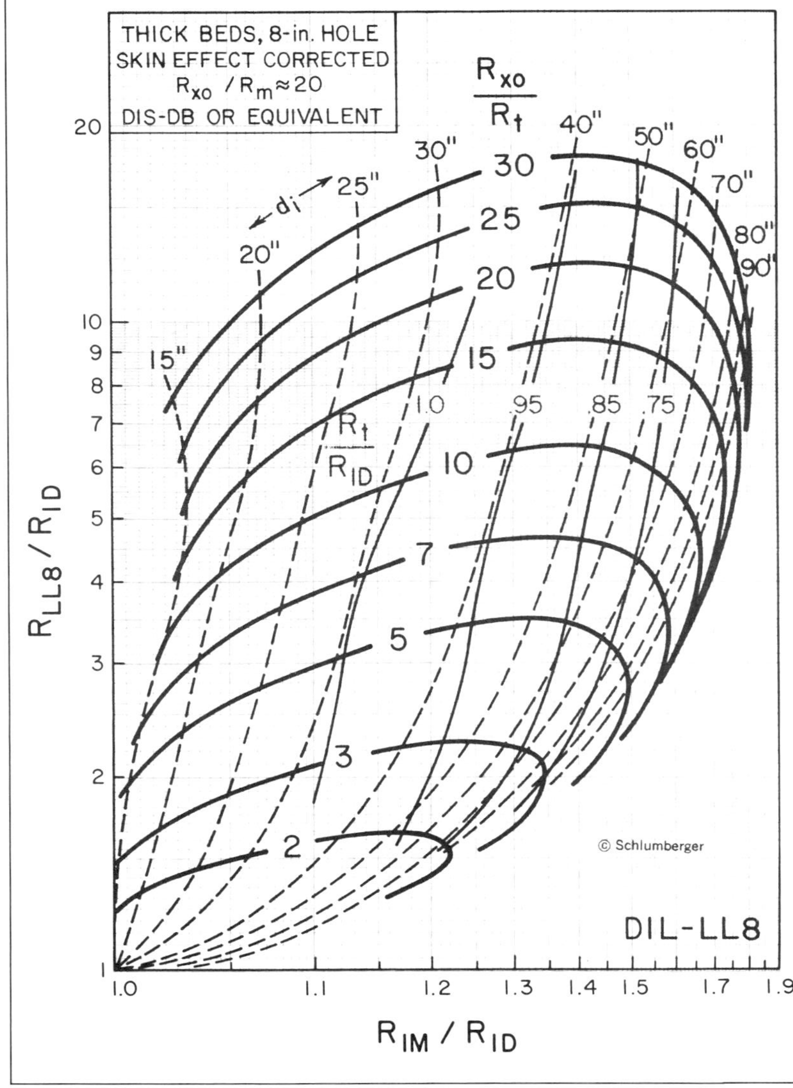


**Figure 3–22.** Resistivity invasion correction chart for Dual Induction-Laterolog 8 for $R_{xo}/R_m \approx 20$ (courtesy of Schlumberger).[4]

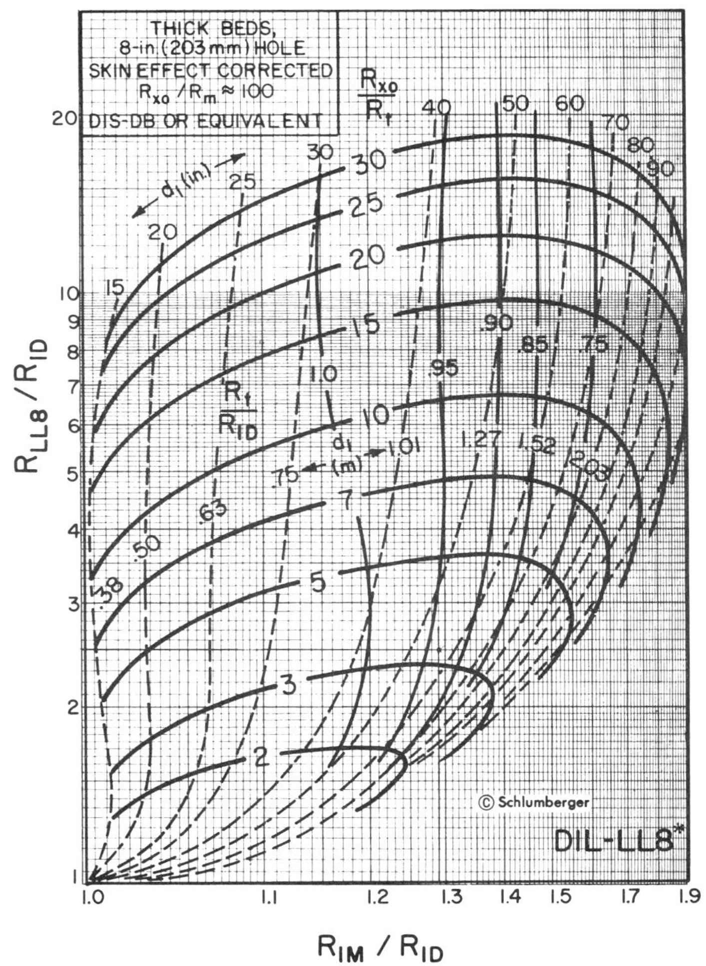


**Figure 3–23.** Resistivity invasion correction chart for Dual Induction-Laterolog 8 for $R_{xo}/R_m \approx 100$ (courtesy of Schlumberger).[4]

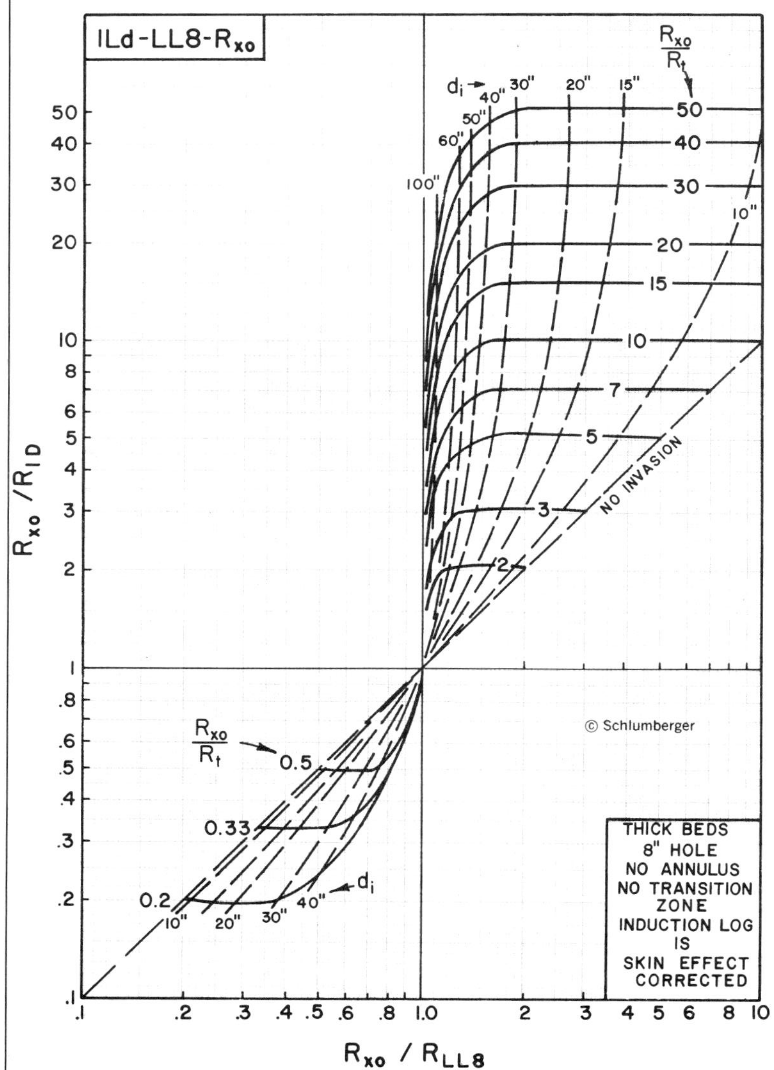


**Figure 3–24.** Resistivity invasion correction chart for Deep Induction-Laterolog 8-$R_{xo}$ (courtesy of Schlumberger).[4]

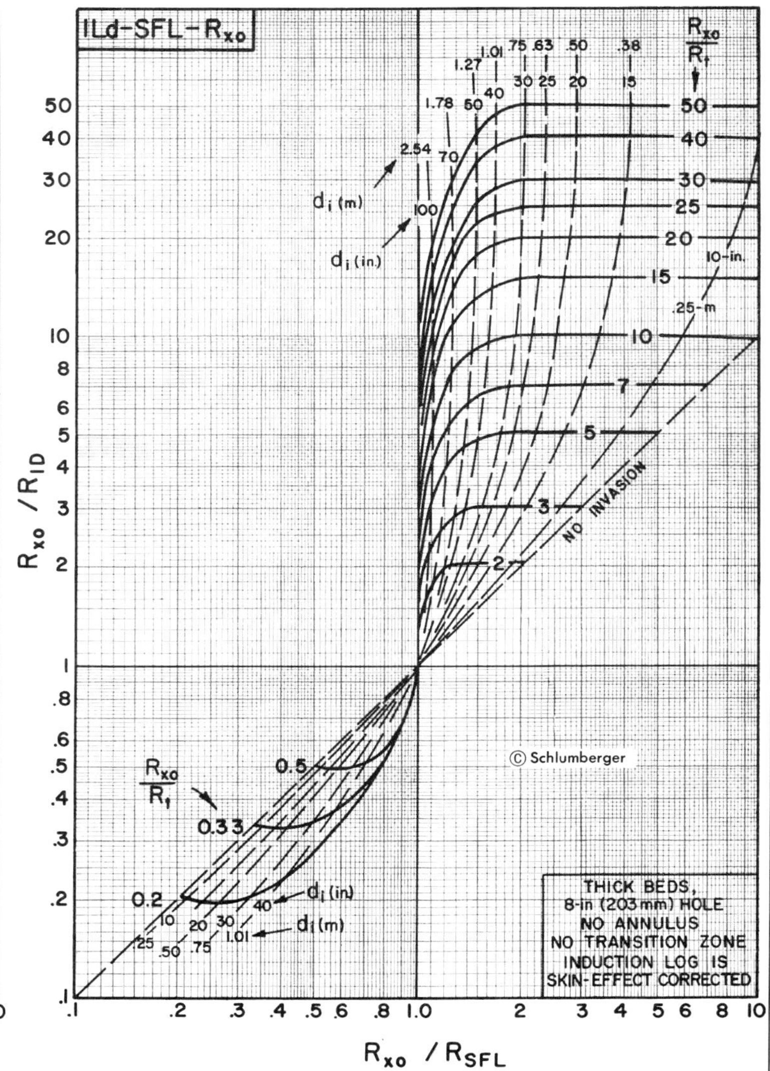


**Figure 3–25.** Resistivity invasion correction chart for Deep Induction-Spherically Focused log-$R_{xo}$ (courtesy of Schlumberger).[4]

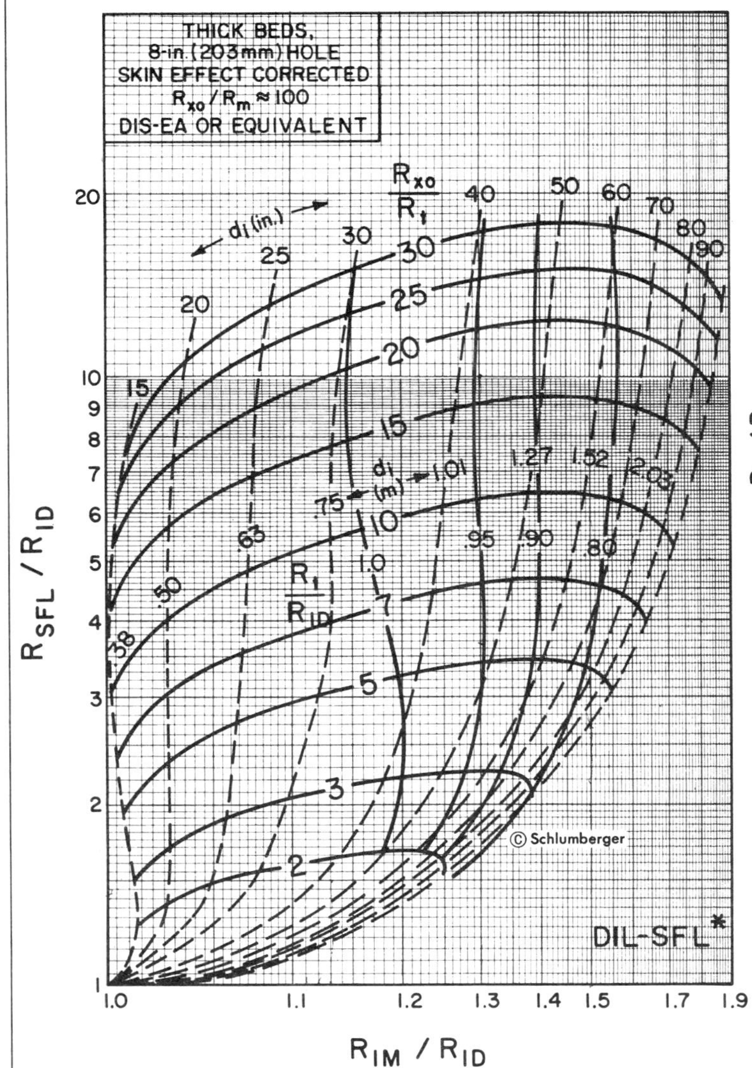


Figure 3–26. Resistivity invasion correction chart for Dual Induction-SFL—$R_{xo}/R_m \approx 100$ (courtesy of Schlumberger).[4]

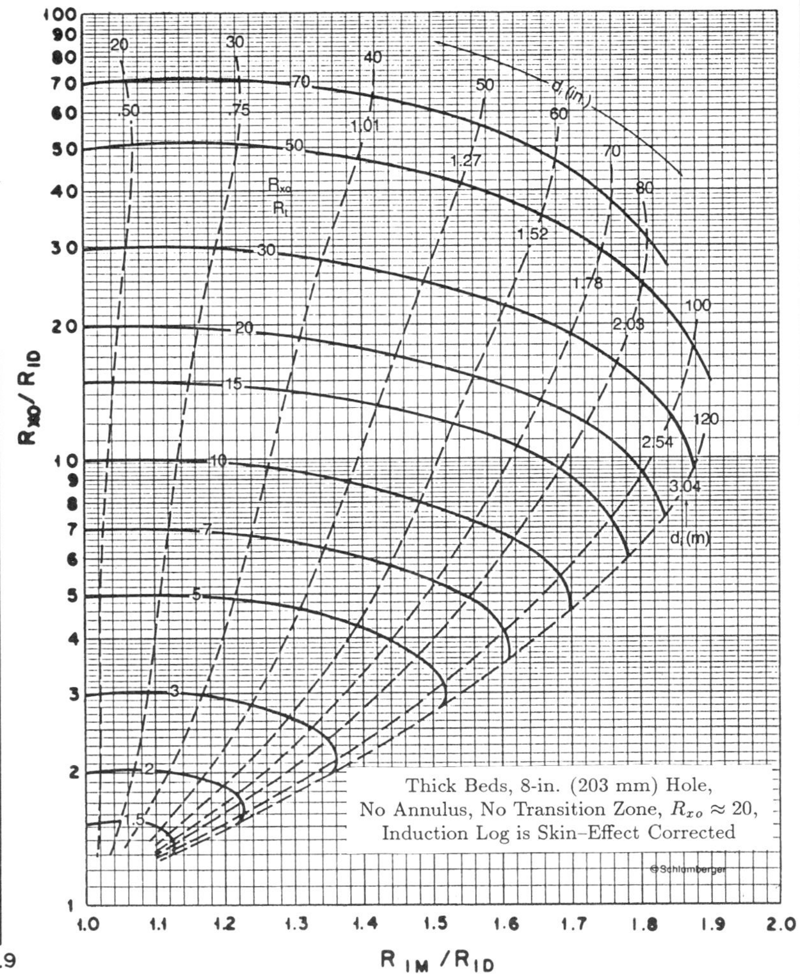


Figure 3–27. Resistivity invasion correction chart for Dual Induction-$R_{xo}$—$R_{xo}/R_m \approx 20$ (courtesy of Schlumberger).[4]

Therefore,

$$
\begin{aligned}
R_t &= R_{ILD}(R_t/R_{ILD}) \\
&= (5)(0.95) \\
&= 4.75 \quad \text{ohm-m} \\
R_{xo} &= R_t(R_{xo}/R_t) \\
&= (4.75)(7.0) \\
&= 33.25 \quad \text{ohm-m}
\end{aligned}
$$

## LATEROLOG

The Laterolog, also referred to as a focused or guard log, is designed to measure $R_t$ in boreholes containing highly conductive drilling mud. The Laterolog is most commonly presented on a four-cycle logarithmic scale ranging from 0.2 to 2,000 ohm-m.

The borehole size and formation thickness affect the Laterolog, but normally the effect is small enough so that Laterolog resistivity can be taken as $R_t$. However, there are conditions where the borehole diameter and bed thickness will affect the Laterolog. The charts presented in Figures 3–28 and 3–29 can be used to correct Laterolog Deep for borehole and bed thickness corrections, respectively.

Relatively shallow invasion will not affect the Laterolog resistivity. As the depth of invasion increases, however, it becomes necessary to correct the apparent resistivity $R_a$ to true resistivity $R_t$. To achieve this objective, resistivity of the flushed zone ($R_{xo}$) and the diameter of invasion are required. The diameter of invasion cannot be determined by the Laterolog, but it can be alternatively estimated from porosity using the relationships given in Table 3–2.[7]

Once the invasion diameter is estimated, and the resistivity corrected for borehole and bed thickness effects, the correction factors given in Table 3–3 can be used to correct $R_a$ for invasion.[7]

**Example 3–9.** The following data were obtained from a well logged with Laterolog deep resistivity log. What is the true resistivity ($R_t$) of the formation?

$R_{LLD} = 5$ ohm-m
$R_{LLM} = 0.05$ ohm-m

Borehole Diameter, $d_h = 16$ in.
$h = 10$ ft
$R_s = 55$ ohm-m

Solution:

$$
\begin{aligned}
R_{LLD}/R_{LLM} &= 5/0.05 \\
&= 100
\end{aligned}
$$

From Figure 3–28a,

$$
\begin{aligned}
R_{LLDcor}/R_{LLD} &= 1.10 \\
R_{LLDcor} &= (5)(1.10) \\
&= 5.50 \quad \text{ohm-m} \\
R_{LLDcor}/R_s &= 5.5/55 \\
&= 0.10
\end{aligned}
$$

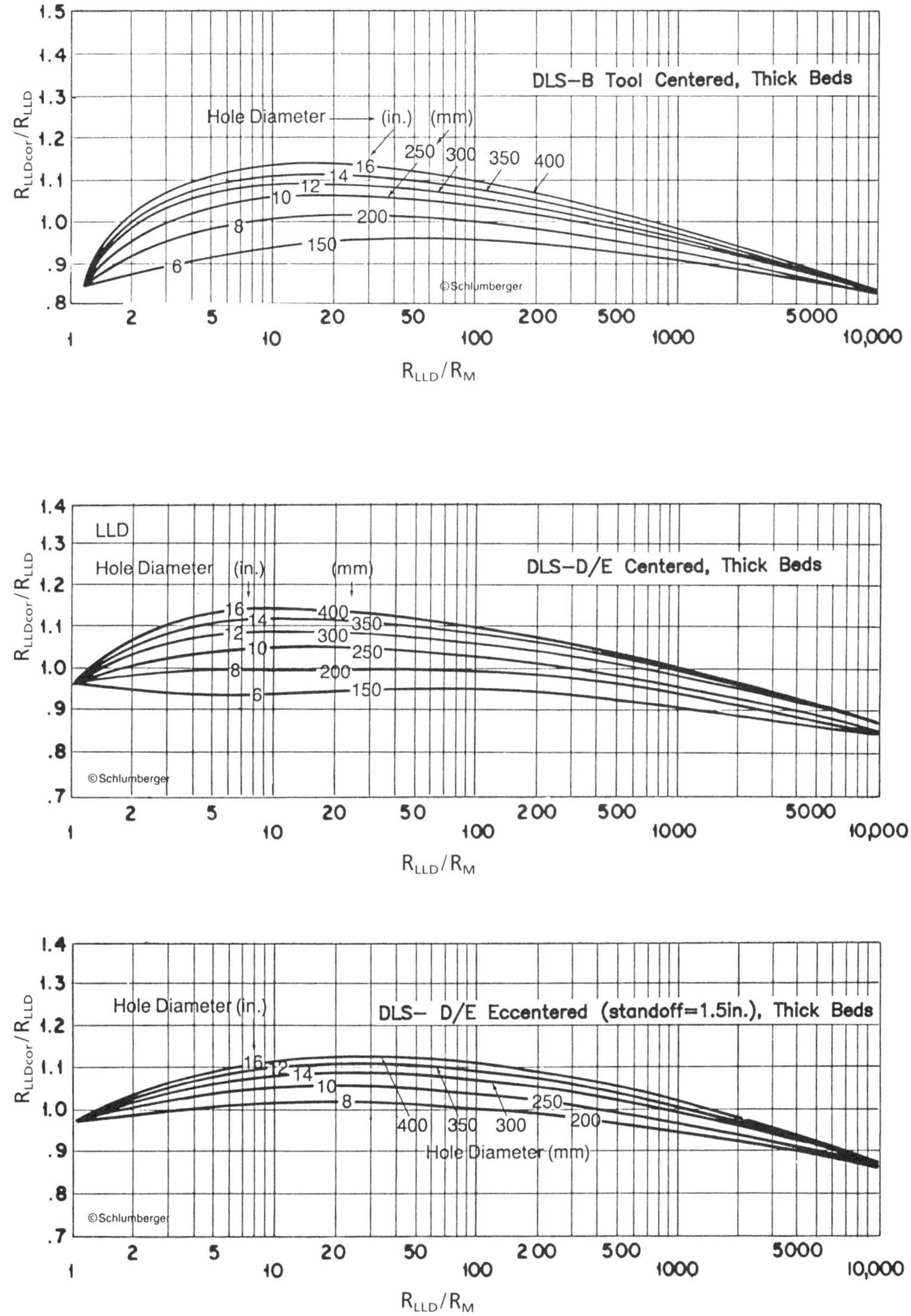


**Figure 3–28.** Charts for correcting Laterolog Deep for borehole effects (courtesy of Schlumberger).[4]

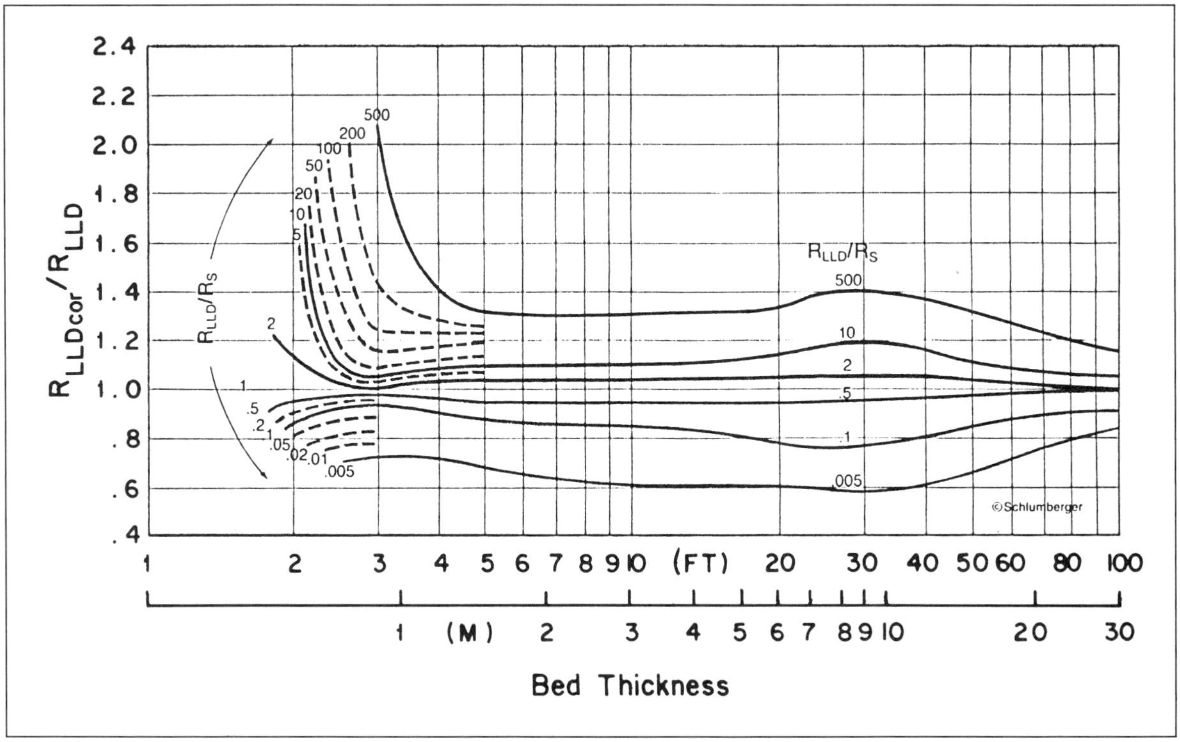


**Figure 3–29.** Chart for correcting Laterolog Deep for bed thickness effects (courtesy of Schlumberger).[4]

From Figure 5–29,

$$R_{LLDcor}/R_{LLD} = 0.88$$

$$R_{LLDcor} = (0.88)(5.5) = 4.84 \quad \text{ohm-m}$$

**TABLE 3–2. Invasion diameter as a function of porosity**

| Porosity (%) | Invasion Diameter |
|---|---|
| 5–10 | $d_i = 10 \times d_h$ |
| 10–15 | $d_i = 5 \times d_h$ |
| 15–20 | $d_i = 2.5 \times d_h$ |

**TABLE 3–3. Invasion correction factors**

| Invasion Diameter | Correction Factor |
|---|---|
| $d_i = 2.5 \times d_h$ | $R_a = 0.2R_{xo} + 0.8R_t$ |
| $d_i = 5.0 \times d_h$ | $R_a = 0.4R_{xo} + 0.6R_t$ |
| $d_i = 10.0 \times d_h$ | $R_a = 0.6R_{xo} + 0.4R_t$ |

## DUAL LATEROLOG-MICROSPHERICALLY FOCUSED LOG

Dual Laterolog consists of two laterologs with different depths of investigation. The shallow Laterolog (LLS) primarily measures $R_i$, and the deep Laterolog (LLD) measures $R_t$. The true resistivity can thus be calculated more precisely by taking into account the effect of the invaded zone in the vicinity of the wellbore.

The Microspherically Focused Log (MSFL) is an electrode log that has a very shallow depth of investigation and measures resistivity of the flushed zone ($R_{xo}$). The Dual Laterolog is accompanied by MSFL or any other $R_{xo}$ measuring log such that three resistivity curves (deep Laterolog, shallow Latrolog, and MSFL) are developed. The three resistivities are then used to correct the deep resistivity for invasion effects.

The shallow Laterolog and the Spherically Focused Log are corrected for borehole effects using the charts in Figure 3–30. The deep Laterolog and the shallow Laterolog are corrected for bed thickness effects using the charts in Figure 3–29 and Figure 3–31, respectively. The MSFL is corrected for mudcake effect using Figure 3–32. The Tornado charts in Figures 3–33 or 3–34 and the corrected values of $R_{LLD}$, $R_{LLS}$, and $R_{xo}$ are then used to correct the deep resistivity for invasion effects. Figures 3–33 and 3–34 are also used to determine the diameter of invasion and the ratio $R_{xo}/R_t$.

## MICROLOG

The microlog is useful in detecting porous and permeable formations, accurately defining the effective pay thickness, recording borehole diameter variations, and estimating the resistivity of the flushed zone ($R_{xo}$).

The resistivities called *micronormal* ($R_2$) and *microinverse* ($R_{1\times 1}$) have different depths of investigation and are recorded simultaneously. Comparison of the two curves readily identifies mudcake, which in turn indicates invaded and therefore permeable formations.

The micronormal measures $R_{xo}$ (investigates 3–4 in. into the formation) and the microinverse measures resistivity of the mudcake ($R_{mc}$). When the micronormal reads higher than the microinverse (*positive separation*), a permeable zone is indicated. Note that positive separation occurs only when $R_{mc} > R_m > R_{mf}$. Shale zones are indicated by no separation or *negative separation* (i.e., micronormal reads less than microinverse).

When formations exhibit resistivities 20 times the resistivity of the drilling mud ($R_m$), the formation, regardless of positive resistivity separation, is usually too dense for production of hydrocarbons. Reliable value of $R_{xo}$ may be obtained from micrologs if the following conditions are met:

1. The ratio $R_{xo}/R_{mc}$ must be less than about 15.
2. Porosity must be greater than at least 15%.
3. Mudcake thickness ($h_{mc}$) must be less than 0.5 in.
4. Depth of invasion must be over 4 in.; otherwise, the microlog readings are affected by $R_t$.

If the above criteria is met, the $R_{xo}$ value can be obtained from microlog by using the chart in Figure 3–35. The value of $R_{mc}$ can be measured directly or estimated from the following equation:

$$R_{mc} = 0.69(R_{mf})\left(\frac{R_m}{R_{mf}}\right)^{2.65} \quad \textbf{(3.10)}$$

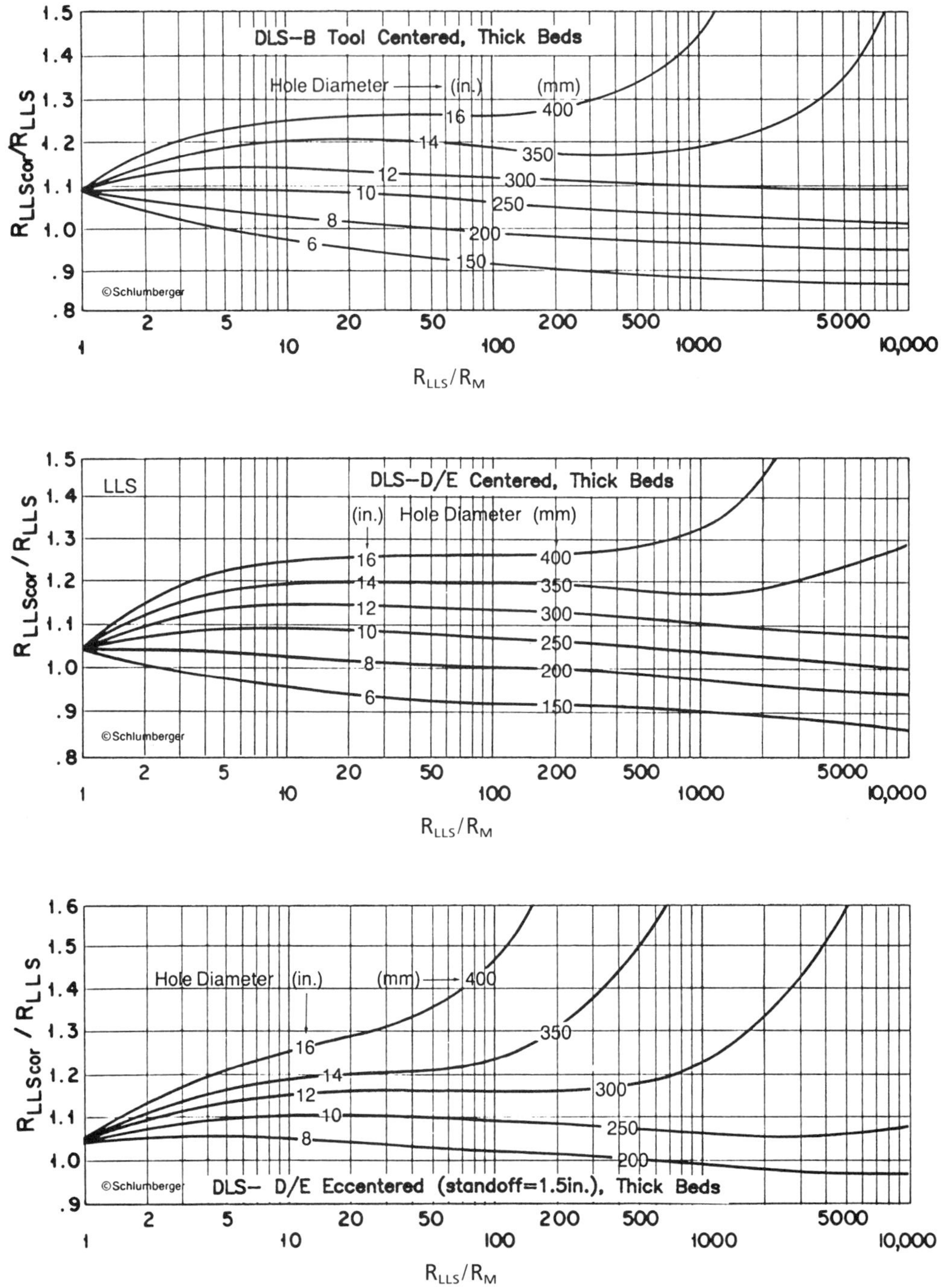


**Figure 3–30.** Charts for correcting shallow Laterolog for borehole effects (courtesy of Schlumberger).[4]

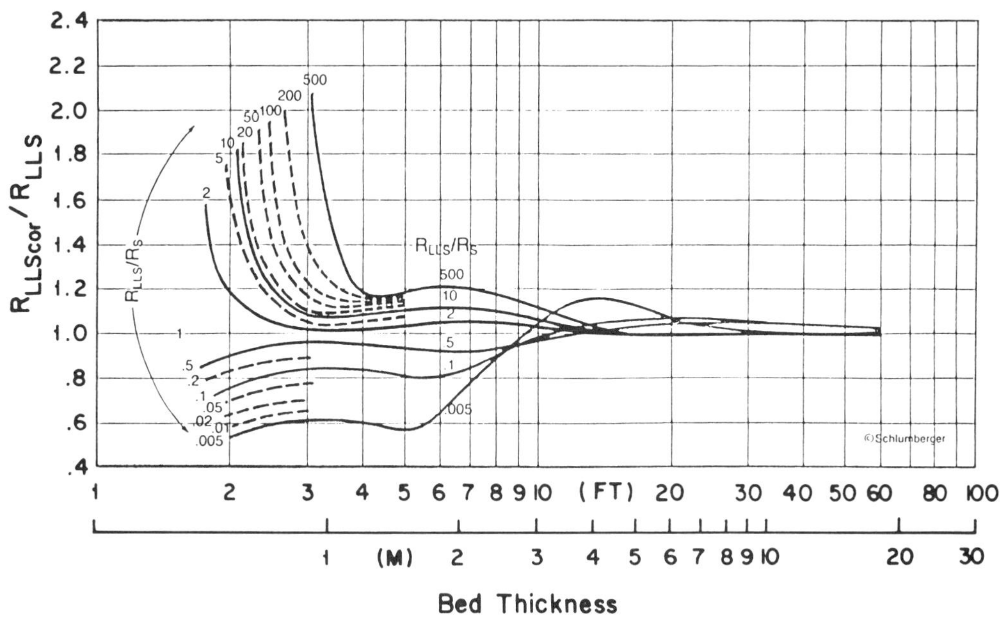


**Figure 3–31.** Chart for correcting shallow Laterolog for bed thickness effects (courtesy of Schlumberger).[4]

Note that the chart in Figure 3–35 is for 8-in. borehole. For hole sizes other than 8 in., multiply $R_{1\times 1}/R_{mc}$ by the following factors before entering the chart.

| Hole Diameter | Factor |
|---|---|
| 4.75 in. | 1.15 |
| 6.00 in. | 1.05 |
| 10.00 in. | 0.93 |

In the chart in Figure 3–35, if the $R_{mc}$ used is incorrect, it will displace the points in the chart along a 45° line. Sometimes this can be recognized when the $h_{mc}$ is different from direct measurement by the microcaliper. This situation can be corrected by moving the plotted point at 45° to intersect the known $h_{mc}$. The $R_{xo}/R_{mc}$ and $R_2/R_{mc}$ are then read from the chart at this new point. The $R_{xo}$ is then estimated using the following equation:

$$R_{xo} = R_2 \left( \frac{R_{xo}/R_{mc}}{R_2/R_{mc}} \right) \tag{3.11}$$

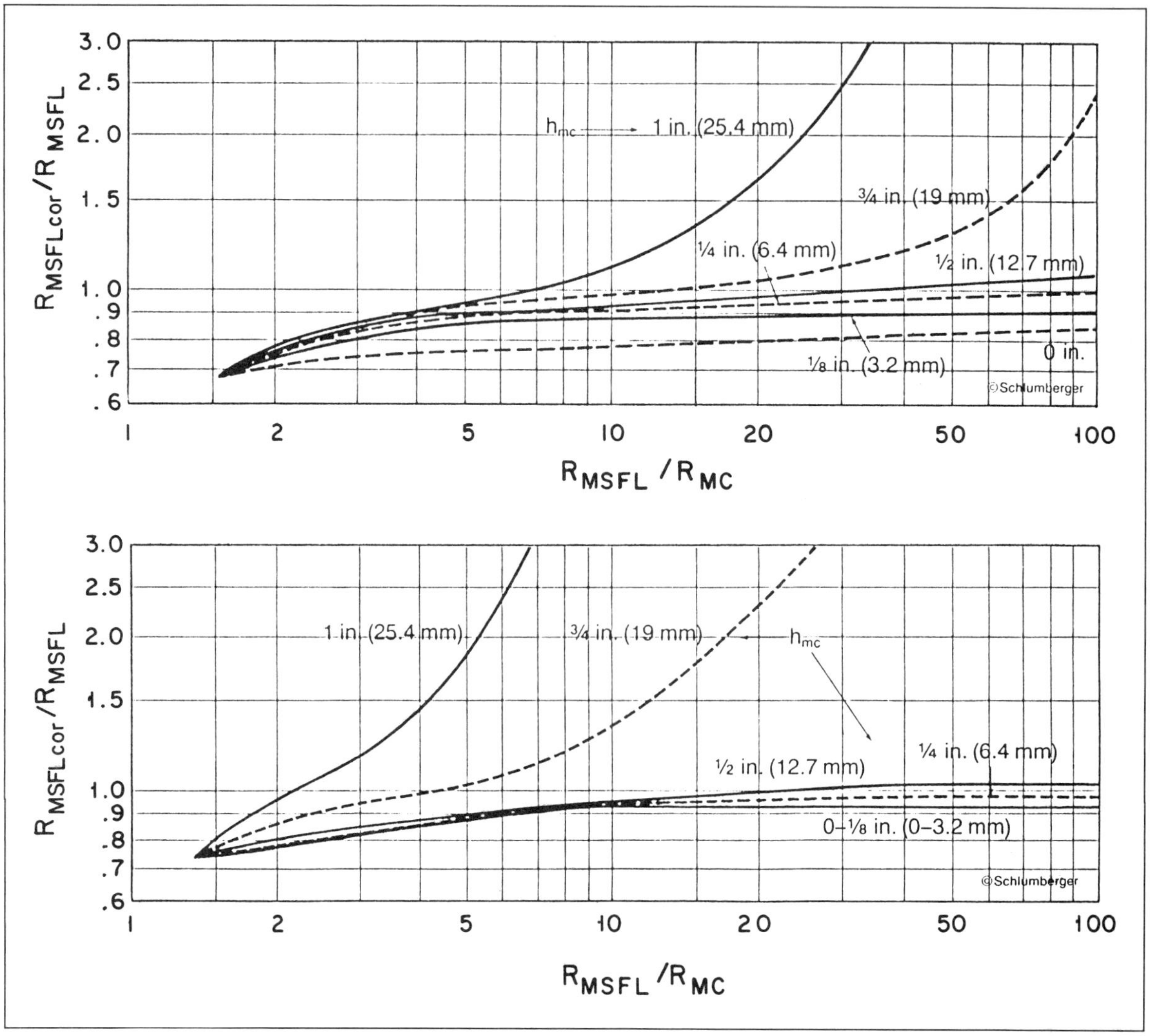


**Figure 3–32.** Charts for correcting MSFL for mudcake effects (courtesy of Schlumberger).[4]

**Example 3–10.** A microlog accompanied by induction electrolog, SP log, and caliper log are shown in Figure 3–36. Analyze the two zones A and B using the microlog and the following data.

$$d_h = 8 \text{ in.} \qquad R_{mf} = 1.3 \text{ ohm-m @}T_f$$
$$R_m = 1.4 \text{ ohm-m @}T_f \qquad R_{mc} = 1.7 \text{ ohm-m @}T_f$$

Solution:

| Interval | $R_{1\times1}$(ohm-m) | $R_2$(ohm-m) | $R_{1\times1}/R_{mc}$ | $R_2/R_{mc}$ |
|---|---|---|---|---|
| A | 4.4 | 7.1 | 2.59 | 4.18 |
| B | 4.0 | 6.3 | 2.35 | 3.71 |

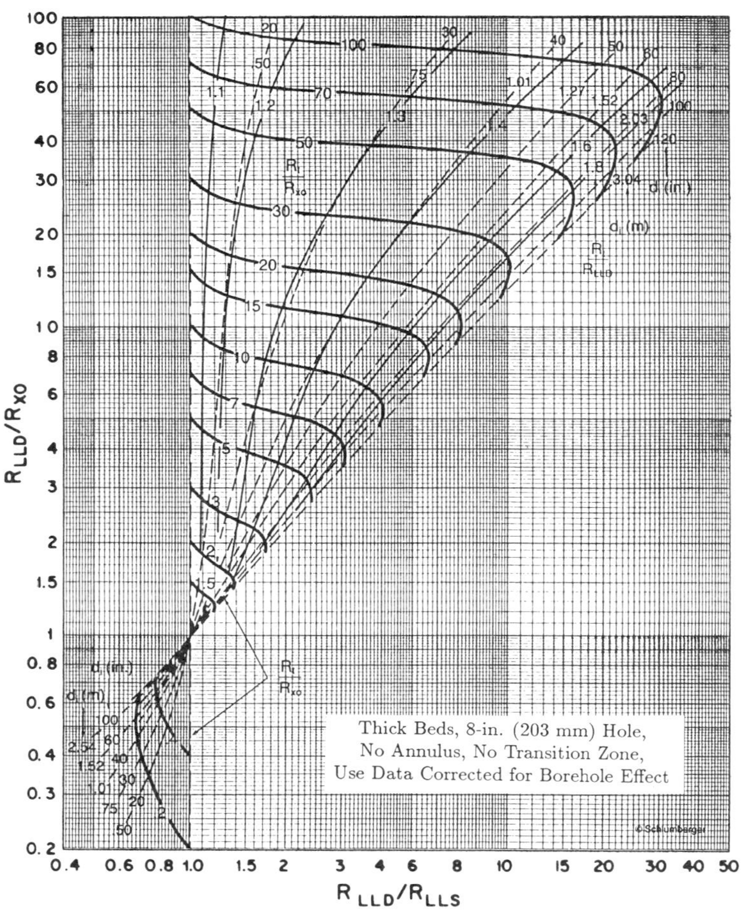


Figure 3–33. Resistivity invasion correction chart for Dual Laterolog-$R_{xo}$ (courtesy of Schlumberger).[4]

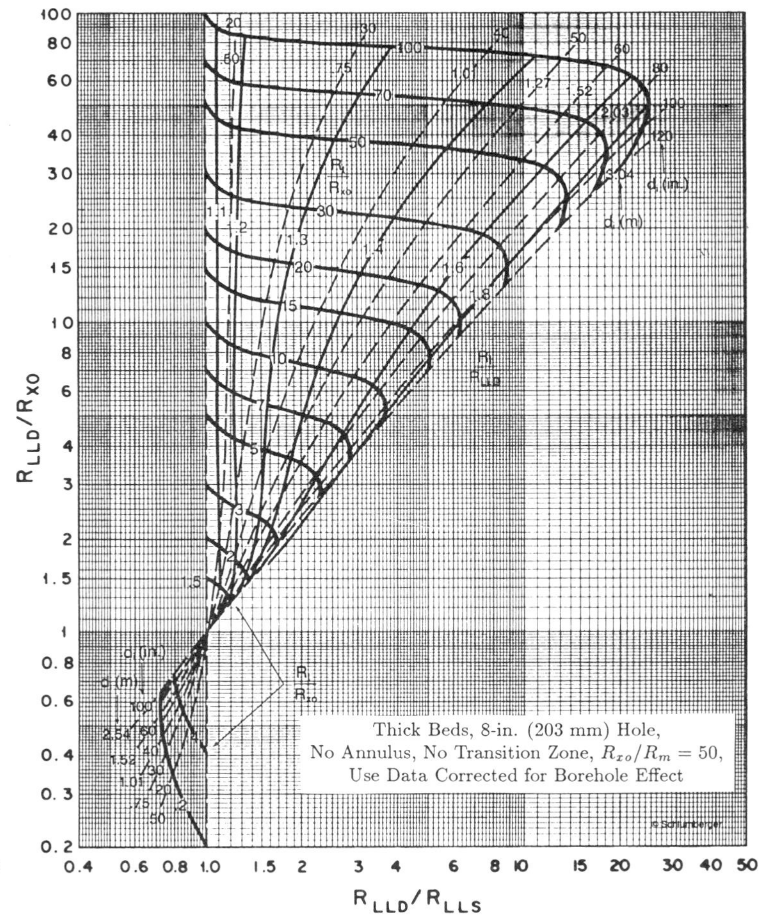


Figure 3–34. Resistivity invasion correction chart for Dual Laterolog-$R_{xo}$—$R_{xo}/R_m$ = 50 (courtesy of Schlumberger).[4]

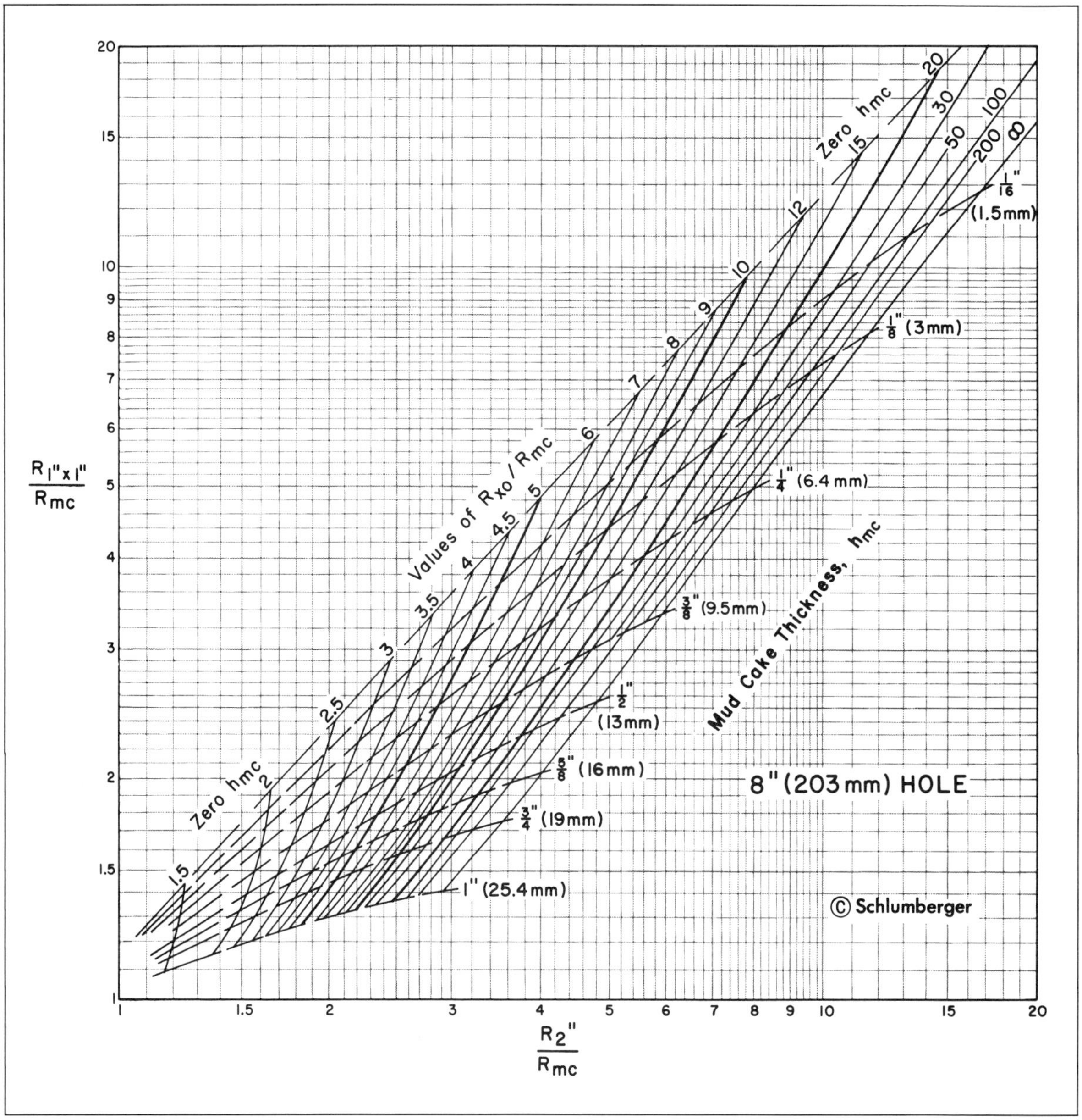


**Figure 3–35.** Microlog interpretation chart (courtesy of Schlumberger).[4]

Using the chart shown in Figure 3–35, $R_{xo}$ for each interval is

| | | |
|---|---|---|
| A | $R_{xo}/R_{mc} = 27$ | $R_{xo} = 45.9$ ohm-m |
| B | $R_{xo}/R_{mc} = 17.5$ | $R_{xo} = 29.8$ ohm-m |

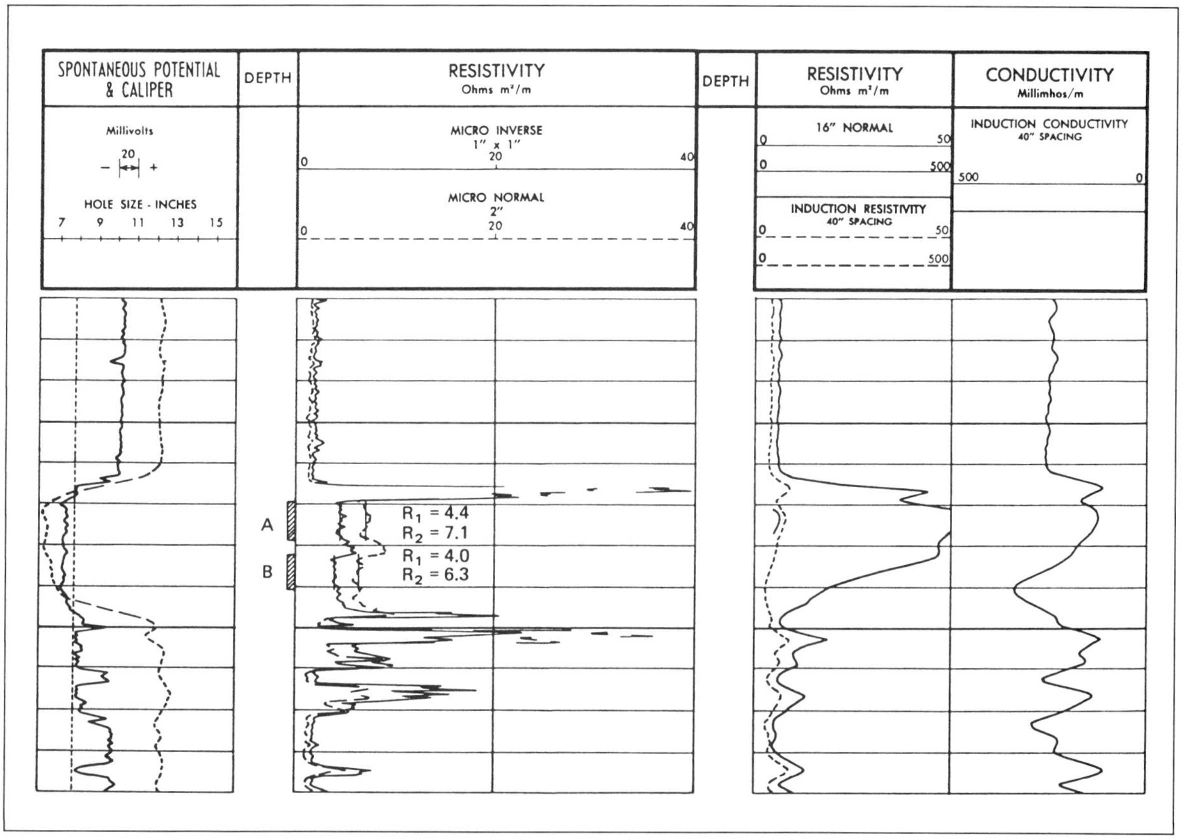


**Figure 3–36.** Log presentation for Example 3–10 (courtesy of Atlas Wireline Services).[7]

## MICROLATEROLOG AND PROXIMITY LOG

The Microlaterolog and Proximity log measure resistivity of the flushed zone ($R_{xo}$). The data provided by these logs are useful for estimating porosity and water saturation. Both these logs are presented on a four-cycle logarithmic scale ranging from 0.20 to 2,000 ohm-m.

The Microlaterolog is used in boreholes with low-resistivity mud and moderately thick mudcakes. On the other hand, Proximity log can be used in boreholes with high-resistivity mud, regardless of mudcake thickness. Both these logs will successfully log low-porosity formations. The Microlaterolog is best suited for formations with high $R_{xo}/R_{mc}$, where the Microlog interpretation lacks resolution.

The Proximity log has a deeper depth of investigation than both the Microlog and the Microlaterolog. Thus the Proximity log measurement may be affected by $R_t$ if the invasion is shallow. The Microlaterolog resistivity should be corrected for mudcake thickness ($h_{mc} > 3/8$ in.) using the chart in Figure 3–37. The Proximity log resistivity should be corrected for mudcake thickness ($h_{mc} > 3/4$ in.) using the chart in Figure 3–38. These logs accompanied by a deep resistivity log and a porosity log can be used to calculate movable hydrocarbons.

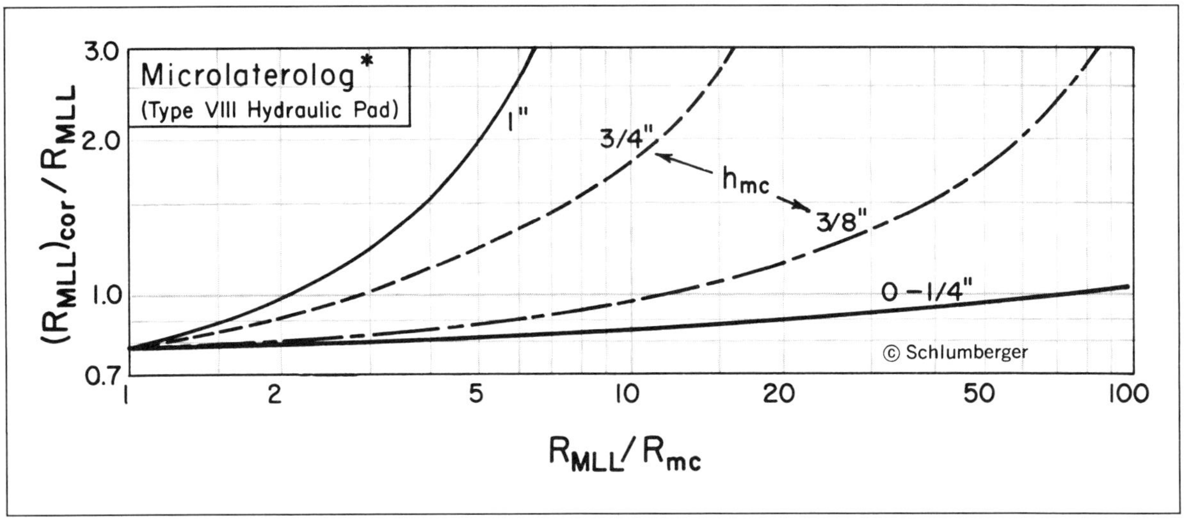


**Figure 3–37.** Chart for correcting Microlaterolog resistivity for mudcake thickness (courtesy of Schlumberger).[4]

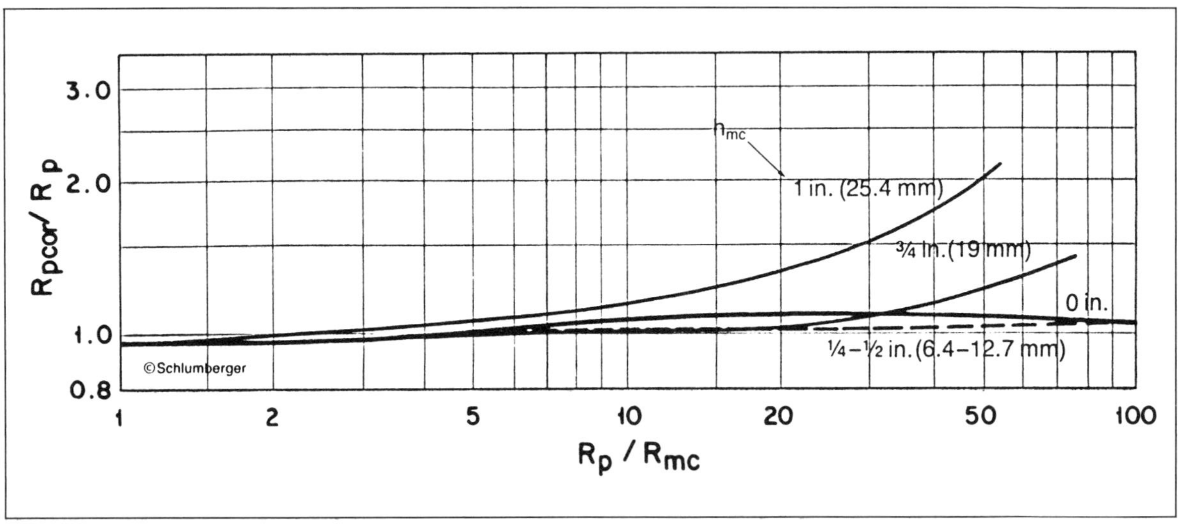


**Figure 3–38.** Chart for correcting Proximity log resistivity for mudcake thickness (courtesy of Schlumberger).[4]

## RESISTIVITY-DERIVED POROSITY

Formations' resistivity measurements can be used to determine porosity of the formation. Porosity in a water-bearing formation can be related to shallow resistivity ($R_{xo}$) and mud filtrate resistivity ($R_{mf}$) by the following equation:

$$\phi = \left( \frac{a \times R_{mf}}{R_{xo}} \right)^{1/m} \qquad \textbf{(3.12)}$$

where $a$ and $m$ are constants.

$a = 1.00$ for carbonates
$a = 0.62$ for unconsolidated sands
$a = 0.81$ for consolidated sands
$m = 2.00$ for consolidated sands and carbonates
$m = 2.15$ for unconsolidated sands



The resistivity-derived porosity will represent true porosity in water-bearing formations; however, in hydrocarbon-bearing formations the resistivity-derived porosity will be lower than the actual porosity. This is because of the higher $R_{xo}$ value due to residual hydrocarbon saturation in the flushed zone.

If the water saturation of the flushed zone ($S_{xo}$) is known or could be estimated, then the true porosity could be calculated using the following equation:

$$\phi = \left[ \frac{a(R_{mf}/R_{xo})}{(S_{xo})^2} \right]^{1/m} \qquad \textbf{(3.13)}$$

Some empirically derived $S_{xo}$ values are given in Table 3–4. If the $S_{xo}$ value is not known, it could be estimated from this table. The resistivity-derived porosity can also be calculated from the deep resistivity ($R_t$); however, in this case the resistivity of formation water ($R_w$) and formation water saturation ($S_w$)

**TABLE 3–4. Percentages of residual hydrocarbon saturation (RHS) as a function of hydrocarbon density and porosity**

| | API° Gravity | RHS (%) | $S_{xo}$%(1 − RHS) |
|---|---|---|---|
| Gas | | 40–5 | 60–95 |
| High Gravity Oil | 40–50 | 10–5 | 90–95 |
| Medium Gravity Oil | 20–40 | 20–10 | 80–90 |
| Low Gravity Oil | 10–20 | 30–20 | 70–80 |
| Porosity (%) | | | |
| 25–35 | | 30 | 70 |
| 15–20 | | 15 | 85 |

must be known. To use $R_t$, $R_w$, and $S_w$, Equations (3.12) and (3.13) are modified (i.e., $R_{mf}$ is replaced by $R_w$, $R_{xo}$ by $R_t$, and $S_{xo}$ by $S_w$).

## POROSITY AND LITHOLOGY DETERMINATION

In this section we will discuss the logs used to calculate formation porosity. The logs discussed in this section are (1) density log, (2) sonic (acoustic) log, and (3) neutron log. All these logs respond to the formation porosity, fluid, and matrix.

The depth of investigation of all these three logs is very shallow—only a few inches or less—and therefore the porosities obtained from these logs will in the majority of cases represent the *flushed zone*. We will also discuss different cross-plotting techniques which helps in identifying the probable lithology. Note that in clean water-filled formations, a single log, either the neutron log, the density log, or (if there is no secondary porosity) the sonic, can be used to determine porosity.

### DENSITY LOG

Density log is primarily used as a porosity log. Other uses include gas detection, determine hydrocarbon density, evaluation of shaly sands and complex lithologies, and evaluation of minerals in evaporite deposits, etc. The log directly measures electron density of a formation with a phenomenon called *Compton scattering* (interaction of gamma ray particles and electrons in the formation). Consequently, electron density is related to formation density in grams/cubic centimeters.

Formation bulk density ($\rho_b$) is a function of matrix density ($\rho_{ma}$) and density of the fluid in the pores ($\rho_f$). To determine density porosity, the matrix density and the type of fluid in the borehole must be known. Matrix densities of common lithologies are given in Table 3–5. The equation for calculating density porosity is then

$$\phi_D = \frac{\rho_{ma} - \rho_b}{\rho_{ma} - \rho_f} \tag{3.14}$$

**TABLE 3–5. Commonly used matrix densities for commonly occurring rock types**

| Mineral | $\rho_{ma}$ Range gms/cc | $\rho_{ma}$ Commonly Used gms/cc |
|---|---|---|
| Sandstone | 2.65–2.70 | 2.65 |
| Limestone | 2.71 | 2.71 |
| Dolomite | 2.83–2.89 | 2.87 |
| Anhydrite | 2.94–3.00 | 2.98 |
| Salt (Halite) | 2.03 | 2.03 |

The fluid density ($\rho_f$) is (1) 1.10 for salt mud, (2) 1.0 for fresh mud, and (3) 0.70 for gas. A typical density log is presented in Figure 3–39 in Track 2 of the log. The log will usually show (1) the bulk density curve, (2) density-derived porosity curve, and (3) the correction curve. The density log has the following characteristics:

1. Gas or light oil in the formation makes the density log show higher apparent porosity (i.e., low bulk density).
2. Shale makes the density log show low apparent porosity or high bulk density. Density porosity in shaly formation is calculated using the following equation:

$$\phi_D = \frac{\rho_{ma} - \rho_b}{\rho_{ma} - \rho_f} - V_{sh}\left(\frac{\rho_{ma} - \rho_{sh}}{\rho_{ma} - \rho_f}\right) \quad \textbf{(3.15)}$$

3. The modern density logs will compensate for borehole irregularities such as mudcake and hole rugosity (compensated density log). The correction curve on the log records the amount of correction that has been applied to the bulk density curve. Whenever the correction curve ($\delta_\rho$) exceeds 0.20 gms/cc, the $\rho_b$ obtained from the log should be considered invalid or used with caution.
4. Under some circumstances, the density log needs to be corrected for borehole size using Figure 3–40.
5. Halite or salt will make the density log read about 40% porosity.

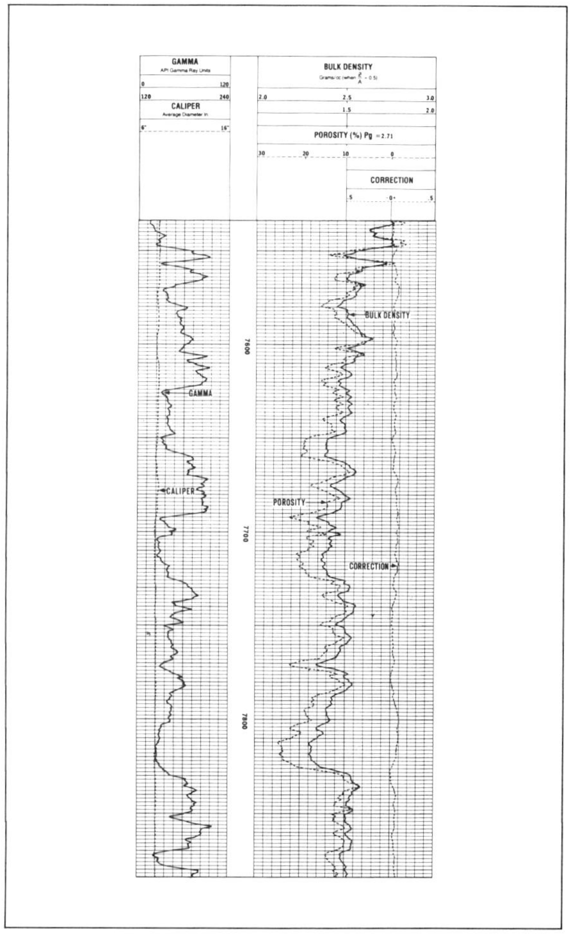


**Figure 3–39.** Typical presentation of density log (courtesy of Welex).[32]

The $\rho_f$ in the invaded zone of a gas/water system often is assumed to be 1.0. This number may be improved for effective porosity determination by taking into consideration $\rho_g$, $S_{xo}$, and $\rho_{mf}$ in the following relationship:

$$\rho_f = (1 - S_{xo})\rho_g + S_{xo}\rho_{mf}$$

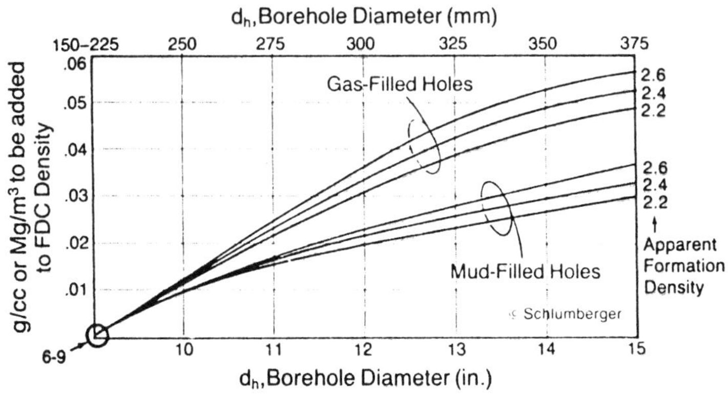


**Figure 3–40.** Chart for correcting density log for borehole diameter (courtesy of Schlumberger).[4]

**Example 3–11.** The following formation properties were obtained from density log. Calculate the density porosity.

$$\rho_b = 2.31 \quad \text{gms/cc}$$
$$\rho_{ma} = 2.71 \quad \text{gms/cc} \quad \text{(limestone matrix)}$$
$$\rho_f = 1.1 \quad \text{gms/cc} \quad \text{(salt mud)}$$
$$h_b = 9 \quad \text{in.} \quad \text{(no correction applicable)}$$

Solution:

$$\phi_D = \frac{\rho_{ma} - \rho_b}{\rho_{ma} - \rho_f} = \frac{2.71 - 2.31}{2.71 - 1.1} \times 100 = 25\%$$

If the density log is the only available log for porosity determination, the bulk density should be corrected for residual hydrocarbons (in the zone of investigation of the log) effects using the following equations:

$$\rho_{LOG} = \rho_b - \Delta\rho_b \qquad \textbf{(3.16)}$$

It has been shown to a good approximation that for oil-bearing formations

$$\Delta\rho_b = 1.07\phi S_{hr}[(1.11 - 0.15P)\rho_{mf} - 1.11\rho_h - 0.03] \qquad \textbf{(3.17)}$$

For gas-bearing formations

$$\Delta\rho_b = 1.07\phi S_{hr}[(1.11 - 0.15P)\rho_{mf} - 1.24\rho_h] \qquad \textbf{(3.18)}$$

where $P$ is the equivalent NaCl concentration in parts per million $\times$ $10^{-6}$ for mud filtrate, $S_{hr}$ is the residual hydrocarbon saturation $(1 - S_{xo})$, and $\rho_h$ is the density of residual hydrocarbons at formation temperature and pressure.

**Example 3–12.** In a clean gas-bearing limestone formation, the density log reads 2.25 gms/cc and the microresistivity log reads $R_{xo}$ = 68 ohm-m. Calculate the formation's porosity (corrected for residual hydrocarbon effect), if $0.15P$ is negligible and

$$\rho_{mf} = 1.0 \text{ gms/cc} \qquad m = 2$$
$$\rho_h = 0.2 \text{ gms/cc} \qquad n = 2$$

Solution: In case of fresh muds, $P$ = 0 and $\rho_{mf}$ = 1.0

$$\phi_D = \frac{\rho_{ma} - \rho_b}{\rho_{ma} - \rho_f} = \frac{2.71 - 2.25}{2.71 - 1.0} = 0.27 \text{ OR } 27\%$$

$$S_{hr} = 1 - S_{xo} = 1 - \left(\frac{R_{mf}}{\phi^2 R_{xo}}\right)^{1/2}$$

The simplified relation for fresh mud would be

$$\phi_D = \frac{\rho_{ma} - \rho_{LOG} - \Delta\rho_b}{\rho_{ma} - 1}$$

Or for gas-bearing formations

$$\phi_D = \frac{\rho_{ma} - \rho_{LOG} + 1.07\sqrt{R_{mf}/R_{xo}}(1.11 - 1.24\rho_h)}{\rho_{ma} - 1 + 1.07(1.11 - 1.24\rho_h)}$$

$$= \frac{2.71 - 2.25 + 1.07\sqrt{1/68}[1.11 - (1.24)(0.20)]}{2.71 - 1 + 1.07[1.11 - (1.24)(0.21)]}$$

$$= \frac{0.57}{2.63}$$

$$= 0.22 \quad \text{OR} \quad 22\%$$



Then the $\Delta\rho_b = 0.09$ gms/cc

## SONIC LOG

The sonic (acoustic) log is a porosity log that records the travel time ($\Delta t$) of a compressional sound wave through one foot of formation. The interval transit time ($\Delta t$) in microseconds per foot (μsec/ft) is the *reciprocal* of the velocity of the compressional sound wave.

The acoustic travel time is a function of the formation lithology and porosity. The following equation is used to calculate sonic porosity in consolidated and compacted formations. A typical sonic log presentation is shown in Figure 3–41.

$$\phi_S = \frac{\Delta t_{LOG} - \Delta t_{ma}}{\Delta t_f - \Delta t_{ma}} \tag{3.19}$$

Some commonly used matrix and fluid velocities and acoustic transit times are given in Table 3–6.

Sonic log has the following characteristics:

1. In carbonates with vuggy or fracture porosity, Equation (3.19) will give very low apparent porosity. This is because the sonic log reads only the matrix porosity. The percentage of vuggy or fracture porosity can be calculated by subtracting *sonic porosity* from the density-neutron porosity.
2. Sonic log is considered as an excellent tool for identifying gas-bearing zones. In uncompacted gas-bearing formations, it will show extremely high apparent porosity.
3. Sonic log will read high apparent porosity in uncompacted formations. Equation (3.19) is modified as given below to account for compaction.

$$\phi_S = \frac{\Delta t_{LOG} - \Delta t_{ma}}{\Delta t_f - \Delta t_{ma}} \times \frac{1}{C_p} \tag{3.20}$$

$$C_p = \frac{\Delta t_{sh} \times C}{100} \tag{3.20a}$$

where $\Delta t_{sh}$ is the interval transit time for the adjacent shale zone and C is a constant which is normally considered as 1.0.[1] The zone of interest is considered uncompacted when $\Delta t_{sh}$ of the adjacent shale zone is more than 100 μsec/ft. The compaction factor can also be determined if density log porosity or resistivity-derived porosity is available in a clean water-bearing zone near the zone of interest. In moderately shaly sands, neutron log can be used advantageously. The $C_p$ in these cases may be determined as follows:

$$C_p = \frac{\phi_S}{\phi_D} \quad \textbf{(3.21)}$$

$$C_p = \frac{\phi_S}{\phi_N} \quad \textbf{(3.22)}$$

4. Shale makes the sonic log read higher apparent porosity. Porosity calculated by using Equation (3.19) or (3.20) is corrected for shale content in laminated shale zones as follows:

$$\phi_{cor} = \phi_S - V_{sh}\phi_{sh} \quad \textbf{(3.23)}$$

or for shaly sand, when only dispersed clay/shale is present:

$$\phi_{cor} = \phi_S - V_{sh} \quad \textbf{(3.24)}$$

5. Hydrocarbons will make sonic log read higher apparent porosity. Hilchie suggests the following empirical corrections for hydrocarbon effect.[1]

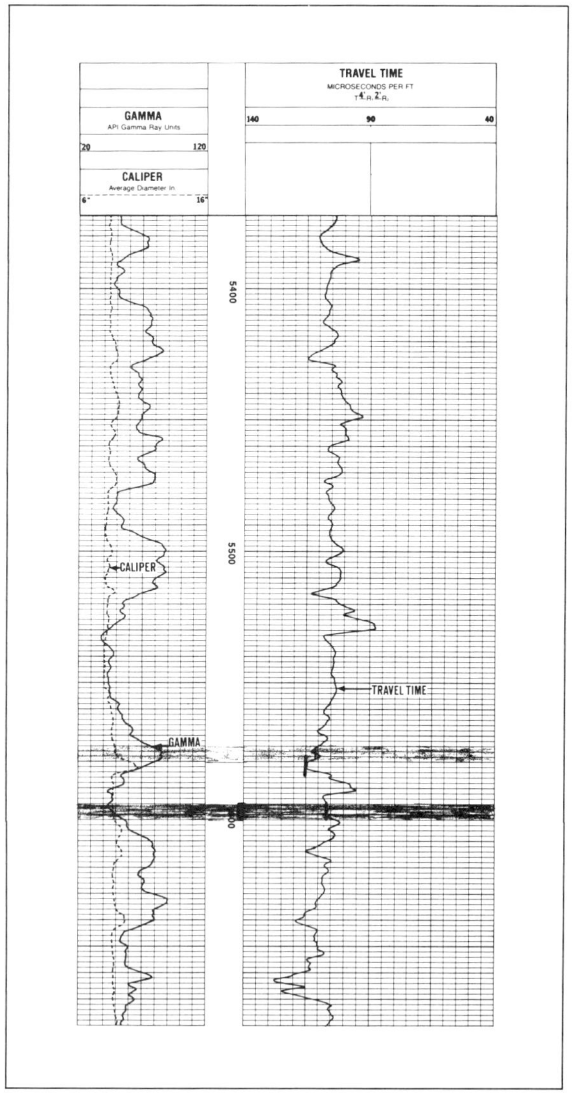


**Figure 3–41.** Typical presentation of sonic log (courtesy of Welex).[32]

For gas:

$$\phi_{cor} = \phi_S \times 0.7 \quad \textbf{(3.25)}$$

For oil:

$$\phi_{cor} = \phi_S \times 0.9 \quad \textbf{(3.26)}$$

**TABLE 3–6. Commonly used matrix and fluid velocities and acoustic transit time**

| Component | $V_{ma}$ (ft/sec) | $\Delta t_{ma}$ (μsec/ft) | Commonly Used $\Delta t_{ma}$ (μsec/ft) |
|---|---|---|---|
| Quartz | 18,000 | 55.6 | 55.6 |
| Sandstone | 18,000–19,500 | 55.5–51.0 | 58.8 |
| Limestones | 21,000–23,000 | 47.6–43.5 | 47.6 |
| Dolomites | 23,000 | 43.5 | 43.5 |
| Anhydrite | 20,000 | 50 | 50.0 |
| Salt | 15,000 | 66.7 | 67.0 |
| **Fluids** | **$V_f$ (ft/sec)** | **$\Delta t_f$ (μsec/ft)** | **Commonly Used $\Delta t_f$ (μsec/ft)** |
| Water (20% NaCl) | 5,300 | 189 | 189 |
| Water (15% NaCl) | 5,000 | 200 | 200 |
| Water (10% NaCl) | 4,800 | 208 | 208 |
| Water (pure) | 4,600 | 218 | 218 |
| Oil | 4,200 | 238 | 238 |
| Methane | 1,600 | 626 | 626 |
| Air | 1,100 | 910 | 910 |

6. The sonic log should always be examined for *cycle skipping*. The cycle skipping results in the acoustic log either spiking (sharp curve movements) or too long a travel time being recorded. The cycle skipping often occurs in unconsolidated formations (particularly in gas-bearing zones), fractured formations, opposite gas-cut mud, or when the borehole is too large (over 12 to 14 in.) and the tool is still centered in the borehole. The cycle skipping may also occur because of bad or weak receiver or transmitter.

**Example 3–13.** Calculate the sonic porosity at a depth of 5,596 ft–5,602 ft from the sonic log given in Figure 3–41. Assume that $\Delta t_f = 189$ μsec/ft and $\Delta t_{ma} = 55.5$ μsec/ft. The shale zone at 5,574–5,580 ft with $\Delta t_{sh} = 111.5$ μsec/ft shows that the formation is uncompacted.

Solution:

$$\Delta t_{LOG} = 107 \quad \mu\text{sec/ft} \quad @5{,}596\text{–}5{,}602 \text{ ft}$$

$$C_p = \frac{\Delta t_{sh}}{100} = \frac{115.5}{100} = 1.115$$

$$\phi_S = \frac{\Delta t_{LOG} - \Delta t_{ma}}{\Delta t_f - \Delta t_{ma}} \times \frac{1}{C_p} = \frac{107 - 55.5}{189 - 55.5} \times \frac{1}{1.115} = 35\%$$

## NEUTRON LOG

Neutron logs are porosity logs that measure the hydrocarbon ion concentration in a formation. In clean, shale-free, and water- or oil-filled porosity, the neutron log measures the liquid-filled porosity.

Neutron logs include sidewall neutron porosity (SNP) logs and compensated neutron porosity logs (CNL). The CNL further includes a dual-spaced neutron log. The CNL and the dual porosity logs are designed to minimize the effects of hole size, and mudcake, etc. In a majority of cases, neutron porosity log and density log are run simultaneously as shown in Figure 3–42. The combination works best in identifying lithology and gas-bearing formations because of their characteristics. The log is recorded in porosity units based on a certain matrix, usually limestone matrix. Neutron log has the following characteristics:

1. Presence of mudcake will make the SNP log read high apparent porosity because of the higher concentration of hydrogen in the mudcake. Figures 3–43a and 3–43b are used to correct SNP porosity for mudcake effect.
2. Both the SNP and CNL logs are recorded in apparent limestone, sandstone, or dolomite porosity units. If a neutron log is recorded in apparent limestone units, and the formation is limestone, the apparent porosity is equal to true porosity. However, if the lithology of a formation is sandstone or dolomite, the apparent limestone porosity must be corrected for lithology using Figure 3–44.
3. Compensated neutron log should be corrected for environmental effects, such as (a) borehole size, (b) mudcake, (c) mud weight, (d) standoff effect, (e) borehole salinity effect, (f) formation salinity effect, and (g) casing and cement effect in cased holes. Figures 3–45, 3–46, and 3–47 are used for correcting compensated neutron porosity for environmental effects.
4. Shale makes the neutron log read higher apparent porosity.
5. Gas-bearing or light oil-bearing formations make a neutron log read lower apparent porosity because of the lower concentration of hydrogen atoms in light hydrocarbons. Note that unlike most other logs, neutron logs must be interpreted from the specific charts designed by a specific logging company. The charts of Dresser Atlas and Schlumberger are presented in this section. Also note that the responses of density log and neutron log in shaly and gas-bearing or light oil-bearing formations are opposite to each other. For this reason, these two logs provide the best combination for identifying and analysis of gas-bearing formations.
6. The neutron logs tend to show anhydrite between 0 and $-1\%$ porosity.
7. Halite or salt will make the neutron log read about 0% porosity.
8. Gypsum, pyrite, and siderite all increase apparent neutron porosity.

Gas-bearing and light oil-bearing formations are identified by a *density-neutron crossover* as shown in Figure 3–42. The magnitude of the crossover depends on gas or light oil and shale concentration in the formation. The degree of separation is also controlled by the density of the gas and the wetness or amount of condensate in the gas. Dry, low-pressure, and high-temperature gases give the larger separation.[1] Shale and some clays in the formation suppress the separation. In some cases,

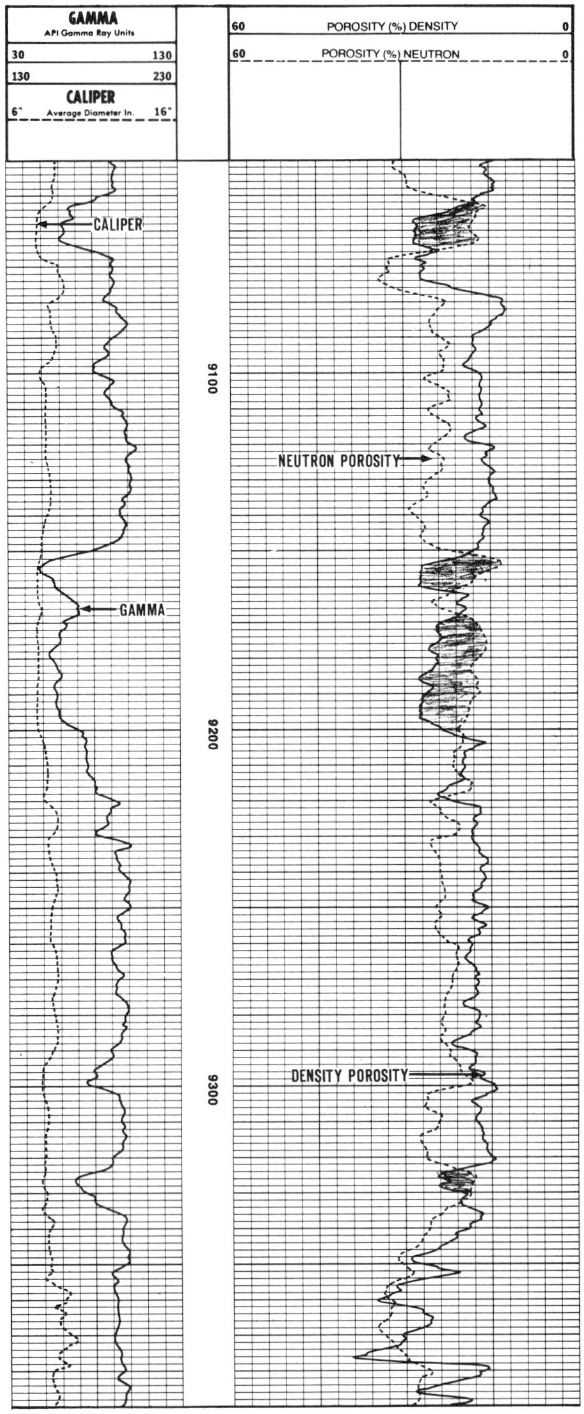


**Figure 3–42.** Combined density porosity and neutron porosity logs (courtesy of Welex).[32]

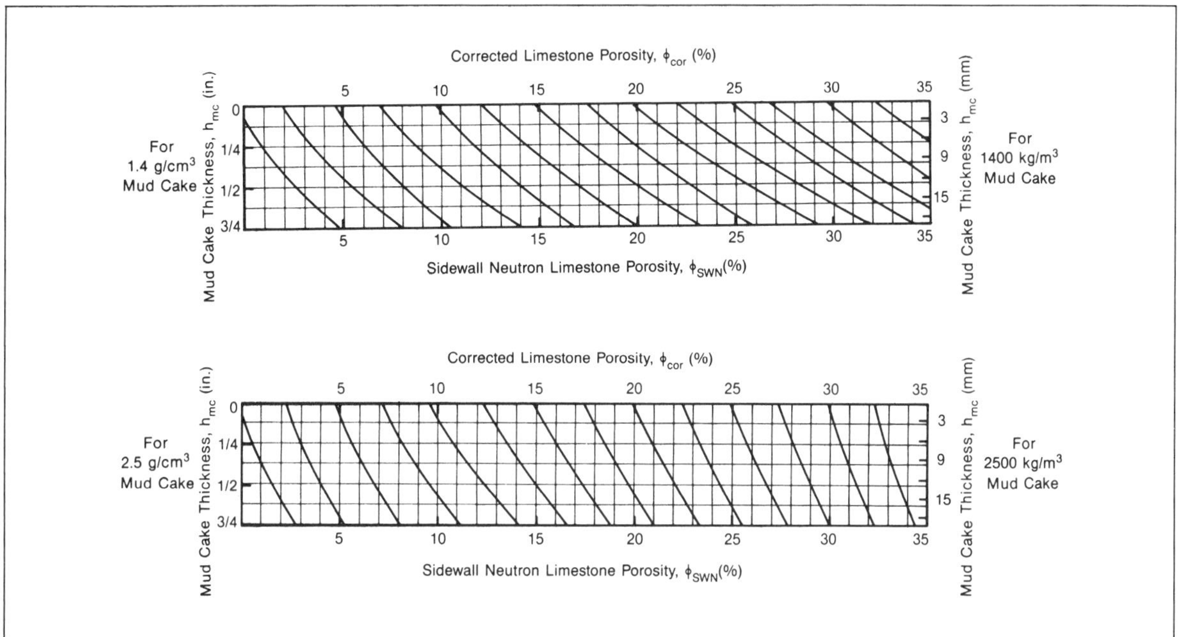


**Figure 3–43a.** AWS Charts for correcting SNP for mudcake effects (courtesy of Atlas Wireline Services).[6]

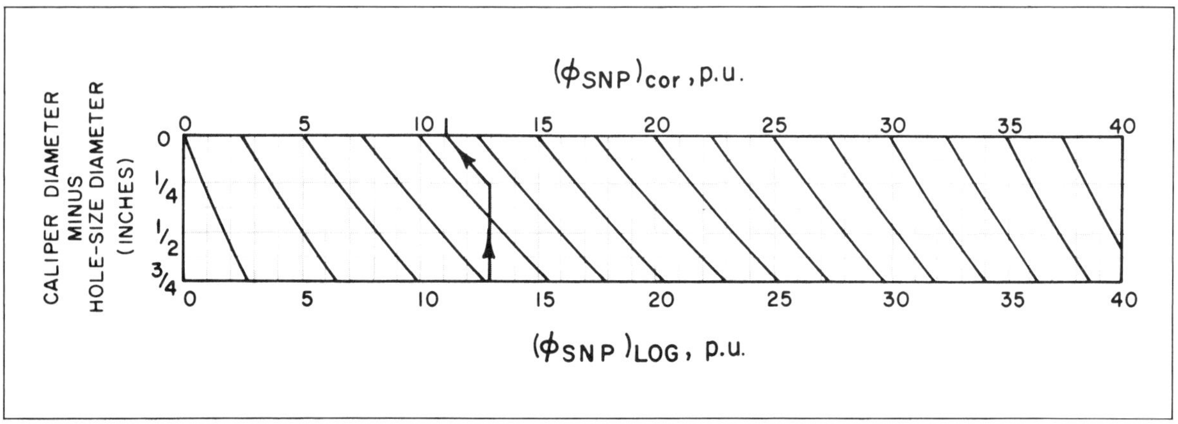


**Figure 3–43b.** Schlumberger chart for correcting SNP for mudcake effects (courtesy of Schlumberger).[4]

depending on hydrocarbon and shale concentration, the logs will not *exactly crossover*, but they will either *tend to crossover*, or they will stack on top of each other.

The true porosity from density-neutron logs is determined using *density-neutron porosity crossplots* (to be discussed in the following pages). The combination is also used to identify lithology. The presence of gas in the zone of investigation of the neutron log further affects the neutron response

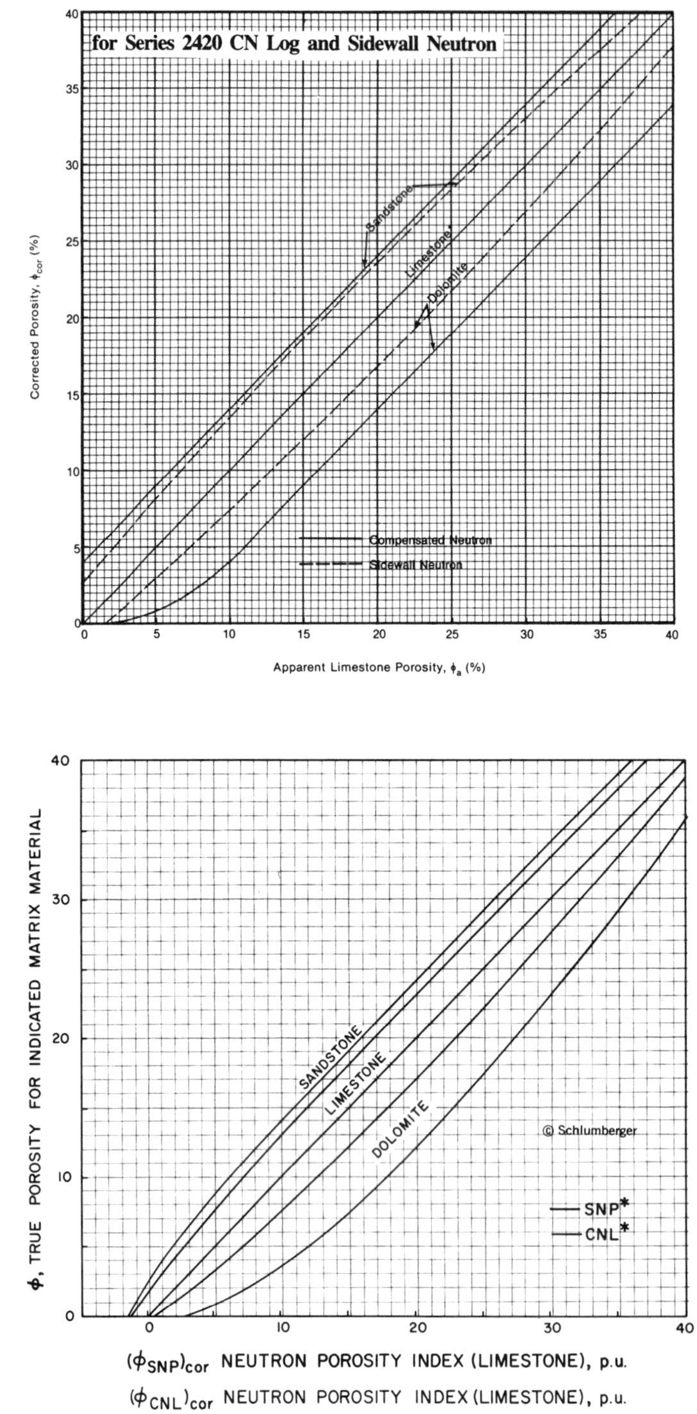


**Figure 3–44.** Neutron porosity equivalence curves (courtesy of Atlas Wireline Services and Schlumberger).[4,6]

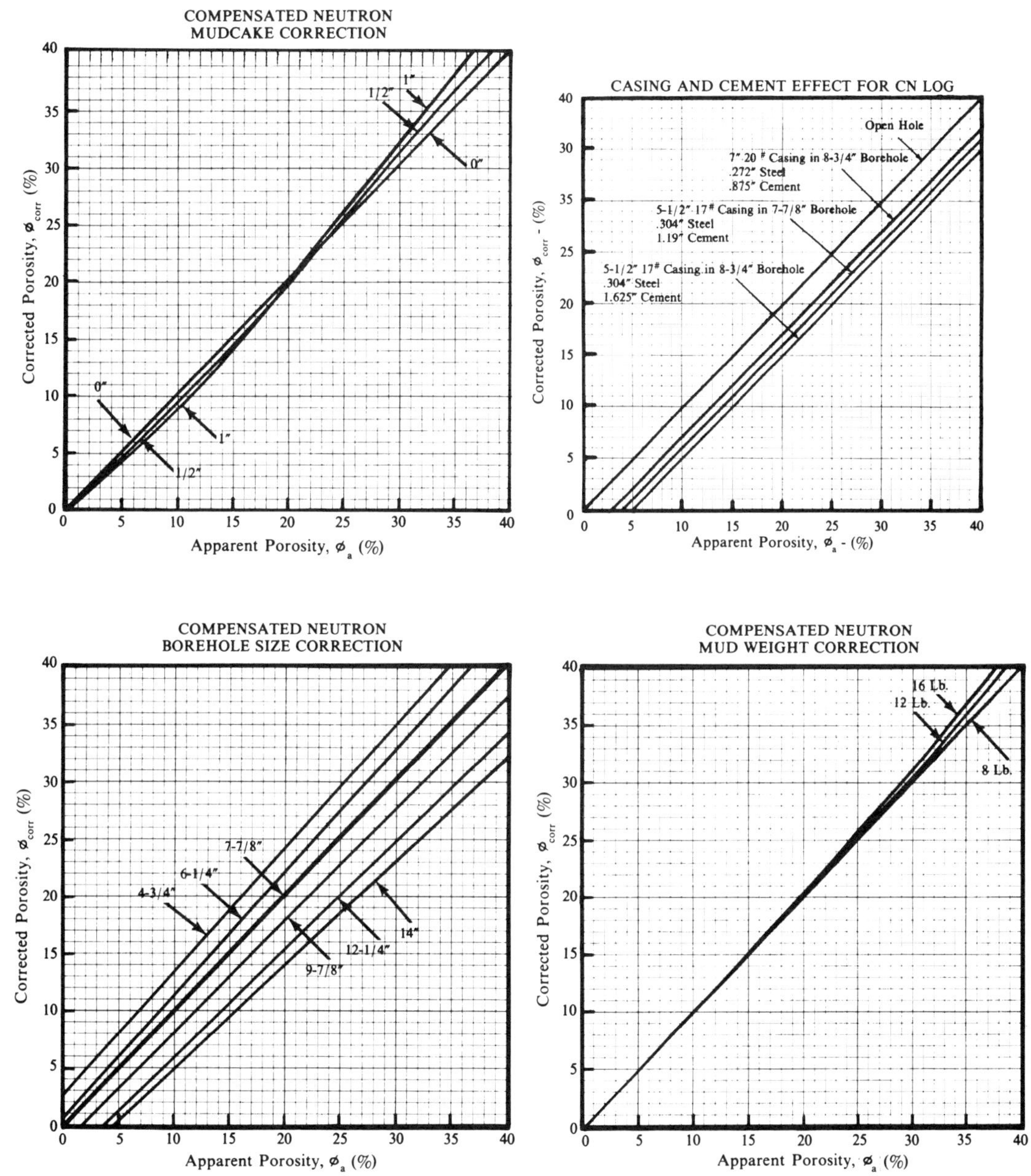


**Figure 3–45.** AWS charts for correcting CNL porosity for environmental effects (courtesy of Atlas Wireline Services).[6]

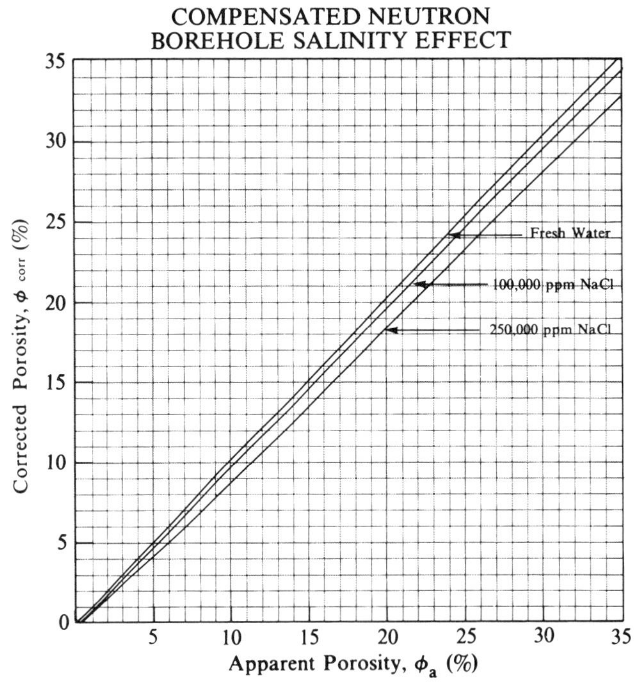


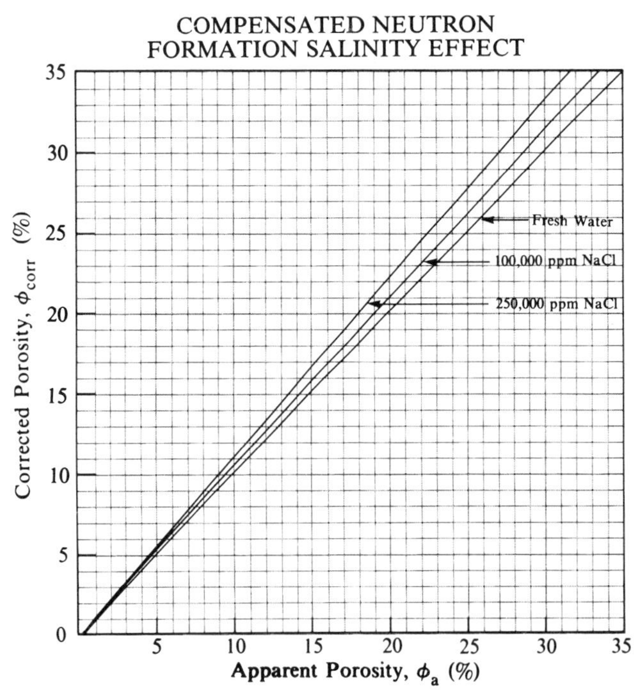


**Figure 3–46.** AWS charts for correcting CNL porosity for environmental effects (courtesy of Atlas Wireline Services).[6]

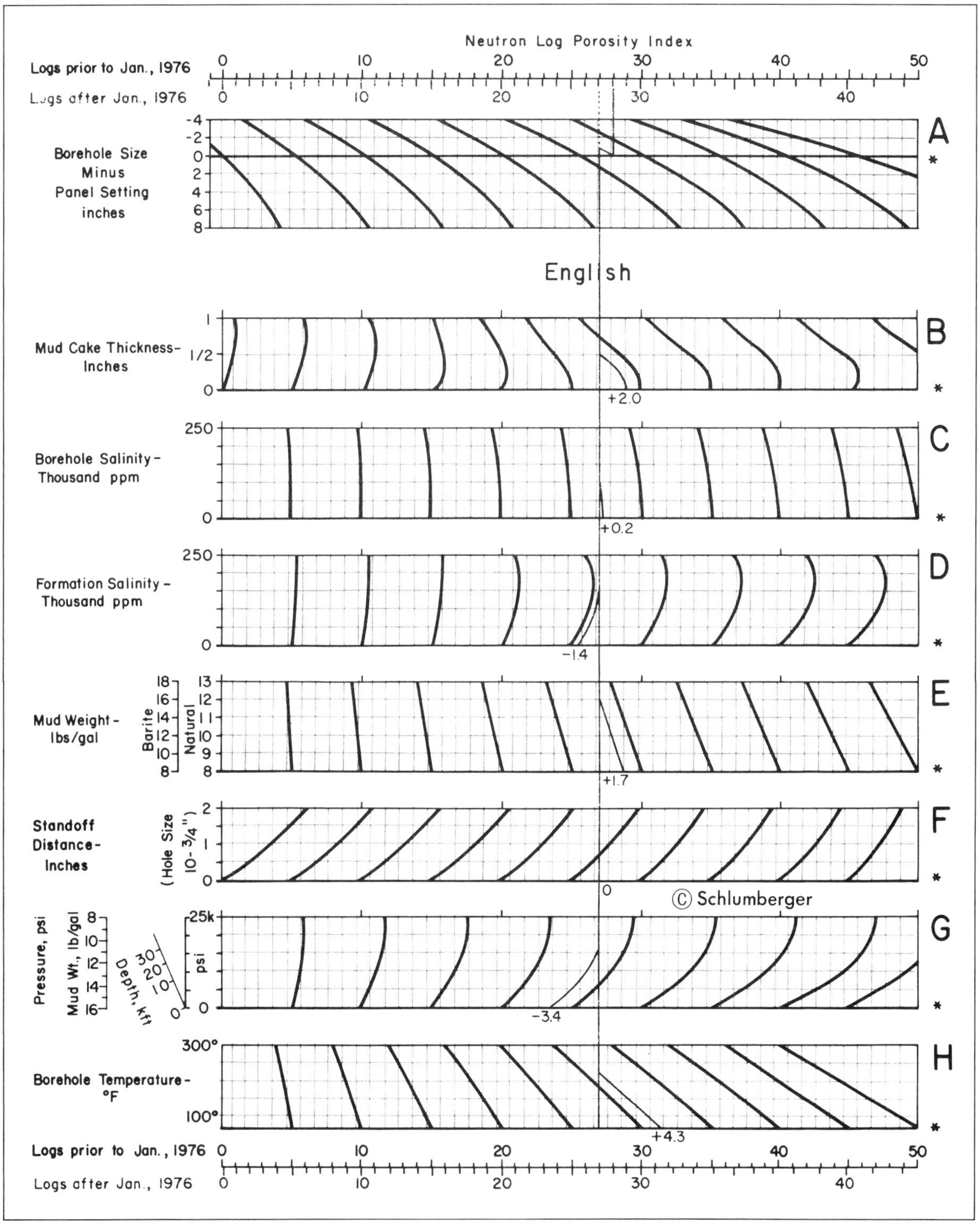


**Figure 3–47.** Schlumberger charts for correcting CNL porosity for environmental effects (courtesy of Schlumberger).[4]

by what is known as the *excavation effect*. Therefore, the quantitative response of the neutron tool to gas or light oil depends primarily on hydrogen index and excavation effect.

Approximate relationships recommended in the case of CNL and SNP log, including the excavation effect and for both oil and gas-bearing ($\rho_h \leq 0.70$) formations

$$\Delta\phi_{SNP} = \phi S_{hr} \frac{\rho_{mf}(1 - P) - 1.67\rho_h + 0.17}{\rho_{mf}(1 - P)} \quad (3.27)$$

$$\Delta\phi_{CNL} = \phi S_{hr} \frac{\rho_{mf}(1 - P) - 2.67\rho_h + 0.87}{\rho_{mf}(1 - P)} \quad (3.28)$$

**Example 3–14.** In a clean gas-bearing sandstone formation, Dresser's CNL reads 10% limestone porosity. The density of gas under reservoir conditions is 0.2 gms/cc. The microresistivity log reads $R_{xo} = 40$ ohm-m, and the well is drilled with fresh mud, $R_{mf}$ at formation temperature being 0.8 ohm-m. Calculate the true porosity of the formation.

Solution: In case of fresh mud $\rho_{mf} = 1.0$ and $P = 0$, also assuming that $a = 1$ and $m = n = 2$.

$$\Delta\phi_{CNL} = \phi S_{hr}(0.87 - 2.67)\rho_h$$

$$\phi = \phi_{CNL} - \frac{\sqrt{R_{mf}/R_{xo}}(0.87 - 2.67)\rho_h}{2.67\rho_h - 0.87}$$

$$= 0.10 - \frac{(0.8/40)^{0.5}[0.87 - (2.67)(0.20)]}{[(2.67)(0.20) - 0.87]}$$

$$= 0.10 - (-0.14)$$

$$= 0.24 \quad \text{OR} \quad 24\% \quad \text{(limestone porosity)}$$

Now using Figure 3–44, correct the above porosity for lithology (sandstone).

$$\phi_{N_{Sandstone}} = 28\%$$

## DENSITY-NEUTRON CROSSPLOT

The density-neutron crossplot is a convenient way to demonstrate how the two logs respond to lithology and porosity. As mentioned before, the density-neutron logs provide the best combination for identifying gas zones and determining their porosity. The combination also provides porosity in complex lithology and volume of shale for shaly formations' evaluation.

Density-neutron crossplots for Schlumberger and Dresser Atlas logs are presented for $\rho_f = 1.0$ gms/cc and $\rho_f = 1.1$ gms/cc in Figures 3–48 through 3–48c. To use these charts, both the density porosity and the neutron porosity must be on a compatible porosity scale (i.e., limestone, sandstone, or dolomite).

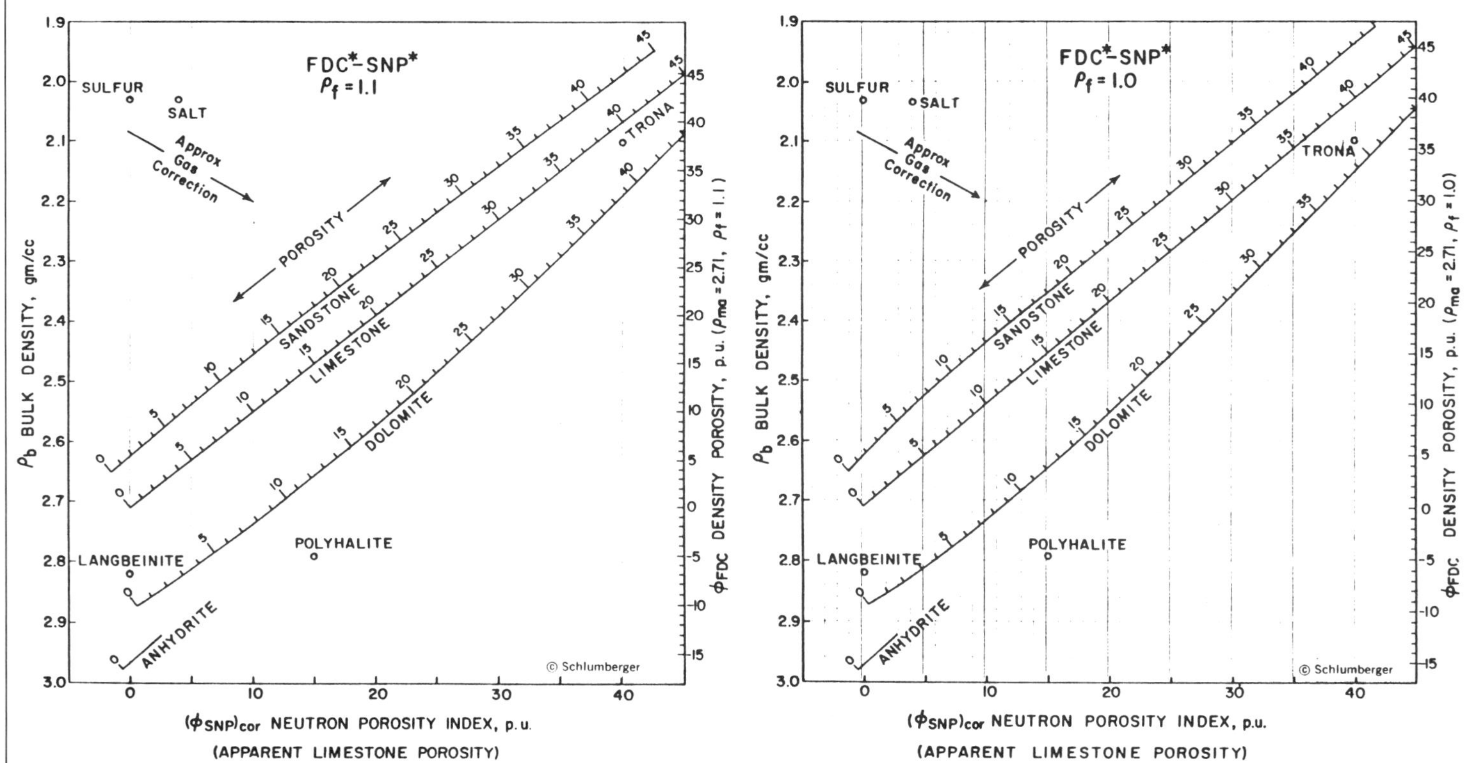


**Figure 3–48.** Density-Neutron (SNP) crossplots for Schlumberger logs (courtesy of Schlumberger).[4]

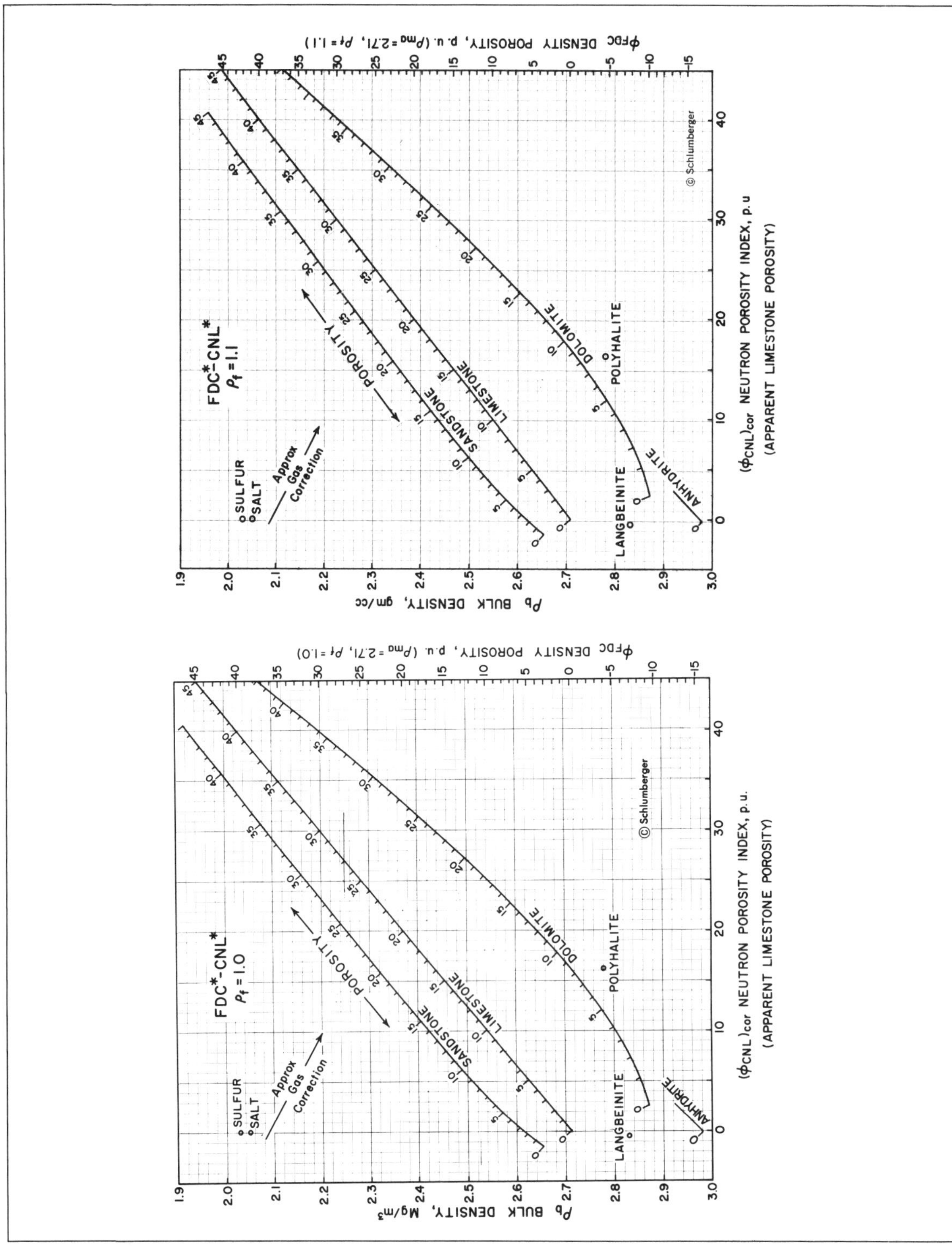


**Figure 3–48a.** Density-Neutron (CNL) crossplots for Schlumberger logs (courtesy of Schlumberger).[4]

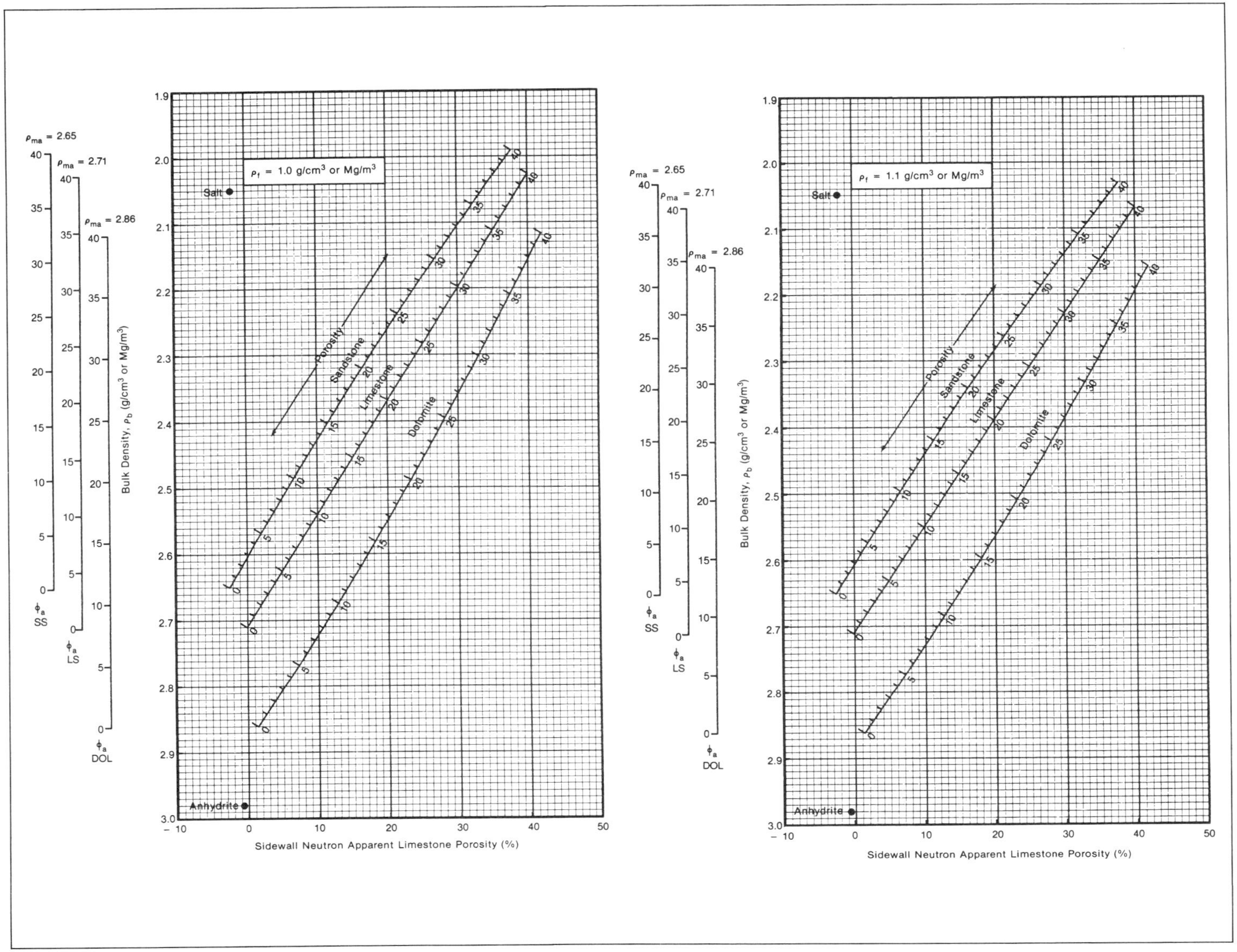


**Figure 3–48b.** Density-Neutron (SNP) crossplots for AWS logs (courtesy of Atlas Wireline Services).[6]

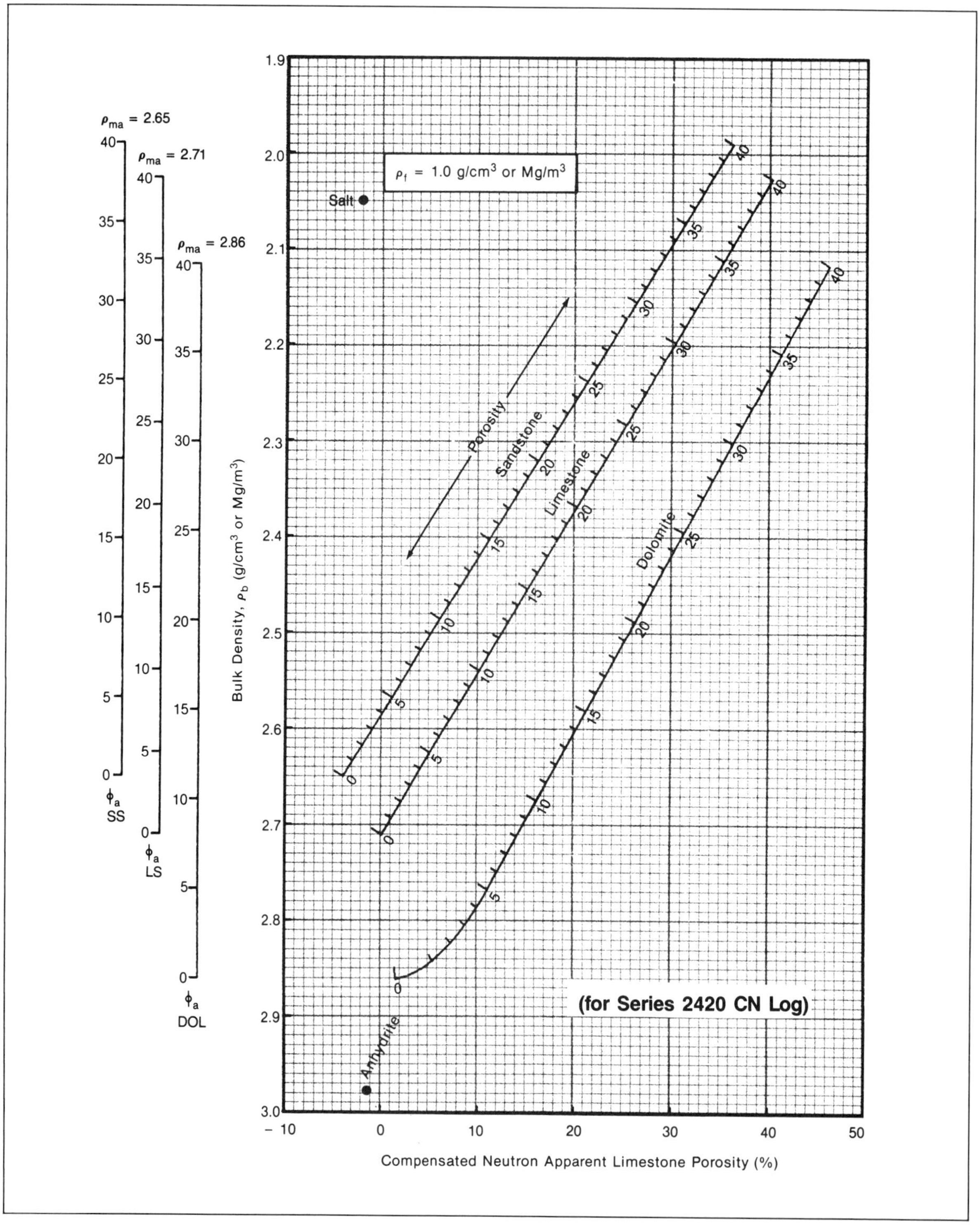


**Figure 3–48c.** Density-Neutron (CNL) crossplots for AWS logs (courtesy of Atlas Wireline Services).[6]

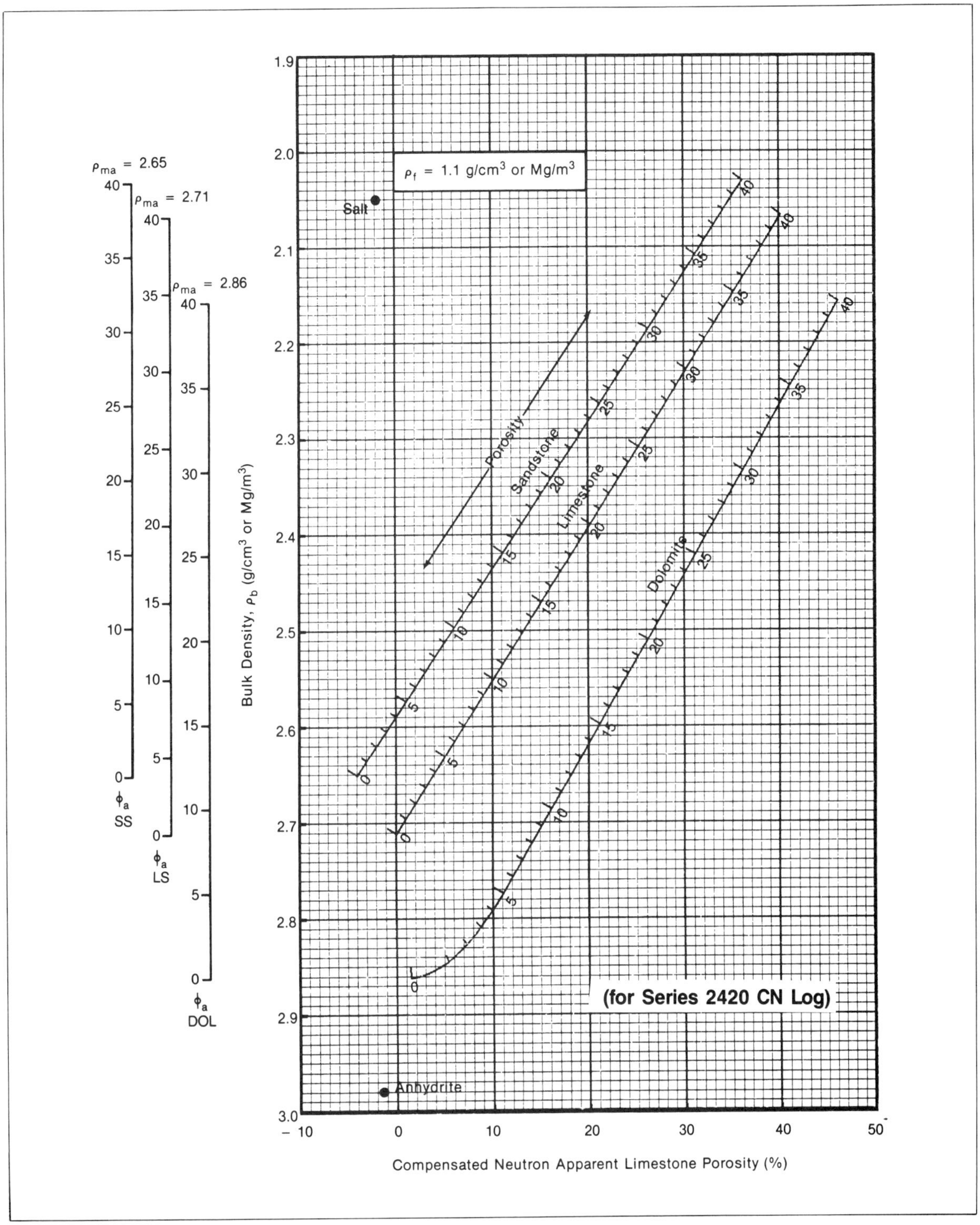


Figure 3–48c (continued)

The best way to use these charts is

1. Make sure that both porosities represent a compatible matrix.
2. Correct both porosities for their respective environmental effects.
3. Enter the appropriate chart starting with the appropriate matrix line, for instance, sandstone.
4. Draw a vertical line at the neutron sandstone porosity ($\phi_{CNL}$ = 20%) and a horizontal line intersecting the density sandstone porosity ($\phi_D$ = 17%). The intersection of the two lines gives a crossplot porosity ($\phi_{CP}$ = 18%) and the possible lithology (limy sandstone in this case).
5. Use the horizontal and vertical axes if (a) both porosities are in limestone matrix or (b) if neutron porosity is in limestone matrix and bulk density is given.

The following qualitative characteristics are worth noting when working with the density-neutron combination. Assume both logs are recorded on limestone porosity.[1]

1. In shale, the neutron reads apparent high porosity and the density reads apparent low porosity.
2. In gas, the neutron reads apparent low porosity and the density reads apparent high porosity.
3. Dolomites show a density-neutron separation of about 15% porosity, with the neutron reading higher porosity.
4. Sandstone or chalk will show a density-neutron separation of about 6%, with neutron reading lower porosity.
5. Halite or salt is detected by a large density-neutron separation, with the density reading about 40% porosity and the neutron about 0%.
6. Limestone with salt in the pore spaces will show up as either pure limestone porosity on both curves, or there will be large separation as noted in Item 5.

## SONIC-DENSITY CROSSPLOT

The sonic-density crossplot ($\Delta t$ vs. $\rho_b$ or $\phi_D$) provides a very poor porosity and reservoir rock resolution in sandstones, limestones, and dolomites. However, the combination is useful in determining some evaporite minerals. A Schlumberger sonic-density crossplot is presented in Figure 3–49.

## SONIC-NEUTRON CROSSPLOT

Sonic-neutron (SNL) and sonic-neutron (CNL) crossplots for Schlumberger logs are presented for $\Delta t_f$ = 189 μsec/ft in Figure 3–50. The sonic-neutron crossplots for Dresser Atlas logs for $\Delta t_f$ = 189 μsec/ft and $\Delta t_f$ = 185 μsec/ft for both SNL and CNL are presented in Figures 3–50a and 3–50b, respectively. The arguments regarding its use are the same as those for the density-neutron crossplot.

With both the density-neutron and sonic-neutron crossplots, the resolution between sandstone, limestone, and dolomite lithologies is good, and therefore errors in choosing the lithology pair will have only a small effect on the porosity value found. However, resolution is lost if evaporites are present in the formation logged.

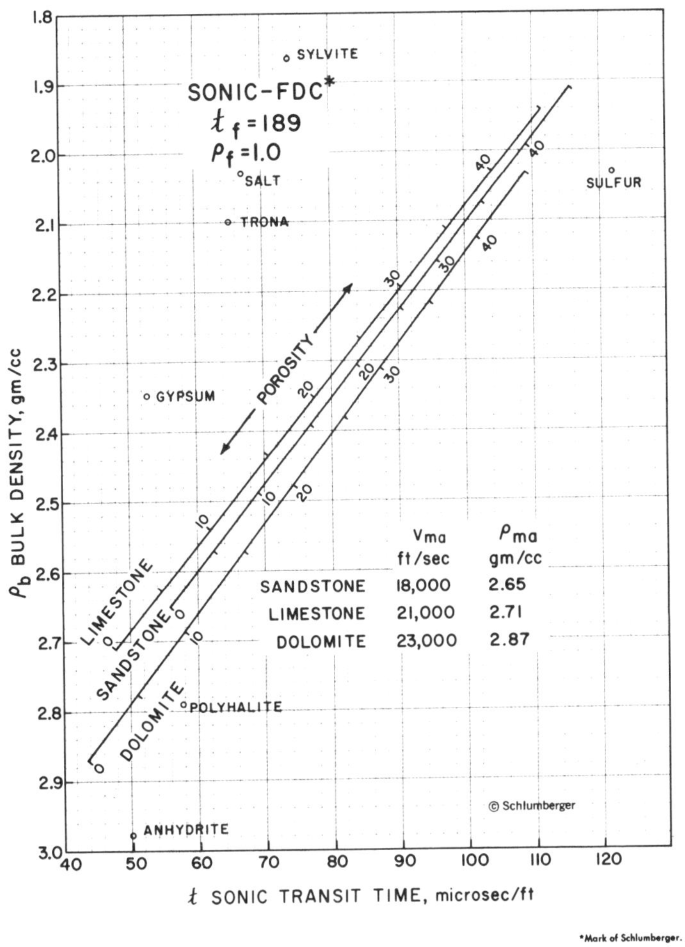


**Figure 3–49.** Lithology identification from sonic-density crossplot (courtesy of Schlumberger).[4]

## GAS DETECTION WITH THE DENSITY-NEUTRON LOG

It has been previously mentioned that the most commonly used gas detection system is the combined density and neutron logs. However, it is important to note the following three situations when detecting gas with this combination:

1. The case where both logs respond equally to the gas saturation.
2. The case where the neutron is seeing more gas due to invasion problems.
3. The case where no crossover takes place because of higher shale concentration.

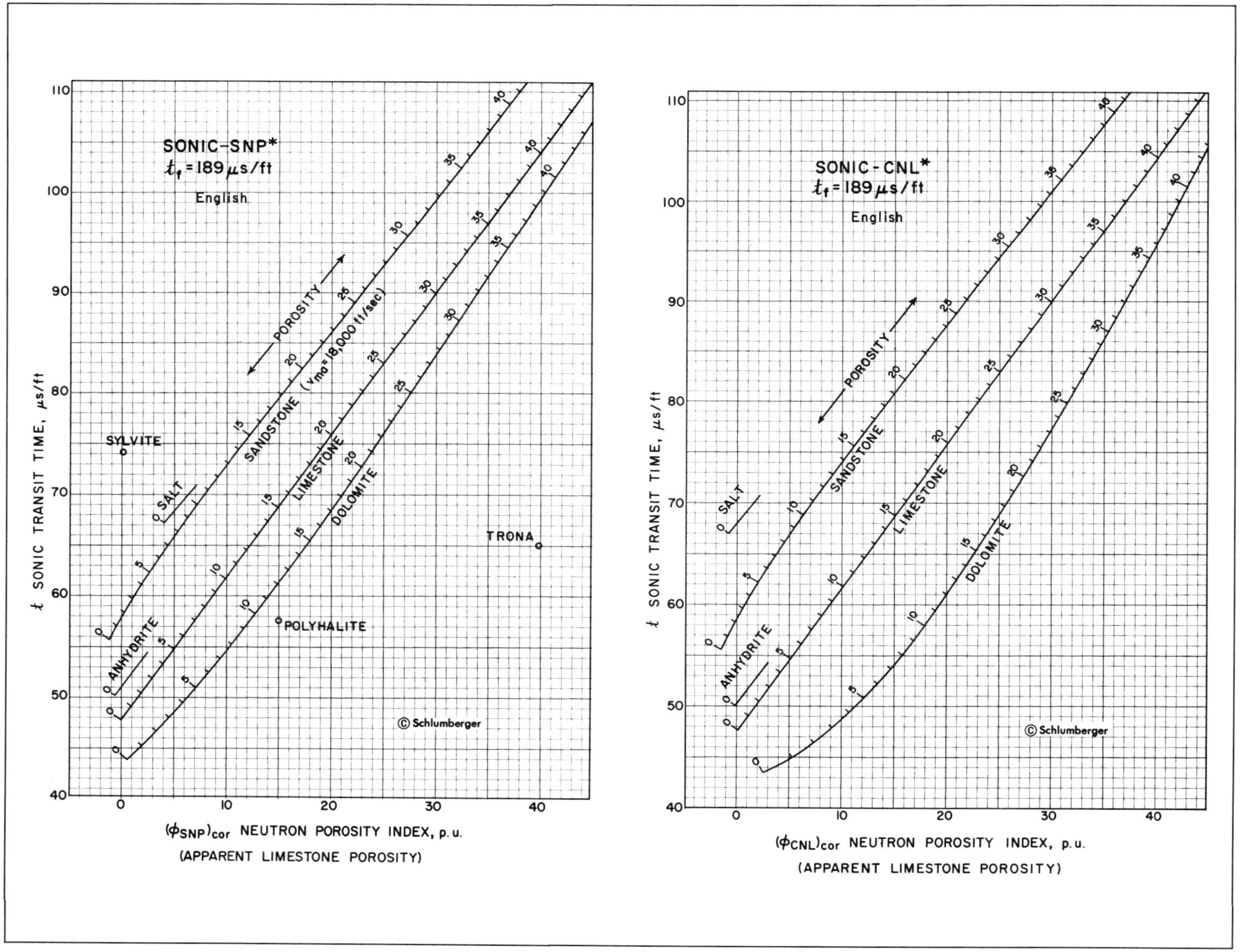


**Figure 3–50.** Schlumberger Sonic-Neutron (SNP) and Sonic-Neutron (CNL) crossplots (courtesy of Schlumberger).[4]

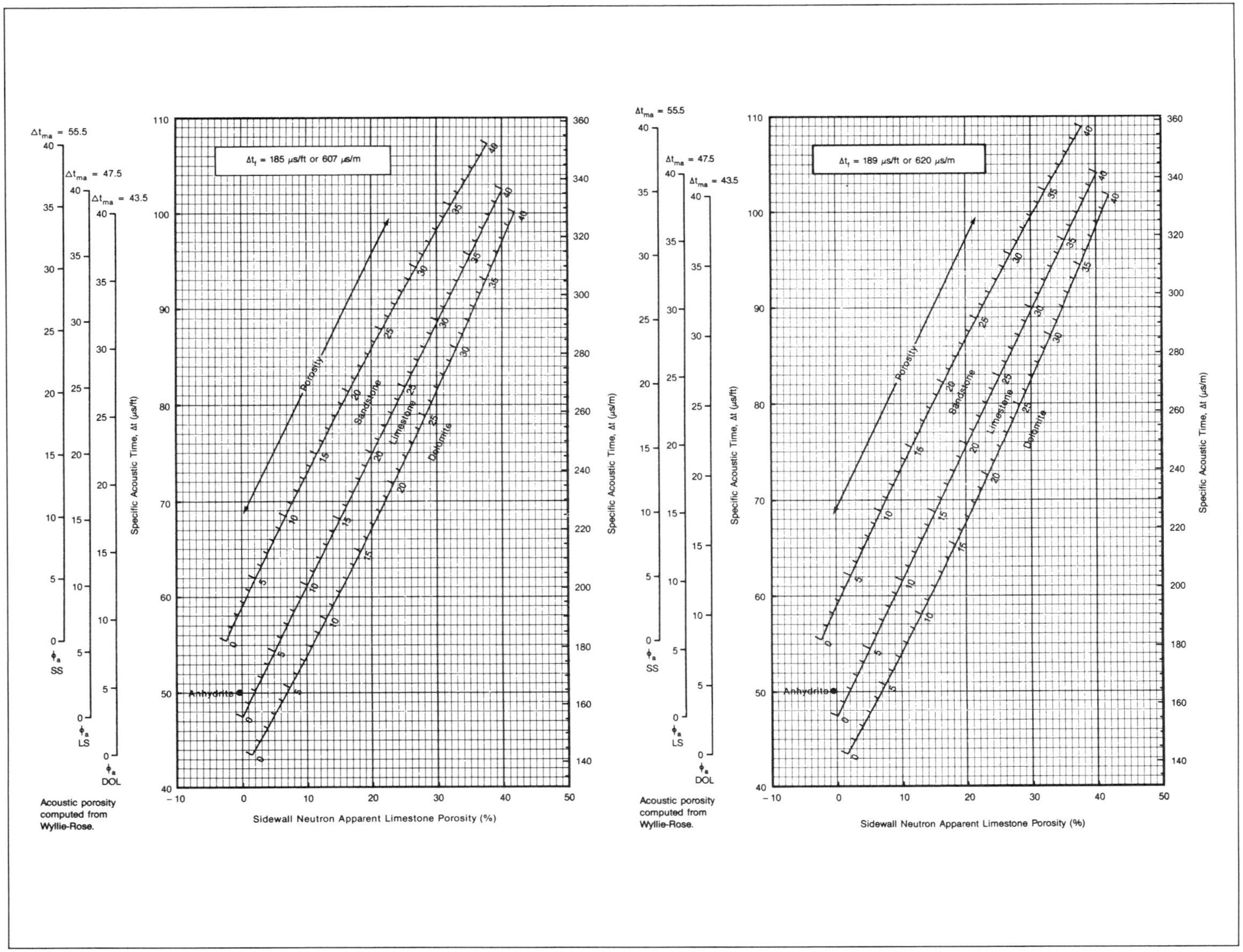


**Figure 3–50a.** AWS Sonic-Neutron (SNP) crossplots (courtesy of Atlas Wireline Services).[6]

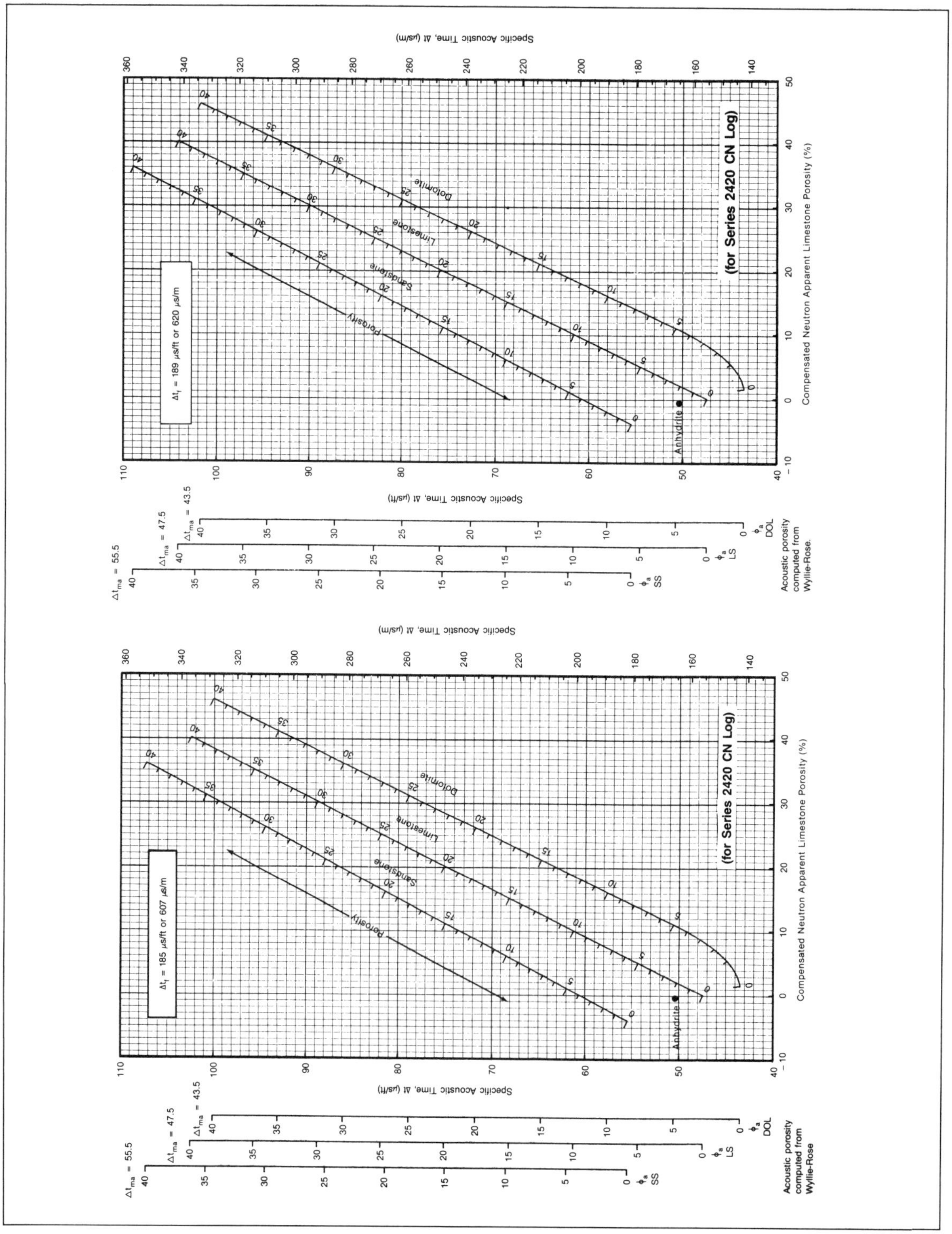


**Figure 3–50b.** AWS Sonic-Neutron (CNL) crossplots (courtesy of Atlas Wireline Services).[6]

In Case I, density-neutron crossover occurs and is recognized by mirror imaging of the two logs. The density-neutron crossplot presented in Figures 3–48 and 3–48a are primarily used for Case 1 correction. A close approximation for the crossplot solution is given by the following equation:

$$\phi_{CP} = \sqrt{\frac{\phi_D^2 + \phi_N^2}{2}} \quad \textbf{(3.29)}$$

A more accurate correction chart is shown in Figure 3–51. The chart grid porosities are for an average depth of about 3,000 ft. Correction is required beyond that depth.

Case II is the situation where the gas crossover occurs but the two logs do not form a mirror image of each other as shown in Figure 3–52. This response is because of the different depth of investigation of each log; i.e., the density log is measuring the invaded zone (in deep invasion), and the neutron log is reading mostly behind the invaded zone. Consequently the two logs see different gas saturation.

For Case II, density log should be used for porosity and neutron log should be ignored. The Case I correction applied to Case II will adversely affect the formation evaluation, such as eliminating a productive zone.

Case III is most commonly encountered in shaly formations where the logs will not show any evidence if the gas is present. In such cases it is recommended to plot ($\phi_N - \phi_D$) versus gamma ray (API units) on Cartesian coordinate paper. If gamma ray log and the density-neutron logs respond appropriately to the shale content of the formation, the plot will essentially be a straight line. Gas points will fall off this straight line region to a position that could be easily distinguished.

## POROSITY ESTIMATION IN HYDROCARBON-BEARING FORMATION

When a combination of density-neutron logs is available, the two porosities could be used to estimate hydrocarbon density and porosity of hydrocarbon-bearing formations.[10–11]

Figure 3–53 presents charts which are used to estimate hydrocarbon density and porosity corrected for hydrocarbon effect. The charts include approximate correction for excavation effects. The charts may lose accuracy in some extreme cases when $\rho_h < 0.25$ (gases).

**Example 3–15.** A hydrocarbon-bearing formation showed $\phi_D = 30\%$ and $\phi_{CNL} = 10\%$, if the $S_{xo}$ calculated from the microlaterolog in the same zone showed $S_{xo} = 60\%$. What is the hydrocarbon density of the hydrocarbons in the formation and the corrected porosity of the formation? Assume that $\phi_D$ and $\phi_N$ are previously corrected for environmental effects.

Solution:

$$S_{hr} = 1 - S_{xo}$$
$$= 1 - 0.60 = 0.40$$
$$\phi_{Ncor}/\phi_{Dcor} = 10/30 = 0.33$$

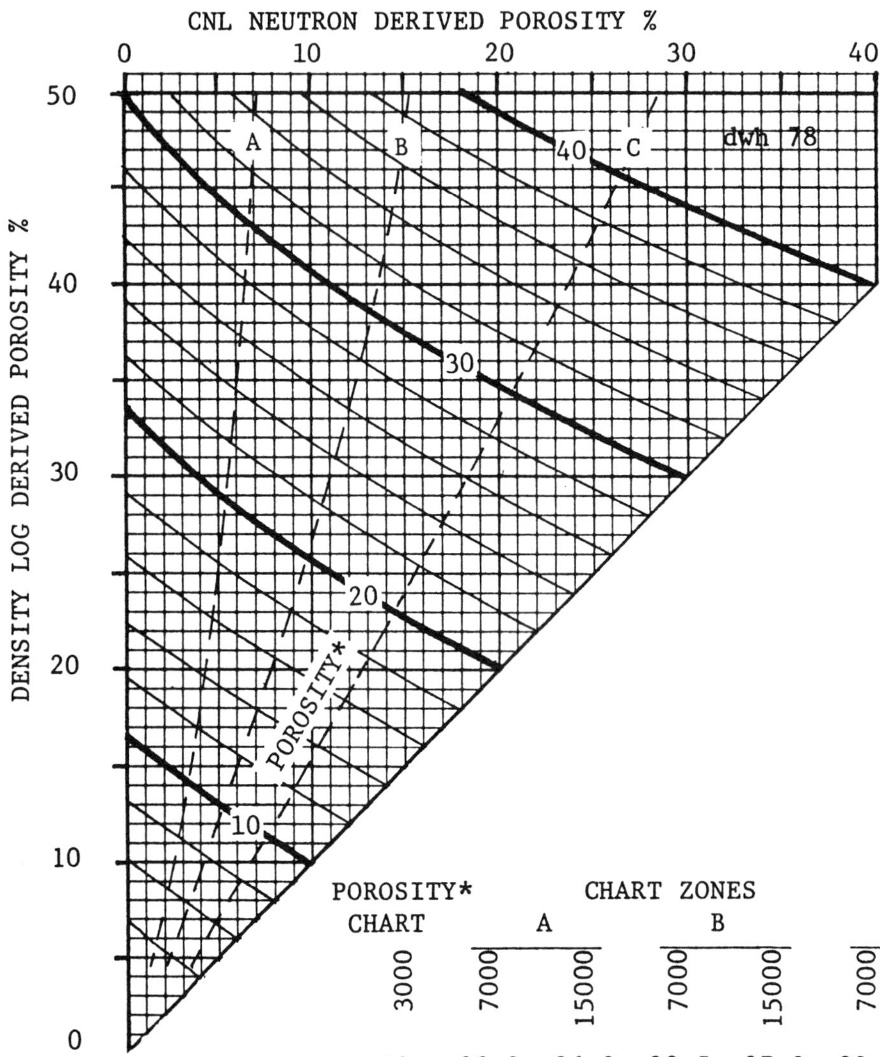


| POROSITY* CHART | CHART ZONES A | | B | | C | |
|---|---|---|---|---|---|---|
| 3000 | 7000 | 15000 | 7000 | 15000 | 7000 | 15000 |
| 30 | 28.0 | 26.0 | 28.5 | 27.0 | 29.3 | 28.5 |
| 20 | 18.8 | 17.5 | 19.0 | 18.0 | 19.5 | 19.0 |
| 10 | 9.5 | 9.0 | 9.5 | 9.0 | 9.3 | 9.5 |
| 0 | 0 | 0 | 0 | 0 | 0 | 0 |

Approximate Correction for Depth

**Figure 3–51.** Gas effect correction chart for Density-Neutron logs (courtesy of D. W. Hilchie, Inc., Boulder, Colorado).[1]

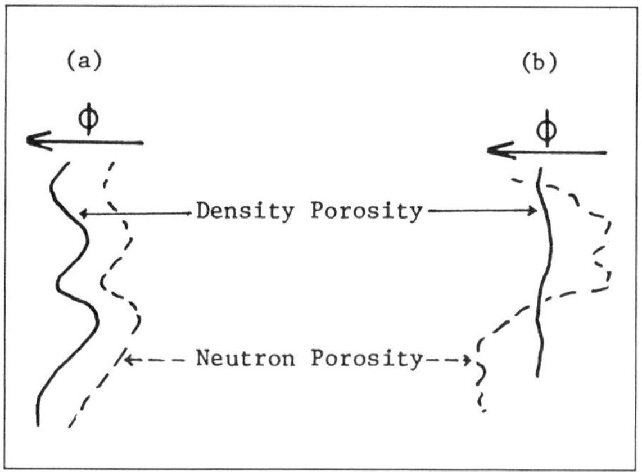


**Figure 3–52.** A schematic of density-neutron logs response in gas zone—Case II.

From Figure 3–53,

$$\rho_h = 0.15$$
$$\phi = 25.6$$
$$\Delta\phi = 1.20$$

Therefore, $\phi_{cor} = 25.6 - 1.2 = 24.40\%$



## FORMATION WATER RESISTIVITY ($R_w$) DETERMINATION

Resistivity of the formation water ($R_w$) is an important parameter required for the calculation of water and/or hydrocarbon saturation. The formation water, also referred to as the connate water or interstitial water, is the water that exists in the uncontaminated portion of the porous formation rock.

In the beginning of this chapter, we discussed the use of SP log for calculating $R_w$. In this section we will discuss several other sources which can be utilized for $R_w$ determination.

One other term that is used extensively in log interpretation is denoted by $R_o$, which is the resistivity of a 100% water-saturated formation. The two resistivities $R_w$ and $R_o$ are related to each other by a *formation resistivity factor*, $F_R$, as shown below.

$$F_R = \frac{R_o}{R_w} \tag{3.30}$$

OR $R_o = F_R \times R_w$

The formation resistivity factor is mathematically related to porosity by the following equation:

$$F_R = \frac{a}{\phi^m} \tag{3.31}$$

The most commonly used versions of this equation are the *Archie Equation* with $a = 1$ and $m = 2$, and the *Humble Equation* with $a = 0.62$ and $m = 2.15$. The Archie equation is used for carbonates and the Humble equation for sandstones. Additional relationships for calculating formation factor used in the industry are listed in Table 3–7. The $a$ and $m$ for a specific formation may also be obtained in the laboratory.

### $R_w$ FROM WATER CATALOG

In many oil-producing regions, water catalogs have been published which list the $R_w$ for many formation waters collected from different fields and different producing horizons of the region. The $R_w$s in the water catalog are accumulated from several different sources. These sources include (1)

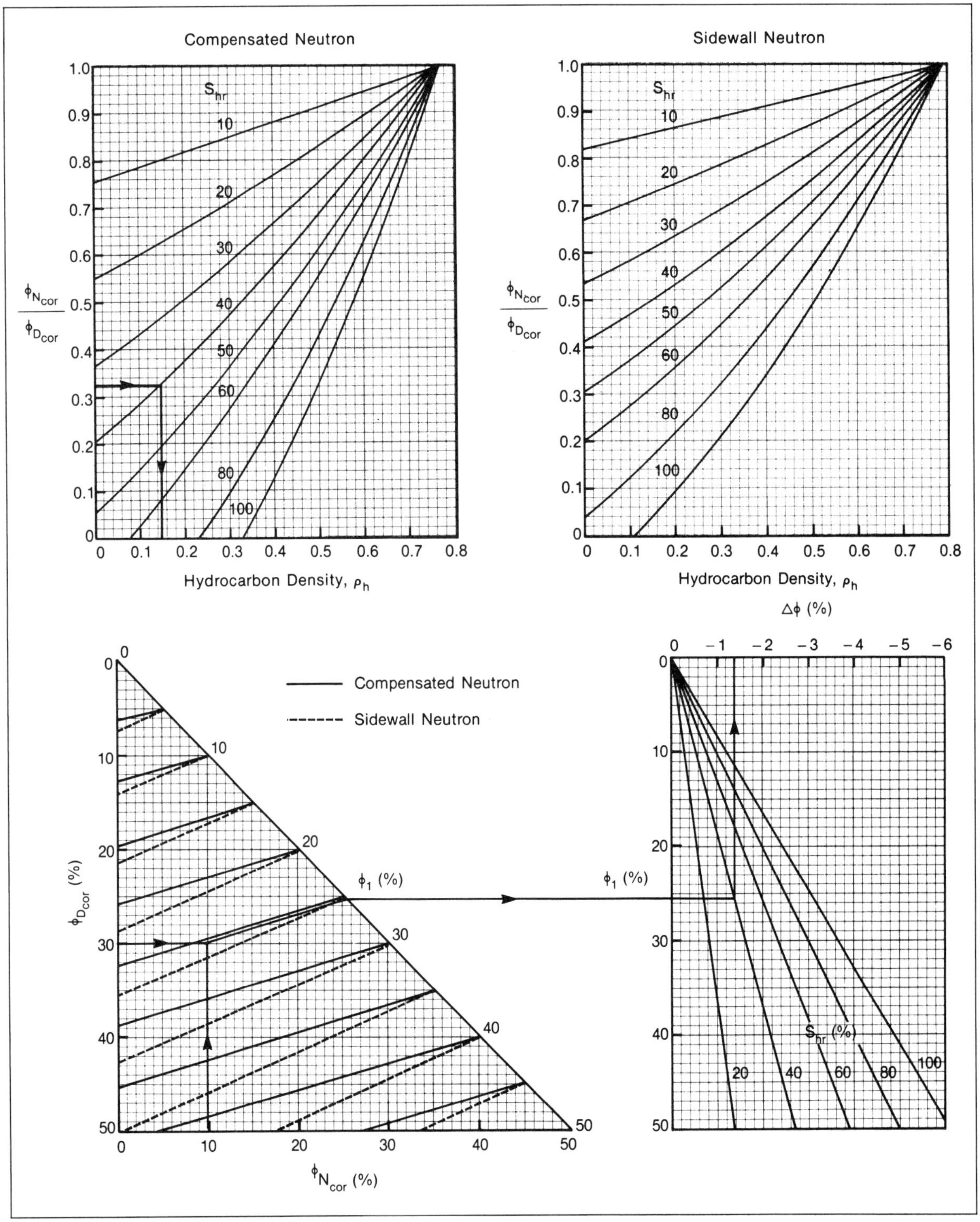


**Figure 3–53.** Charts for estimating hydrocarbon density in clean sands and porosity in hydrocarbon-bearing formations (courtesy of Atlas Wireline Services).[6]

**TABLE 3–7. Different coefficients and exponents used to calculate $F_R$.[12–14]**

| Equation | Formation |
|---|---|
| $F_R = 1.45/\phi^{1.54}$ | for average sands |
| $F_R = 1.65/\phi^{1.33}$ | for shaly sands |
| $F_R = 1.45/\phi^{1.7}$ | for calcareous sands |
| $F_R = 0.85/\phi^{2.14}$ | for carbonates |
| $F_R = 2.45/\phi^{1.08}$ | for Pliocene sands, CA |
| $F_R = 1.97/\phi^{1.29}$ | for Miocene sands, TX, LA, GC |
| $F_R = 1/\phi^{(2.05-\phi)}$ | for clean granular formations |

water samples obtained from actual production, (2) production test, (3) drillstem test, or (4) well logs.

These catalogs are compiled by U.S.G.S., by oil companies, by other government entities, or by educational groups. The $R_w$s obtained from other sources may be compared with those from the water catalog if one is available for the area.

## $R_w$ FROM CHEMICAL ANALYSIS

Formation resistivity can be measured directly (preferred method), or it can be estimated from analysis of the water sample. The latter method uses weighting coefficients to convert concentrations of individual ions into equivalent NaCl concentrations.

Figure 3–54 provides multipliers for converting various ions to electrically equivalent sodium chloride (NaCl) concentration. The electrically equivalent NaCl concentration is the weighted sum of individual ion concentration. Once the equivalent NaCl concentration is obtained, the $R_w$ can be obtained using Figure 3–6 or the following equation:

$$R_{w\ @75°F} = 0.0123 + \frac{3{,}647.5}{[\text{NaCl}\ \ (\text{ppm})]^{0.955}} \tag{3.32}$$

**Example 3–16.** A brine with 50,000 ppm total solids includes 10,000 ppm Na, 16,000 ppm Cl, 7,000 ppm Mg, 5,000 ppm Ca, and 12,000 ppm $SO_4$. Determine the equivalent NaCl concentration and $R_w$ at 75°F for this sample.

| Ion (1) | Concentration (ppm) (2) | Multiplier (Fig. 3–54) (3) | Eq. NaCl Concentration (4) = (2 × 3) |
|---|---|---|---|
| Na | 10,000.00 | 1.000 | 10,000.00 |
| CL | 16,000.00 | 1.000 | 16,000.00 |
| Mg | 7,000.00 | 0.922 | 6,454.00 |
| Ca | 5,000.00 | 0.778 | 3,890.00 |
| $SO_4$ | 12,000.00 | 0.386 | 4,632.00 |
| Total | 50,000.00 | | 40,976.00 |

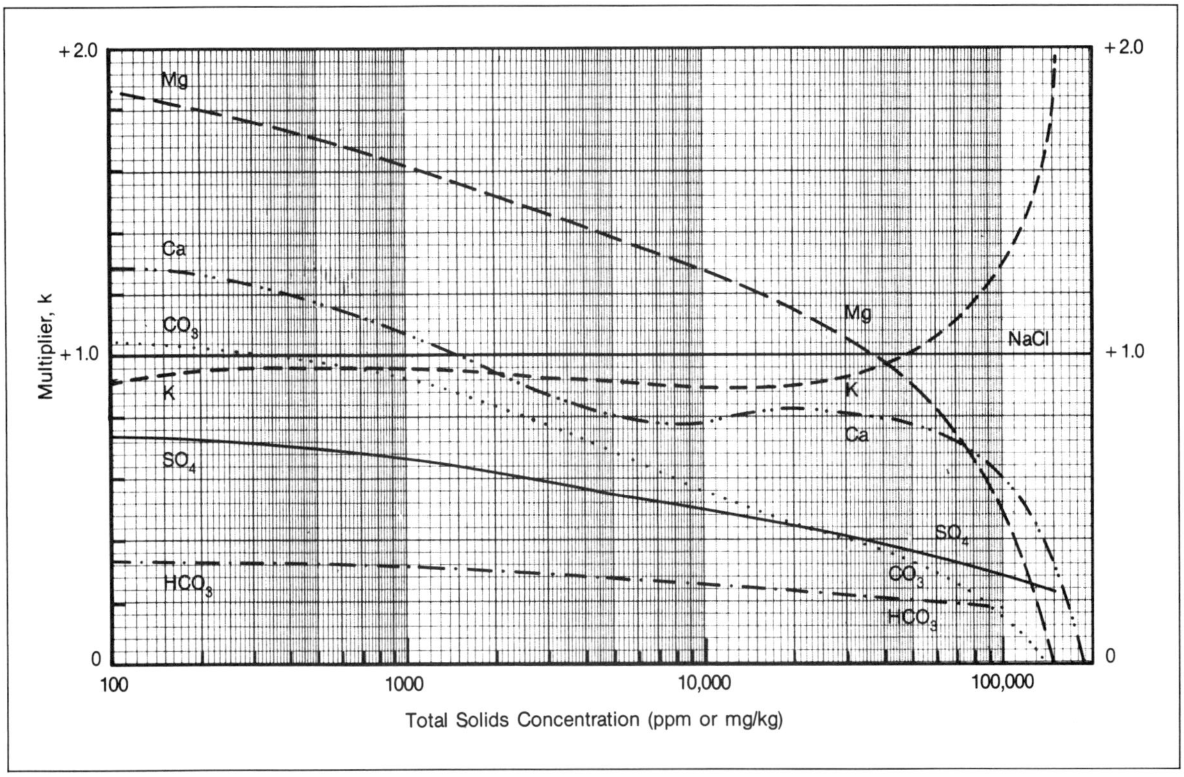


**Figure 3–54.** Equivalent NaCl concentrations from ionic concentrations (courtesy of Atlas Wireline Services).[6]

Solution: Using Equation (3.32),

$$
\begin{aligned}
R_{w\ @75^\circ\mathrm{F}} &= 0.0123 + \frac{3{,}647.5}{[\mathrm{NaCl}\ \ (\mathrm{ppm})]^{0.955}} \\
&= 0.0123 + \frac{3{,}647.5}{40{,}976^{0.955}} \\
&= 0.156 \quad \text{ohm-m}
\end{aligned}
$$

## $R_w$ FROM RESISTIVITY-POROSITY CROSSPLOT

Two different resistivity-porosity crossplots (RPC) generally used in log interpretation are presented in this section. The two techniques solve for different unknowns. In the industry these techniques are referred to as (1) the Pickett Crossplot method and (2) the Hingle Crossplot.[16]

**The Pickett Crossplot.** This crossplot involves plotting $\phi$ versus resistivity $R_t$ or $R_{deep}$ on a log-log graph paper. The technique is used to determine (1) formation water resistivity ($R_w$), (2) cementation factor ($m$), (3) matrix parameters for porosity logs ($\Delta t_{ma}$ and $\rho_{ma}$), and (4) water saturations.

In addition to plotting true porosity versus ($R_{ILD}$ or $R_{LLD}$), $\Delta t_{LOG} - \Delta t_{ma}$ or $\rho_{ma} - \rho_b$ may also be plotted on the y-axis. This combination helps in determining a value for formation matrix ($\Delta t_{ma}$ or $\rho_{ma}$).

The Pickett plot utilizes the Archie equation to determine water saturation ($S_w$). The Archie equation is

$$S_w^n = \frac{aR_w}{\phi^m R_t} = \frac{R_o}{R_t} \tag{3.33}$$

Rearrangement of Equation (3.33) gives a straight-line equation (on log-log graph paper) as given below.

$$\log R_t = -m \log \phi + \log (aR_w) - n \log S_w \tag{3.34}$$

If $S_w = 100\%$, Equation (3.34) reduces to

$$\log R_t = -m \log \phi + \log (aR_w) \tag{3.35}$$

The Pickett plot requires the following steps:

1. Tabulate the porosity and the corresponding resistivity values for the expected hydrocarbon-bearing formations and the possible nearby 100% water-saturated formations. Use crossplot porosities if two porosity logs are run.
2. Plot the data on a log-log (two cycle by three cycle preferred) graph paper.
3. On the plot, a zone with constant $R_w$, $m$, and $S_w = 100\%$ will have data points plotted along a single straight-line trend. Draw a best-fit straight line through these points. This straight-line trend represents the $R_o$ line.
4. Determine the slope of the $R_o$ line. The slope of this line is equal to the cementation factor ($m$). The $R_t$ at which this line intersects $\phi = 100\%$ represents $aR_w$.
5. Data plotted above the $R_o$ line represent water saturation values less than 100%. This is true only when $R_w$ and $m$ are both constant.
6. A quick look assessment of a zone's water saturation can be made by drawing various water saturation lines parallel to the $R_o$ line. The position of the various water saturation lines is determined using Equation (3.33).
7. To construct additional water saturation lines on the graph (a) draw a horizontal line to the right, beginning at the point where the $R_o$ line crosses the porosity scale ($\phi = 0.1, 1.0, 10.0$, etc.) (i.e., $R_t = R_o = 0.1, 1.0, 10.0$, etc.); (b) on this line mark the $R_t$s which correspond to the indicated water saturations. Remember that for $R_o = 1.0$, $R_t = 1/S_w^n$ or $R_t = R_o/S_w^n$.

When $\Delta t_{LOG} - \Delta t_{ma}$ or $\rho_{ma} - \rho_b$ are plotted versus $R_t$ using incorrect matrix value, the $R_o$ line for $\Delta t_{LOG} - \Delta t_{ma}$ or $\rho_{ma} - \rho_b$ versus $R_t$ plot will not plot as a straight line, but will curve. In such cases several matrix values ($\Delta t_{ma}$ or $\rho_{ma}$) are tried until the $R_o$ line is a straight line. This trial and error helps us determine the correct matrix value for the formation.

**Example 3–17.** Construct a Pickett plot for the zones shown on the log in Figure 3–55. Determine $m$ and $R_w$ from the plot and $S_w$ for the nine zones. Assume $\Delta t_{ma}$ = 48 μsec/ft, and $\Delta t_f$ = 189 μsec/ft.



Solution: The log data for the nine zones in Figure 3–55 are given in Table 3–8. The log-log crossplot for the data in Table 3–8 is shown in Figure 3–56. From Figure 3–56,

$$m = 2.02$$
$$aR_w = 0.0225 \quad \text{ohm-m}$$

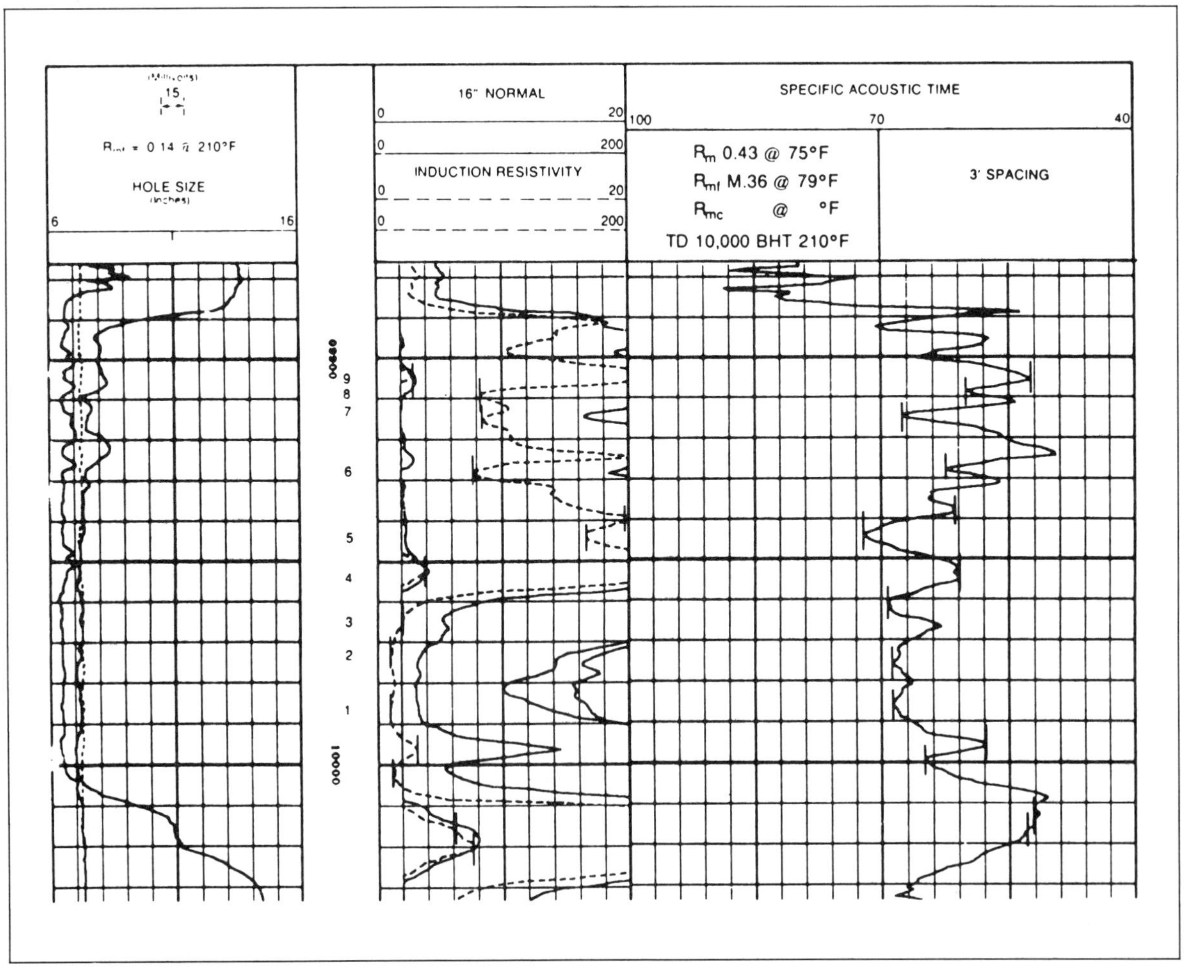


**Figure 3–55.** Log section for Example 3–17 (courtesy of Atlas Wireline Services).[5]

**TABLE 3–8. Log data for the log in Figure 3–55**

| Zone | $\Delta t_{LOG}$ | $R_{ILD}$ | $\Delta t_{LOG} - \Delta t_{ma}$ | $\phi$ |
|---|---|---|---|---|
| 1 | 68.8 | 1.00 | 21.3 | 0.15 |
| 2 | 68.5 | 1.10 | 21.0 | 0.15 |
| 3 | 69.5 | 2.10 | 22.0 | 0.16 |
| 4 | 61.0 | 4.00 | 13.5 | 0.10 |
| 5 | 72.3 | 16.5 | 24.8 | 0.18 |
| 6 | 62.5 | 7.50 | 15.0 | 0.11 |
| 7 | 67.8 | 8.30 | 20.3 | 0.14 |
| 8 | 60.0 | 8.20 | 12.5 | 0.09 |
| 9 | 52.0 | 28.0 | 4.50 | 0.03 |



To find the $S_w$ for the nine zones, we construct water saturation lines for $S_w$ = 80%, 60%, 40%, 20%, and 10% on Figure 3–56, using $\phi$ = 10% and $R_t$ as given below.

| $\phi$ | $S_w$ | $R_o$ | $R_t = R_o/S_w^2$ |
|---|---|---|---|
| 0.10 | 80 | 2.4 | 3.75 |
| 0.10 | 60 | 2.4 | 6.67 |
| 0.10 | 40 | 2.4 | 15.00 |
| 0.10 | 20 | 2.4 | 60.00 |
| 0.10 | 10 | 2.4 | 240.00 |

**The Hingle Crossplot.** The Hingle crossplotting technique is a powerful technique allowing a long section of well logs to be analyzed in a minimum of time. The advantage of Hingle plot over the Pickett plot is that a matrix value does not have to be known or assumed. It is determined from the crossplot.

If very little is known about a formation to be analyzed, both the Pickett and Hingle crossplots can be used together. The Pickett plot gives us $m$ and the Hingle plot gives us the matrix value and $R_w$. However, the Hingle plot requires a special graph paper. Figures 3–57 and 3–58 show two graph papers for $F_R = 0.62/\phi^{2.15}$ (sandstone) and $F_R = 1/\phi^2$ (carbonates) respectively.

The following steps will help in constructing the Hingle plot and determine $S_w$, $R_w$, and appropriate matrix value:

1. Select the appropriate graph paper. A special graph paper could be constructed if $a$ and $m$ are different than those in Figures 3–57 and 3–58.
2. Scale the $x$-axis on a linear scale, using $\Delta t_{LOG}$, $\rho_b$, $\phi_N$, $\phi_D$, or $\phi_{CP}$.
3. Plot deep resistivity values on $y$-axis corresponding to the respective porosity values. The resistivity and conductivity scale may be changed to facilitate plotting of any particular set of data. The scale can be changed by dividing or multiplying it by a constant.

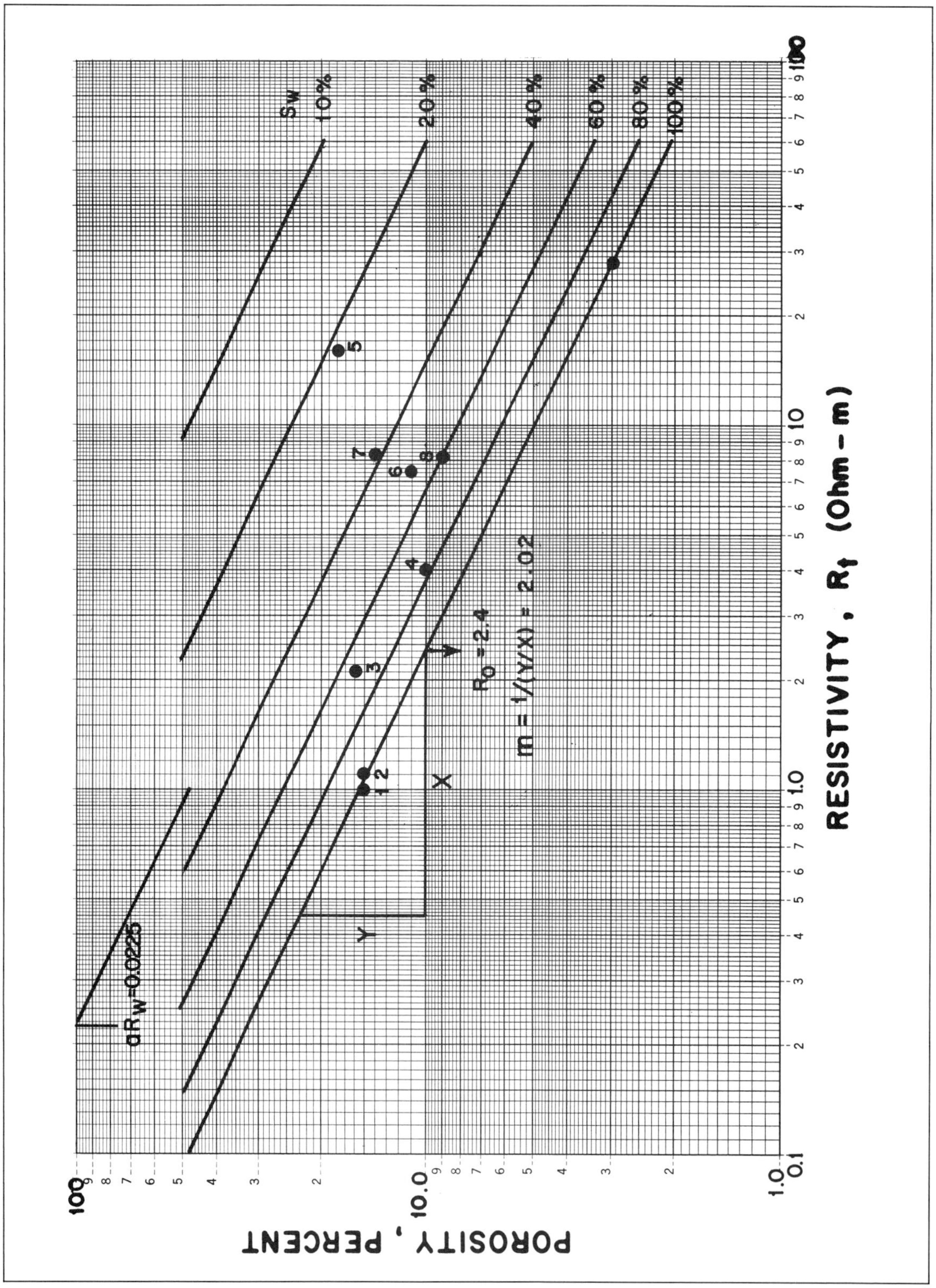


**Figure 3–56.** Resistivity-porosity crossplot (Pickett) for the log in Figure 3–55.

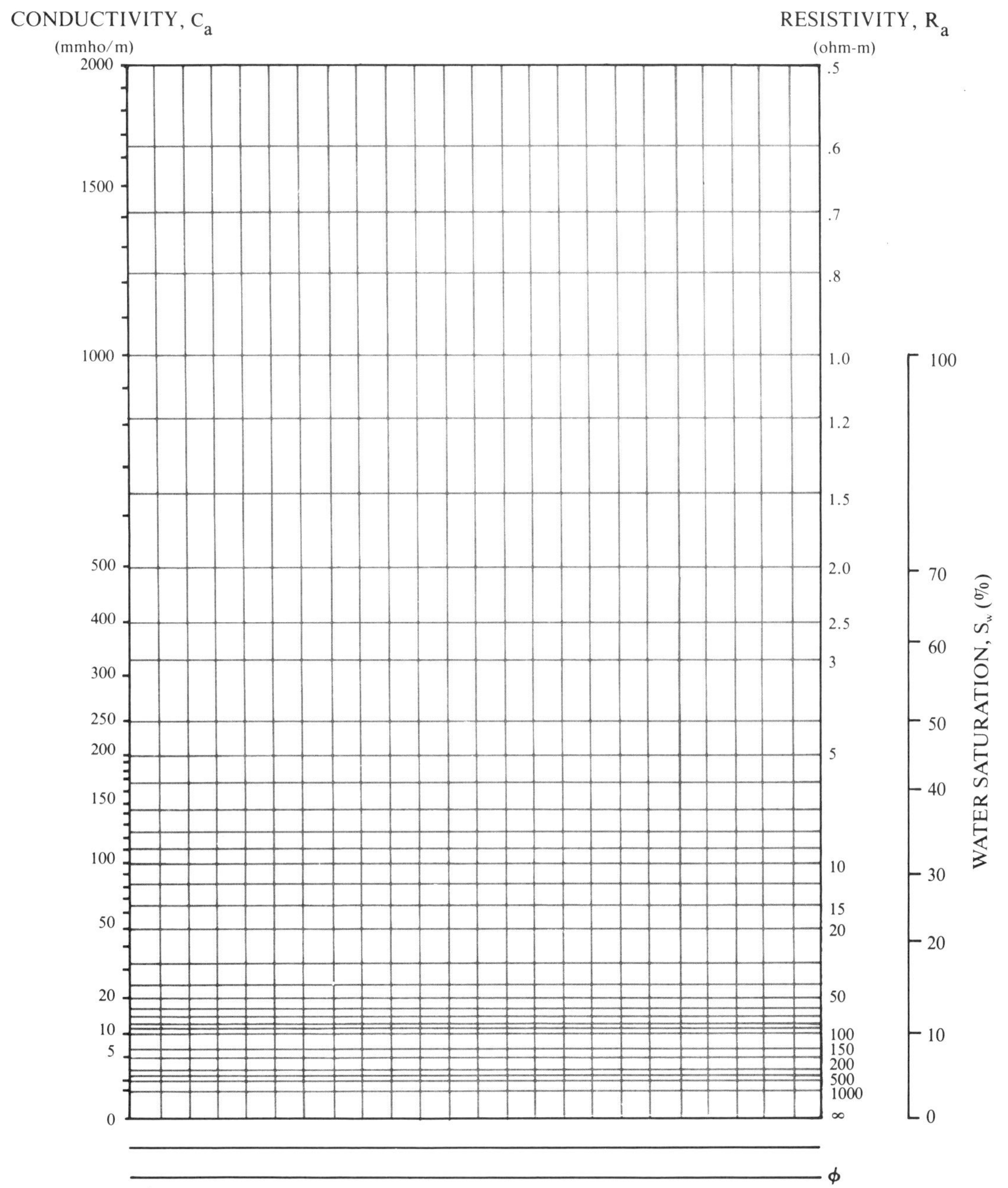


**Figure 3–57.** Resistivity-porosity crossplot graph paper for sandstones (courtesy of Atlas Wireline Services).[6]

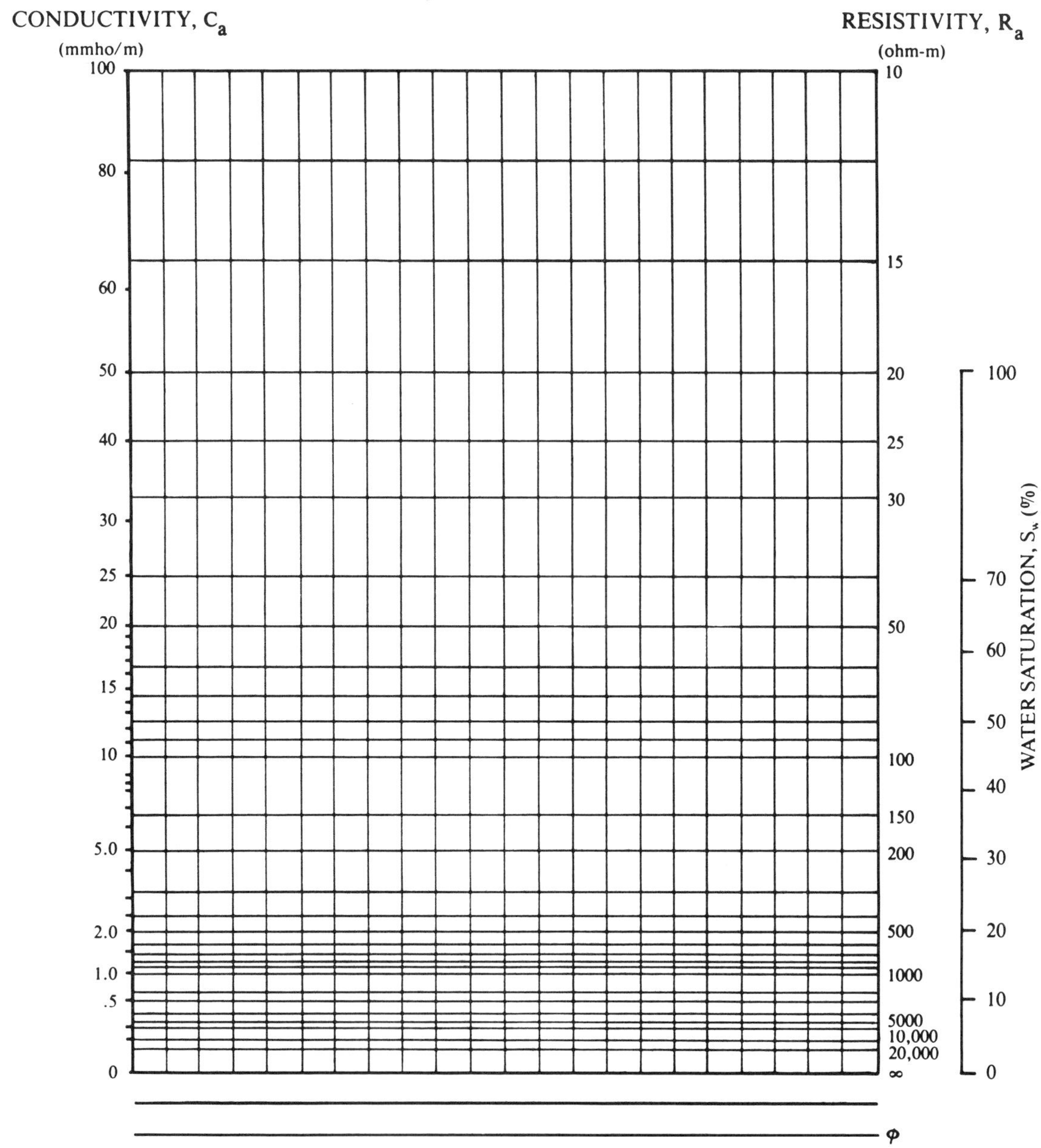


**Figure 3–58.** Resistivity-porosity crossplot graph paper for carbonates (courtesy of Atlas Wireline Services).[6]

4. Construct a straight line through the most northwesterly points, and extrapolate the line to infinite resistivity ($R_t = \infty$). This line represents the $R_o$ line or the $S_w = 100\%$ line.
5. The intersection point of the *x*-axis and the $R_o$ line represent $\phi = 0\%$ and the correct formation matrix value. The scale should then be readjusted accordingly.
6. Using the formula $R_w = R_o/F_R$ and any corresponding set of $\phi$ and $R_o$ values, calculate $R_w$.
7. Construct the lines of constant $S_w$ using $S_w = \sqrt{R_o/R_t}$ for any given porosity value. Note that all $S_w$ lines will converge at the matrix point ($\phi = 0$ and $R_t = \infty$). These lines are valid only if $R_w$ is constant throughout the interval whose resistivity and porosity values are plotted.

**Example 3–18.** For the logs in Example 3–17, plot the data from the nine zones on appropriate Hingle crossplot graph paper. Determine $R_w$ and the matrix travel time $\Delta t_{ma}$. Also construct the $S_w$ lines for $S_w = 80\%, 60\%, 40\%, 20\%$, and $10\%$.

Solution: We will use the graph paper for carbonates as given in Figure 3–58.

The data plotted on the Hingle crossplot are shown in Figure 3–59. Note that the $R_o$ line drawn through Points 1, 2, and 9 intersects the *x*-axis at $\Delta t_{ma} = 48$ μsec/ft.

$R_o$ (at $\phi = 7.5\%$) $= 3.0$ ohm-m

$$
\begin{aligned}
F_R &= 1/\phi^2 \\
&= (0.075)^{-2} \\
&= 177.78 \\
R_w &= R_o/F_R \\
&= 3.0/177.78 \\
&= 0.02 \quad \text{ohm-m}
\end{aligned}
$$

The $R_w$ is in close agreement with the $R_w$ determined in Example 3–17. The $S_w$ lines are drawn using the point at $\phi = 0$ and ($\phi = 7.5$ and $R_t = R_o/S_w^2$).

| $S_w$ | $\phi$ | $R_o$ | $R_t = R_o/S_w^2$ |
|---|---|---|---|
| 80 | 7.5 | 3.0 | 4.69 |
| 60 | 7.5 | 3.0 | 8.33 |
| 40 | 7.5 | 3.0 | 18.75 |
| 20 | 7.5 | 3.0 | 75.00 |
| 10 | 7.5 | 3.0 | 300.00 |

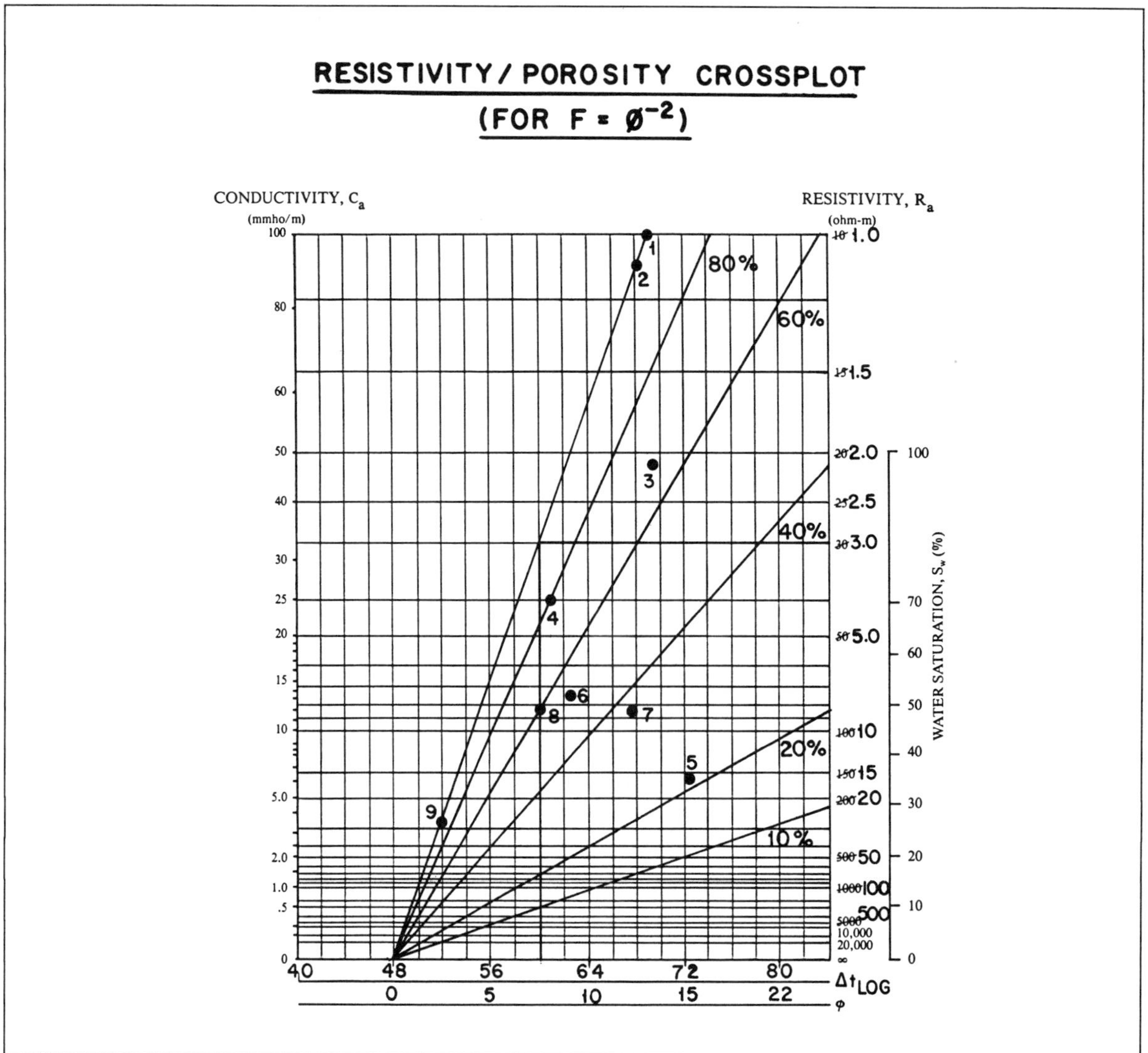


**Figure 3–59.** Resistivity-porosity crossplot (Hingle) for data in Table 3–8.

## $R_w$ FROM $R_{wa}$–SP PLOT[3]

A plot of $R_{wa}$ ($R_{wa} = R_t/F_R$) on logarithmic scale versus SP deflection on linear scale can be used to estimate $R_w$. This plot is much valuable at shallow depths when $R_w$ can change rapidly. The rapidly changing $R_w$ makes it difficult to determine $R_w$ from SP log.

Figure 3–60 is a plot of $R_{wa}$ versus SP from an interval where $R_w$ changes rapidly. The $R_{wa}$ and the corresponding SP values when plotted on semilogarithmic graph paper will show a trend parallel to and just above the $R_w$ line, $R_w$ as computed from the SP curve. This indicates slightly greater than average amounts of salts other than NaCl. The hydrocarbon-bearing formations are identified by data points falling above the main trend, such as the Points 5 and 11 in Figure 3–60.

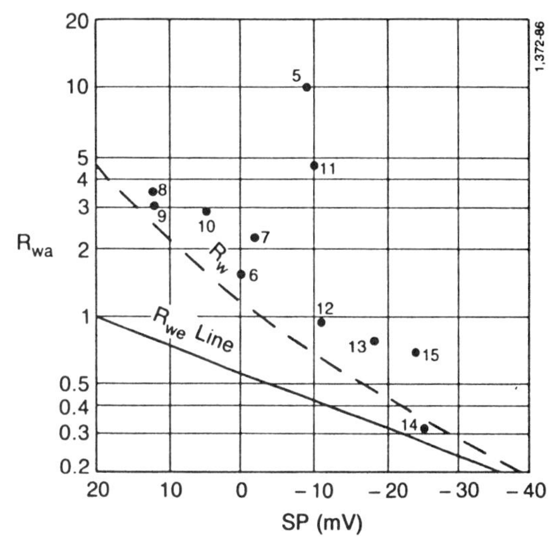


**Figure 3–60.** $R_{wa}$ vs. SP for determining $R_w$ and distinguishing hydrocarbon-bearing zones in intervals of changing $R_w$ (courtesy of Schlumberger).[3]

### $R_w$ FROM $R_{xo}$ AND $R_t$[3]

The $R_w$, $R_{xo}$, and $R_t$ are related to each other by the following equation:

$$\frac{R_{xo}}{R_t} = \left(\frac{S_w}{S_{xo}}\right)^2 \frac{R_{mf}}{R_w} \qquad (3.36)$$

In 100% water saturated zones, $S_w = S_{xo} = 1.0$ (i.e., 100%), and $R_{xo}/R_t$ is equal to $R_{mf}/R_w$. Thus by computing $R_{xo}/R_t$ over an interval containing clean, invaded-water sands, the value of $R_{mf}/R_w$ can be found. Then, knowing $R_{mf}$, $R_w$ can be determined.

A similar result can be obtained by plotting $R_{xo}$ values (from microspherically focused log or Proximity log) versus the corresponding $R_t$ values (from a deep resistivity log) over an interval containing some water sands. A line through the points corresponding to the highest values of $R_{xo}/R_t$ provides an estimate of $R_{mf}/R_w$.

## OPENHOLE LOG INTERPRETATION—CLEAN FORMATIONS

In this section we will discuss the most commonly used method of quantitative log interpretation, called the $R_{wa}$ method. Real-life logs and their step-by-step interpretation will be provided. Also included in this section are some qualitative techniques referred to as the *Quick Look* interpretation techniques. These techniques are helpful because they provide *flags* which point to possible hydrocarbon zones requiring further investigation.

Resistivity ratio methods are also presented. When no porosity or formation factor data are available, the resistivity ratios are used to calculate water saturations. The use of ratio methods is dependent upon good $R_{xo}$ and $R_t$ values (i.e., invasion must be deep enough to allow a shallow investigating resistivity device to measure $R_{xo}$ but not so deep that a deep resistivity device cannot measure $R_t$).

### $R_{wa}$ METHOD

The $R_{wa}$ method is probably the single most used analysis approach for analyzing openhole well logs. The Archie equation is used, but it is divided into several steps for convenience and systematic analysis.

The Archie equation for calculating water saturation ($S_w$) in the uninvaded zone is

$$S_w = \left(\frac{a}{\phi^m} \times \frac{R_w}{R_t}\right)^{1/n} \tag{3.37}$$

and for calculating water saturation ($S_{xo}$) in the invaded zone is



$$S_{xo} = \left(\frac{a}{\phi^m} \times \frac{R_{mf}}{R_{xo}}\right)^{1/n} \tag{3.38}$$

As previously mentioned, $a$ is the tortuosity factor, $m$ is the cementation factor, and $n$ is the saturation exponent which varies from 1.8 to 2.5 but is normally equal to 2.0. Water saturation of the flushed zone ($S_{xo}$) can be used as an indicator of hydrocarbon movability. If $S_{xo} >> S_w$, then hydrocarbons in the flushed zone have probably been moved or flushed out of the zone in the vicinity of the borehole by the drilling mud.

The following steps provide a systematic approach for the analysis of openhole well logs in clean formations:

1. Correlate the porosity and resistivity logs. This is necessary because the logs may be off depth. Many times, failure to put both logs on the same depth reference will completely invalidate the interpretation.
2. Pick out the permeable beds. This can be done using SP log, gamma ray log, separation of $R_i$ and $R_t$ logs, separation of microlog, or a combination of all these logs.
3. Mark the zones to be analyzed on the logs. Read the resistivity and porosity (or $\Delta t_{LOG}$, $\rho_b$, etc.) for each zone to be analyzed. Make the necessary corrections to the resistivity and porosity values to obtain $\phi_{cor}$ or $\phi_{CP}$ and $R_t$.
4. Note that if there is a wide variation in the porosity and resistivity values in a particular zone, read the values every 2 ft. Avoid too much averaging of the log responses.
5. Calculate $R_w$ from SP log if possible, or obtain it from other sources previously discussed.
6. Calculate formation resistivity factor $F_R = 1/\phi^2$ (carbonates) and $F_R = 0.62/\phi^{2.15}$ (sandstones). If other values for $a$ and $m$ for the specific formation are available, use those. Other relationships are given in Table 3–7.
7. Calculate apparent water resistivity ($R_{wa}$) by dividing $R_t$ by $F_R$ (i.e., $R_{wa} = R_t/R_R$ for each interval for which $\phi$ and $R_t$ are obtained).
8. Compare the lowest $R_{wa}$ (lowest $R_{wa} = R_w$) in Step 7 with the $R_w$s in Step 5, use the lowest $R_w$ of the two.
9. Calculate water saturation $S_w$ for each interval; $S_w = \sqrt{R_w/R_{wa}}$.
10. Calculate the average $\phi$ and $S_w$ using the averaging equations given in Chapter 2.

In some cases $R_{wa}$ curve can be developed and presented with SP or GR log. This curve will help in quickly identifying the zones (with $R_{wa} \geq R_w$) for further investigation. The development of this curve requires that $R_w$ should be consistent or vary in a consistent and recognizable manner over the interpreted interval.

The following examples will clarify the steps involved in log analysis and the $R_{wa}$ technique.

**Example 3–19.** A set of logs presented in Figure 3–61 have several zones marked on the logs. Using the $R_{wa}$ technique, calculate water saturation for each zone. The well was drilled with nonsalt mud.



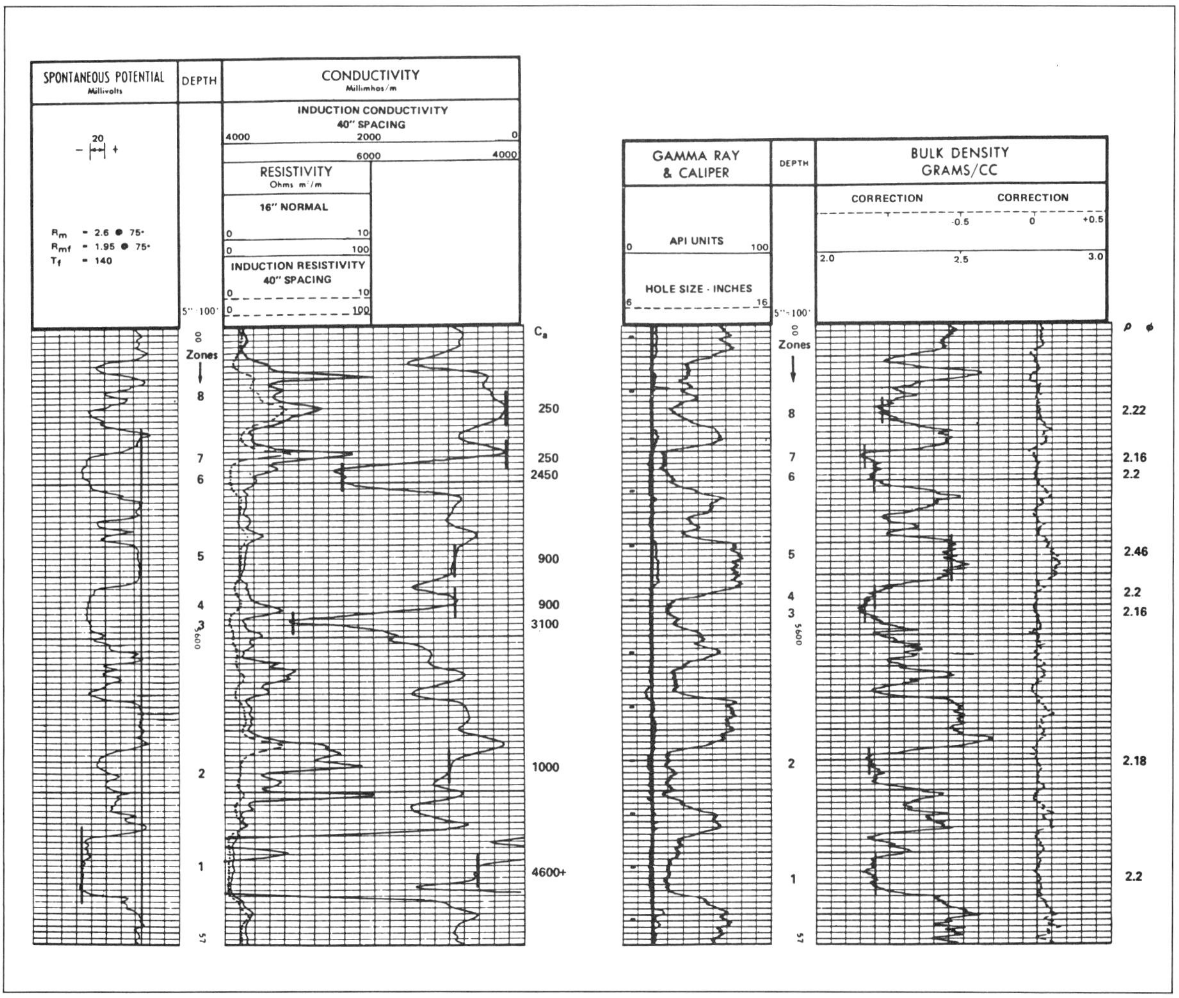


**Figure 3–61.** Openhole well logs for Example 3–19 (courtesy of D. W. Hilchie, Inc., Boulder, Colorado).[1]

Solution: Calculate $R_w$ from SP, SSP = −82 mv.

$$R_{mf} = 1.95 \quad @75°F$$

$$R_{mf} \quad @T_f = R_{mf@75}\left(\frac{75 + 6.77}{140 + 6.77}\right)$$

$$= 1.95\left(\frac{81.77}{146.77}\right)$$

$$= 1.09 > 1.0 = R_{mfeq}$$

Using Equations (3.3) and (3.4),

$$K = 60 + (0.133 \times T_f)$$

$$= 60 + (0.133)(140)$$

$$= 78.62$$

$$\log R_{weq} = \left(\frac{SSP + K \log R_{mfeq}}{K}\right)$$

$$= \frac{-82 + 78.62 \log (1.09)}{78.62}$$

$$= -1.01$$

$$R_{weq} = 0.10$$

Since $R_{weq} \geq 0.10$, therefore, $R_w = R_{weq} = 0.10$

The log analyses are shown in Table 3–9. $R_t$ = 1,000/$C_a$ and $\phi_D$ is calculated using Equation (3.14).

Note the difference in $R_w$ from SP log and $R_{wa}$ calculations. The difference may be because (1) the zone is shaly or (2) the $R_{mf}$ value is poor.

**TABLE 3–9. Log analysis of the logs in Figure 3–61, $\rho_f$ = 1.0 and $\rho_m$ = 2.65**

| Zone | $\rho_b$ gms/cc | $\phi_D$ | $C_a$ mmhos/m | $R_t$ ohm-m | $F_R$ | $R_{wa}$ ohm-m | $S_w = \sqrt{R_w/R_{wa}}$ (%) |
|---|---|---|---|---|---|---|---|
| 1 | 2.20 | 0.27 | 4,600 | 0.22 | 10.35 | 0.021 | 100 |
| 2 | 2.18 | 0.28 | 1,000 | 1.00 | 9.57 | 0.104 | 45 |
| 3 | 2.16 | 0.30 | 3,100 | 0.32 | 8.25 | 0.039 | 73 |
| 4 | 2.20 | 0.27 | 900 | 1.11 | 10.35 | 0.107 | 44 |
| 5 | 2.46 | | Shale Zone | | | | |
| 6 | 2.20 | 0.27 | 2,450 | 0.41 | 10.35 | 0.040 | 72 |
| 7 | 2.16 | 0.30 | 250 | 4.00 | 8.25 | 0.485 | 21 |
| 8 | 2.22 | 0.26 | 250 | 4.00 | 11.23 | 0.358 | 24 |

**Example 3–20.** Figure 3–62 shows gamma ray log, caliper log, Dual Laterolog with Microlaterolog, SP log, and a combination of density porosity and compensated neutron porosity logs (Dresser Atlas) on a well in Kimball County, Nebraska. We need to analyze the zone (Wolfcamp formation) from 6,728 to 6,738 ft. The zone was drillstem tested from 6,733 to 6,743 ft; water sample from the DST showed $R_w = 0.20$ ohm-m at 65°F. The formation temperature from the DST was 192°F. The well was drilled with saturated salt mud.

Solution: Correct $R_w$ for $T_f = 192°\text{F}$.

$$
\begin{aligned}
R_w \quad T_f &= R_w \quad {}_{@65}\left(\frac{65 + 6.77}{T_f + 6.77}\right) \\
&= 0.20\left(\frac{71.77}{198.77}\right) \\
&= 0.072 \quad \text{ohm-m}
\end{aligned}
$$

The log analyses are shown in Table 3–10.

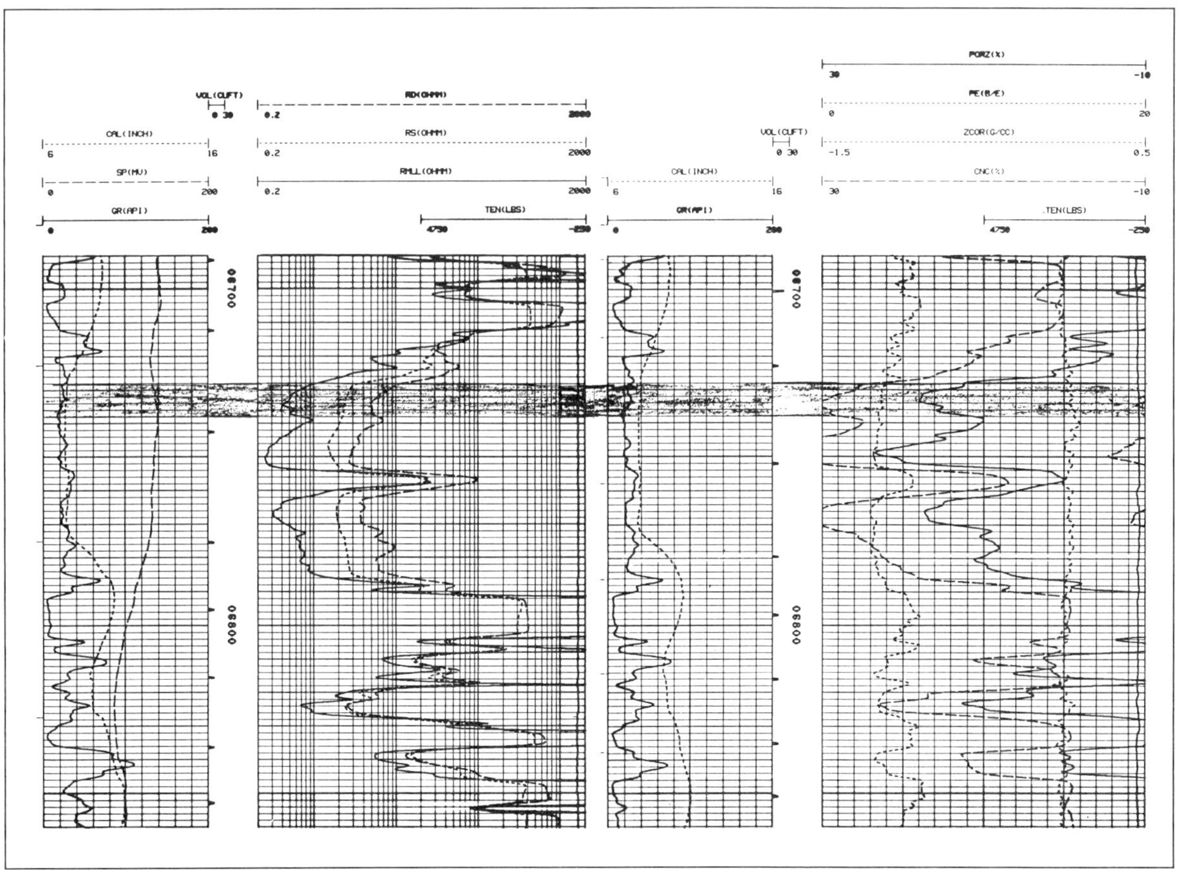


**Figure 3–62.** Openhole well logs for Example 3–20 (courtesy of Miller Oil Co., Ft. Collins, Colorado).

**TABLE 3–10. Log analysis of the logs in Figure 3–62**

| Depth | $\phi_D$ | $\phi_N$ | $\phi_{CP}$ | $R_{LLD}$ | $R_{LLS}$ | $R_{MLL}$ |
|---|---|---|---|---|---|---|
| 6,728–6,730 | 13 | 26 | 21 | 9.5 | 3.0 | 0.8 |
| 6,730–6,734 | 18 | 28 | 24 | 6.0 | 2.4 | 0.5 |
| 6,734–6,736 | 15 | 31 | 24 | 5.0 | 2.2 | 0.5 |
| 6,736–6,738 | 14 | 29 | 23 | 5.0 | 2.2 | 0.5 |

| $R_{LLD}/R_{xo}$ | $R_{LLD}/R_{LLS}$ | $R_t/R_{LLD}$ | $R_t$ | $F_R = 1/\phi^2$ | $R_{wa}$ | $S_w = \sqrt{R_w/R_{wa}}$ |
|---|---|---|---|---|---|---|
| 11.9 | 3.2 | 1.4 | 13.3 | 22.7 | 0.586 | 35 |
| 12.0 | 2.5 | 1.36 | 8.2 | 17.4 | 0.471 | 39 |
| 10.0 | 2.3 | 1.36 | 6.8 | 17.4 | 0.391 | 42 |
| 10.0 | 2.3 | 1.36 | 6.8 | 18.9 | 0.360 | 44 |

$R_t$ corrected for invasion using Figure 3–33. Figure 3–48c used to determine crossplot porosity.

Note (1) the SP deflection on the log (featureless SP) because of saturated salt mud, and (2) the density-neutron crossplot shows that the lithology is dolomite.

**Example 3–21.** Calculate porosity and water saturation for the interval from 1,654 ft to 1,660 ft, shown in the logs in Figure 3–63. Other relevant information is as follows:

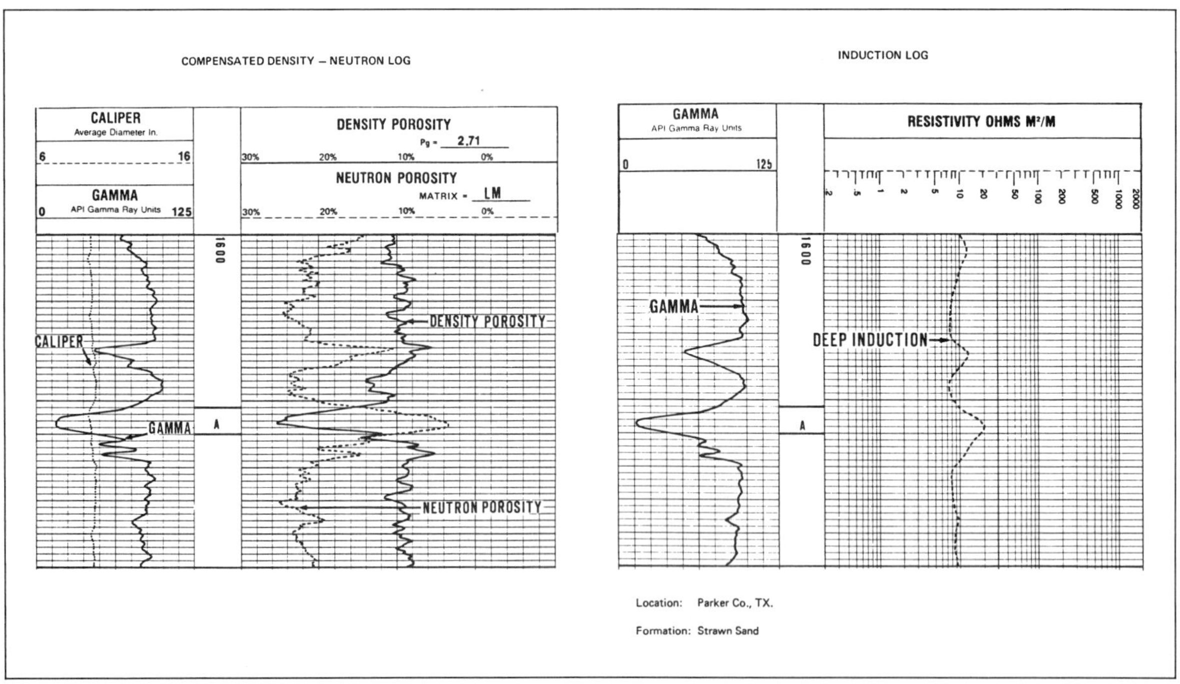


**Figure 3–63.** Openhole well logs for Example 3–21 (courtesy of Welex).[32]

$$R_w = 0.10 \quad \text{ohm-m}$$
$$m = 1.90$$
$$a = 1.0$$
$$\rho_{ma} = 2.65 \quad \text{gms/cc}$$

Solution: From the logs, $\phi_D = 0.25$, $\phi_N = 0.04$, and $R_t = 20$ ohm-m.

$$\phi_{CP} = \sqrt{\frac{\phi_D^2 + \phi_N^2}{2}}$$
$$= \sqrt{\frac{(0.25)^2 + (0.04)^2}{2}}$$
$$= 0.18$$
$$F_R = \frac{a}{\phi^m}$$
$$= \frac{1}{(0.18)^{1.9}}$$
$$= 26$$
$$R_{wa} = \frac{R_t}{F_R} = \frac{20}{26}$$
$$= 0.769$$
$$S_w = \sqrt{\frac{R_w}{R_{wa}}}$$
$$= \sqrt{\frac{0.10}{0.769}}$$
$$= 36\%$$

Note the density-neutron crossover (mirror image) shows presence of gas. The presence of gas is also confirmed by plotting $\phi_D$ and $\phi_N$ on Figure 3–48c. Check this observation yourself.

## LOG F–LOG $R_{deep}$ OVERLAY[3]

Instead of developing $R_wa$ curve and displaying it on the log, the formation resistivity factor ($F$) can be recorded directly on logarithmic scale on the resistivity log. The deep resistivity curve and the $F$ curve are presented in Figure 3–64. The log $R_{wa}$ on a logarithmic scale is simply the algebraic difference between $R_{deep}$ and $F$ as shown below.

The $R_{wa}$ value can be read directly from the overlay by using a transparent overlay with a

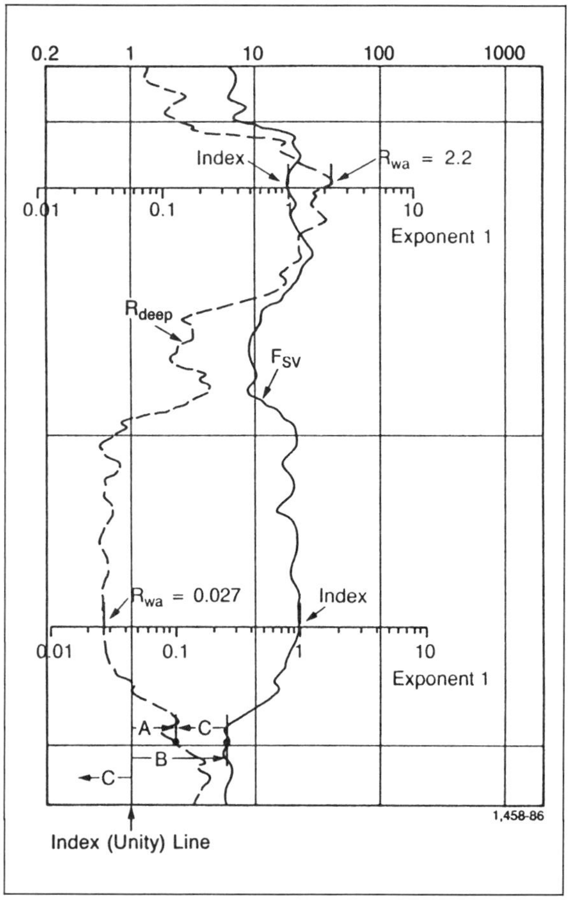


**Figure 3–64.** The log $R_{deep}$-log F overlay (courtesy of Schlumberger).[3]

logarithmic scale, the same as the resistivity log scale. To find $R_{wa}$ at any given level on the log, the *unity index line* of the transparent overlay is placed on the F curve. Then $R_{wa}$ is read directly from the position of the $R_{deep}$ curve relative to the scale. Figure 3–64 shows how $R_{wa}$ is read from the log $F$–$R_{deep}$ overlay.

The F curve can be developed using any porosity log; however, crossplot porosity will be preferred because of the conflicting response of porosity logs to shale and hydrocarbons. The technique will work well only when invasion is shallow because in deep invasion $R_{deep}$ has to be corrected for invasion effects.

## $R_o$ OVERLAY AND $F_R$ OVERLAY[3]

The $R_o$ curve can be developed by adding the logarithm of $R_w$ to the logarithm of $F$ ($R_o = FR_w$) and plotted on the resistivity grid for comparison with $R_{deep}$. The separation between log $R_o$ and log $R_{deep}$ will provide water saturation ($S_w$). The separation between the curves (on logarithmic scale) will be approximately twice the logarithm of $S_w$ (i.e., $2 \log S_w \approx \log R_o - \log R_{deep}$).

The $R_o$ overlay as plotted on the resistivity grid is shown in Figure 3–65a. The $S_w$ is obtained by using a transparent logarithmic overlay with a scale twice as long as the resistivity scale (i.e., the scale using exponent 0.50). To find the $S_w$ at a certain interval, the unity (1.0) index line of the transparent overlay is placed on $R_{deep}$ curve; $S_w$ is read from the resulting position of the $R_o$ curve on the scale as shown in Figure 3–65a.

The $F_R$ overlay is similar to the F overlay and is a variation of the $R_o$ overlay. When both F and $F_R$ curves are plotted on the logarithmic scale, the $S_w$ can be determined by the separation between F and $F_R$. The F curve was developed using porosity, while the $F_R$ curve is developed using resistivity:

$$F_R = \frac{R_{deep}}{R_w} \approx \frac{R_t}{R_w} \tag{3.39}$$

In terms of logarithms, $2 \log S_w = \log F - \log F_R$. Thus the log $F_R$ curve is found by shifting the logarithmic $R_{deep}$ curve by a distance equal to log $R_w$ as shown in Figure 3–65b. If $R_w$ is not known,

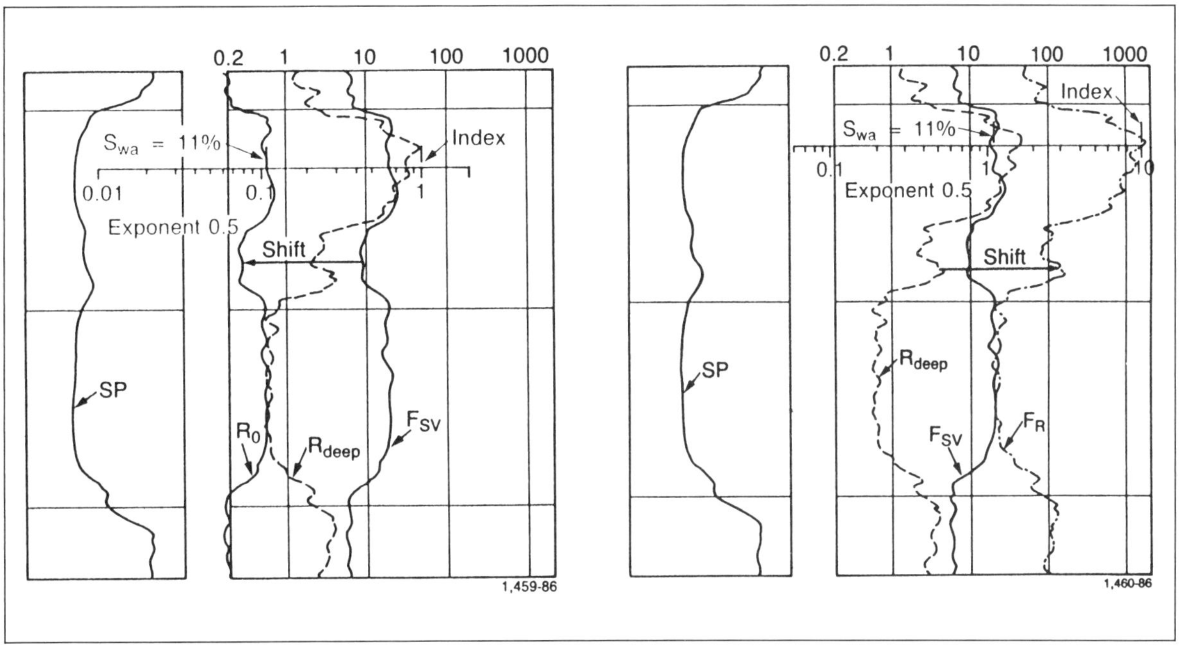


**Figure 3–65.** The $R_o$ overlay and $F_R$ overlay (courtesy of Schlumberger).[3]



the shift could also be determined by making the shift such that the $F_R$ *and F curves overlay in water-bearing zones.* The transparent overlay used to find $S_w$ in Figure 3–65a can be used here too.

## FLUSHED ZONE METHOD (RESISTIVITY RATIO METHOD)

This method is used to determine $S_w$ by using the ratios of resistivities in the uninvaded and the flushed zones. This method does not require any knowledge of porosity and thus the formation factor. Rearrangement of Equation (3.36) gives the ratio of $S_w$ to $S_{xo}$ as follows:

$$\left(\frac{S_w}{S_{xo}}\right)^2 = \frac{R_{xo}/R_t}{R_{mf}/R_w} \tag{3.40}$$

$R_{xo}$ may be found from a MicroSFL, $R_t$ from an induction or Laterolog, and $R_{mf}/R_w$ from measured values, or from the SP curve, or from the $R_w$ methods previously discussed.

The ratio $S_w/S_{xo}$ can be termed as *hydrocarbon movability index.* When $S_w/S_{xo} = 1$ in a permeable zone, the zone will produce water or be nonproductive, regardless of water saturation. The $S_w/S_{xo}$ less than 1.0 indicates that the zone is permeable and contains some hydrocarbons, and that the hydrocarbons have been flushed by invasion. Thus the zone contains producible hydrocarbons.

An empirical relationship between $S_w$ and $S_{xo}$ given below can be used to determine water saturation ($S_w$) in the uninvaded zone. The equation will provide useful results for moderate invasion and average residual oil saturation. If $S_{xo}$ is known, then $S_w = (S_{xo})$.[5]

$$S_w = \left(\frac{R_{xo}/R_t}{R_{mf}/R_w}\right)^{5/8} \tag{3.41}$$

**Example 3–22.** Using the ratio method, calculate the ratio $S_w/S_{xo}$ and $S_w$ for the zones analyzed in the logs presented in Figure 3–62 (Example 3–20). For each interval indicate if hydrocarbons have been moved by invasion. The log heading shows $R_{mf} = 0.04$ ohm-m at 97°F.



Solution:

$$R_{mf} \quad @T_f = R_{mf@97}\left(\frac{97 + 6.77}{T_f + 6.77}\right)$$

$$= 0.04\left(\frac{97 + 6.77}{192 + 6.77}\right)$$

$$= 0.021 \quad \text{ohm-m}$$

The $S_w/S_{xo}$ and $S_w$ values for each interval are given in Table 3–11.

Note that the average water saturation determined in Example 3–20 is within $S_w = 1\%$ of the average $S_w$ calculated using the ratio method. This also confirms that use of $F_R = 1/\phi^2$ in Example 3–20 was appropriate.

**TABLE 3–11. Analysis of openhole well log using ratio method (for Example 3–22)**

| Depth | $R_{xo}$ | $R_t$ | $S_w/S_{xo}$ | $S_w$ (%) | Movable Hydrocarbons |
|---|---|---|---|---|---|
| 6,728–6,730 | 0.80 | 13.3 | 0.45 | 37 | Yes |
| 6,730–6,734 | 0.50 | 8.2 | 0.46 | 38 | Yes |
| 6,734–6,736 | 0.50 | 6.8 | 0.50 | 42 | Yes |
| 6,736–6,738 | 0.50 | 6.8 | 0.50 | 42 | Yes |

## INVADED ZONE METHOD (RESISTIVITY RATIO METHOD)[3]

This method is useful for the determination of water saturation ($S_w$) when only an ES, IES, or other early resistivity log is available. Once again, the porosity log and the formation factor ($F$) are not required. The method requires that $R_i/R_m$ must be at least 10. The water saturation in the invaded zone ($S_{wi}$) is given by

$$S_{wi}^2 = \frac{FR_z}{R_t} \tag{3.42}$$

where $R_z$ is the resistivity of the water in the invaded zone, which is a mixture of the mud filtrate and the formation water. Studies of many logs suggest that $S_{wi}$ and $S_w$ are related by[3]

$$S_{wi} = (S_w)^{0.50} \tag{3.42a}$$

The water saturation of the uninvaded zone can be determined using the following equation:

$$S_w = \frac{R_i/R_t}{R_z/R_w} \tag{3.43}$$

where $R_t$ is taken from deep resistivity log (induction or laterolog), $R_i$ is taken from shallow resistivity log (LL8, 16-in. normal, or SFL), and $R_z$ is given by the following equation:

$$\frac{1}{R_z} = \frac{Z}{R_w} + \frac{1 - Z}{R_{mf}} \tag{3.44}$$

The $Z$ in the above equation is the fraction of the invaded zone pore water, which is formation water, and the remaining fraction $(1 - Z)$ is the mud filtrate. The value of $Z$ varies from 0.075 in case of normal invasion to 0.035 in cases of deep invasion or vuggy formations. The $R_i/R_t \approx 1$ will show that the formation may be (1) hydrocarbon-bearing, (2) dense and impermeable, or (3) extremely invaded or there may be little invasion.

Rodermund et al. developed a chart that can be used instead of Equation (3.43).[17] The chart is presented in Figure 3–66.

## POROSITY BALANCE[3]

The porosity balance method can be used to check on the $S_w$ calculated from the resistivity ratio method. The method involves comparing porosity ($\phi_t$) calculated from well logs or determined from core analysis or other methods with porosity ($\phi_c$) back calculated from the $S_w$ calculated from the ratio methods.

This *porosity check* can verify the applicability of the value of $S_w$ from the ratio methods. The same check can be obtained by calculating $S_w$ using the Archie equation (using independent $\phi$) and comparing it with $S_w$ calculated from the ratio methods. If the porosity check indicates that the ratio method results are in error, the porosity balance will indicate how the error can be corrected.

1. If $\phi_c = \phi_t$, the ratio method solution is correct and $S_{wc} = S_w$.
2. If $\phi_c > \phi_t$, then $F_c$ is too low and $S_w$ is too high. This may be because invasion is either deeper or shallower than one of the resistivity devices can accurately read. The shallow resistivity ($R_{xo}$ or $R_i$) is too low because of shallow invasion or the deep resistivity ($R_t$) is too high because of deep invasion.

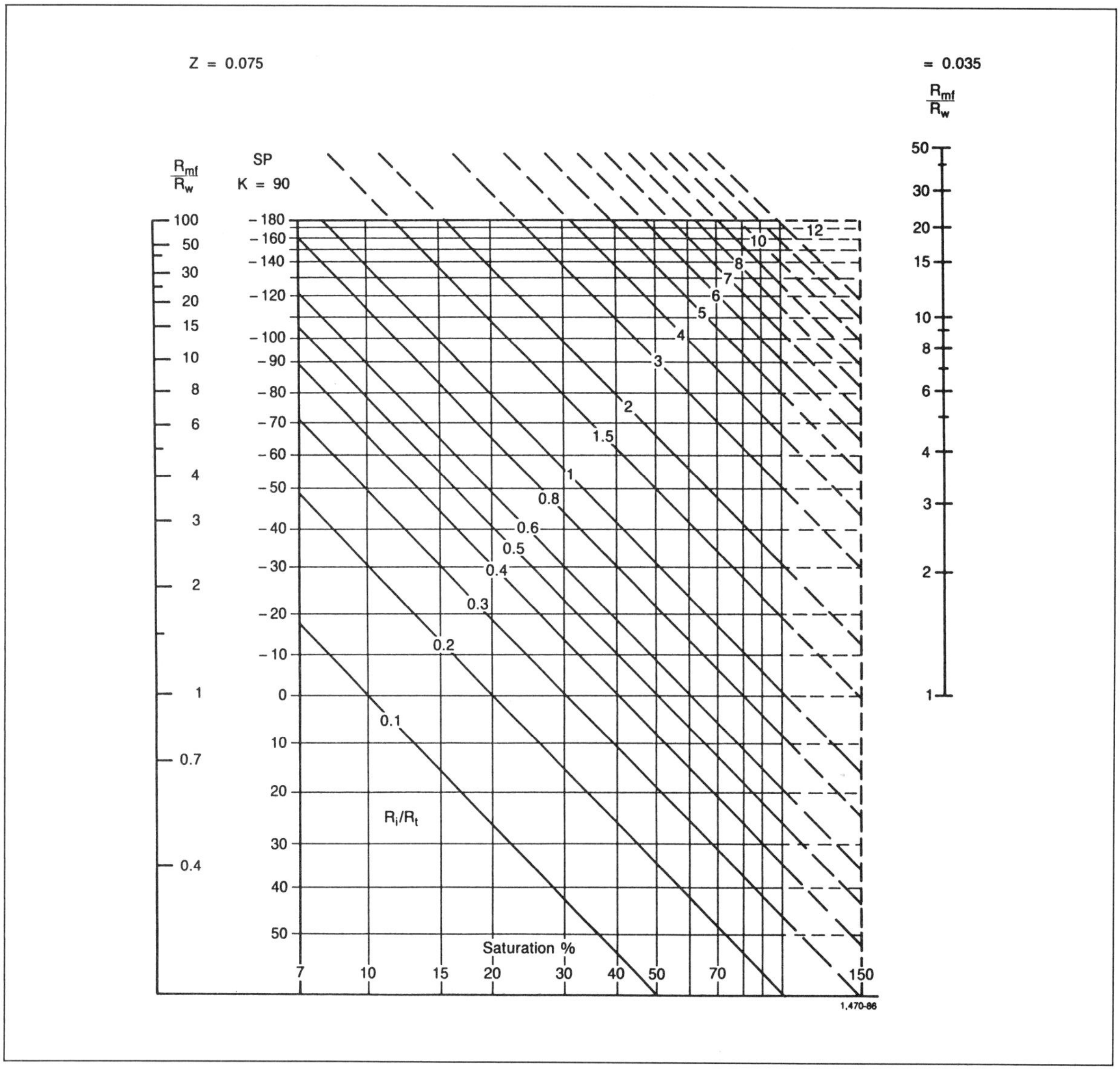


**Figure 3–66.** Empirical resistivity-ratio method (courtesy of Oil & Gas Journal).[17]

a. If $R_{shallow}/R_{deep} < 1.4$ and the induction log was used for $R_t$, use $R_{ILD} = R_t$ in the Archie equation to find $S_w$.

b. If $R_{shallow}/R_{deep} > 1.4$ and $FR_z > R_{shallow}$, invasion is shallow. Use $R_{deep} = R_t$ in the Archie equation to calculate $S_w$.

c. If $R_{shallow}/R_{deep} > 1.4$ but $FR_z < R_{shallow}$, invasion is deep. Equation (3.42a) is solved for $S_w$ with $S_w = S_i^2$.

3. If $\phi_c > \phi_t$, then $S_{wc}$ is too high. This occurs when $R_{xo}/R_t$ or $R_i/R_t$ is too high, as might happen in the case of an annulus. Equation (3.42a) is solved for $S_w$ with $S_w = S_i^2$.

If the water saturation from the Archie ($S_{wA}$) method is compared with the water saturation from the ratio methods ($S_{wR}$), then note the following:

1. If $S_{wA} \approx S_{wR}$, then assumption of a step contact invasion profile is indicated to be correct, and all values determined ($S_w$, $R_t$, $R_{mf}$, $R_{xo}$, $R_i$, and $d_i$) are correct.
2. If $S_{wA} > S_{wR}$, then $R_{xo}/R_t$ is too low, a transition type invasion profile may be indicated and $S_{wA}$ is considered a good value for $S_w$.
3. If $S_{wA} < S_{wR}$, then $R_{xo}/R_t$ is too high, an annulus type of invasion profile may be indicated and/or $S_{xo} < (S_w)^{1/5}$. In this case a more accurate value for water saturation can be estimated using the following equation:[3]

$$(S_w)_{cor} = S_{wA}\left(\frac{S_{wA}}{S_{wR}}\right)^{0.25} \tag{3.45}$$

**Note:** This author has experienced that the difference in $S_{wA}$ and $S_{wR}$ is because of the values used for $m$ and $a$ for calculating $F$ used in the Archie equation. Some clays in the formation (even in small amount) will affect the value of $m$.

## $R_{xo}/R_t$ QUICKLOOK

The $R_{xo}/R_t$ presentation as shown in Figure 3–67, obtained from Dual Induction Log data or Dual Laterolog data, is an effective hydrocarbon indicator that does not require a porosity log for qualitative interpretation.

On the log, the $R_{xo}/R_t$ curve is plotted as a dashed line in Track #1. The dashed line tracks the SP curve in wet zones, but deflects to the right, away from the SP curve, in hydrocarbon-bearing zones. In Figure 3–67, hydrocarbon-bearing zone is indicated at 6,110–6,132 ft, which is confirmed by higher resistivity.

In zones with very high hydrocarbon saturation (i.e., $S_w < 25\%$), the $R_{xo}/R_t$ curve may move in the *opposite direction* from the SP curve. The $R_{xo}/R_t$ quicklook technique is applicable to fresh mud conditions ($R_{xo} > R_t$) in formations where invasions falls within the limits demanded by the $R_{xo}/R_t$ computation. The technique efficiently handles variations in formation water resistivity ($R_w$) and in shaliness. Shaliness and $R_w$ affect the SP and $R_{xo}/R_t$ amplitude in a similar way. Shaliness reduces the $R_{xo}/R_t$ value and the SP amplitude.

## CONDUCTIVITY DERIVED POROSITY CURVE (CDP QUICK LOOK)

We have discussed how to use resistivity or conductivity to calculate porosity. Equations (3.12) and (3.13) are used to calculate resistivity-derived porosity. A conductivity-derived porosity curve plotted on a crossplot porosity curve helps in identifying hydrocarbon-bearing zones. Figure 3–68 shows that CDP and the Acoustilog porosity are essentially the same for the two lower sands (B, C, and D), while in the upper sand (A), the CDP is considerably less than the acoustilog porosity.

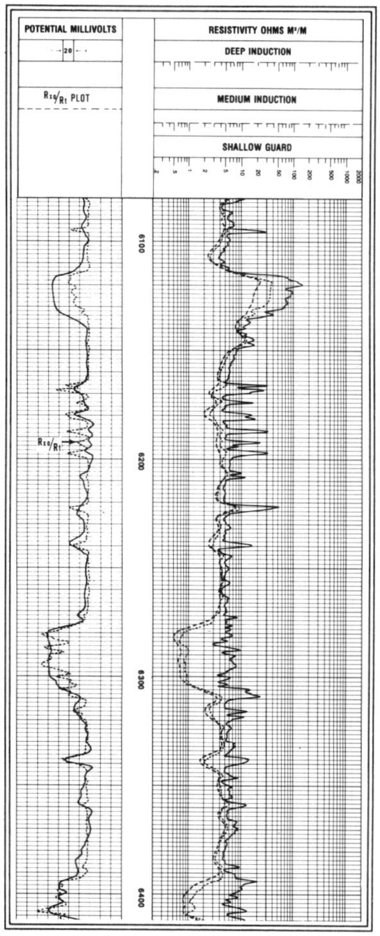


**Figure 3–67.** $R_{xo}/R_t$ plot (courtesy of Welex).[32]

The lower sands are water-bearing and the upper sand is hydrocarbon-bearing. This finding is also confirmed by the $R_{wa}$ curve plotted in Figure 3–68.

### F-MOP MOVABLE OIL PLOT

The F-MOP plot is used as an interpretation technique. The plot uses two resistivity curves and a porosity curve recorded on logarithmic scale to show hydrocarbon saturation and movability. The recorded curves are $F$, $F_{deep}$, and $F_{xo}$. $F$ is the formation factor calculated using appropriate $a$ and $m$, and the $F_{deep}$ and $F_{xo}$ values are calculated as follows:

$$F_{deep} = \frac{R_{deep}}{R_w} \approx FS_w^2 \tag{3.46}$$

$$F_{xo} = \frac{R_{xo}}{R_{mf}} \approx FS_{xo}^2 \tag{3.47}$$

In water-bearing zones, all three formation factors will yield similar values. When hydrocarbons are present, since $R_{deep}$ and $R_{xo}$ are affected to some degree, higher values of $F_{deep}$ and $F_{xo}$ will result as shown in Figure 3–69. The following two equations can be used to calculate $S_w$ and $S_{xo}$.

$$S_w = \sqrt{\frac{F}{F_{deep}}} \tag{3.48}$$

$$S_{xo} = \sqrt{\frac{F}{F_{xo}}} \tag{3.49}$$

The $S_w$ and $S_{xo}$ measurements can be used to calculate the amount of oil remaining in the invaded zone and the amount of movable oil. For example, if $S_{xo}$ is 60% and $S_w$ is 20%, then there is 80% oil in the uninvaded portion of the formation, but half of that oil (40%) has remained as residual oil in the invaded zone.

The three values of $F$ plotted in Figure 3–69 help in easily indentifying hydrocarbon-bearing zones and give some idea of where the movable oil exists. The $S_w$ and $S_{xo}$ can be read directly from the curves by using a transparent logarithmic scale of exponent 0.5. The scale applied to $F$ and $F_{deep}$

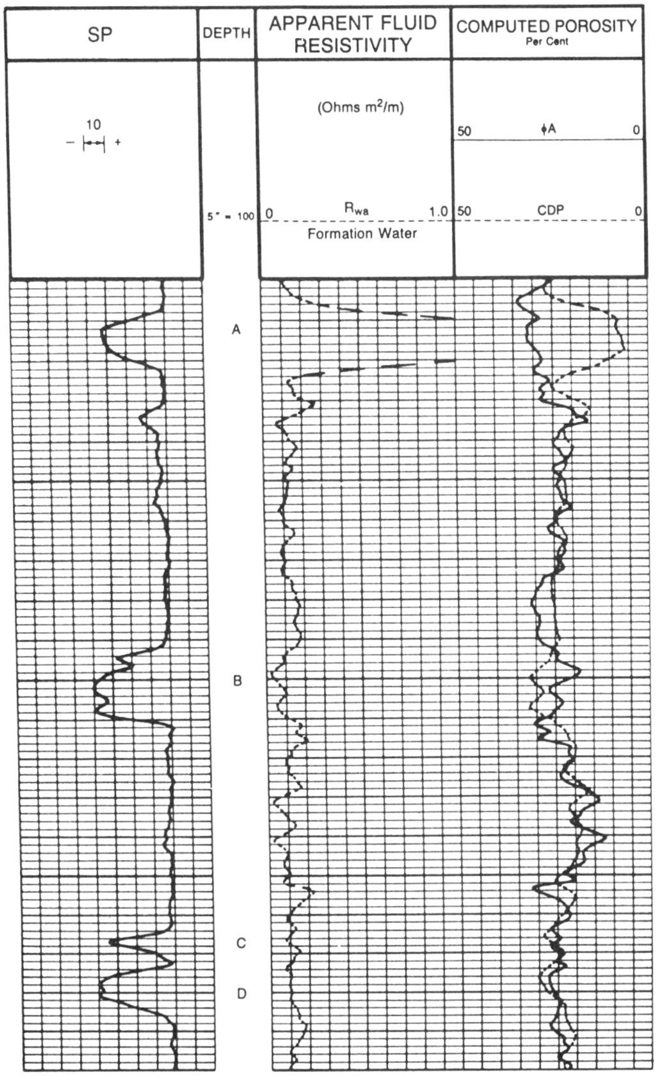


**Figure 3–68.** CDP and Acoustilog porosity (courtesy of Atlas Wireline Services).[8]

will give $S_w$, and $F$ and $F_{xo}$ will give $S_{xo}$. In each case the unity index line of the overlay scale is placed on the $F_{deep}$ or $F_{xo}$ curves.

## OPENHOLE LOG INTERPRETATION—SHALY FORMATIONS

The presence of shale (i.e., clay minerals) in a reservoir rock can cause erroneous values for water saturation and porosity derived from logs. The effect of shales on various logs is discussed when

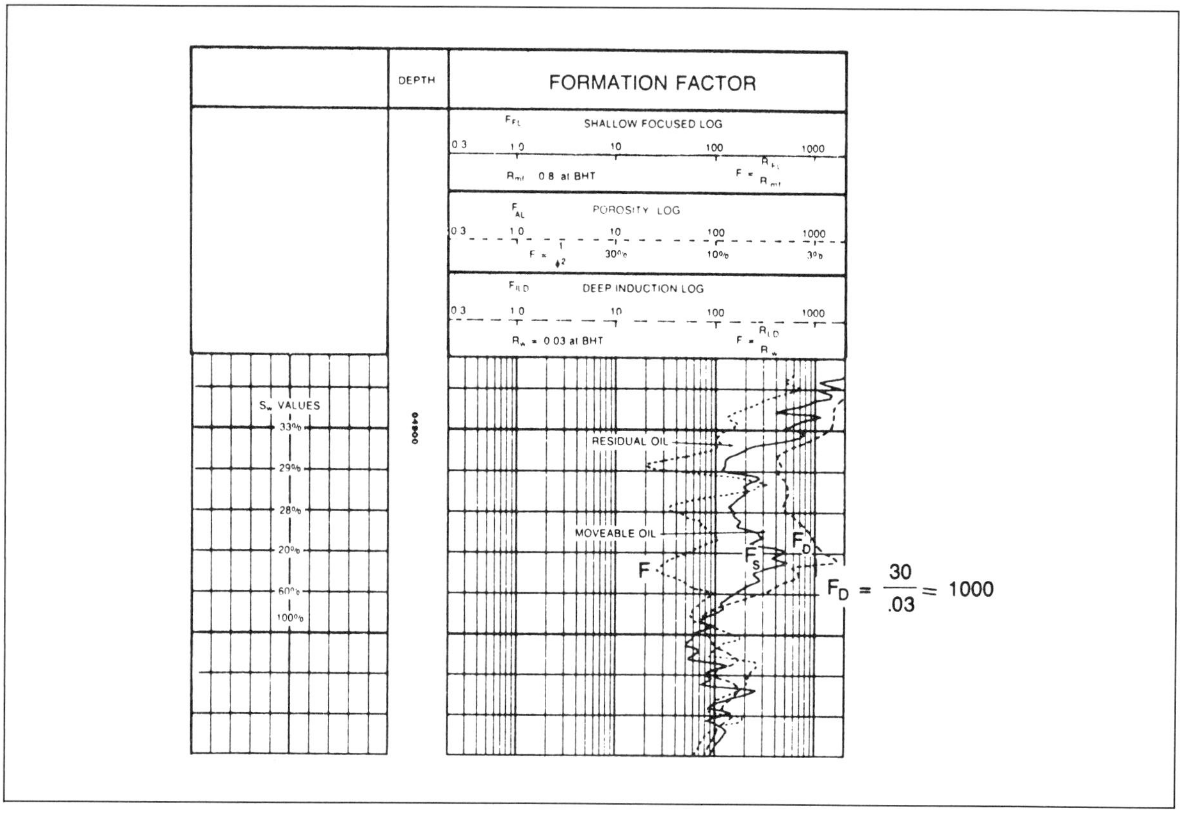


**Figure 3–69.** Movable oil plot F-MOP (courtesy of Atlas Wireline Services).[8]

we discussed the different logging devices. The most common shale problem is that associated with sandstones. The shale effects are summarized below.

1. Shale makes the density log read lower apparent porosity, except when $\rho_{sh} = \rho_{ma}$.
2. Shale makes the neutron log read higher apparent porosity.
3. Increasing amounts of shale in a formation reduce the resistivity ($R_t$) of hydrocarbon-bearing formations and increase the resistivity of water-bearing formations. This makes the visual detection of pay zones very difficult.
4. In some cases the resistivity and acoustic travel time, or neutron porosity, tend to be self-compensating for water saturation calculation. Thus these combinations could be used without corrections for shale content.
5. Shale reduces the gamma ray and SP deflection. The reduction in most cases is proportional to the percentage of shale in the formation.
6. The shale effect is reduced in formations containing very high salinity brine.

The first step in shaly formation evaluation is to obtain the *volume of shale*, $V_{sh}$. The $V_{sh}$ can be obtained from gamma ray log, SP log, or a combination of density-neutron porosity. Note that if the volume of shale is over estimated, a water-bearing zone may calculate like a hydrocarbon-bearing zone. A summary of the equations used to calculate $V_{sh}$ ($V_{cl}$) and the conditions in which they can be used are given in Table 3–12.

The second step is to correct the porosity logs for shale content (refer to discussion on porosity logs). If a combination of density-neutron logs are run, the following equations are used to correct $\phi_D$ and $\phi_N$ for shale:

$$\phi_{N\text{-}corr} = \phi_N - \left[\left(\frac{\phi_{N\text{-}clay}}{0.45}\right) \times 0.30 \times V_{sh}\right] \quad \textbf{(3.50)}$$



$$\phi_{D\text{-}corr} = \phi_D - \left[\left(\frac{\phi_{D\text{-}clay}}{0.45}\right) \times 0.13 \times V_{sh}\right] \quad \textbf{(3.51)}$$

The density-neutron crossplot porosity is then determined using Equation (3.29) or Figure 3–48.

One other useful technique to determine $V_{sh}$ and shale-corrected porosity is to use *density-neutron crossplot* ($\phi_D$ vs. $\phi_N$), as shown in Figure 3–70. The *shale point* is the density and neutron porosity in the adjacent shale bed (assuming the properties of shale in the formation are the same as the adjacent shale bed).

In a similar way, porosity and salt content can be evaluated in salt and limestone matrix. To obtain satisfactory porosity values in such formations, the zone must have been drilled with salt-saturated mud so that the salt has not been washed away from the matrix surrounding the borehole.

A possibly better technique of $V_{sh}$ estimation is through the use of gamma ray. This technique, however, is not suited for sections in which clean zones are absent. To overcome this problem, it is suggested to crossplot $V_{sh}$ from a neutron-density method against gamma. The intersection of the line defined by the plot with the 0% neutron-density $V_{sh}$ line gives an approximate value for clean sand.

Note the $\phi_D$ vs. $\phi_N$ crossplot technique can be used with acceptable accuracy in oil-bearing formations. The presence of gas or light oil, however, will cause the point to shift in the northwesterly direction. In such cases $V_{sh}$ should be evaluated from GR or SP, in order to evaluate the amount of the shift.

After the volume of shale and the corrected porosity have been determined, the water saturation can be calculated. Water saturation calculations require the use of shaly sand equations to compensate for clay-related resistivity decrease. A summary of other water saturation equations which may be used to evaluate shaly clastic reservoir rocks is given in Table 3–13.

Three of the most commonly used shaly formation equations for water saturation calculation are:

Simandoux Equation:

$$S_w = \frac{0.4R_w}{\phi^2} \times \left[\frac{-V_{sh}}{R_{sh}} + \sqrt{\left(\frac{V_{sh}}{R_{sh}}\right)^2 + \frac{5\phi^2}{(R_tR_w)}}\right] \quad \textbf{(3.52)}$$

**TABLE 3–12. Log-derived clay volume indicators[18]**

| Logging Curve | Mathematical Relationship | Favorable Conditions | Unfavorable Conditions* |
|---|---|---|---|
| Spontaneous Potential (SP curve) | $V_{cl} = 1.0 - (PSP/SSP) = 1.0 - \alpha$ | Water-bearing, laminated shaly sands ($<R_t$). | $R_{mf}/R_w \pm 1.0$. Thin, $\gg R_t$ zones. Hydrocarbon-bearing. Large electrokinetic and/or invasion effects. |
| | $V_{cl} = (PSP - SP_{min})/(SSP - SP_{min})$ | $c < 1.0$ as function of clay type. | |
| | $V_{cl} = 1.0 - c\alpha$<br>$1.0 - \alpha = \log A/\log [(A - V_{cl}B)/(1 - V_{cl}B)]$,<br>where $A = R_t/R_{xo}$,<br>$B = R_t/R_{cl}$ | Knowledge of several parameters required, including $\alpha$, $R_t$, $R_{xo}$, and $R_{cl}$. Similar limitations as for straightforward SP equations. | |
| | $1.0 - \alpha = (KV_{cl}W)/(KV_{cl}W + \phi S_{xo})$ | $K$ = log-derived coefficient; $W$ = clay porosity from bulk and matrix $\rho_{cl}$; $S_{xo}$ = flushed zone water saturation; laboratory-derived; too many requirements. | |
| Gamma Ray (GR) | $V_{cl} = (GR - GR_{min})/(GR_{max} - GR_{min})$ | Only clay minerals are radioactive. | Radioactive minerals other than clays (mica, feldspar, and silt). |
| | $V_{cl} = C(GR - GR_{min})/(GR_{max} - GR_{min})$ | $C < 1.0$, frequently approximately 0.5 when $V_{cl} < 40\%$. | Only potassium-deficient kaolinite present. Uranium enrichment in permeable, fractured zones. |
| | $V_{cl} = (GR - W)/Z$, where $W$, $Z$ = geologic area coefficient. | | Radiobarite scales on casing. Severe washouts ($\ll$ GR). |
| | $V_{cl} = 0.33(2^{2V_{cl}} - 1.0)$** | Highly consolidated and Mesozoic rocks. | Younger, unconsolidated rocks. |
| | $V_{cl} = 0.083(2^{3.7V_{cl}} - 1.0)$** | Tertiary clastics. | Older, consolidated rocks. |

| | | | |
|---|---|---|---|
| Spectralog<br>Gamma ray spectral logging provides individual measurements of potassium (K, %) and thorium (Th, ppm) content | $V_{cl} = (A - A_{min})/(A_{max} - A_{min})$<br>$V_{cl} = C(A - A_{min})/(A_{max} - A_{min})$<br>$V_{cl} = 0.33(2^{2V_{cl}} - 1.0)$†<br>$V_{cl} = 0.083(2^{3.7V_{cl}} - 1.0)$† | Conditions similar to gamma ray discussion. A = Spectralog readings (K in %, Th in ppm). $A_{min}$ = minimum value (K or Th) in clean zones. $A_{max}$ = maximum values (K, Th) in essentially pure shales. | Similar to gamma ray discussion. However, uranium enrichment in permeable, fractured zones and radiobarite buildup are no limitations.<br>If Th curve is used, localized bentonite streaks should be ignored. |
| Resistivity<br>If several resistivity logs are available, use the one that exhibits highest resistivity values in subject well | $V_{cl} = (R_{cl}/R_t)^{1/b}$,<br>where $b = 1.0$<br>$b = 2.0$ | Low-porosity zones (carbonate, marls); pay zones with low $(S_w - S_{wir})$.<br>$R_{cl}/R_t$ from 0.5 to 1.0.<br>$R_{cl}$ approaches $R_t$. | High-porosity water sand; high $R_{cl}$ values. |
| | $V_{cl} = \{R_{cl}(R_{max} - R_t)/[R_t(R_{max} - R_{cl})]\}^{1/b}$<br>where<br>$(1/b) = 1.0$ when $R_{cl}/R_t \geq 0.5$<br>$(1/b) = 0.5/(1 - R_{cl}/R_t)$ when $R_{cl}/R_t < 0.5$ | In clean hydrocarbon-bearing zones, $V_{cl} = 0$. | |
| Neutron | $V_{cl} = \phi_N/\phi_{Ncl}$ | High gas saturation or very low reservoir porosity. | $\phi_{Ncl}$ is low. |
| | $V_{cl} = (\phi_N - \phi_{min})/(\phi_{Ncl} - \phi_{min})$ | $\phi_{min}$ can be varied. | |
| Pulsed Neutron | $V_{cl} = (\Sigma - \Sigma_{min})/(\Sigma_{max} - \Sigma_{min})$ | Freshwater environment; low-porosity and gas-bearing zones. | |
| | $V_{cl} = (\Sigma_{cl}/\Sigma)(\Sigma - \Sigma_{min})/(\Sigma_{max} - \Sigma_{min})$ | | |
| Density/Neutron | $V_{cl} = \dfrac{\rho_B(\phi_{Nma} - 1.0) - \phi_N(\rho_{ma} - \rho_f) - \rho_f\phi_{Nma} + \rho_{ma}}{(\rho_{sh} - \rho_f)(\phi_{Nma} - 1.0) - (\phi_{Nsh} - 1.0)(\rho_{ma} - \rho_f)}$ | | Too low $V_{cl}$ in prolific gas zones. Not for use in severe hole conditions.<br>Lithology affected. |

**TABLE 3–12. Continued**

| Logging Curve | Mathematical Relationship | Favorable Conditions | Unfavorable Conditions* |
|---|---|---|---|
| Density/Acoustic | $V_{cl} = \dfrac{\rho_B(\Delta t_{ma} - \Delta t_f) - \Delta t(\rho_{ma} - \rho_f) - \rho_f \Delta t_f \Delta t_{ma} + \rho_{ma}\Delta t_f}{(\Delta t_{ma} - \Delta t_f)(\rho_{sh} - \rho_f) - (\rho_{ma} - \rho_f)(\Delta t_{sh} - \Delta t_f)}$ | Less dependent on lithology and fluid conditions than density/neutron crossplot. Use in gauge boreholes. | Badly washed out wellbores. Highly undercompacted formations (shallow, overpressures). |
| Neutron/Acoustic | $V_{cl} = \dfrac{\phi_N(\Delta t_{ma} - \Delta t_f) - \Delta t(\phi_{Nma} - 1.0) - \Delta t_{ma} + \phi_{Nma}\Delta t_f}{(\Delta t_{ma} - \Delta t_f)(\phi_{Nsh} - 1.0) - (\phi_{Nma} - 1.0)(\Delta t_{sh} - \Delta t_f)}$ | Use only in gas-bearing zones with low $S_w$. | Similar effects because of shaliness on both logs. |

* For further definitions, see Ref. 4.

** $V_{cl} = (GR - GR_{min})/(GR_{max} - GR_{min})$

† $V_{cl} = (A - A_{min})/(A_{max} - A_{min})$

Fertl Equation, where $a = 0.25$ for Gulf Coast and $a = 0.35$ for Rocky Mountains:

$$S_w = \frac{1}{\phi} \times \left[ \sqrt{\frac{R_w}{R_t} + \left(\frac{a \times V_{sh}}{2}\right)^2} - \frac{a \times V_{sh}}{2} \right] \quad \textbf{(3.53)}$$

Schlumberger Equation:

$$S_w = \frac{\dfrac{-V_{sh}}{R_{sh}} + \sqrt{\left(\dfrac{V_{sh}}{R_{sh}}\right)^2 + \dfrac{\phi^2}{0.2 \times R_w (1 - V_{sh}) R_t}}}{\dfrac{\phi^2}{0.4 \times R_w (1 - V_{sh})}} \quad \textbf{(3.54)}$$

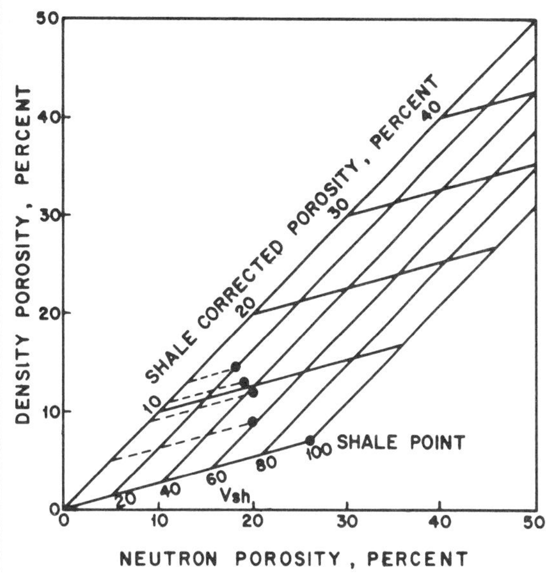


**Figure 3–70.** Density-neutron crossplot showing shale point, $V_{sh}$ (%), and shale-corrected porosity.

**Example 3–23.** Figure 3–71 shows dual induction focused log, compensated density log, compensated neutron log, and gamma ray log for Codell-Niobrara formation in Weld County, Colorado. Using shaly formation analysis, calculate water saturation for the zone at 6,586–6,598 ft (use Schlumberger Equation). Other data from the log heading are

| | | |
|---|---|---|
| Depth, Logger = 6,716 ft | | $R_m$ @BHT = 1.23 ohm-m |
| $R_m$ @101°F = 2.0 ohm-m | | Bit Size = 7.875 in. |
| $R_{mf}$ @101°F = 1.5 ohm-m | | BHT = 164°F |
| $R_{mc}$ @101°F = 2.5 ohm-m | | |

Solution: The density-neutron crossplot for this example is shown in Figure 3–70. The shale point is $\phi_{D-sh} = 7\%$ and $\phi_{N-sh} = 26\%$. Resistivity of the adjacent shale bed is $R_{sh} = 3.5$ ohm-m. Table 3–14 shows the log analysis.

Recently Aguilera presented methods which can be used to analyze laminated, dispersed, and total shale models.[19] The techniques involve the use of log-log crossplots of effective porosity

**TABLE 3–13. Classification of water saturation equations in shaly clastic reservoir rocks[6]**

| Reference | Equation* | Type** | Comments† |
|---|---|---|---|
| Poupon *et al.* | $C_t = \frac{(1 - V_{sh})C_w S_w^2}{F} + V_{sh}C_{sh}$ | 1 | Laminated shale model.<br>$F$ = formation factor of clean-sand streaks.<br>$S_w$ relates to total interconnected pore space of clean sand streaks. |
| Hossin | $C_t = \frac{C_w}{F} S_w^2 + V_{sh}^2 C_{sh}$ | 1 | |
| Simandoux | $C_t = \frac{C_w}{F} S_w^2 + \epsilon V_{sh} C_{sh}$ | 1 | $\epsilon = 1$ for high $S_w$;<br>$\epsilon < 1$ for low $S_w$. |
| L. de Witte | $C_t = \frac{2.15 k m_w}{F} S_w^2 + \frac{k m_{sh}}{F} S_w$ | 2 | $m_w$ = molal concentration of exchangeable cations in formation water.<br>$m_{sh}$ = molal concentration of exchangeable cations associated with shale.<br>$k$ = conversion from $m_{sh,w}$ to conductivity.<br>$F$ relates to total interconnected porosity.<br>$S_w$ relates to total interconnected pore space. |
| A. J. de Witte | $C_t = \frac{C_w}{F} S_w^2 + A S_w$ | 2 | $F$ = maximum formation factor.<br>$FA$ = shaliness factor.<br>$S_w$ relates to total interconnected pore space. |

| | | | |
|---|---|---|---|
| Patchett and Rausch | $C_t = \frac{C_w}{F} S_w^2 + C_s S_w$ | 2 | $C_s$ = conductivity due to shale ($\neq C_{sh}$). $F$ relates to total interconnected porosity. $S_w$ relates to total interconnected pore space. |
| Waxman and Smits | $C_t = \frac{C_w}{F^*} S_w^2 + \frac{BQ_V}{F^*} S_w$ | 2 | $F^*$ relates to total interconnected porosity. $S_w$ relates to total interconnected pore space. |
| Bardon and Pied | $C_t = \frac{C_w}{F} S_w^2 + V_{sh} C_{sh} S_w$ | 2 | Modified Simandoux equation. |
| Schlumberger | $C_t = \frac{C_w}{F(1 - V_{sh})} S_w^2 + V_{sh} C_{sh} S_w$ | 2 | $F$ relates to the free fluid porosity of the total rock volume, inclusive of intraformational (laminated) shales. |
| Clavier *et al.* | $C_t = \frac{C_w}{F_o} S_w^2 + \frac{(C_{bw} - C_w) V_Q Q_V}{F_o} S_w$ | 2 | Dual-water model. $F_o$ relates to total interconnected porosity. $S_w$ relates to the total interconnected pore space. |
| Juhasz | $C_t = \frac{C_w}{F} S_w^2 + \left(\frac{C_{sh}}{F_{sh}} - C_w\right) \frac{V_{sh} \phi_{sh} S_w}{\phi}$ | 2 | Normalized Waxman-Smits equation. $F = 1/\phi^m$ where $\phi$ is the porosity derived from the density log and corrected for hydrocarbon effects. $F_{sh} = 1/\phi_{sh}^m$ where $\phi_{sh}$ is the shale porosity derived from the density log. $S_w$ relates to total interconnected pore space. |

**TABLE 3–13. Continued**

| Reference | Equation* | Type** | Comments† |
|---|---|---|---|
| Doll | $C_t = \frac{C_w}{F} S_w^2 + 2V_{sh} \sqrt{\frac{C_w C_{sh}}{F}} + V_{sh}^2 C_{sh}$ | 3 | |
| Alger *et al.* | $C_t = \frac{C_w(1 - q)^2}{F} S_w^2 + \frac{q(1 - q)(C_{sh} + C_w)}{F} S_w + \frac{q^2 C_{sh}}{F}$ | 3 | Clay slurry model.<br>$F$ relates to total volume occupied by fluid and clay.<br>$S_w$ relates to fluid-filled pore space. |
| Husten and Anton | $C_t = \frac{C_w}{F} S_w^2 + 2V_{sh} \sqrt{\frac{C_w C_{sh}}{F}} \left(1 - \sqrt{\frac{C_w}{C_{bw}}}\right) S_w + V_{sh}^2 C_{sh} \left(1 - \sqrt{\frac{C_w}{C_{bw}}}\right)^2$ | 3 | $F = 1/\phi_t^2$ where $\phi_t$, is the total interconnected porosity.<br>$C_w = C_{bw}$.<br>$S_w$ relates to total interconnected pore space. |
| Patchett and Herrick | $C_t = \frac{(1 - V_{sh})C_w}{F} S_w^2 + \frac{(1 - V_{sh})}{F} BQ_v S_w + V_{sh} C_{sh}$ | 3 | Laminated sand-shale model.<br>$V_{sh}$ = volume fraction of laminated shales only.<br>$F$ relates to total interconnected porosity within shaly-sand streaks.<br>$S_w$ relates to total interconnected pore space within shaly-sand streaks. |
| Poupon and Leveaux | $C_t = \frac{C_w}{F} S_w^2 + 2 \sqrt{\frac{C_w V_{sh}^{2-V_{sh}} C_{sh}}{F}} S_w^2 + V_{sh}^{2-V_{sh}} C_{sh} S_w^2$ | 4 | "Indonesia" formula. |
| Poupon and Leveaux | $C_t = \frac{C_w}{F} S_w^2 + 2 \sqrt{\frac{C_w V_{sh} C_{sh}}{F}} S_w^2 + V_{sh}^2 C_{sh} S_w^2$ | 4 | Simplified Indonesia formula for $V_{sh} \leq 0.5$. |

| | | | |
|---|---|---|---|
| Woodhouse | $C_t = \frac{C_w}{F} S_w^2 + 2 \sqrt{\frac{C_w V_{sh}^{2-V_{sh}} C_{sh}}{F}} S_w^2 + V_{sh}^{2-V_{sh}} C_{sh} S_w^2$ | 4 | Modification of Poupon and Leveaux's equation for tar sands. |
| Raiga-Clemenceau *et al.* | $C_t = \frac{C_w}{F_t} S_w^2 + 2 \sqrt{\frac{C_w \phi_{bw}^{1.72} C_{bw}}{F}} S_w^{1.5} + \phi_{bw}^{1.72} C_{bw} S_w$ | 4 | "Dual-porosity" model; $\phi_{bw} = \phi_t - \phi_e$. |

* $F$ relates to free-fluid porosity unless otherwise stated; $S_w$ relates to free-fluid pore space unless otherwise stated; equations are written with $n = 2$

** TYPE EQUATIONS FOR $S_w$ RELATIONSHIPS

Type 1: $C_t = \alpha S_w^n + \gamma$
No interactive term; $S_w$ does not appear in both terms.

Type 2: $C_t = \alpha S_w^n + \gamma S_w^s$
No interactive term; $S_w$ appears in both terms.

Type 3: $C_t = \alpha S_w^n + \beta S_w^t + \gamma$
Interactive terms, $S_w$ does not appear in all terms.

Type 4: $C_t = \alpha S_w^n + \beta S_w^t + \gamma S_w^s$
Interactive term; $S_w$ appears in all terms.

$\alpha$ = predominant sand term; $\beta$ = predominant interactive term; $\gamma$ = predominant shale term; $t$, the saturation exponent, is an interactive term; $s$, the saturation exponent, is a shale term.

† For further definitions, see Ref. 3.

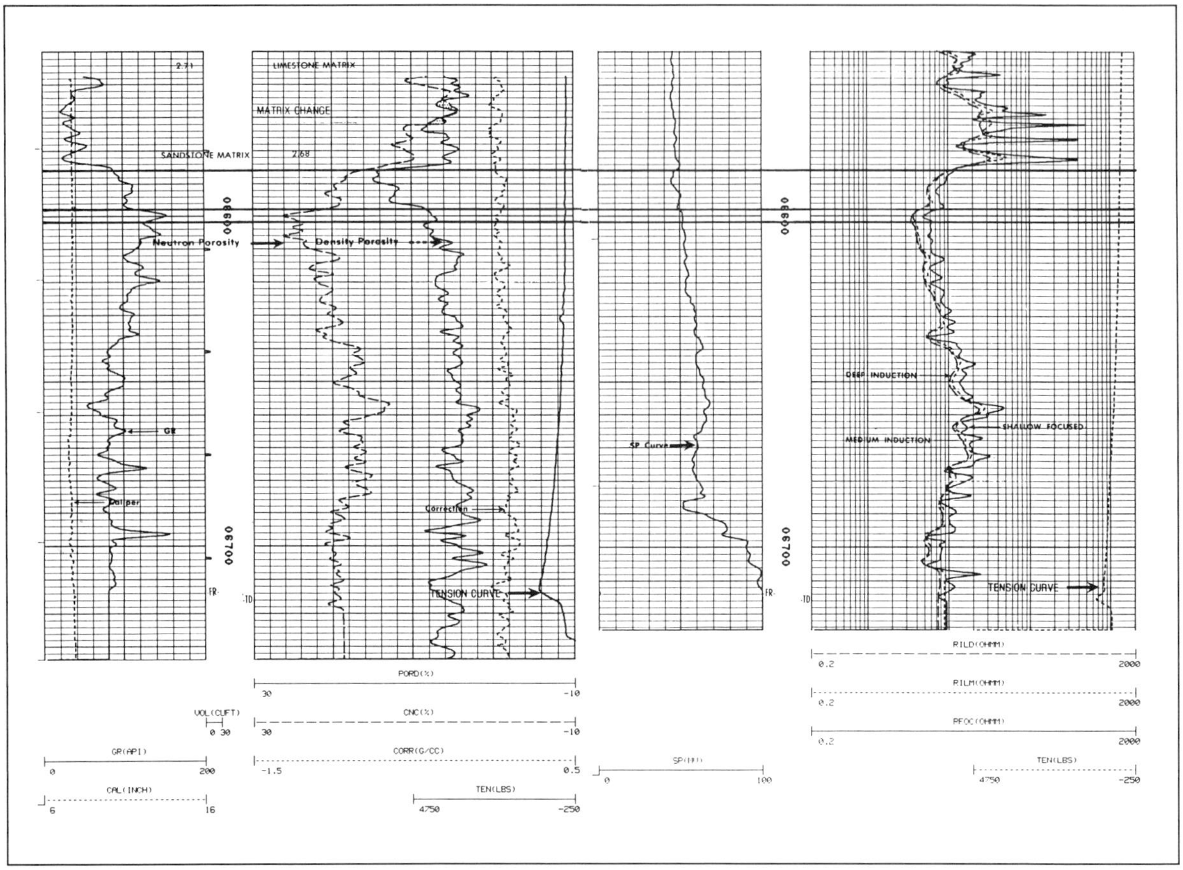


**Figure 3–71.** Dual induction focused log, compensated density, compensated neutron, and GR log—Codell formation, Weld County, Colorado (courtesy of Miller Oil Co., Ft. Collins, Colorado).

versus true resistivity as affected by a shale group ($A_{sh}$). The key advantages of the proposed techniques are:

1. The value of $m$ does not have to be assumed; in fact, it can be determined from this analysis by trial and error.
2. Water resistivity does not have to be known in advance provided that the reservoir contains some water-bearing intervals.

**TABLE 3–14. Shaly formation evaluation for Example 3–23**

| Depth | $\phi_D$ | $\phi_N$ | $\phi_{cor}$ | $R_t$ | $V_{sh}$ | $S_w$ |
|---|---|---|---|---|---|---|
| 6,586–6,590 | 0.15 | 0.18 | 0.13 | 7.5 | 0.20 | 0.37 |
| 6,590–6,594 | 0.12 | 0.20 | 0.09 | 5.6 | 0.42 | 0.49 |
| 6,594–6,596 | 0.13 | 0.19 | 0.11 | 6.0 | 0.32 | 0.43 |
| 6,596–6,598 | 0.09 | 0.20 | 0.05 | 5.2 | 0.59 | 0.84 |

3. All the analyzed intervals can be displayed in a single log-log plot even if shale volumes are changing continuously.

The equations that handle the laminated, dispersed, and total shale models, respectively, have been written in such a way as to allow the generation of log-log plots from which values of water-saturation can be readily calculated.

**Laminated Shale Model (Poupon et al.[20]):**



$$S_w = \sqrt{\frac{a(1 - V_{lam})}{\phi^2}\left(\frac{1}{R_t} - \frac{V_{lam}}{R_{sh}}\right)} \tag{3.55}$$

Equation (3.55) can be written in general terms as follows:

$$\frac{R_t}{A_{lam}} = aR_w\phi^{-m}S_w^{-n} \tag{3.56}$$

where

$$A_{lam} = \frac{(R_{sh} - R_t V_{lam})(1 - V_{lam})}{R_{sh}}$$

Equation (3.56) shows that a log-log plot of $(R_t/A_{lam})$ vs. $\phi$ should result in a straight line with a slope equal to $-m$, the value of $aR_w$ can be read at a point obtained by extrapolating the 100% water saturation line to a porosity of 100%.

If there are intervals at irreducible water saturation which meet the criteria $S_{iw}\phi = C$ (constant), a plot of $(R_t/A_{lam})$ vs. $\phi$ on log-log graph paper should result in a straight line with a slope equal to $(n - m)$ for intervals at irreducible water saturation. Since the value of $m$ has already been determined, the water saturation exponent can be readily obtained. The following steps are followed for calculating $n$:

1. Determine the value of $n$ from $(n - m)$ slope.
2. Extrapolate the irreducible water saturation line to $\phi = 100\%$. Read $R_t/A_{lam}$ at this point, which is equal to $aR_w/C^n$. Calculate C.
3. Extrapolate the irreducible water saturation line until it intercepts the 100% water saturation line. If the porosity read at the intercept is equal to C in Step 2, the selected $S_{iw}$ line is correct, otherwise shift the line and repeat the procedure.

Water saturation, $S_w$, can be calculated from the log-log plot by using the following relationship:

$$S_w = \left(\frac{\phi_w}{\phi_{hy}}\right)^{m/n}$$

where $\phi_w$ is read at the 100% water saturation line and $\phi_{hy}$ is the porosity of the hydrocarbon-bearing interval; both $\phi_w$ and $\phi_{hy}$ are read at the same value of $R_t/A_{lam}$.

**Dispersed Shale Model (Schlumberger Model):**

$$S_w = \sqrt{\frac{aR_w}{\phi^2 R_t} + \left[\frac{V_{dis}(R_{dis} - R_w)}{2\phi R_{dis}}\right]^2} - \frac{V_{dis}}{\phi}\left(\frac{R_{dis} + R_w}{2R_{dis}}\right) \quad \textbf{(3.57)}$$

Equation (3.57) can be written in general terms as follows:

$$\frac{R_t}{A_{dis}} = aR_w\phi^{-m}S_w^{-2} \quad \textbf{(3.58)}$$

where

$$A_{dis} = 1 + \frac{\phi^m R_t}{aR_w}\left[B_{dis} + C_{dis}^2 - \sqrt{2\frac{aR_w}{\phi^m R_t} + B_{dis}C_{dis}}\right]$$

$$B_{dis} = \left[\frac{V_{dis}(R_{dis} - R_w)}{2\phi R_{dis}}\right]^2$$

and

$$C_{dis} = \frac{V_{dis}(R_{dis} + R_w)}{2\phi R_{dis}}$$

Thus a log-log plot of $R_t/A_{dis}$ vs. $\phi$ should result in a straight line with a slope equal to $-m$ for intervals with constant $aR_w$ and constant $S_w$. The water saturation is calculated from the log-log plot using the following relationship:

$$S_w = \left[\frac{(R_t/A_{dis})_{hy}}{(R_t/A_{dis})_w}\right]^{-1/2} = I_{dis}^{-1/n}$$

where both numerator and denominator can be read directly from log-log plot at the porosity value of the interval being analyzed, and $n$ is assumed to be equal to 2.0.

Note that the method is a trial and error since calculation of $A_{dis}$ requires $aR_w$ and $m$ in order to prepare the log-log crossplot. Thus reasonable values of $aR_w$ and $m$ are assumed, and a plot of $R_t/A_{dis}$ vs. $\phi$ prepared. This should result in a straight line for intervals of constant $S_w$. The slope of the line gives $m$, and the intercept at $\phi = 100\%$ gives the product $aR_w$. If the $aR_w$ and $m$ values agree with the input values to calculate $R_t/A_{dis}$, the assumed straight line is correct; otherwise the procedure is repeated until a satisfactory agreement is obtained.

**Total Shale Model (Simandoux[21] and Poupon et al.[20]):**

$$S_w = \sqrt{\frac{aR_w}{\phi^2 R_t} + \left(\frac{aR_w}{2\phi^2}\frac{V_{tsh}}{R_{tsh}}\right)^2} - \left(\frac{aR_w}{2\phi^2}\frac{V_{tsh}}{R_{tsh}}\right) \tag{3.59}$$

Equation (3.59) can be written in general terms as follows:

$$\frac{R_t}{A_{tsh}} = aR_w\phi^{-m}S_w^{-2} \tag{3.60}$$

where

$$A_{tsh} = 1 + \frac{\phi^m R_t}{aR_w}\left(2B_{tsh}^2 - 2B_{tsh}\sqrt{\frac{aR_w}{\phi^m} + B_{tsh}^2}\right)$$

and

$$B_{tsh} = \frac{aR_w}{2\phi^m}\frac{V_{tsh}}{R_{tsh}}$$

The procedure to calculate $m$, $aR_w$, and $S_w$ is the same as discussed in the case of dispersed shale model.

## NEUTRON LIFETIME LOG

The Neutron Lifetime Log (NLL) corresponds to the newer version by Atlas Wireline Services called PDK-100 and the Thermal Decay Time (TDT) log by Schlumberger, and is a measure of the time required for thermal neutrons to be captured versus depth. The thermal neutron capture cross section ($\Sigma$) is a basic property of a formation and can be correlated with the usual openhole electrical resistivity log.[22]

The NLL is used to distinguish between gas, oil, and salt water-bearing formations in *cased holes*. This is applicable to finding of new hydrocarbon reserves or verifying the presence of hydrocarbons that were initially in place.

In this section we will discuss how to calculate water saturation from NLL. We will also discuss the use of NLL to detect gas/oil/water contacts, hydrocarbon migrations between zones, and recovery efficiency.

### CAPTURE CROSS SECTION ($\Sigma$)

Capture cross section of a reservoir rock is dependent upon the porosity, formation water salinity, and the quantity and type of hydrocarbons contained in the pore spaces. $\Sigma$ of a reservoir rock decreases with increased hydrocarbon content, increases with increasing porosity, and increases with the increasing formation water salinity.

A typical NLL consists of four curves as shown in Figure 3–72: (1) the capture gamma rays measured during Gate 1 (400–600 microseconds), (2) the capture gamma rays measured during Gate 2 (700–900 microseconds), (3) a monitor of the source output, and (4) the calculated $\Sigma$.

Note that the calculated $\Sigma$ curve is developed using the following equation:

$$\Sigma = \frac{10{,}000}{\Delta t} \log_{10} \frac{N_1}{N_2} \tag{3.61}$$

where $\Delta t$ is the separation between the two gates which is usually 300 microseconds. Table 3–15 shows $\Sigma$ values for common materials encountered around wellbores. The practical values listed

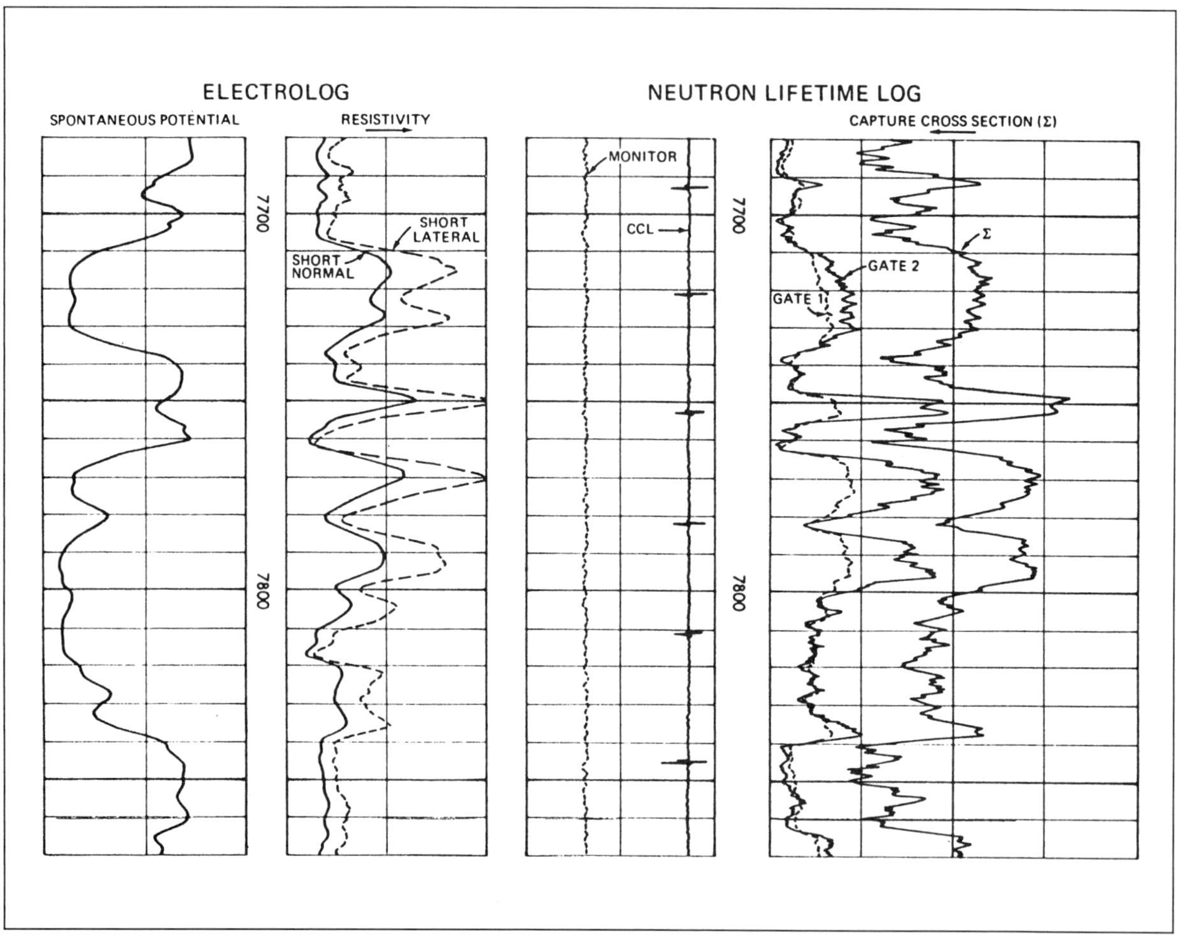


**Figure 3–72.** Typical Neutron Lifetime Log correlated with an electrolog (courtesy of Atlas Wireline Services).[7]

**TABLE 3–15. $\Sigma$ values of various materials**

| Material | Calculated $\Sigma_{ma}$ | Practical $\Sigma_{ma}$ Values |
|---|---|---|
| Sandstone ($SiO_2$) | 4.3 | 8–13 |
| Dolomite ($CaMg(CO_3)_2$) | 4.7 | 8–12 |
| Limestone ($CaCO_3$) | 7.1 | 8–12 |
| Anhydrite | 12.5 | 18–21 |
| Gypsum | 18.0 | |
| Portland Cement | 13.1 | |
| Pozzolamic Cement | 17 or > | |
| Fresh Water | 22.2 | 22 |
| Oil (Dead Crude) | 22.2 | 22 |
| Shale | | 20–40 |
| Iron | 200 | |



represent the $\Sigma$ values normally encountered where the matrix or solid rock values are obtained in a water sand of known porosity and water salinity.[7]

## INTERPRETATION BASICS

The Thermal Neutron Lifetime ($L$) in microseconds is given by the following equation:

$$L = \frac{3{,}150}{\Sigma} \tag{3.62}$$

The water saturation ($S_w$) calculation, which is one of the prime uses of NLL, is based on the following equation:

$$S_w = \frac{\Sigma_t - \Sigma_{ma} + \phi(\Sigma_{ma} - \Sigma_{hc})}{\phi(\Sigma_w - \Sigma_{hc})} \tag{3.63}$$

In shaly formations $S_w$ is given by

$$S_w = \frac{\Sigma_t - \Sigma_{ma} + \phi(\Sigma_{ma} - \Sigma_{hc}) + V_{sh}(\Sigma_{ma} - \Sigma_{sh})}{\phi(\Sigma_w - \Sigma_{hc})} \tag{3.64}$$

where $\Sigma_{ma}$ is given by the following equation:

$$\Sigma_{ma} = \frac{\Sigma_t - \phi\Sigma_w}{1 - \phi} \tag{3.65}$$

The interpretation of NLL in terms of water or hydrocarbon saturation requires knowledge of $\Sigma$ of various materials in the formation. Table 3–15 provides a list of some frequently used $\Sigma$ values. Figures 3–73 through 3–76 provide charts which may be used to determine $\Sigma$ values for various materials.

If reservoir gas specific gravity ($\gamma_g$) is available, then $\Sigma_{hc}$ is given by

$$\Sigma_{hc} = \Sigma_{Methane} \times (0.23 + 1.4\gamma_g) \quad \textbf{(3.66)}$$

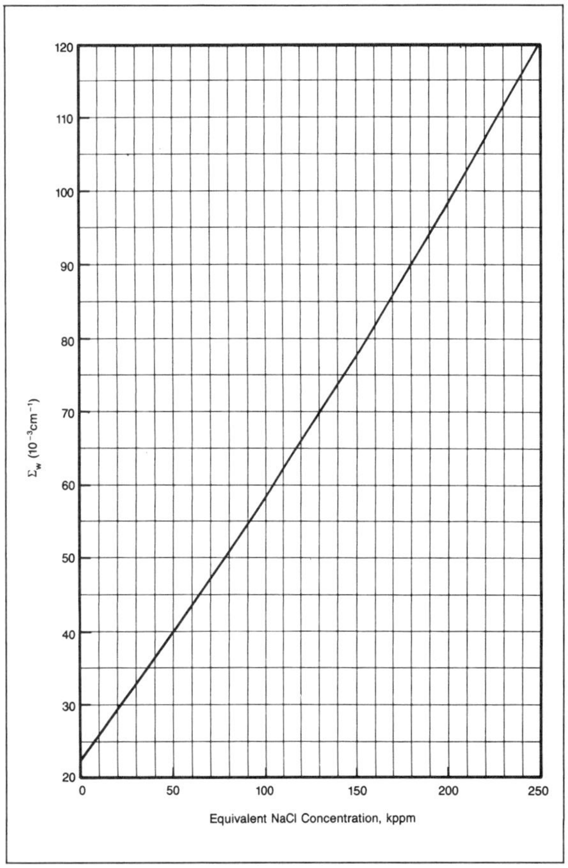


**Figure 3–73.** Chart for determining $\Sigma_w$ when water salinity is known (courtesy of Atlas Wireline Services).[6]

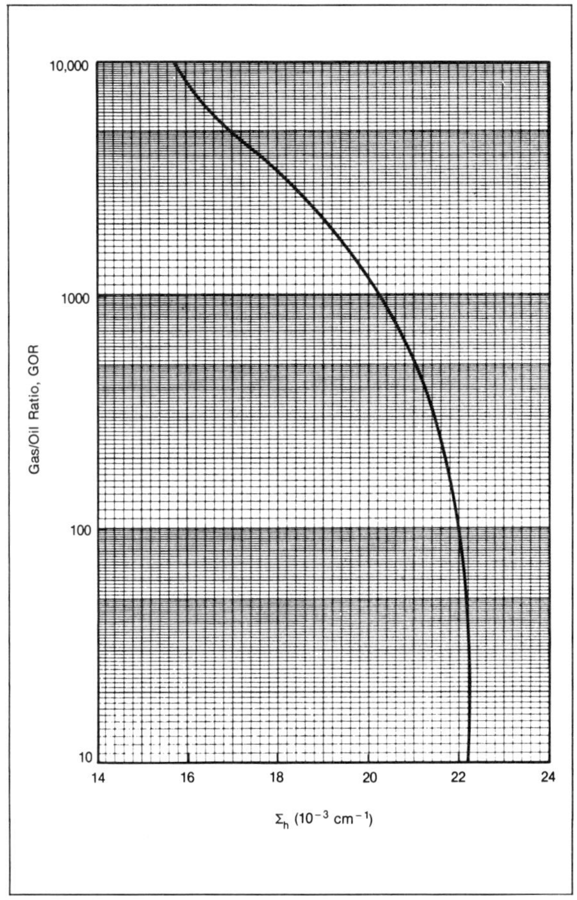


**Figure 3–74.** Average reservoir crude oil cross section (courtesy of Atlas Wireline Services).[6]

## QUALITATIVE INTERPRETATION

In this section we will discuss two techniques which can be used to quickly scan the NLL log for zones that appear to be hydrocarbon-bearing. The following count-rate features should be noted while eyeballing NLL.

1. Shales are characterized by low count rates.
2. Saltwater reservoirs will show low count rates.
3. Oil reservoirs will show higher count rates.
4. Gas-bearing reservoirs will show even higher count rates.

The two techniques discussed here are (1) the shale reference method and (2) the Gulf Coast method.

**The Shale Reference Method.** This method involves establishing a sort of a shale base line (shale reference line). This is done by averaging the highest $\Sigma$s of the shales. The shale reference line is shown in Figure 3–77. The $\Sigma$ value for the 100% water sands has been found in many areas to be 10 sigma units less than the $\Sigma_{shale}$. Therefore, draw the $\Sigma_o$ ($S_w$ = 100%) line 10 sigma units from the shale reference line as shown in Figure 3–77.

All the sands in this case must have about the same porosity and water salinity. The next step is to establish $\Sigma_{os}$ (100% oil saturation) line. Thus any zones with $\Sigma$ value between $\Sigma_o$ and $\Sigma_{os}$ are considered to be potential hydrocarbon-bearing zones, and any zones with $\Sigma$ values less than $\Sigma_{os}$ are considered gas-bearing zones.

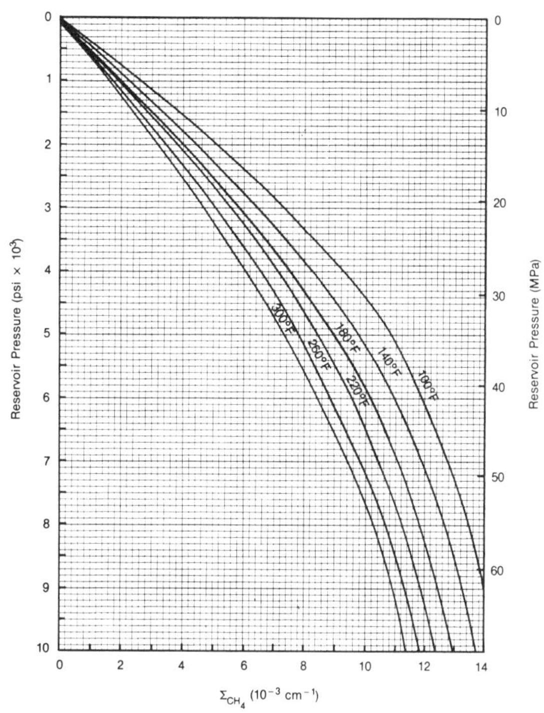


**Figure 3–75.** Neutron capture cross section of Methane at reservoir temperature and pressure (courtesy of Atlas Wireline Services).[6]

Note that this method should only be used where there are no obvious water sands available.

**The Gulf Coast Method.** In Gulf Coast where the water salinity, porosity, and rock type are relatively constant, this technique is applicable. The chart presented in Figure 3–78 is utilized in this method.

The $\Sigma_{os}$ and $\Sigma_{gs}$ (the 100% oil and gas saturation values of sigma) are almost constant for constant porosity and lithology. These $\Sigma$ values are shown in Figure 3–78. If porosity and $\Sigma_{ma}$ are known, $\Sigma_{gs}$ and $\Sigma_{os}$ can be obtained from Table 3–16. Also needed is the $\Sigma_o$ value. This is obtained from the log and plotted on the graph. The $\Sigma_o$ and the appropriate $\Sigma_{gs}$ or $\Sigma_{os}$ are joined by a straight line. This line is then used to establish relationship between $S_w$ and $\Sigma$ from the log.

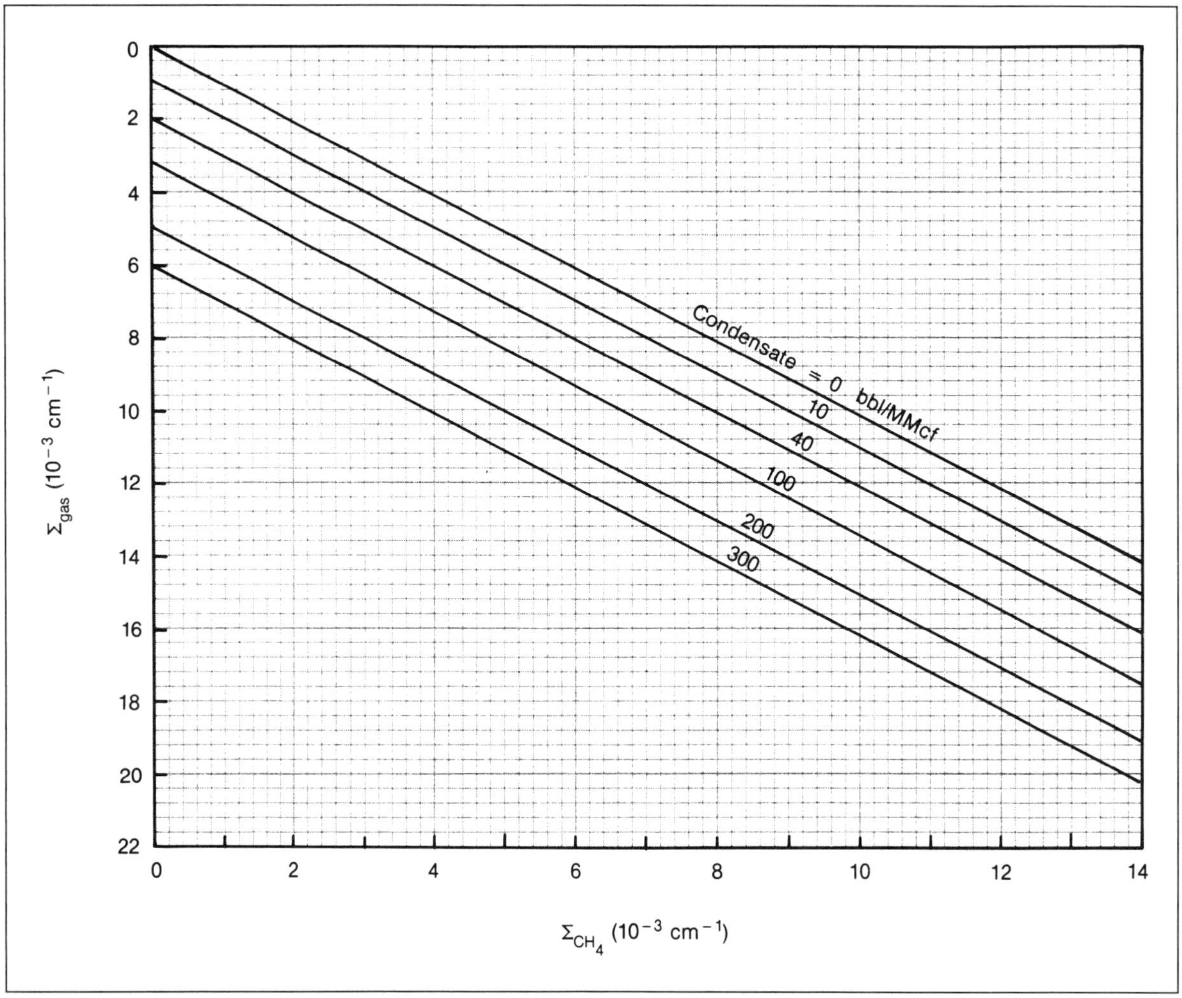


**Figure 3–76.** Chart for determining $\Sigma_{hc}$ when $\Sigma_{Methane}$ and condensate yield are known (courtesy of Atlas Wireline Services).[6]

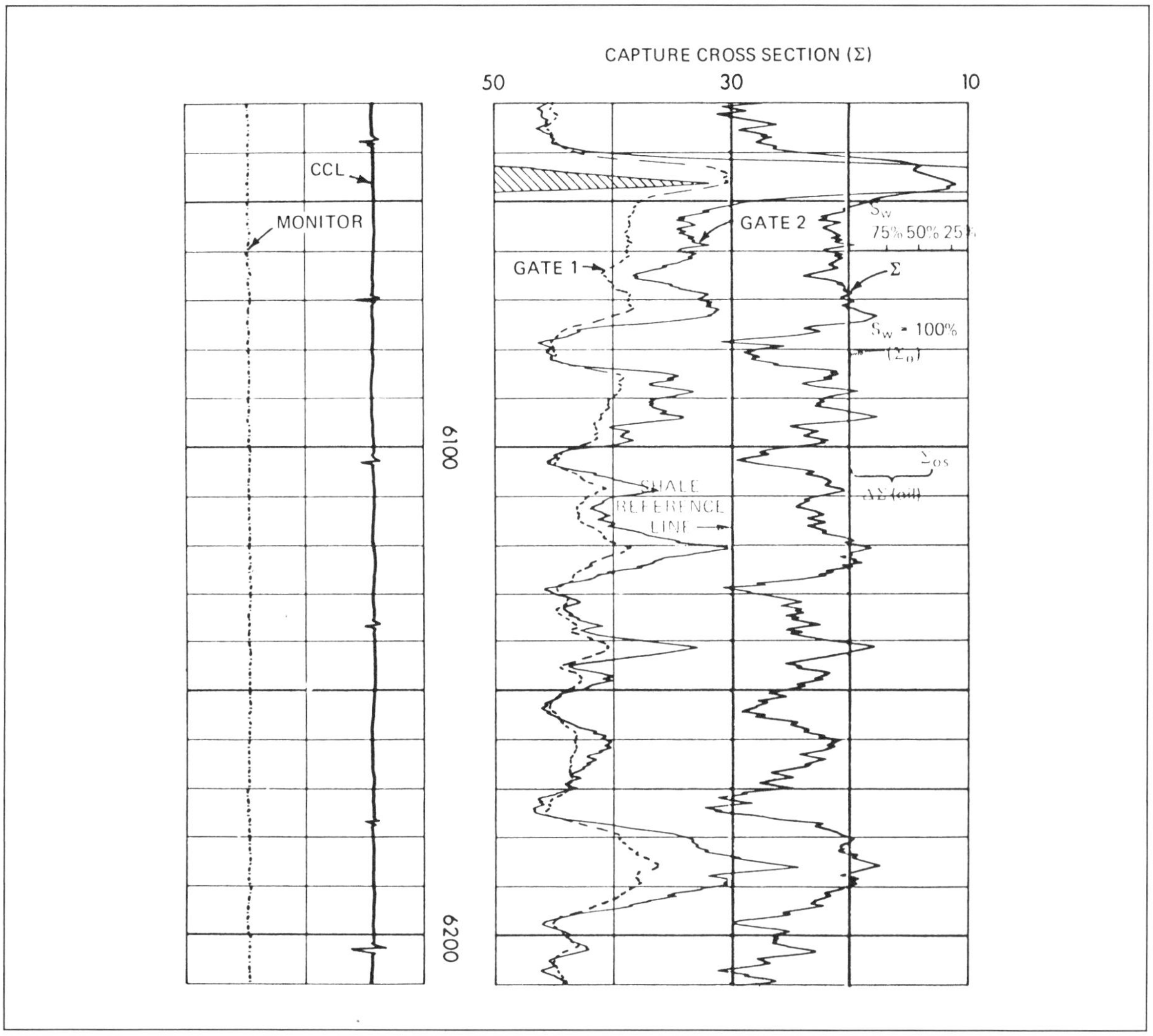


**Figure 3–77.** Presentation of the Shale Reference method (courtesy of Atlas Wireline Services).[22]

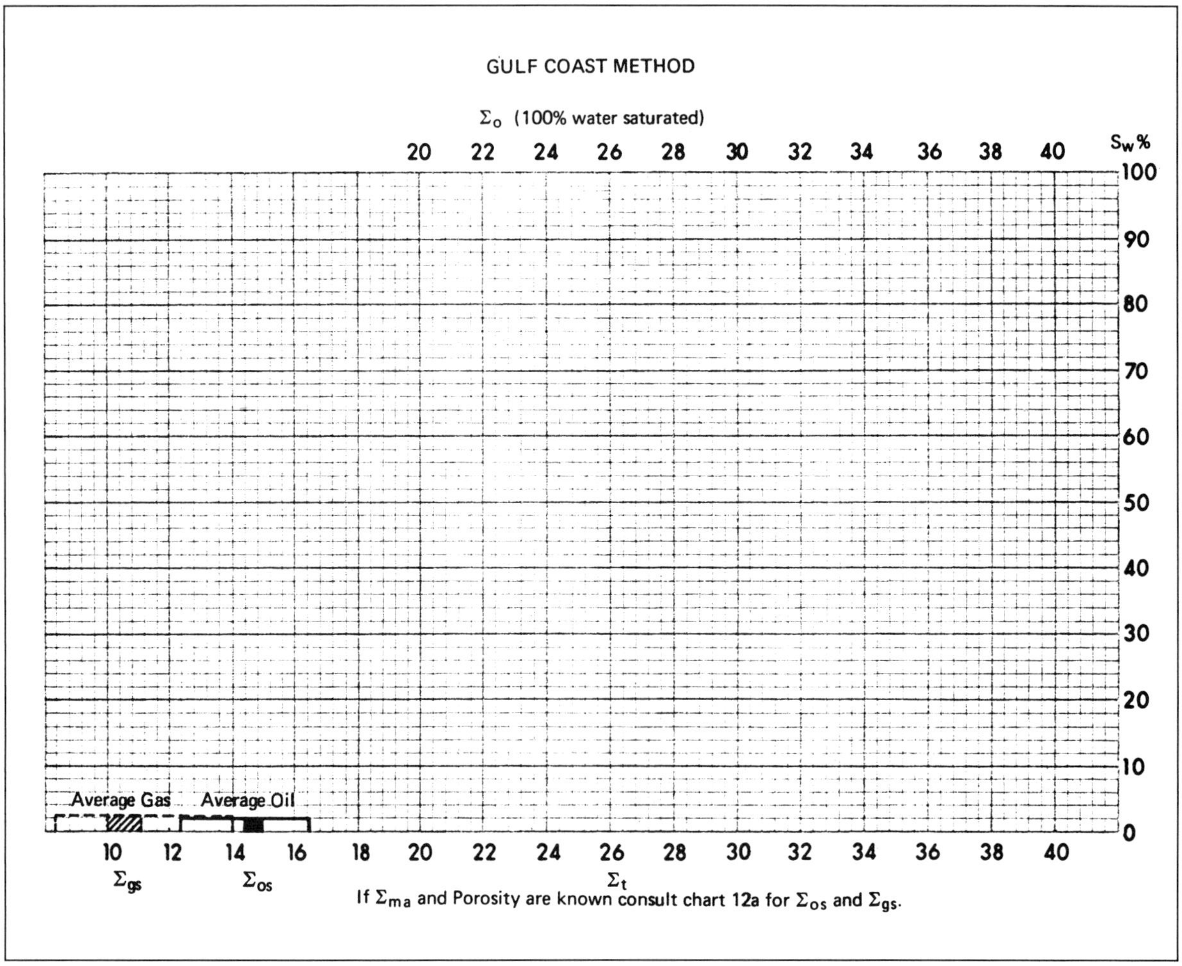


**Figure 3–78.** Chart for qualitative analysis of NLL using the Gulf Coast method (courtesy of Atlas Wireline Services).[22]

**TABLE 3–16. Neutron lifetime sigma, calculated for 100% oil and gas (courtesy Dresser Atlas)[22]**

| | | | | $\Sigma_{os}(\Sigma_{hc} = 22)$ Porosity Percent | | | |
|---|---|---|---|---|---|---|---|
| $\Sigma_{ma}$ | 5 | 10 | 15 | 20 | 25 | 30 | 35 |
| 8 | 8.7 | 9.4 | 10.1 | 10.8 | 11.5 | 12.2 | 12.9 |
| 9 | 9.6 | 10.3 | 10.9 | 11.6 | 12.2 | 12.9 | 13.5 |
| 10 | 10.6 | 11.2 | 11.8 | 12.4 | 13.0 | 13.6 | 14.2 |
| 11 | 11.5 | 12.1 | 12.6 | 13.2 | 13.7 | 14.3 | 14.8 |
| 12 | 12.5 | 13.0 | 13.5 | 14.0 | 14.5 | 15.0 | 15.5 |
| 13 | 13.5 | 13.9 | 14.3 | 14.8 | 15.2 | 15.7 | 16.1 |
| 14 | 14.4 | 14.8 | 15.2 | 15.6 | 16.0 | 16.4 | 16.8 |
| | | | | $\Sigma_{gs}(\Sigma_{hc} = 7.5)$ | | | |
| 8 | 8.0 | 8.0 | 7.9 | 7.9 | 7.9 | 7.9 | 7.8 |
| 9 | 8.9 | 8.9 | 8.8 | 8.7 | 8.7 | 8.6 | 8.5 |
| 10 | 9.9 | 9.8 | 9.6 | 9.5 | 9.4 | 9.3 | 9.1 |
| 11 | 10.8 | 10.7 | 10.5 | 10.3 | 10.2 | 10.0 | 9.8 |
| 12 | 11.8 | 11.5 | 11.2 | 11.1 | 10.9 | 10.6 | 10.3 |
| 13 | 12.7 | 12.5 | 12.2 | 11.9 | 11.7 | 11.4 | 11.1 |
| 14 | 13.7 | 13.4 | 13.0 | 12.7 | 12.4 | 12.1 | 11.7 |

## QUANTITATIVE ANALYSIS—CLEAN FORMATIONS

The determination of water saturation in clean formations using NLL requires the use of Equation (3.63). In this equation $\Sigma_t$ is obtained from the NLL, and $\Sigma_{hc}$ is either assumed to be oil (i.e., $\Sigma_{hc} = 22$) or is obtained from Figures 3–74 through 3–76. If $\Sigma_{hc} = 22$ is assumed and it resulted in $S_w < 0$, then hydrocarbon is probably gas and a recalculation must be performed using appropriate $\Sigma_{hc}$.

The formation water cross section ($\Sigma_w$) may be obtained from catalog, or $R_w$ may be obtained using the methods previously discussed. The $R_w$ is then converted to salinity and then to $\Sigma_w$ using Figure 3–73.

The $\Sigma_{ma}$ must be determined in the water zone or from previous experience. In general, cleaner the formation, the smaller is the $\Sigma_{ma}$. Equation (3.65) may be used to determine $\Sigma_{ma}$ in obvious water-bearing formations where $\phi$ and $\Sigma_w$ are known. The $\Sigma_{ma}$ may also be selected from previous experience in the area or from Table 3–15.

The variation in $\Sigma$ observed between completely water-bearing and hydrocarbon-bearing formations is given by the following equation:

$$\Delta\Sigma = \phi(\Sigma_w - \Sigma_{hc}) \qquad \textbf{(3.62)}$$

As with other logging programs (Pickett and Hingle plots) it is often desirable to plot $\Sigma_t$ versus $\phi$ on Cartesian coordinate paper. The plot provides $S_w = 100\%$ line, $\Sigma_{ma}$, and $\Sigma_w$ (slope of $\Sigma_o$ line).

## QUANTITATIVE ANALYSIS—SHALY FORMATIONS

The $\Sigma$s of shale are normally high due to interstitial water and concentrations of boron, iron, aluminum, and other high $\Sigma$ materials. The water saturation for shaly formations is calculated using Equation (3.64).

The $\Sigma_{sh}$ is the capture cross section of the adjacent shale. The volume of shale ($V_{sh}$) may be determined from gamma ray log, SP log (should be considered a maximum shaliness indicator due to the presence of hydrocarbons), or density-neutron crossplot. Refer to the shaly formation evaluation of openhole well logs.

Figure 3–79 provides an empirical relationship which can be used to correct NLL for shale content and to calculate $S_w$. The right two-thirds of this monograph also can be used to calculate water saturation for clean (nonshaly) formations.

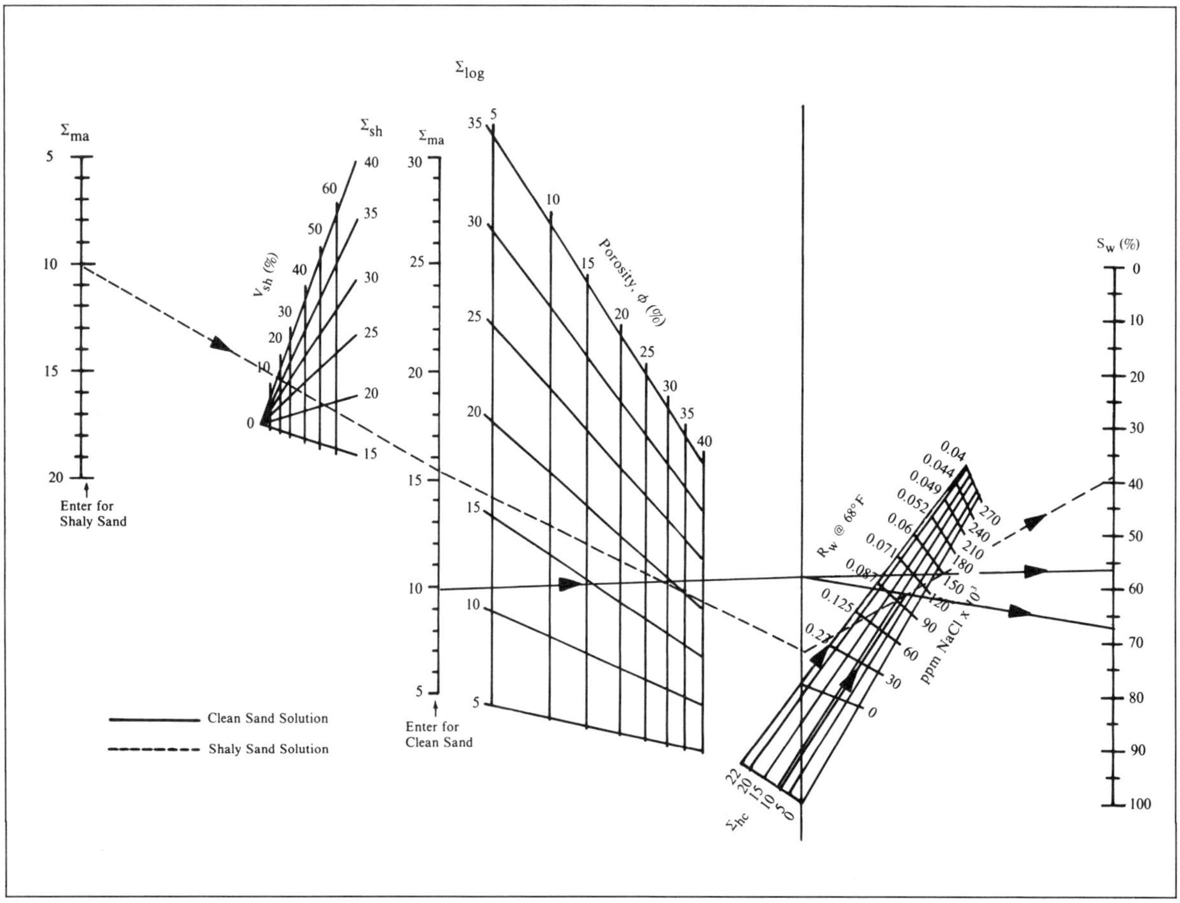


**Figure 3–79.** NLL shale correction and $S_w$ nomograph (courtesy of Atlas Wireline Services).[6]

**Example 3–24.** Figure 3–80 is an early NLL over a zone of high-porosity (28%) and high-salinity (165,000 ppm) water. Calculate water saturation for the zones marked on the log from 6,996 to 7,002 ft. The ratio of Gate 1 to Gate 2 scales is 6, and time separation of gates is 300 microseconds.

Solution: Since no $\Sigma$ curve is available, we have to calculate $\Sigma$ using Equation (3.61).

$$\Sigma_w = 84 \qquad \phi = 0.28$$



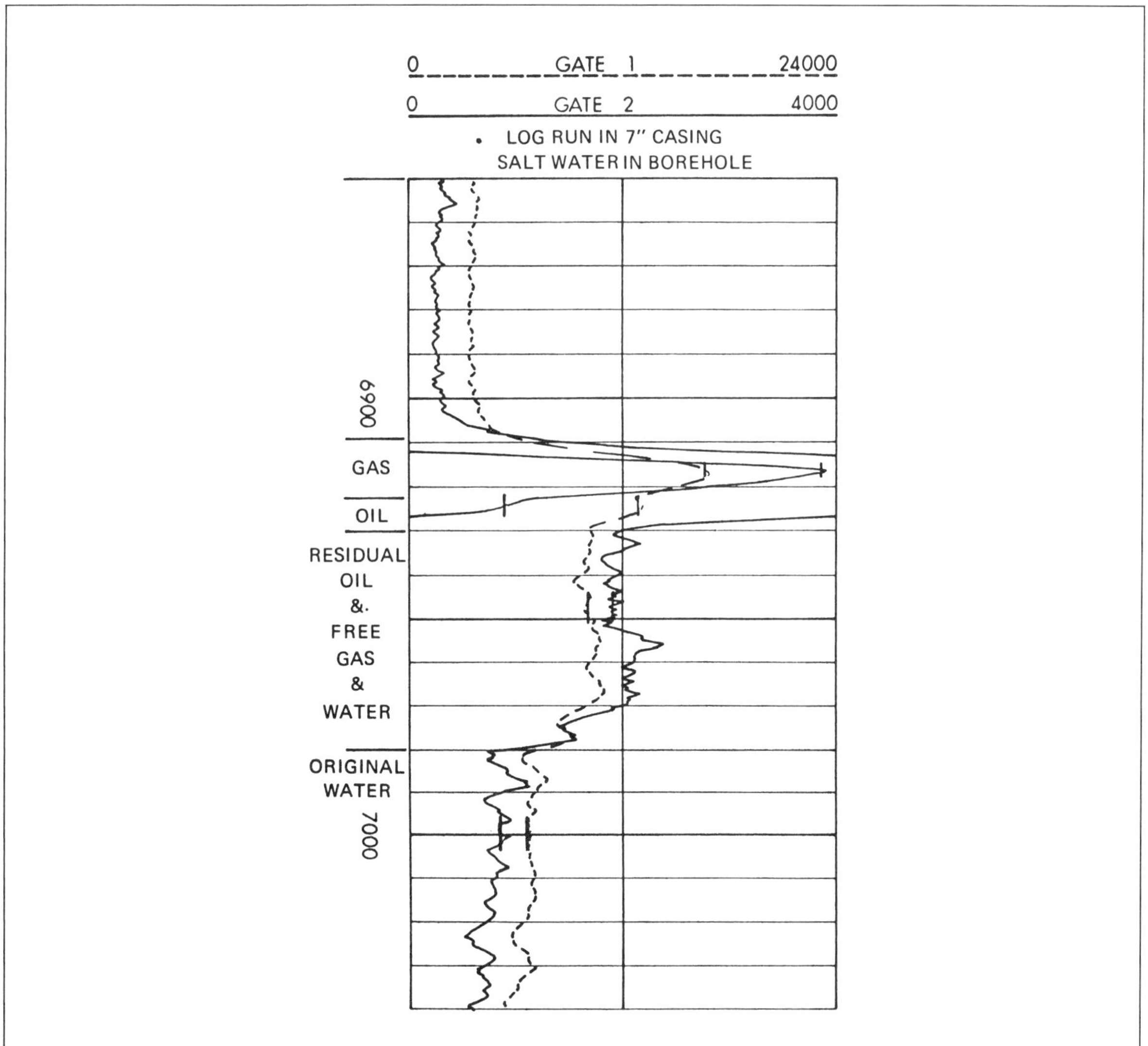


**Figure 3–80.** Neutron lifetime log for Example 3–24 (courtesy of Atlas Wireline Services).[22]

Calculate $\Sigma_o$ from water zone at 6,996 to 7,002 ft, $D_1 = 5.5$ and $D_2 = 4.5$ divisions.

$$\Sigma_o = \frac{10{,}500}{\Delta t} \log \frac{6D_1}{D_2}$$
$$= \frac{10{,}500}{300} \log \left(6 \times \frac{5.5}{4.5}\right) = 30.29 \text{ su}$$

Using Equation (3.65),

$$\Sigma_{ma} = \frac{\Sigma_o - \phi\Sigma_w}{1 - \phi}$$
$$= \frac{30.29 - (0.28)(84)}{1 - 0.28}$$
$$= 9.40 \text{ su}$$

Now calculate $S_w$ for the zone at 6,944 to 6,950 feet. Assume $\Sigma_{hc} = 22$ su. The zone shows $D_1 = 8.4$ and $D_2 = 9.6$. Therefore, $\Sigma_t$ is

$$\Sigma_t = 35 \log \frac{6D_1}{D_2}$$
$$= 35 \log \left(6 \times \frac{8.4}{9.6}\right)$$
$$= 25.21$$
$$S_w = \frac{\Sigma_t - \Sigma_{ma} + \phi(\Sigma_{ma} - \Sigma_{hc})}{\phi(\Sigma_w - \Sigma_{hc})}$$
$$= \frac{25.21 - 9.40 + 0.28(9.4 - 22)}{0.28(84 - 22)}$$
$$= 71\%$$

It is an oil zone which has been displaced by movement of water into the zone. Residual oil saturation is 29% ($1 - 0.71 = 0.29$). Similarly for the other two zones:

| | | | | |
|---|---|---|---|---|
| 6,923–6,926 | $D_1 = 10.8$ | $D_2 = 24.7$ | $\Sigma_t = 14.66$ | $S_w = 0.10$ |
| 6,917 | $D_1 = 14.0$ | $D_2 = 39.2$ | $\Sigma_t = 11.58$ | $S_w < 0\%$ |

The zone at 6,923 to 6,926 ft is an oil zone. Low water saturation is probably due to some gas saturation around the wellbore. The zone at 6,917 ft calculates $S_w < 0\%$; therefore, it must be a gas zone.

Let us assume that $T_f = 210°F$ and reservoir pressure at 6,917 ft is approximately 3,500 psi. From Figure 3–75, $\Sigma_{hc} = 6.4$; therefore, $S_w = 13\%$.

**Example 3–25.** Using the shaly formation analysis technique of NLL analysis, calculate water saturation $S_w$ for the three intervals shown in Figure 3–81. Other relevant data are as follows:

$$\Sigma_w = 60 \qquad \Sigma_{hr} = 22 \quad \text{if oil}$$
$$\Sigma_{ma} = 10 \quad \text{(assumed)} \qquad \Sigma_{hc} = 9 \quad \text{if gas}$$
$$T_f = 185°\text{F} \qquad p_R = 5{,}000 \quad \text{psi}$$

Solution: For Interval I @11,261–11,264 ft

$$GR_{max} = 8.8 \qquad GR_{(1)} = 5.3$$
$$GR_{min} = 2.3 \qquad \Sigma_t = 19$$
$$I_{GR} = \frac{GR_{(1)} - GR_{min}}{GR_{max} - GR_{min}} = \frac{5.3 - 2.3}{8.8 - 2.3} = 0.46$$

From Figure 3–13, assuming older rocks, $V_{sh} = 0.31$.

Using Equation (3.64), $S_w$ is

$$S_w = \frac{19 - 10 + 0.27(10 - 9) + 0.31(10 - 37)}{0.27(60 - 9)} \approx 7\%$$

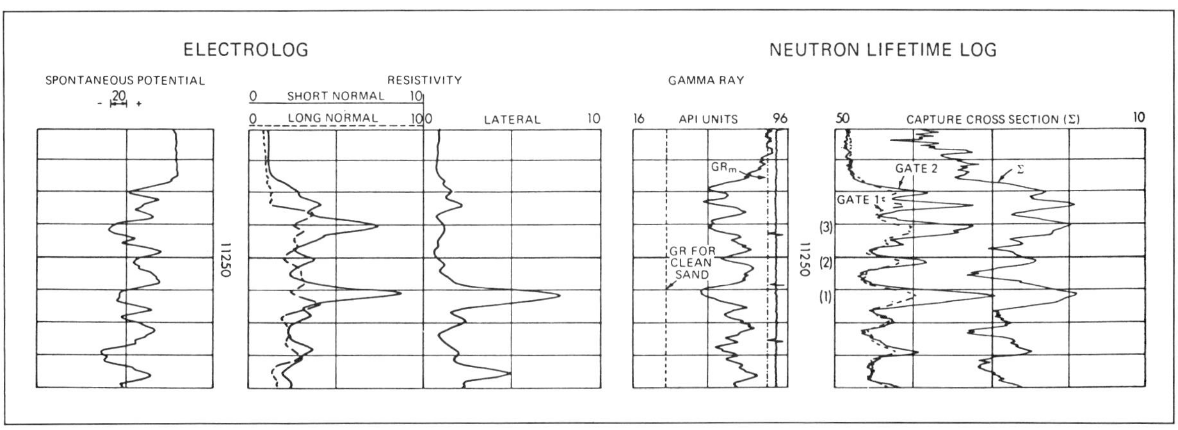


**Figure 3–81.** Neutron lifetime log for Example 3–25 (courtesy of Atlas Wireline Services).[22]

Similarly for the other two zones:

| Depth | $\Sigma_t$ | $GR_{LOG}$ | $I_{GR}$ | $V_{sh}$ | $S_w$ |
|---|---|---|---|---|---|
| 11,250–11,253 | 24 | 5.9 | 0.55 | 0.42 | 21 |
| 11,240–11,249 | 20 | 5.2 | 0.45 | 0.32 | 12 |

## NATURALLY FRACTURED RESERVOIRS

In this section some qualitative and quantitative interpretation techniques for evaluating well logs in naturally fractured reservoirs are presented. Naturally fractured reservoirs hold large hydrocarbon reservoirs and represent potentially attractive economic targets in exploratory ventures. Evaluation techniques in these reservoirs, however, have lagged far behind the advanced technology applied to conventional reservoirs.

Although fractures cannot be detected with absolute certainty, their presence may be deduced from some of the qualitative techniques presented in this section. Aguilera has presented very comprehensive work on the subject.[25–27]

### QUALITATIVE ANALYSIS

Logging tools are designed to respond to different characteristics of the wellbore environment. Some tools respond to lithology, some to porosity, and others to fluid saturation. However, no logs respond primarily to fractures; only open fractures may affect the response of some logging tools. The qualitative interpretation of fractures from some logs is presented in the following paragraphs.

**Sonic Amplitude Logs.** These logs have been extensively used in efforts to detect fractures. Four wave types can be identified from acoustic velocity logs as (1) compressional wave, (2) shear wave, (3) fluid or water wave, and (4) low-velocity wave. The compressional wave is generally more attenuated by high angle and vertical fractures, while the shear wave is more sensitive to low-angle and horizontal fractures.

In fractured zones, the appearance of the wavetrain as recorded on the Variable Density log (VDL) shows sudden changes—vague or blurred zones, chevron patterns, and others as illustrated in Figure 3–82. However, experience indicates that sonic amplitude logs are not universally applicable because changes in amplitude as large as those due to fractures can also be produced by variations in lithology, hole size, tool centralization, borehole rugosity, and porosity variation. Also, solid-to-solid contact across the fractures would diminish the degree of acoustic discontinuity.

Cycle skipping on acoustic logs can also be used to qualitatively identify fractures. If travel times are larger than the times generally recorded, cycle skipping has occurred. Cycle skipping at the fracture interface is because the shear energy is reflected, some is converted to other modes of wave propagation, and some is refracted. As a result, the amount of shear energy reaching the receiver is significantly reduced.

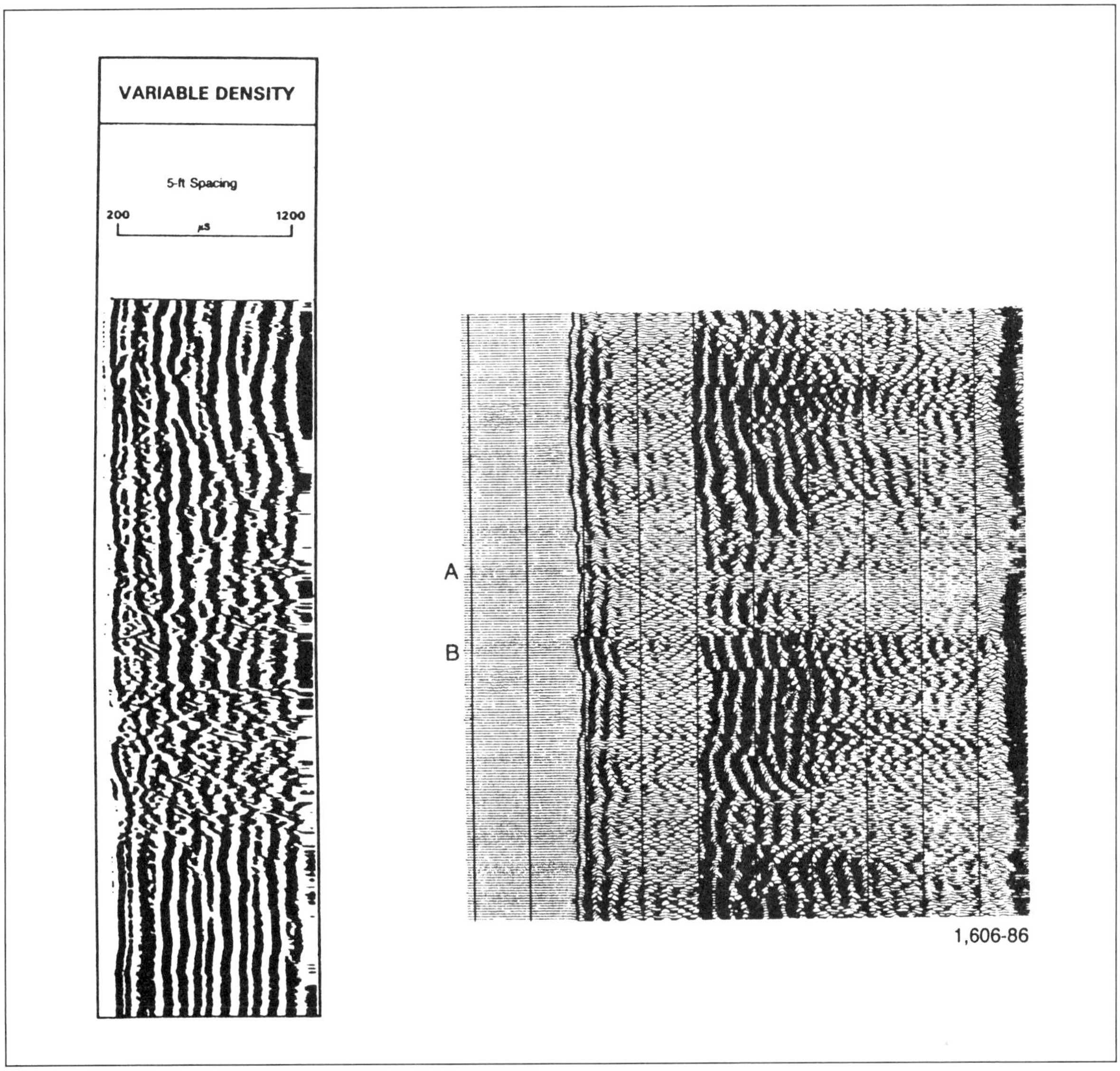


**Figure 3–82.** (a) Fracture identification with VDL display and (b) Sonic wavetrain response over a large horizontal fracture (courtesy of Schlumberger).[3]

**Spontaneous Potential (SP) Log.** The SP curve can be used qualitatively to determine the presence of fractures. Due to the mud filtrate invasion into the fractures, a streaming potential effect takes place giving a hechured SP curve. Each pip on the curve most likely corresponds to a streaming potential effect.

**Sonic and Neutron or Density Logs.** Fractures may be detected with the use of a sonic log and either a density or a neutron log. It is assumed that the sonic log sees matrix porosity while the neutron or density log reads the total fluid-filled porosity. The difference between the logs is interpreted as fracture porosity. The following difficulties may, however, create some interpretation uncertainties:

1. Since total porosity is usually estimated from a measure in only one side of the borehole, it may be underestimated.
2. Variations in shaliness and lithology may indicate the presence of nonexisting fractures.
3. Hole irregularities may lead to optimistic estimates of total porosity, even if fractures are not present.
4. Secondary porosity, such as vugs, may be interpreted as fracture porosity.

**Core Porosity and Neutron Logs.** Comparison of core porosities with neutron log response and of neutron log response with other porosity logs may be helpful in determining the total number of zones that may be fractured in a reservoir.[24] This technique, however, depends on the presence of a normal distribution around the average correction of the responses when no fractures are present.

**Caliper Measurements.** A multiarm, multidirectional caliper, such as the high-resolution dipmeter (HDT) or Dual Dipmeter (SHDT) can be used to detect fractures. When drilling a fractured zone, rock edges of the fractures are often chipped away, thereby enlarging the borehole in the plane of the fracture system. This borehole enlargement and elongation in a formation, which will otherwise indicate a gauged and circular borehole, can indicate fractures.

Using caliper for fracture detection, however, should be used with caution, since directional drilling, deviated boreholes, drilling through steeply dipping beds, an oriented pore structure, and other events can also cause an elongated borehole.[3] If lost circulation material is used during drilling, fractures can sometimes have the opposite effect on the calipers.

**Density Measurements.** The correction curve on a density log, $\Delta\rho$, can be used as a fracture indicator. An active, erratic $\Delta\rho$ curve (spikes) may indicate fractures when the hole is in gauge.[3] A disadvantage of this curve is that it sees only a small portion of the borehole and therefore miss the fracture. Also, the "chipping-out" of the fracture during drilling may slightly increase the $\phi_D$. The $\Delta\rho$ will also respond to washouts not associated with fractures. Therefore, care should be exercised.

**Resistivity Measurements.** Induction log resistivity anomalies might suggest fracture, provided that sufficient contrast with the resistivities of adjacent beds is present. The generation of such anomaly depends on invasion of vertical fractures with nonconductive fluid (oil-base mud). On a dual induction log, fractures may be indicated when the SFL or LL8 curves exhibit spurious low-resistivity readings not evident on the deep or medium induction readings. The larger the separation, the stronger the fracture intensity, other things being equal. Figure 3–83 is an example of fracture detection with the DLL-MSFL logs.

Some of the other logs used for fracture identification are dipmeter measurements, Borehole Televiewer tool, and the most recent introduction of Formation Microscanner (FMS) tool. As can be deduced from the preceding paragraphs, no one log will furnish unequivocal proof of the presence of fractures. Several logs or techniques should be used simultaneously in order to effectively confirm the identification of fractures.

## QUANTITATIVE ANALYSIS

In this section techniques used for quantitative analysis of naturally fractured reservoirs from conventional well logs are discussed. The methods are not claimed to be perfect; however, they have provided reasonable results in evaluating orders of magnitude of fracture, total, and matrix porosities and in water saturations in matrix, fractures, and the composite matrix-fractures system.[25]

The literature suggests that the porosity exponent $m$ of a fracture-matrix system is smaller (ranging from about 1.1 to 1.3) than the porosity exponent $m_{ma}$ of the matrix if the fracture is not healed and that $m = n$ in the Archie equation. It has been shown that the value of $m$ in the fracture-matrix system may range somewhere between about 1.1 and $m_{ma}$ determined in the laboratory. Thus a comparison of $m$ (obtained from logs) and $m_{ma}$ (determined in the laboratory) gives a reliable way to detect naturally fractured systems.

Pirson presented the following relationship for the case in which the matrix and fractures are 100% saturated with water:[28]

$$\frac{1}{R_{fo}} = \frac{\nu\phi_t}{R_w} + \frac{1 - \nu}{R_o} \tag{3.67}$$

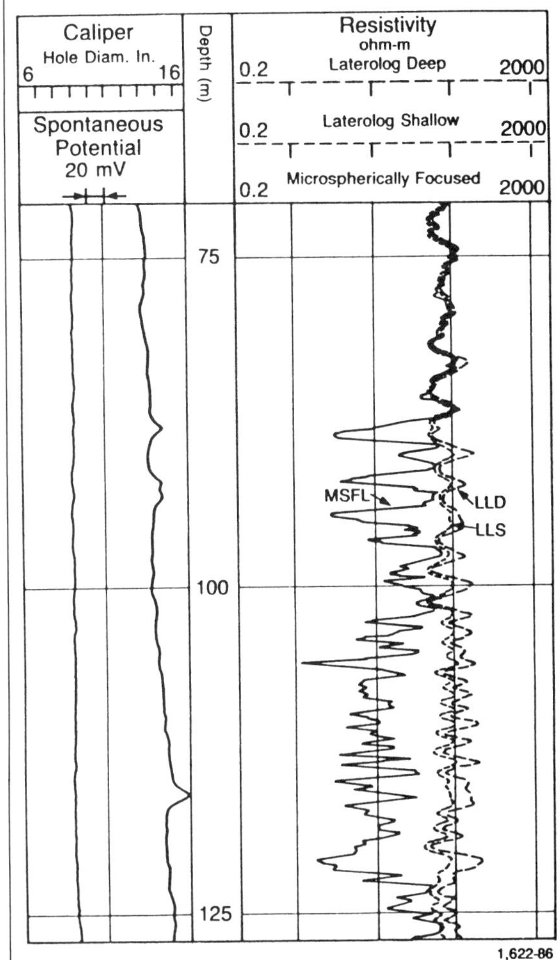


**Figure 3–83.** Fracture identification with DLL-MSFL logs (courtesy of Schlumberger).[3]

where $\nu$ is the partitioning coefficient representing the fraction of total pore volume made up of fractures and is given by[29]

$$\nu = \frac{\phi_t - \phi_{ma}}{\phi_t(1.0 - \phi_{ma})} \tag{3.68}$$

OR

$$\phi_t = \frac{-\phi_{ma}}{\nu(1 - \phi_{ma}) - 1} \tag{3.68a}$$

and

$$R_{fo} = \frac{R_w R_o}{\nu \phi_t R_o + (1 - \nu) R_w} \tag{3.69}$$



The system formation factor $F_t$ is then given by

$$F_t = \left[ \frac{R_w R_o}{\nu \phi_t R_o + (1 - \nu) R_w} \right] / R_w \tag{3.70}$$

OR

$$F_t = \frac{1}{\nu \phi_t + (1 - \nu)/F} \tag{3.70a}$$

Finally,

$$\phi_t^{-m} = \frac{1}{\nu \phi_t + (1 - \nu)/\phi_{ma}^{-m_{ma}}} \tag{3.71}$$

OR

$$m = \frac{-\log \left[ \frac{1}{\nu \phi_t + (1 - \nu)/\phi_{ma}^{-ma}} \right]}{\log \phi_t} \tag{3.71a}$$

Equations (3.68) and (3.71) can be used to estimate the following:

1. Total ($\phi_t$) and fracture ($\phi_f$) porosities and the partitioning coefficient ($\nu$) as a function of $\phi_{ma}$, $m_t$, and $m_{ma}$.
2. Matrix ($\phi_{ma}$) and fracture ($\phi_f$) porosities and the partitioning coefficient ($\nu$) as a function of $\phi_t$, $m_t$, and $m_{ma}$.
3. The double-porosity exponent ($m_t$) as a function of $\phi_t$, $\phi_{ma}$, and $m_{ma}$. The values of $m_{ma}$ and $\phi_{ma}$ can be obtained in the laboratory, and total porosity can be determined from density and neutron logs.

The value of porosity exponent $m_t$ of the double-porosity system can be obtained from a log-log plot of $(\Delta t - \Delta t_{ma})$ vs. $R_t$. Such a plot should result in a straight line with a slope $-m$ for zones of $S_w = 100\%$, constant $R_w$, $m$, and $\Delta t_{ma}$ (refer to resistivity-porosity crossplots in this chapter), provided correct value of $\Delta t_{ma}$ is used. If the value of $\Delta t_{ma}$ is not correct, an upward or downward curved line will result. Similar type of analyses can be made for different porosity tools by crossplotting

1. Density: $\log (\rho_{ma} - \rho_b)$ vs. $\log R_t$
2. Epithermal neutron (SNP): $\log \phi_{SNP}$ vs. $\log R_t$
3. Pulsed Neutron Log: $\log(\Sigma - \Sigma_{ma})$ vs. $\log R_t$

The value of $aR_w$ can be determined directly from the log-log plot by extrapolating the 100% water-bearing trend to a porosity of $\phi = 100\%$. The hydrocarbon-bearing zones in such a plot will be easily identified.

Finally, the value of $S_w = I^{-1/n}$ can be calculated for each hydrocarbon point by calculating the resistivity index (I) for the hydrocarbon zone by dividing its resistivity ($R_t$) by the resistivity of the water-bearing trend ($R_o$) at the same level of porosity.

Sometimes it is difficult to establish the position of the 100% water-bearing trend. In such cases a statistical analysis using a parameter ($P$) becomes a valuable tool. The parameter $P$ is found to have a square-root-normal distribution for zones with 100% water saturation as

$$P^{1/2} = [R_t(\Delta t - \Delta t_{ma})^m]^{1/2} \tag{3.72}$$

Similarly,

$$P^{1/2} = [R_t(\rho_{ma} - \rho_b)^m]^{1/2} \tag{3.72a}$$

and

$$P^{1/2} = [R_t(\Sigma - \Sigma_{ma})^m]^{1/2} \tag{3.72b}$$

These relationships show that a plot of $P^{1/2}$ vs. cumulative frequency on probability graph paper should result in an approximate straight line with the hydrocarbon zones deviated from this line in the north-east direction. Note that this method was initially introduced by Porter et al. for the analysis of homogeneous formations.[30]

Once the hydrocarbon zones have been recognized, value of water saturation for these zones can be calculated as follows:

1. Consider the 100% water saturation zones as a single distribution. Replot $P^{1/2}$ vs. the cumulative frequency for these data on probability graph paper.
2. Draw the best-fit line through the data points and determine the value of $P$ at a cumulative frequency of 50%.

3. Calculate the resistivity index $I = P_H/P_{100}$, where $P_H$ is the value of $P$ for hydrocarbon-bearing zone and $P_{100}$ is the mean value of $P$.
4. Calculate water saturation $S_w = I^{-1/n}$ for each zone, assuming that $m_t = n$.

**Example 3–26.**[25] This hypothetical example provides a working knowledge of the quantitative analysis presented in this section. Using the data given in Table 3–17, determine:

1. The porosity exponent $m$
2. Water saturation $S_w$ for zones 10 and 11.

Solution:

1. Log-log plot of ($\Delta t - 55$) vs. $R_t$ is shown in Figure 3–84, from the plot $m = 1.44$ and Zones 10 and 11 are identified as hydrocarbon-bearing zones.
2. To calculate water saturation, $P^{1/2}$ as calculated in Table 3–17 using $m = 1.44$ is plotted versus cumulative frequency distribution for all zones as shown in Figure 3–85. The plot confirms the hydrocarbon zones as identified in Figure 3–84.

Frequency distribution for the water-bearing zones only is calculated as shown in Table 3–18 and plotted in Figure 3–86. The mean value of $P^{1/2}$ for water-bearing zones is 16.94 at a cumulative frequency of 50%; consequently $P_{100} = 296.964$.

**TABLE 3–17. Data for the analysis of fractured reservoir ($\Delta t_m = 55$ μsec/ft)[25]**

| Zone | $\Delta t$ μsec/ft | $R_t$ Ohm-m | $P^{1/2} = [R_t(\Delta t - \Delta t_{ma})^m]^{1/2}$ |
|---|---|---|---|
| 1 | 58.0 | 55 | 16.357 |
| 2 | 60.0 | 27 | 16.555 |
| 3 | 64.2 | 11 | 16.392 |
| 4 | 61.0 | 29 | 19.564 |
| 5 | 59.0 | 49 | 18.992 |
| 6 | 60.0 | 35 | 18.849 |
| 7 | 61.0 | 23 | 17.423 |
| 8 | 62.0 | 19 | 17.695 |
| 9 | 63.0 | 15 | 17.309 |
| 10 | 63.0 | 100 | 44.691 |
| 11 | 62.0 | 115 | 43.533 |

**TABLE 3–18. Frequency distribution of $P^{1/2}$ calculated in Table 3–17**

| Range of $P^{1/2}$ | No. of Samples | All Zones Frequency | All Zones Cum. Freq. | Water Zones Frequency | Water Zones Cum. Freq. |
|---|---|---|---|---|---|
| 16–17 | 3 | 0.273 | 0.273 | 0.333 | 0.333 |
| 17–18 | 3 | 0.273 | 0.546 | 0.333 | 0.666 |
| 18–19 | 2 | 0.182 | 0.728 | 0.222 | 0.888 |
| 19–20 | 1 | 0.091 | 0.819 | 0.111 | 0.999 |
| 43–44 | 1 | 0.091 | 0.910 | | |
| 44–45 | 1 | 0.091 | 1.001 | | |

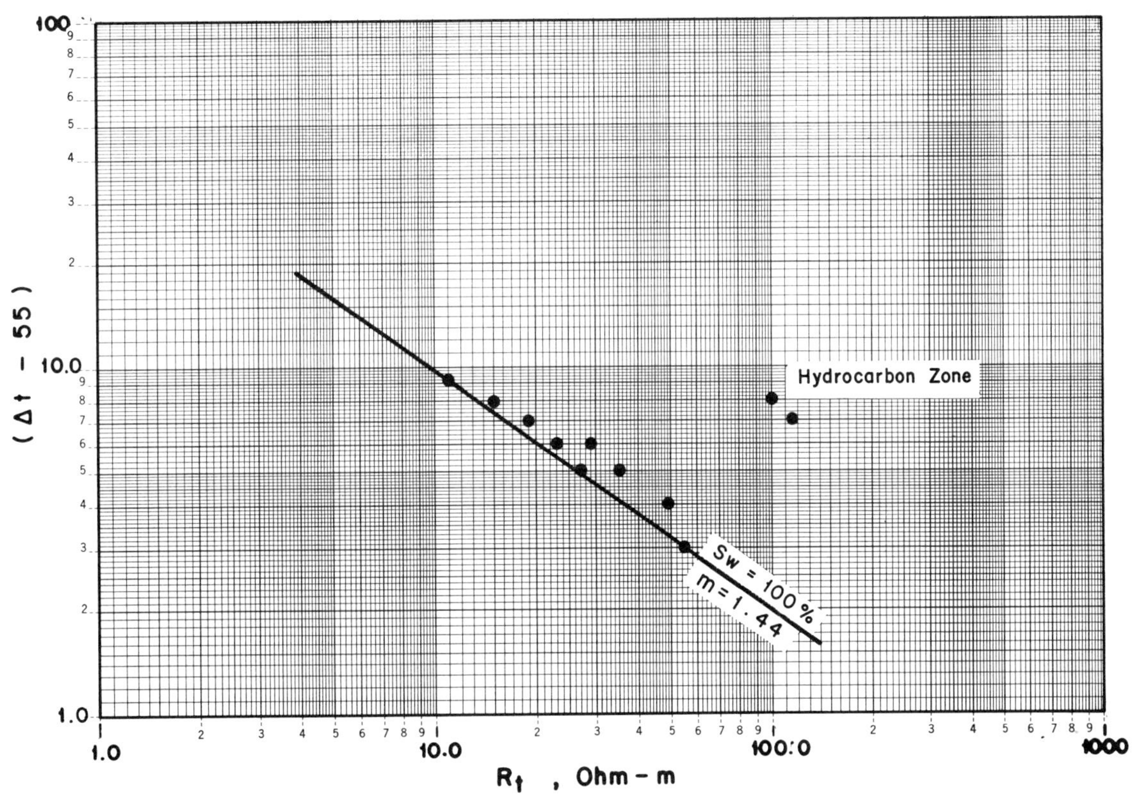


**Figure 3–84.** ($\Delta t$ − 55) vs. $R_t$ allows estimation of $m$ value to be used in the $P^{1/2}$ statistical analysis.

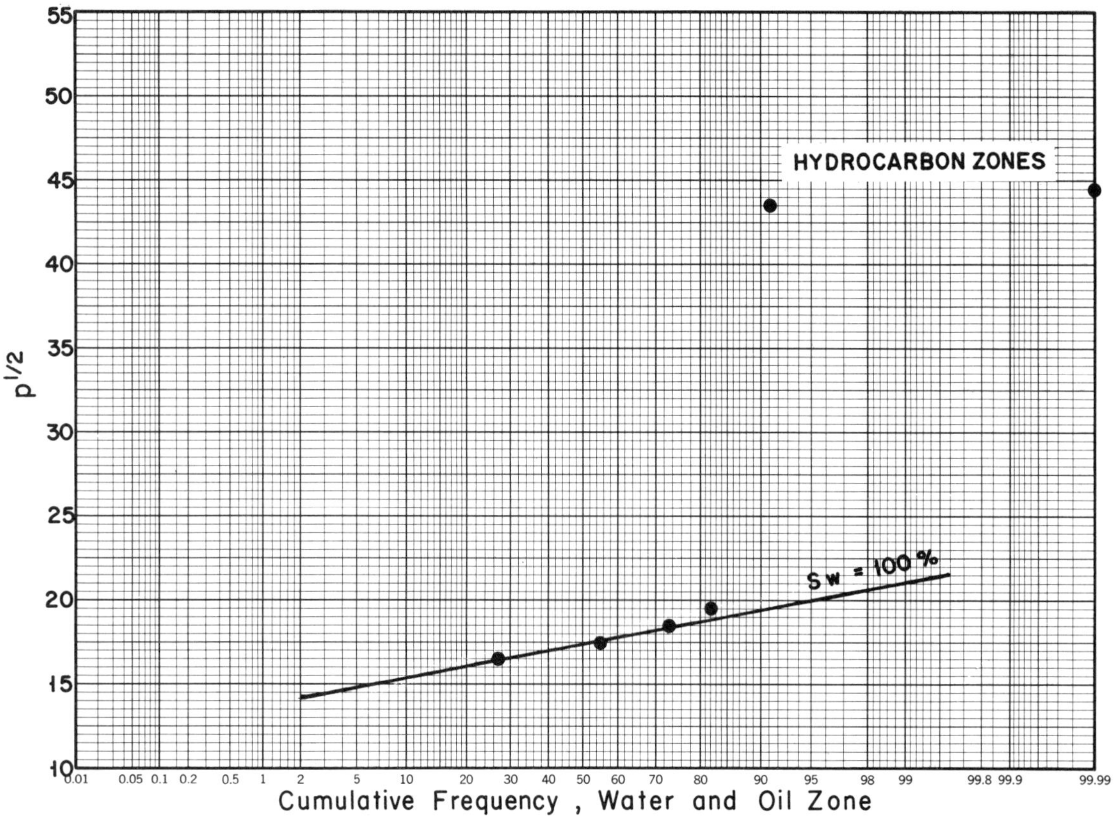


**Figure 3–85.** $P^{1/2}$ vs. cumulative frequency for oil and water zones.

Values of resistivity index ($I$) for the hydrocarbon zones can be determined from the relationship $I = P_H/P_{100}$ as follows:

| Zone | $P^{1/2}$ | $P_H$ | $I = P_H/P_{100}$ | $S_w = I^{-1/n}$ |
|---|---|---|---|---|
| 10 | 44.691 | 1997.285 | 6.960 | 0.26 |
| 11 | 43.533 | 1895.122 | 6.604 | 0.27 |

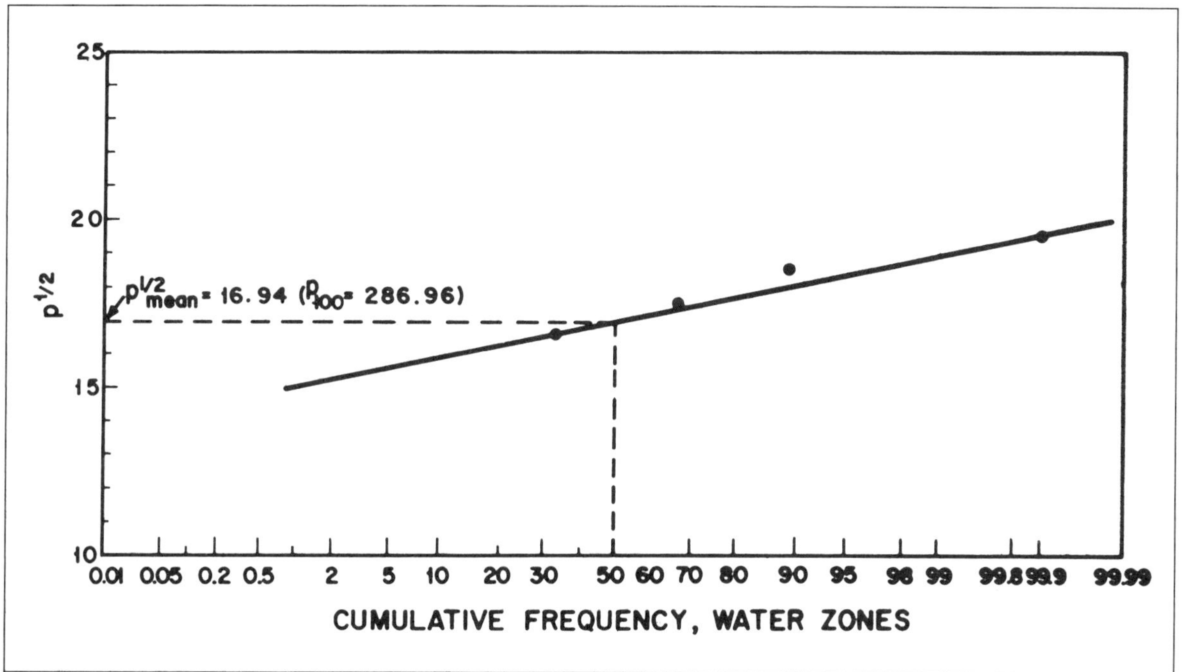


**Figure 3–86.** $P^{1/2}$ vs. cumulative frequency for water zones only.

## ELECTROMAGNETIC PROPAGATION TOOL (EPT)

In this section a relatively newer logging device referred to as the *Electromagnetic Propagation Tool (EPT)* and the techniques used to interpret the data obtained with EPT are discussed.

The EPT is a subsurface logging device which measures the propagation travel time ($t_{pl}$) and signal attenuation ($A_{LOG}$), which are the basic measurements that are displayed on the log. The log responds primarily to the bulk volume of water in the formation. This includes the bulk volume of *bound water* associated with shales ($\phi_B$) and bulk volume of *free water* ($\phi_F$). The free water consists of formation connate water plus mud filtrate if the zone is invaded.

The EPT has a relatively shallow depth of investigation (ranging from 1 to 6 in., depending on conductivity) and therefore the measurements respond primarily to the flushed zone. Comparison of the bulk volume water from EPT measurement with the total porosity measured with neutron and density logs provides water saturation of the flushed zone without requiring previous knowledge of $R_w$ and $R_t$.

The EPT tool is designed to operate under these conditions: (1) $R_{xo} \geq 0.3$ ohm-m, (2) formation pressures $\leq$ 20,000 psi, (3) formation temperatures $\leq$ 350°F, (4) borehole diameter in the range of $6\frac{1}{4}$ to 22 in., and (5) mudcake thickness $\leq \frac{3}{8}$ in. The hole size has no significant effect on the EPT measurement as long as the hole diameter is within the tool's operational limits. However, hole rugosity can result in erratic and erroneously high $t_{pl}$ measurements. The EPT will not give accurate measurements in holes with air- or oil-based mud systems.

**TABLE 3–19. Relative dielectric constants and propagation time for various minerals and fluids[3]**

| Mineral | Relative Dielectric Permitivity ($\epsilon$) | Propagation Time $t_{pl}$ (ns/m) |
|---|---|---|
| Sandstone | 4.65 | 7.2 |
| Dolomite | 6.80 | 8.7 |
| Limestone | 7.5–9.2 | 9.1–10.2 |
| Anhydrite | 6.35 | 8.4 |
| Halite | 5.6–6.35 | 7.9–8.4 |
| Gypsum | 4.16 | 6.8 |
| Dry Colloids | 5.76 | 8.0 |
| Shale | 5.25 | 7.45–16.6 |
| Oil | 2.0–2.4 | 4.7–5.2 |
| Gas | 1.0 | 3.3 |
| Water | 56–80 | 25–30 |
| Fresh Water | 78.3 | 29.5 |

Table 3–19 lists the relative dielectric constants and corresponding propagation times of commonly occurring rocks and fluids. A typical EPT log is shown in Figure 3–87.

Three interpretation methods for computation of water-filled porosity from EPT are presented in this section. These are referred to as (1) complex refractive index method (CRIM), (2) $t_{pl}$ and $A_c$ method, and (3) $t_{po}$ method.

## COMPLEX REFRACTIVE INDEX METHOD (CRIM)

This technique is used to determine water-filled porosity and the salinity of the water in the region investigated by the EPT. The technique involves an iterative process in which the water-filled porosity and the saturating water electromagnetic wave propagation are determined by iteration. The technique is only appropriate for computer solution.

$$\frac{\sqrt{\epsilon'}}{\sqrt{1 - \tan^2 \dfrac{\delta}{2}}} = \frac{\phi_{EPT}\sqrt{\epsilon_w'}}{\sqrt{1 - \tan^2 \dfrac{\delta_w}{2}}} + (1 - \phi_{EPT})\sqrt{\epsilon_w'} \tag{3.73}$$

and

$$\frac{\sqrt{\epsilon' \tan \dfrac{\delta}{2}}}{\sqrt{1 - \tan^2 \dfrac{\delta}{2}}} = \frac{\phi_{EPT}\sqrt{\epsilon_w' \tan \dfrac{\delta_w}{2}}}{\sqrt{1 - \tan^2 \dfrac{\delta_w}{2}}} \tag{3.73a}$$

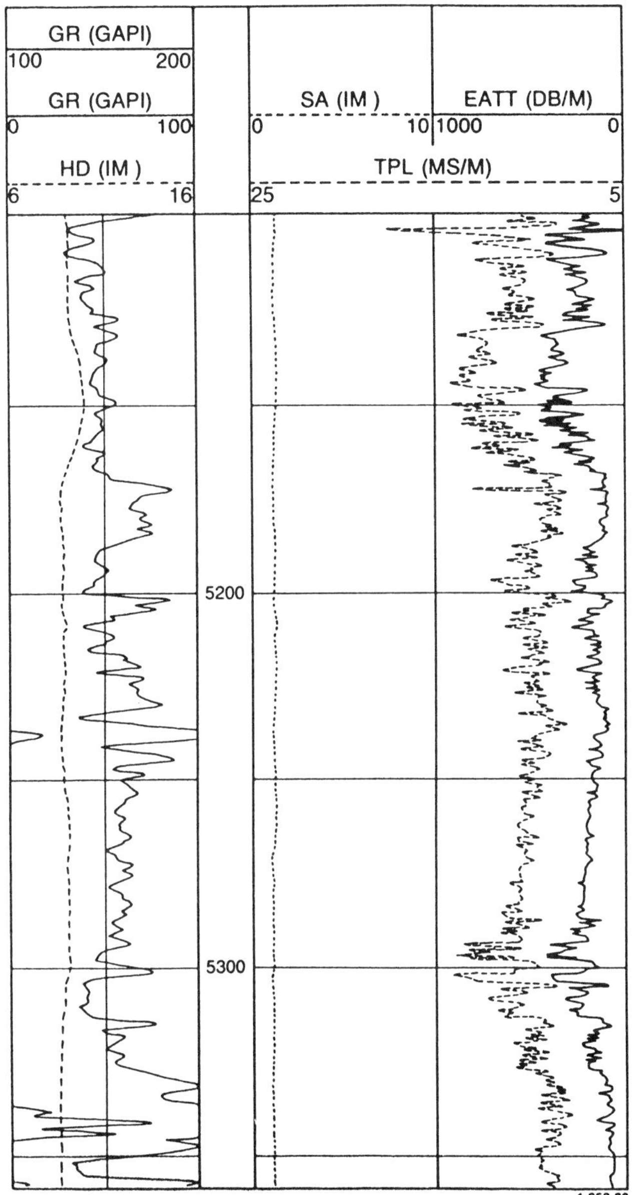


**Figure 3–87.** Typical Electromagnetic Propagation Log (courtesy of Schlumberger).[3]

where

$$\epsilon' = 94.88 - 0.2317T_f + 0.000217T_f - 0.1556 - 0.413(\text{Kppm}) + 0.00158(\text{Kppm})^2 \quad \textbf{(3.74)}$$

and

$$\tan \delta_w = \frac{5.66 + 2.65(\text{Kppm}) - 0.0045(\text{Kppm})^2}{\epsilon'_w} \quad \textbf{(3.74a)}$$

where $T_f$ is the formation temperature in °F, Kppm is the salinity in thousand parts per million, $\epsilon'$ and $\epsilon'_w$ are the real parts of the dielecric permitivity of the composite formation and waters, and tanδ and tan$\delta_w$ are their respective loss tangents. The $\epsilon'$ and tanδ/2 are computed using $t_{pl}$ and A(EATT in Figure 3–87) from EPT log as

$$\epsilon' = 0.09t_{pl}^2 - 2.4972 \times 10^{-5}\,A_c^2 \quad \textbf{(3.75)}$$

$$\epsilon'' = 1.832 \times 10^{-4}\,A_c t_{pl} \quad \textbf{(3.75a)}$$

and

$$\tan \delta = \frac{\epsilon''}{\epsilon'} \quad \textbf{(3.76)}$$

where $A_c$ is the corrected attenuation given by

$$A_c = A_{LOG} - (45 + 1.3t_{pi} + 0.18t_{pl}^2) \quad \textbf{(3.77)}$$

The following iterative steps are required to solve for $\phi_{EPT}$ and salinity:

1. Assume initial value of salinity in the invaded zone. Usually mud filtrate salinity is a good starting point.
2. Calculate $\epsilon'_w$ using $T_f$, Kppm, and Equation (3.74).
3. Calculate tan $\delta_w$ using Equation (3.74a).
4. Compute $\epsilon'$ and tan δ using Equations (3.75) and (3.76).
5. Calculate $\phi_{EPT}$ by iteration using Equation (3.73).
6. Using the $\phi_{EPT}$ calculated in Step 5, a corrected tan $\delta_w$ is determined from Equation (3.73a).
7. The value of tan $\delta_w$ is then used to determine a new value for the salinity, and a new $\phi_{EPT}$ and salinity calculated.

The iteration process is continued until the $\phi_{EPT}$ and salinity converge to constant values.

## $t_{pl}$ METHOD AND $A_c$ METHOD

The following equations are used for the calculation of $\phi_{EPT}$ individually using the $t_{pl}$ and $A_c$ measurements:

$$t_{pl} = \phi_{EPT} t_{pw} + (1 - \phi_{EPT}) t_{pma} \quad \textbf{(3.78)}$$

and for the attenuation, it is

$$A_c = \phi_{EPT} A_w \quad \textbf{(3.79)}$$

where $t_{pw}$ and $t_{pma}$ are estimated from Figure 3–88. The $t_{pw}$ estimate is based on laboratory measurements and $t_{pma}$ is estimated from knowledge of the apparent matrix density ($\rho_{maa}$). Figure 3–89 is used to estimate $A_w$ for Equation (3.79).

Comparing $\phi_{EPT}$ with actual porosity of the formation gives an estimate of water saturation of the zone investigated, most likely the flushed zone. The water saturation is roughly given by

$$S_{xo} = \frac{\phi_{EPT}}{\phi}$$

and

$$\phi_{hc} = \phi - \phi_{EPT}$$

## $t_{po}$ METHOD

This method of computing $\phi_{EPT}$ currently is used in wellsite computations. The measurement of the lossless travel time $t_{pl}$ and the correct attenuation $A_c$ is used to correct $t_{pl}$ to the lossless travel time $t_{po}$ using the following equation:

$$t_{po} = \left( t_{pl}^2 - \frac{A_c^2}{3{,}604} \right)^{1/2} \quad \textbf{(3.80)}$$

The $\phi_{EPT}$ is then given by

$$\phi_{EPT} = \frac{t_{po} - t_{pma}}{t_{pwo} - t_{pma}} \quad \textbf{(3.81)}$$

where $t_{pwo}$ is the lossless propagation time of water calculated from the following temperature-related equation and $t_{pma}$ is taken from Table 3–19.

$$t_{pwo} = 20 \left( \frac{710 - T_f/3}{444 - T_f/3} \right) \quad \textbf{(3.82)}$$

where $T_f$ is the formation temperature in °F.

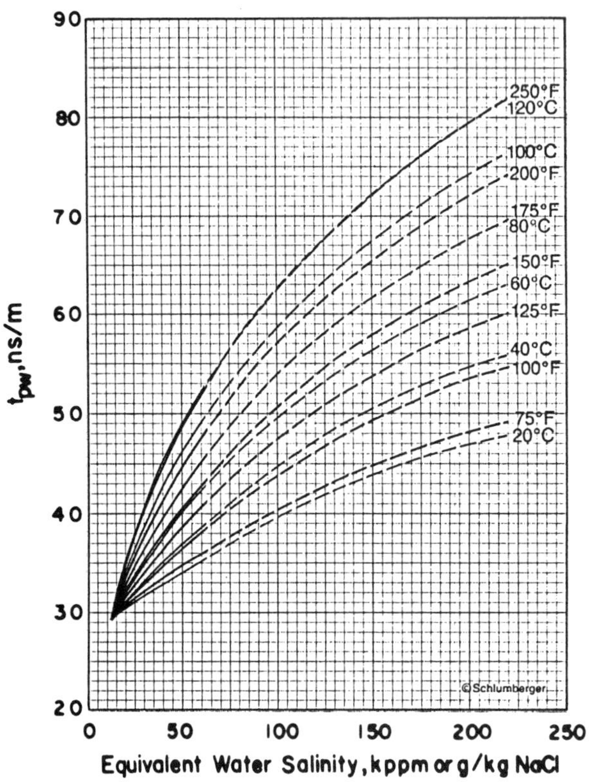


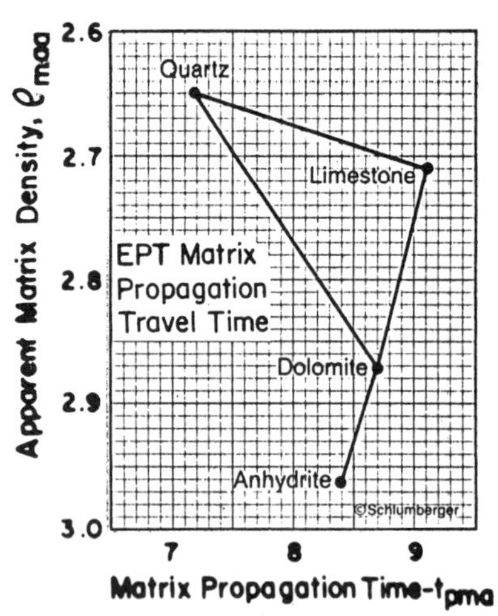


**Figure 3–88.** EPT propagation time for NaCl solutions and apparent matrix density (courtesy of Schlumberger).[4]

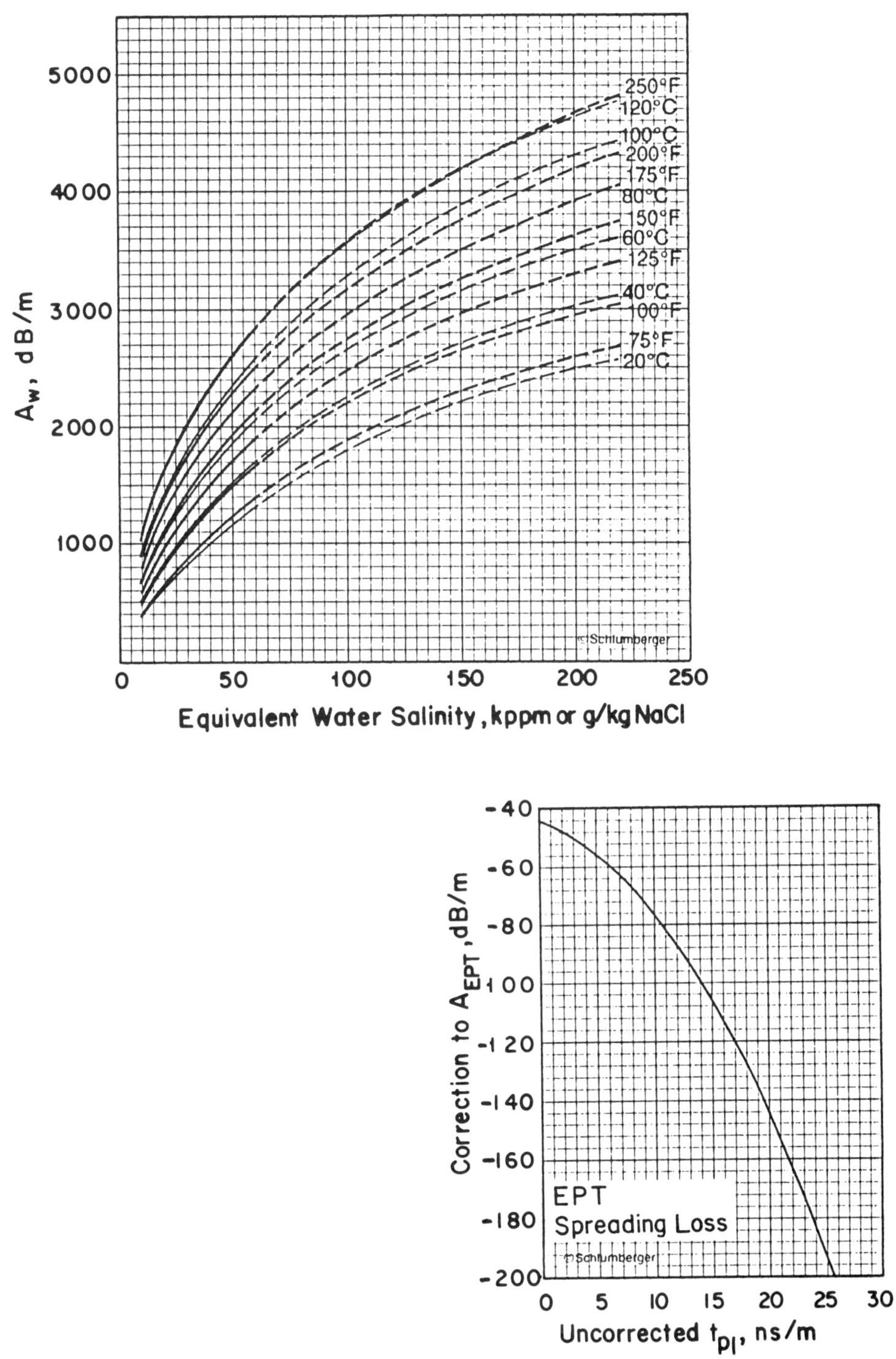


**Figure 3–89.** EPT attenuation for NaCl solutions (courtesy of Schlumberger).[4]

The water saturation $S_{xo}$ is given by

$$S_{xo} = \frac{(t_{po} - t_{pma}) + \phi_t(t_{pma} - t_{ph})}{\phi_t(t_{pwo} - t_{ph})} \tag{3.83}$$

where $t_{ph}$ is the propagation time for hydrocarbon and $\phi_t$ is the total porosity of the formation. If the $t_{pma} \approx t_{ph}$, then $s_{xo} = \phi_{EPT}/\phi_t$.

The EPT derived apparent formation water resistivity ($E_{Rfa}$) as an estimation for $R_w$ is given by

$$E_{Rfa} = (\phi_{EPT}^2)(R_t) \tag{3.84}$$

The value of $E_{Rfa}$ is a very good estimate of $R_w$ when there is no flushing of hydrocarbons. If there is flushing of hydrocarbons, $E_{Rfa}$ will represent an upper limit for the value of $R_w$.[31]

Figure 3–90 shows some typical comparison of $\phi_{EPT}$ with $\phi_N$ and $\phi_D$. In Figure 3–90, Zone A

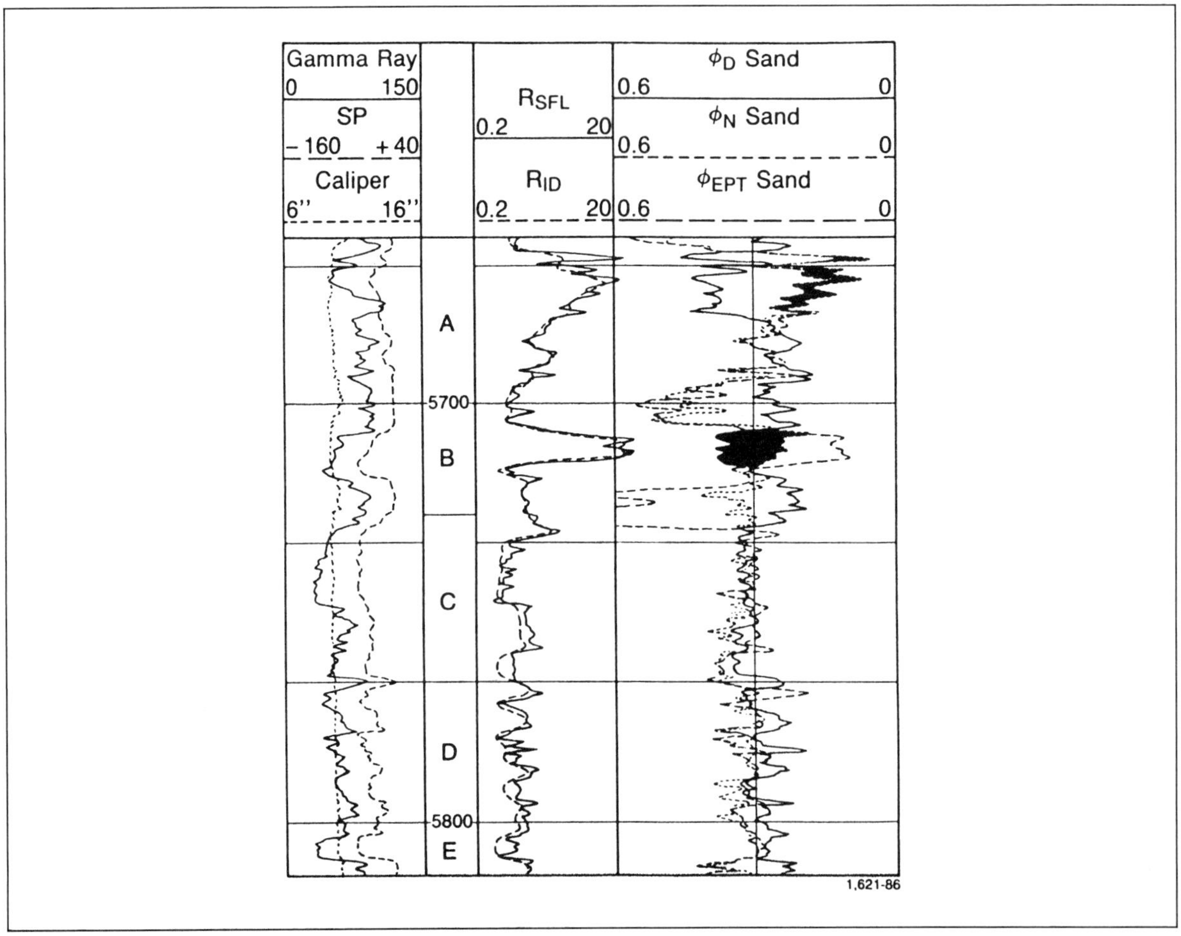


**Figure 3–90.** ISF/EPT/CNL/FDC logs (courtesy of Schlumberger).[3]

is obviously gas bearing as evidenced by the density-neutron crossover. This finding is confirmed by $\phi_{EPT}$ which is reading closer to $\phi_N$.

Zone *B* exhibits a different porosity profile. Once again $\phi_N < \phi_D$, indicating the presence of some light hydrocarbons; however, $\phi_{EPT}$ is now much less than both $\phi_N$ and $\phi_D$. This shows that Zone *B* contains more oil or condensate than Zone A. The three porosity logs read similar values in Zone C which is indicative of a water-bearing zone. Water-bearing zones are identified when $\phi_{EPT}$ is about the same or higher than $\phi_N$ and $\phi_D$.

The very top of Zone *D* contains oil since the $\phi_{EPT}$ is much lower than the $\phi_N$ and $\phi_D$ which read about the same porosity.



## SUMMARY OF LOGS' RESPONSE

1. Sandstones are classified as clastic rocks and are usually composed of quartz.
2. Shales have low effective porosity and very low permeability.
3. Resistivity of the formation is a function of the resistivity of the formation fluids and porosity.
4. A step profile of invasion assumes that invading fluids are never mixed with the formation fluids.
5. When an annulus invasion profile exists and $R_{xo}/R_t$ ratios are greater than five, the deep curve will record a resistivity approximately equal to the medium induction curve or less than the medium induction curve.
6. Higher salt concentrations tend to increase the normal pressure gradient, whereas increasing amounts of gases in solution would decrease the normal pressure gradient.
7. The measured bottomhole temperature during the logging operations is expected to be lower than the true static formation temperature.
8. The laterolog devices measure the drop in *potential* created by the flow of current through the formation to a current return electrode.
9. If a laterolog device is recorded in a 100% saline water-saturated formation which is deeply invaded with fresh mud filtrate, the apparent resistivity will be low.
10. The Minilog is a microresistivity measuring device with high vertical resolution and therefore provides a good indication of mudcake along the borehole.
11. The information available from minilog is affected by rugose borehole, air-filled borehole, and when $R_{mc}/R_{xo} = 1$.
12. The Microlaterolog is most useful in formations drilled with salt mud.
13. Shale in a reservoir containing *saline* formation water and drilled with a fresh mud system reduces the SP amplitude.
14. Gamma ray log is a shale indicator; shale reduces the GR amplitude (i.e., show high GR counts in API units).
15. In gas reservoirs with little or no invasion, the apparent porosity calculated using $\rho_f = 1.0$ is too high.
16. If there is no invasion in a water zone filled with very salty water, the apparent porosity using $\rho_f = 1.0$ is too low.

17. Shale makes the density log read lower apparent porosity.
18. Gas makes the density log read higher apparent porosity.
19. In hydrocarbon-bearing unconsolidated sands, the calculated sonic porosities are generally high.
20. On the acoustic log in a zone of constant compacted lithology (rock type) curve movements to the left are higher porosity.
21. The neutron log sees dry gas as low apparent porosity.
22. The neutron log sees shale as high apparent porosity.
23. Density versus neutron porosity provides a quick identification of gas-bearing sands in sand-shale sequences.
24. In a limestone formation, which contains gypsum, the apparent limestone porosity as seen by the density and neutron logs will be too high.
25. In extremely shaly, compacted sandstones, a combination of neutron versus acoustic porosity is most applicable to detect gas.
26. The presence of mica is easily detected by comparing the neutron and density porosity response.
27. If anhydrite is present in a limestone formation, the apparent density limestone porosity will be too low.
28. The best method of determining effective porosity in sand-shale sequences is to use the density-neutron crossplot.
29. If $\rho_{sh} = \rho_{ma}$, then $\phi_D$ should be equal to density-neutron crossplot porosity in shaly formations.
30. If acoustic porosity is recorded in sandstone units, the $R_{wa}$ in a clean carbonate will be too low.
31. In shaly formations, neutron porosity and resistivity logs provide true $S_w$.
32. In hydrocarbon-bearing formations, $F_R$ will be greater than the $F$ determined from porosity.
33. If $F < F_R = F_{xo}$, then the zone has no movable hydrocarbons.
34. If $S_w/S_{xo} = 1.0$, then the zone has no movable hydrocarbons.
35. The Pickett crossplot is typically employed more than the Hingle crossplot in clean carbonates.

## REFERENCES

[1] D. W. Hilchie, *Applied Openhole Log Interpretation for Geologists and Engineers,* Douglas W. Hilchie, Inc., Golden, Colorado, 1987.

[2] G. Asquith, *Basic Well Log Analysis For Geologists,* The American Association of Petroleum Geologists, Tulsa, Oklahoma, 1982.

[3] *Schlumberger Log Interpretation Principles/Applications,* Schlumberger Educational Services, 1331 Lamar, Suite 1175, Houston, Texas 77010, 1987.

[4] *Schlumberger Log Interpretation Charts,* Schlumberger Well Services, Inc., Houston, Texas, 1986.

[5] *Log Interpretation Fundamentals,* Dresser Atlas Division, Dresser Industries, Inc., Houston, Texas, 1975.

[6] *Log Interpretation Charts,* Atlas Wireline Services, Western Atlas International, Inc., Houston, Texas, 1985.

[7] *Log Review 1,* Dresser Atlas Division, Dresser Industries, Inc., Houston, Texas, 1974.

[8] *Well Logging and Interpretation Techniques,* The Course for Home Study, Third Edition, Dresser Atlas Division, Dresser Industries, Inc., Houston, Texas, 1982.

[9] *Welex Log Interpretation Charts,* Halliburton Company, 1985.

[10] R. Gaymard and A. Poupon, "Response of Neutron and Formation Density Logs in Hydrocarbon-Bearing Formations," *The Log Analysts,* Sept.–Oct. 1968.

[11] F. Segesman and O. Liu, "The Excavation Effect," *Trans.,* SPWLA Annual Logging Symposium, 1971.

[12] J. E. Carothers, "A Statistical Study of Formation Factor Relation to Porosity," *The Log Analyst,* Vol. 9, 1968, pp. 38–52.

[13] J. E. Carothers and C. R. Porter, "Formation Factor-Porosity Relation from Well Log Data," *Society of Professional Well Log Analysts,* 11th Annual Logging Symposium Transactions, Paper D, 1970.

[14] D. K. Sethi, "Some Considerations about the Formation Resistivity Factor-Porosity Relationship," *Society of Professional Well Log Analysts,* 20th Annual Logging Symposium Transactions, Paper L, 1979.

[15] M. A. Mian and D. W. Hilchie, "Comparison of Results from Three Cation Exchange Capacity (CEC) Analysis Techniques," *The Log Analyst,* Vol. XXIII, No. 5, Sept.–Oct. 1982.

[16] A. T. Hingle, "The Use of Logs in Exploration Problems," *Society of Exploration Geophysicists,* 29th Meeting, Los Angeles, 1959.

[17] C. G. Rodermund, R. P. Alger, and J. Tittman, "Logging Empty Holes," *Oil and Gas Journal,* June 2, 1961.

[18] W. H. Fertl, "Log-Derived Evaluation of Shaly Clastic Reservoirs," *Journal of Petroleum Technology,* Feb. 1987, pp. 175–194.

[19] R. Aguilera, "New Methods for Evaluation of Shaly Formation by Well Logs," *Paper SPE 17517,* presented at the SPE Rocky Mountain Regional Meeting, held in Casper, Wyoming, May 11–13, 1988.

[20] A. Poupon, M. E. Loy, and M. P. Tixier, "A Contribution to Electrical Log Interpretation in Shaly Sands," *Journal of Petroleum Technology,* June 1954.

[21] P. Simandoux, "Mesures Dielectriques en Milieu Poreux, Application a Mesure des Saturation en Eau, Etude du Comportement des Massifs Argileux," Reveu de L'Institute Français du Petrols, Supplementary Issue, 1963.

[22] *Neutron Lifetime Log Interpretation,* Dresser Atlas Division, Dresser Industries, Inc., Houston, Texas, 1974.

[23] G. R. Pickett, "Acoustic Character Logs and Their Applications in Formation Evaluation," *Trans.,* AIME, **228,** 1963, pp. 659–667.

[24] G. R. Pickett and E. B. Reynolds, "Evaluation of Fractured Reservoirs," *Soc. Pet. Eng. J.,* March 1969, pp. 28–38.

[25] R. Aguilera, *Naturally Fractured Reservoirs,* The PennWell Publishing Co., Tulsa, Oklahoma, 1980.

[26] R. Aguilera, "Analysis of Naturally Fractured Reservoirs From Sonic and Resistivity Logs," *Journal of Petroleum Technology,* Nov. 1974, pp. 1233–1238.

[27] R. Aguilera, "Analysis of Naturally Fractured Reservoirs From Conventional Well Logs," *Journal of Petroleum Technology,* July 1976, pp. 764–772.

[28] S. J. Pirson, "Log Interpretation in Rocks With Multiple Porosity Types—Water or Oil Wet," *World Oil,* June 1957, p. 196.

[29] S. J. Pirson, "Petrophysical Interpretation of Formation Tester Pressure Buildup Records," Paper presented at the *Third Annual Logging Symposium,* SPWLA, May 17–18, 1962.

[30] C. R. Pickett, G. R. Pickett, and W. W. Whitman, "A Statistical Method for Determination of Water Saturation from Logs," Paper presented at the 10th *Annual Logging Symposium,* SPWLA, May 25–28, 1969.

[31] J. M. Delano Jr. and R. P. Wharton, "An EPT Interpretation Procedure and Application in Freshwater, Shaly Oil Sands, *Journal of Petroleum Technology,* Oct. 1984, pp. 1763–1772.

[32] "An Introduction to Well Log Analysis," Welex—A Haliburton Company, Houston, Texas, 1978.

CHAPTER FOUR

# RESERVOIR ENGINEERING AND EVALUATION



In this chapter we will discuss the techniques used in estimating oil and gas reserves and the pattern in which these reserves will be produced in the future years.

An oil and/or gas property goes through several development stages, and at each stage, most likely, a different reserve estimate is developed with a different degree of confidence. The first stage in the development of any oil and/or gas property is the exploratory stage when a wildcat well is drilled. The reserve estimates at this stage are based on analogy with other similar prospects and geologic settings. If the estimates are favorable, the drilling of a wildcat is approved.

After the first well is drilled and a productive horizon identified, more physical information about the rock and fluid properties is obtained with the help of well logs, core analysis, PVT analysis, and transient well test analysis, etc. Based on this information, the reserve estimates are revised appropriately.

After the first well starts producing oil and/or gas, reservoir pressure and performance become available. This information may further warrant the revision of the reserve estimates. With the passage of time, the data on the property accumulates and the estimates revised accordingly. Also at different stages, more appropriate techniques are utilized to obtain better estimates.

The drilling of every new well into the same productive horizon (reservoir) provides added information such as the aerial extent of the reservoir, the type of oil and/or gas accumulation, and the reservoir producing mechanism, etc. Unfortunately, the reserve estimates are no longer estimates when the property is abandoned.

Three techniques are commonly used in estimating oil and/or gas reserves. These techniques include (1) analogy, (2) volumetrics, and (3) performance. The performance-based techniques depending on the data availability and age of the property are further divided into three subcategories as: material balance calculations, simulation studies, and decline trend (pressure or production) analysis.

In this chapter we will discuss the volumetric methods, material balance techniques and their variations, and the decline trend analysis techniques. The techniques will be utilized accounting for the different reservoir drive mechanisms and type of fluids in the reservoir. Included also is a section on reservoir economics which shows how the reserve estimates are converted into their associated monetary value.

In dealing with oil and/or gas reserve estimates, we will utilize the following most commonly used terminology:

1. Original or initial oil and/or gas in place
2. Ultimate oil and/or gas recovery or reserves
3. Cumulative oil and/or gas production
4. Remaining oil and/or gas reserves

Original oil and/or gas in place (OOIP or OGIP) and stock-tank oil initially in place (STOIIP) are the hydrocarbon volumes in the reservoir at the time it is encountered through drilling. Having determined the original oil and/or gas in place, some empirical correlations and/or experience are applied in order to predict the percentage of the total hydrocarbons volume that will be economically and technically producible. This producible volume is referred to as the *ultimate* oil and/or gas recovery or reserves, also referred to as *recoverable* oil and/or gas reserves. The recovery efficiency depends on the reservoir drive mechanism and the rock and fluid properties. When a property starts producing, the production of oil, gas, and/or water are accumulated and referred to as the *cumulative* production (oil, gas, and/or water) at a specific point in time. The reserves estimated to be produced after that point in time are referred to as the *remaining* oil and/or gas reserves as of that time. The sum of the cumulative production and the remaining reserves is again referred to as the *ultimate* oil and/or gas recovery or reserves.

## ESTIMATING OIL AND GAS RESERVES—VOLUMETRICS

Volumetric estimates of oil and gas are generally used only at early times in the life of a field for reserve purposes. These estimates are considered as preliminary estimates as compared to the estimates obtained by performance. Reservoir heterogeneities are a commonly overlooked factor which makes the volumetric estimates often different than those obtained by performance.

Most reservoir engineers will remember the volumetric equations by heart, and therefore, calculating oil and/or gas in place volumetrically is a relatively simple chore. In this section we will discuss the calculations of original oil in place (OOIP), original gas in place (OGIP), and estimating fieldwide oil and gas reserves volumetrically.

## ORIGINAL OIL IN PLACE AND RECOVERABLE OIL

The data required for estimating OOIP are (1) formation thickness ($h$, feet), (2) drainage area (A, acres), (3) porosity ($\phi$, fraction), (4) formation oil saturation ($S_o$, fraction), and (5) oil formation volume factor ($B_o$, RB/STB). The following simple equation is then used to estimate OOIP (N) in stock-tank barrels (STB) as follows:

$$N = \frac{7{,}758\phi(1 - S_{iw})hA}{B_{oi}} \tag{4.1}$$



where porosity and water saturations are obtained from well logs or core analysis or both. The formation thickness is estimated from resistivity logs or from geologic maps if the well is in a developed reservoir. The drainage area is estimated based on experience, type of reservoir producing mechanism, analogy to wells producing from similar horizons in other areas, and from geologic maps. The oil formation volume factor is either determined in the laboratory upon fluid analysis, or it is estimated from empirical correlations presented in Chapter 2. The constant 7,758 in Equation (4.1) is reservoir barrels per 1 acre-foot.

Once the original oil in place is estimated, the engineer has to estimate as to how much of that oil is recoverable. This is done by applying a recovery factor, also referred to as recovery efficiency ($E_R$) to the original oil in place. The $E_R$ is based on experience, reservoir drive mechanism, analogy, and rock and fluid properties. The American Petroleum Institute presented a statistical study of oil recovery factors. The study was based on 226 sandstone and 86 carbonate reservoirs classified by their predominant drive mechanism. The API subcommittee established the following empirical relationship for the $E_{Ro}$:[1]

For waterdrive reservoirs:

$$E_{ow} = 54.898\left[\frac{\phi(1 - S_{iw})}{B_{oi}}\right]^{0.0422}\left(\frac{k\mu_{wi}}{\mu_{oi}}\right)^{0.0770}(S_{iw})^{-0.1903}\left(\frac{p_i}{p_a}\right)^{-0.2159} \tag{4.2}$$

For solution gas drive reservoirs below the bubblepoint:

$$E_{og} = 41.815\left[\frac{\phi(1 - S_{iw})}{B_{ob}}\right]^{0.1611}\left(\frac{k}{\mu_{ob}}\right)^{0.0979}(S_{iw})^{0.3722}\left(\frac{p_b}{p_a}\right)^{0.1744} \tag{4.3}$$

where the permeability $k$ is in darcies.

Note that in Equation (4.2), oil properties are calculated at the *initial reservoir pressure,* while in Equation (4.3), oil properties are calculated at the *bubblepoint pressure.* The two equations can be of value in the estimation of *ultimate recovery* ($N_{ul}$) from *carbonate reservoirs* where the data required for more detailed studies are lacking. The API correlations for waterdrive reservoirs should not be misused as estimates for waterflood reservoirs.

Guthrie and Greenberger found the following correlation between waterdrive recovery and five variables which affect recovery in sandstone reservoirs.[2] A test of the equation showed that 50% of the fields studied showed recovery factor within ±6.2 percentage points of that predicted, 75% were within ±9.0 recovery percents, and 100% were within ±19.0 recovery percent.

$$E_{ow} = 0.144 + 0.272 \log k + 0.256 S_{iw} - 0.136 \log \mu_o - 1.538\phi - 0.00035h \quad \textbf{(4.4)}$$

As a rule of thumb, the recovery factor in dissolved-gas or depletion-type reservoir ranges from 5% to 20% of the original oil in place. The recovery range in gas cap drive reservoirs ranges from 10% to 30%, provided the oil reservoir is properly produced. In waterdrive reservoirs the recovery factors will range from low of 30% to as high as 80%.

The following relationships can also be used to calculate *unit ultimate* (STB/ac-ft) oil recovery in (1) saturated depletion-type oil reservoirs, (2) undersaturated oil reservoirs without waterdrive above the bubblepoint, and (3) oil reservoirs with waterdrive.

Saturated depletion-type oil reservoirs:

$$N_{ug} = 7{,}758\phi\left(\frac{1 - S_{iw}}{B_{oi}} - \frac{1 - S_{iw} - S_{gr}}{B_{oa}}\right) \quad \textbf{(4.5)}$$

Undersaturated oil reservoirs without waterdrive, above the bubblepoint:

$$N_p = \frac{7{,}758\phi(p_i - p_b)[c_o + c_f - S_{iw}(c_o - c_w)]}{B_{oi}[1 + c_o(p_i - p_b)]} \quad \textbf{(4.6)}$$

Oil reservoir with waterdrive:

$$N_{uw} = 7{,}758\phi\left(\frac{1 - S_w}{B_{oi}} - \frac{S_{or}}{B_{oa}}\right) \quad \textbf{(4.7)}$$

The residual gas saturation ($S_{gr}$, fraction) in Equation (4.5) can be calculated using an empirical correlation as $S_{gr} = (0.625 - 1.3125\phi)$. A typical $S_{gr}$ value for average consolidated sand, a medium solution GOR of 400–500 ft³/bbl, and a crude oil gravity of 30–40° API is 0.25.

The typical $S_{gr}$ value of 0.25 may be reduced by a factor of about 0.05 due to either (1) a high degree of cementation, (2) a high shale content of sand, or (3) a 50% reduction in solution GOR. With clean, loose, unconsolidated sands or doubling of solution GOR, the $S_{gr}$ value may increase by as much as 0.10. The $S_{gr}$ value will also increase or decrease by about 0.01 for every 3°API gravity.

The $S_{or}$ value in Equation (4.7) play a role similar to the final (residual) gas saturation ($S_{gr}$)

in the depletion-type reservoirs. The $S_{or}$ value is typically determined in the laboratory using *flood-pot tests.*

**Example 4–1.** An oil well has been drilled and completed. The productive zone has been encountered at a depth of 7,815–7,830 ft. The log analysis showed an average porosity of 15% and an average water saturation of 35%. The oil formation volume factor was determined in the laboratory to be 1.215 RB/STB. Experience has shown that other reservoirs of about the same properties will drain 80 ac with a recovery factor of 12%. Compute the OOIP and the ultimate oil recovery.



Solution:

$$\begin{array}{llll} \phi = 0.15 & h = 15 \text{ ft} & B_o = 1.215 & \text{RB/STB} \\ S_{iw} = 0.35 & A = 80 \text{ ac} & E_R = 0.12 & \end{array}$$

Using Equation (4.1),

$$\begin{aligned} N &= \frac{7{,}758(0.15)(1 - 0.35)(15)(80)}{1.215} \\ &\approx 747{,}000 \text{ STB} \\ N_{ul} &= N \times E_R \\ &= 747{,}000 \times 0.12 \\ &\approx 89{,}600 \text{ STB} \end{aligned}$$

## ORIGINAL GAS IN PLACE AND RECOVERABLE GAS

The volumetric calculations of gas reserves require about the same information as that used in the volumetric calculations of oil reserves. The reservoir pay thickness, drainage area, and recovery factor still have the same limitations. The recovery factor for gas wells may be based on experience or it may be calculated from an estimated abandonment pressure. Therefore the recovery factor in gas wells may be calculated relatively more accurately.

Since gas flows more readily than oil, usually a higher percentage (70–90%) of the original gas in place is recoverable. At the same time a larger area could be drained. For this reason, the spacing of oil wells is usually denser than the spacing of gas wells. The following equation is used for original-gas-in-place calculations:

$$G = 43{,}560\phi(1 - S_{iw})hAB_{gi} \qquad \textbf{(4.8)}$$

where the constant 43,560 converts acres into cubic feet and gas formation volume factor ($B_{gi}$, SCF/ft$^3$) converts reservoir volume (cubic feet) to standard conditions (standard cubic feet, SCF). Equation (2.13) is used to calculate $B_{gi}$ at initial reservoir pressure.

The recovery factor from a gas reservoir is primarily a function of the abandonment pressure and permeability. The abandonment pressure depends on various factors, such as price of gas, the productivity index of the wells, the size of the field, its location with respect to market, the type of market, and the operating pressure of the transmission pipeline. Some gas pipeline companies use an abandonment pressure of 100 psi/1,000 ft of depth. In large fields where the expense of compressor installation is economically feasible, the abandonment pressure could be lowered, thus yielding more recoverable reserves.

The recovery factor in waterdrive gas reservoirs is usually less than the closed (depletion-type) reservoirs. This is because of the high abandonment pressure in waterdrive gas reservoirs due to water encroachment into the gas reservoirs. Higher permeability also aids in increasing recovery factor because of higher deliverability for a given pressure drop. For strong waterdrives where residual gas is trapped at high pressures, recovery factor may be 50–60% compared to 70–80% for partial waterdrive and 80–90% for volumetric reservoirs.

The following equations are used to calculate recovery factors in gas reservoirs. These recovery factors are valid provided the unit neither drains nor is drained by adjacent units.

For depletion-type gas reservoirs:

$$E_g = \left(1 - \frac{B_{gi}}{B_{ga}}\right) = \left(1 - \frac{p_a z_i}{p_i z_a}\right) \quad \textbf{(4.9)}$$

For waterdrive gas reservoirs:

$$\begin{aligned} E_{gw} &= \frac{(S_{gi}B_{ga} - S_{gr}B_{gi})}{S_{gi}B_{ga}} \\ &= \frac{(S_{gi}p_i/z_i) - (S_{gr}p_a/z_a)}{(S_{gi}p_i/z_i)} \end{aligned} \quad \textbf{(4.10)}$$

The gas saturation at abandonment pressure ($S_{gr}$, fraction) can be estimated from the following equation:

$$S_{gr} = 0.625 - 1.3125(\phi) \quad \textbf{(4.11)}$$

**Example 4–2.** A gas well is completed at a depth of 8,550 ft. The log analysis showed a total formation thickness of 12 ft of 16% porosity and 30% water saturation. On potential test, the well produced dry gas with a specific gravity of 0.75 (air = 1). The reservoir pressure was determined from a drillstem test (DST) to be 3,850 psi and have a temperature of 155°F. The gas well will be produced at the surface where the standard temperature and pressure would be 60°F and 14.65 psi, respectively. Calculate the original gas in place based on a 160-ac drainage. What will be the recovery factor and the recoverable gas at abandonment pressure of 855 psi?

Solution:

$$\phi = 0.16 \qquad A = 160 \text{ ac} \qquad \gamma_g = 0.75 \text{ (air} = 1.0)$$
$$h = 12 \text{ ft} \qquad p_R = 3{,}850 \text{ psi} \qquad p_a = 855 \text{ psi}$$
$$S_{iw} = 0.30 \qquad T_R = 155°\text{F}$$

$$T_{pc} = 405°\text{R} \qquad \text{Equation (2.3)}$$
$$p_{pc} = 667 \text{ psi} \qquad \text{Equation (2.4)}$$
$$T_{pr} = \frac{T_R}{T_{pc}} = \frac{155 + 460}{405} = 1.52$$
$$p_{pc} = \frac{p_R}{p_{pc}} = \frac{3{,}850}{667} = 5.77$$

Gas deviation factor $z = 0.855$ from Figure 2–3.

Using Equation (2.13),

$$\begin{aligned} B_{gi} &= \frac{(60 + 460)(3{,}850)}{(14.65)(0.855)(155 + 460)} \\ &= 259.89 \text{ SCF/CF} \\ G &= 43{,}560\phi(1 - S_{iw})hAB_{gi} \\ &= 43{,}560(0.16)(1 - 0.30)(12)(160)(259.89) \\ &= 2{,}434{,}426{,}638 \text{ SCF} \\ &= 2{,}434 \text{ MMSCF} \end{aligned}$$

To calculate the recovery factor, we need to find gas deviation factor ($z_a$) at abandonment pressure ($p_a$).

$$p_{pra} = \frac{p_a}{p_{pc}} = \frac{855}{667} = 1.3$$

At $T_{pr} = 1.52$ and $p_{pra} = 1.3$, $z_a = 0.883$.

$$\begin{aligned} E_g &= \left(1 - \frac{p_a z_i}{p_i z_a}\right) \times 100 \\ &= 1 - \frac{(855)(0.855)}{(3{,}850)(0.883)} \\ &= 78.5\% \\ G_{ul} &= G \times E_g \\ &= 2{,}434 \times 0.785 \\ &= 1{,}911 \text{ MMSCF} \end{aligned}$$

**Example 4–3.** Assuming the reservoir in Example 4–2 is a waterdrive gas reservoir, what would be the recovery factor at abandonment pressure of 855 psi?

Solution: Using Equation (4.11),

$$S_{gr} = 0.625 - 1.3125(0.16)$$
$$= 0.415$$

Using Equation (4.10),

$$E_{gw} = \frac{(0.70)(3{,}850/0.855) - (0.415)(855/0.883)}{(0.70)(3{,}850/0.855)}$$
$$= \frac{3{,}152.05 - 401.84}{3{,}152.05}$$
$$= 0.873 = 87.3\%$$

## ORIGINAL OIL AND GAS IN PLACE—FIELDWIDE

Fieldwide reserve estimates cannot be prepared until sufficient wells have been drilled to delineate its areal geometry and thickness. Once enough wells are drilled so that sufficient subsurface well control is available to enable geologic mapping, geologic maps are then used to compute the reservoir volume. The reservoir rock volume is calculated using *isopach* map and the reservoir hydrocarbon volume is calculated using *isovolume* map, also referred to as the *hydrocarbon thickness* map. Isovolume map is a contour map based on the net pay thickness determined at each well multiplied by the porosity and hydrocarbon saturation at that location [isovol $= \phi h(1 - S_{iw})$].

A sample of a subsurface geologic map contoured on the subsea depth of the top of the sand (solid lines) and on the subsea depth of the base of the sand (dashed lines) is presented in Figure 4–1. The total area enclosed by each contour is planimetered and plotted as the abscissa on an acre-feet diagram as shown in Figure 4–2, against the corresponding subsea depth as the ordinate. After connecting the points in Figure 4–2, several methods may be used to determine volume of oil- and/or gas-bearing zone.[3]

The formation volume can be either calculated directly by planimetering the acre-feet diagram in Figure 4–2 or, if the number of contours are even, volume can be calculated by Simpson's rule, Trapezoidal rule, or Pyramidal rule using the relationships given below.

Simpson's rule (if the number of contour intervals is even):

$$V_B = \frac{h}{3}[(y_o + y_n) + 4(y_1 + y_3 + \cdots + y_{n-1}) + 2(y_2 + y_4 + \cdots + y_{n-2})] \quad \textbf{(4.12)}$$

Trapezoidal rule (with somewhat less accuracy):

$$V_B = h[1/2(y_o + y_n) + y_1 + y_2 + \cdots + y_{n-1}] \quad \textbf{(4.13)}$$

Pyramidal rule (somewhat more complicated):

$$V_B = \frac{h}{3}(A_n + A_{n+1} + \sqrt{A_n A_{n+1}}) \qquad \textbf{(4.14)}$$

where $V_B$ is the reservoir bulk volume (acre-feet), $h$ is the contour interval (feet), $y_o$ is the area on top of sand less area on base of sand at highest contour, and $y_n$ is the area on top of sand less area on base of sand at lowest contour. In Equation (4.14), $A_n$ is the area enclosed by the lower isopach lines in acres, $A_{n+1}$ is the area enclosed by the upper isopach line in acres, and $h$ is the contour interval in feet.

For best accuracy, Equation (4.12) or Equation (4.14) should be used; however, since the trapezoidal equation, Equation (4.13), is simpler, it is commonly used. As a rule of thumb, whenever the ratio of the areas of any two successive isopach lines is found to be larger than 0.50, the trapezoidal rule is applied. Whenever the ratio is smaller than 0.50, Simpson's or the pyramidal rule is applied.

Using Equations (4.12), (4.13), and (4.14), the volume of the acre-feet diagram in Figure 4–2 can be calculated as follows:

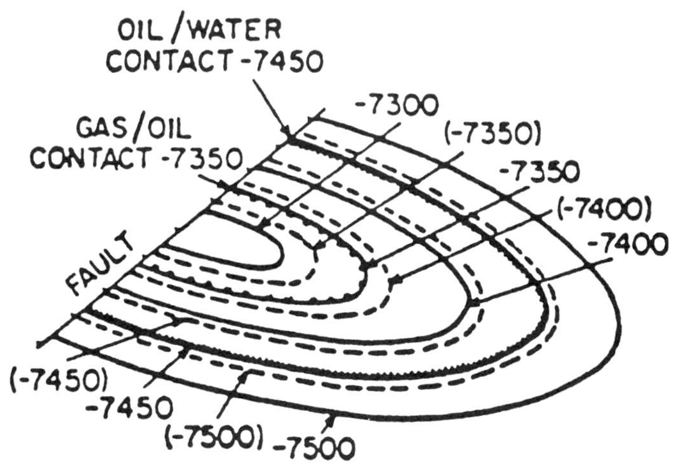


**Figure 4–1.** Geological map on top and base of a reservoir (© SPE, Petroleum Engineering Handbook, 1987).[3]

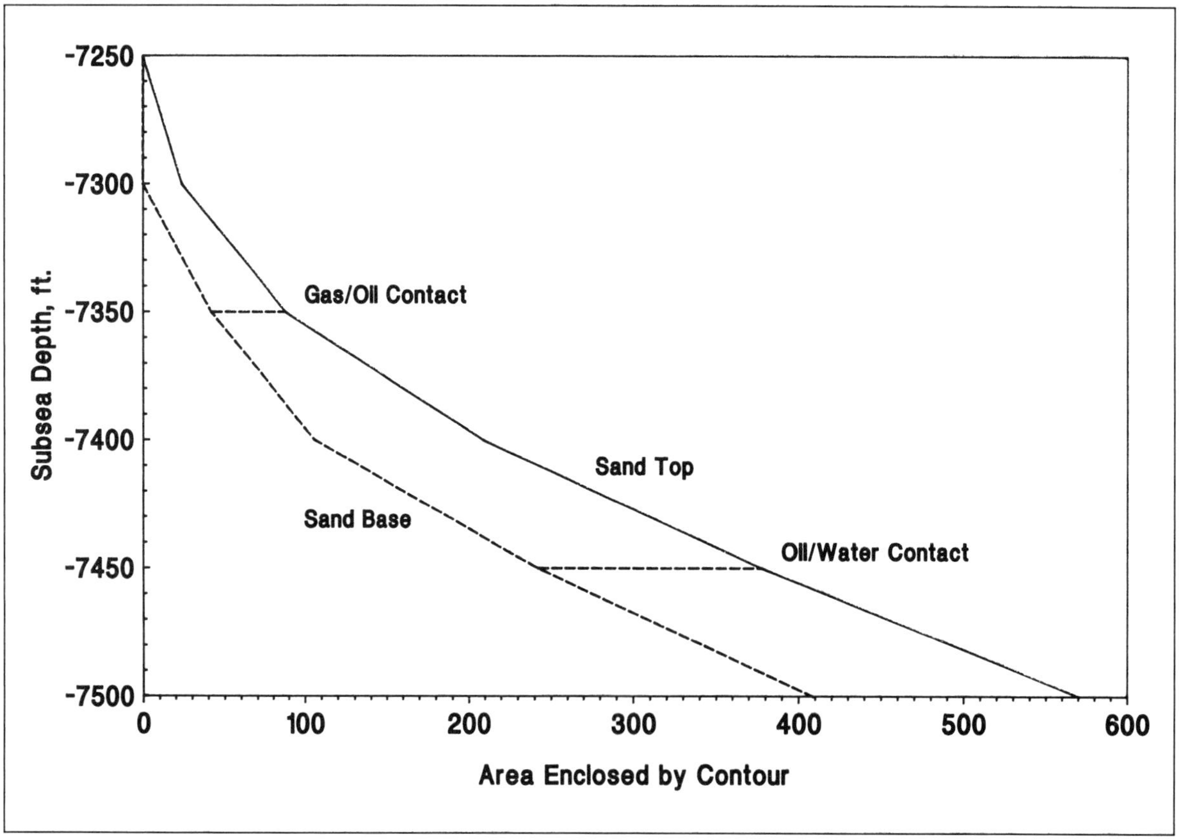


**Figure 4–2.** Acre-feet diagram for the map in Fig. 4–1.

Simpson's rule:

$$V_B = \frac{50}{3}[0 + (378 - 242) + 4(24 - 0 + 209 - 106) + 2(88 - 42)]$$

$$= \frac{50}{3}[(136) + 4(24 + 103) + (2 \times 46)]$$

$$= 12{,}267 \quad \text{ac-ft}$$

Trapezoidal rule:

$$V_B = 50[1/2(0 + 136) + (24 + 46 + 103)]$$

$$= 12{,}050 \quad \text{ac-ft}$$

Pyramidal rule:

$$V_B = \frac{50}{3}[(0 + 136) + 2(24 + 46 + 103) + \sqrt{24 \times 88} + \sqrt{88 \times 209} + \sqrt{209 \times 378}$$

$$- \sqrt{42 \times 106} - \sqrt{106 \times 242}]$$

$$= 11{,}963 \quad \text{ac-ft}$$

These acre-feet volumes are then multiplied by average porosity ($\phi_{av}$) and average hydrocarbon saturation ($1 - S_{iw}$) to determine reservoir hydrocarbon volume. If porosity and water saturation for a zone vary substantially from well to well, the hydrocarbon volume should be determined from isovolume map.

**Example 4–4.** An oil reservoir is represented by an isovolume map as shown in Figure 4–3. The map is contoured on a 0.1-ft contour interval. The zero thickness line is constructed as a best estimate beyond which no economical hydrocarbon production is expected. The planimetered area covered by each contour and the hydrocarbon thickness for each well are as follows:

| Contour | Area (acres) | Well | $\phi h(1 - S_{iw})$ |
|---|---|---|---|
| 0 | 1,776 | C | 0.15 |
| 0.1 | 1,021 | D | 0.10 |
| 0.2 | 434 | F | 0.10 |
| 0.3 | 302 | G | 0.73 |
| 0.4 | 158 | I | 0.05 |
| 0.5 | 83 | J | 0.40 |
| 0.6 | 45 | K | 0.22 |
| 0.7 | 10 | M | 0.18 |

Determine the reservoir oil volume in acre-feet and reservoir barrels.

Solution:

Using the trapezoidal rule,

$$V_B = 0.10\left[\frac{1}{2}(1{,}776) + 1{,}021 + 434 + 302 + 158 + 83 + 45 + \frac{10}{2}\right]$$

$$= 293.6$$

$$= 293.6 \times 7{,}758 \simeq 2{,}278{,}000 \quad \text{RB}$$

## MATERIAL BALANCE CALCULATIONS, OIL RESERVOIRS

The amount of oil or gas in place in a reservoir may be estimated by the material balance method. The material balance calculations are applicable only when sufficient pressure production performance data are recorded and PVT data describing the reservoir fluid behavior are available. The material balance could also be used to evaluate the water influx into the reservoir, or the size of the gas cap, etc.

In its simplest form, the material balance equation can be written as *Initial volume = volume remaining + volume recovered.* The material balance calculations require significant amount of good

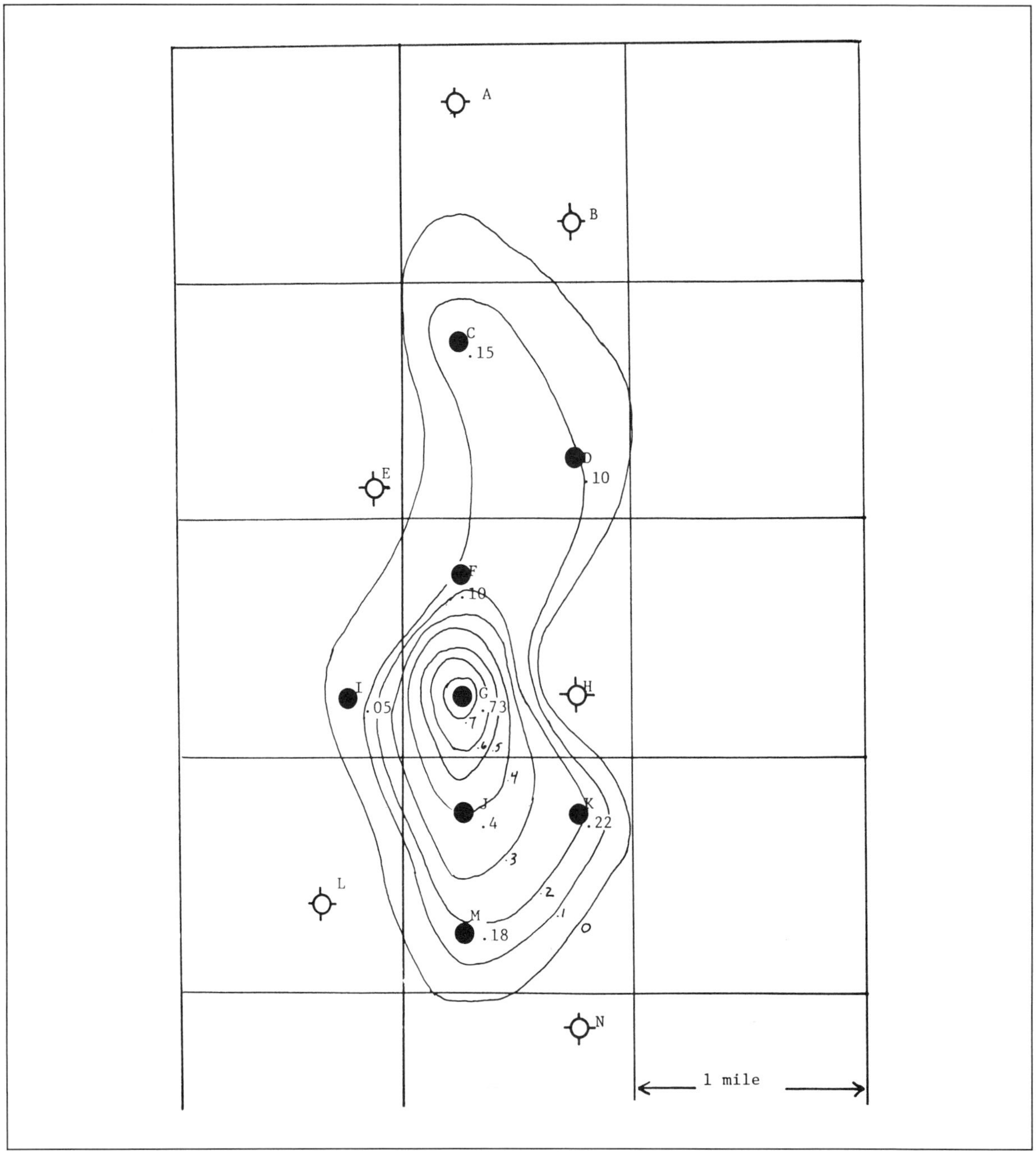


**Figure 4–3.** Isovolume map of lower cretaceous oil field (courtesy of D. W. Hilchie, Inc., Boulder, Colorado).

reservoir rock and fluid data. Field production data including cumulative oil, water, and gas production; reservoir pressure as a function of time and fluid production from the reservoir; and appropriate laboratory PVT data or empirical relationships for the physical properties of oil, water, and gas are all required for material balance calculations.

In addition to these data, it is helpful to know the type of reservoir drive mechanism which is operative in order to expedite estimation of the volume of the initial hydrocarbons in the reservoir. Material balance calculations also may be designed to extend the observed performance of a reservoir into the future. These predictive tools assume a homogeneous reservoir and require application of relative permeability concepts.

The Schilthuis material balance equation has long been regarded as one of the basic tools of reservoir engineers for interpreting and predicting reservoir performance.[4] In this chapter we will use mainly the interpretive techniques of Havlena and Odeh, who introduced the material balance equation as an equation of straight line, and some other commonly used methods.[5] The general material balance equation is reduced to account for specific reservoir drive mechanisms. The reservoir drive mechanisms which will be studied are (1) solution gas drive, (2) gas cap drive, (3) natural waterdrive, and (4) combination of solution gas drive, gas cap drive, and waterdrive.

## GENERAL FORM OF THE MATERIAL BALANCE EQUATION

The general form of the material balance equation is given below. All of the terms of this equation are never significant at any one time. The equation results from equating the cumulative observed production, expressed as underground withdrawal, to the expansion of the fluids in the reservoir resulting from a finite pressure drop.

$$N = \frac{N_pB_o + (G_p - N_pR_s)B_g + (W_p - W_i - W_e)B_w - G_iB_g}{(B_o - B_{oi}) + (R_{si} - R_s)B_g + mB_{oi}\left(\dfrac{B_g - B_{gi}}{B_{gi}}\right) + \dfrac{B_{oi}}{1 - S_{iw}}(1 + m)(S_{iw}c_w + c_f)(\Delta p_R)} \quad \textbf{(4.15)}$$

where the terms

$$N(B_o - B_{oi}) = \text{expansion of initial oil}$$

$$N(R_{si} - R_s)B_g = \text{volume occupied by liberated solution gas}$$

$$mNB_{oi}\left(\frac{B_g - B_{gi}}{B_{gi}}\right) = \text{gas cap gas expansion}$$

$$\frac{NB_{oi}(1 + m)}{1 - S_{iw}} S_{iw}c_w(p_{iR} - p_R) = \text{water expansion}$$

$$\frac{NB_{oi}(1 + m)}{1 - S_{iw}} c_f(p_{iR} - p_R) = \text{rock expansion}$$

$$N_pB_o = \text{oil production} + (G_{ps} - N_pR_s)B_g(\text{liberated solution gas production})$$

$$G_{pc}B_g = \text{gas cap gas production}$$

$G_iB_g$ = gas injection
$W_pB_w$ = water production, RB
$W_iB_w$ = water injection
$W_eB_w$ = water influx, RB

and

$R_s$ = solution GOR, SCF/STB
$R_{si}$ = value of $R_s$ at initial pressure, SCF/STB
$m$ = PV of gas cap/PV of oil zone
$G_p = G_{ps} + G_{pc}$

## OOIP—SOLUTION GAS DRIVE RESERVOIR

For the solution gas drive reservoir, the general material balance equation, Equation (4.15), reduces to the following forms with only one unknown, N (assuming no gas cap, $m = 0$, and negligible water influx, $W_e = 0$).

Initially undersaturated oil reservoirs

Above bubblepoint pressure:

$$N = \frac{\left[N_p(1 + \Delta p_R c_o) + \frac{B_w W_p}{B_{oi}}\right](1 - S_{iw})}{\Delta p_R[c_o + c_f - S_{iw}(c_o - c_w)]} \tag{4.16}$$

OR

$$N = \frac{N_pB_o + W_pB_w}{B_{oi}c_e(p_{iR} - p_R)} \tag{4.16a}$$

where

$$c_e = c_o + \frac{S_{iw}c_w}{1 - S_{iw}} + \frac{c_f}{1 - S_{iw}}$$

Below bubblepoint pressure:

$$N = \frac{N_p[B_o + B_g(R_p - R_s)] + B_wW_p}{B_o - B_{oi} + B_g(R_{si} - R_s)} \tag{4.17}$$

Note that $B_g$ in Equation (4.17) is in RB/SCF. Since the compressibilities $c_w$ and $c_f$ are generally of the same order of magnitude as the compressibility of oil ($c_o$), *the compressibilities in Equation (4.16) may not be neglected.* The cumulative produced gas/oil ratio $R_p = G_p/N_p$.

Havlena and Odeh showed that a plot of $Q_p = N_p[B_t + B_g(R_p - R_{si})]$ vs. $\Delta B_t = B_t - B_{ti}$ on Cartesian coordinate graph paper will result in a straight line passing through the origin, with its slope equal to the value of the initial OIP (N).[5] If the plot of $Q_p$ vs. $\Delta B_t$ is not a straight line, possible reasons include (1) erroneous average and/or fluid properties, (2) water influx, (3) gas cap expansion, and (4) gravity drainage that is affecting reservoir performance ($R_p$ lower than for solution gas drive alone).

Note that $Q_p$ is the net fluid production in RB, $\Delta B_t$ is the oil expansion per STB of initial OIP in RB/STB, and $B_g$ is in RB/SCF. $B_t$ is the two-phase formation volume factor, $B_t = B_o + B_g(R_{si} - R_s)$, in reservoir barrels occupied by one barrel of stock-tank oil plus the gas that was initially dissolved in that oil at reservoir conditions.

**Example 4–5.**[6] Calculation of initial oil in place in a volumetric ($W_e = 0$, $m = 0$), undersaturated reservoir. Connate water salinity is 20,000 ppm.

$$\begin{aligned} B_{ob} &= 1.391 \quad \text{RB/STB} & S_{iw} &= 0.20 \\ p_{av} &= 3{,}600 \quad \text{psi} & \phi &= 0.09 \\ p_i &= 5{,}000 \quad \text{psi} & N_p &= 1.25 \quad \text{MMSTB} \\ T_R &= 220°\text{F} & W_p &= 32{,}000 \quad \text{STB} \end{aligned}$$

Solution:

$$\begin{aligned} c_w &= 3.6 \times 10^{-6} \quad 1/\text{psi} \\ B_w &= 1.04 \quad \text{RB/STB @3,600 psi} \\ B_o &= 1.375 \quad \text{RB/STB @3,600 psi} \end{aligned}$$

$$\begin{aligned} B_{oi} &= 1.355 \quad \text{RB/STB @5,000 psi} \\ c_f &= 5.0 \times 10^{-6} \quad 1/\text{psi for } \phi = 0.09 \end{aligned}$$

$$\begin{aligned} c_o &= \frac{B_o - B_{oi}}{B_{oi}(p_{iR} - p_R)} \\ &= \frac{1.375 - 1.355}{(1.355)(5{,}000 - 3{,}600)} \\ &= 10.54 \times 10^{-6} \quad 1/\text{psi} \end{aligned}$$

Substituting in Equation (4.16),

$$\begin{aligned} N &= \frac{\left[1.25 \times 10^6(1 + 1{,}400 \times 10.54 \times 10^{-6}) + \dfrac{1.04(32{,}000)}{1.355}\right](1 - 0.20)}{1{,}400[10.54 \times 10^{-6} + 5.0 \times 10^{-6} - 0.20(10.54 - 3.6) \times 10^{-6}]} \\ &= \left(\frac{1{,}034{,}404.71}{0.0198}\right)/10^6 \\ &= 52.243 \quad \text{MMSTB} \end{aligned}$$

If rock and water compressibilities are neglected in this example, the initial oil in place will calculate to approximately 86 MMSTB. Therefore, it is important to add all the terms in the material balance calculations applied to expansion type reservoirs producing above the bubblepoint pressure.

The recovery factors at any pressures may be calculated using the following equations:

Above the bubblepoint pressure:

$$E_{Rp} = \frac{B_{oi}}{B_o} c_e \Delta p_R \tag{4.18}$$

Below the bubblepoint pressure:

$$E_{Rp} = \frac{(B_o - B_{oi}) + (R_{si} - R_s)B_g}{B_o + (R_p - R_s)B_g} \tag{4.19}$$

where $R_p$ is the *net cumulative produced gas/oil ratio* and is the quotient of all the gas produced from the reservoir $G_p$ and all the oil produced $N_p$. Since $R_p$ occurs in the denominator of Equation (4.19), the recovery factor will depend on $R_p$. To obtain a high primary recovery, as much gas as possible should be kept in the reservoir, which means that cumulative gas/oil ratio should be maintained as low as possible.

The gas saturation in the reservoir at abandonment pressure can then be calculated using the following equations:

$$S_g = 1 - \left(1 - \frac{N_p}{N}\frac{B_o}{B_{oi}}\right)(1 - S_{iw}) \tag{4.20}$$

$$S_o = \frac{(N - N_p)B_o(1 - S_{iw})}{NB_{oi}} \tag{4.21}$$

From initial reservoir pressure to the bubblepoint, the reservoir oil volume remains constant because oil is produced by liquid expansion and the following equation results.

$$N = \frac{N_p B_o}{(B_o - B_{oi})} \tag{4.22}$$

It is assumed that from initial pressure to the bubblepoint pressure, the initial dissolved GOR is equal to the producing GOR. When the pressure drops below the bubblepoint pressure, the *free gas* ($G_f$) phase develops in the reservoir. If the hydrocarbon pore volume remains constant, the following relationship results:

$$V_{oi} = V_o + V_g \tag{4.23}$$

The volume of the free gas at a pressure below the bubblepoint pressure will be given by the following equation:

$$\text{Free Gas} = \text{Initial Gas} - \text{Solution Gas} - \text{Produced Gas}$$
$$G_f = NR_{si} - (N - N_p)R_s - N_pR_p \tag{4.24}$$

**Example 4–6.**[6] The following data are taken from an oil field which had no original gas cap ($m = 0$) and no waterdrive ($W_e = 0$).

Oil pore volume of reservoir = 75 MM ft$^3$
Solubility of gas in crude = 0.42 SCF/STB/psi
Initial bottomhole pressure = 3,500 psi
Bottomhole temperature = 140°F
Saturation pressure of the reservoir = 2,400 psi
Formation volume factor at 3,500 psi = 1.333 RB/STB
Compressibility factor of the gas at 1,500 psi and 140°F = 0.95
Oil produced when pressure is 1,500 psi = 1 MMSTB
Net cumulative produced GOR = 2,800 SCF/STB

Calculate:

(a) The initial STB of oil in the reservoir
(b) The initial SCF of gas in the reservoir
(c) The initial dissolved GOR of the reservoir
(d) The SCF of gas remaining in the reservoir at 1,500 psi
(e) The gas volume factor of the escaped gas at 1,500 psi and at standard conditions of 14.7 psi and 60°F
(f) The reservoir volume of the free gas at 1,500 psi.

Solution:

(a)
$$N = \frac{V_o}{B_{oi}} = \frac{75/5.615}{1.333} = 10 \quad \text{MMSTB}$$

(b)
$$G = NR_{si} = (10 \times 10^6)(0.42 \times 2{,}400) = 10 \quad \text{MMMSCF}$$

(c)
$$R_{si} = \frac{G}{N} = \frac{10 \times 10^9}{10 \times 10^6} = 1{,}000 \quad \text{SCF/STB}$$

(d)
$$\text{Remaining Gas} = G - N_pR_p = 10 \times 10^9 - [(1 \times 10^6)(2{,}800)] = 7.20 \quad \text{MMMSCF}$$

(e)
$$B_g = \frac{p_{sc}z(T_R + 460)}{(T_{sc} + 460)p_R}$$
$$= \frac{(14.7)(0.95)(140 + 460)}{(60 + 460)(1{,}500)}$$
$$= 0.0107 \quad \text{ft}^3/\text{SCF}$$

(f)
$$G_f = NR_{si} - (N - N_p)R_s - N_pR_p$$
$$= 10 \times 10^9 - (10 \times 10^6 - 1 \times 10^6)630 - (1 \times 10^6 \times 2{,}800)$$
$$= 1.53 \times 10^9 \quad \text{SCF}$$
$$= (1.53 \times 10^9)(0.0107) = 16.37 \quad \text{MMft}^3$$

**Example 4–7.**[6] The "R" Sand is a volumetric oil reservoir whose PVT properties are shown in Figure 4–4. When the reservoir pressure dropped from an initial pressure of 2,500 psi to an average pressure of 1,600 psi, a total of 26.0 MMSTB of oil had been produced. The cumulative GOR at 1,600 psi was 954 SCF/STB and the current GOR was 2,250 SCF/STB. The average porosity for field is 18% and average connate water 18%. No appreciable amount of water was produced and standard conditions are 14.7 psi and 60 °F.

Calculate:

(a) The initial oil in place, N
(b) The SCF of evolved gas remaining in the reservoir at 1,600 psi
(c) The average gas saturation in the reservoir at 1,600 psi

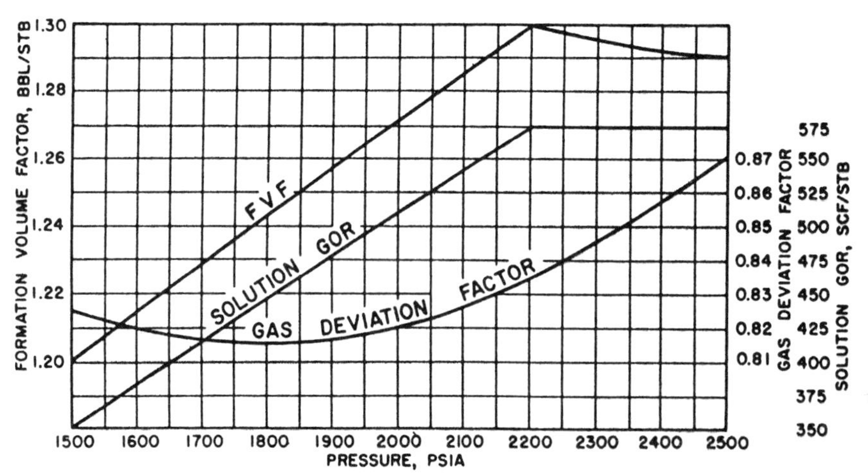


**Figure 4–4.** PVT data for the "R" sand reservoir at 150° F for Example 4–7 (courtesy of Prentice-Hall, Inc.).[6]

(d) The barrels of oil which would have been recovered at 1,600 psi if all the produced gas had been returned to the reservoir
(e) The initial free gas in the reservoir at 2,500 psi

Solution: From Figure 4–4,

$$T_R = 190°\text{F} \quad \text{(assume)} \qquad z = 0.82$$
$$p_b = 2{,}200 \quad \text{psi} \qquad R_{si} = 575 \quad \text{SCF/STB @2,500 psi}$$
$$B_{oi} = 1.29 \quad \text{RB/STB @2,500 psi} \qquad R_s = 385 \quad \text{SCF/STB @1,600 psi}$$
$$B_o = 1.215 \quad \text{RB/STB @1,600 psi}$$

$$\begin{aligned} B_g &= \frac{p_{sc} z (T_R + 460)}{(T_{sc} + 460) p_R} \\ &= \frac{(14.7)(0.82)(190 + 460)}{(60 + 460)(1{,}600)} \\ &= 0.0094 \quad \text{ft}^3\text{/SCF} = 0.0017 \quad \text{RB/SCF} \end{aligned}$$

Using Equation (4.17),

(a)
$$\begin{aligned} N &= \frac{26 \times 10^6[1.215 + 0.0017(954 - 385)]}{1.215 - 1.29 + 0.0017(575 - 385)} \\ &= 229 \quad \text{MMSTB} \end{aligned}$$

(b)
$$\begin{aligned} G_f &= NR_{si} - (N - N_p)R_s - N_pR_p \\ &= \{(229)(575) - [(229 - 26)(385)] - (26)(954)\} \times 10^6 \\ &= (131{,}675 - 78{,}155 - 24{,}804) \times 10^6 \\ &= 28.72 \quad \text{MMMSCF} \end{aligned}$$

(c)
$$\begin{aligned} S_g &= 1 - \left(1 - \frac{N_p}{N}\frac{B_o}{B_{oi}}\right)(1 - S_{iw}) \\ &= 1 - \left(1 - \frac{26}{229}\frac{1.215}{1.29}\right)(1 - 0.18) \\ &= 26.77\% \end{aligned}$$

(d) In this part we assume that $R_p = 0$ and using Equation (4.19),

$$\begin{aligned} E_{Rp} &= \frac{(1.215 - 1.29) + (575 - 385)(0.0017)}{1.215 + (0 - 385)(0.0017)} \\ &= 44.25\% \end{aligned}$$

$$\begin{aligned} N_{p=1600} @R_p = 0 &= 229 \times 0.4425 \\ &= 101.33 \quad \text{MMSTB} \end{aligned}$$

(e) Since at 2,500 psi the reservoir pressure is above the bubblepoint pressure of $p_b = 2{,}200$ psi, there is no free gas in the reservoir.

## OOIP—GAS CAP DRIVE RESERVOIR

In this section we will present a method to evaluate original oil in place in a solution gas drive plus a gas cap drive oil reservoir. The general material balance equation is modified for a solution gas–gas cap drive reservoir by assuming that (1) there is no water encroachment and injection, (2) the change in connate water and pore volume is small enough to be treated as zero, and (3) the fluid properties of dissolved gas and gas cap gas are the same. The material balance equation thus derived is as follows:

$$N_p[(B_o + (R_p - R_s)B_g] = NB_{oi}\left[\frac{(B_o - B_{oi}) + (R_{si} - R_s)B_g}{B_{oi}} + m\left(\frac{B_g}{B_{gi}} - 1\right)\right] \quad \textbf{(4.25)}$$

Rearranging the above equation in terms of $B_t$, we get

$$N = \frac{N_p[B_t + 0.1781B_g(R_p - R_{si})] + B_wW_p}{mB_{oi}\left(\dfrac{B_g}{B_{gi}} - 1\right) + (B_t - B_{oi})} \quad \textbf{(4.26)}$$

There are two unknowns in these equations: the original oil in place (N) and the ratio of the initial hydrocarbon pore volume of the gas cap to that of the oil column, designated by $m$. In this type of reservoir drive, the intent is not to produce gas cap gas but rather to let it expand and displace oil towards the producing oil wells.

Havlena and Odeh presented a graphical technique which can be used in order to estimate $m$ and at the same time N.[5] This technique involves plotting a variable F versus $(E_o + mE_g)$ for an assumed value of $m$. The correct value of $m$ will result in a straight line passing through the origin. If the value of the selected $m$ is small or large, the plot will curve upward or downward, respectively. The slope of the straight line will be N. Remember that the *straight line must pass through the origin.* Havlena and Odeh suggested finding the correct $m$ by trial-and-error; however, it could be easily found by using the computer method given in Chapter 1. The variables F and $(E_o + mE_g)$ are defined by the following equations:

$$F = N(E_o + mE_g) \quad \textbf{(4.27)}$$

where $F$ is the underground withdrawal term, $E_o$ represents the expansion of the oil and its originally dissolved gas, and $E_g$ represents the expansion of the gas cap gas. These terms are related to the PVT properties by the following equations:

$$F = N_p[B_o + (R_p - R_s)B_g] + W_pB_w \quad \textbf{(4.28)}$$

$$E_o = (B_o - B_{oi}) + (R_{si} - R_s)B_g \quad \textbf{(4.29)}$$

$$E_g = B_{oi}\left(\frac{B_g}{B_{gi}} - 1\right) \quad \textbf{(4.30)}$$

**TABLE 4–1.[7] Reservoir performance and PVT data for Example 4–8**

| Pressure (psi) | $N_p$ (MMSTB) | $R_p$ (SCF/STB) | $B_o$ (RB/STB) | $R_s$ (SCF/STB) | $B_g$ (RB/SCF) |
|---|---|---|---|---|---|
| 3,330 | | | 1.2511 | 510 | 0.00087 |
| 3,150 | 3.295 | 1,050 | 1.2353 | 477 | 0.00092 |
| 3,000 | 5.903 | 1,060 | 1.2222 | 450 | 0.00096 |
| 2,850 | 8.852 | 1,160 | 1.2122 | 425 | 0.00101 |
| 2,700 | 11.503 | 1,235 | 1.2022 | 401 | 0.00107 |
| 2,550 | 14.513 | 1,265 | 1.1922 | 375 | 0.00113 |
| 2,400 | 17.730 | 1,300 | 1.1822 | 352 | 0.00120 |

**Example 4–8.**[7] Cumulative oil production $N_p$, cumulative gas/oil ratio $R_p$, and PVT properties at several pressure increments for a gas cap drive reservoir are given in Table 4–1. The geologists have estimated (from individual wells in the reservoir) the original oil in place to be 115 million STB. Calculate the OOIP using the data in Table 4–1; also estimate $m$.

Solution: The variables $F$, $E_o$, $E_g$, and some trials of $m$ for the data given in Table 4–1 are given in Table 4–2. The plot of $F$ vs. $(E_o + mE_g)$ is shown in Figure 4–5. Note that $m = 0.5$ gives relatively straight line with a slope of $114 \times 10^6$; therefore, $N = 114 \times 10^6$ STB.

Havlena and Odeh also recommended that second method also be used as a check, even though the method described above "is a more powerful method" because "it specifies that the line must pass through the origin."[5] The second method requires a plot of $F/E_o$ vs. $E_g/E_o$ on Cartesian coordinate graph paper. The plot resulting in a straight line is based on the following relationship:

$$\frac{F}{E_o} = N + mN\frac{E_g}{E_o} \quad (4.31)$$

**TABLE 4–2. Calculated values of F, $E_o$, $E_g$, and $m$ for data in Table 4–1**

| F (MMRB) | $E_o$ (RB/STB) | $E_g$ (RB/STB) | $E_o + mE_g$: $m = 0.40$ | $E_o + mE_g$: $m = 0.50$ | $E_o + mE_g$: $m = 0.60$ |
|---|---|---|---|---|---|
| 5.807 | 0.01456 | 0.07190 | 0.0433 | 0.0505 | 0.0577 |
| 10.671 | 0.02870 | 0.12942 | 0.0805 | 0.0934 | 0.1064 |
| 17.302 | 0.04695 | 0.20133 | 0.1275 | 0.1476 | 0.1677 |
| 24.094 | 0.06773 | 0.28761 | 0.1828 | 0.2115 | 0.2403 |
| 31.898 | 0.09365 | 0.37389 | 0.2432 | 0.2806 | 0.3180 |
| 41.130 | 0.12070 | 0.47456 | 0.3105 | 0.3580 | 0.4054 |

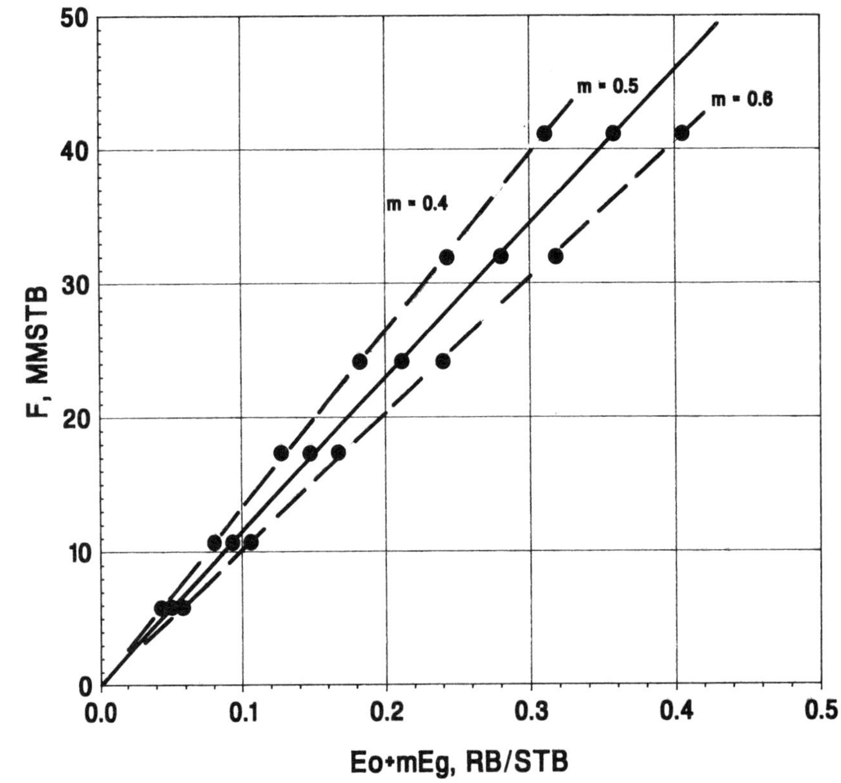


**Figure 4–5.** $F$ vs. $(E_o + mE_g)$ plot for the data in Table 4–2.

Figure 4–6 shows the plot of $F/E_o$ vs. $E_g/E_o$ for the data in Table 4–2, regression analysis of the data points give $N \approx 109$ MMSTB (*y* intercept) and $m = 0.54$ (slope of the straight line). Comparison of these values with those determined by the first method is a desirable check and also may aid selection of the best value of *m* for use in the first method.

## OOIP—NATURAL WATERDRIVE RESERVOIR

If the reduction in reservoir pressure leads to an expansion of the adjacent aquifer water and consequent influx into the reservoir, the general material balance equation must be modified as follows:

Initially undersaturated oil reservoir with active water drive and no gas cap:

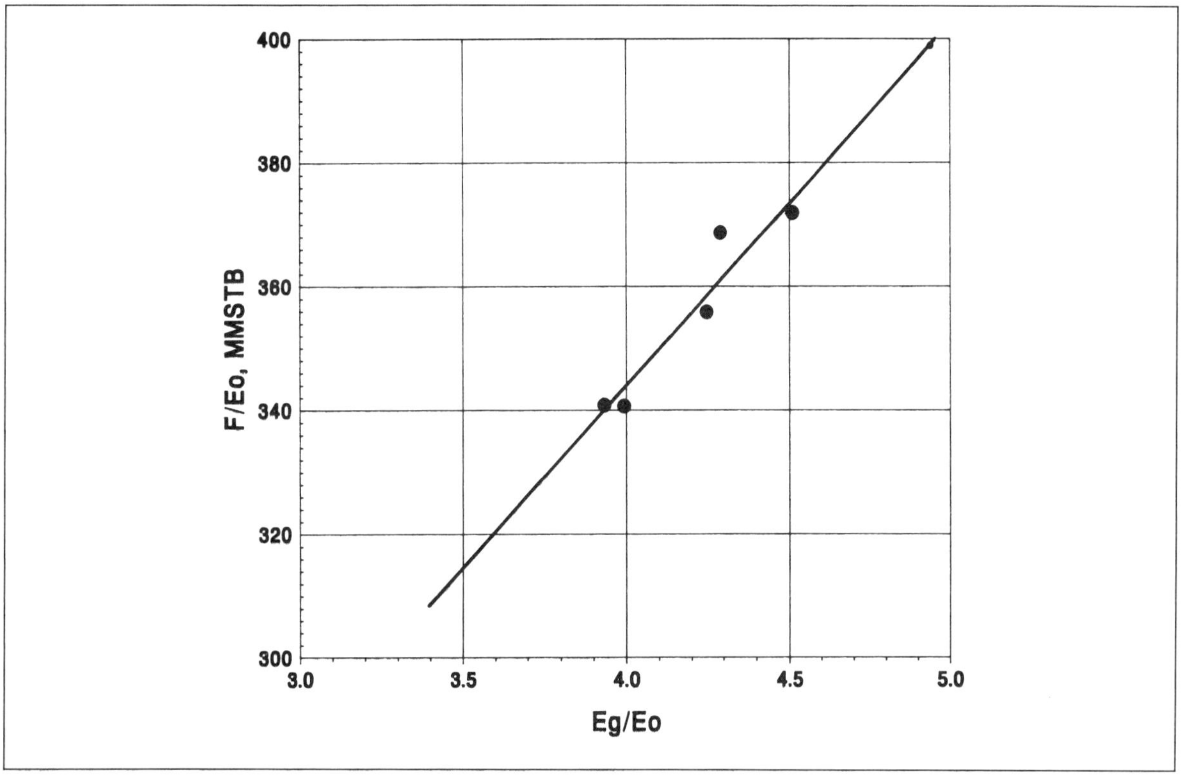


**Figure 4–6.** Alternative graphical method for determining $m$ and N according to Havlena and Odeh.



Above the bubblepoint pressure:

$$N = \frac{\left[N_p(1 + \Delta p_R c_o) - \dfrac{W_e - B_w W_p}{B_{oi}}\right](1 - S_{iw})}{\Delta p_R[c_o + c_f - S_{iw}(c_o - c_w)]} \tag{4.32}$$

Below the bubblepoint pressure:

$$N = \frac{N_p[B_t + 0.01781 B_g(R_p - R_{si})] - (W_e - B_w W_p)}{B_t - B_{oi}} \tag{4.33}$$

where $B_g$ is in ft$^3$/SCF and the water influx term $W_e$ and N are unknown in the equations. The best way to calculate $W_e$ is by using the unsteady-state equation which we will discuss in this chapter. However, Havlena and Odeh proposed the following relationship, according to which $W_e$ can be determined by trial-and-error or the method given in Chapter 1.[5]

$$\frac{V_R}{E_N} = \frac{N + m_p \sum_{j=1}^{n} \Delta p_{n-1-j} W_{edj}}{E_N} \tag{4.34}$$

where

$$V_R = N_p[B_t + B_g(R_p - R_{si})] + (W_p - W_i)B_w - G_iB_{gi}$$

$$E_N = B_t - B_{ti} + \frac{B_{ti}}{1 - S_{iw}}(c_f + S_{iw}c_w)(p_i - p_n) + m\frac{B_{ti}}{B_{gi}}(B_g - B_{gi})$$

$V_R$ is the cumulative voidage at the end of interval $n$ in RB, $E_N$ is the cumulative expansion per stock-tank barrel OOIP in RB, and $G_i$ is the cumulative gas injected in SCF.

Equation (4.34) is the equation of a straight line with a slope $m_p$ and a $y$ intercept of N. Assume an acquifer with $r_D = r_e/r_w$ and a dimensionless time $\Delta t_D$, calculate the summation term in Equation (4.34), and plot $V_R/E_N$ vs. $\Sigma\Delta pW_{edj}/E_N$. If the assumed acquifer and dimensionless time are correct, the plot will be a straight line (Fig. 4–7) with N being the $y$ intercept and $m_p$ being the slope.

If incorrect acquifer model is used, an upward or downward curving line will result. An increasing slope will indicate that the summation terms are too small, while a decreasing slope will indicate that the summation terms are too large. In such cases the procedure is repeated, using different estimates of $r_D$ and/or $\Delta t_D$ until a straight line plot is obtained. Note that more than one combination of $r_D$ and $\Delta t_D$ may yield a reasonable straight line.

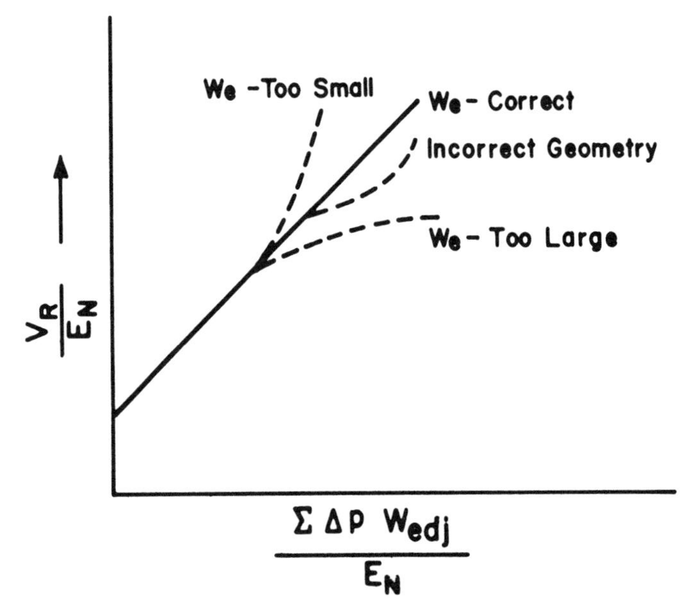


**Figure 4–7.** Trial-and-error method of determining the correct aquifer model.

**Example 4–9.** A waterdrive reservoir produced 103,000 STB of oil and 9,420 RB of water for pressure drop from 2,850 psi to 2,475 psi. The bubblepoint pressure of the reservoir is 2,514 psi. Other PVT properties are given below. If the cumulative water influx is calculated to be 50,000 RB, what would be the original oil in place?

$$
\begin{aligned}
R_{si} &= 510 \quad \text{SCF/STB} & T_R &= 155°\text{F} \\
R_s &= 500.5 \quad \text{SCF/STB} & \phi &= 0.125 \\
B_{oi} &= 1.279 \quad \text{RB/STB} & S_{iw} &= 0.40 \\
B_t &= 1.289 \quad \text{RB/STB} & \gamma_o &= 31.7°\text{API} \\
B_g &= 0.0056 \quad \text{ft}^3/\text{SCF} & R_p &= 512 \quad \text{SCF/STB}
\end{aligned}
$$

Solution: Using Equation (4.33),

$$
\begin{aligned}
N &= \frac{103{,}300[1.289 + 0.1781(0.0056)(512 - 510)] - (50{,}000 - 9{,}420)}{1.289 - 1.279} \\
&= \frac{92{,}780}{0.01} = 9.3 \quad \text{MMSTB}
\end{aligned}
$$

**Example 4–10.**[6] Assume the reservoir of Example 4–7 is a waterdrive reservoir in which $25 \times 10^6$ barrels of water encroached into the reservoir when the pressure fell to 1,600 psi. Calculate the initial oil in place. Use the same current and cumulative GORs, the same PVT data, and assume no water production.

Solution:

$$
\begin{aligned}
B_t &= B_o + (R_{si} - R_s)B_g \\
&= 1.215 + (575 - 385)0.0017 \\
&= 1.538 \quad \text{RB/STB} \\
N &= \frac{26 \times 10^6[1.538 + 0.1781(0.0094)(954 - 575)] - (25 \times 10^6 - 0)}{1.538 - 1.29} \\
&= \frac{31.48 \times 10^6}{0.248} \approx 127 \quad \text{MMSTB}
\end{aligned}
$$

## COMBINATION DRIVE RESERVOIRS

The material balance methods to calculate OOIP outlined in the previous sections are applicable only when the reservoir is being produced exclusively by the specific drive mechanism. Many reservoirs, however, are produced by a combination of two or more driving mechanisms. Techniques for predicting behavior of these *combination drive reservoirs* must be altered to fit the particular reservoir being studied.

A combination of some of the following factors will help in recognizing combination drive reservoirs:[8]

1. Relatively rapid pressure decline: water encroachment and/or external gas cap expansion are insufficient to maintain reservoir pressure.
2. Water encroaching slowly into the lower part of the reservoir: structurally lower completions will show increasing water cut.
3. As a substantial percentage of the OOIP is recovered due to the depletion drive mechanism, the gas/oil ratio of structurally lower wells will continue to increase due to evolution of gas from solution from oil as pressure is reduced.
4. The presence of small and *expanding* gas cap will exhibit increasing gas/oil ratio in the structurally higher completions. A *decreasing* gas/oil ratio exhibited by structurally higher completions will be indicative of gas cap shrinkage due to excess free gas production. This situation must be controlled as much as possible because large volume of oil will be lost.



In most combination drive reservoirs, some type of pressure maintenance operation, either gas injection, water injection, or both gas and water injection will increase ultimate oil recovery.

In the study of combination drive reservoirs (i.e., a reservoir producing simultaneously under depletion drive, gas cap drive, and waterdrive), it is of practical interest to determine the relative magnitude of each of these mechanisms which contribute to the production. The material balance equation is rearranged so as to give three fractions whose sum is one. The three fractions are (1) depletion drive index (DDI), (2) the segregation (gas cap) index (SDI), and (3) the waterdrive index (WDI).

The following form of material balance equation is applied to fields under the simultaneous influence of dissolved gas drive, waterdrive, and gas cap drive:

$$N = \frac{N_p[B_t + (R_p - R_{si})B_g] - (W_e - B_w W_p)}{B_t - B_{ti} + \frac{mB_{ti}}{B_{gi}}(B_g - B_{gi})} \quad \textbf{(4.35)}$$

The three drive indices are given by

$$\text{DDI} = \frac{N(B_t - B_{ti})}{N_p[B_t + (R_p - R_{si})B_g} \quad \textbf{(4.36)}$$

$$\text{SDI} = \frac{N_m B_{ti}/B_{gi}(B_g - B_{gi})}{N_p[B_t + (R_p - R_{si})B_g]} \quad \textbf{(4.37)}$$

$$\text{WDI} = \frac{(W_e - B_w W_p)}{N_p[B_t + (R_p - R_{si})B_g]} \quad \textbf{(4.38)}$$

If the three indices can be calculated, the overall recovery can be estimated using these indices. An increase in the depletion drive and gas drive indices would be reflected by declining pressures and increasing gas/oil ratios, and might indicate the need for water injection to supplement the natural water influx and to turn the recovery mechanism more towards waterdrive.

When all three driving mechanisms simultaneously exist, the depletion drive index should be maintained as low as possible. Maximum advantage should always be taken of the most efficient drive available.

**Example 4–11.**[6] Calculate the stock-tank barrels of oil initially in place in a combination drive reservoir. The following reservoir and PVT data are used:

| | | |
|---|---|---|
| $p_i = 2{,}710$ psi | $N_p = 20 \times 10^6$ STB | $W_e = 11.58 \times 10^6$ bbl |
| $B_g = 0.008479$ ft³/SCF | $p_R = 2{,}000$ psi | $W_p = 1.05 \times 10^6$ STB |
| $B_{gi} = 0.006266$ ft³/SCF | $R_p = 700$ SCF/STB | $B_w = 1.028$ RB/STB |
| $R_{si} = 562$ SCF/STB | $B_t = 1.4954$ RB/STB | $B_{ti} = 1.340$ RB/STB |

Assuming the ratio of the volume of bulk gas cap to the volume of the bulk oil zone, $m = 0.175$.

Solution: Using Equation (4.34),

$$N = \frac{20 \times 10^6[1.4954 + 0.1781(700 - 562)(0.008479)] - (11.58 - 1.05 \times 1.028)}{1.4954 - 1.340 + \dfrac{(0.175)(1.340)}{0.006266/5.615}(0.008479 - 0.006266)/5.615}$$

$$= \frac{23.576 \times 10^6}{0.2382}$$

$$= 98.98 \times 10^6 \text{ STB}$$

## SUMMARY OF MATERIAL BALANCE EQUATIONS

Summary of the various versions of the general material balance equations are given below:

Oil reservoir with gas cap and active waterdrive:

$$N = \frac{N_p[B_t + 0.1781B_g(R_p - R_{si})] - (W_e - W_p)}{mB_{oi}\left(\dfrac{B_g}{B_{gi}} - 1\right) + (B_t - B_{oi})}$$

Oil reservoir with gas cap, no active waterdrive ($W_e = 0$):

$$N = \frac{N_p[B_t + 0.1781B_g(R_p - R_{si})] + W_p}{mB_{oi}\left(\dfrac{B_g}{B_{gi}} - 1\right) + (B_t - B_{oi})}$$

Initially undersaturated oil reservoir with active waterdrive ($m = 0$):

1. Above the bubblepoint pressure:

$$N = \frac{\left[N_p(1 + \Delta p_R c_o) - \dfrac{W_e - W_p}{B_{oi}}\right](1 - S_{iw})}{\Delta p[(c_o + c_f - S_{iw}(c_o - c_w)]}$$

2. Below the bubblepoint pressure:

$$N = \frac{N_p[B_t + 0.1781B_g(R_p - R_{si})] - (W_e - W_p)}{B_t - B_{oi}}$$

Initially undersaturated oil reservoir, no active waterdrive ($m = 0$), ($W_e = 0$):

1. Above the bubblepoint pressure:

$$N = \frac{\left[N_p(1 + \Delta p_R c_o) - \dfrac{W_p}{B_{oi}}\right](1 - S_{iw})}{\Delta p_R[c_o + c_f - S_{iw}(c_o - c_w)]}$$

2. Below the bubblepoint pressure:

$$N = \frac{N_p B_t + 0.1781B_g(R_p - R_{si})] + W_p}{B_t - B_{oi}}$$

where

$N_p$ = cumulative oil produced, STB
$R_p$ = cumulative GOR, SCF/STB
$R_{si}$ = initial solution GOR, SCF/STB
$W_e$ = cumulative water influx, bbl
$W_p$ = cumulative water produced, bbl
$\Delta p_R$ = change in reservoir pressure, psi
$B_{gi}$ = initial gas FVF, ft$^3$/SCF

## BOTTOMHOLE PRESSURE IN GAS WELLS

The static or flowing bottomhole pressure of a gas well is required for gas deliverability calculations, material balance calculations, and volumetric estimates. The preferred method of determining the bottomhole pressure is, of course, to measure the pressure with a bottomhole pressure bomb. However, it is often impractical or too expensive to measure in this manner. Alternatively, in most cases, the bottomhole pressures could be estimated from the surface pressures with sufficient accuracy.

Many authors have presented mathematical relationships for converting the surface shut-in or flowing pressure to the bottomhole shut-in or flowing pressure. Most techniques require an iterative procedure because of the variation of gas properties with pressure and temperature.

Two methods are presented in this section. The first method is called the average temperature and average $z$-factor method, and the second method is referred to as the Cullender and Smith method.[9] Both these techniques are iterative; however, the Cullender and Smith technique is widely accepted in the industry, it gives the most accurate results, and it is also easily programmable for computer calculations.

## STATIC BOTTOMHOLE PRESSURE FROM WELLHEAD SHUT-IN PRESSURE

Calculation of bottomhole shut-in pressure amounts to evaluating the pressure difference equal to the weight of the gas column. The following equations can be used to calculate the bottomhole shut-in pressure from wellhead shut-in pressure:

For average temperature and $z$-factor method:

$$p_{ws} = p_{ts} \exp\left[\frac{0.01875\gamma_g H}{\bar{z}(\bar{T}_f + 460)}\right] \quad \textbf{(4.39)}$$

where $p_{ws}$ is the static or shut-in bottomhole pressure (psi), $H$ is the depth of well (feet), $p_{ts}$ is the static wellhead pressure (psi), and $\bar{z}$ is the $z$-factor at $\bar{T}_f$ and $\bar{p}_{ws}$.

For shallow depths and low-pressured reservoirs, Equation (4.39) will give sufficiently accurate results for many purposes. Otherwise, a value for $p_{ws}$ is assumed at which $\bar{z}$ is calculated and the BHP is calculated and compared to the assumed value; if different, the arithmetic average pressure is calculated and the process repeated. The iterations continue until there is no further change in the average pressure. Normally the solution converges to an acceptable accuracy very quickly.

**Example 4–12.**[10] Calculate the static BHP of a gas well having a depth of 7,500 ft. The gas gravity is 0.744, and the pressure at the wellhead is 2,600 psi. The average temperature of the flow string is 152.5°F. The pseudocritical pressure and temperature are 663.8 psi and 385.6°R, respectively.

Solution: **First Trial:** Assume $p_{ws} = 3{,}100$ psi.

$$\bar{p} = \frac{p_{ws} + p_{ts}}{2}$$
$$= \frac{3{,}100 + 2{,}600}{2} = 2{,}850 \quad \text{psi}$$
$$p_{pr} = 2{,}850/663.8 = 4.29$$
$$T_{pr} = (152.5 + 460)/385.6 = 1.59$$
$$\bar{z} = 0.826$$

Therefore, using Equation (4.39),

$$p_{ws1} = 2{,}600 \exp\left[\frac{0.01877(0.744)(7{,}500)}{(0.826)(152.5 + 460)}\right]$$
$$= 2{,}600e^{0.2070}$$
$$\approx 3{,}198 \quad \text{psi}$$

**Second Trial:**

$$\bar{p} = \frac{3{,}198 + 2{,}600}{2} = 2{,}899 \quad \text{psi}$$
$$\bar{z} = 0.828 \quad @\ \bar{p} = 2{,}899 \text{ and } \bar{T} = 152.5$$
$$p_{ws2} = 2{,}600e^{0.2065}$$
$$\approx 3{,}196 \quad \text{psi}$$

Since $p_{ws2} = 3{,}196$ psi is very close to $p_{ws1} = 3{,}198$ psi, therefore, the static BHP is 3,196 psi. Measured pressure at 7,500 ft equals 3,193 psi.

The Cullender and Smith method involves dividing the wellbore into two halves. This method is widely used in the industry since its adoption by various state regulatory agencies. The method avoids the assumption of a constant average temperature. The procedure is as follows:

For upper half of wellbore:

1. Calculate the integrand $T_{ts}$ at the wellhead using the following equation:

$$I = \frac{1{,}000}{p/(T + 460)z} \tag{4.40}$$

2. As a first approximation, assume $I_{ms}$ at the midpoint equals $I_{ts}$.
3. Calculate the pressure at the midpoint, $p_{ms}$, using the following equation:

$$(p_{ms} - p_{ts})(I_{ms} + I_{ts}) = 18.75\gamma_g H \tag{4.41}$$

4. Use the new $p_{ms}$ to determine $I_{ms}$.
5. Recalculate $p_{ms}$ and iterate.

Repeat Steps 1 through 5 for the bottom half of the wellbore using the following equation:

$$(p_{ws} - p_{ms})(I_{ws} + I_{ms}) = 18.75\gamma_g H \tag{4.42}$$

Once the bottomhole shut-in pressure ($p_{ws}$) has been determined, the accuracy may be improved by using the Simpson's rule (parabolic interpolation) as follows:

$$p_{ws} = p_{ts} + \frac{112.5\gamma_g H}{I_{ts} + 4I_{ms} + I_{ws}} \tag{4.43}$$

## FLOWING BOTTOMHOLE PRESSURE FROM WELLHEAD FLOWING PRESSURE

The bottomhole flowing pressure can be calculated from the wellhead flowing pressure by calculating the difference in pressure due to the weight of the gas as well as frictional pressure drop through the production string.

Major problems in bottomhole pressure calculations from surface measurements (shut-in or flowing) occur because of (1) slugs of unknown amounts of liquid hydrocarbons or water in the tubing, (2) unusual temperature distributions, (3) variations in gas $z$-factor from the correlations, and (4) changes in fluid composition with depth for condensate systems.[11]

Using the average temperature and average $z$-factor method, the flowing bottomhole pressure is given by the following equation:

$$p_{wf}^2 = p_{tf}^2 e^s + \frac{25\gamma_g q_g^2(\bar{T} + 460)\bar{z} f H(e^s - 1)}{s d_i^5} \tag{4.44}$$

where

$\bar{z}$ = $z$-factor at $\bar{p}$ and $\bar{T}$
$s$ = $0.0375\gamma_g H/(\bar{T} + 460)\bar{z}$
$q_g$ = gas flow rate, MMSCF/d
$d_i$ = inside tubing diameter, in.
$f$ = friction factor as a function of $N_{Re}$ and $\epsilon/d$
$N_{Re}$ = Reynolds number
$\epsilon/d$ = relative roughness of pipe in the same units as $d$

The Reynolds number can be calculated from:

$$N_{Re} = \frac{1{,}488\rho_g(\text{lbm/ft}^3)v(\text{ft/sec})d(\text{ft})}{\mu_g(\text{cp})} \quad \textbf{(4.45)}$$

OR

$$N_{Re} \approx \frac{20q_g(\text{MSCF/D})\gamma_g}{\mu_g(\text{cp})d(\text{in.})} \quad \textbf{(4.45a)}$$

Values of the friction factor $f$ can be obtained from the Moody friction factor diagram in Figure 4–8 or, for turbulent flow ($N_{Re} > 2{,}100$), by the following equation:

$$\frac{1}{\sqrt{f}} = 1.14 - 2\log\left[\frac{\epsilon}{d} + \frac{21.25}{(N_{Re})^{0.9}}\right] \quad \textbf{(4.46)}$$

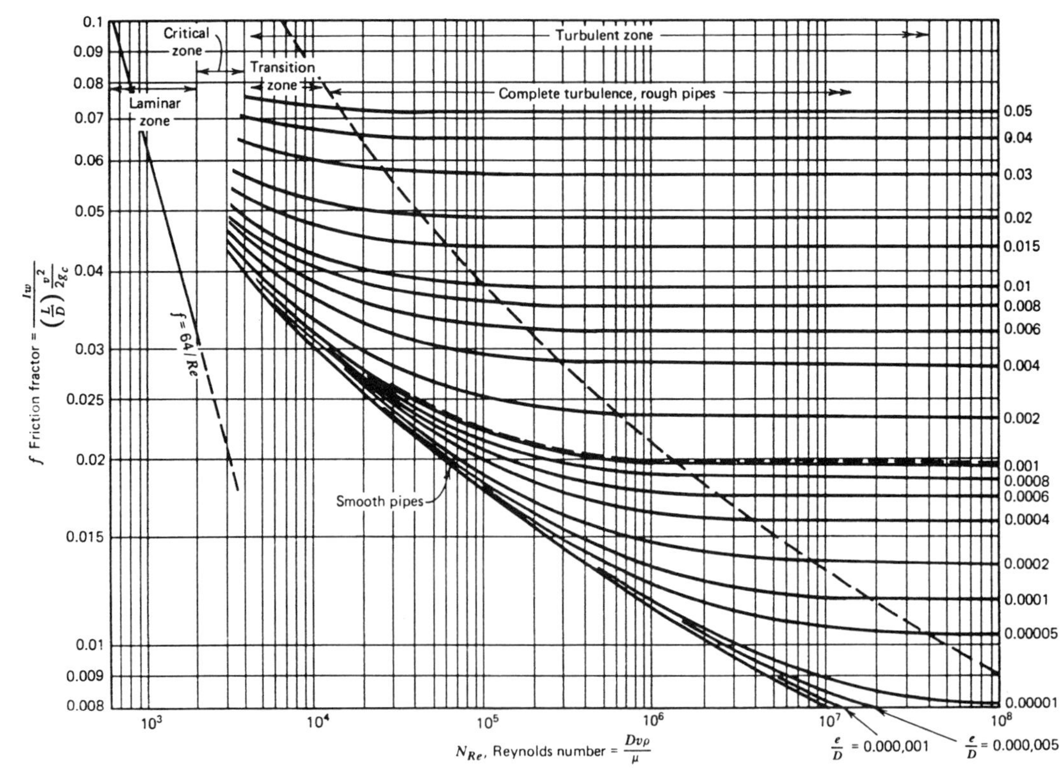


**Figure 4–8.** Friction factor $f$ versus Reynolds number $N_{Re}$ (© SPE, Petroleum Engineering Handbook, 1987).[3]

The calculation of bottomhole flowing pressure using the Cullender and Smith method is obtained as follows:

$$I = \frac{p/(T + 460)z}{0.001[p/(T + 460)z]^2 + F^2} \quad \textbf{(4.47)}$$

where $d < 4.277$ in.

$$F = F_r q_g = \frac{0.10797 q_g}{d^{2.612}} \quad \textbf{(4.48)}$$

and $d \geq 4.277$ in.

$$F = F_r q_g = \frac{0.10337 q_g}{d^{2.582}} \quad \textbf{(4.48a)}$$

where $q_g$ is in MMSCF/D.

The rest of the procedure is the same as given in the previous section. Equations (4.48) and (4.48a) presume an absolute pipe roughness of 0.0006 in.

A modified version of the Cullender and Smith equation given below (which requires iterations on Zavg) is in common use in many computer programs. The bottomhole pressure is solved iteratively from the surface pressure (or vice versa) as follows:

$$p_{wf}^2 = p_{tf}^2 e^A + \frac{\gamma_g \bar{T} F L (e^A - 1) q_g^2}{40{,}000 A d^5} \quad \textbf{(4.49)}$$

where

$q_g$ = gas flow rate, MSCF/D
$A = \gamma_g TVD/26.67(\bar{T} + 460)\bar{z}$
$\bar{T} = (T_{bh} + T_s)/2$, °F
$\bar{z} = (z_{wf} + z_{tf})/2$
$L$ = measured depth

The bottomhole flowing or shut-in pressures are initially estimated using the following relationships:[11] If flowing tubing pressure is given,

$$p_{wf} = p_{tf} + \frac{p_{tf} H}{40{,}000} \quad \textbf{(4.50)}$$

If shut-in tubing pressure is given,

$$p_{ws} = p_{ts} + \frac{p_{ts}H}{40{,}000} \tag{4.50a}$$

**Example 4–13.**[12] Calculate the flowing bottomhole pressure by the method of Cullender and Smith from the following well data:

$$\gamma_g = 0.75 \quad (\text{air} = 1.0) \qquad p_{tf} = 2{,}000 \quad \text{psi}$$
$$L = \text{TVD} = 10{,}000 \quad \text{ft} \qquad q_g = 4.915 \times 10^6 \quad \text{SCF/D}$$
$$T_{wh} = 570°\text{R} \qquad d_i = 2.441 \quad \text{in.}$$
$$T_{bh} = 705°\text{R} \qquad T_{pc} = 408°\text{R}$$
$$p_{pc} = 667 \quad \text{psi}$$

Solution:

$$\bar{T} = \frac{T_{wh} + T_{bh}}{2} = \frac{570 + 705}{2} = 637.50°\text{R}$$

$$\text{Wellhead } T_{pr} = \frac{T_{wh}}{T_{pc}} = \frac{570}{408} = 1.397°\text{R}$$

$$\text{Midpoint } T_{pr} = \frac{T_{mp}}{T_{pc}} = \frac{637.5}{408} = 1.563°\text{R}$$

$$\text{Bottom } T_{pr} = \frac{T_{bh}}{T_{pc}} = \frac{705}{408} = 1.728°\text{R}$$

$$\text{Wellhead } p_{pr} = \frac{p_{tf}}{p_{pc}} = \frac{2{,}000}{667} = 2.999 \quad \text{psi}$$

$$z = 0.7073 \quad @p_{pr} = 2.999 \text{ and } T_{pr} = 1.397°\text{R}$$

Using Equation (4.48),

$$F = \frac{(0.10797)(4.915)}{(2.441)^{2.612}}$$
$$= 0.0516$$
$$F^2 = 0.0027$$

**First Trial**

Using Equation (4.47),

$$\text{assuming } I_{tf} = I_{mf} = \frac{2{,}000/(570)(0.7073)}{0.001[2{,}000/(570)(0.7073)]^2 + 0.0027}$$
$$= 181.714$$

Using Equation (4.41),

$$18.75\gamma_g L = (p_{mf} - p_{tf})(I_{mf} - I_{tf})$$
$$18.75(0.75)(10{,}000) = (p_{mf} - 2{,}000)(181.714 + 181.714)$$
$$p_{mf} = 2{,}387 \quad \text{psi}$$

**Second Trial:**

$$p_{pr} = \frac{p_{mf}}{p_{pc}} = \frac{2{,}387}{667} = 3.579$$
$$z_{mf} = 0.8055 \quad @T_{pr} = 1.563,\ p_{pr} = 3.579$$
$$\frac{p_{mf}}{T_{mf} z_{mf}} = \frac{2{,}387}{(637.5)(0.8055)} = 4.648$$
$$I_{mf} = \frac{4.648}{0.001(4.648)^2 + 0.0027} = 191.245$$

Solving Equation (4.41) for $p_{wf}$,

$$140{,}625 = (p_{mf} - 2{,}000)(191.245 + 181.714)$$
$$p_{mf} = 2{,}377 \quad \text{psi}$$

**Third Trial:**

$$p_{pr} = \frac{2{,}377}{667} = 3.564$$
$$z_{mf} = 0.8054 \quad @T_{pr} = 1.563,\ p_{pr} = 3.564$$
$$\frac{p_{mf}}{T_{mf} z_{mf}} = \frac{2{,}377}{(637.5)(0.8054)} = 4.630$$
$$I_{mf} = \frac{4.630}{0.001(4.630)^2 + 0.0027} = 191.822$$

Solving Equation (4.41) for $p_{mf}$,

$$140{,}625 = (p_{mf} - 2{,}000)(191.822 + 181.714)$$
$$p_{mf} = 2{,}376 \quad \text{psi}$$

Since the $p_{mf} = 2{,}377$ from second trial is close to $p_{mf} = 2{,}376$ from third trial, therefore, $p_{mf}$ for the first half is 2,376 psi.

For the lower half of the flow string assume $I_{wf} = I_{mf} = 191.822$. Solving Equation (4.42) for $p_{wf}$,

**First Trial:**

$$18.75\gamma_g d = (p_{wf} - p_{mf})(I_{wf} + I_{mf})$$
$$140{,}625 = (p_{wf} - 2{,}376)(191.822 + 191.822)$$
$$p_{wf} = 2{,}743 \quad \text{psi}$$

**Second Trial:**

$$p_{pr} = \frac{2{,}743}{667} = 4.112$$
$$z_{wf} = 0.877 \quad @T_{pr} = 1.728,\ p_{pr} = 4.112$$
$$\frac{p_{wf}}{T_{wf}z_{wf}} = \frac{2{,}743}{(705)(0.8771)} = 4.436$$
$$I_{wf} = \frac{4.436}{0.001(4.436)^2 + 0.0027}$$
$$= 198.230$$

Solving Equation (4.42) for $p_{wf}$,

$$140{,}625 = (p_{wf} - 2{,}376)(198.23 + 191.822)$$
$$p_{wf} = 2{,}737 \quad \text{psi}$$

**Third Trial:**

$$p_{pr} = \frac{2{,}737}{667} = 4.103$$
$$z_{wf} = 0.8769 \quad @T_{pr} = 1.728,\ p_{pr} = 4.103$$
$$\frac{p_{wf}}{T_{wf}z_{wf}} = \frac{2{,}737}{(705)(0.8769)} = 4.427$$
$$I_{wf} = \frac{4.427}{0.001(4.427)^2 + 0.0027}$$
$$= 198.535$$
$$140{,}625 = (p_{wf} - 2{,}376)(198.535 + 191.822)$$
$$p_{wf} = 2{,}736 \quad \text{psi}$$

Using Simpson's rule from Equation (4.43),

$$p_{wf} = 2{,}000 + \frac{112.5(0.75)(10{,}000)}{181.714 + 4(191.822) + 198.535}$$
$$= 2{,}735 \quad \text{psi}$$

## BOTTOMHOLE PRESSURE CALCULATIONS—GAS/WATER AND GAS-CONDENSATE FLOW

The Cullender and Smith method of calculating bottomhole pressure in gas wells can be easily modified to account for gas/water mist flow and gas-condensate wells. To include the effect of water in the Cullender and Smith calculations, the integrand $I$ is modified as follows: 

$$I = \frac{[p/(T + 460)z](\bar{\rho}/\rho_g)}{F^2\left(\dfrac{q_g + q_w}{q_g}\right)^2 + 0.001[p/(T + 460)z]^2(\bar{\rho}/\rho_g)^2} \tag{4.51}$$

where the volumetric average density $\bar{\rho}$ at flowing conditions can be calculated as

$$\bar{\rho} = \frac{q_g\rho_g + q_w\rho_w}{q_g + q_w} \tag{4.52}$$

where $q$ is the volumetric flow rate at flowing conditions.

Calculations of BHP in gas-condensate wells are based on equations previously presented for dry gas wells. The application of these equations may be limited somewhat by the amount of liquid present in the flow string.

In the flow equations for gas, the gas gravity $\gamma_g$ is replaced by the flow stream gravity. This is calculated using Equation (2.28) or the following equations:

$$\gamma_{ff} = \frac{(\gamma_g)_{sp} + (4{,}591\gamma_L/R_{gL})}{1 + (1.123/R_{gL})} \tag{4.53}$$

where

$(\gamma_g)_{sp}$ = separator gas gravity (air = 1.0)
$\gamma_L$ = specific gravity of condensate
$R_{gL}$ = gas/liquid ratio, ft$^3$/bbl

Another equation for computing the specific gravity of the flowing fluid is given below.

$$\gamma_{ff} = \frac{R_{gL}\gamma_g + 4{,}603\gamma_o}{R_{gL} + V_L} \tag{4.54}$$

where

$V_L$ = vapor volume equivalent of 1 bbl (60°F) of hydrocarbon liquid, $ft^3/bbl$

$$V_L = 369 + 5\gamma_{API} + 0.04(\gamma_{API})^2 \quad \textbf{(4.55)}$$

## CALCULATION OF GAS INJECTION BOTTOMHOLE PRESSURE

Cullender and Smith's Integrand in Equation (4.47) can be rearranged to calculate the BHP for the gas injection as follows:

$$I = \frac{p/(T + 460)z}{0.001[p/(T + 460)z]^2 - F^2} \quad \textbf{(4.56)}$$

The solution of this equation is identical to that previously described for flowing gas wells (upward flow).

Similarly, by considering the downward flow of gas, the simplified equation developed by Smith for upward flow can be rearranged so that the pressure for vertical flow downward can be calculated.[13]

$$e^s p_{th}^2 - p_{bh}^2 = \frac{25 f q_g^2 \bar{T}^2 \bar{z}^2 (e^s - 1)}{0.0375 d_i^5} \quad \textbf{(4.57)}$$

If gas injection is down the annulus of a well, $d_i^5$ of Equation (4.57) is replaced as follows:

$$d_i^5 = (d_{ci} + d_{to})^2 (d_{ci} - d_{to})^3 \quad \textbf{(4.58)}$$

where $d_{ci}$ is the inside diameter of casing (inches) and $d_{to}$ is the outside diameter of tubing (inches).

Cullender and Binckley derived the following expression for the friction factor as defined by Figure 4–8 that can be used in both upward gas production and downward gas injection.[14]

$$f = 30.9208 \times 10^{-3} \frac{q_g^{-0.065} d_i^{-0.058} \gamma_g^{-0.065}}{\mu_g^{-0.065}} \quad \textbf{(4.59)}$$

where $q_g$ = MMSCF/D, $d_i$ is in feet, and gas viscosity $\mu_g$ is in lbm/ft-sec.

## MATERIAL BALANCE CALCULATIONS—GAS RESERVOIRS

In the previous sections we discussed techniques used to calculate original oil and gas in place using the volumetric method. We also discussed material balance calculations as applied to oil reservoirs. In this section we will discuss material balance calculations as applied to gas reservoirs.

However, the gas material balance calculations are relatively simple as compared to oil material balance and therefore, they are used much more frequently. The relative simplicity is due to the fact that unlike saturated crude oils and condensates, natural gas does not undergo phase changes upon reduction in reservoir pressure.

The gas material balance is of interest specifically in situations when gas production is severely curtailed due to market demand and production performance can not be relied upon for calculating gas reserves. Using material balance, gas in place, reserves, and water influx may be estimated from pressure performance history.



The gas material balance calculations are based on the following assumptions:

1. No large pressure gradients exist across the reservoir at any given time (i.e., pressure equilibrium exists throughout the reservoir).
2. Reliable production, injection, and reservoir pressure measurements are available.
3. Laboratory PVT data apply to the reservoir gas at the average pressures used.
4. The change in volume of interstitial water and the change in porosity with pressure are negligible.

## OGIP—DEPLETION-TYPE RESERVOIR

The term depletion, also referred to as the volumetric depletion, means that when reservoir pressure declines due to production, there is no water influx into the reservoir. The material balance equation for depletion-type gas reservoir is given below.

$$G = \frac{G_p B_g + 5.615 W_p}{B_g - B_{gi}} = \frac{G_p(p_i/z_i) + 5.615 W_p}{p_i/z_i - p/z} \tag{4.60}$$

Early in the producing life of a reservoir, the denominator of the right-hand side of Equation (4.60) is very small, therefore a small change in the denominator will result in a large discrepancy in the calculated value of original gas in place (G).

To confirm the validity of any material balance calculations, the calculations should be performed using *at least three* pressure points. If the original hydrocarbons in place calculated at three pressure points are in good agreement, the average of the three values should be the resultant in-place volume, and it confirms the depletion-type reservoir.

**Example 4–14.** A depletion-type gas reservoir produced 1.0 BSCF with a pressure drop of 386 psi from initial reservoir pressure of 3,250 psi. Calculate the original gas in place (G). Other relevant information is

$T_{sc} = 60°F$ $\quad z\,@3{,}250 \text{ psi} = 0.910$

$T_R = 213°F$ $\quad z\,@2{,}864 \text{ psi} = 0.888$

$p_{sc} = 15.025$ psi

Solution: Calculate $B_{gi}$ and $B_g$ at 2,864 psi as

$$B_{gi} = \frac{T_{sc} p_{Ri}}{p_{sc} z_i T_R}$$

$$= \frac{(60 + 460)(3{,}250)}{(15.025)(0.910)(213 + 460)}$$

$$= 183.66 \quad \text{SCF/ft}^3$$

$$B_g \quad @2{,}864 \text{ psi} = \frac{(60 + 460)(2{,}864)}{(15.025)(0.888)(213 + 460)}$$

$$= 165.86 \quad \text{SCF/ft}^3$$

$$G = G_p \frac{B_{gi}}{B_{gi} - B_g}$$

$$= 1.0 \times 10^9 \left( \frac{183.66}{183.66 - 165.86} \right)$$

$$= 10.32 \quad \text{BSCF}$$

In practice, the most common way in which gas material balance calculations are performed is to plot $p/z$ versus $G_p$ on Cartesian coordinate graph paper. Theoretically the plot should be a straight line as shown in Figure 4–9, resulting in the following equation:

$$p/z = a + b(G_p) \tag{4.61}$$

where

$$a = \frac{p_i V_i T_{sc}}{z_i p_{sc} T_R}$$

$$b = \frac{V_i T_{sc}}{p_{sc} T_R}$$

In Equation (4.61), $p/z$ is the dependent variable, $G_p$ is the cumulative at $p_i/z_i$, and $a$ and $b$ are the coefficients of the equation, corresponding to the intercept and slope parameters, respectively. The coefficients $a$ and $b$ may be found by using the method of least squares (regression). The following equations are utilized:

$$b = \frac{n\Sigma(p/z)(G_p) - \Sigma p/z \Sigma G_p}{n\Sigma G_p^2 - (\Sigma G_p)^2} \tag{4.62}$$

$$a = \overline{p/z} - b\overline{G}_p \tag{4.63}$$

$$G = \frac{a}{-b} \tag{4.64}$$

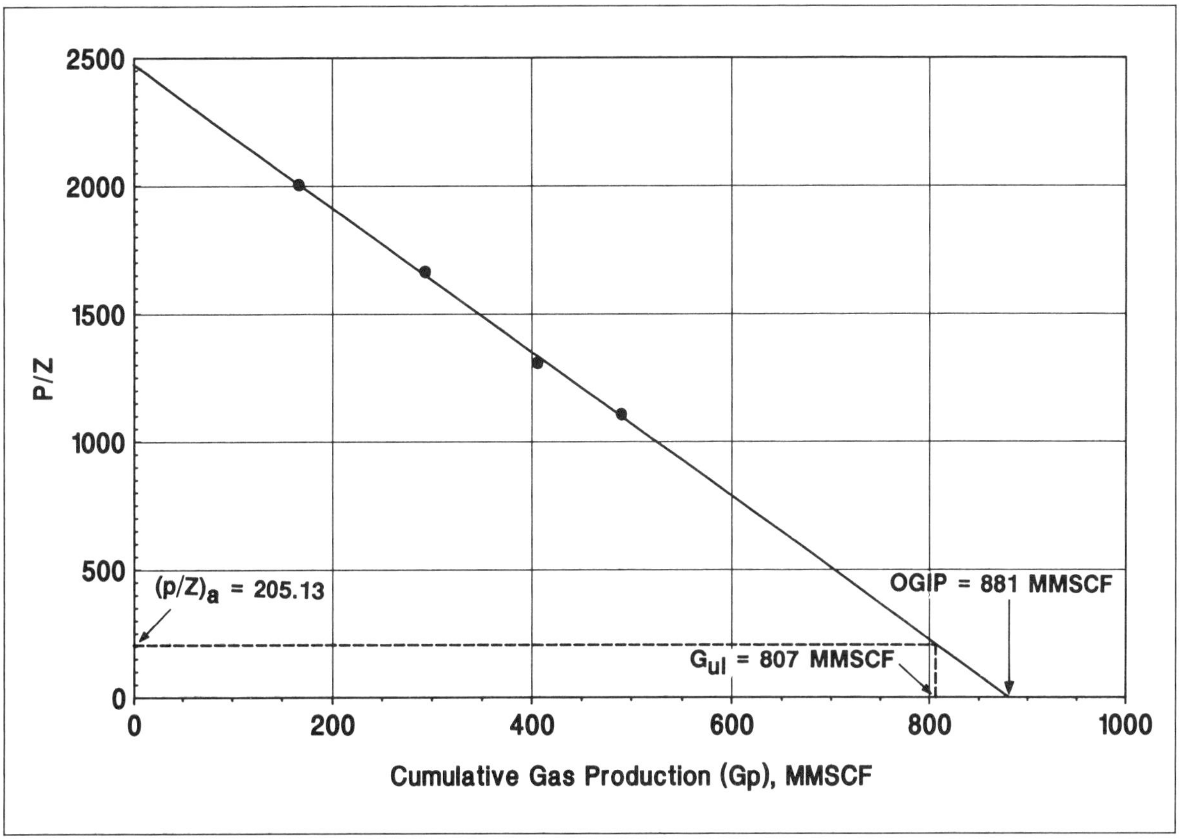


**Figure 4–9.** Pressure-production graph for gas reservoir.

$$\text{Recoverable gas at } p_x/z_x = \frac{a - p_x/z_x}{-b} \tag{4.64a}$$

where

$n$ = the number of $p/z$ vs. $G_p$ pairs
$\overline{p/z}$ = average $p/z$ value, $(\Sigma p/z)/n$
$\overline{G_p}$ = average $G_p$ value, $(\Sigma G_p)/n$

The original gas in place can be found at $p/z = 0$ directly from the $p/z$ plot (commonly identified in the industry). Also, recoverable gas reserves at any abandonment pressure ($p_a/z_a$) may be directly obtained from the plot. Note that the extrapolation of the straight line to $G_p = 0$ will not *always* give $p_i/z_i$ value.

Also, a plot of $G_pB_g$ versus $B_g - B_{gi}$ on Cartesian coordinate graph paper should result in a straight line going through the origin, with G being the slope of this straight line.

**Example 4–15.** Given the following well test information on a Morrow formation gas well in Clark County, Kansas, compute (a) the original gas in place, (b) the recoverable gas ($G_{ul}$) at an abandonment $p_a = 200$ psi, and (c) the remaining gas reserves at $p/z = 800$. Other pertinent reservoir gas PVT data are given below.

$$T_{pc} = 370.88°R \qquad \gamma_g = 0.64 \quad (\text{air} = 1.0)$$
$$p_{pc} = 671.24 \text{ psi} \qquad T_R = 123°F$$

| Date | $p_{ts}$, psi | $G_p$, MMSCF |
|---|---|---|
| 1/1/85 | 1,422 | 166.406 |
| 1/1/86 | 1,208 | 293.967 |
| 1/1/87 | 980 | 406.317 |
| 1/1/88 | 847 | 490.430 |

Solution: The following steps are required:

1. Convert wellhead shut-in pressure to bottomhole shut-in pressure.
2. Determine gas deviation factor $z$ at each bottomhole shut-in pressure.
3. Compute $p/z$ value for each pressure.
4. Plot $p/z$ versus $G_p$.

The calculations are tabulated in Table 4–3 and plotted in Figure 4–9. From Figure 4–9 $p_i/z_i$ is approximately equal to 2,460. The original gas in place G is also calculated individually at each pressure and presented in Table 4–3. The calculated Gs are in good agreement.

(a) The original gas in place at $p/z = 0$ in Figure 4–9 is G ≈ 881 MMSCF.
(b) To find recoverable gas at $p_a = 200$ psi, first find $z$ at 200 psi.

$$p_{pr} = \frac{p_a}{p_{pc}} = \frac{200}{671.24} = 0.298 \quad \text{psi}$$

$$T_{pr} = \frac{T_R + 460}{T_{pc}} = \frac{123 + 460}{370.88} = 1.572°R$$

**TABLE 4–3. $p/z$ data for Example 4–15**

| Date | $p_{ts}$ (psi) | $p_{ws}$ (psi) | $z$ | $p/z$ | $G_p$ (MMSCF) | $G = \dfrac{G_p p_i/z_i}{p_i/z_i - p/z}$ (MMSCF) |
|---|---|---|---|---|---|---|
| 1/1/85 | 1,422 | 1,661 | 0.8287 | 2,004 | 166.406 | 897.717 |
| 1/1/86 | 1,208 | 1,406 | 0.8456 | 1,663 | 293.967 | 907.351 |
| 1/1/87 | 980 | 1,135 | 0.8687 | 1,307 | 406.317 | 866.904 |
| 1/1/88 | 847 | 978 | 0.8841 | 1,106 | 490.430 | 891.032 |

Therefore, $z = 0.975$ and $p_a/z_a = 205.13$.

$$\begin{aligned} G_p \ @\ p_a &= 200 \text{ psi} \approx 807 \quad \text{MMSCF} \qquad \text{(From Fig. 4–9)} \\ G_p \ @\ p/z &= 800 \approx 600 \quad \text{MMSCF} \qquad \text{(From Fig. 4–9)} \\ \text{(c)} \qquad G_{ul} \ @\ p/z &= 800 = G_{p205} - G_{p800} \\ &= 807 - 600 \\ &= 207 \quad \text{MMSCF} \end{aligned}$$

**Example 4–16.** Rework the data in Example 4–15 to calculate OGIP and recoverable gas at $p/z = 800$, using regression analysis.

Solution: The data for regression analysis are tabulated in Table 4–4.

Using Equation (4.62),

$$\begin{aligned} b &= \frac{4(1{,}895{,}816.644) - (6{,}080)(1{,}375.120)}{4(519{,}722.644) - (1{,}357.120)^2} \\ &= -2.817 \end{aligned}$$

$$\overline{p/z} = \frac{6{,}080}{4} = 1{,}520$$

$$\overline{G}_p = \frac{1{,}357.120}{4} = 339.280$$

Using Equation (4.63),

$$\begin{aligned} a &= 1{,}520 - (-2.817)(339.280) \\ &= 2{,}475.752 \end{aligned}$$

Using Equation (4.64),

**TABLE 4–4. Regression analysis of $p/z$ data**

| $p/z$ | $G_p$ | $(p/z)(G_p)$ | $G_p^2$ |
|---|---|---|---|
| 2,004 | 166.406 | 333,477.624 | 27,690.957 |
| 1,663 | 293.967 | 488,867.121 | 86,416.597 |
| 1,307 | 406.317 | 531,056.319 | 165,093.505 |
| 1,106 | 490.430 | 542,415.580 | 240,521.585 |
| 6,080 | 1,357.120 | 1,895,816.644 | 519,722.644 |

$$G = \frac{2{,}475.752}{-(-2.817)}$$
$$\approx 879 \quad \text{MMSCF}$$
$$G_p\ @\ p/z = 800 = \frac{2{,}475.752 - 800}{-(-2.817)}$$
$$\approx 595 \quad \text{MMSCF}$$



## OGIP—WATERDRIVE RESERVOIR

In a depletion-type gas reservoir, a plot of G versus $G_p$ should result in a horizontal straight line, or the data points will evenly scatter around a horizontal line. However, if water (or other fluid) influx is occurring, a plot of G versus $G_p$ (assuming a depletion-type reservoir) will result in a straight line showing increasing values of G as shown in Figure 4–10. If there is gas leakage to another zone due to bad cement job or casing leaks, the straight line will show decreasing G values as shown in Figure 4–10.[15]

For a waterdrive gas reservoir, the following material balance equation is presented. Note that

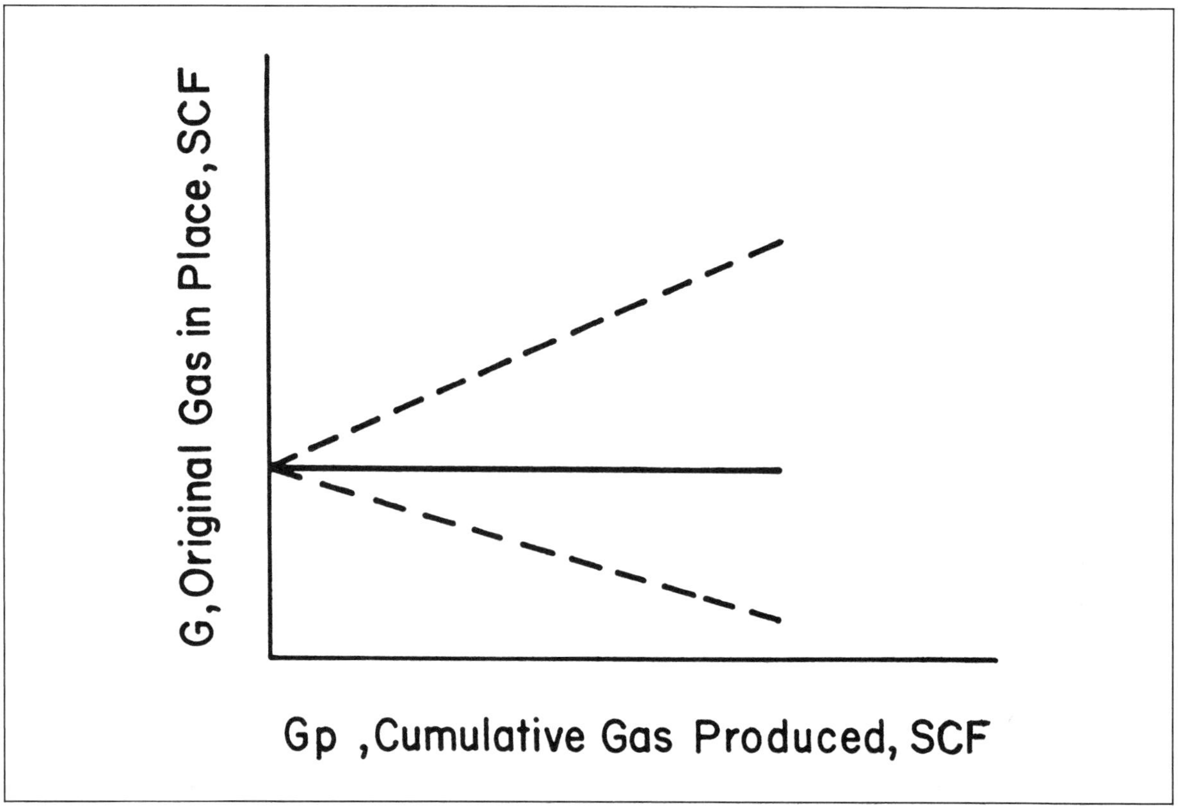


**Figure 4–10.** Variation of G with $G_p$ in water drive gas reservoir.

there are two unknowns: the original gas in place (G) and the cumulative water influx ($W_e$). The $W_e$ may be independently estimated using the unsteady-state equation or it may be estimated from the material balance equation by trial-and-error.

$$G = \frac{G_p B_g - (W_e - W_p B_w)5.615}{B_g - B_{gi}} \tag{4.65}$$

If the value of $W_e$ is known from independent sources, then Equation (4.65) can be used directly to calculate G. If, however, $W_e$ is not known, then a method suggested by Bruns et al. may be used to calculate G.[16] According to this method apparent gas in place $G_a = G_p/(1 - B_g/B_{gi})$ is calculated, assuming a depletion-type gas reservoir. If there is an active waterdrive, the value of $G_a$ calculated for known values of $B_g$ and $G_p$ will not be unique. Successive $G_a$ values will increase as the deviation of $p/z$ above the depletion material balance line increases.

The calculated values of $G_a$ are then plotted versus $W_e B_g/(1 - B_g/B_{gi})$. The result should be a straight line, provided the correct $W_e$ value has been used. If $W_e$ is not correct, the line will *curve upward* for $W_e$ being too small or *downward* for $W_e$ being too large as shown in Figure 4–11. Once the straight line is obtained, the proper value of G can be determined by its linear extrapolation to the ordinate.

The periodic water influx corresponding to a particular pressure and $G_p$ may be calculated using the following rearranged equation:

$$W_e = G_p B_g - G(B_g - B_{gi}) \tag{4.66}$$

A plot of $[G + (W_e/B_g - B_{gi})]$ versus $G_p$ on Cartesian coordinate graph paper is illustrated in Figure 4–12. Extrapolation of a best-fit line drawn through the data points to $G_p = 0$ will give the true value of G, because when $G_p = 0$, the $W_e/(B_g - B_{gi})$ is also zero. At any time the difference between the horizontal line and the sloping line will give the value of $(W_e/B_g - B_{gi})$. This technique will, therefore, help in forecasting the future water influx $W_e$, and it will provide some guidance for scheduling production and expected cash flow.

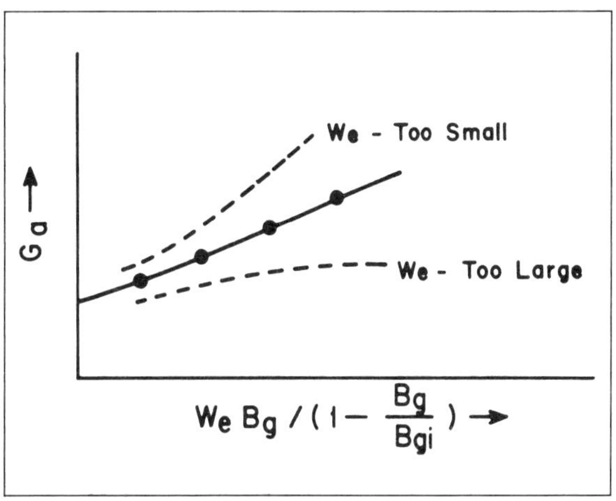


**Figure 4–11.** Determination of G in a waterdrive gas reservoir (after Bruns et. al.).[16]

**Example 4–17.**[6] The "M" Sand is a small gas reservoir with an initial bottomhole pressure of 3,200 psi and bottomhole temperature of 220°F. It is desired to inventory the gas in place at three production intervals. The pressure-production history and gas volume factors in SCF/ft$^3$ at standard conditions of 14.7 psi and 60°F are

| Pressure (psi) | Cumulative Gas Production (MMSCF) | Gas Volume Factor (SCF/ft³) | Gas Volume Factor (ft³/SCF) |
|---|---|---|---|
| 3,200 | 0 | 190.035 | 0.0052622 |
| 2,925 | 79 | 175.426 | 0.0057004 |
| 2,525 | 221 | 153.114 | 0.0065311 |
| 2,125 | 452 | 129.266 | 0.007360 |

(a) Calculate the initial gas in place at each pressure interval, assuming volumetric behavior.
(b) Show why the calculations of G in Part (a) indicate waterdrive.
(c) Show that a waterdrive exists by plotting $p/z$ vs. $G_p$.
(d) If the original gas in place is G = 950 MMSCF, and if the sand is under waterdrive, what is the volume of water encroached into the reservoir at the end of each interval. Assume water production is negligible.

Solution:

(a)

$$G_{2,925} = \frac{G_{p=2,925} \times B_{g=3,200}}{B_{g=3,200} - B_{g=2,925}} = \frac{(79)(190.035)}{(190.035 - 175.426)} = 1{,}028 \text{ MMSCF}$$

Similarly,

$$G_{2,525} = 1{,}138 \text{ MMSCF}$$

$$G_{2,125} = 1{,}413 \text{ MMSCF}$$

(b) Plot G versus $G_p$ on Cartesian coordinate graph paper as shown in Figure 4–13. Since the line drawn through the data points is increasing from the horizontal line, water influx is indicated.

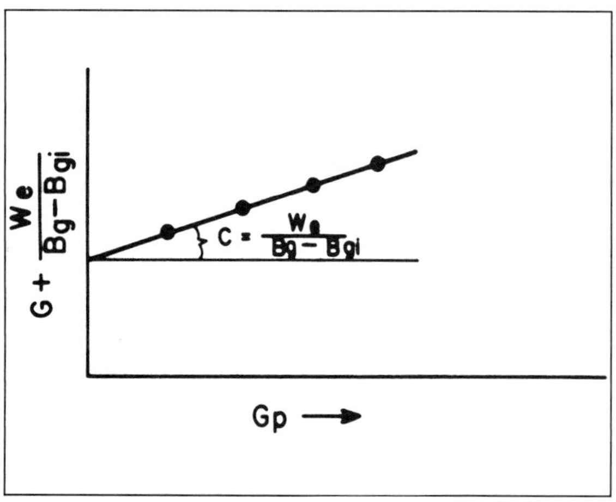


**Figure 4–12.** Effect of water influx on gas material balance calculations.

(c) To plot $p/z$ versus $G_p$, $p/z$ values are back-calculated (in this particular case) from $B_g$ values as given below.

$$p/z = \frac{B_g p_{sc}(T_R + 460)}{(T_{sc} + 460)}$$

$$(p/z)_{3,200} = \frac{(190.35)(14.7)(220 + 460)}{(60 + 460)} = 3{,}653$$

Similarly,

$$(p/z)_{2,925} = 3{,}372$$

$$(p/z)_{2,525} = 2{,}943$$

$$(p/z)_{2,125} = 2{,}485$$

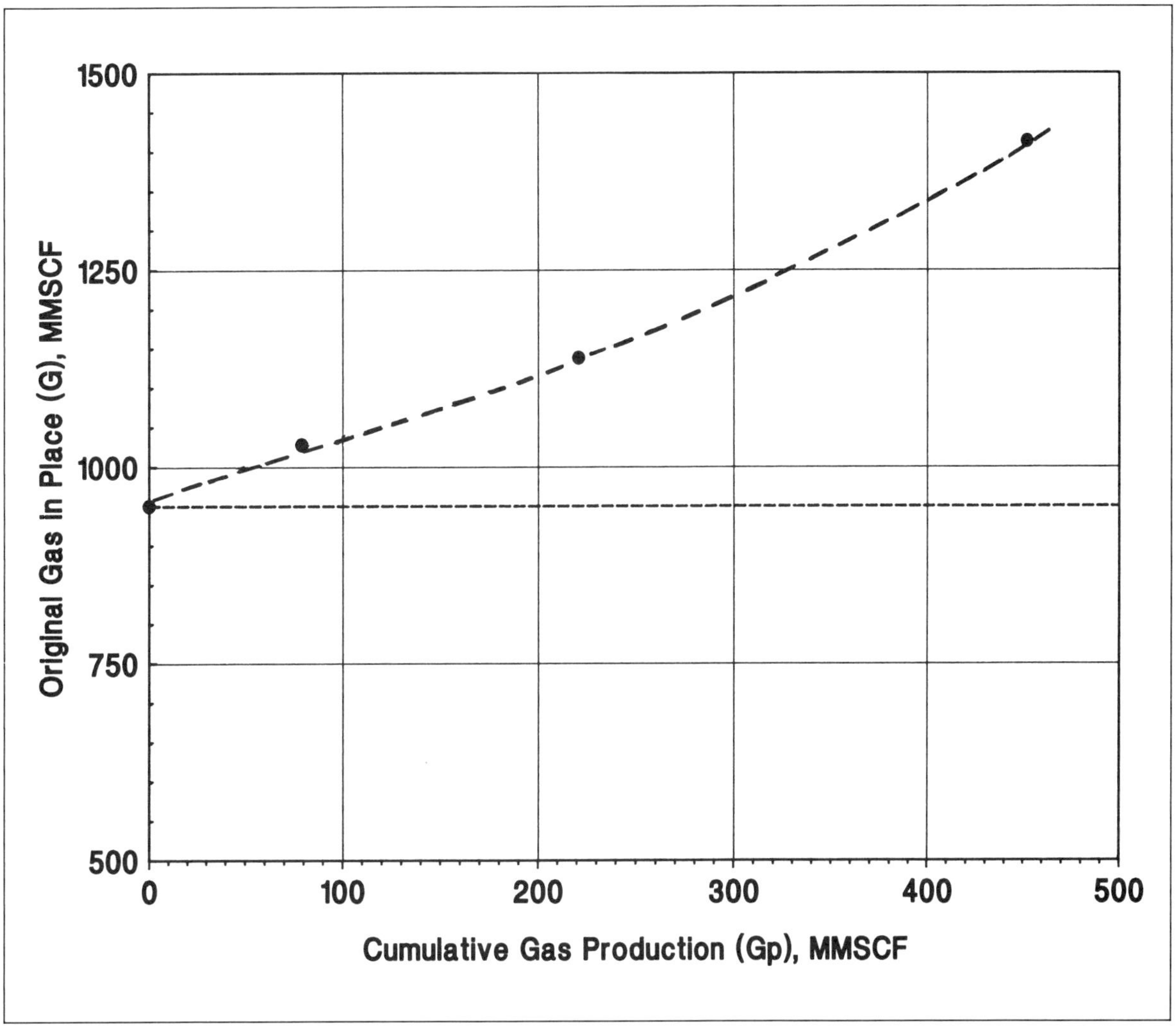


**Figure 4–13.** Plot of G versus $G_p$ for data in Example 4–17.

The $p/z$ plot for this data is shown in Figure 4–14.

(d)
$$W_e = [G_pB_g - G(B_g - B_{gi})]$$

$$W_{e2,925} = \left[\left(79 \times 10^6 \frac{0.0057004}{5.615}\right) - (950 \times 10^6)\left(\frac{(0.0057004 - 0.0052622)}{5.615}\right)\right]$$

$$= (0.0802 - 0.0741) \times 10^6$$

$$= 6{,}100 \quad \text{bbls}$$

Similarly, $W_{e2,525} = 42{,}372 \quad \text{bbls}$

$$W_{e2,125} = 204{,}197 \quad \text{bbls}$$

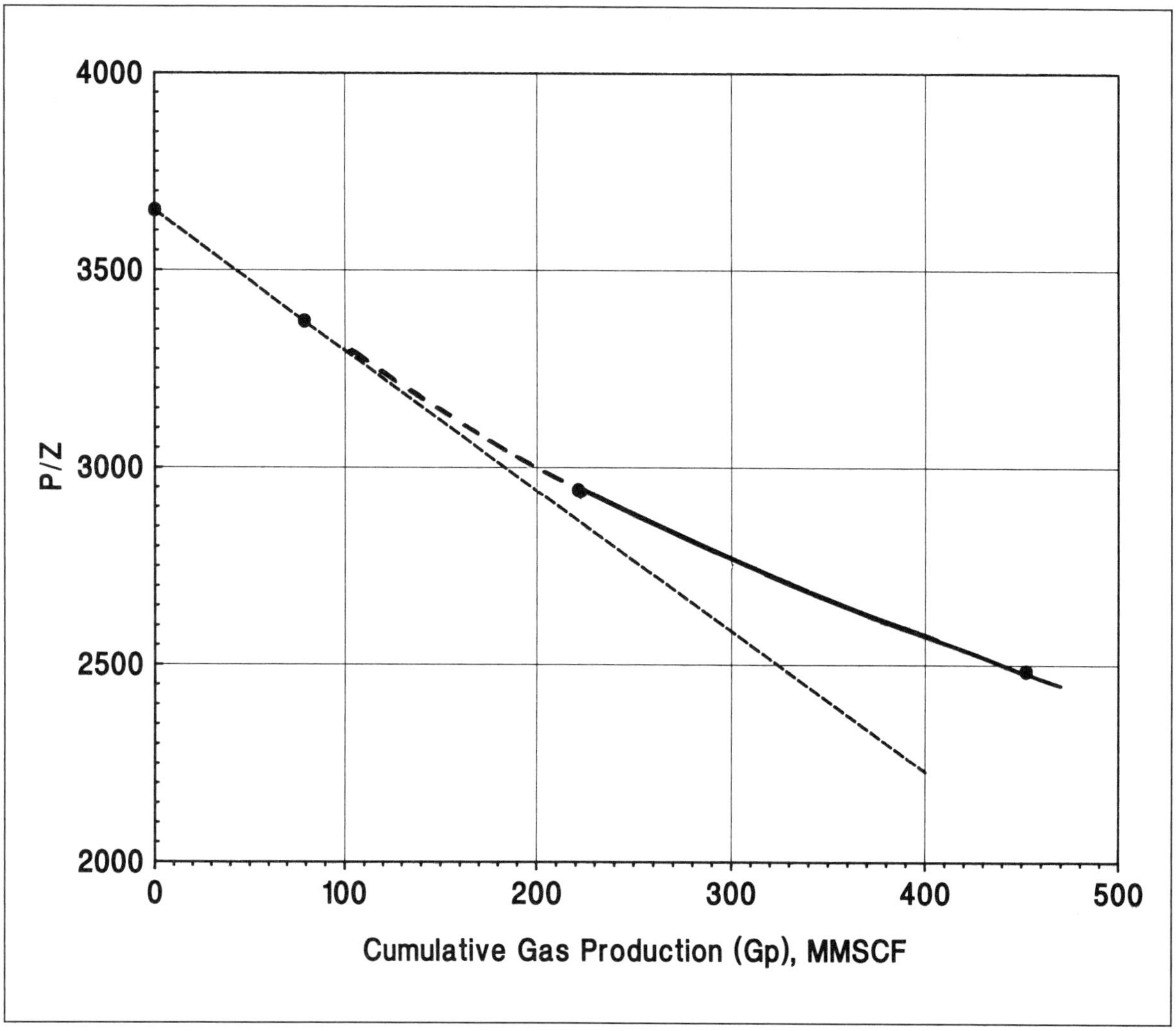


**Figure 4–14.** $p/z$ vs. $G_p$ plot for Example 4–17.

## OGIP—WATERDRIVE (INFINITE LINEAR AQUIFER CASE)

For a linear aquifer, water influx is proportional to the square root of time and

$$W_e = C'\Sigma(\Delta p_n \sqrt{t - t_n}) \qquad \textbf{(4.67)}$$

For other aquifer geometries, a dimensionless time $t_D$ must be estimated. Then the dimensionless water influx $Q_D$ is determined, and $W_e = C\Sigma\Delta p Q_D$. Water influx determination is discussed in more detail in the next section.

If gas material balance calculations indicate water influx, the infinite linear aquifer case should first be tried since it does not require the estimation of dimensionless time and is easy to perform.

For the linear aquifer case, $G_pB_g/(B_g - B_{gi})$ is plotted versus $[\Sigma\Delta p_n\sqrt{t - t_n}/(B_g - B_{gi})]$ on a Cartesian coordinate graph. The plot should result in a straight line with G being the y intercept and C′ (water influx constant) being the slope of the straight line.

**Example 4–18.**[17] Production, pressures and pertinent expansion factors for a dry gas reservoir are presented in Table 4–5. Calculate the OGIP (G).



Solution: A summary of data and calculations is presented in Table 4–5. The depletion-type material balance, Equation (4.60), was tried first. A plot of $G_pB_g$ vs. $(B_g - B_{gi})$ showed upward curvature, indicating water influx. Because of this condition, the MBE with waterdrive (linear aquifer case) was tried next.

**TABLE 4–5.**[17] **Reservoir data for Example 4–18**

| Time (months) | Average Reservoir Pressure (psi) | $(B_g - B_{gi}) \times 10^{-6}$ (ft³/SCF) | $(G_pB_g) \times 10^6$ (ft³) | $\frac{\Sigma\Delta p_n\sqrt{t - t_n}}{B_g - B_{gi}}$ ($10^6$) | $\frac{G_pB_g}{B_g - B_{gi}}$ ($10^{12}$) |
|---|---|---|---|---|---|
| 0 | 2,883 | 0.0 | — | — | — |
| 2 | 2,881 | 4.0 | 5.5340 | 0.3536 | 1.3835 |
| 4 | 2,874 | 18.0 | 24.5967 | 0.4647 | 1.3665 |
| 6 | 2,866 | 34.0 | 51.1776 | 0.6487 | 1.5052 |
| 8 | 2,857 | 52.0 | 76.9246 | 0.7860 | 1.4793 |
| 10 | 2,849 | 68.0 | 103.3184 | 0.9306 | 1.5194 |
| 12 | 2,841 | 85.0 | 131.5371 | 1.0358 | 1.5475 |
| 14 | 2,826 | 116.5 | 180.0178 | 1.0315 | 1.5452 |
| 16 | 2,808 | 154.5 | 240.7764 | 1.0594 | 1.5584 |
| 18 | 2,794 | 185.5 | 291.3014 | 1.1485 | 1.5703 |
| 20 | 2,782 | 212.0 | 336.6281 | 1.2426 | 1.5879 |
| 22 | 2,767 | 246.0 | 392.8592 | 1.2905 | 1.5970 |
| 24 | 2,755 | 273.5 | 441.3134 | 1.3702 | 1.6136 |
| 26 | 2,741 | 305.5 | 497.2907 | 1.4219 | 1.6278 |
| 28 | 2,726 | 340.0 | 556.1110 | 1.4672 | 1.6356 |
| 30 | 2,712 | 373.5 | 613.6513 | 1.5174 | 1.6430 |
| 32 | 2,699 | 405.0 | 672.5969 | 1.5714 | 1.6607 |
| 34 | 2,688 | 432.5 | 723.0868 | 1.6332 | 1.6719 |
| 36 | 2,667 | 455.5 | 771.4902 | 1.7016 | 1.6937 |

$\Delta p_n = \frac{p_{n-1} - p_{n+1}}{2}$, $t_n = t_n$ to $t = 0$

As evident from Figure 4–15, the necessary straight line relationship was regarded as a satisfactory evidence of the presence of linear aquifer. The original gas in place was 1.325 TSCF and $C' = 212.7 \times 10^3$ ft$^3$/psi.

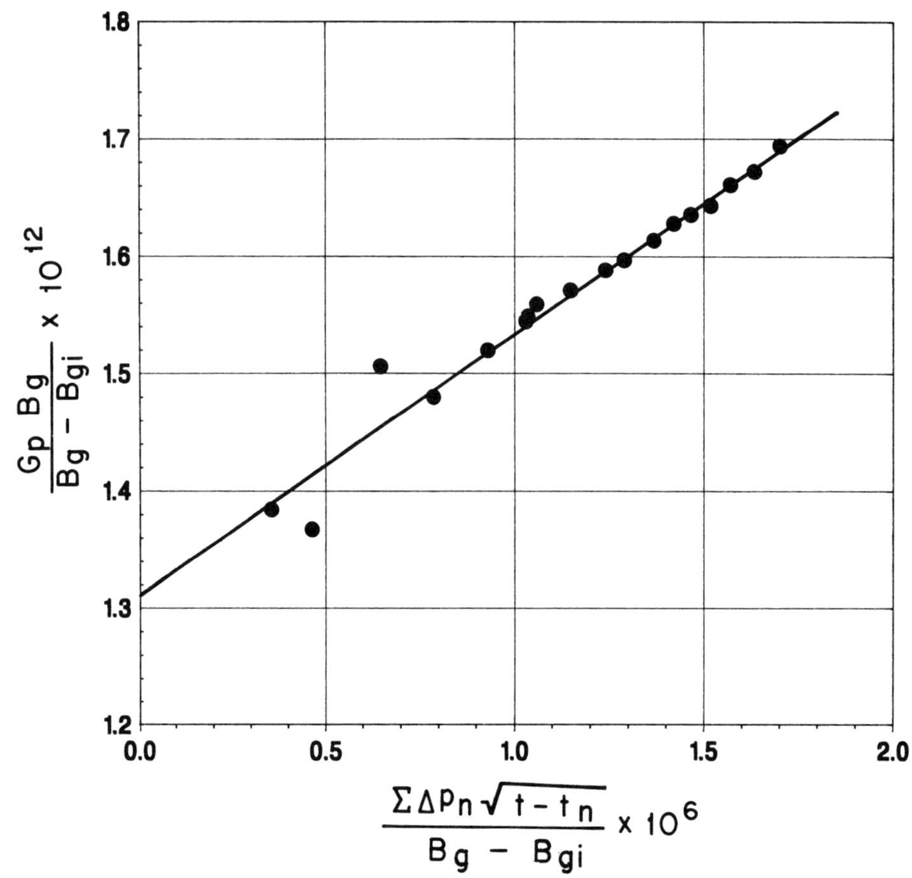


**Figure 4–15.** Determination of OGIP in waterdrive dry gas reservoir—linear aquifer case.

## EFFECT OF WATER AND PORE COMPRESSIBILITIES

The water and pore compressibilities ($c_w$ and $c_f$) are usually omitted from the gas material balance equation, Equation (4.60). The reason for its omission is because $c_w$ and $c_f$ are usually, although not always, insignificant in comparison to the gas compressibility. Note that the pore compressibility can sometimes be very large in shallow unconsolidated reservoirs.

If the formation compressibility is not zero, then the $p/z$ plot should not be a straight line.

When the formation and connate water volume changes are omitted, the error in the resultant OGIP is given by

$$\text{Percent Error} = \frac{G_{app} - G_{act}}{G_{act}} \times 100 \tag{4.68}$$

where

$$G_{app} = \frac{G_p}{\left(1 - \frac{pz_i}{p_i z}\right)}$$

$$G_{act} = \frac{G_p}{1 - \left(\frac{pz_i}{p_i z}\right)\left[1 - (c_w S_{iw} + c_f)\left(\frac{p_i - p}{1 - S_{iw}}\right)\right]}$$

OR

$$\text{Percent Error} = \frac{(c_w S_{iw} + c_f)(p_i - p)/(1 - S_{iw})}{(p_i z/p z_i) - 1} \tag{4.69}$$

## OGIP—ABNORMALLY HIGH PRESSURED GAS RESERVOIRS

The material presented in this section was originally published by Hammerlindl.[18] In abnormally high pressured gas reservoirs, the effective compressibility ($c_e$) may equal two or more times that of gas compressibility ($c_g$). Since the effective compressibility is higher than the gas compressibility, gas production is due to (1) gas expansion, (2) formation compaction due to formation compressibility, (3) crystal expansion, and (4) water expansion until the compaction is complete at approximately the normal pressure. At the normal pressure, the gas compressibility approaches the effective compressibility, and the gas is produced due to gas expansion.

When $p/z$ is plotted versus $G_p$, two distinct slopes are evident as shown in Figure 4–16. One slope occurs during the early life of the well, which is due to compressibility differences and hence pressure maintenance; the second slope starts at normal pressure, which is due to gas expansion. In most cases, the majority of the gas is produced before the reservoir pressure reaches normal pressure. Consequently, reserve estimates based on the early life portion of the curve are erroneously high. More realistic estimate is possible once the second-slope line has started. Two methods are presented for correcting the reserve estimates from the early life data (assuming no water influx).

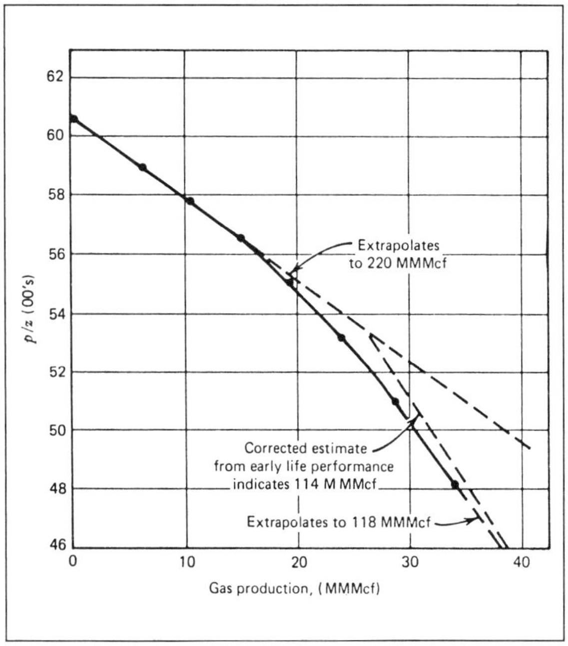


**Figure 4–16.** $p/z$ versus $G_p$ plot for North Osmun Field, Lafayette Parish, Louisiana NS2B reservoir (© SPE, paper presented at the 64th annual fall meeting, Oct. 1971).[18]

**Using $p/z$-Plot:** To obtain the actual gas in place, the following method is used:

1. Using Figure 4–17, determine $c_f$ from appropriate curve.
2. Calculate $c_{gi}$ and $c_{g2}$ ($p_2$ should be less than or equal to 0.5 psi/ft of depth).
3. Calculate effective compressibility at $p_i$ and $p_2$ using Equation (2.66c).

$$c_e = \frac{c_{gi}S_{iw} + c_{wi}S_{iw} + c_f}{S_{gi}}$$

4. Find the $(c_e/c_g)_{avg}$, which is approximately equal to

$$\left(\frac{\dfrac{c_{ei}}{c_{gi}} + \dfrac{c_{e2}}{c_{g2}}}{2}\right) \tag{4.70}$$

5. Extrapolate the first-slope line to $p/z = 0$ to estimate *apparent* original gas in place. Calculate the *corrected* gas in place using the following equation:

$$G_{cor} = \frac{G_{app}}{\left(\dfrac{c_e}{c_g}\right)_{avg}} \tag{4.71}$$

6. To predict the recoverable gas reserves at any abandonment pressure:
   a. Determine the point of normal gradient on the first-slope line (0.45 to 0.5 psi/ft in $p/z$ value).
   b. Plot the $G_{cor}$ value at $p/z = 0$.
   c. Draw a straight line between $G_{cor}$ and the normal gradient pressure point.
   d. Use the second-slope line to determine the reserves at any abandonment pressure.

**Using Material Balance:** In this method the original gas in place is calculated in the same way as for the depletion-type gas reservoirs (discussed in the previous sections). The corrected gas in place ($G_{cor}$) is then the product of $G_{app}$ and $G_{pr}$. The term $G_{pr}$ is calculated using the following equation:

$$G_{pr} = \frac{G_{g2} - B_{gi}}{B_{g2} - B_{gi} + \dfrac{B_{gi}\Delta p(c_f + c_w S_{iw})}{1 - S_{iw}}} \tag{4.72}$$

where $B_g$ is in $ft^3/SCF$. The following steps will help in using this method:

1. Determine $G_{app}$ using the material balance for depletion type reservoir (first-slope line of $p/z$-plot may also be used).

2. Determine $B_{gi}$ and $B_{g2}$; pressures should be above the normal gradient.
3. Determine $c_f$ from Figure 4–17.
4. Calculate $G_{pr}$ using Equation (4.72).
5. Determine the corrected original gas in place ($G_{cor}$), $G_{cor} = G_{app} \times G_{pr}$.

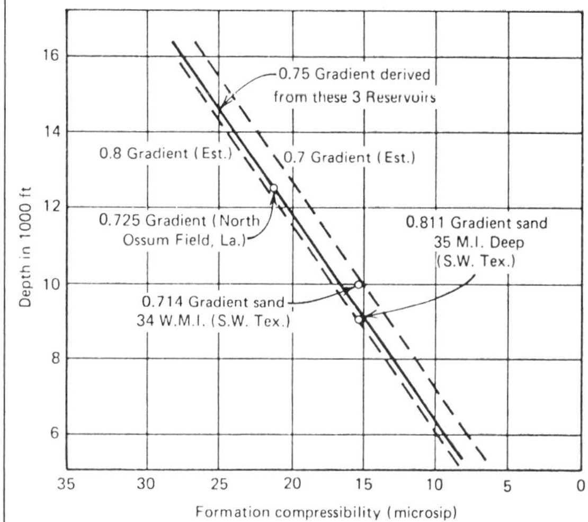


**Figure 4–17.** Depth vs. $c_f$ in abnormally high pressured reservoir (© SPE, paper presented at the 64th annual fall meeting, Oct. 1971).[18]

**Example 4–19.** Reservoir and fluid data for the well in North Osmun Field (Fig. 4–16) are given below. Using the material balance method, calculate the original gas in place. Standard temperature and pressure are 60°F and 14.7 psi, respectively.

$$\text{Depth, ft} = 12{,}500$$
$$\text{Pressure, psi} = 8{,}921$$
$$\text{Pressure gradient, psi/ft} = 0.725$$
$$\text{Temperature, °F} = 248$$
$$\text{Formation Thickness, ft} = 100$$
$$\text{Water Saturation, fraction} = 0.34$$
$$p_{pc} = 609 \quad \text{psi}$$
$$T_{pc} = 475°\text{R}$$
$$z_i = 1/472$$
$$G_p @ p = 6{,}921 \text{ psi} = 15 \quad \text{MMSCF}$$
$$c_{gi} = 30 \quad \text{microsip} \qquad c_f = 21.9 \quad \text{microsip}$$

Solution: Gas formation volume factors calculated using HP–41 Petroleum Fluids Pac as

$$B_{gi} = 304.266 \quad \text{SCF/ft}^3$$
$$B_{g6,921} = 281.699 \quad \text{SCF/ft}^3$$

Using Equation (4.60),

$$G_{app} = \frac{15(304.266)}{(304.266 - 281.699)} = 202 \quad \text{MMSCF}$$

Calculate $G_{pr}$ using Equation (4.72).

$$G_{pr} = \frac{0.00355 - 0.00329}{0.00355 - 0.00329 + \dfrac{0.00329(2{,}000)}{1 - 0.34}[21.9 - (0.34)(3) \times 10^{-6}]}$$
$$= \frac{0.00026}{0.00026 + 0.00021}$$
$$= 0.553$$
$$G_{cor} = G_{app} \times G_{pr} = (202)(0.553)$$
$$\approx 112 \quad \text{MMSCF}$$

Another method suggested by Roach is a direct graphical technique for analyzing abnormally high pressured reservoirs.[19] According to this method the material balance equation for gas reservoirs may be written as

$$(p/z)c_t = \left(\frac{p_i}{z_i}\right)\left[1 - \frac{G_p}{G}\right] \tag{4.73}$$

where

$$c_t = 1 - \frac{(c_f + c_w S_{iw})(p_i - p)}{1 - S_{iw}}$$

Plotting $(p/z)c_t$ vs. $G_p$ on Cartesian coordinate graph paper results in a straight line with an $x$-intercept at the original gas in place and a $y$-intercept at the initial $p/z$. The $c_t$ value must be found by trial-and-error resulting in the best straight-line fit.

## GAS RESERVOIR SIZE

It is always of interest to know how much area each well will drain in order to arrive at total field development plan. A combination of material balance equation and volumetric equation is used to calculate the drainage area of a well.

The best way to do this is to first calculate the ultimate gas recovery ($G_{ul}$) at a selected abandonment pressure using appropriate material balance equation. Then use Equation (4.8) to calculate drainage area. The rearranged version of Equation (4.8) would be

$$A = \frac{G_{ul}}{43{,}560\phi(1 - S_{iw})hB_{gi}} \tag{4.74}$$

where $B_{gi}$ is in SCF/ft$^3$.

**Example 4–20.** Calculate the drainage area of the well in Example 4–14 if the log calculated porosity and water saturation are 15% and 30%, respectively, and formation thickness is 20 ft.

Solution:

$$\begin{aligned} G &= 10.32 \quad \text{MMSCF} \qquad \text{(from Example 4–14.)} \\ B_{gi} &= 187.721 \quad \text{SCF/ft}^3 \qquad \text{(from Example 4–14.)} \end{aligned}$$

Using Equation (4.74),

$$\begin{aligned} A &= \frac{10.32 \times 10^9}{(43{,}560)(0.15)(1 - 0.30)(20)(187.721)} \\ &\approx 601 \quad \text{ac} \end{aligned}$$

## MATERIAL BALANCE GAS PRODUCTION

High-pressure gas in the reservoir will generally drop out some of the heavier components upon production as condensate. In this case the $G_p$ values in the material balance equations should be the gas equivalent of produced condensate.

Since the gas in the reservoir is saturated with water at reservoir conditions, when gas is produced at the surface, all of this water is removed prior to gas sales. Thus the material balance calculations should also include the water vapor that was a part of the original gas in place. However, the water produced as a result of condensation from the gas should not be included as $W_p$.



The material presented in this section should not be confused with retrograde condensate reservoirs which will be presented in a separate section. In this section we deal with situations when the reservoir fluid remains in a single (gas) phase. If a hydrocarbon liquid phase develops in the reservoir, these methods are then not applicable.

The difference between the total produced water and the original water vapor in the produced gas must be treated as produced water ($W_p$). To obtain the produced gas value ($G_p$), the following simple relationship is applicable. Note that the weight of 1 bbl of fresh water is 350.5 lbs.

$$\begin{aligned} G_p = &\ (\text{Dry Gas Sold}) + (\text{Gas Equivalent of Condensate Produced}) \\ &+ (\text{Original Water Vapor in Produced Gas}) \\ &+ (\text{Vent Gas from Condensate Storage}) \end{aligned} \tag{4.75}$$

The gas equivalent (GE) of one stock-tank barrel of condensate liquid is given by the following equation:

$$GE = \frac{nRT_{sc}}{p_{sc}} = \frac{350.5\gamma_o RT_{sc}}{M_o p_{sc}} \tag{4.76}$$

The molecular weight of the tank oil ($M_o$) may be estimated using the following equation:

$$M_o = \frac{6{,}084}{°API - 5.9} \tag{4.77}$$

If the analysis of the separator liquid is available, a more precise determination of its gas equivalent can be made.

**Example 4–21.**[20] Calculate the total daily reservoir gas production ($G_p$) and water production ($W_p$) for material balance calculations using the following data. Standard temperature and pressure are 60°F and 14.7 psi, respectively.

Separator gas production = 10.0 MMSCF/D
Condensate production = 150.0 STB/D
Condensate Gravity = 55°API

$$
\begin{aligned}
\text{Stock-tank gas production (vent gas)} &= 30 \quad \text{MSCF/D} \\
\text{Fresh water production} &= 15 \quad \text{RB/D} \\
\text{Initial reservoir pressure} &= 4{,}000 \quad \text{psi} \\
\text{Reservoir temperature} &= 200°\text{F} \\
\text{Formation water salinity} &= 150{,}000 \quad \text{ppm}
\end{aligned}
$$

Solution:

$$
\begin{aligned}
\gamma_o &= \frac{141.5}{55 + 131.5} \\
&= 0.759 \\
M_o &= \frac{6{,}084}{55 - 5.9} = 123.91 \\
GE_o &= \frac{350.5\gamma_o RT_{sc}}{M_o p_{sc}} \\
&= \frac{(350.5)(0.759)(10.73)(60 + 460)}{(123.91)(14.7)} \\
&\approx 815 \quad \text{SCF/STB} \\
\text{Condensate Gas} &= (815)(150) = 122{,}250 \quad \text{SCF/STB} \\
GE_w &= \frac{350.5\gamma_w RT_{sc}}{M_w P_{sc}} \\
&= \frac{(350.5)(1.0)(10.73)(60 + 460)}{(18)(14.7)} \\
&= 7{,}391 \quad \text{SCF/RB}
\end{aligned}
$$

Water vapor in original gas = (350.5)(0.89) = 312 lb/MMSCF

$$
\begin{aligned}
\text{Water Production} &= \frac{W_p \times 350.5}{G_p + GE_o} \\
&= \frac{(15)(350.5)}{(10.0 + 0.122)} \approx 519 \quad \text{lb/MMSCF} \\
\text{SCF of water vapor} &= \left(\frac{312}{350.5}\right)(7{,}391)(10.122) \\
&= 66{,}594 \quad \text{SCF} \\
G_p/d &= 10.0 + 0.0666 + 0.122 + 0.03 \\
&= 10.219 \quad \text{MMSCF/D} \\
W_p/d &= \frac{(519 - 312)(10.122)}{350.5} \\
&= 5.98 \quad \text{RB/D}
\end{aligned}
$$

## TO CALCULATE PRESSURE FROM $p/z$

In most cases it will be desirable to know the reservoir pressure at which a certain volume of gas may be recovered, or the reservoir pressure after a certain volume of gas is produced.

Two techniques are presented in this section which will help in calculating reservoir pressure from $p/z$. The $p/z$ is obtained from the $p/z$ plot at a desired gas volume. Theoretically the initial reservoir pressure may also be calculated from $p_i/z_i$ at $G_p = 0$ (extrapolating $p/z$ to $G_p = 0$).

The reservoir pressure may be calculated by trial-and-error using the following equation:



$$p_{i+1} = p_i - \frac{p_i - (p/z - z_i)}{(p_i/p_{pc})(c_{ri})} \quad \textbf{(4.78)}$$

where $z_i$ is $z$ at $p_i$ and $c_{ri}$ is the pseudoreduced compressibility at $p_i$. The ratio $p/z$ is also used as the initial guess. Alternatively, a plot of $(p_r/z)$ versus $z$ for various reduced temperatures as shown in Figure 4–18 may be used to calculate reservoir pressure from $p/z$. The $p_r/z$ versus $z$ plot will facilitate the direct evaluation of $z$-factor, from the ratio $p_r/z$ ($p_r/z = (p/z)/p_{pc}$), and the immediate calculation of the pressure by using $z$ and the critical pressure.

**Example 4–22.** Using the plot in Figure 4–18, evaluate the initial reservoir pressure from the initial $p_i/z_i = 4{,}348$. Other pertinent data are as follows:

$$\gamma_g = 0.72 \quad (\text{air} = 1.0)$$
$$T_{pc} = 395.52°\text{R}$$
$$p_{pc} = 668.36 \quad \text{psi}$$

Solution:

$$\frac{p_r}{z} = \frac{\frac{p_i}{z_i}}{p_{pc}}$$
$$= \frac{4{,}348}{668.36} = 6.505$$
$$T_{pr} = \frac{T + 460}{T_{pc}}$$
$$= \frac{210 + 460}{359.52} = 1.694$$

From Figure 4–18, $z = 0.92$.

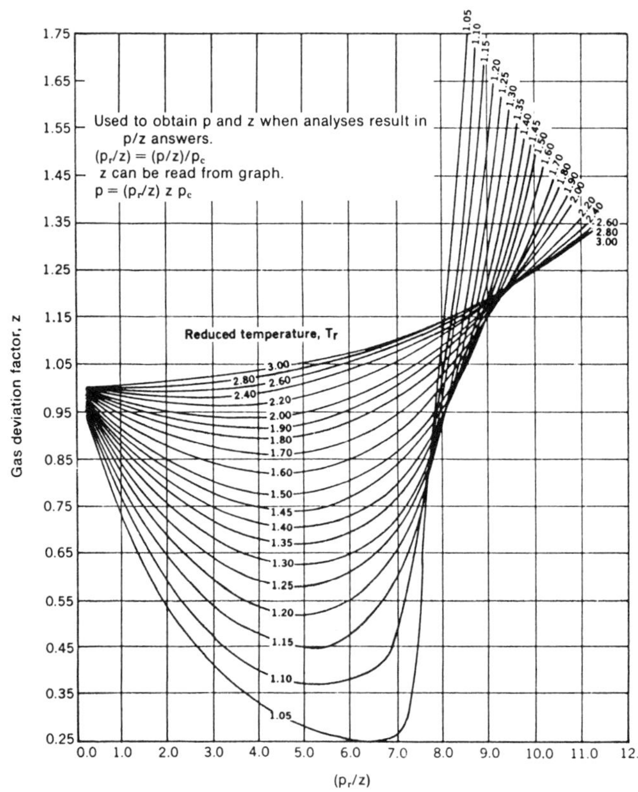


**Figure 4–18.** Gas deviation factor ($z$) versus ($p_r/z$) (from Slider, courtesy of PennWell Books).[20]

$$
\begin{aligned}
p_i &= (p_r/z) \times z \times p_{pc} \\
&= (6.505)(0.92)(668.36) \\
&= 3{,}999.87 \quad \approx 4{,}000 \quad \text{psi}
\end{aligned}
$$

## RESERVOIRS UNDER NATURAL WATERDRIVE

Production of oil by a water-displacing process is usually the most efficient displacing agent naturally available in the reservoir. The following reservoir characteristics will help in identifying whether a reservoir is under a natural waterdrive or not.

1. In waterdrive reservoirs, the pressure decline is very gradual. This is because the oil and gas withdrawals from the reservoir are replaced almost volume-for-volume by water influx into the producing formation. To achieve maximum possible hydrocarbon recovery, the gas and water

production has to be minimized. This may be accomplished by shutting in wells producing large quantities of water.

2. Hydrocarbon production from structurally low wells will be accompanied by excessive amount of water production. In such cases it may be economically feasible to shut in the structurally low wells, and this oil may be produced by the structurally higher wells.
3. If there is no initial free gas cap, there will be very little change in the producing gas/oil ratio during the life of a reservoir.
4. The water/oil ratio will increase with time as more and more water encroaches into the reservoir.

Natural water influx into oil reservoirs is usually from the edge inward parallel to the bedding planes (edgewater drive) or upward from below (bottomwater drive). Bottomwater drive occurs when (1) the reservoir thickness exceeds the thickness of the oil column, so that the oil/water contact underlies the entire oil reservoir, or (2) when the vertical permeabilities are high and there is little or no horizontal stratification.

In this section we will discuss a technique to predict water encroachment independent of material balance. Water encroachment is not only a function of the aquifer flow characteristics but is also a function of pressure history and time.

## CUMULATIVE WATER INFLUX—EDGEWATER DRIVE RESERVOIRS

In a waterdrive reservoir, pressure drop in the reservoir is responded (offset or retard pressure drop) by a water influx or encroachment from the adjoining aquifer by (1) expansion of the water in the aquifer, (2) expansion of other known or unknown hydrocarbon accumulations in the aquifer rock, (3) compressibility of the aquifer rock, and/or (4) artesian flow (i.e., the rise of aquifer to a level above the reservoir).

The volume of water an aquifer of radius $r_e$ will deliver to a reservoir of radius $r_w$ by rock and water compressibility under a pressure drop ($\Delta p$) throughout the aquifer is given by the following equation:

$$\Delta V = c_e \pi (r_e^2 - r_w^2) \phi h \Delta p \qquad \textbf{(4.79)}$$

where $c_e$ is the effective water compressibility, which for an aquifer will be the sum of $c_f$ and $c_w$. Equation (4.79), however, does not indicate the rate at which the influx will occur. This rate will depend on the permeability of the aquifer rock.

The best method of predicting water influx is by the unsteady-state compressible fluid flow theory; the process often referred to as the Van Everdingen and Hurst method.[21] Van Everdingen and Hurst solved the radial diffusivity equation for the aquifer-reservoir system by applying the Laplace transformation to the equation. The practical form of their equation is frequently expressed as given below.

$$W_{e_n} = m_p \sum_{j=1}^{n} (\Delta \overline{p}_{(n+1-j)} W_{eDj}) \qquad \textbf{(4.80)}$$

where

$W_{e_n}$ = cumulative water influx at end of $n$th period
$W_{eD}$ = Dimensionless water influx term
$m_p$ = Water influx constant in barrels/psi
$\Delta\bar{p}_{(n+1-j)}$ = Average pressure drop in period $n + 1 - j$.



Theoretically, the term $m_p$ can be calculated using the following equations:

For radial aquifer geometry:

$$m_p = 1.119\phi(c_w + c_f)r_w^2 h \frac{\theta}{360} \quad \textbf{(4.81)}$$

For linear aquifer geometry:

$$m_p = 0.17811 wLh\phi(c_w + c_f) \quad \textbf{(4.81a)}$$

where $\theta$ is the angle subtended by the reservoir circumference. For a full circle $\theta = 360°$ and for a semicircular reservoir against a fault, $\theta = 180°$. In Equation (4.81a), $w$ is the width and $L$ the length of the water aquifer as shown in Figure 4–19.

The dimensionless water influx term $W_{eD}$ is given as a function of dimensionless time $t_D$, aquifer geometry, and aquifer size ($t_D$ for radial aquifers).

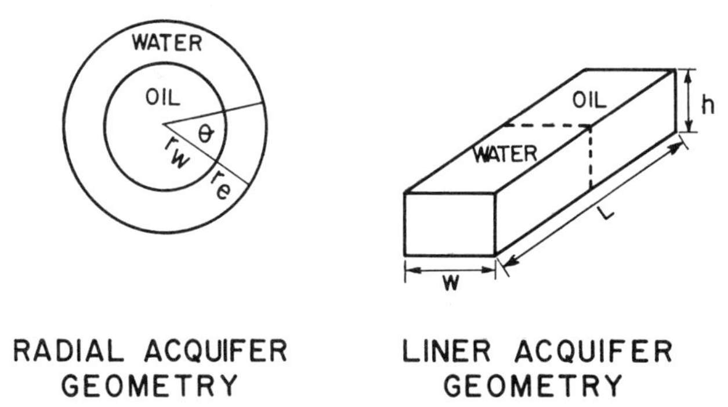


**Figure 4–19.** Schematic of radial and linear aquifer geometries.

For radial geometry:

$$t_D = \frac{0.000264kt}{\phi\mu_w(c_w + c_f)r_w^2} \quad \textbf{(4.82)}$$

For linear geometry:

$$t_D = \frac{0.000264kt}{\phi\mu_w(c_w + c_f)L^2} \quad \textbf{(4.82a)}$$

where $t$ is in hours.

The calculation of $m_p$ from Equation (4.81) requires input of aquifer properties for porosity, compressibility, radius, angle subtended by the aquifer, and aquifer thickness. These properties, however, are unavailable in most cases, since no wells are intentionally drilled into the aquifer or are too erratic for valid interpretation. It is recommended to obtain $m_p$ from performance if it is available.

The calculation of $m_p$ using performance data is based on the calculation of $W_e$ at each historical point using the appropriate material balance equation. Equation (4.79) is then used to calculate the implied value of $m_p$ for each time-pressure water influx point. This process is repeated for different values of $r_e/r_w$ to give a better estimate of the area of the aquifer. Note that a plot of $W_{eD}$ at various times versus the summation term for these times should give a straight line plot whose slope is the constant $m_p$.

The dimensionless water influx term $W_{eD}$ in Equation (4.79) is frequently presented in tabular form as shown in Tables 4–6 and 4–7. Plots of $W_{eD}$ versus $t_D$ for both radial and linear geometry and, in the former case, for selected values of $r_{eD}$ are included in Figures 4–20 through 4–24.

The cumulative water influx for an *infinite linear aquifer* can be calculated directly using the following equation:

$$W_e = 3.26 \times 10^{-3}hw\sqrt{\frac{\phi k(c_w + c_f)t}{\mu_w}} \times \Delta p \quad \textbf{(4.83)}$$

A modification presented by Carter and Tracy permits calculations of $W_e$ that approximate the $W_e$ calculated using Equation (4.80).[22] This method is more appropriate for hand calculations. The following equation can be used to calculate water influx at time $t_n$ directly from the previous value.

$$W_{e_n} = W_{e_{(n-1)}} + \frac{[m_p\Delta p_n t_{Dn} - W_{e(n-1)}\, p'_{Dn}][t_{Dn} - t_{D(n-1)}]}{p_{Dn} - t_{D_{(n-1)}}p'_{Dn}} \quad \textbf{(4.84)}$$

where

$$p'_{Dn} = \frac{p_{Dn} - p_{D(n-1)}}{t_{Dn} - t_{D(n-1)}}$$

**TABLE 4–6. Dimensionless water influx and dimensionless pressures for infinite radial aquifers (courtesy of SPE)[3]**

| $t_D$ | $W_{eD}$ | $p_D$ | $t_D$ | $W_{eD}$ | $t_D$ | $W_{eD}$ | $t_D$ | $W_{eD}$ |
|---|---|---|---|---|---|---|---|---|
| $1.0 \times 10^{-2}$ | 0.112 | 0.112 | $1.5 \times 10^{3}$ | $4.136 \times 10^{2}$ | $1.5 \times 10^{7}$ | $1.828 \times 10^{6}$ | $1.5 \times 10^{11}$ | $1.17 \times 10^{10}$ |
| $5.0 \times 10^{-2}$ | 0.278 | 0.229 | $2.0 \times 10^{3}$ | $5.315 \times 10^{2}$ | $2.0 \times 10^{7}$ | $2.398 \times 10^{6}$ | $2.0 \times 10^{11}$ | $1.55 \times 10^{10}$ |
| $1.0 \times 10^{-1}$ | 0.404 | 0.315 | $2.5 \times 10^{3}$ | $6.466 \times 10^{2}$ | $2.5 \times 10^{7}$ | $2.961 \times 10^{6}$ | $2.5 \times 10^{11}$ | $1.92 \times 10^{10}$ |
| $1.5 \times 10^{-1}$ | 0.520 | 0.376 | $3.0 \times 10^{3}$ | $7.590 \times 10^{2}$ | $3.0 \times 10^{7}$ | $3.517 \times 10^{6}$ | $3.0 \times 10^{11}$ | $2.29 \times 10^{10}$ |
| $2.0 \times 10^{-1}$ | 0.606 | 0.424 | $4.0 \times 10^{3}$ | $9.757 \times 10^{2}$ | $4.0 \times 10^{7}$ | $4.610 \times 10^{6}$ | $4.0 \times 10^{11}$ | $3.02 \times 10^{10}$ |
| $2.5 \times 10^{-1}$ | 0.689 | 0.469 | $5.0 \times 10^{3}$ | $11.88 \times 10^{2}$ | $5.0 \times 10^{7}$ | $5.689 \times 10^{6}$ | $5.0 \times 10^{11}$ | $3.75 \times 10^{10}$ |
| $3.0 \times 10^{-1}$ | 0.758 | 0.503 | $6.0 \times 10^{3}$ | $13.95 \times 10^{3}$ | $6.0 \times 10^{7}$ | $6.758 \times 10^{6}$ | $6.0 \times 10^{11}$ | $4.47 \times 10^{11}$ |
| $4.0 \times 10^{-1}$ | 0.898 | 0.564 | $7.0 \times 10^{3}$ | $15.99 \times 10^{3}$ | $7.0 \times 10^{7}$ | $7.816 \times 10^{6}$ | $7.0 \times 10^{11}$ | $5.19 \times 10^{10}$ |
| $5.0 \times 10^{-1}$ | 1.020 | 0.616 | $8.0 \times 10^{3}$ | $18.00 \times 10^{3}$ | $8.0 \times 10^{7}$ | $8.866 \times 10^{6}$ | $8.0 \times 10^{11}$ | $5.89 \times 10^{10}$ |
| $6.0 \times 10^{-1}$ | 1.140 | 0.659 | $9.0 \times 10^{3}$ | $19.99 \times 10^{3}$ | $9.0 \times 10^{7}$ | $9.911 \times 10^{6}$ | $9.0 \times 10^{11}$ | $6.58 \times 10^{10}$ |
| $7.0 \times 10^{-1}$ | 1.251 | 0.702 | $1.0 \times 10^{4}$ | $21.96 \times 10^{2}$ | $1.0 \times 10^{8}$ | $10.95 \times 10^{6}$ | $1.0 \times 10^{12}$ | $7.28 \times 10^{10}$ |
| $8.0 \times 10^{-1}$ | 1.359 | 0.735 | $1.5 \times 10^{4}$ | $3.146 \times 10^{3}$ | $1.5 \times 10^{8}$ | $1.604 \times 10^{7}$ | $1.5 \times 10^{12}$ | $1.08 \times 10^{11}$ |
| $9.0 \times 10^{-1}$ | 1.469 | 0.772 | $2.0 \times 10^{4}$ | $4.679 \times 10^{3}$ | $2.0 \times 10^{8}$ | $2.108 \times 10^{7}$ | $2.0 \times 10^{12}$ | $1.42 \times 10^{11}$ |
| 1.0 | 1.570 | 0.802 | $2.5 \times 10^{4}$ | $4.991 \times 10^{3}$ | $2.5 \times 10^{8}$ | $2.607 \times 10^{7}$ | | |
| 1.5 | 2.032 | 0.927 | $3.0 \times 10^{4}$ | $5.891 \times 10^{3}$ | $3.0 \times 10^{8}$ | $3.100 \times 10^{7}$ | | |
| 2.0 | 2.442 | 1.020 | $4.0 \times 10^{4}$ | $7.634 \times 10^{3}$ | $4.0 \times 10^{8}$ | $4.071 \times 10^{7}$ | | |
| 2.5 | 2.838 | 1.101 | $5.0 \times 10^{4}$ | $9.342 \times 10^{3}$ | $5.0 \times 10^{8}$ | $5.032 \times 10^{7}$ | | |
| 3.0 | 3.209 | 1.169 | $6.0 \times 10^{4}$ | $11.03 \times 10^{4}$ | $6.0 \times 10^{8}$ | $5.984 \times 10^{7}$ | | |
| 4.0 | 3.897 | 1.275 | $7.0 \times 10^{4}$ | $12.69 \times 10^{4}$ | $7.0 \times 10^{8}$ | $6.928 \times 10^{7}$ | | |
| 5.0 | 4.541 | 1.362 | $8.0 \times 10^{4}$ | $14.33 \times 10^{4}$ | $8.0 \times 10^{8}$ | $7.865 \times 10^{7}$ | | |
| 6.0 | 5.148 | 1.436 | $9.0 \times 10^{4}$ | $15.95 \times 10^{4}$ | $9.0 \times 10^{8}$ | $8.797 \times 10^{7}$ | | |
| 7.0 | 5.749 | 1.500 | $1.0 \times 10^{5}$ | $17.56 \times 10^{4}$ | $1.0 \times 10^{9}$ | $9.725 \times 10^{7}$ | | |
| 8.0 | 6.314 | 1.556 | $1.5 \times 10^{5}$ | $2.538 \times 10^{4}$ | $1.5 \times 10^{9}$ | $1.429 \times 10^{8}$ | | |
| 9.0 | 6.861 | 1.604 | $2.0 \times 10^{5}$ | $3.308 \times 10^{4}$ | $2.0 \times 10^{9}$ | $1.880 \times 10^{8}$ | | |
| $1.0 \times 10^{1}$ | 7.417 | 1.651 | $2.5 \times 10^{5}$ | $4.066 \times 10^{4}$ | $2.5 \times 10^{9}$ | $2.328 \times 10^{8}$ | | |
| $1.5 \times 10^{1}$ | 9.965 | 1.829 | $3.0 \times 10^{5}$ | $4.817 \times 10^{4}$ | $3.0 \times 10^{9}$ | $2.771 \times 10^{8}$ | | |
| $2.0 \times 10^{1}$ | $1.229 \times 10^{1}$ | 1.960 | $4.0 \times 10^{5}$ | $6.267 \times 10^{4}$ | $4.0 \times 10^{9}$ | $3.645 \times 10^{8}$ | | |
| $2.5 \times 10^{1}$ | $1.455 \times 10^{1}$ | 2.067 | $5.0 \times 10^{5}$ | $7.699 \times 10^{4}$ | $5.0 \times 10^{4}$ | $4.510 \times 10^{8}$ | | |
| $3.0 \times 10^{1}$ | $1.681 \times 10^{1}$ | 2.147 | $6.0 \times 10^{5}$ | $9.113 \times 10^{4}$ | $6.0 \times 10^{9}$ | $5.368 \times 10^{8}$ | | |
| $4.0 \times 10^{1}$ | $2.088 \times 10^{1}$ | 2.282 | $7.0 \times 10^{5}$ | $10.51 \times 10^{5}$ | $7.0 \times 10^{9}$ | $6.220 \times 10^{8}$ | | |
| $5.0 \times 10^{1}$ | $2.482 \times 10^{1}$ | 2.388 | $8.0 \times 10^{5}$ | $11.89 \times 10^{5}$ | $8.0 \times 10^{9}$ | $7.066 \times 10^{8}$ | | |
| $6.0 \times 10^{1}$ | $2.860 \times 10^{1}$ | 2.476 | $9.0 \times 10^{5}$ | $13.26 \times 10^{5}$ | $9.0 \times 10^{9}$ | $7.909 \times 10^{8}$ | | |
| $7.0 \times 10^{1}$ | $3.228 \times 10^{1}$ | 2.550 | $1.0 \times 10^{6}$ | $14.62 \times 10^{5}$ | $1.0 \times 10^{10}$ | $8.747 \times 10^{8}$ | | |
| $8.0 \times 10^{1}$ | $3.599 \times 10^{1}$ | 2.615 | $1.5 \times 10^{6}$ | $2.126 \times 10^{5}$ | $1.5 \times 10^{10}$ | $1.288 \times 10^{9}$ | | |
| $9.0 \times 10^{1}$ | $3.942 \times 10^{1}$ | 2.672 | $2.0 \times 10^{6}$ | $2.781 \times 10^{5}$ | $2.0 \times 10^{10}$ | $1.697 \times 10^{9}$ | | |
| $1.0 \times 10^{2}$ | $4.301 \times 10^{1}$ | 2.723 | $2.5 \times 10^{6}$ | $3.427 \times 10^{5}$ | $2.5 \times 10^{10}$ | $2.103 \times 10^{9}$ | | |
| $1.5 \times 10^{2}$ | $5.980 \times 10^{1}$ | 2.921 | $3.0 \times 10^{6}$ | $4.064 \times 10^{5}$ | $3.0 \times 10^{10}$ | $2.505 \times 10^{9}$ | | |
| $2.0 \times 10^{2}$ | $7.586 \times 10^{1}$ | 3.064 | $4.0 \times 10^{6}$ | $5.313 \times 10^{5}$ | $4.0 \times 10^{10}$ | $3.299 \times 10^{9}$ | | |
| $2.5 \times 10^{2}$ | $9.120 \times 10^{1}$ | 3.173 | $5.0 \times 10^{6}$ | $6.544 \times 10^{5}$ | $5.0 \times 10^{10}$ | $4.087 \times 10^{9}$ | | |
| $3.0 \times 10^{2}$ | $10.58 \times 10^{1}$ | 3.263 | $6.0 \times 10^{6}$ | $7.761 \times 10^{5}$ | $6.0 \times 10^{10}$ | $4.868 \times 10^{9}$ | | |
| $4.0 \times 10^{2}$ | $13.48 \times 10^{1}$ | 3.406 | $7.0 \times 10^{6}$ | $8.965 \times 10^{5}$ | $7.0 \times 10^{10}$ | $5.643 \times 10^{9}$ | | |
| $5.0 \times 10^{2}$ | $16.24 \times 10^{1}$ | 3.516 | $8.0 \times 10^{6}$ | $10.16 \times 10^{6}$ | $8.0 \times 10^{10}$ | $6.414 \times 10^{9}$ | | |
| $6.0 \times 10^{2}$ | $18.97 \times 10^{1}$ | 3.608 | $9.0 \times 10^{6}$ | $11.34 \times 10^{6}$ | $9.0 \times 10^{10}$ | $7.183 \times 10^{9}$ | | |
| $7.0 \times 10^{2}$ | $21.60 \times 10^{1}$ | 3.684 | $1.0 \times 10^{7}$ | $12.52 \times 10^{6}$ | $1.0 \times 10^{11}$ | $7.948 \times 10^{9}$ | | |
| $8.0 \times 10^{2}$ | $24.23 \times 10^{1}$ | 3.750 | | | | | | |
| $9.0 \times 10^{2}$ | $26.77 \times 10^{1}$ | 3.809 | | | | | | |
| $1.0 \times 10^{3}$ | $29.31 \times 10^{1}$ | 3.860 | | | | | | |

and

$$\Delta p_n = p_i - p_n$$

**Example 4–23.** Calculate the volume of water an aquifer of 25,000-ft radius can deliver to a 3,500-ft radius reservoir by rock and water compressibility under a 1,500 psi pressure drop throughout the aquifer. Assume $\phi = 0.18$, $c_w = 3 \times 10^{-6}$ 1/psi, $c_f = 4.6 \times 10^{-6}$ 1/psi, and $h = 32$ ft.

Solution:

$$\begin{aligned} c_e &= c_w + c_f \\ &= (3 + 4.6) \times 10^{-6} = 7.6 \times 10^{-6} \end{aligned}$$

Using Equation (4.79),

$$\begin{aligned} \Delta V &= 7.6 \times 10^{-6}\pi(25{,}000^2 - 3{,}500^2)(0.18)(32)(1{,}500) \\ &= 126.4 \times 10^6 \quad \text{ft}^3 \end{aligned}$$

**Example 4–24.**[6] Calculate the water influx after 100, 200, 400, and 800 days into a reservoir whose boundary pressure is suddenly lowered and held at 2,724 psi from initial pressure of 2,734 psi. Assume $\phi = 0.20$, $k = 83$ md, $c_e = 8 \times 10^{-6}$ 1/psi, $r_w = 3{,}000$ ft, $r_e = 30{,}000$ ft, $\mu_w = 0.62$ cp, $\theta = 360°$, and $h = 40$ ft. What would be the water influx if the same pressure drop is transmitted instantaneously throughout the aquifer?

Solution: Using Equation (4.82),

$$\begin{aligned} t_D &= \frac{0.000264(83)(24t)}{(0.20)(0.62)(8 \times 10^{-6})(3{,}000)^2} \\ &= 0.0589t \end{aligned}$$

For $t = 100$ days, $t_D = (0.0589)(100) = 5.89$. For this example $r_e/r_w = 10$. In Table 4–7 and column for $r_e/r_w = 10$, the lowest $t_D$ is 15. Since the $t_D$ at 100 days is 5.89 (which is less than 15), the aquifer behaves essentially as if it were infinite. Therefore, using Table 4–6, we find $W_{eD}$ at $t_D = 5.89$.

$$W_{eD} = 4.539 \quad \text{at } t_D = 5$$
$$W_{eD} = 5.153 \quad \text{at } t_D = 6$$

By linear interpolation,

$$\frac{6 - 5.89}{6 - 5} = \frac{5.153 - W_{eD}}{5.153 - 4.539}$$
$$W_{eD} = 5.086 \quad \text{at } t_D = 5.89$$

## TABLE 4–7. Dimensionless water influx for finite outcroping radial aquifer (courtesy of SPE)[3]

| $r_D = 1.5$ | | $r_D = 2.0$ | | $r_D = 2.5$ | | $r_D = 3.0$ | | $r_D = 3.5$ | | $r_D = 4.0$ | | $r_D = 4.5$ | |
|---|---|---|---|---|---|---|---|---|---|---|---|---|---|
| $t_D$ | $W_{eD}$ | $t_D$ | $W_{eD}$ | $t_D$ | $W_{eD}$ | $t_D$ | $W_{eD}$ | $t_D$ | $W_{eD}$ | $t_D$ | $W_{eD}$ | $t_D$ | $W_{eD}$ |
| $5.0 \times 10^{-2}$ | 0.276 | $5.0 \times 10^{-2}$ | 0.278 | $1.0 \times 10^{-1}$ | 0.408 | $3.0 \times 10^{-1}$ | 0.755 | 1.00 | 1.571 | 2.00 | 2.442 | 2.5 | 2.835 |
| $6.0 \times 10^{-2}$ | 0.304 | $7.5 \times 10^{-2}$ | 0.345 | $1.5 \times 10^{-1}$ | 0.509 | $4.0 \times 10^{-1}$ | 0.895 | 1.20 | 1.761 | 2.20 | 2.598 | 3.0 | 3.196 |
| $7.0 \times 10^{-2}$ | 0.330 | $1.0 \times 10^{-1}$ | 0.404 | $2.0 \times 10^{-1}$ | 0.599 | $5.0 \times 10^{-1}$ | 1.023 | 1.40 | 1.940 | 2.40 | 2.748 | 3.5 | 3.537 |
| $8.0 \times 10^{-2}$ | 0.354 | $1.25 \times 10^{-1}$ | 0.458 | $2.5 \times 10^{-1}$ | 0.681 | $6.0 \times 10^{-1}$ | 1.143 | 1.60 | 2.111 | 2.60 | 2.893 | 4.0 | 3.859 |
| $9.0 \times 10^{-2}$ | 0.375 | $1.50 \times 10^{-1}$ | 0.507 | $3.0 \times 10^{-1}$ | 0.758 | $7.0 \times 10^{-1}$ | 1.256 | 1.80 | 2.273 | 2.80 | 3.034 | 4.5 | 4.165 |
| $1.0 \times 10^{-1}$ | 0.395 | $1.75 \times 10^{-1}$ | 0.553 | $3.5 \times 10^{-1}$ | 0.829 | $8.0 \times 10^{-1}$ | 1.363 | 2.00 | 2.427 | 3.00 | 3.170 | 5.0 | 4.454 |
| $1.1 \times 10^{-1}$ | 0.414 | $2.00 \times 10^{-1}$ | 0.597 | $4.0 \times 10^{-1}$ | 0.897 | $9.0 \times 10^{-1}$ | 1.465 | 2.20 | 2.574 | 3.25 | 3.334 | 5.5 | 4.727 |
| $1.2 \times 10^{-1}$ | 0.431 | $2.25 \times 10^{-1}$ | 0.638 | $4.5 \times 10^{-1}$ | 0.962 | 1.00 | 1.563 | 2.40 | 2.715 | 3.50 | 3.493 | 6.0 | 4.986 |
| $1.3 \times 10^{-1}$ | 0.446 | $2.50 \times 10^{-1}$ | 0.678 | $5.0 \times 10^{-1}$ | 1.024 | 1.25 | 1.791 | 2.60 | 2.849 | 3.75 | 3.645 | 6.5 | 5.231 |
| $1.4 \times 10^{-1}$ | 0.461 | $2.75 \times 10^{-1}$ | 0.715 | $5.5 \times 10^{-1}$ | 1.083 | 1.50 | 1.997 | 2.80 | 2.976 | 4.00 | 3.792 | 7.0 | 5.464 |
| $1.5 \times 10^{-1}$ | 0.474 | $3.00 \times 10^{-1}$ | 0.751 | $6.0 \times 10^{-1}$ | 1.140 | 1.75 | 2.184 | 3.00 | 3.098 | 4.25 | 3.932 | 7.5 | 5.684 |
| $1.6 \times 10^{-1}$ | 0.486 | $3.25 \times 10^{-1}$ | 0.785 | $6.5 \times 10^{-1}$ | 1.195 | 2.00 | 2.353 | 3.25 | 3.242 | 4.50 | 4.068 | 8.0 | 5.892 |
| $1.7 \times 10^{-1}$ | 0.497 | $3.50 \times 10^{-1}$ | 0.817 | $7.0 \times 10^{-1}$ | 1.248 | 2.25 | 2.507 | 3.50 | 3.379 | 4.75 | 4.198 | 8.5 | 6.089 |
| $1.8 \times 10^{-1}$ | 0.507 | $3.75 \times 10^{-1}$ | 0.848 | $7.5 \times 10^{-1}$ | 1.229 | 2.50 | 2.646 | 3.75 | 3.507 | 5.00 | 4.323 | 9.0 | 6.276 |
| $1.9 \times 10^{-1}$ | 0.517 | $4.00 \times 10^{-1}$ | 0.877 | $8.0 \times 10^{-1}$ | 1.348 | 2.75 | 2.772 | 4.00 | 3.628 | 5.50 | 4.560 | 9.5 | 6.453 |
| $2.0 \times 10^{-1}$ | 0.525 | $4.25 \times 10^{-1}$ | 0.905 | $8.5 \times 10^{-1}$ | 1.395 | 3.00 | 2.886 | 4.25 | 3.742 | 6.00 | 4.779 | 10 | 6.621 |
| $2.1 \times 10^{-1}$ | 0.533 | $4.50 \times 10^{-1}$ | 0.932 | $9.0 \times 10^{-1}$ | 1.440 | 3.25 | 2.990 | 4.50 | 3.850 | 6.50 | 4.982 | 11 | 6.930 |
| $2.2 \times 10^{-1}$ | 0.541 | $4.75 \times 10^{-1}$ | 0.958 | $9.5 \times 10^{-1}$ | 1.484 | 3.50 | 3.084 | 4.75 | 3.951 | 7.00 | 5.169 | 12 | 7.208 |
| $2.3 \times 10^{-1}$ | 0.548 | $5.00 \times 10^{-1}$ | 0.982 | 1.0 | 1.526 | 3.75 | 3.170 | 5.00 | 4.047 | 7.50 | 5.343 | 13 | 7.457 |
| $2.4 \times 10^{-1}$ | 0.554 | $5.50 \times 10^{-1}$ | 1.028 | 1.1 | 1.605 | 4.00 | 3.247 | 5.50 | 4.222 | 8.00 | 5.504 | 14 | 7.680 |
| $2.5 \times 10^{-1}$ | 0.559 | $6.00 \times 10^{-1}$ | 1.070 | 1.2 | 1.679 | 4.25 | 3.317 | 6.00 | 4.378 | 8.50 | 5.653 | 15 | 7.880 |
| $2.6 \times 10^{-1}$ | 0.565 | $6.50 \times 10^{-1}$ | 1.108 | 1.3 | 1.747 | 4.50 | 3.381 | 6.50 | 4.516 | 9.00 | 5.790 | 16 | 8.060 |
| $2.8 \times 10^{-1}$ | 0.574 | $7.00 \times 10^{-1}$ | 1.143 | 1.4 | 1.811 | 4.75 | 3.439 | 7.00 | 4.639 | 9.50 | 5.917 | 18 | 8.365 |
| $3.0 \times 10^{-1}$ | 0.582 | $7.50 \times 10^{-1}$ | 1.174 | 1.5 | 1.870 | 5.00 | 3.491 | 7.50 | 4.749 | 10 | 6.035 | 20 | 8.611 |
| $3.2 \times 10^{-1}$ | 0.588 | $8.00 \times 10^{-1}$ | 1.203 | 1.6 | 1.924 | 5.50 | 3.581 | 8.00 | 4.846 | 11 | 6.246 | 22 | 8.809 |
| $3.4 \times 10^{-1}$ | 0.594 | $9.00 \times 10^{-1}$ | 1.253 | 1.7 | 1.975 | 6.00 | 3.656 | 8.50 | 4.932 | 12 | 6.425 | 24 | 8.968 |
| $3.6 \times 10^{-1}$ | 0.599 | 1.00 | 1.295 | 1.8 | 2.022 | 6.50 | 3.717 | 9.00 | 5.009 | 13 | 6.580 | 26 | 9.097 |
| $3.8 \times 10^{-1}$ | 0.603 | 1.1 | 1.330 | 2.0 | 2.106 | 7.00 | 3.767 | 9.50 | 5.078 | 14 | 6.712 | 28 | 9.200 |
| $4.0 \times 10^{-1}$ | 0.606 | 1.2 | 1.358 | 2.2 | 2.178 | 7.50 | 3.809 | 10.00 | 5.138 | 15 | 6.825 | 30 | 9.283 |
| $4.5 \times 10^{-1}$ | 0.613 | 1.3 | 1.382 | 2.4 | 2.241 | 8.00 | 3.843 | 11 | 5.241 | 16 | 6.922 | 34 | 9.404 |
| $5.0 \times 10^{-1}$ | 0.617 | 1.4 | 1.402 | 2.6 | 2.294 | 9.00 | 3.894 | 12 | 5.321 | 17 | 7.004 | 38 | 9.481 |
| $6.0 \times 10^{-1}$ | 0.621 | 1.6 | 1.432 | 2.8 | 2.340 | 10.00 | 3.928 | 13 | 5.385 | 18 | 7.076 | 42 | 9.532 |
| $7.0 \times 10^{-1}$ | 0.623 | 1.7 | 1.444 | 3.0 | 2.380 | 11.00 | 3.951 | 14 | 5.435 | 20 | 7.189 | 46 | 9.565 |
| $8.0 \times 10^{-1}$ | 0.624 | 1.8 | 1.453 | 3.4 | 2.444 | 12.00 | 3.967 | 15 | 5.476 | 22 | 7.272 | 50 | 9.586 |
| | | 2.0 | 1.468 | 3.8 | 2.491 | 14.00 | 3.985 | 16 | 5.506 | 24 | 7.332 | 60 | 9.612 |
| | | 2.5 | 1.487 | 4.2 | 2.525 | 16.00 | 3.993 | 17 | 5.531 | 26 | 7.377 | 70 | 9.621 |
| | | 3.0 | 1.495 | 4.6 | 2.551 | 18.00 | 3.997 | 18 | 5.551 | 30 | 7.434 | 80 | 9.623 |
| | | 4.0 | 1.499 | 5.0 | 2.570 | 20.00 | 3.999 | 20 | 5.579 | 34 | 7.464 | 90 | 9.624 |
| | | 5.0 | 1.500 | 6.0 | 2.599 | 22.00 | 3.999 | 25 | 5.611 | 38 | 7.481 | 100 | 9.625 |
| | | | | 7.0 | 2.613 | 24.00 | 4.000 | 30 | 5.621 | 42 | 7.490 | | |
| | | | | 8.0 | 2.619 | | | 35 | 5.624 | 46 | 7.494 | | |
| | | | | 9.0 | 2.622 | | | 40 | 5.625 | 50 | 7.497 | | |
| | | | | 10.0 | 2.624 | | | | | | | | |

### TABLE 4–7.[3] Continued

| $r_D = 5.0$ | | $r_D = 6.0$ | | $r_D = 7.0$ | | $r_D = 8.0$ | | $r_D = 9.0$ | | $r_D = 10.0$ | |
|---|---|---|---|---|---|---|---|---|---|---|---|
| $t_D$ | $W_{eD}$ | $t_D$ | $W_{eD}$ | $t_D$ | $W_{eD}$ | $t_D$ | $W_{eD}$ | $t_D$ | $W_{eD}$ | $t_D$ | $W_{eD}$ |
| 3.0 | 3.195 | 6.0 | 5.148 | 9.00 | 6.861 | 9 | 6.861 | 10 | 7.417 | 15 | 9.965 |
| 3.5 | 3.542 | 6.5 | 5.440 | 9.50 | 7.127 | 10 | 7.398 | 15 | 9.945 | 20 | 12.32 |
| 4.0 | 3.875 | 7.0 | 5.724 | 10 | 7.389 | 11 | 7.920 | 20 | 12.26 | 22 | 13.22 |
| 4.5 | 4.193 | 7.5 | 6.002 | 11 | 7.902 | 12 | 8.431 | 22 | 13.13 | 24 | 14.09 |
| 5.0 | 4.499 | 8.0 | 6.273 | 12 | 8.397 | 13 | 8.930 | 24 | 13.98 | 26 | 14.95 |
| 5.5 | 4.792 | 8.5 | 6.537 | 13 | 8.876 | 14 | 9.418 | 26 | 14.79 | 28 | 15.78 |
| 6.0 | 5.074 | 9.0 | 6.795 | 14 | 9.341 | 15 | 9.895 | 28 | 15.59 | 30 | 16.59 |
| 6.5 | 5.345 | 9.5 | 7.047 | 15 | 9.791 | 16 | 10.361 | 30 | 16.35 | 32 | 17.38 |
| 7.0 | 5.605 | 10.0 | 7.293 | 16 | 10.23 | 17 | 10.82 | 32 | 17.10 | 34 | 18.16 |
| 7.5 | 5.854 | 10.5 | 7.533 | 17 | 10.65 | 18 | 11.26 | 34 | 17.82 | 36 | 18.91 |
| 8.0 | 6.094 | 11 | 7.767 | 18 | 11.06 | 19 | 11.70 | 36 | 18.52 | 38 | 19.65 |
| 8.5 | 6.325 | 12 | 8.220 | 19 | 11.46 | 20 | 12.13 | 38 | 19.19 | 40 | 20.37 |
| 9.0 | 6.547 | 13 | 8.651 | 20 | 11.85 | 22 | 12.95 | 40 | 19.85 | 42 | 21.07 |
| 9.5 | 6.760 | 14 | 9.063 | 22 | 12.58 | 24 | 13.74 | 42 | 20.48 | 44 | 21.76 |
| 10 | 6.965 | 15 | 9.456 | 24 | 13.27 | 26 | 14.50 | 44 | 21.09 | 46 | 22.42 |
| 11 | 7.350 | 16 | 9.829 | 26 | 13.92 | 28 | 15.23 | 46 | 21.69 | 48 | 23.07 |
| 12 | 7.706 | 17 | 10.19 | 28 | 14.53 | 30 | 15.92 | 48 | 22.26 | 50 | 23.71 |
| 13 | 8.035 | 18 | 10.53 | 30 | 15.11 | 34 | 17.22 | 50 | 22.82 | 52 | 24.33 |
| 14 | 8.339 | 19 | 10.85 | 35 | 16.39 | 38 | 18.41 | 52 | 23.36 | 54 | 24.94 |
| 15 | 8.620 | 20 | 11.16 | 40 | 17.49 | 40 | 18.97 | 54 | 23.89 | 56 | 25.53 |
| 16 | 8.879 | 22 | 11.74 | 45 | 18.43 | 45 | 20.26 | 56 | 24.39 | 58 | 26.11 |
| 18 | 9.338 | 24 | 12.16 | 50 | 19.24 | 50 | 21.42 | 58 | 24.88 | 60 | 26.67 |
| 20 | 9.731 | 25 | 12.50 | 60 | 20.51 | 55 | 22.46 | 60 | 25.36 | 65 | 28.02 |
| 22 | 10.07 | 31 | 13.74 | 70 | 21.45 | 60 | 23.40 | 65 | 26.48 | 70 | 29.29 |
| 24 | 10.35 | 35 | 14.40 | 80 | 22.13 | 70 | 24.98 | 70 | 27.52 | 75 | 30.49 |
| 26 | 10.59 | 39 | 14.93 | 90 | 22.63 | 80 | 26.26 | 75 | 28.48 | 80 | 31.61 |
| 28 | 10.80 | 51 | 16.05 | 100 | 23.00 | 90 | 27.28 | 80 | 29.36 | 85 | 32.67 |
| 30 | 10.89 | 60 | 16.56 | 120 | 23.47 | 100 | 28.11 | 85 | 30.18 | 90 | 33.66 |
| 34 | 11.26 | 70 | 16.91 | 140 | 23.71 | 120 | 29.31 | 90 | 30.93 | 95 | 34.60 |
| 38 | 11.46 | 80 | 17.14 | 160 | 23.85 | 140 | 30.08 | 95 | 31.63 | 100 | 35.48 |
| 42 | 11.61 | 90 | 17.27 | 180 | 23.92 | 160 | 30.58 | 100 | 32.27 | 120 | 38.51 |
| 46 | 11.71 | 100 | 17.36 | 200 | 23.96 | 180 | 30.91 | 120 | 34.39 | 140 | 40.89 |
| 50 | 11.79 | 110 | 17.41 | 500 | 24.00 | 200 | 31.12 | 140 | 35.92 | 160 | 42.75 |
| 60 | 11.91 | 120 | 17.45 | | | 240 | 31.34 | 160 | 37.04 | 180 | 44.21 |
| 70 | 11.96 | 130 | 17.46 | | | 280 | 31.43 | 180 | 37.85 | 200 | 45.36 |
| 80 | 11.98 | 140 | 17.48 | | | 320 | 31.47 | 200 | 38.44 | 240 | 46.95 |
| 90 | 11.99 | 150 | 17.49 | | | 360 | 31.49 | 240 | 39.17 | 280 | 47.94 |
| 100 | 12.00 | 160 | 17.49 | | | 400 | 31.50 | 280 | 39.56 | 320 | 48.54 |
| 120 | 12.0 | 180 | 17.50 | | | 500 | 31.50 | 320 | 39.77 | 360 | 48.91 |
| | | 200 | 17.50 | | | | | 360 | 39.88 | 400 | 49.14 |
| | | 220 | 17.50 | | | | | 400 | 39.94 | 440 | 49.28 |
| | | | | | | | | 440 | 39.97 | 480 | 49.36 |
| | | | | | | | | 480 | 39.98 | | |

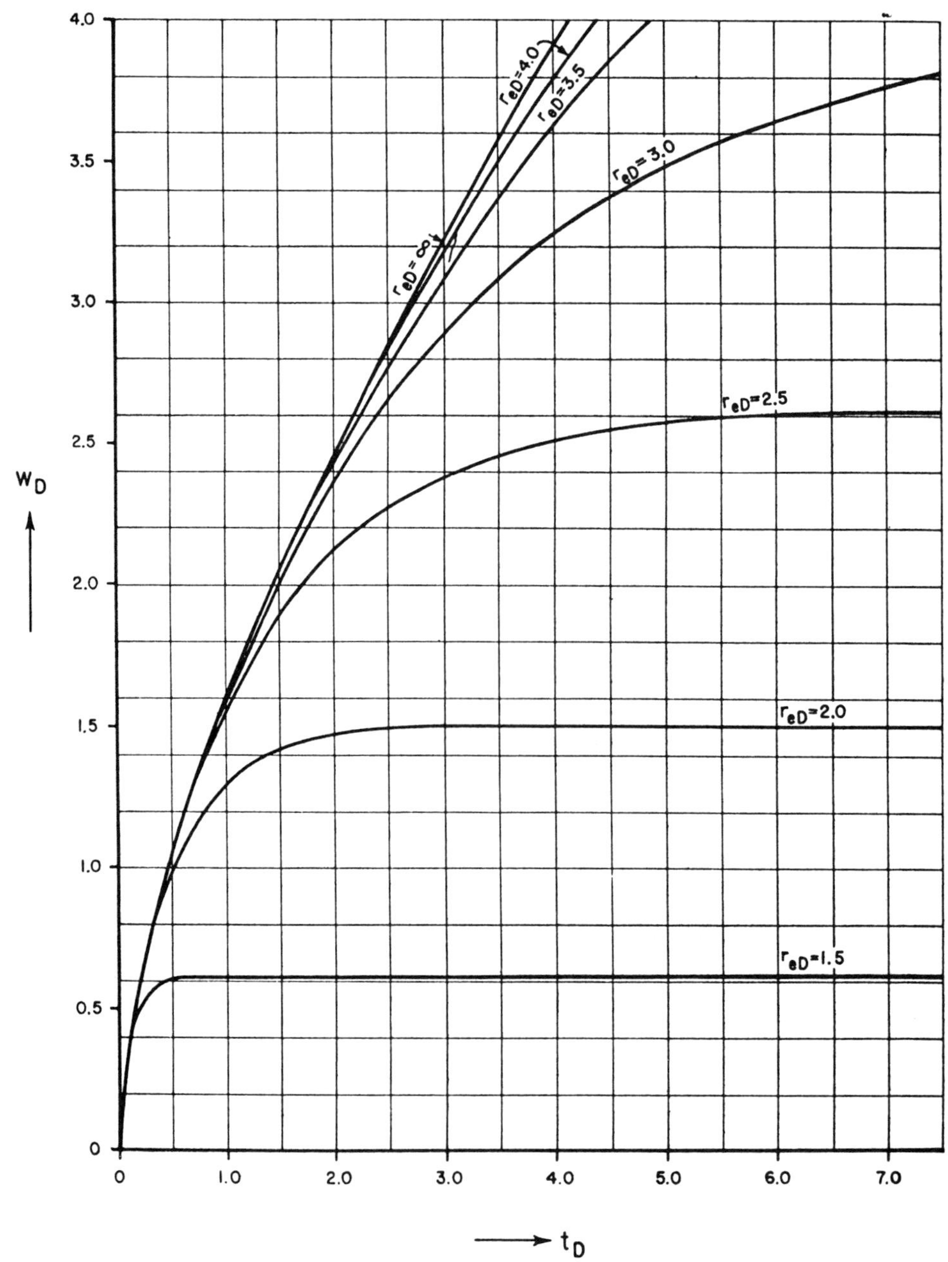


**Figure 4–20.** Dimensionless water influx, constant terminal pressure case, radial flow (courtesy of Elsevier).[7]

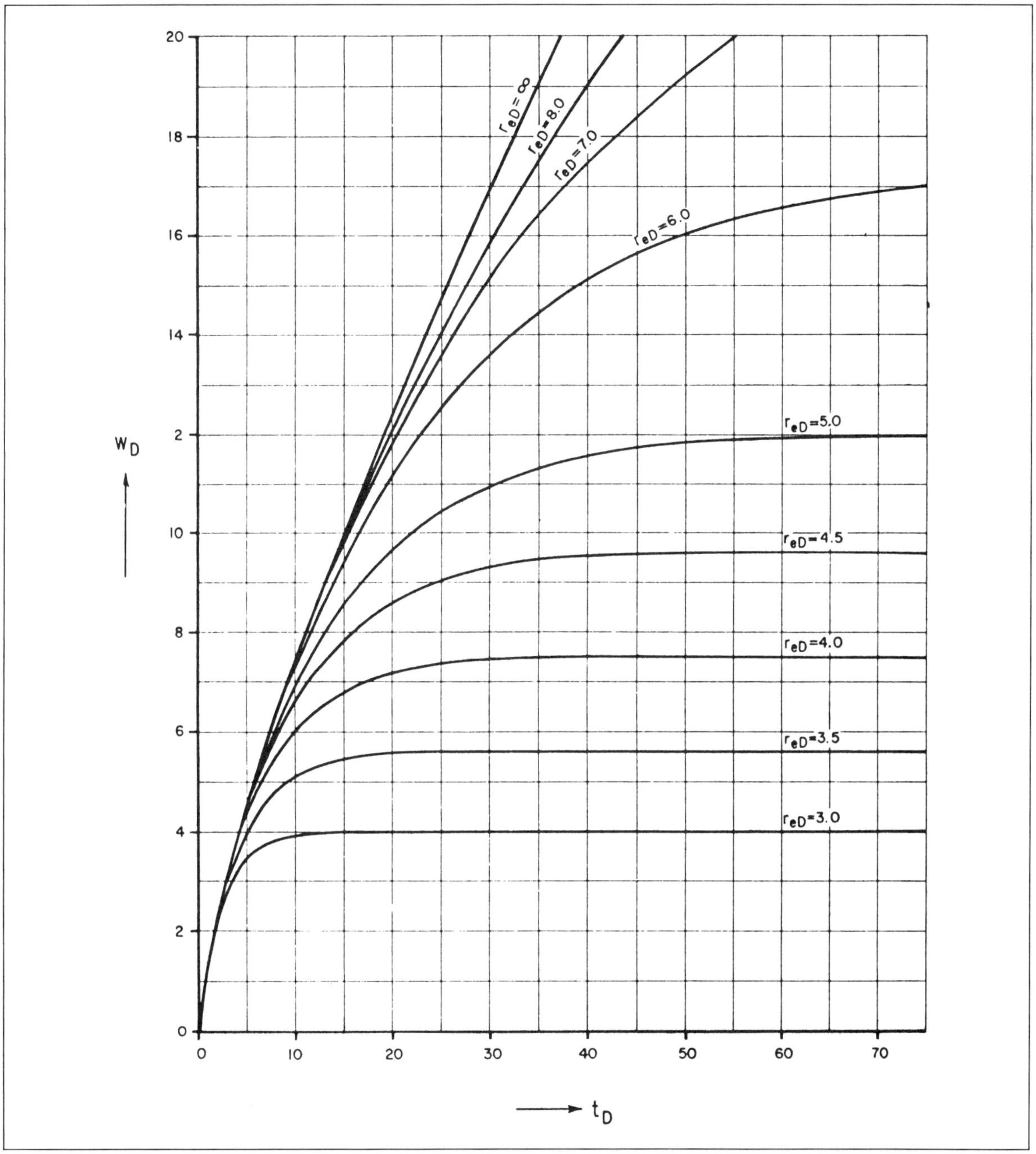


**Figure 4–21.** Dimensionless water influx, constant terminal pressure case, radial flow (courtesy of Elsevier).[7]

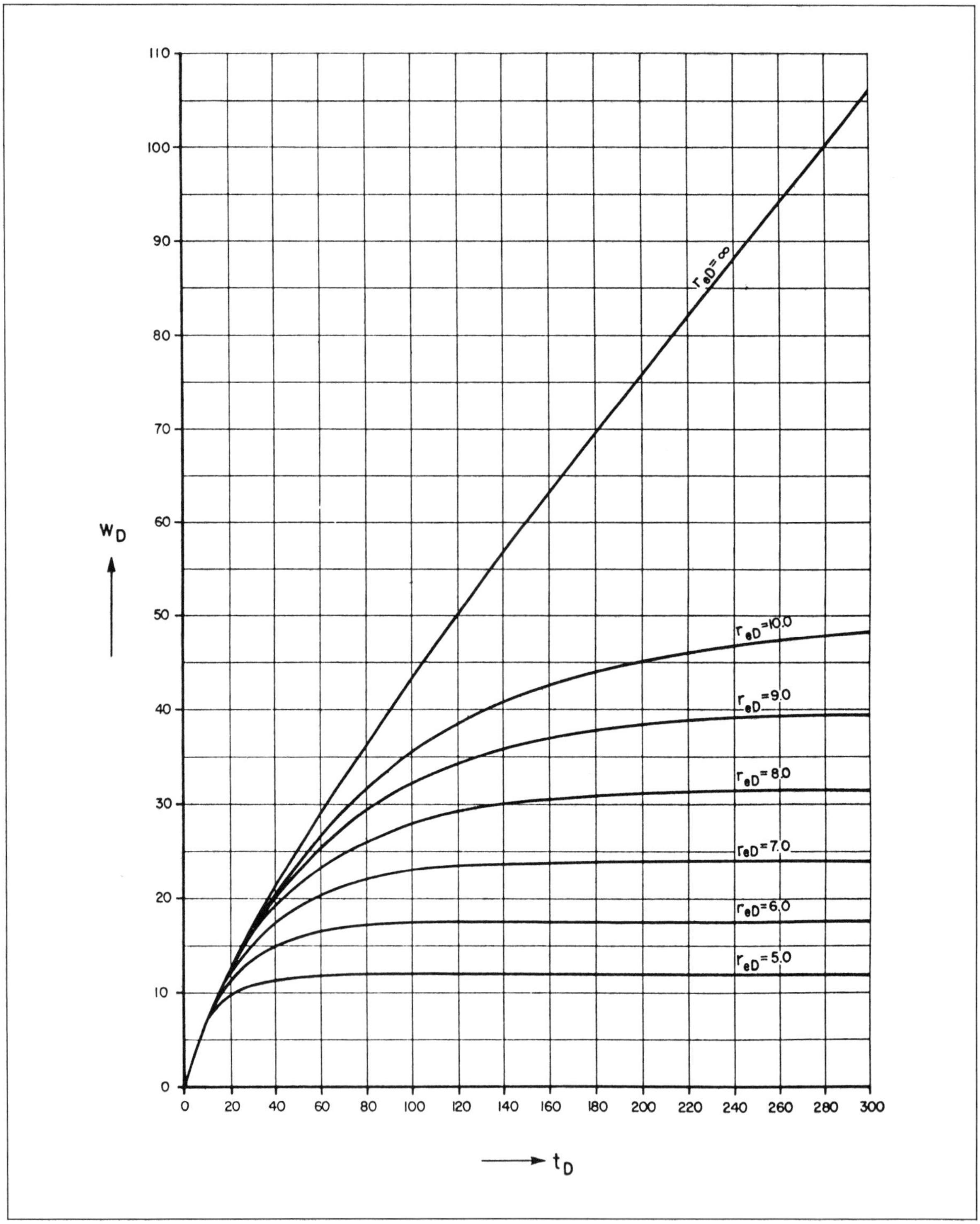


**Figure 4–22.** Dimensionless water influx, constant terminal pressure case, radial flow (courtesy of Elsevier).[7]

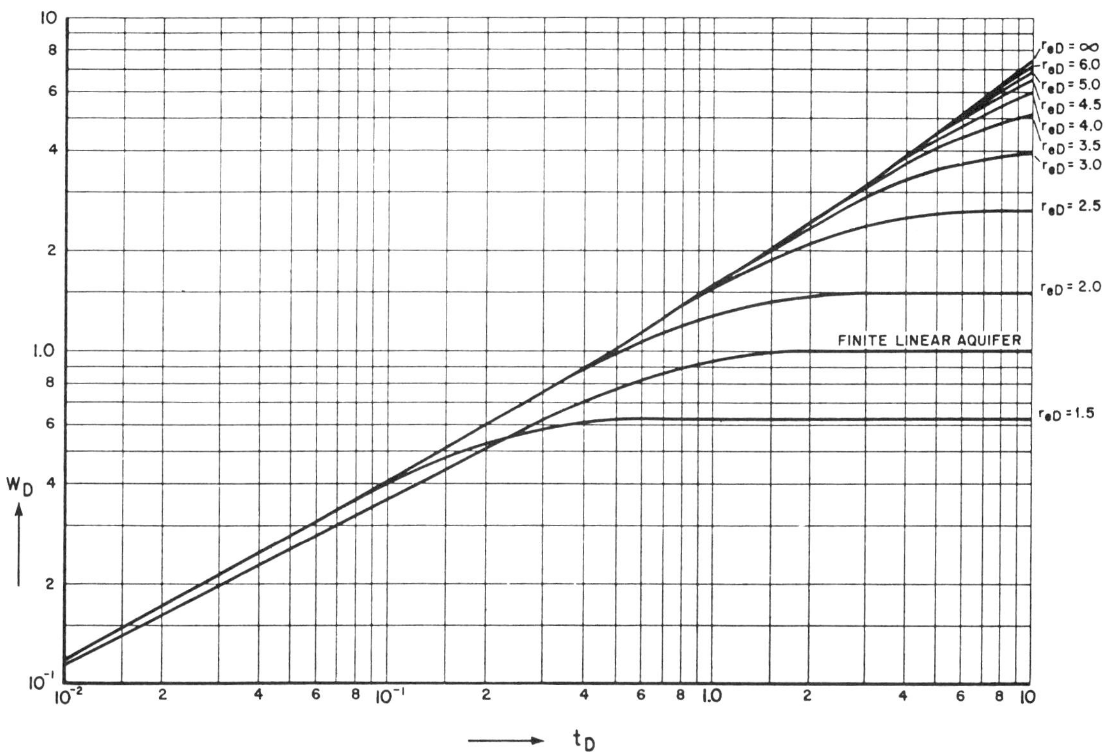


**Figure 4–23.** Dimensionless water influx, constant terminal pressure case, radial and linear flow (courtesy of Elsevier).[7]

Using Equation (4.81), calculate water influx constant, $m_p$.

$$m_p = 1.119(0.2)(8 \times 10^{-6})(3{,}000)^2(40)\left(\frac{360}{360}\right)$$
$$= 644.5 \quad \text{bbls/psi}$$

Water influx at 100 days is

$$W_{e100} = m_p \Delta p W_{eD}$$
$$= (644.5)(2{,}734 - 2{,}724)(5.086)$$
$$= 32{,}779 \quad \text{bbls}$$

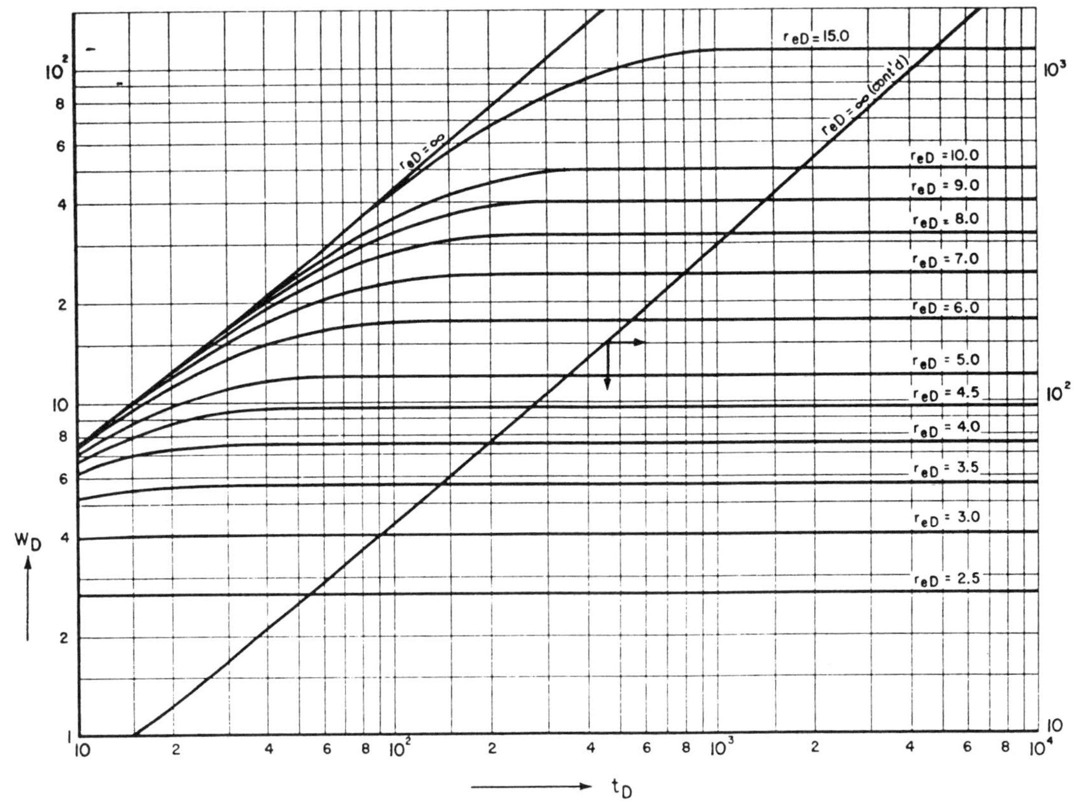


**Figure 4–24.** Dimensionless water influx, constant terminal pressure case, radial flow (courtesy of Elsevier).[7]

Similarly at 200, 400, and 800 days,

| | $t_D$ | $W_{eD}$ | $W_e$ |
|---|---|---|---|
| 200 days | 11.76 | 8.43 | 54,330 |
| 400 days | 23.52 | 13.90 | 89,590 |
| 800 days | 47.04 | 22.75 | 146,600 |

If the pressure drop is transmitted instantaneously throughout the aquifer, then

$$W_e = c_e \frac{\theta}{360} \pi(r_e^2 - r_w^2)h\phi\Delta p/5.615$$

$$= \left[\frac{(8 \times 10^{-6})(1.0)(\pi)(30{,}000^2 - 3{,}000^2)(40)(0.2)(10)}{5.615}\right]$$

$$= 319{,}049 \quad \text{bbls}$$

In Figure 4–24, note that for dimensionless times greater than $t_D \approx 330$, $W_{eD} = 50$, and it remains constant at this value, indicating that the maximum amount of water influx due to the 10 psi pressure drop is 

$$W_e = (644.5)(2{,}734 - 2{,}724)(50)$$

$$= 322{,}250 \quad \text{bbls}$$

which is the same as 319,049 bbls calculated above. This situation is reached at approximately $t$ = 50/0.0589 = 849 days.

## CUMULATIVE WATER INFLUX—BOTTOMWATER DRIVE RESERVOIRS

In the previous section an aquifer influx model for edgewater drive developed by van Everdingen and Hurst was presented. Although the assumptions made in deriving this model are not strictly valid for bottomwater drive systems, water influx in this case can sometimes be closely approximated by radial flow. Since the results are often quite adequate, it has been a common practice to apply the van Everdingen and Hurst method to both bottomwater and edgewater drive systems.

Coats originally developed a model accounting for vertical flow effects, and he showed that these effects are fairly significant.[23] The model was applied only to *terminal-rate* case allowing the user to calculate pressure from known influx and it was applicable only to infinite aquifers.

Allard and Chen recently presented a bottomwater drive model based on the solution to the *terminal-pressure* case and applies to both finite and infinite aquifers.[24] The results are presented in the form of dimensionless group as tabulated in Tables 4–8 through 4–12. Equations (4.80), (4.81), and (4.82) are still applicable with $W_{eD}$ as a function of not only $t_D$ but also the reservoir/aquifer geometry. To describe the system geometry, the following dimensionless constants are introduced:

$$r'_D = r_e/r_D \qquad \textbf{(4.85)}$$

and

$$z'_D = h/(r_R F_k^{0.5}) \qquad \textbf{(4.86)}$$

where $r_e$ is the aquifer radius, $r_R$ is the reservoir radius, $F_k$ is the ratio of vertical to horizontal permeability, $h$ is aquifer thickness, and $z'_D$ is dimensionless thickness constant. For fixed values of these two parameters, $W_{eD}$ is a function of only $t_D$.

## TABLE 4–8. Dimensionless influx, $W_{eD}$, for infinite aquifers (courtesy of SPE)[24]

| | $r'_D$ | | | | | | |
|---|---|---|---|---|---|---|---|
| $t_D$ | 0.05 | 0.1 | 0.3 | 0.5 | 0.7 | 0.9 | 1.0 |
| 0.1 | 0.700 | 0.677 | 0.508 | 0.349 | 0.251 | 0.195 | 0.176 |
| 0.2 | 0.793 | 0.786 | 0.696 | 0.547 | 0.416 | 0.328 | 0.295 |
| 0.3 | 0.936 | 0.926 | 0.834 | 0.692 | 0.548 | 0.440 | 0.396 |
| 0.4 | 1.051 | 1.041 | 0.952 | 0.812 | 0.662 | 0.540 | 0.486 |
| 0.5 | 1.158 | 1.155 | 1.059 | 0.918 | 0.764 | 0.631 | 0.569 |
| 0.6 | 1.270 | 1.268 | 1.167 | 1.021 | 0.862 | 0.721 | 0.651 |
| 0.7 | 1.384 | 1.380 | 1.270 | 1.116 | 0.953 | 0.806 | 0.729 |
| 0.8 | 1.503 | 1.499 | 1.373 | 1.205 | 1.039 | 0.886 | 0.803 |
| 0.9 | 1.621 | 1.612 | 1.477 | 1.286 | 1.117 | 0.959 | 0.872 |
| 1 | 1.743 | 1.726 | 1.581 | 1.347 | 1.181 | 1.020 | 0.932 |
| 2 | 2.402 | 2.393 | 2.288 | 2.034 | 1.827 | 1.622 | 1.509 |
| 3 | 3.031 | 3.018 | 2.895 | 2.650 | 2.408 | 2.164 | 2.026 |
| 4 | 3.629 | 3.615 | 3.477 | 3.223 | 2.949 | 2.669 | 2.510 |
| 5 | 4.217 | 4.201 | 4.048 | 3.766 | 3.462 | 3.150 | 2.971 |
| 6 | 4.784 | 4.766 | 4.601 | 4.288 | 3.956 | 3.614 | 3.416 |
| 7 | 5.323 | 5.303 | 5.128 | 4.792 | 4.434 | 4.063 | 3.847 |
| 8 | 5.829 | 5.808 | 5.625 | 5.283 | 4.900 | 4.501 | 4.268 |
| 9 | 6.306 | 6.283 | 6.094 | 5.762 | 5.355 | 4.929 | 4.680 |
| 10 | 6.837 | 6.816 | 6.583 | 6.214 | 5.792 | 5.344 | 5.080 |
| 11 | 7.263 | 7.242 | 7.040 | 6.664 | 6.217 | 5.745 | 5.468 |
| 12 | 7.742 | 7.718 | 7.495 | 7.104 | 6.638 | 6.143 | 5.852 |
| 13 | 8.196 | 8.172 | 7.943 | 7.539 | 7.052 | 6.536 | 6.231 |
| 14 | 8.648 | 8.623 | 8.385 | 7.967 | 7.461 | 6.923 | 6.604 |
| 15 | 9.094 | 9.068 | 8.821 | 8.389 | 7.864 | 7.305 | 6.973 |
| 16 | 9.534 | 9.507 | 9.253 | 8.806 | 8.262 | 7.682 | 7.338 |
| 17 | 9.969 | 9.942 | 9.679 | 9.218 | 8.656 | 8.056 | 7.699 |
| 18 | 10.399 | 10.371 | 10.100 | 9.626 | 9.046 | 8.426 | 8.057 |
| 19 | 10.823 | 10.794 | 10.516 | 10.029 | 9.432 | 8.793 | 8.411 |
| 20 | 11.241 | 11.211 | 10.929 | 10.430 | 9.815 | 9.156 | 8.763 |
| 21 | 11.664 | 11.633 | 11.339 | 10.826 | 10.194 | 9.516 | 9.111 |
| 22 | 12.075 | 12.045 | 11.744 | 11.219 | 10.571 | 9.874 | 9.457 |
| 23 | 12.486 | 12.454 | 12.147 | 11.609 | 10.944 | 10.229 | 9.801 |
| 24 | 12.893 | 12.861 | 12.546 | 11.996 | 11.315 | 10.581 | 10.142 |
| 25 | 13.297 | 13.264 | 12.942 | 12.380 | 11.683 | 10.931 | 10.481 |
| 26 | 13.698 | 13.665 | 13.336 | 12.761 | 12.048 | 11.279 | 10.817 |
| 27 | 14.097 | 14.062 | 13.726 | 13.140 | 12.411 | 11.625 | 11.152 |
| 28 | 14.493 | 14.458 | 14.115 | 13.517 | 12.772 | 11.968 | 11.485 |
| 29 | 14.886 | 14.850 | 14.501 | 13.891 | 13.131 | 12.310 | 11.816 |
| 30 | 15.277 | 15.241 | 14.884 | 14.263 | 13.488 | 12.650 | 12.145 |
| 31 | 15.666 | 15.628 | 15.266 | 14.634 | 13.843 | 12.990 | 12.473 |
| 32 | 16.053 | 16.015 | 15.645 | 15.002 | 14.196 | 13.324 | 12.799 |
| 33 | 16.437 | 16.398 | 16.023 | 15.368 | 14.548 | 13.659 | 13.123 |
| 34 | 16.819 | 16.780 | 16.398 | 15.732 | 14.897 | 13.992 | 13.446 |
| 35 | 17.200 | 17.160 | 16.772 | 16.095 | 15.245 | 14.324 | 13.767 |
| 36 | 17.579 | 17.538 | 17.143 | 16.456 | 15.592 | 14.654 | 14.088 |
| 37 | 17.956 | 17.915 | 17.513 | 16.815 | 15.937 | 14.983 | 14.406 |
| 38 | 18.331 | 18.289 | 17.882 | 17.173 | 16.280 | 15.311 | 14.724 |
| 39 | 18.704 | 18.662 | 18.249 | 17.529 | 16.622 | 15.637 | 15.040 |
| 40 | 19.088 | 19.045 | 18.620 | 17.886 | 16.964 | 15.963 | 15.356 |
| 41 | 19.450 | 19.407 | 18.982 | 18.240 | 17.305 | 16.288 | 15.671 |
| 42 | 19.821 | 19.777 | 19.344 | 18.592 | 17.644 | 16.611 | 15.985 |
| 43 | 20.188 | 20.144 | 19.706 | 18.943 | 17.981 | 16.933 | 16.297 |
| 44 | 20.555 | 20.510 | 20.065 | 19.293 | 18.317 | 17.253 | 16.608 |
| 45 | 20.920 | 20.874 | 20.424 | 19.641 | 18.651 | 17.573 | 16.918 |
| 46 | 21.283 | 21.237 | 20.781 | 19.988 | 18.985 | 17.891 | 17.227 |
| 47 | 21.645 | 21.598 | 21.137 | 20.333 | 19.317 | 18.208 | 17.535 |
| 48 | 22.006 | 21.958 | 21.491 | 20.678 | 19.648 | 18.524 | 17.841 |
| 49 | 22.365 | 22.317 | 21.844 | 21.021 | 19.978 | 18.840 | 18.147 |
| 50 | 22.722 | 22.674 | 22.196 | 21.363 | 20.307 | 19.154 | 18.452 |
| 51 | 23.081 | 23.032 | 22.547 | 21.704 | 20.635 | 19.467 | 18.757 |

## TABLE 4–8. Continued

| | $\xi'_D$ | | | | | | |
|---|---|---|---|---|---|---|---|
| $t_D$ | 0.05 | 0.1 | 0.3 | 0.5 | 0.7 | 0.9 | 1.0 |
| 52 | 23.436 | 23.387 | 22.897 | 22.044 | 20.962 | 19.779 | 19.060 |
| 53 | 23.791 | 23.741 | 23.245 | 22.383 | 21.288 | 20.091 | 19.362 |
| 54 | 24.145 | 24.094 | 23.593 | 22.721 | 21.613 | 20.401 | 19.664 |
| 55 | 24.498 | 24.446 | 23.939 | 23.058 | 21.937 | 20.711 | 19.965 |
| 56 | 24.849 | 24.797 | 24.285 | 23.393 | 22.260 | 21.020 | 20.265 |
| 57 | 25.200 | 25.147 | 24.629 | 23.728 | 22.583 | 21.328 | 20.564 |
| 58 | 25.549 | 25.496 | 24.973 | 24.062 | 22.904 | 21.636 | 20.862 |
| 59 | 25.898 | 25.844 | 25.315 | 24.395 | 23.225 | 21.942 | 21.160 |
| 60 | 26.246 | 26.191 | 25.657 | 24.728 | 23.545 | 22.248 | 21.457 |
| 61 | 26.592 | 26.537 | 25.998 | 25.059 | 23.864 | 22.553 | 21.754 |
| 62 | 26.938 | 26.883 | 26.337 | 25.390 | 24.182 | 22.857 | 22.049 |
| 63 | 27.283 | 27.227 | 26.676 | 25.719 | 24.499 | 23.161 | 22.344 |
| 64 | 27.627 | 27.570 | 27.015 | 26.048 | 24.816 | 23.464 | 22.639 |
| 65 | 27.970 | 27.913 | 27.352 | 26.376 | 25.132 | 23.766 | 22.932 |
| 66 | 28.312 | 28.255 | 27.688 | 26.704 | 25.447 | 24.068 | 23.225 |
| 67 | 28.653 | 28.596 | 28.024 | 27.030 | 25.762 | 24.369 | 23.518 |
| 68 | 28.994 | 28.936 | 28.359 | 27.356 | 26.075 | 24.669 | 23.810 |
| 69 | 29.334 | 29.275 | 28.693 | 27.681 | 26.389 | 24.969 | 24.101 |
| 70 | 29.673 | 29.614 | 29.026 | 28.006 | 26.701 | 25.268 | 24.391 |
| 71 | 30.011 | 29.951 | 29.359 | 28.329 | 27.013 | 25.566 | 24.681 |
| 72 | 30.349 | 30.288 | 29.691 | 28.652 | 27.324 | 25.864 | 24.971 |
| 73 | 30.686 | 30.625 | 30.022 | 28.974 | 27.634 | 26.161 | 25.260 |
| 74 | 31.022 | 30.960 | 30.353 | 29.296 | 27.944 | 26.458 | 25.548 |
| 75 | 31.357 | 31.295 | 30.682 | 29.617 | 28.254 | 26.754 | 25.836 |
| 76 | 31.692 | 31.629 | 31.012 | 29.937 | 28.562 | 27.049 | 26.124 |
| 77 | 32.026 | 31.963 | 31.340 | 30.257 | 28.870 | 27.344 | 26.410 |
| 78 | 32.359 | 32.296 | 31.668 | 30.576 | 29.178 | 27.639 | 26.697 |
| 79 | 32.692 | 32.628 | 31.995 | 30.895 | 29.485 | 27.933 | 26.983 |
| 80 | 33.024 | 32.959 | 32.322 | 31.212 | 29.791 | 28.226 | 27.268 |
| 81 | 33.355 | 33.290 | 32.647 | 31.530 | 30.097 | 28.519 | 27.553 |
| 82 | 33.686 | 33.621 | 32.973 | 31.846 | 30.402 | 28.812 | 27.837 |
| 83 | 34.016 | 33.950 | 33.297 | 32.163 | 30.707 | 29.104 | 28.121 |
| 84 | 34.345 | 34.279 | 33.622 | 32.478 | 31.011 | 29.395 | 28.404 |
| 85 | 34.674 | 34.608 | 33.945 | 32.793 | 31.315 | 29.686 | 28.687 |
| 86 | 35.003 | 34.935 | 34.268 | 33.107 | 31.618 | 29.976 | 28.970 |
| 87 | 35.330 | 35.263 | 34.590 | 33.421 | 31.921 | 30.266 | 29.252 |
| 88 | 35.657 | 35.589 | 34.912 | 33.735 | 32.223 | 30.556 | 29.534 |
| 89 | 35.984 | 35.915 | 35.233 | 34.048 | 32.525 | 30.845 | 29.815 |
| 90 | 36.310 | 36.241 | 35.554 | 34.360 | 32.826 | 31.134 | 30.096 |
| 91 | 36.636 | 36.566 | 35.874 | 34.672 | 33.127 | 31.422 | 30.376 |
| 92 | 36.960 | 36.890 | 36.194 | 34.983 | 33.427 | 31.710 | 30.656 |
| 93 | 37.285 | 37.214 | 36.513 | 35.294 | 33.727 | 31.997 | 30.935 |
| 94 | 37.609 | 37.538 | 36.832 | 35.604 | 34.026 | 32.284 | 31.215 |
| 95 | 37.932 | 37.861 | 37.150 | 35.914 | 34.325 | 32.570 | 31.493 |
| 96 | 38.255 | 38.183 | 37.467 | 36.223 | 34.623 | 32.857 | 31.772 |
| 97 | 38.577 | 38.505 | 37.785 | 36.532 | 34.921 | 33.142 | 32.050 |
| 98 | 38.899 | 38.826 | 38.101 | 36.841 | 35.219 | 33.427 | 32.327 |
| 99 | 39.220 | 39.147 | 38.417 | 37.149 | 35.516 | 33.712 | 32.605 |
| 100 | 39.541 | 39.467 | 38.733 | 37.456 | 35.813 | 33.997 | 32.881 |
| 105 | 41.138 | 41.062 | 40.305 | 38.987 | 37.290 | 35.414 | 34.260 |
| 110 | 42.724 | 42.645 | 41.865 | 40.508 | 38.758 | 36.821 | 35.630 |
| 115 | 44.299 | 44.218 | 43.415 | 42.018 | 40.216 | 38.221 | 36.993 |
| 120 | 45.864 | 45.781 | 44.956 | 43.520 | 41.666 | 39.612 | 38.347 |
| 125 | 47.420 | 47.334 | 46.487 | 45.012 | 43.107 | 40.995 | 39.694 |
| 130 | 48.966 | 48.879 | 48.009 | 46.497 | 44.541 | 42.372 | 41.035 |
| 135 | 50.504 | 50.414 | 49.523 | 47.973 | 45.967 | 43.741 | 42.368 |
| 140 | 52.033 | 51.942 | 51.029 | 49.441 | 47.386 | 45.104 | 43.696 |
| 145 | 53.555 | 53.462 | 52.528 | 50.903 | 48.798 | 46.460 | 45.017 |
| 150 | 55.070 | 54.974 | 54.019 | 52.357 | 50.204 | 47.810 | 46.333 |
| 155 | 56.577 | 56.479 | 55.503 | 53.805 | 51.603 | 49.155 | 47.643 |

## TABLE 4–8. Continued

| | $z_D'$ | | | | | | |
|---|---|---|---|---|---|---|---|
| $t_D$ | 0.05 | 0.1 | 0.3 | 0.5 | 0.7 | 0.9 | 1.0 |
| 160 | 58.077 | 57.977 | 56.981 | 55.246 | 52.996 | 50.494 | 48.947 |
| 165 | 59.570 | 59.469 | 58.452 | 56.681 | 54.384 | 51.827 | 50.247 |
| 170 | 61.058 | 60.954 | 59.916 | 58.110 | 55.766 | 53.156 | 51.542 |
| 175 | 62.539 | 62.433 | 61.375 | 59.534 | 57.143 | 54.479 | 52.832 |
| 180 | 64.014 | 63.906 | 62.829 | 60.952 | 58.514 | 55.798 | 54.118 |
| 185 | 65.484 | 65.374 | 64.276 | 62.365 | 59.881 | 57.112 | 55.399 |
| 190 | 66.948 | 66.836 | 65.718 | 63.773 | 61.243 | 58.422 | 56.676 |
| 195 | 68.406 | 68.293 | 67.156 | 65.175 | 62.600 | 59.727 | 57.949 |
| 200 | 69.860 | 69.744 | 68.588 | 66.573 | 63.952 | 61.028 | 59.217 |
| 205 | 71.309 | 71.191 | 70.015 | 67.967 | 65.301 | 62.326 | 60.482 |
| 210 | 72.752 | 72.633 | 71.437 | 69.355 | 66.645 | 63.619 | 61.744 |
| 215 | 74.191 | 74.070 | 72.855 | 70.740 | 67.985 | 64.908 | 63.001 |
| 220 | 75.626 | 75.503 | 74.269 | 72.120 | 69.321 | 66.194 | 64.255 |
| 225 | 77.056 | 76.931 | 75.678 | 73.496 | 70.653 | 67.476 | 65.506 |
| 230 | 78.482 | 78.355 | 77.083 | 74.868 | 71.981 | 68.755 | 66.753 |
| 235 | 79.903 | 79.774 | 78.484 | 76.236 | 73.306 | 70.030 | 67.997 |
| 240 | 81.321 | 81.190 | 79.881 | 77.601 | 74.627 | 71.302 | 69.238 |
| 245 | 82.734 | 82.602 | 81.275 | 78.962 | 75.945 | 72.570 | 70.476 |
| 250 | 84.144 | 84.010 | 82.664 | 80.319 | 77.259 | 73.736 | 71.711 |
| 255 | 85.550 | 85.414 | 84.050 | 81.672 | 78.570 | 75.098 | 72.943 |
| 260 | 86.952 | 86.814 | 85.432 | 83.023 | 79.878 | 76.358 | 74.172 |
| 265 | 88.351 | 88.211 | 86.811 | 84.369 | 81.182 | 77.614 | 75.398 |
| 270 | 89.746 | 89.604 | 88.186 | 85.713 | 82.484 | 78.868 | 76.621 |
| 275 | 91.138 | 90.994 | 89.558 | 87.053 | 83.782 | 80.119 | 77.842 |
| 280 | 92.526 | 92.381 | 90.926 | 88.391 | 85.078 | 81.367 | 79.060 |
| 285 | 93.911 | 93.764 | 92.292 | 89.725 | 86.371 | 82.612 | 80.276 |
| 290 | 95.293 | 95.144 | 93.654 | 91.056 | 87.660 | 83.855 | 81.489 |
| 295 | 96.672 | 96.521 | 95.014 | 92.385 | 88.948 | 85.095 | 82.700 |
| 300 | 98.048 | 97.895 | 96.370 | 93.710 | 90.232 | 86.333 | 83.908 |
| 305 | 99.420 | 99.266 | 97.724 | 95.033 | 91.514 | 87.568 | 85.114 |
| 310 | 100.79 | 100.64 | 99.07 | 96.35 | 92.79 | 88.80 | 86.32 |
| 315 | 102.16 | 102.00 | 100.42 | 97.67 | 94.07 | 90.03 | 87.52 |
| 320 | 103.52 | 103.36 | 101.77 | 98.99 | 95.34 | 91.26 | 88.72 |
| 325 | 104.88 | 104.72 | 103.11 | 100.30 | 96.62 | 92.49 | 89.92 |
| 330 | 106.24 | 106.08 | 104.45 | 101.61 | 97.89 | 93.71 | 91.11 |
| 335 | 107.60 | 107.43 | 105.79 | 102.91 | 99.15 | 94.93 | 92.30 |
| 340 | 108.95 | 108.79 | 107.12 | 104.22 | 100.42 | 96.15 | 93.49 |
| 345 | 110.30 | 110.13 | 108.45 | 105.52 | 101.68 | 97.37 | 94.68 |
| 350 | 111.65 | 111.48 | 109.78 | 106.82 | 102.94 | 98.58 | 95.87 |
| 355 | 113.00 | 112.82 | 111.11 | 108.12 | 104.20 | 99.80 | 97.06 |
| 360 | 114.34 | 114.17 | 112.43 | 109.41 | 105.45 | 101.01 | 98.24 |
| 365 | 115.68 | 115.51 | 113.76 | 110.71 | 106.71 | 102.22 | 99.42 |
| 370 | 117.02 | 116.84 | 115.08 | 112.00 | 107.96 | 103.42 | 100.60 |
| 375 | 118.36 | 118.18 | 116.40 | 113.29 | 109.21 | 104.63 | 101.78 |
| 380 | 119.69 | 119.51 | 117.71 | 114.57 | 110.46 | 105.83 | 102.95 |
| 385 | 121.02 | 120.84 | 119.02 | 115.86 | 111.70 | 107.04 | 104.13 |
| 390 | 122.35 | 122.17 | 120.34 | 117.14 | 112.95 | 108.24 | 105.30 |
| 395 | 123.68 | 123.49 | 121.65 | 118.42 | 114.19 | 109.43 | 106.47 |
| 400 | 125.00 | 124.82 | 122.94 | 119.70 | 115.43 | 110.63 | 107.64 |
| 405 | 126.33 | 126.14 | 124.26 | 120.97 | 116.67 | 111.82 | 108.80 |
| 410 | 127.65 | 127.46 | 125.56 | 122.25 | 117.90 | 113.02 | 109.97 |
| 415 | 128.97 | 128.78 | 126.86 | 123.52 | 119.14 | 114.21 | 111.13 |
| 420 | 130.28 | 130.09 | 128.16 | 124.79 | 120.37 | 115.40 | 112.30 |
| 425 | 131.60 | 131.40 | 129.46 | 126.06 | 121.60 | 116.59 | 113.46 |
| 430 | 132.91 | 132.72 | 130.75 | 127.33 | 122.83 | 117.77 | 114.62 |
| 435 | 134.22 | 134.03 | 132.05 | 128.59 | 124.06 | 118.96 | 115.77 |
| 440 | 135.53 | 135.33 | 133.34 | 129.86 | 125.29 | 120.14 | 116.93 |
| 445 | 136.84 | 136.64 | 134.63 | 131.12 | 126.51 | 121.32 | 118.08 |
| 450 | 138.15 | 137.94 | 135.92 | 132.38 | 127.73 | 122.50 | 119.24 |
| 455 | 139.45 | 139.25 | 137.20 | 133.64 | 128.96 | 123.68 | 120.39 |

TABLE 4–8. Continued

| | $z'_D$ | | | | | | |
|---|---|---|---|---|---|---|---|
| $t_D$ | 0.05 | 0.1 | 0.3 | 0.5 | 0.7 | 0.9 | 1.0 |
| 460 | 140.75 | 140.55 | 138.49 | 134.90 | 130.18 | 124.86 | 121.54 |
| 465 | 142.05 | 141.85 | 139.77 | 136.15 | 131.39 | 126.04 | 122.69 |
| 470 | 143.35 | 143.14 | 141.05 | 137.40 | 132.61 | 127.21 | 123.84 |
| 475 | 144.65 | 144.44 | 142.33 | 138.66 | 133.82 | 128.38 | 124.98 |
| 480 | 145.94 | 145.73 | 143.61 | 139.91 | 135.04 | 129.55 | 126.13 |
| 485 | 147.24 | 147.02 | 144.89 | 141.15 | 136.25 | 130.72 | 127.27 |
| 490 | 148.53 | 148.31 | 146.16 | 142.40 | 137.46 | 131.89 | 128.41 |
| 495 | 149.82 | 149.60 | 147.43 | 143.65 | 138.67 | 133.06 | 129.56 |
| 500 | 151.11 | 150.89 | 148.71 | 144.89 | 139.88 | 134.23 | 130.70 |
| 510 | 153.68 | 153.46 | 151.24 | 147.38 | 142.29 | 136.56 | 132.97 |
| 520 | 156.25 | 156.02 | 153.78 | 149.85 | 144.70 | 138.88 | 135.24 |
| 530 | 158.81 | 158.58 | 156.30 | 152.33 | 147.10 | 141.20 | 137.51 |
| 540 | 161.36 | 161.13 | 158.82 | 154.79 | 149.49 | 143.51 | 139.77 |
| 550 | 163.91 | 163.68 | 161.34 | 157.25 | 151.88 | 145.82 | 142.03 |
| 560 | 166.45 | 166.22 | 163.85 | 159.71 | 154.27 | 148.12 | 144.28 |
| 570 | 168.99 | 168.75 | 166.35 | 162.16 | 156.65 | 150.42 | 146.53 |
| 580 | 171.52 | 171.28 | 168.85 | 164.61 | 159.02 | 152.72 | 148.77 |
| 590 | 174.05 | 173.80 | 171.34 | 167.05 | 161.39 | 155.01 | 151.01 |
| 600 | 176.57 | 176.32 | 173.83 | 169.48 | 163.76 | 157.29 | 153.25 |
| 610 | 179.09 | 178.83 | 176.32 | 171.92 | 166.12 | 159.58 | 155.48 |
| 620 | 181.60 | 181.34 | 178.80 | 174.34 | 168.48 | 161.85 | 157.71 |
| 630 | 184.10 | 183.85 | 181.27 | 176.76 | 170.83 | 164.13 | 159.93 |
| 640 | 186.60 | 186.35 | 183.74 | 179.18 | 173.18 | 166.40 | 162.15 |
| 650 | 189.10 | 188.84 | 186.20 | 181.60 | 175.52 | 168.66 | 164.37 |
| 660 | 191.59 | 191.33 | 188.66 | 184.00 | 177.86 | 170.92 | 166.58 |
| 670 | 194.08 | 193.81 | 191.12 | 186.41 | 180.20 | 173.18 | 168.79 |
| 680 | 196.57 | 196.29 | 193.57 | 188.81 | 182.53 | 175.44 | 170.99 |
| 690 | 199.04 | 198.77 | 196.02 | 191.21 | 184.86 | 177.69 | 173.20 |
| 700 | 201.52 | 201.24 | 198.46 | 193.60 | 187.19 | 179.94 | 175.39 |
| 710 | 203.99 | 203.71 | 200.90 | 195.99 | 189.51 | 182.18 | 177.59 |
| 720 | 206.46 | 206.17 | 203.34 | 198.37 | 191.83 | 184.42 | 179.78 |
| 730 | 208.92 | 208.63 | 205.77 | 200.75 | 194.14 | 186.66 | 181.97 |
| 740 | 211.38 | 211.09 | 208.19 | 203.13 | 196.45 | 188.89 | 184.15 |
| 750 | 213.83 | 213.54 | 210.62 | 205.50 | 198.76 | 191.12 | 186.34 |
| 760 | 216.28 | 215.99 | 213.04 | 207.87 | 201.06 | 193.35 | 188.52 |
| 770 | 218.73 | 218.43 | 215.45 | 210.24 | 203.36 | 195.57 | 190.69 |
| 780 | 221.17 | 220.87 | 217.86 | 212.60 | 205.66 | 197.80 | 192.87 |
| 790 | 223.61 | 223.31 | 220.27 | 214.96 | 207.95 | 200.01 | 195.04 |
| 800 | 226.05 | 225.74 | 222.68 | 217.32 | 210.24 | 202.23 | 197.20 |
| 810 | 228.48 | 228.17 | 225.08 | 219.67 | 212.53 | 204.44 | 199.37 |
| 820 | 230.91 | 230.60 | 227.48 | 222.02 | 214.81 | 206.65 | 201.53 |
| 830 | 233.33 | 233.02 | 229.87 | 224.36 | 217.09 | 208.86 | 203.69 |
| 840 | 235.76 | 235.44 | 232.26 | 226.71 | 219.37 | 211.06 | 205.85 |
| 850 | 238.18 | 237.86 | 234.65 | 229.05 | 221.64 | 213.26 | 208.00 |
| 860 | 240.59 | 240.27 | 237.04 | 231.38 | 223.92 | 215.46 | 210.15 |
| 870 | 243.00 | 242.68 | 239.42 | 233.72 | 226.19 | 217.65 | 212.30 |
| 880 | 245.41 | 245.08 | 241.80 | 236.05 | 228.45 | 219.85 | 214.44 |
| 890 | 247.82 | 247.49 | 244.17 | 238.37 | 230.72 | 222.04 | 216.59 |
| 900 | 250.22 | 249.89 | 246.55 | 240.70 | 232.98 | 224.22 | 218.73 |
| 910 | 252.62 | 252.28 | 248.92 | 243.02 | 235.23 | 226.41 | 220.87 |
| 920 | 255.01 | 254.68 | 251.28 | 245.34 | 237.49 | 228.59 | 223.00 |
| 930 | 257.41 | 257.07 | 253.65 | 247.66 | 239.74 | 230.77 | 225.14 |
| 940 | 259.80 | 259.46 | 256.01 | 249.97 | 241.99 | 232.95 | 227.27 |
| 950 | 262.19 | 261.84 | 258.36 | 252.28 | 244.24 | 235.12 | 229.39 |
| 960 | 264.57 | 264.22 | 260.72 | 254.59 | 246.48 | 237.29 | 231.52 |
| 970 | 266.95 | 266.60 | 263.07 | 256.89 | 248.72 | 239.46 | 233.65 |
| 980 | 269.33 | 268.98 | 265.42 | 259.19 | 250.96 | 241.63 | 235.77 |
| 990 | 271.71 | 271.35 | 267.77 | 261.49 | 253.20 | 243.80 | 237.89 |
| 1,000 | 274.08 | 273.72 | 270.11 | 263.79 | 255.44 | 245.96 | 240.00 |
| 1,010 | 276.35 | 275.99 | 272.35 | 265.99 | 257.58 | 248.04 | 242.04 |

## TABLE 4–8. Continued

| $t_D$ | $z'_D$ | | | | | | |
|---|---|---|---|---|---|---|---|
| | 0.05 | 0.1 | 0.3 | 0.5 | 0.7 | 0.9 | 1.0 |
| 1,020 | 278.72 | 278.35 | 274.69 | 268.29 | 259.81 | 250.19 | 244.15 |
| 1,030 | 281.08 | 280.72 | 277.03 | 270.57 | 262.04 | 252.35 | 246.26 |
| 1,040 | 283.44 | 283.08 | 279.36 | 272.86 | 264.26 | 254.50 | 248.37 |
| 1,050 | 285.81 | 285.43 | 281.69 | 275.15 | 266.49 | 256.66 | 250.48 |
| 1,060 | 288.16 | 287.79 | 284.02 | 277.43 | 268.71 | 258.81 | 252.58 |
| 1,070 | 290.52 | 290.14 | 286.35 | 279.71 | 270.92 | 260.95 | 254.69 |
| 1,080 | 292.87 | 292.49 | 288.67 | 281.99 | 273.14 | 263.10 | 256.79 |
| 1,090 | 295.22 | 294.84 | 290.99 | 284.26 | 275.35 | 265.24 | 258.89 |
| 1,100 | 297.57 | 297.18 | 293.31 | 286.54 | 277.57 | 267.38 | 260.98 |
| 1,110 | 299.91 | 299.53 | 295.63 | 288.81 | 279.78 | 269.52 | 263.08 |
| 1,120 | 302.26 | 301.87 | 297.94 | 291.07 | 281.98 | 271.66 | 265.17 |
| 1,130 | 304.60 | 304.20 | 300.25 | 293.34 | 284.19 | 273.80 | 267.26 |
| 1,140 | 306.93 | 306.54 | 302.56 | 295.61 | 286.39 | 275.93 | 269.35 |
| 1,150 | 309.27 | 308.87 | 304.87 | 297.87 | 288.59 | 278.06 | 271.44 |
| 1,160 | 311.60 | 311.20 | 307.18 | 300.13 | 290.79 | 280.19 | 273.52 |
| 1,170 | 313.94 | 313.53 | 309.48 | 302.38 | 292.99 | 282.32 | 275.61 |
| 1,180 | 316.26 | 315.86 | 311.78 | 304.64 | 295.19 | 284.44 | 277.69 |
| 1,190 | 318.59 | 318.18 | 314.08 | 306.89 | 297.38 | 286.57 | 279.77 |
| 1,200 | 320.92 | 320.51 | 316.38 | 309.15 | 299.57 | 288.69 | 281.85 |
| 1,210 | 323.24 | 322.83 | 318.67 | 311.39 | 301.76 | 290.81 | 283.92 |
| 1,220 | 325.56 | 325.14 | 320.96 | 313.64 | 303.95 | 292.93 | 286.00 |
| 1,230 | 327.88 | 327.46 | 323.25 | 315.89 | 306.13 | 295.05 | 288.07 |
| 1,240 | 330.19 | 329.77 | 325.54 | 318.13 | 308.32 | 297.16 | 290.14 |
| 1,250 | 332.51 | 332.08 | 327.83 | 320.37 | 310.50 | 299.27 | 292.21 |
| 1,260 | 334.82 | 334.39 | 330.11 | 322.61 | 312.68 | 301.38 | 294.28 |
| 1,270 | 337.13 | 336.70 | 332.39 | 324.85 | 314.85 | 303.49 | 296.35 |
| 1,280 | 339.44 | 339.01 | 334.67 | 327.08 | 317.03 | 305.60 | 298.41 |
| 1,290 | 341.74 | 341.31 | 336.95 | 329.32 | 319.21 | 307.71 | 300.47 |
| 1,300 | 344.05 | 343.61 | 339.23 | 331.55 | 321.38 | 309.81 | 302.54 |
| 1,310 | 346.35 | 345.91 | 341.50 | 333.78 | 323.55 | 311.92 | 304.60 |
| 1,320 | 348.65 | 348.21 | 343.77 | 336.01 | 325.72 | 314.02 | 306.65 |
| 1,330 | 350.95 | 350.50 | 346.04 | 338.23 | 327.89 | 316.12 | 308.71 |
| 1,340 | 353.24 | 352.80 | 348.31 | 340.46 | 330.05 | 318.22 | 310.77 |
| 1,350 | 355.54 | 355.09 | 350.58 | 342.68 | 332.21 | 320.31 | 312.82 |
| 1,360 | 357.83 | 357.38 | 352.84 | 344.90 | 334.38 | 322.41 | 314.87 |
| 1,370 | 360.12 | 359.67 | 355.11 | 347.12 | 336.54 | 324.50 | 316.92 |
| 1,380 | 362.41 | 361.95 | 357.37 | 349.34 | 338.70 | 326.59 | 318.97 |
| 1,390 | 364.69 | 364.24 | 359.63 | 351.56 | 340.85 | 328.68 | 321.02 |
| 1,400 | 366.98 | 366.52 | 361.88 | 353.77 | 343.01 | 330.77 | 323.06 |
| 1,410 | 369.26 | 368.80 | 364.14 | 355.98 | 345.16 | 332.86 | 325.11 |
| 1,420 | 371.54 | 371.08 | 366.40 | 358.19 | 347.32 | 334.94 | 327.15 |
| 1,430 | 373.82 | 373.35 | 368.65 | 360.40 | 349.47 | 337.03 | 329.19 |
| 1,440 | 376.10 | 375.63 | 370.90 | 362.61 | 351.62 | 339.11 | 331.23 |
| 1,450 | 378.38 | 377.90 | 373.15 | 364.81 | 353.76 | 341.19 | 333.27 |
| 1,460 | 380.65 | 380.17 | 375.39 | 367.02 | 355.91 | 343.27 | 335.31 |
| 1,470 | 382.92 | 382.44 | 377.64 | 369.22 | 358.06 | 345.35 | 337.35 |
| 1,480 | 385.19 | 384.71 | 379.88 | 371.42 | 360.20 | 347.43 | 339.38 |
| 1,490 | 387.46 | 386.98 | 382.13 | 373.62 | 362.34 | 349.50 | 341.42 |
| 1,500 | 389.73 | 389.25 | 384.37 | 375.82 | 364.48 | 351.58 | 343.45 |
| 1,525 | 395.39 | 394.90 | 389.96 | 381.31 | 369.82 | 356.76 | 348.52 |
| 1,550 | 401.04 | 400.55 | 395.55 | 386.78 | 375.16 | 361.93 | 353.59 |
| 1,575 | 406.68 | 406.18 | 401.12 | 392.25 | 380.49 | 367.09 | 358.65 |
| 1,600 | 412.32 | 411.81 | 406.69 | 397.71 | 385.80 | 372.24 | 363.70 |
| 1,625 | 417.94 | 417.42 | 412.24 | 403.16 | 391.11 | 377.39 | 368.74 |
| 1,650 | 423.55 | 423.03 | 417.79 | 408.60 | 396.41 | 382.53 | 373.77 |
| 1,675 | 429.15 | 428.63 | 423.33 | 414.04 | 401.70 | 387.66 | 378.80 |
| 1,700 | 434.75 | 434.22 | 428.85 | 419.46 | 406.99 | 392.78 | 383.82 |
| 1,725 | 440.33 | 439.79 | 434.37 | 424.87 | 412.26 | 397.89 | 388.83 |
| 1,750 | 445.91 | 445.37 | 439.89 | 430.28 | 417.53 | 403.00 | 393.84 |
| 1,775 | 451.48 | 450.93 | 445.39 | 435.68 | 422.79 | 408.10 | 398.84 |

## TABLE 4–8. Continued

| | $z'_D$ | | | | | | |
|---|---|---|---|---|---|---|---|
| $t_D$ | 0.05 | 0.1 | 0.3 | 0.5 | 0.7 | 0.9 | 1.0 |
| 1,880 | 457.04 | 456.48 | 450.88 | 441.07 | 428.04 | 413.20 | 403.83 |
| 1,825 | 462.59 | 462.03 | 456.37 | 446.46 | 433.29 | 418.28 | 408.82 |
| 1,850 | 468.13 | 467.56 | 461.85 | 451.83 | 438.53 | 423.36 | 413.80 |
| 1,875 | 473.67 | 473.09 | 467.32 | 457.20 | 443.76 | 428.43 | 418.77 |
| 1,900 | 479.19 | 478.61 | 472.78 | 462.56 | 448.98 | 433.50 | 423.73 |
| 1,925 | 484.71 | 484.13 | 478.24 | 467.92 | 454.20 | 438.56 | 428.69 |
| 1,950 | 490.22 | 489.63 | 483.69 | 473.26 | 459.41 | 443.61 | 433.64 |
| 1,975 | 495.73 | 495.13 | 489.13 | 478.60 | 464.61 | 448.66 | 438.59 |
| 2,000 | 501.22 | 500.62 | 494.56 | 483.93 | 469.81 | 453.70 | 443.53 |
| 2,025 | 506.71 | 506.11 | 499.99 | 489.26 | 475.00 | 458.73 | 448.47 |
| 2,050 | 512.20 | 511.58 | 505.41 | 494.58 | 480.18 | 463.76 | 453.40 |
| 2,075 | 517.67 | 517.05 | 510.82 | 499.89 | 485.36 | 468.78 | 458.32 |
| 2,100 | 523.14 | 522.52 | 516.22 | 505.19 | 490.53 | 473.80 | 463.24 |
| 2,125 | 528.60 | 527.97 | 521.62 | 510.49 | 495.69 | 478.81 | 468.15 |
| 2,150 | 534.05 | 533.42 | 527.02 | 515.78 | 500.85 | 483.81 | 473.06 |
| 2,175 | 539.50 | 538.86 | 532.40 | 521.07 | 506.01 | 488.81 | 477.96 |
| 2,200 | 544.94 | 544.30 | 537.78 | 526.35 | 511.15 | 493.81 | 482.85 |
| 2,225 | 550.38 | 549.73 | 543.15 | 531.62 | 516.29 | 498.79 | 487.74 |
| 2,250 | 555.81 | 555.15 | 548.52 | 536.89 | 521.43 | 503.78 | 492.63 |
| 2,275 | 561.23 | 560.56 | 553.88 | 542.15 | 526.56 | 508.75 | 497.51 |
| 2,300 | 566.64 | 565.97 | 559.23 | 547.41 | 531.68 | 513.72 | 502.38 |
| 2,325 | 572.05 | 571.38 | 564.58 | 552.66 | 536.80 | 518.69 | 507.25 |
| 2,350 | 577.46 | 576.78 | 569.92 | 557.90 | 541.91 | 523.65 | 512.12 |
| 2,375 | 582.85 | 582.17 | 575.26 | 563.14 | 547.02 | 528.61 | 516.98 |
| 2,400 | 588.24 | 587.55 | 580.59 | 568.37 | 552.12 | 533.56 | 521.83 |
| 2,425 | 593.63 | 592.93 | 585.91 | 573.60 | 557.22 | 538.50 | 526.68 |
| 2,450 | 599.01 | 598.31 | 591.23 | 578.82 | 562.31 | 543.45 | 531.53 |
| 2,475 | 604.38 | 603.68 | 596.55 | 584.04 | 567.39 | 548.38 | 536.37 |
| 2,500 | 609.75 | 609.04 | 601.85 | 589.25 | 572.47 | 553.31 | 541.20 |
| 2,550 | 620.47 | 619.75 | 612.45 | 599.65 | 582.62 | 563.16 | 550.86 |
| 2,600 | 631.17 | 630.43 | 623.03 | 610.04 | 592.75 | 572.99 | 560.50 |
| 2,650 | 641.84 | 641.10 | 633.59 | 620.40 | 602.86 | 582.80 | 570.13 |
| 2,700 | 652.50 | 651.74 | 644.12 | 630.75 | 612.95 | 592.60 | 579.73 |
| 2,750 | 663.13 | 662.37 | 654.64 | 641.07 | 623.02 | 602.37 | 589.32 |
| 2,800 | 673.75 | 672.97 | 665.14 | 651.38 | 633.07 | 612.13 | 598.90 |
| 2,850 | 684.34 | 683.56 | 675.61 | 661.67 | 643.11 | 621.88 | 608.45 |
| 2,900 | 694.92 | 694.12 | 686.07 | 671.94 | 653.12 | 631.60 | 617.99 |
| 2,950 | 705.48 | 704.67 | 696.51 | 682.19 | 663.13 | 641.32 | 627.52 |
| 3,000 | 716.02 | 715.20 | 706.94 | 692.43 | 673.11 | 651.01 | 637.03 |
| 3,050 | 726.54 | 725.71 | 717.34 | 702.65 | 683.08 | 660.69 | 646.53 |
| 3,100 | 737.04 | 736.20 | 727.73 | 712.85 | 693.03 | 670.36 | 656.01 |
| 3,150 | 747.53 | 746.68 | 738.10 | 723.04 | 702.97 | 680.01 | 665.48 |
| 3,200 | 758.00 | 757.14 | 748.45 | 733.21 | 712.89 | 689.64 | 674.93 |
| 3,250 | 768.45 | 767.58 | 758.79 | 743.36 | 722.80 | 699.27 | 684.37 |
| 3,300 | 778.89 | 778.01 | 769.11 | 753.50 | 732.69 | 708.87 | 693.80 |
| 3,350 | 789.31 | 788.42 | 779.42 | 763.62 | 742.57 | 718.47 | 703.21 |
| 3,400 | 799.71 | 798.81 | 789.71 | 773.73 | 752.43 | 728.05 | 712.62 |
| 3,450 | 810.10 | 809.19 | 799.99 | 783.82 | 762.28 | 737.62 | 722.00 |
| 3,500 | 820.48 | 819.55 | 810.25 | 793.90 | 772.12 | 747.17 | 731.38 |
| 3,550 | 830.83 | 829.90 | 820.49 | 803.97 | 781.94 | 756.72 | 740.74 |
| 3,600 | 841.18 | 840.24 | 830.73 | 814.02 | 791.75 | 766.24 | 750.09 |
| 3,650 | 851.51 | 850.56 | 840.94 | 824.06 | 801.55 | 775.76 | 759.43 |
| 3,700 | 861.83 | 860.86 | 851.15 | 834.08 | 811.33 | 785.27 | 768.76 |
| 3,750 | 872.13 | 871.15 | 861.34 | 844.09 | 821.10 | 794.76 | 778.08 |
| 3,800 | 882.41 | 881.43 | 871.51 | 854.09 | 830.86 | 804.24 | 787.38 |
| 3,850 | 892.69 | 891.70 | 881.68 | 864.08 | 840.61 | 813.71 | 796.68 |
| 3,900 | 902.95 | 901.95 | 891.83 | 874.05 | 850.34 | 823.17 | 805.96 |
| 3,950 | 913.20 | 912.19 | 901.96 | 884.01 | 860.06 | 832.62 | 815.23 |
| 4,000 | 923.43 | 922.41 | 912.09 | 893.96 | 869.77 | 842.06 | 824.49 |
| 4,050 | 933.65 | 932.62 | 922.20 | 903.89 | 879.47 | 851.48 | 833.74 |

## TABLE 4–8. Continued

| | $z'_D$ | | | | | | |
|---|---|---|---|---|---|---|---|
| $t_D$ | 0.05 | 0.1 | 0.3 | 0.5 | 0.7 | 0.9 | 1.0 |
| 4,100 | 943.86 | 942.82 | 932.30 | 913.82 | 889.16 | 860.90 | 842.99 |
| 4,150 | 954.06 | 953.01 | 942.39 | 923.73 | 898.84 | 870.30 | 852.22 |
| 4,200 | 964.25 | 963.19 | 952.47 | 933.63 | 908.50 | 879.69 | 861.44 |
| 4,250 | 974.42 | 973.35 | 962.53 | 943.52 | 918.16 | 889.08 | 870.65 |
| 4,300 | 984.58 | 983.50 | 972.58 | 953.40 | 927.80 | 898.45 | 879.85 |
| 4,350 | 994.73 | 993.64 | 982.62 | 963.27 | 937.43 | 907.81 | 889.04 |
| 4,400 | 1,004.9 | 1,003.8 | 992.7 | 973.1 | 947.1 | 917.2 | 898.2 |
| 4,450 | 1,015.0 | 1,013.9 | 1,002.7 | 983.0 | 956.7 | 926.5 | 907.4 |
| 4,500 | 1,025.1 | 1,024.0 | 1,012.7 | 992.8 | 966.3 | 935.9 | 916.6 |
| 4,550 | 1,035.2 | 1,034.1 | 1,022.7 | 1,002.6 | 975.9 | 945.2 | 925.7 |
| 4,600 | 1,045.3 | 1,044.2 | 1,032.7 | 1,012.4 | 985.5 | 954.5 | 934.9 |
| 4,650 | 1,055.4 | 1,054.2 | 1,042.6 | 1,022.2 | 995.0 | 963.8 | 944.0 |
| 4,700 | 1,065.5 | 1,064.3 | 1,052.6 | 1,032.0 | 1,004.6 | 973.1 | 953.1 |
| 4,750 | 1,075.5 | 1,074.4 | 1,062.6 | 1,041.8 | 1,014.1 | 982.4 | 962.2 |
| 4,800 | 1,085.6 | 1,084.4 | 1,072.5 | 1,051.6 | 1,023.7 | 991.7 | 971.4 |
| 4,850 | 1,095.6 | 1,094.4 | 1,082.4 | 1,061.4 | 1,033.2 | 1,000.9 | 980.5 |
| 4,900 | 1,105.6 | 1,104.5 | 1,092.4 | 1,071.1 | 1,042.8 | 1,010.2 | 989.5 |
| 4,950 | 1,115.7 | 1,114.5 | 1,102.3 | 1,080.9 | 1,052.3 | 1,019.4 | 998.6 |
| 5,000 | 1,125.7 | 1,124.5 | 1,112.2 | 1,090.6 | 1,061.8 | 1,028.7 | 1,007.7 |
| 5,100 | 1,145.7 | 1,144.4 | 1,132.0 | 1,110.0 | 1,080.8 | 1,047.2 | 1,025.8 |
| 5,200 | 1,165.6 | 1,164.4 | 1,151.7 | 1,129.4 | 1,099.7 | 1,065.6 | 1,043.9 |
| 5,300 | 1,185.5 | 1,184.3 | 1,171.4 | 1,148.8 | 1,118.6 | 1,084.0 | 1,062.0 |
| 5,400 | 1,205.4 | 1,204.1 | 1,191.1 | 1,168.2 | 1,137.5 | 1,102.4 | 1,080.0 |
| 5,500 | 1,225.3 | 1,224.0 | 1,210.7 | 1,187.5 | 1,156.4 | 1,120.7 | 1,098.0 |
| 5,600 | 1,245.1 | 1,243.7 | 1,230.3 | 1,206.7 | 1,175.2 | 1,139.0 | 1,116.0 |
| 5,700 | 1,264.9 | 1,263.5 | 1,249.9 | 1,226.0 | 1,194.0 | 1,157.3 | 1,134.0 |
| 5,800 | 1,284.6 | 1,283.2 | 1,269.4 | 1,245.2 | 1,212.8 | 1,175.5 | 1,151.9 |
| 5,900 | 1,304.3 | 1,302.9 | 1,288.9 | 1,264.4 | 1,231.5 | 1,193.8 | 1,169.8 |
| 6,000 | 1,324.0 | 1,322.6 | 1,308.4 | 1,283.5 | 1,250.2 | 1,211.9 | 1,187.7 |
| 6,100 | 1,343.6 | 1,342.2 | 1,327.9 | 1,302.6 | 1,268.9 | 1,230.1 | 1,205.5 |
| 6,200 | 1,363.2 | 1,361.8 | 1,347.3 | 1,321.7 | 1,287.5 | 1,248.3 | 1,223.3 |
| 6,300 | 1,382.8 | 1,381.4 | 1,366.7 | 1,340.8 | 1,306.2 | 1,266.4 | 1,241.1 |
| 6,400 | 1,402.4 | 1,400.9 | 1,386.0 | 1,359.8 | 1,324.7 | 1,284.5 | 1,258.9 |
| 6,500 | 1,421.9 | 1,420.4 | 1,405.3 | 1,378.8 | 1,343.3 | 1,302.5 | 1,276.6 |
| 6,600 | 1,441.4 | 1,439.9 | 1,424.6 | 1,397.8 | 1,361.9 | 1,320.6 | 1,294.3 |
| 6,700 | 1,460.9 | 1,459.4 | 1,443.9 | 1,416.7 | 1,380.4 | 1,338.6 | 1,312.0 |
| 6,800 | 1,480.3 | 1,478.8 | 1,463.1 | 1,435.6 | 1,398.9 | 1,356.6 | 1,329.7 |
| 6,900 | 1,499.7 | 1,498.2 | 1,482.4 | 1,454.5 | 1,417.3 | 1,374.5 | 1,347.4 |
| 7,000 | 1,519.1 | 1,517.5 | 1,501.5 | 1,473.4 | 1,435.8 | 1,392.5 | 1,365.0 |
| 7,100 | 1,538.5 | 1,536.9 | 1,520.7 | 1,492.3 | 1,454.2 | 1,410.4 | 1,382.6 |
| 7,200 | 1,557.8 | 1,556.2 | 1,539.8 | 1,511.1 | 1,472.6 | 1,428.3 | 1,400.2 |
| 7,300 | 1,577.1 | 1,575.5 | 1,559.0 | 1,529.9 | 1,491.0 | 1,446.2 | 1,417.8 |
| 7,400 | 1,596.4 | 1,594.8 | 1,578.1 | 1,548.6 | 1,509.3 | 1,464.1 | 1,435.3 |
| 7,500 | 1,615.7 | 1,614.0 | 1,597.1 | 1,567.4 | 1,527.6 | 1,481.9 | 1,452.8 |
| 7,600 | 1,634.9 | 1,633.2 | 1,616.2 | 1,586.1 | 1,545.9 | 1,499.7 | 1,470.3 |
| 7,700 | 1,654.1 | 1,652.4 | 1,635.2 | 1,604.8 | 1,564.2 | 1,517.5 | 1,487.8 |
| 7,800 | 1,673.3 | 1,671.6 | 1,654.2 | 1,623.5 | 1,582.5 | 1,535.3 | 1,505.3 |
| 7,900 | 1,692.5 | 1,690.7 | 1,673.1 | 1,642.2 | 1,600.7 | 1,553.0 | 1,522.7 |
| 8,000 | 1,711.6 | 1,709.9 | 1,692.1 | 1,660.8 | 1,619.0 | 1,570.8 | 1,540.1 |
| 8,100 | 1,730.8 | 1,729.0 | 1,711.0 | 1,679.4 | 1,637.2 | 1,588.5 | 1,557.6 |
| 8,200 | 1,749.9 | 1,748.1 | 1,729.9 | 1,698.0 | 1,655.3 | 1,606.2 | 1,574.9 |
| 8,300 | 1,768.9 | 1,767.1 | 1,748.8 | 1,716.6 | 1,673.5 | 1,623.9 | 1,592.3 |
| 8,400 | 1,788.0 | 1,786.2 | 1,767.7 | 1,735.2 | 1,691.6 | 1,641.5 | 1,609.7 |
| 8,500 | 1,807.0 | 1,805.2 | 1,786.5 | 1,753.7 | 1,709.8 | 1,659.2 | 1,627.0 |
| 8,600 | 1,826.0 | 1,824.2 | 1,805.4 | 1,772.2 | 1,727.9 | 1,676.8 | 1,644.3 |
| 8,700 | 1,845.0 | 1,843.2 | 1,824.2 | 1,790.7 | 1,746.0 | 1,694.4 | 1,661.6 |
| 8,800 | 1,864.0 | 1,862.1 | 1,842.9 | 1,809.2 | 1,764.0 | 1,712.0 | 1,678.9 |
| 8,900 | 1,883.0 | 1,881.1 | 1,861.7 | 1,827.7 | 1,782.1 | 1,729.6 | 1,696.2 |
| 9,000 | 1,901.9 | 1,900.0 | 1,880.5 | 1,846.1 | 1,800.1 | 1,747.1 | 1,713.4 |
| 9,100 | 1,920.8 | 1,918.9 | 1,899.2 | 1,864.5 | 1,818.1 | 1,764.7 | 1,730.7 |

## TABLE 4–8. Continued

| | $z'_D$ | | | | | | |
|---|---|---|---|---|---|---|---|
| $t_D$ | 0.05 | 0.1 | 0.3 | 0.5 | 0.7 | 0.9 | 1.0 |
| 9,200 | 1,939.7 | 1,937.4 | 1,917.9 | 1,882.9 | 1,836.1 | 1,782.2 | 1,747.9 |
| 9,300 | 1,958.6 | 1,956.6 | 1,936.6 | 1,901.3 | 1,854.1 | 1,799.7 | 1,765.1 |
| 9,400 | 1,977.4 | 1,975.4 | 1,955.2 | 1,919.7 | 1,872.0 | 1,817.2 | 1,782.3 |
| 9,500 | 1,996.3 | 1,994.3 | 1,973.9 | 1,938.0 | 1,890.0 | 1,834.7 | 1,799.4 |
| 9,600 | 2,015.1 | 2,013.1 | 1,992.5 | 1,956.4 | 1,907.9 | 1,852.1 | 1,816.6 |
| 9,700 | 2,033.9 | 2,031.9 | 2,011.1 | 1,974.7 | 1,925.8 | 1,869.6 | 1,833.7 |
| 9,800 | 2,052.7 | 2,050.6 | 2,029.7 | 1,993.0 | 1,943.7 | 1,887.0 | 1,850.9 |
| 9,900 | 2,071.5 | 2,069.4 | 2,048.3 | 2,011.3 | 1,961.6 | 1,904.4 | 1,868.0 |
| $1.00 \times 10^{4}$ | $2.090 \times 10^{3}$ | $2.088 \times 10^{3}$ | $2.067 \times 10^{3}$ | $2.029 \times 10^{3}$ | $1.979 \times 10^{3}$ | $1.922 \times 10^{3}$ | $1.885 \times 10^{3}$ |
| $1.25 \times 10^{4}$ | $2.553 \times 10^{3}$ | $2.551 \times 10^{3}$ | $2.526 \times 10^{3}$ | $2.481 \times 10^{3}$ | $2.421 \times 10^{3}$ | $2.352 \times 10^{3}$ | $2.308 \times 10^{3}$ |
| $1.50 \times 10^{4}$ | $3.009 \times 10^{3}$ | $3.006 \times 10^{3}$ | $2.977 \times 10^{3}$ | $2.925 \times 10^{3}$ | $2.855 \times 10^{3}$ | $2.775 \times 10^{3}$ | $2.724 \times 10^{3}$ |
| $1.75 \times 10^{4}$ | $3.457 \times 10^{3}$ | $3.454 \times 10^{3}$ | $3.421 \times 10^{3}$ | $3.362 \times 10^{3}$ | $3.284 \times 10^{3}$ | $3.193 \times 10^{3}$ | $3.135 \times 10^{3}$ |
| $2.00 \times 10^{4}$ | $3.900 \times 10^{3}$ | $3.897 \times 10^{3}$ | $3.860 \times 10^{3}$ | $3.794 \times 10^{3}$ | $3.707 \times 10^{3}$ | $3.605 \times 10^{3}$ | $3.541 \times 10^{3}$ |
| $2.50 \times 10^{4}$ | $4.773 \times 10^{3}$ | $4.768 \times 10^{3}$ | $4.724 \times 10^{3}$ | $4.646 \times 10^{3}$ | $4.541 \times 10^{3}$ | $4.419 \times 10^{3}$ | $4.341 \times 10^{3}$ |
| $3.00 \times 10^{4}$ | $5.630 \times 10^{3}$ | $5.625 \times 10^{3}$ | $5.574 \times 10^{3}$ | $5.483 \times 10^{3}$ | $5.361 \times 10^{3}$ | $5.219 \times 10^{3}$ | $5.129 \times 10^{3}$ |
| $3.50 \times 10^{4}$ | $6.476 \times 10^{3}$ | $6.470 \times 10^{3}$ | $6.412 \times 10^{3}$ | $6.309 \times 10^{3}$ | $6.170 \times 10^{3}$ | $6.009 \times 10^{3}$ | $5.906 \times 10^{3}$ |
| $4.00 \times 10^{4}$ | $7.312 \times 10^{3}$ | $7.305 \times 10^{3}$ | $7.240 \times 10^{3}$ | $7.125 \times 10^{3}$ | $6.970 \times 10^{3}$ | $6.790 \times 10^{3}$ | $6.675 \times 10^{3}$ |
| $4.50 \times 10^{4}$ | $8.139 \times 10^{3}$ | $8.132 \times 10^{3}$ | $8.060 \times 10^{3}$ | $7.933 \times 10^{3}$ | $7.762 \times 10^{3}$ | $7.564 \times 10^{3}$ | $7.437 \times 10^{3}$ |
| $5.00 \times 10^{4}$ | $8.959 \times 10^{3}$ | $8.951 \times 10^{3}$ | $8.872 \times 10^{3}$ | $8.734 \times 10^{3}$ | $8.548 \times 10^{3}$ | $8.331 \times 10^{3}$ | $8.193 \times 10^{3}$ |
| $6.00 \times 10^{4}$ | $1.057 \times 10^{3}$ | $1.057 \times 10^{3}$ | $1.047 \times 10^{3}$ | $1.031 \times 10^{3}$ | $1.010 \times 10^{3}$ | $9.846 \times 10^{3}$ | $9.684 \times 10^{3}$ |
| $7.00 \times 10^{4}$ | $1.217 \times 10^{4}$ | $1.217 \times 10^{4}$ | $1.206 \times 10^{4}$ | $1.188 \times 10^{4}$ | $1.163 \times 10^{4}$ | $1.134 \times 10^{4}$ | $1.116 \times 10^{4}$ |
| $8.00 \times 10^{4}$ | $1.375 \times 10^{4}$ | $1.375 \times 10^{4}$ | $1.363 \times 10^{4}$ | $1.342 \times 10^{4}$ | $1.315 \times 10^{4}$ | $1.283 \times 10^{4}$ | $1.262 \times 10^{4}$ |
| $9.00 \times 10^{4}$ | $1.532 \times 10^{4}$ | $1.531 \times 10^{4}$ | $1.518 \times 10^{4}$ | $1.496 \times 10^{4}$ | $1.465 \times 10^{4}$ | $1.430 \times 10^{4}$ | $1.407 \times 10^{4}$ |
| $1.00 \times 10^{5}$ | $1.687 \times 10^{4}$ | $1.686 \times 10^{4}$ | $1.672 \times 10^{4}$ | $1.647 \times 10^{4}$ | $1.614 \times 10^{4}$ | $1.576 \times 10^{4}$ | $1.551 \times 10^{4}$ |
| $1.25 \times 10^{5}$ | $2.071 \times 10^{4}$ | $2.069 \times 10^{4}$ | $2.052 \times 10^{4}$ | $2.023 \times 10^{4}$ | $1.982 \times 10^{4}$ | $1.936 \times 10^{4}$ | $1.906 \times 10^{4}$ |
| $1.50 \times 10^{5}$ | $2.448 \times 10^{4}$ | $2.446 \times 10^{4}$ | $2.427 \times 10^{4}$ | $2.392 \times 10^{4}$ | $2.345 \times 10^{4}$ | $2.291 \times 10^{4}$ | $2.256 \times 10^{4}$ |
| $2.00 \times 10^{5}$ | $3.190 \times 10^{4}$ | $3.188 \times 10^{4}$ | $3.163 \times 10^{4}$ | $3.119 \times 10^{4}$ | $3.059 \times 10^{4}$ | $2.989 \times 10^{4}$ | $2.945 \times 10^{4}$ |
| $2.50 \times 10^{5}$ | $3.918 \times 10^{4}$ | $3.916 \times 10^{4}$ | $3.885 \times 10^{4}$ | $3.832 \times 10^{4}$ | $3.760 \times 10^{4}$ | $3.676 \times 10^{4}$ | $3.622 \times 10^{4}$ |
| $3.00 \times 10^{5}$ | $4.636 \times 10^{4}$ | $4.633 \times 10^{4}$ | $4.598 \times 10^{4}$ | $4.536 \times 10^{4}$ | $4.452 \times 10^{4}$ | $4.353 \times 10^{4}$ | $4.290 \times 10^{4}$ |
| $4.00 \times 10^{5}$ | $6.048 \times 10^{4}$ | $6.044 \times 10^{4}$ | $5.999 \times 10^{4}$ | $5.920 \times 10^{4}$ | $5.812 \times 10^{4}$ | $5.687 \times 10^{4}$ | $5.606 \times 10^{4}$ |
| $5.00 \times 10^{5}$ | $7.436 \times 10^{4}$ | $7.431 \times 10^{4}$ | $7.376 \times 10^{4}$ | $7.280 \times 10^{4}$ | $7.150 \times 10^{4}$ | $6.998 \times 10^{4}$ | $6.900 \times 10^{4}$ |
| $6.00 \times 10^{5}$ | $8.805 \times 10^{4}$ | $8.798 \times 10^{4}$ | $8.735 \times 10^{4}$ | $8.623 \times 10^{4}$ | $8.471 \times 10^{4}$ | $8.293 \times 10^{4}$ | $8.178 \times 10^{4}$ |
| $7.00 \times 10^{5}$ | $1.016 \times 10^{5}$ | $1.015 \times 10^{5}$ | $1.008 \times 10^{5}$ | $9.951 \times 10^{4}$ | $9.777 \times 10^{4}$ | $9.573 \times 10^{4}$ | $9.442 \times 10^{4}$ |
| $8.00 \times 10^{5}$ | $1.150 \times 10^{5}$ | $1.149 \times 10^{5}$ | $1.141 \times 10^{5}$ | $1.127 \times 10^{5}$ | $1.107 \times 10^{5}$ | $1.084 \times 10^{5}$ | $1.070 \times 10^{5}$ |
| $9.00 \times 10^{5}$ | $1.283 \times 10^{5}$ | $1.282 \times 10^{5}$ | $1.273 \times 10^{5}$ | $1.257 \times 10^{5}$ | $1.235 \times 10^{5}$ | $1.210 \times 10^{5}$ | $1.194 \times 10^{5}$ |
| $1.00 \times 10^{6}$ | $1.415 \times 10^{5}$ | $1.412 \times 10^{5}$ | $1.404 \times 10^{5}$ | $1.387 \times 10^{5}$ | $1.363 \times 10^{5}$ | $1.335 \times 10^{5}$ | $1.317 \times 10^{5}$ |
| $1.50 \times 10^{6}$ | $2.059 \times 10^{5}$ | $2.060 \times 10^{5}$ | $2.041 \times 10^{5}$ | $2.016 \times 10^{5}$ | $1.982 \times 10^{5}$ | $1.943 \times 10^{5}$ | $1.918 \times 10^{5}$ |
| $2.00 \times 10^{6}$ | $2.695 \times 10^{5}$ | $2.695 \times 10^{5}$ | $2.676 \times 10^{5}$ | $2.644 \times 10^{5}$ | $2.601 \times 10^{5}$ | $2.551 \times 10^{5}$ | $2.518 \times 10^{5}$ |
| $2.50 \times 10^{6}$ | $3.320 \times 10^{5}$ | $3.319 \times 10^{5}$ | $3.296 \times 10^{5}$ | $3.254 \times 10^{5}$ | $3.202 \times 10^{5}$ | $3.141 \times 10^{5}$ | $3.101 \times 10^{5}$ |
| $3.00 \times 10^{6}$ | $3.937 \times 10^{5}$ | $3.936 \times 10^{5}$ | $3.909 \times 10^{5}$ | $3.864 \times 10^{5}$ | $3.803 \times 10^{5}$ | $3.731 \times 10^{5}$ | $3.684 \times 10^{5}$ |
| $4.00 \times 10^{6}$ | $5.154 \times 10^{5}$ | $5.152 \times 10^{5}$ | $5.118 \times 10^{5}$ | $5.060 \times 10^{5}$ | $4.981 \times 10^{5}$ | $4.888 \times 10^{5}$ | $4.828 \times 10^{5}$ |
| $5.00 \times 10^{6}$ | $6.352 \times 10^{5}$ | $6.349 \times 10^{5}$ | $6.308 \times 10^{5}$ | $6.238 \times 10^{5}$ | $6.142 \times 10^{5}$ | $6.029 \times 10^{5}$ | $5.956 \times 10^{5}$ |
| $6.00 \times 10^{6}$ | $7.536 \times 10^{5}$ | $7.533 \times 10^{5}$ | $7.485 \times 10^{5}$ | $7.402 \times 10^{5}$ | $7.290 \times 10^{5}$ | $7.157 \times 10^{5}$ | $7.072 \times 10^{5}$ |
| $7.00 \times 10^{6}$ | $8.709 \times 10^{5}$ | $8.705 \times 10^{5}$ | $8.650 \times 10^{5}$ | $8.556 \times 10^{5}$ | $8.427 \times 10^{5}$ | $8.275 \times 10^{5}$ | $8.177 \times 10^{5}$ |
| $8.00 \times 10^{6}$ | $9.972 \times 10^{5}$ | $9.867 \times 10^{5}$ | $9.806 \times 10^{5}$ | $9.699 \times 10^{5}$ | $9.555 \times 10^{5}$ | $9.384 \times 10^{5}$ | $9.273 \times 10^{5}$ |
| $9.00 \times 10^{6}$ | $1.103 \times 10^{6}$ | $1.102 \times 10^{6}$ | $1.095 \times 10^{6}$ | $1.084 \times 10^{6}$ | $1.067 \times 10^{6}$ | $1.049 \times 10^{6}$ | $1.036 \times 10^{6}$ |
| $1.00 \times 10^{7}$ | $1.217 \times 10^{6}$ | $1.217 \times 10^{6}$ | $1.209 \times 10^{6}$ | $1.196 \times 10^{6}$ | $1.179 \times 10^{6}$ | $1.158 \times 10^{6}$ | $1.144 \times 10^{6}$ |
| $1.50 \times 10^{7}$ | $1.782 \times 10^{6}$ | $1.781 \times 10^{6}$ | $1.771 \times 10^{6}$ | $1.752 \times 10^{6}$ | $1.727 \times 10^{6}$ | $1.697 \times 10^{6}$ | $1.678 \times 10^{6}$ |
| $2.00 \times 10^{7}$ | $2.337 \times 10^{6}$ | $2.336 \times 10^{6}$ | $2.322 \times 10^{6}$ | $2.298 \times 10^{6}$ | $2.266 \times 10^{6}$ | $2.227 \times 10^{6}$ | $2.202 \times 10^{6}$ |
| $2.50 \times 10^{7}$ | $2.884 \times 10^{6}$ | $2.882 \times 10^{6}$ | $2.866 \times 10^{6}$ | $2.837 \times 10^{6}$ | $2.797 \times 10^{6}$ | $2.750 \times 10^{6}$ | $2.720 \times 10^{6}$ |
| $3.00 \times 10^{7}$ | $3.425 \times 10^{6}$ | $3.423 \times 10^{6}$ | $3.404 \times 10^{6}$ | $3.369 \times 10^{6}$ | $3.323 \times 10^{6}$ | $3.268 \times 10^{6}$ | $3.232 \times 10^{6}$ |
| $4.00 \times 10^{7}$ | $4.493 \times 10^{6}$ | $4.491 \times 10^{6}$ | $4.466 \times 10^{6}$ | $4.422 \times 10^{6}$ | $4.361 \times 10^{6}$ | $4.290 \times 10^{6}$ | $4.244 \times 10^{6}$ |
| $5.00 \times 10^{7}$ | $5.547 \times 10^{6}$ | $5.544 \times 10^{6}$ | $5.514 \times 10^{6}$ | $5.460 \times 10^{6}$ | $5.386 \times 10^{6}$ | $5.299 \times 10^{6}$ | $5.243 \times 10^{6}$ |
| $6.00 \times 10^{7}$ | $6.590 \times 10^{6}$ | $6.587 \times 10^{6}$ | $6.551 \times 10^{6}$ | $6.488 \times 10^{6}$ | $6.401 \times 10^{6}$ | $6.299 \times 10^{6}$ | $6.232 \times 10^{6}$ |
| $7.00 \times 10^{7}$ | $7.624 \times 10^{6}$ | $7.620 \times 10^{6}$ | $7.579 \times 10^{6}$ | $7.507 \times 10^{6}$ | $7.407 \times 10^{6}$ | $7.290 \times 10^{6}$ | $7.213 \times 10^{6}$ |
| $8.00 \times 10^{7}$ | $8.651 \times 10^{6}$ | $8.647 \times 10^{6}$ | $8.600 \times 10^{6}$ | $8.519 \times 10^{6}$ | $8.407 \times 10^{6}$ | $8.274 \times 10^{6}$ | $8.188 \times 10^{6}$ |
| $9.00 \times 10^{7}$ | $9.671 \times 10^{6}$ | $9.666 \times 10^{6}$ | $9.615 \times 10^{6}$ | $9.524 \times 10^{6}$ | $9.400 \times 10^{6}$ | $9.252 \times 10^{6}$ | $9.156 \times 10^{6}$ |
| $1.00 \times 10^{8}$ | $1.069 \times 10^{7}$ | $1.067 \times 10^{7}$ | $1.062 \times 10^{7}$ | $1.052 \times 10^{7}$ | $1.039 \times 10^{7}$ | $1.023 \times 10^{7}$ | $1.012 \times 10^{7}$ |
| $1.50 \times 10^{8}$ | $1.567 \times 10^{7}$ | $1.567 \times 10^{7}$ | $1.555 \times 10^{7}$ | $1.541 \times 10^{7}$ | $1.522 \times 10^{7}$ | $1.499 \times 10^{7}$ | $1.483 \times 10^{7}$ |
| $2.00 \times 10^{8}$ | $2.059 \times 10^{7}$ | $2.059 \times 10^{7}$ | $2.048 \times 10^{7}$ | $2.029 \times 10^{7}$ | $2.004 \times 10^{7}$ | $1.974 \times 10^{7}$ | $1.954 \times 10^{7}$ |

**TABLE 4–8. Continued**

| | $z_D'$ | | | | | | |
|---|---|---|---|---|---|---|---|
| $t_D$ | 0.05 | 0.1 | 0.3 | 0.5 | 0.7 | 0.9 | 1.0 |
| $2.50 \times 10^{8}$ | $2.546 \times 10^{7}$ | $2.545 \times 10^{7}$ | $2.531 \times 10^{7}$ | $2.507 \times 10^{7}$ | $2.476 \times 10^{7}$ | $2.439 \times 10^{7}$ | $2.415 \times 10^{7}$ |
| $3.00 \times 10^{8}$ | $3.027 \times 10^{7}$ | $3.026 \times 10^{7}$ | $3.010 \times 10^{7}$ | $2.984 \times 10^{7}$ | $2.947 \times 10^{7}$ | $2.904 \times 10^{7}$ | $2.875 \times 10^{7}$ |
| $4.00 \times 10^{8}$ | $3.979 \times 10^{7}$ | $3.978 \times 10^{7}$ | $3.958 \times 10^{7}$ | $3.923 \times 10^{7}$ | $3.875 \times 10^{7}$ | $3.819 \times 10^{7}$ | $3.782 \times 10^{7}$ |
| $5.00 \times 10^{8}$ | $4.920 \times 10^{7}$ | $4.918 \times 10^{7}$ | $4.894 \times 10^{7}$ | $4.851 \times 10^{7}$ | $4.793 \times 10^{7}$ | $4.724 \times 10^{7}$ | $4.679 \times 10^{7}$ |
| $6.00 \times 10^{8}$ | $5.852 \times 10^{7}$ | $5.850 \times 10^{7}$ | $5.821 \times 10^{7}$ | $5.771 \times 10^{7}$ | $5.702 \times 10^{7}$ | $5.621 \times 10^{7}$ | $5.568 \times 10^{7}$ |
| $7.00 \times 10^{8}$ | $6.777 \times 10^{7}$ | $6.774 \times 10^{7}$ | $6.741 \times 10^{7}$ | $6.684 \times 10^{7}$ | $6.605 \times 10^{7}$ | $6.511 \times 10^{7}$ | $6.450 \times 10^{7}$ |
| $8.00 \times 10^{8}$ | $7.700 \times 10^{7}$ | $7.693 \times 10^{7}$ | $7.655 \times 10^{7}$ | $7.590 \times 10^{7}$ | $7.501 \times 10^{7}$ | $7.396 \times 10^{7}$ | $7.327 \times 10^{7}$ |
| $9.00 \times 10^{8}$ | $8.609 \times 10^{7}$ | $8.606 \times 10^{7}$ | $8.564 \times 10^{7}$ | $8.492 \times 10^{7}$ | $8.393 \times 10^{7}$ | $8.275 \times 10^{7}$ | $8.199 \times 10^{7}$ |
| $1.00 \times 10^{9}$ | $9.518 \times 10^{7}$ | $9.515 \times 10^{7}$ | $9.469 \times 10^{7}$ | $9.390 \times 10^{7}$ | $9.281 \times 10^{7}$ | $9.151 \times 10^{7}$ | $9.066 \times 10^{7}$ |
| $1.50 \times 10^{9}$ | $1.401 \times 10^{8}$ | $1.400 \times 10^{8}$ | $1.394 \times 10^{8}$ | $1.382 \times 10^{8}$ | $1.367 \times 10^{8}$ | $1.348 \times 10^{8}$ | $1.336 \times 10^{8}$ |
| $2.00 \times 10^{9}$ | $1.843 \times 10^{8}$ | $1.843 \times 10^{8}$ | $1.834 \times 10^{8}$ | $1.819 \times 10^{8}$ | $1.799 \times 10^{8}$ | $1.774 \times 10^{8}$ | $1.758 \times 10^{8}$ |
| $2.50 \times 10^{9}$ | $2.281 \times 10^{8}$ | $2.280 \times 10^{8}$ | $2.269 \times 10^{8}$ | $2.251 \times 10^{8}$ | $2.226 \times 10^{8}$ | $2.196 \times 10^{8}$ | $2.177 \times 10^{8}$ |
| $3.00 \times 10^{9}$ | $2.714 \times 10^{8}$ | $2.713 \times 10^{8}$ | $2.701 \times 10^{8}$ | $2.680 \times 10^{8}$ | $2.650 \times 10^{8}$ | $2.615 \times 10^{8}$ | $2.592 \times 10^{8}$ |
| $4.00 \times 10^{9}$ | $3.573 \times 10^{8}$ | $3.572 \times 10^{8}$ | $3.556 \times 10^{8}$ | $3.528 \times 10^{8}$ | $3.489 \times 10^{8}$ | $3.443 \times 10^{8}$ | $3.413 \times 10^{8}$ |
| $5.00 \times 10^{9}$ | $4.422 \times 10^{8}$ | $4.421 \times 10^{8}$ | $4.401 \times 10^{8}$ | $4.367 \times 10^{8}$ | $4.320 \times 10^{8}$ | $4.263 \times 10^{8}$ | $4.227 \times 10^{8}$ |
| $6.00 \times 10^{9}$ | $5.265 \times 10^{8}$ | $5.262 \times 10^{8}$ | $5.240 \times 10^{8}$ | $5.199 \times 10^{8}$ | $5.143 \times 10^{8}$ | $5.077 \times 10^{8}$ | $5.033 \times 10^{8}$ |
| $7.00 \times 10^{9}$ | $6.101 \times 10^{8}$ | $6.098 \times 10^{8}$ | $6.072 \times 10^{8}$ | $6.025 \times 10^{8}$ | $5.961 \times 10^{8}$ | $5.885 \times 10^{8}$ | $5.835 \times 10^{8}$ |
| $8.00 \times 10^{9}$ | $6.932 \times 10^{8}$ | $6.930 \times 10^{8}$ | $6.900 \times 10^{8}$ | $6.847 \times 10^{8}$ | $6.775 \times 10^{8}$ | $6.688 \times 10^{8}$ | $6.632 \times 10^{8}$ |
| $9.00 \times 10^{9}$ | $7.760 \times 10^{8}$ | $7.756 \times 10^{8}$ | $7.723 \times 10^{8}$ | $7.664 \times 10^{8}$ | $7.584 \times 10^{8}$ | $7.487 \times 10^{8}$ | $7.424 \times 10^{8}$ |
| $1.00 \times 10^{10}$ | $8.583 \times 10^{8}$ | $8.574 \times 10^{8}$ | $8.543 \times 10^{8}$ | $8.478 \times 10^{8}$ | $8.389 \times 10^{8}$ | $8.283 \times 10^{8}$ | $8.214 \times 10^{8}$ |
| $1.50 \times 10^{10}$ | $1.263 \times 10^{9}$ | $1.264 \times 10^{9}$ | $1.257 \times 10^{9}$ | $1.247 \times 10^{9}$ | $1.235 \times 10^{9}$ | $1.219 \times 10^{9}$ | $1.209 \times 10^{9}$ |
| $2.00 \times 10^{10}$ | $1.666 \times 10^{9}$ | $1.666 \times 10^{9}$ | $1.659 \times 10^{9}$ | $1.646 \times 10^{9}$ | $1.630 \times 10^{9}$ | $1.610 \times 10^{9}$ | $1.596 \times 10^{9}$ |
| $2.50 \times 10^{10}$ | $2.065 \times 10^{9}$ | $2.063 \times 10^{9}$ | $2.055 \times 10^{9}$ | $2.038 \times 10^{9}$ | $2.018 \times 10^{9}$ | $1.993 \times 10^{9}$ | $1.977 \times 10^{9}$ |
| $3.00 \times 10^{10}$ | $2.458 \times 10^{9}$ | $2.458 \times 10^{9}$ | $2.447 \times 10^{9}$ | $2.430 \times 10^{9}$ | $2.405 \times 10^{9}$ | $2.376 \times 10^{9}$ | $2.357 \times 10^{9}$ |
| $4.00 \times 10^{10}$ | $3.240 \times 10^{9}$ | $3.239 \times 10^{9}$ | $3.226 \times 10^{9}$ | $3.203 \times 10^{9}$ | $3.171 \times 10^{9}$ | $3.133 \times 10^{9}$ | $3.108 \times 10^{9}$ |
| $5.00 \times 10^{10}$ | $4.014 \times 10^{9}$ | $4.013 \times 10^{9}$ | $3.997 \times 10^{9}$ | $3.968 \times 10^{9}$ | $3.929 \times 10^{9}$ | $3.883 \times 10^{9}$ | $3.852 \times 10^{9}$ |
| $6.00 \times 10^{10}$ | $4.782 \times 10^{9}$ | $4.781 \times 10^{9}$ | $4.762 \times 10^{9}$ | $4.728 \times 10^{9}$ | $4.682 \times 10^{9}$ | $4.627 \times 10^{9}$ | $4.591 \times 10^{9}$ |
| $7.00 \times 10^{10}$ | $5.546 \times 10^{9}$ | $5.544 \times 10^{9}$ | $5.522 \times 10^{9}$ | $5.483 \times 10^{9}$ | $5.430 \times 10^{9}$ | $5.366 \times 10^{9}$ | $5.345 \times 10^{9}$ |
| $8.00 \times 10^{10}$ | $6.305 \times 10^{9}$ | $6.303 \times 10^{9}$ | $6.278 \times 10^{9}$ | $6.234 \times 10^{9}$ | $6.174 \times 10^{9}$ | $6.102 \times 10^{9}$ | $6.055 \times 10^{9}$ |
| $9.00 \times 10^{10}$ | $7.060 \times 10^{9}$ | $7.058 \times 10^{9}$ | $7.030 \times 10^{9}$ | $6.982 \times 10^{9}$ | $6.914 \times 10^{9}$ | $6.834 \times 10^{9}$ | $6.782 \times 10^{9}$ |
| $1.00 \times 10^{11}$ | $7.813 \times 10^{9}$ | $7.810 \times 10^{9}$ | $7.780 \times 10^{10}$ | $7.726 \times 10^{9}$ | $7.652 \times 10^{10}$ | $7.564 \times 10^{9}$ | $7.506 \times 10^{9}$ |
| $1.50 \times 10^{11}$ | $1.154 \times 10^{10}$ | $1.153 \times 10^{10}$ | $1.149 \times 10^{10}$ | $1.141 \times 10^{10}$ | $1.130 \times 10^{10}$ | $1.118 \times 10^{10}$ | $1.109 \times 10^{10}$ |
| $2.00 \times 10^{11}$ | $1.522 \times 10^{10}$ | $1.521 \times 10^{10}$ | $1.515 \times 10^{10}$ | $1.505 \times 10^{10}$ | $1.491 \times 10^{10}$ | $1.474 \times 10^{10}$ | $1.463 \times 10^{10}$ |
| $2.50 \times 10^{11}$ | $1.886 \times 10^{10}$ | $1.885 \times 10^{10}$ | $1.878 \times 10^{10}$ | $1.866 \times 10^{10}$ | $1.849 \times 10^{10}$ | $1.828 \times 10^{10}$ | $1.814 \times 10^{10}$ |
| $3.00 \times 10^{11}$ | $2.248 \times 10^{10}$ | $2.247 \times 10^{10}$ | $2.239 \times 10^{10}$ | $2.224 \times 10^{10}$ | $2.204 \times 10^{10}$ | $2.179 \times 10^{10}$ | $2.163 \times 10^{10}$ |
| $4.00 \times 10^{11}$ | $2.965 \times 10^{10}$ | $2.964 \times 10^{10}$ | $2.953 \times 10^{10}$ | $2.934 \times 10^{10}$ | $2.907 \times 10^{10}$ | $2.876 \times 10^{10}$ | $2.855 \times 10^{10}$ |
| $5.00 \times 10^{11}$ | $3.677 \times 10^{10}$ | $3.675 \times 10^{10}$ | $3.662 \times 10^{10}$ | $3.638 \times 10^{10}$ | $3.605 \times 10^{10}$ | $3.566 \times 10^{10}$ | $3.540 \times 10^{10}$ |
| $6.00 \times 10^{11}$ | $4.383 \times 10^{10}$ | $4.381 \times 10^{10}$ | $4.365 \times 10^{10}$ | $4.337 \times 10^{10}$ | $4.298 \times 10^{10}$ | $4.252 \times 10^{10}$ | $4.221 \times 10^{10}$ |
| $7.00 \times 10^{11}$ | $5.085 \times 10^{10}$ | $5.082 \times 10^{10}$ | $5.064 \times 10^{10}$ | $5.032 \times 10^{10}$ | $4.987 \times 10^{10}$ | $4.933 \times 10^{10}$ | $4.898 \times 10^{10}$ |
| $8.00 \times 10^{11}$ | $5.783 \times 10^{10}$ | $5.781 \times 10^{10}$ | $5.760 \times 10^{10}$ | $5.723 \times 10^{10}$ | $5.673 \times 10^{10}$ | $5.612 \times 10^{10}$ | $5.572 \times 10^{10}$ |
| $9.00 \times 10^{11}$ | $6.478 \times 10^{10}$ | $6.476 \times 10^{10}$ | $6.453 \times 10^{10}$ | $6.412 \times 10^{10}$ | $6.355 \times 10^{10}$ | $6.288 \times 10^{10}$ | $6.243 \times 10^{10}$ |
| $1.00 \times 10^{12}$ | $7.171 \times 10^{10}$ | $7.168 \times 10^{10}$ | $7.143 \times 10^{10}$ | $7.098 \times 10^{10}$ | $7.035 \times 10^{10}$ | $6.961 \times 10^{10}$ | $6.912 \times 10^{10}$ |
| $1.50 \times 10^{12}$ | $1.060 \times 10^{11}$ | $1.060 \times 10^{11}$ | $1.056 \times 10^{11}$ | $1.050 \times 10^{11}$ | $1.041 \times 10^{11}$ | $1.030 \times 10^{11}$ | $1.022 \times 10^{11}$ |
| $2.00 \times 10^{12}$ | $1.400 \times 10^{11}$ | $1.399 \times 10^{11}$ | $1.394 \times 10^{11}$ | $1.386 \times 10^{11}$ | $1.374 \times 10^{11}$ | $1.359 \times 10^{11}$ | $1.350 \times 10^{11}$ |

## SIMULTANEOUS EVALUATION OF OOIP AND WATER INFLUX

Occasionally it may be necessary to estimate OOIP and cumulative water influx $W_e$ simultaneously. In general, the methods available are very sensitive to errors in basic data, so a large amount of accurate data is required. The data obtained after reservoir pressure has dropped below bubblepoint will be significant in evaluating the OOIP.

Two methods are in common use: (1) the van Everdingen, Timmerman, and McMahon Method[25] and (2) the Havlena-Odeh Method. For both methods, the aquifer is assumed to be infinite

**TABLE 4–9. Dimensionless influx, $W_{eD}$, for $r'_D = 4$ (courtesy of SPE)[24]**

| $t_D$ | $z'_D$ | | | | | | |
|---|---|---|---|---|---|---|---|
| | 0.05 | 0.1 | 0.3 | 0.5 | 0.7 | 0.9 | 1.0 |
| 2 | 2.398 | 2.389 | 2.284 | 2.031 | 1.824 | 1.620 | 1.507 |
| 3 | 3.006 | 2.993 | 2.874 | 2.629 | 2.390 | 2.149 | 2.012 |
| 4 | 3.552 | 3.528 | 3.404 | 3.158 | 2.893 | 2.620 | 2.466 |
| 5 | 4.053 | 4.017 | 3.893 | 3.627 | 3.341 | 3.045 | 2.876 |
| 6 | 4.490 | 4.452 | 4.332 | 4.047 | 3.744 | 3.430 | 3.249 |
| 7 | 4.867 | 4.829 | 4.715 | 4.420 | 4.107 | 3.778 | 3.587 |
| 8 | 5.191 | 5.157 | 5.043 | 4.757 | 4.437 | 4.096 | 3.898 |
| 9 | 5.464 | 5.434 | 5.322 | 5.060 | 4.735 | 4.385 | 4.184 |
| 10 | 5.767 | 5.739 | 5.598 | 5.319 | 5.000 | 4.647 | 4.443 |
| 11 | 5.964 | 5.935 | 5.829 | 5.561 | 5.240 | 4.884 | 4.681 |
| 12 | 6.188 | 6.158 | 6.044 | 5.780 | 5.463 | 5.107 | 4.903 |
| 13 | 6.380 | 6.350 | 6.240 | 5.983 | 5.670 | 5.316 | 5.113 |
| 14 | 6.559 | 6.529 | 6.421 | 6.171 | 5.863 | 5.511 | 5.309 |
| 15 | 6.725 | 6.694 | 6.589 | 6.345 | 6.044 | 5.695 | 5.495 |
| 16 | 6.876 | 6.844 | 6.743 | 6.506 | 6.213 | 5.867 | 5.671 |
| 17 | 7.014 | 6.983 | 6.885 | 6.656 | 6.371 | 6.030 | 5.838 |
| 18 | 7.140 | 7.113 | 7.019 | 6.792 | 6.523 | 6.187 | 5.999 |
| 19 | 7.261 | 7.240 | 7.140 | 6.913 | 6.663 | 6.334 | 6.153 |
| 20 | 7.376 | 7.344 | 7.261 | 7.028 | 6.785 | 6.479 | 6.302 |
| 22 | 7.518 | 7.507 | 7.451 | 7.227 | 6.982 | 6.691 | 6.524 |
| 24 | 7.618 | 7.607 | 7.518 | 7.361 | 7.149 | 6.870 | 6.714 |
| 26 | 7.697 | 7.685 | 7.607 | 7.473 | 7.283 | 7.026 | 6.881 |
| 28 | 7.752 | 7.752 | 7.674 | 7.563 | 7.395 | 7.160 | 7.026 |
| 30 | 7.808 | 7.797 | 7.741 | 7.641 | 7.484 | 7.283 | 7.160 |
| 34 | 7.864 | 7.864 | 7.819 | 7.741 | 7.618 | 7.451 | 7.350 |
| 38 | 7.909 | 7.909 | 7.875 | 7.808 | 7.719 | 7.585 | 7.496 |
| 42 | 7.931 | 7.931 | 7.909 | 7.864 | 7.797 | 7.685 | 7.618 |
| 46 | 7.942 | 7.942 | 7.920 | 7.898 | 7.842 | 7.752 | 7.697 |
| 50 | 7.954 | 7.954 | 7.942 | 7.920 | 7.875 | 7.808 | 7.764 |
| 60 | 7.968 | 7.968 | 7.965 | 7.954 | 7.931 | 7.898 | 7.864 |
| 70 | 7.976 | 7.976 | 7.976 | 7.968 | 7.965 | 7.942 | 7.920 |
| 80 | 7.982 | 7.982 | 7.987 | 7.976 | 7.976 | 7.965 | 7.954 |
| 90 | 7.987 | 7.987 | 7.987 | 7.984 | 7.983 | 7.976 | 7.965 |
| 100 | 7.987 | 7.987 | 7.987 | 7.987 | 7.987 | 7.983 | 7.976 |
| 120 | 7.987 | 7.987 | 7.987 | 7.987 | 7.987 | 7.987 | 7.987 |

**TABLE 4–10. Dimensionless influx, $W_{eD}$, for $r'_D = 6$ (courtesy of SPE)[24]**

| $t_D$ | $z'_D$ | | | | | | |
|---|---|---|---|---|---|---|---|
| | 0.05 | 0.1 | 0.3 | 0.5 | 0.7 | 0.9 | 1.0 |
| 6 | 4.780 | 4.762 | 4.597 | 4.285 | 3.953 | 3.611 | 3.414 |
| 7 | 5.309 | 5.289 | 5.114 | 4.779 | 4.422 | 4.053 | 3.837 |
| 8 | 5.799 | 5.778 | 5.595 | 5.256 | 4.875 | 4.478 | 4.247 |
| 9 | 6.252 | 6.229 | 6.041 | 5.712 | 5.310 | 4.888 | 4.642 |
| 10 | 6.750 | 6.729 | 6.498 | 6.135 | 5.719 | 5.278 | 5.019 |
| 11 | 7.137 | 7.116 | 6.916 | 6.548 | 6.110 | 5.648 | 5.378 |
| 12 | 7.569 | 7.545 | 7.325 | 6.945 | 6.491 | 6.009 | 5.728 |
| 13 | 7.967 | 7.916 | 7.719 | 7.329 | 6.858 | 6.359 | 6.067 |
| 14 | 8.357 | 8.334 | 8.099 | 7.699 | 7.214 | 6.697 | 6.395 |
| 15 | 8.734 | 8.709 | 8.467 | 8.057 | 7.557 | 7.024 | 6.713 |
| 16 | 9.093 | 9.067 | 8.819 | 8.398 | 7.884 | 7.336 | 7.017 |
| 17 | 9.442 | 9.416 | 9.160 | 8.730 | 8.204 | 7.641 | 7.315 |
| 18 | 9.775 | 9.749 | 9.485 | 9.047 | 8.510 | 7.934 | 7.601 |
| 19 | 10.09 | 10.06 | 9.794 | 9.443 | 8.802 | 8.214 | 7.874 |
| 20 | 10.40 | 10.37 | 10.10 | 9.646 | 9.087 | 8.487 | 8.142 |
| 22 | 10.99 | 10.96 | 10.67 | 10.21 | 9.631 | 9.009 | 8.653 |
| 24 | 11.53 | 11.50 | 11.20 | 10.73 | 10.13 | 9.493 | 9.130 |
| 26 | 12.06 | 12.03 | 11.72 | 11.23 | 10.62 | 9.964 | 9.594 |
| 28 | 12.52 | 12.49 | 12.17 | 11.68 | 11.06 | 10.39 | 10.01 |
| 30 | 12.95 | 12.92 | 12.59 | 12.09 | 11.46 | 10.78 | 10.40 |
| 35 | 13.96 | 13.93 | 13.57 | 13.06 | 12.41 | 11.70 | 11.32 |
| 40 | 14.69 | 14.66 | 14.33 | 13.84 | 13.23 | 12.53 | 12.15 |
| 45 | 15.27 | 15.24 | 14.94 | 14.48 | 13.90 | 13.23 | 12.87 |
| 50 | 15.74 | 15.71 | 15.44 | 15.01 | 14.47 | 13.84 | 13.49 |
| 60 | 16.40 | 16.38 | 16.15 | 15.81 | 15.34 | 14.78 | 14.47 |
| 70 | 16.87 | 16.83 | 16.67 | 16.38 | 15.99 | 15.50 | 15.24 |
| 80 | 17.20 | 17.18 | 17.04 | 16.80 | 16.48 | 16.06 | 15.83 |
| 90 | 17.43 | 17.42 | 17.30 | 17.10 | 16.85 | 16.50 | 16.29 |
| 100 | 17.58 | 17.58 | 17.49 | 17.34 | 17.12 | 16.83 | 16.66 |
| 110 | 17.71 | 17.69 | 17.63 | 17.50 | 17.34 | 17.09 | 16.93 |
| 120 | 17.78 | 17.78 | 17.73 | 17.63 | 17.49 | 17.29 | 17.17 |
| 130 | 17.84 | 17.84 | 17.79 | 17.73 | 17.62 | 17.45 | 17.34 |
| 140 | 17.88 | 17.88 | 17.85 | 17.79 | 17.71 | 17.57 | 17.48 |
| 150 | 17.92 | 17.91 | 17.88 | 17.84 | 17.77 | 17.66 | 17.58 |
| 175 | 17.95 | 17.95 | 17.94 | 17.92 | 17.87 | 17.81 | 17.76 |
| 200 | 17.97 | 17.97 | 17.86 | 17.95 | 17.93 | 17.98 | 17.86 |
| 225 | 17.97 | 17.97 | 17.97 | 17.96 | 17.95 | 17.93 | 17.91 |
| 250 | 17.98 | 17.98 | 17.98 | 17.97 | 17.96 | 17.95 | 17.95 |
| 300 | 17.98 | 17.98 | 17.98 | 17.98 | 17.98 | 17.97 | 17.97 |
| 350 | 17.98 | 17.98 | 17.98 | 17.98 | 17.98 | 17.98 | 17.98 |
| 400 | 17.98 | 17.98 | 17.98 | 17.98 | 17.98 | 17.98 | 17.98 |
| 450 | 17.98 | 17.98 | 17.98 | 17.98 | 17.98 | 17.98 | 17.98 |
| 500 | 17.98 | 17.98 | 17.98 | 17.98 | 17.98 | 17.98 | 17.98 |

and radial. The Havlena-Odeh method has been discussed in the section on OOIP-natural water-drive reservoirs using Equation (4.34). The van Everdingen, Timmerman, and McMahon method is as follows:

**van Everdingen, Timmerman, and McMahon Method.[25]** This method assumes that OOIP, aquifer conductivity $kh\alpha/\mu$, and diffusivity $k/(\phi\mu c_t)$ unknown. Solving the material balance equation and Equation (4.80) simultaneously for OOIP gives

$$N = A + m_p F(t) \tag{4.87}$$

**TABLE 4–11. Dimensionless influx, $W_{eD}$, for $r'_D$ = 8 (courtesy of SPE)[24]**

| | $\tau'_D$ | | | | | | |
|---|---|---|---|---|---|---|---|
| $t_D$ | 0.05 | 0.1 | 0.3 | 0.5 | 0.7 | 0.9 | 1.0 |
| 9 | 6.301 | 6.278 | 6.088 | 5.756 | 5.350 | 4.924 | 4.675 |
| 10 | 6.828 | 6.807 | 6.574 | 6.205 | 5.783 | 5.336 | 5.072 |
| 11 | 7.250 | 7.229 | 7.026 | 6.650 | 6.204 | 5.732 | 5.456 |
| 12 | 7.725 | 7.700 | 7.477 | 7.086 | 6.621 | 6.126 | 5.836 |
| 13 | 8.173 | 8.149 | 7.919 | 7.515 | 7.029 | 6.514 | 6.210 |
| 14 | 8.619 | 8.594 | 8.355 | 7.937 | 7.432 | 6.895 | 6.578 |
| 15 | 9.058 | 9.032 | 8.783 | 8.351 | 7.828 | 7.270 | 6.940 |
| 16 | 9.485 | 9.458 | 9.202 | 8.755 | 8.213 | 7.634 | 7.293 |
| 17 | 9.907 | 9.879 | 9.613 | 9.153 | 8.594 | 7.997 | 7.642 |
| 18 | 10.32 | 10.29 | 10.01 | 9.537 | 8.961 | 8.343 | 7.979 |
| 19 | 10.72 | 10.69 | 10.41 | 9.920 | 9.328 | 8.691 | 8.315 |
| 20 | 11.12 | 11.08 | 10.80 | 10.30 | 9.687 | 9.031 | 8.645 |
| 22 | 11.89 | 11.86 | 11.55 | 11.02 | 10.38 | 9.686 | 9.280 |
| 24 | 12.63 | 12.60 | 12.27 | 11.72 | 11.05 | 10.32 | 9.896 |
| 26 | 13.36 | 13.32 | 12.97 | 12.40 | 11.70 | 10.94 | 10.49 |
| 28 | 14.06 | 14.02 | 13.65 | 13.06 | 12.33 | 11.53 | 11.07 |
| 30 | 14.73 | 14.69 | 14.30 | 13.68 | 12.93 | 12.10 | 11.62 |
| 34 | 16.01 | 15.97 | 15.54 | 14.88 | 14.07 | 13.18 | 12.67 |
| 38 | 17.21 | 17.17 | 16.70 | 15.99 | 15.13 | 14.18 | 13.65 |
| 40 | 17.80 | 17.75 | 17.26 | 16.52 | 15.64 | 14.66 | 14.12 |
| 45 | 19.15 | 19.10 | 18.56 | 17.76 | 16.83 | 15.77 | 15.21 |
| 50 | 20.42 | 20.36 | 19.76 | 18.91 | 17.93 | 16.80 | 16.24 |
| 55 | 21.46 | 21.39 | 20.80 | 19.96 | 18.97 | 17.83 | 17.24 |
| 60 | 22.40 | 22.34 | 21.75 | 20.91 | 19.93 | 18.78 | 18.19 |
| 70 | 23.97 | 23.92 | 23.36 | 22.55 | 21.58 | 20.44 | 19.86 |
| 80 | 25.29 | 25.23 | 24.71 | 23.94 | 23.01 | 21.91 | 21.32 |
| 90 | 26.39 | 26.33 | 25.85 | 25.12 | 24.24 | 23.18 | 22.61 |
| 100 | 27.30 | 27.25 | 26.81 | 26.13 | 25.29 | 24.29 | 23.74 |
| 120 | 28.61 | 28.57 | 28.19 | 27.63 | 26.90 | 26.01 | 25.51 |
| 140 | 29.55 | 29.51 | 29.21 | 28.74 | 28.12 | 27.33 | 26.90 |
| 160 | 30.23 | 30.21 | 29.96 | 29.57 | 29.04 | 28.37 | 27.99 |
| 180 | 30.73 | 30.71 | 30.51 | 30.18 | 29.75 | 29.18 | 28.84 |
| 200 | 31.07 | 31.04 | 30.90 | 30.63 | 30.26 | 29.79 | 29.51 |
| 240 | 31.50 | 31.49 | 31.39 | 31.22 | 30.98 | 30.65 | 30.45 |
| 280 | 31.72 | 31.71 | 31.66 | 31.56 | 31.39 | 31.17 | 31.03 |
| 320 | 31.85 | 31.84 | 31.80 | 31.74 | 31.64 | 31.49 | 31.39 |
| 360 | 31.90 | 31.90 | 31.88 | 31.85 | 31.78 | 31.68 | 31.61 |
| 400 | 31.94 | 31.94 | 31.93 | 31.90 | 31.86 | 31.79 | 31.75 |
| 450 | 31.96 | 31.96 | 31.95 | 31.94 | 31.91 | 31.88 | 31.85 |
| 500 | 31.97 | 31.97 | 31.96 | 31.96 | 31.95 | 31.93 | 31.90 |
| 550 | 31.97 | 31.97 | 31.97 | 31.96 | 31.96 | 31.95 | 31.94 |
| 600 | 31.97 | 31.97 | 31.97 | 31.96 | 31.97 | 31.96 | 31.95 |
| 700 | 31.97 | 31.97 | 31.97 | 31.97 | 31.97 | 31.97 | 31.97 |
| 800 | 31.97 | 31.97 | 31.97 | 31.97 | 31.97 | 31.97 | 31.97 |

where

$$A = \frac{1}{(F_v - 1)B_{oi}}[N_pF_vB_o + N_p(R_p - R_s)B_g + W_p] \quad \textbf{(4.88)}$$

$$F(t) = \frac{1}{(F_v - 1)B_{oi}}\left[\sum_{j=1}^{n} \Delta p_{(n+1-j)} W_{eDj}\right] \quad \textbf{(4.89)}$$

$$F_v = \frac{p_b - p}{pY} + 1 \quad \textbf{(4.90)}$$

$$Y = \frac{p_b - p}{p(F_v - 1)} \quad \textbf{(4.91)}$$

$F_v$ = ratio of volume of oil and its dissolved original gas at a given pressure to its volume at initial pressure. Generally, Y is calculated with laboratory-determined values of $F_v - 1$. Smooth values of Y can be calculated from the straight-line relationship of Y vs. $p$ plot.

The equations for obtaining the least-squares fit to Equations (4.88) and (4.89) for a given dimensionless time interval, $\Delta t_D$, and $n$ data points are

$$nN = \sum_{j=1}^{n} A_j - m_p \sum_{j=1}^{n} F(t)_j \quad \textbf{(4.92)}$$

and

$$N \sum_{j=1}^{n} F(t)_j = \sum_{j=1}^{n} A_j[F(t)]_j - m_p \sum_{j=1}^{n} [F(t)]_j^2 \quad \textbf{(4.93)}$$

The variance of this fit from field data can be calculated by

$$\sigma^2 = \frac{1}{n}\sum_{j=1}^{n} \{A_j - N + m_p[F(t)]_j\}^2 \quad \textbf{(4.94)}$$

The variance is then plotted versus the various assumed values of $\Delta t_D$, and the $\Delta t_D$ at the minimum variance is then used in Equations (4.92) and (4.93) to solve for the best estimate of N and $m_p$.

## APPLICATION OF WATER INFLUX THEORY IN HISTORY MATCHING

In history matching the observed reservoir pressure, it is necessary to calculate the water influx corresponding to a continuous pressure decline at the reservoir/aquifer boundary (original hydrocarbon/water contact). To perform such calculations, the continuous decline is divided into a series of discrete pressure steps. The water influx is then calculated for a pressure drop ($\Delta p$) between each pressure step. The cumulative water influx is then obtained by the superposition of the influxes at each $\Delta p$. Van Everdingen, Timmerman, and McMahon suggested a technique to approximate the continuous pressure decline by a series of discrete steps, which is illustrated in Figure 4–25.[25] 

For a pressure drop $\Delta p_j$ at time $t_j$ and $Q(t_D - t_{Dj})$ as the dimensionless water influx for dimensionless times $(t_D - t_{Dj})$, the cumulative water influx at some arbitrary time $t$ is given by

$$W_{e_t} = m_p \sum_{j=0}^{n} [\Delta p_j W_{eD}(t_D - t_{Dj})] \tag{4.95}$$

OR

$$W_{e_t} = m_p[\Delta p_o W_{eD}(t_D) p_1 W_{eD}(t_D - t_{D1}) + \Delta p_2 W_{eD}(t_D - t_{D2}) + \cdots \\ \Delta p_j W_{eD}(t_D - t_{Dj}) + \cdots \Delta p_{n-1} W_{eD}(t_D - t_{D_{n-1}})] \tag{4.95}$$

In case of an infinite linear aquifer, the cumulative water influx at time $t$ due to step-like pressure decline at the original hydrocarbon/water contact can be calculated using the following equation:

$$W_{e_t} = 3.26 \times 10^{-3} hw \sqrt{\frac{\phi h c_e}{\pi \mu_w}} \sum_{j=0}^{n-1} [\Delta p_j \sqrt{t - t_j}] \tag{4.96}$$

The following steps will help in performing acquifer fitting (modelling) using the unsteady-state theory of van Everdingen and Hurst:

1. Plot the pressure versus time on Cartesian coordinate graph paper, approximate the continuous pressure decline by a series of discrete pressure steps as shown in Figure 4–25.
2. Calculate the water influx using the unsteady-state water influx theory of van Everdingen and Hurst at an assumed $r_{eD} = r_e/r_w$.
3. Using the material balance technique of Havlena and Odeh (previously discussed), prepare a plot of $V_R/E_N$ vs. $\Sigma \Delta p W_{eDj}/E_N$. For the correct model ($r_e/r_w$), the plot should be linear with intercept as N and slope as $m_p$.
4. If the plot in Step 3 is nonlinear for the $r_e/r_w$ used, incorrect aquifer size is indicated. Repeat Steps 2 and 3, varying the value of $r_e/r_w$ until a straight line is obtained.

**Example 4–25.**[7] A 10-year pressure history and the corresponding PVT properties on a waterdrive oil reservoir are given in Table 4–13. The estimated OOIP is $312 \times 10^6$ STB. Other reservoir and aquifer properties are given below. Compute aquifer size.

| | | |
|---|---|---|
| $h = 100$ ft | $\phi = 0.25$ | $c_f = 4.0 \times 10^{-6}$ 1/psi |
| $k = 200$ md | $\theta = 140°$ | $B_w = 1.0$ RB/STB |
| $\mu_w = 0.55$ cp | $c_w = 3.0 \times 10^{-6}$ 1/psi | $r_w = 9{,}200$ ft |

**TABLE 4–12. Dimensionless influx, $W_{eD}$, for $r'_D$ = 10 (courtesy of SPE)[24]**

| | $z'_D$ | | | | | | |
|---|---|---|---|---|---|---|---|
| $t_D$ | 0.05 | 0.1 | 0.3 | 0.5 | 0.7 | 0.9 | 1.0 |
| 22 | 12.07 | 12.04 | 11.74 | 11.21 | 10.56 | 9.865 | 9.449 |
| 24 | 12.86 | 12.83 | 12.52 | 11.97 | 11.29 | 10.55 | 10.12 |
| 26 | 13.65 | 13.62 | 13.29 | 12.72 | 12.01 | 11.24 | 10.78 |
| 28 | 14.42 | 14.39 | 14.04 | 13.44 | 12.70 | 11.90 | 11.42 |
| 30 | 15.17 | 15.13 | 14.77 | 14.15 | 13.38 | 12.55 | 12.05 |
| 32 | 15.91 | 15.87 | 15.49 | 14.85 | 14.05 | 13.18 | 12.67 |
| 34 | 16.63 | 16.59 | 16.20 | 15.54 | 14.71 | 13.81 | 13.28 |
| 36 | 17.33 | 17.29 | 16.89 | 16.21 | 15.35 | 14.42 | 13.87 |
| 38 | 18.03 | 17.99 | 17.57 | 16.86 | 15.98 | 15.02 | 14.45 |
| 40 | 18.72 | 18.68 | 18.24 | 17.51 | 16.60 | 15.61 | 15.02 |
| 42 | 19.38 | 19.33 | 18.89 | 18.14 | 17.21 | 16.19 | 15.58 |
| 44 | 20.03 | 19.99 | 19.53 | 18.76 | 17.80 | 16.75 | 16.14 |
| 46 | 20.67 | 20.62 | 20.15 | 19.36 | 18.38 | 17.30 | 16.67 |
| 48 | 21.30 | 21.25 | 20.76 | 19.95 | 18.95 | 17.84 | 17.20 |
| 50 | 21.92 | 21.87 | 21.36 | 20.53 | 19.51 | 18.38 | 17.72 |
| 52 | 22.52 | 22.47 | 21.95 | 21.10 | 20.05 | 18.89 | 18.22 |
| 54 | 23.11 | 23.06 | 22.53 | 21.66 | 20.59 | 19.40 | 18.72 |
| 56 | 23.70 | 23.64 | 23.09 | 22.20 | 21.11 | 19.89 | 19.21 |
| 58 | 24.26 | 24.21 | 23.65 | 22.74 | 21.63 | 20.39 | 19.68 |
| 60 | 24.82 | 24.77 | 24.19 | 23.26 | 22.13 | 20.87 | 20.15 |
| 65 | 26.18 | 26.12 | 25.50 | 24.53 | 23.34 | 22.02 | 21.28 |
| 70 | 27.47 | 27.41 | 26.75 | 25.73 | 24.50 | 23.12 | 22.36 |
| 75 | 28.71 | 28.55 | 27.94 | 26.88 | 25.60 | 24.17 | 23.39 |
| 80 | 29.89 | 29.82 | 29.08 | 27.97 | 26.65 | 25.16 | 24.36 |
| 85 | 31.02 | 30.95 | 30.17 | 29.01 | 27.65 | 26.10 | 25.31 |
| 90 | 32.10 | 32.03 | 31.20 | 30.00 | 28.60 | 27.03 | 26.25 |
| 95 | 33.04 | 32.96 | 32.14 | 30.95 | 29.54 | 27.93 | 27.10 |
| 100 | 33.94 | 33.85 | 33.03 | 31.85 | 30.44 | 28.82 | 27.98 |
| 110 | 35.55 | 35.46 | 34.65 | 33.49 | 32.08 | 30.47 | 29.62 |
| 120 | 36.97 | 36.90 | 36.11 | 34.98 | 33.58 | 31.98 | 31.14 |
| 130 | 38.28 | 38.19 | 37.44 | 36.33 | 34.96 | 33.38 | 32.55 |
| 140 | 39.44 | 39.37 | 38.64 | 37.56 | 36.23 | 34.67 | 33.85 |
| 150 | 40.49 | 40.42 | 39.71 | 38.67 | 37.38 | 35.86 | 35.04 |
| 170 | 42.21 | 42.15 | 41.51 | 40.54 | 39.33 | 37.89 | 37.11 |
| 190 | 43.62 | 43.55 | 42.98 | 42.10 | 40.97 | 39.62 | 38.90 |
| 210 | 44.77 | 44.72 | 44.19 | 43.40 | 42.36 | 41.11 | 40.42 |
| 230 | 45.71 | 45.67 | 45.20 | 44.48 | 43.54 | 42.38 | 41.74 |
| 250 | 46.48 | 46.44 | 46.01 | 45.38 | 44.53 | 43.47 | 42.87 |
| 270 | 47.11 | 47.06 | 46.70 | 46.13 | 45.36 | 44.40 | 43.84 |
| 290 | 47.61 | 47.58 | 47.25 | 46.75 | 46.07 | 45.19 | 44.68 |
| 310 | 48.03 | 48.00 | 47.72 | 47.26 | 46.66 | 45.87 | 45.41 |
| 330 | 48.38 | 48.35 | 48.10 | 47.71 | 47.16 | 46.45 | 46.03 |
| 350 | 48.66 | 48.64 | 48.42 | 48.08 | 47.59 | 46.95 | 46.57 |
| 400 | 49.15 | 49.14 | 48.99 | 48.74 | 48.38 | 47.89 | 47.60 |
| 450 | 49.46 | 49.45 | 49.35 | 49.17 | 48.91 | 48.55 | 48.31 |
| 500 | 49.65 | 49.64 | 49.58 | 49.45 | 49.26 | 48.98 | 48.82 |
| 600 | 49.84 | 49.84 | 49.81 | 49.74 | 49.65 | 49.50 | 49.41 |
| 700 | 49.91 | 49.91 | 49.90 | 49.87 | 49.82 | 49.74 | 49.69 |
| 800 | 49.94 | 49.94 | 49.93 | 49.92 | 49.90 | 49.85 | 49.83 |
| 900 | 49.96 | 49.96 | 49.94 | 49.94 | 49.93 | 49.91 | 49.90 |
| 1,000 | 49.96 | 49.96 | 49.96 | 49.96 | 49.94 | 49.93 | 49.93 |
| 1,200 | 49.96 | 49.96 | 49.96 | 49.96 | 49.96 | 49.96 | 49.96 |

Solution:

$$t_D = \frac{0.000264kt}{\phi\mu_w c_e r_w^2}$$

$$= \frac{0.000264(200)(t \times 24 \times 365)}{(0.25)(0.55)(3 + 4) \times 10^{-6}(9{,}200)^2}$$

$$= 5.67t$$

$$m_p = 1.119\phi c_e r_w^2 h \frac{\theta}{360}$$

$$= (1.119)(0.25)(4.0 + 3.0)(10^{-6})(9{,}200)^2(100)\left(\frac{140}{360}\right)$$

$$\approx 6{,}446 \quad \text{bbls/psi}$$

Table 4–14 shows the $W_{eD}(t_D - t_{Dj})$ values for $r_{ed}$ = 10 and $r_{eD}$ = 5 and the corresponding water influx values for each pressure drop. A sample calculation of $W_{eD}$ at $r_{eD}$ = 10 is given below.

Sample calculations for some $W_e$ values at $r_{eD}$ = 10 are

Year

1 $W_e = 6{,}446(120 \times 4.95) = 3.829$

2 $W_e = 6{,}446[(120 \times 8.12) + (225 \times 4.95)] = 13.460$

3 $W_e = 6{,}446[(120 \times 10.9) + (225 \times 8.12) + (196 \times 4.95)] = 26.462$

4 $W_e = 6{,}446[(120 \times 13.5) + (225 \times 10.9) + (196 \times 8.12) + (170 \times 4.95)] = 41.935$

and so on.

Table 4–15 shows the parameters calculated for use in the Havlena and Odeh material balance plot. The plot of $F/E_o$ versus $W_e/E_o$ for both $r_{eD}$ = 10 and $r_{eD}$ = 5 is shown in Figure 4–26. The line for $r_{eD}$ = 10 deviates from the unit-slope line indicating that the correct value of $r_{eD}$ should be somewhat less than $r_{eD}$ = 10. The line for $r_{eD}$ = 5 shows linear relationship and a unit-slope line; it also confirms the OOIP at 312 MMSTB. Therefore, $r_{eD}$ = 5 is the correct value of the ratio of the aquifer/reservoir radius ($r_e/r_w$).

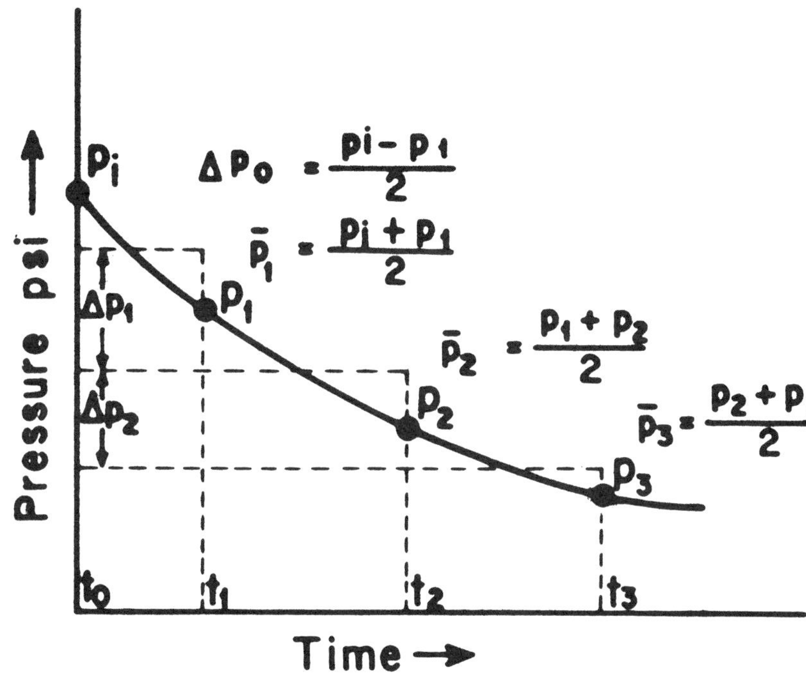


**Figure 4–25.** Approximating a continuous pressure decline at aquifer-reservoir boundary by a series of discrete pressure steps.

**TABLE 4–13. Ten-year pressure history and corresponding PVT properties for reservoir in Example 4–25[7]**

| (Years) | Pressure @ W.O.C. (psi) | Plateau Pressure Levels (psi) | $\Delta p$ (psi) | $N_p$ $10^6$ (STB) | $R_p$ (SCF/STB) | $B_o$ (RB/STB) | $R_s$ (SCF/STB) | $B_g$ (RB/SCF) |
|---|---|---|---|---|---|---|---|---|
| 0 | 2,740 | — | 120 | — | 650 | 1.404 | 650 | 0.00093 |
| 1 | 2,500 | 2,620 | 225 | 7.88 | 760 | 1.374 | 592 | 0.00098 |
| 2 | 2,290 | 2,395 | 196 | 18.42 | 845 | 1.349 | 545 | 0.00107 |
| 3 | 2,109 | 2,199 | 170 | 29.15 | 920 | 1.329 | 507 | 0.00117 |
| 4 | 1,949 | 2,029 | 146 | 40.69 | 975 | 1.316 | 471 | 0.00128 |
| 5 | 1,818 | 1,883 | 123 | 50.14 | 1,025 | 1.303 | 442 | 0.00139 |
| 6 | 1,702 | 1,760 | 105 | 58.42 | 1,065 | 1.294 | 418 | 0.00150 |
| 7 | 1,608 | 1,655 | 84 | 65.39 | 1,095 | 1.287 | 398 | 0.00160 |
| 8 | 1,535 | 1,571 | 64 | 70.74 | 1,120 | 1.280 | 383 | 0.00170 |
| 9 | 1,480 | 1,507 | 47 | 74.54 | 1,145 | 1.276 | 371 | 0.00176 |
| 10 | 1,440 | 1,460 | | 77.43 | 1,160 | 1.273 | 364 | 0.00182 |

TABLE 4–14. **Calculated values of $t_D$, $W_{eD}(t_D - t_{Dj})$, and $W_e$ for Example 4–25**

| Time, $t$ (Years) | $t_D$ (5.67 $t$) | $\Delta p_i$ (psi) | $W_{eD}$ ($r_{eD} = 10$) | $W_{eD}$ ($r_{eD} = 5$) | $W_e$ (MMRB) $r_{eD} = 10$ | $W_e$ (MMRB) $r_{eD} = 5$ |
|---|---|---|---|---|---|---|
| 0 | 0 | 120 | — | — | — | — |
| 1 | 5.67 | 225 | 4.95 | 4.88 | 3.829 | 3.775 |
| 2 | 11.34 | 196 | 8.12 | 7.46 | 13.460 | 12.848 |
| 3 | 17.01 | 170 | 10.90 | 9.10 | 26.462 | 24.024 |
| 4 | 22.68 | 146 | 13.50 | 10.09 | 41.935 | 35.775 |
| 5 | 28.35 | 123 | 15.90 | 10.83 | 59.207 | 47.276 |
| 6 | 34.02 | 105 | 18.10 | 11.27 | 77.628 | 58.035 |
| 7 | 39.69 | 84 | 20.20 | 11.52 | 96.805 | 67.778 |
| 8 | 45.36 | 64 | 22.20 | 11.69 | 116.284 | 76.259 |
| 9 | 51.03 | 47 | 24.00 | 11.81 | 135.601 | 83.398 |
| 10 | 56.70 | — | 25.70 | 11.89 | 154.401 | 89.225 |

TABLE 4–15. **Material balance parameters for Example 4–25 (courtesy of Elsevier)[7]**

$$F = N_p[B_o + (R_p - R_s)B_g]$$
$$E_o = [(B_o - B_{oi}) + (R_{si} - R_s)B_g]$$

| Time (Years) | F (MMRB) | $E_o$ (RB/STB) | $F/E_o$ (MMSTB) | $W_e/E_o$ ($r_{eD} = 10$) | $W_e/E_o$ ($r_{eD} = 5$) |
|---|---|---|---|---|---|
| 1 | 12.124 | 0.0268 | 452.4 | 142.9 | 140.9 |
| 2 | 30.761 | 0.0574 | 535.9 | 234.5 | 223.8 |
| 3 | 52.826 | 0.0923 | 572.3 | 286.7 | 260.3 |
| 4 | 79.798 | 0.1411 | 565.5 | 297.2 | 253.5 |
| 5 | 105.964 | 0.1881 | 563.3 | 314.8 | 251.3 |
| 6 | 132.292 | 0.2380 | 555.8 | 326.2 | 243.8 |
| 7 | 157.080 | 0.2862 | 548.8 | 338.2 | 236.8 |
| 8 | 179.177 | 0.3299 | 543.1 | 352.5 | 231.2 |
| 9 | 196.654 | 0.3630 | 541.7 | 373.6 | 229.7 |
| 10 | 210.743 | 0.3895 | 541.1 | 396.4 | 229.1 |

## WATER INFLUX THEORY OF FETKOVITCH FOR FINITE AQUIFER

Although the unsteady-state water influx theory of van Everdingen and Hurst is extensively used in the industry for water influx calculations, Fetkovitch introduced a technique that works equally well in majority of cases.[26] The advantage of Fetkovitch method over the unsteady-state method is that it does not require trial-and-error calculations and the use of dimensionless time.

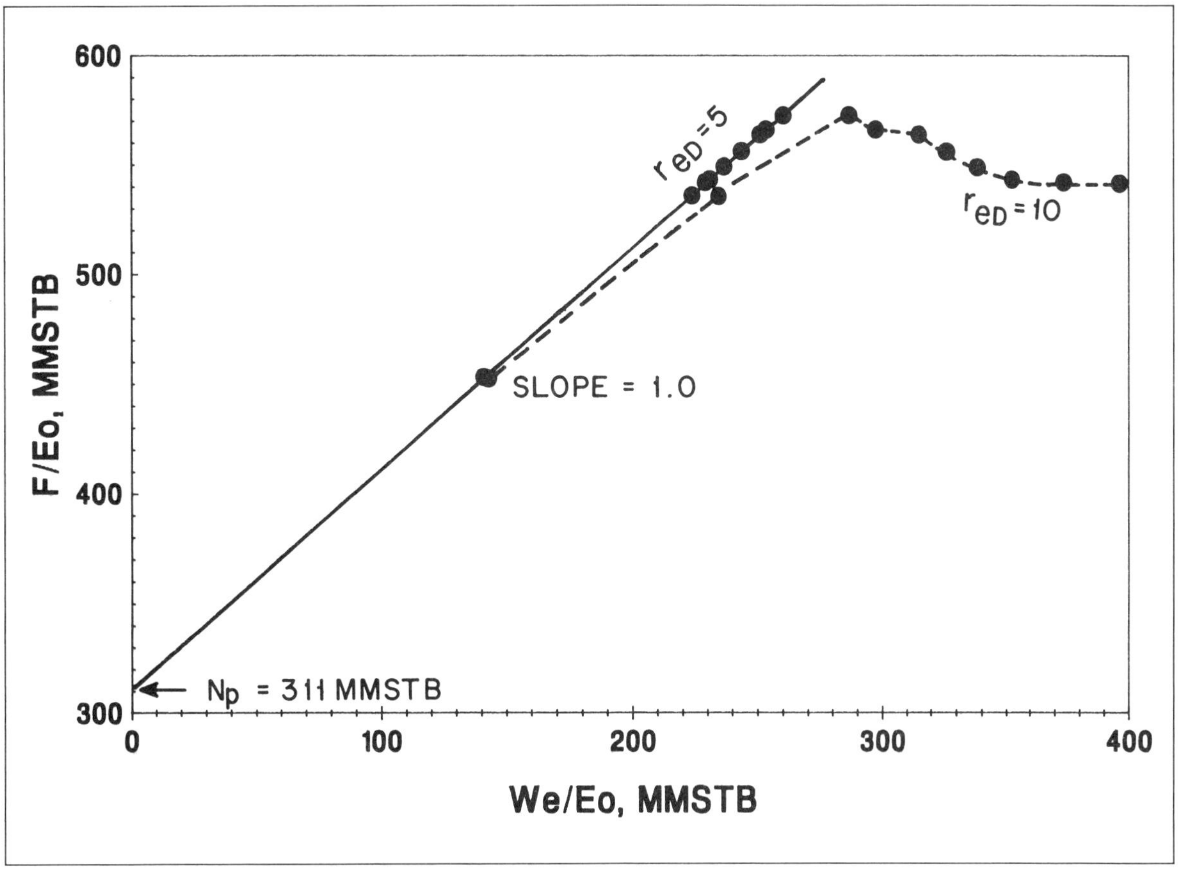


**Figure 4–26.** Aquifer fitting using the interpretation technique of Havlena and Odeh.

When dealing with large aquifer, $r_{eD} = 10$, in which, for the first few years, the water influx occurs under transient flow conditions, Fetkovitch's method fails to provide good match with the results obtained using unsteady-state method. In such cases unsteady-state method is used for the first few time steps with a switch to Fetkovitch's method. Fetkovitch's method cannot be used for influx calculations from an infinite aquifer. Using the method of Fetkovitch, the following equations are required:

$$\bar{p}_{a_{n-1}} = p_i\left(1 - \frac{\sum_{j=1}^{n-1} \Delta W_{ej}}{W_{ej}}\right) \quad \textbf{(4.97)}$$

$$\Delta W_{e_n} = \frac{W_{e_i}}{p_i}(\bar{p}_{a_{n-1}} - \bar{p}_n)(1 - e^{-Jp_i\Delta t_n/Wei}) \quad \textbf{(4.98)}$$

where

$\bar{p}_{a_n}$ = average pressure in the aquifer at the end of the $(n - 1)$th time interval

$\bar{p}_n$ = average reservoir-aquifer boundary pressure during the $n$th time interval (obtained from discrete pressure steps).

$W_{ei} = \frac{\theta}{360} c_e \pi (r_e^2 - r_w^2) h \phi p_i / 5.615$ RB

$\Delta t_n$ = time steps in days

For the first time step $\bar{p}_{a_{n-1}} = p_i$ (the initial reservoir and aquifer pressure). The aquifer productivity index ($J$) is given by the following equations in barrels/day/psi.

For semisteady state (drawdown expressed as $\bar{p}_a - p$):

Radial aquifers

$$J = \frac{0.00708 \frac{\theta}{360} kh}{\mu_w \left( \ln \frac{r_e}{r_w} - 0.75 \right)} \tag{4.99}$$

Linear aquifers

$$J = \frac{0.001127khw}{\mu_w L} \tag{4.99a}$$

For steady state (drawdown expressed as $p_i - p$):

Radial aquifer

$$J = \frac{0.00708 \frac{\theta}{360} kh}{\mu_w \ln \frac{r_e}{r_w}} \tag{4.100}$$

Linear aquifers

$$J = \frac{0.001127khw}{\mu_w L} \tag{4.100a}$$

If a reservoir is asymmetrically situated within a noncircular shaped aquifer, the Dietz shape factors introduced in Chapter 6 may be incorporated in the radial PI equation as follows:

$$J = \frac{0.00708 \dfrac{\theta}{360} kh}{\dfrac{\mu_w}{2} \ln \dfrac{4A}{\gamma C_A r_w^2}} \tag{4.101}$$

**Example 4–26.** Using Fetkovitch method, calculate water influx for the data given in Example 4–25 and an aquifer size of $r_{eD} = 5$.

Solution:

**Step 1.** Calculate $W_{ei}$.

$$\begin{aligned} W_{ei} &= c_e \frac{\theta}{360} \pi (r_e^2 - r_w^2) h \phi p_i / 5.615 \\ &= (7 \times 10^{-6}) \left(\frac{140}{360}\right) \pi (46{,}000^2 - 9{,}200^2)(100)(0.25)(2{,}740)/5.615 \\ &= 211.9 \times 10^6 \quad \text{RB} \end{aligned}$$

**Step 2.** Calculate PI for a finite radial aquifer.

$$\begin{aligned} J &= \frac{0.00708\theta/360kh}{\mu_w \left(\ln \dfrac{r_e}{r_w} - 0.75\right)} \\ &= \frac{0.00708(140/360)(200)(100)}{(0.55)\left(\ln \dfrac{46{,}000}{9{,}200} - 0.75\right)} \\ &= \frac{55.067}{0.473} = 116.42 \quad \text{B/D/psi} \end{aligned}$$

**Step 3.** Calculate $1 - e^{-Jp_i\Delta t/W_{ei}}$ for time steps of one year.

$$\begin{aligned} 1 - e^{-Jp_i\Delta t/W_{ei}} &= 1 - e^{[(-116.42)(2{,}740)(365)/(211.9 \times 10^6)]} \\ &= 0.4227 \end{aligned}$$

**Step 4.** Calculate $\Delta W_{e_n}$ in times of $(\bar{p}_{a_{n-1}} - \bar{p})$ using Equation (4.98).

Also calculate $\Delta W_{e1}$ for the first time step.

$$\Delta W_{e_n} = \frac{211.9 \times 10^6}{2{,}740}(\bar{p}_{a_{n-1}} - \bar{p}_n)(0.4227)$$
$$= 32{,}690(\bar{p}_{a_{n-1}} - \bar{p}_n)$$
$$\Delta W_{e1} = 32{,}690(2{,}740 - 2{,}620)$$
$$= 3.922 \times 10^6 \quad \text{RB}$$



**Step 5.** Calculate $\bar{p}_{a_n} = p_i\left(1 - \frac{\Delta W_{e_n}}{W_{ei}}\right)$, and calculate $\bar{p}_{a_{n-1}} - \bar{p}_n$.

$$\bar{p}_{a_1} = p_i\left(1 - \frac{\Delta W_{e1} + \Delta W_{e0}}{W_{ei}}\right)$$
$$= 2{,}740\left(1 - \frac{3.922}{211.9}\right)$$
$$= 2{,}689$$

Therefore,

$$\bar{p}_{a_{n-1}} - \bar{p}_n = 2{,}689 - 2{,}395 = 294$$

**Step 6.** Calculate $W_{e_n} = \Delta W_{e_n} + \Delta W_{e_{n-1}}$; therefore, for time step one, $W_{e_1}$ is

$$W_{e_1} = \Delta W_{e_1} + \Delta W_{e_{1-1}}$$
$$= 3.922 \times 10^6 + 0 = 3.922 \times 10^6 \quad \text{RB}$$

**Step 7.** Repeat Steps 4 through 6 for successive time steps. Thus for the second time step

$$\Delta W_{e_2} = 32{,}690(2{,}689 - 2{,}395)$$
$$= 9.611 \times 10^6 \quad \text{RB}$$
$$\bar{p}_{a_2} = p_i\left(1 - \frac{\Delta W_{e_1} + \Delta W_{e_2}}{W_{e_i}}\right)$$
$$= 2{,}740\left(1 - \frac{3.922 + 9.611}{211.9}\right)$$
$$= 2{,}565$$

Therefore,

$$\bar{p}_{a_{n-1}} - \bar{p}_n = (2{,}565 - 2{,}199)$$
$$= 366$$
$$W_{e_2} = \Delta W_{e_2} + \Delta W_{e_{2-1}}$$
$$= 9.611 + 3.922$$
$$= 13.533 \times 10^6 \quad \text{RB}$$

In the same way, $W_e$ at each pressure and the corresponding time step for the data in Example 4–25 is calculated. The values for each parameter are shown in Table 4–16.

**TABLE 4–16. Calculated parameters using Fetkovitch water influx method**

| Time (Years) | $\bar{p}_n$ (psi) | $\bar{p}_{a_n}$ (psi) | $\bar{p}_{a_{n-1}} - \bar{p}_n$ (psi) | $\Delta W_{e_n}$ (MMRB) | $W_{e_n}$ (MMRB) |
|---|---|---|---|---|---|
| 0 | 2,740 | — | — | — | — |
| 1 | 2,620 | 2,689 | 120 | 3.922 | 3.922 |
| 2 | 2,395 | 2,565 | 294 | 9.611 | 13.533 |
| 3 | 2,199 | 2,410 | 366 | 11.965 | 25.498 |
| 4 | 2,029 | 2,249 | 381 | 12.455 | 37.953 |
| 5 | 1,883 | 2,095 | 366 | 11.965 | 49.918 |
| 6 | 1,760 | 1,953 | 335 | 10.951 | 60.869 |
| 7 | 1,655 | 1,827 | 298 | 9.742 | 70.611 |
| 8 | 1,571 | 1,719 | 256 | 8.369 | 78.980 |
| 9 | 1,507 | 1,628 | 212 | 6.988 | 85.968 |
| 10 | 1,460 | 1,558 | 168 | 5.492 | 91.460 |

## PREDICTING THE AMOUNT OF WATER INFLUX

In the previous sections we discussed techniques to calculate the amount of water influx into a reservoir and to develop a mathematical aquifer model through which aquifer response is matched with reservoir production and pressure history. Once a satisfactory aquifer model is developed, the next step is to use this model to predict the future pressure of the reservoir and its corresponding aquifer response.

The material balance and the water influx equations presented in the previous sections are solved simultaneously by an iterative process to predict future reservoir pressure, water influx, and cumulative hydrocarbon production.

The prediction procedure involves (1) predicting an estimated future reservoir pressure from a historic pressure-time plot, (2) calculating water influx $W_e$ using the $W_{eD}$ function, Equation (4.80), and (3) calculating $W_e$ using material balance ($W_{e_{MBE}}$). When a pressure is found that will make $W_e = W_{e_{MBE}}$, it is assumed that a correct pressure and cumulative water influx is predicted.

Since the material balance calculations involve cumulative oil and/or gas production ($N_p$ and/or $G_p$), these variables are iteratively determined at the estimated predicted pressure. The water influx prediction procedure is based on the assumptions that

1. No free gas exists in the water bank
2. A uniform saturation throughout the unaffected zone
3. Producing wells are shut in whenever they start producing water, i.e., all production is from the uninvaded zone
4. The reservoir acts in a piston-like displacement
5. All bypassed hydrocarbons in the water zone can be represented by constant oil saturation, $S_{oby}$

Since simultaneous satisfaction of several iterations are required in order to predict behavior of a waterdrive reservoir, Slider in *Practical Petroleum Reservoir Engineering Methods* (published by PennWell Books) presented the following stepwise procedure which can be followed for the calculations:[20]

1. Estimate the average reservoir pressure in the uninvaded portion of the reservoir using a historic pressure-time plot. Determine the corresponding pressure at the original water/oil contact at this time.
2. Calculate $W_e$ using Equation (4.80). The aquifer constant $m_p$ must have been previously determined by model fitting.
3. Assume wnicn wells will be producing during the next prediction period. This can be based on the observation of the water encroachment area at the start of the period and its advance as a result of the water encroachment $W_e$ calculated in Step 2 and the bypassed oil in the invaded zone.
4. Estimate $N_p$ for the prediction period, say $N_{pn}$, using the historic rate-time extrapolation.
5. Evaluate oil saturation in the uninvaded portion ($S_{o_{UN}}$) of the reservoir at the end and the beginning of the period. The following equation can be used to calculate $S_{o_{UN}}$:

$$S_{o_{UN}} = \frac{\dfrac{(N - N_p)B_o - S_{oby}(W_e - W_p)}{(1 - S_{oby} - S_w)}}{\dfrac{[N\,B_{oi}/(1 - S_w)] - (W_e - W_p)}{(1 - S_{oby} - S_w)}} = \frac{\text{Reservoir oil in the uninvaded zone}}{\text{Pore volume in uninvaded zone}} \tag{4.102}$$

6. Using the following equation, calculate the average productivity index for the prediction period for the wells that will be producing during this period.

$$\frac{J_1}{J_2} = \frac{\left(\frac{k_o}{B_o\mu_o}\right)_1}{\left(\frac{k_o}{\quad}\right)} \qquad \textbf{(4.103)}$$

7. Based on the result of Step 6 and the number of wells producing, calculate $N_{pn}$. If $N_{pn}$ calculated does not agree sufficiently well with $N_{pn}$ estimated in Step 4, repeat Step 4 through 7 until an acceptable agreement is obtained.
8. Calculate $G_{pn}$ based on $S_{o_{UN}}$ at the end of the period. Use the following equations:

$$R = R_s + \frac{k_g\mu_o B_o}{k_o\mu_g B_g} \qquad \textbf{(4.104)}$$

$$G_{pn} = G_{p(n-1)} + [(R_n + R_{(n-1)})/2](N_{pn} - N_{p(n-1)}) \qquad \textbf{(4.105)}$$

9. Calculate $W_{e_{MBE}}$ by material balance.
10. If $W_e = W_{e_{MBE}}$, proceed to the prediction for the next time period. If $W_e > W_{e_{MBE}}$, repeat Step 1 through Step 10 assuming a larger pressure in Step 1. If $W_e < W_{e_{MBE}}$, repeat Step 1 through Step 10 assuming a smaller pressure in Step 1.



The use of the above procedure is not restricted to the outlined assumptions; however, modifications to the procedure and equations used are required; for example, for a situation when wells will be produced to some water cut greater than zero.

Since there are several possible combinations of various material balance and waterdrive equations, there are several methods of solutions. Another method used for future performance of waterdrive reservoirs is outlined by Bradley.[3] For this method it is necessary to know (1) the saturations behind the front from core analysis or other sources, (2) the water production as a function of frontal advance, and (3) the pressure gradient in the flooded portion of the reservoir.

The pressure gradient between the new and original front positions is given by

$$\Delta p_{o_n} - \Delta p_{w_n} = \frac{\mu_w W_{e_n}}{k_w} F_G = p_{w_n} - p_{o_n} \qquad \textbf{(4.106)}$$

where the reservoir geometry factor $F_G$ is

For the linear frontal advance:

$$F_G = \frac{L_f}{0.001127hb} \qquad \textbf{(4.107)}$$

For the radial frontal advance:

$$F_G = \frac{2\pi \ln (r_w/r_f)}{0.00708h\alpha} \tag{4.107a}$$

where

$L_f$ = linear penetration of water front into reservoir, ft
$r_f$ = radius of water front after penetration, ft
$\alpha$ = angle subtended by reservoir, radians
$b$ = aquifer width, ft
$\Delta p_{O_n}$ = total reservoir pressure drop from $p_i$ to $p_n$.
$W_{e_n}$ = water influx rate, B/D

If historical pressure gradients between the reservoir and the original reservoir boundary are known, correlation may be developed between $F_G$ and the frontal advance which then can be used to estimate future values of $F_G$.

The water influx rate ($W_{e_n}$) for a reservoir above the bubblepoint pressure can be calculated directly however, a trial-and-error procedure is required for the reservoir producing below the bubblepoint pressure.

**Reservoir above the Bubblepoint Pressure:**

$$W_{e_n} = \frac{\Delta p_{O(n-1)} + (\Delta t q_{t_n}/V_p c_{ot}) - m_r \sum_{j=2}^{n} W_{e(n+1-j)} \Delta p_{D_j}}{m_r \Delta p_{D_j} + (\Delta t/V_p c_{ot}) + (\mu_w F_G/k_w)} \tag{4.108}$$

where

For radial aquifers:

$$m_r = \frac{\mu_w}{0.001127kh\alpha}$$

For infinite linear aquifers:

$$m_r = \frac{\mu_w}{0.001127kh}$$

For finite linear aquifers:

$$m_r = \frac{\mu_w L}{0.001127khb}$$

The calculated $W_{e_n}$ now can be used to calculate $\Delta p_{o_n}$ for the next time interval as

$$\Delta p_{o_n} = \frac{(q_{t_n} - W_{e_n})\Delta t}{V_p c_{ot}} + \Delta p_{o(n-1)} \tag{4.109}$$

where

$L$ = aquifer length, ft
$V_p$ = total reservoir pore volume, bbl
$c_{ot}$ = total reservoir compressibility, 1/psi
$\Delta p_{D_j}$ = dimensionless pressure drop to time period $j$

**Reservoir below Bubblepoint Pressure:** To simplify the calculation procedure, it was assumed that (1) uniform saturations exist ahead of and behind the front, (2) the saturations do not change as any portion of the reservoir is bypassed, and (3) the changes in pressure are selected small enough so that the changes in $B_o$ are very small. The following iterative procedure is to be used:

1. Estimate the pressure drop during the next time interval.
2. Calculate the water influx rate as follows:

$$W_{e_n} = \frac{\Delta p_{o(n-1)} - m_r \sum_{j=2}^{n} W_{e(n+1-j)} \Delta p_{D_j}}{m_r \Delta p_{D_j} - (\mu_w F_G k_w)} \tag{4.110}$$

3. Calculate $\Delta V_n$ and $V_n$ using the following equations:

$$\Delta V_n = \frac{(W_{e_n} - q_{w_n})\Delta t_n}{f_R(1 - S_{iw} - S_{or} - S_{gr})_{n-1}} \tag{4.111}$$

$$V_n = V_{n-1} - \Delta V_n \tag{4.112}$$

where $f_R$ is the fraction of reservoir swept.

4. Calculate the oil saturation in $V_n$ for the predicted oil production during interval $n$ with the following equation:

$$S_{o_n} = \frac{B_{o_n}}{V_n}\left\{\frac{S_{o(n-1)}V_n}{B_{o(n-1)}} + \frac{f_R \Delta V_n \, [S_{o(n-1)} - S_{or_n}]}{B_{o_n}} - q_{o_n}\Delta t_n\right\} \tag{4.113}$$

5. Calculate gas production using

$$\Delta G_{p_n} = \frac{V_n\,[S_{g(n-1)} - S_{gn}]}{B_{g(n-1)}} + \frac{f_R \Delta V_n\,[S_{g(n-1)} - S_{gr_n}]}{B_{gn}} + q_{o_n} \Delta t_n \overline{R}_{s_n} \tag{4.114}$$

6. Calculate the GOR using the following equation:

$$R_n = \left( \frac{k_{rg} \mu_o\, \overline{B}_o}{k_{ro} \mu_g\, \overline{B}_g} \right) + \overline{R}_{sn} \tag{4.115}$$

7. Calculate the GOR using the following equation for average values of pressure and saturation:

$$R_n = \frac{\Delta G_n}{q_{o_n} \Delta t_n} \tag{4.116}$$

8. If the GORs obtained in Steps 6 and 7 are in good agreement, proceed to the next interval. If they do not agree, estimate a new pressure drop and repeat Steps 2 through 8.

## PREDICTING SOLUTION GAS DRIVE BEHAVIOR

In the previous section, we discussed how to predict the behavior of a waterdrive reservoir. In this section we will present techniques used to predict the behavior of solution gas drive reservoirs. Due to the difficulty and uncertainty of calculating some reservoir parameters, such as gas/oil relative permeability, the theoretical prediction of reservoir behavior may not be any more accurate than using statistical forecasting techniques (decline curve analysis) applied to historic performance. However, these methods could have considerable value, and they will provide accurate prediction when the past production is restricted. In this case, reservoir producing characteristics would not be evident from just the past production performance.

The prediction techniques require the same data as used in material balance calculations. The OOIP and OGIP have to be known which are also calculated using appropriate material balance equations. Four techniques are presented; however, the Tarner method is recommended for use because it has a more direct approach to the problem and it is self-correcting.

### BEHAVIOR OF THE PRODUCED GAS/OIL RATIO

A key factor in predicting solution gas drive reservoir performance is to predict the behavior of the produced gas/oil ratio. Above the bubblepoint, the produced gas/oil ratio ($R_s$) is equal to the dissolved gas/oil ratio ($R_s = R_{si}$). As the pressure drops below the bubblepoint pressure and after enough free gas has formed in the reservoir, an additional component will be added due to free gas flow in the reservoir. Thus,

$$R = R_s + R_{flow} \tag{4.117}$$

$$R = R_s + k_g/k_o \frac{\mu_{ob}}{\mu_g} \frac{B_{ob}}{B_g} \tag{4.104}$$

where $B_g$ is in bbls/SCF.

The effective permeabilities in Equation (4.104) are the same as relative permeabilities, $k_{rg}/k_{ro}$, which is a function of the liquid saturation. In solution gas drive reservoirs (no waterdrive), the saturation ($S_{iw}$) is assumed to be constant; however, the oil saturation ($S_o$) is changing continually as more and more oil is produced. Thus it is necessary to calculate $S_o$ corresponding to each cumulative production ($N_p$) point. This can be done using the following equation:

$$S_o = \frac{\text{Reservoir Barrels of Oil}}{\text{Barrels of Pore Volume}} = \frac{B_{ob}(N - N_{p_n})(1 - S_{iw} - S_g)}{NB_{oi}} \qquad \textbf{(4.118)}$$



A typical GOR history for a solution gas drive reservoir is shown in Figure 4–27. As noted previously, GOR is constant until the reservoir pressure reaches the bubblepoint pressure, and free gas begins to form in the reservoir. At this point the GOR will theoretically decline slightly (may not be noticeable in practice) until gas saturation becomes high enough (equilibrium gas saturation) to prevent flow.

Once gas begins to flow, the GOR increases rapidly due to the increase in relative permeability ratio ($k_{rg}/k_{ro}$) with a change in saturation and the lower gas viscosity as compared to viscosity of oil. During the late life of the reservoir the GOR may decline due to the continuous increase in gas formation volume factor ($B_g$) which is inversely proportional to pressure.

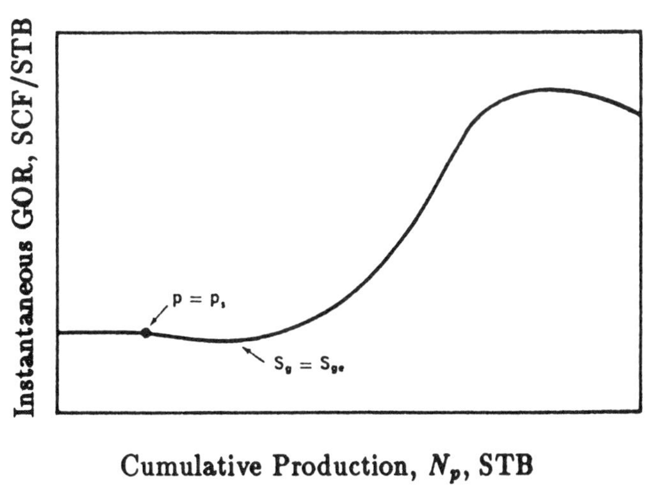


**Figure 4–27.** Typical GOR behavior for a solution gas drive reservoir (from Slider, courtesy of PennWell Books).[20]

### THE SCHILTHUIS METHOD[4]

Once again the prediction is a result of using material balance equation for solution gas drive reservoir, Equation (4.17), rearranged as given here, and an iterative process.

$$N_p = \frac{N[B_o B_{oi} + (R_{si} - R_s)B_g] + G(B_g - B_{gi}) - G_p B_g}{B_o - B_g R_s} \quad \textbf{(4.119)}$$

where $B_g$ is in bbls/SCF.

To predict the cumulative oil production at a particular stage of depletion, it is necessary to know the original oil and gas in the reservoir, the original reservoir pressure, and the PVT data ($B_{oi}$, $R_{si}$, and $B_{gi}$) at original pressure.



The total cumulative gas production $G_p$ is the total sum of the product of an increment of oil production ($\Delta N_p$) and the average produced GOR during the production of that increment. The following equation is used:

$$G_p = \sum_{j=1}^{j=n} R_j \Delta N_{p_j} \quad \textbf{(4.120)}$$

The following steps will help in predicting solution gas drive reservoir behavior:

1. Calculate $k_g/k_o$ and $S_o$ for as much reservoir production and pressure history as possible; use Equations (4.104) and (4.118).
2. Plot ($k_g/k_o$) versus $S_L$ (total liquid saturation, $S_L = S_o + S_w$) on semilogarithmic graph paper. If the laboratory relative permeability data are available, plot that data on the same graph paper and extend the field-calculated permeability data parallel to the laboratory data.
3. Set the reservoir pressure.
4. Assume $G_p$.
5. Using Equation (4.119), calculate $N_p$.
6. Using Equation (4.118), calculate $S_o$.
7. Determine the relative permeability ratio based on the liquid saturation from the $k_g/k_o$ vs. $S_L$ plot in Step 2.
8. Calculate $G_p$ using Equations (4.120) and (4.104).
9. If the $G_p$ assumed in Step 4 agrees satisfactorily with the $G_p$ calculated in Step 8, the prediction is complete. Go to Step 3 to set another pressure.
10. If the assumed and calculated $G_p$ are not in agreement, go to Step 4 and assume another $G_p$.

## THE TARNER METHOD

Tarner suggested to iterate produced GOR at the time when $N_P$ barrels of oil have been produced.[27] The pressure at which cumulative oil and gas ($N_P$ and $G_P$) are desired and the corresponding GOR for iteration are obtained from a plot of instantaneous GOR versus pressure. Data for the plot may be actual historic data or calculated data. The following equation is then used to calculate $N_{p_n}$:

$$N_{p_n} = \frac{N[B_o - B_{oi} + (R_{si} - R_s)B_g] + G(B_g - B_{gi})}{B_o - B_g R_s + (R_n - R_{n-1})B_g/2} + \frac{B_g[G_{p_{n-1}} - (R_n + R_{n-1})N_{p_{n-1}}/2]}{B_o - B_g R_s + (R_n + R_{n-1})B_g/2} \quad \textbf{(4.121)}$$

Once $N_{p_n}$ is calculated, oil saturation ($S_{o_n}$) is then calculated using Equation (4.118). Using liquid saturation ($S_L = S_{o_n} + S_w$) and a previously prepared plot of $k_g/k_o$ versus $S_L$, $k_g/k_o$ at $S_{o_n}$ is obtained. Then, using Equation (4.104), $R_n$ is calculated. If the assumed $R_n$ and the calculated $R_{n_{cal}}$ are in good agreement, proceed to the next pressure; otherwise repeat the procedure using a revised assumed $R_n$. After a correct $R_n$ is achieved, adjust the GOR versus pressure plot accordingly. This will help in assuming a better value for $R_n$ at the next desired pressure.



**Example 4–27.**[20] The following data applies to a solution gas drive reservoir. Calculate (a) $S_o$ and $k_g/k_o$ when the reservoir pressure is 1,302 psi, cumulative production is 1,179,000 STB and 1,123 MMSCF of gas and (b) the cumulative oil and gas production when the reservoir pressure reaches 1,200 psi. Base your $R_{1,200}$ guess on extrapolation of the observed GOR in Figure 4–28 and your $k_g/k_o$ values on the trend established by the field data in Figure 4–29.

$N = 10{,}025{,}000$ STB $\quad p_i = 3{,}013$ psi

$S_w = 0.22 \quad p_b = 2{,}496$ psi

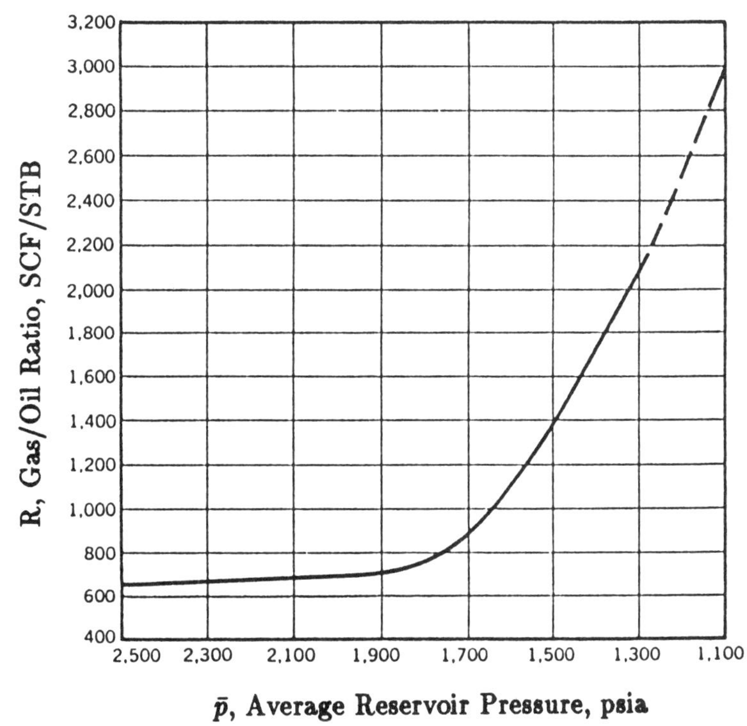


**Figure 4–28.** GOR performance for Example 4–27 (from Slider, courtesy of PennWell Books).[20]

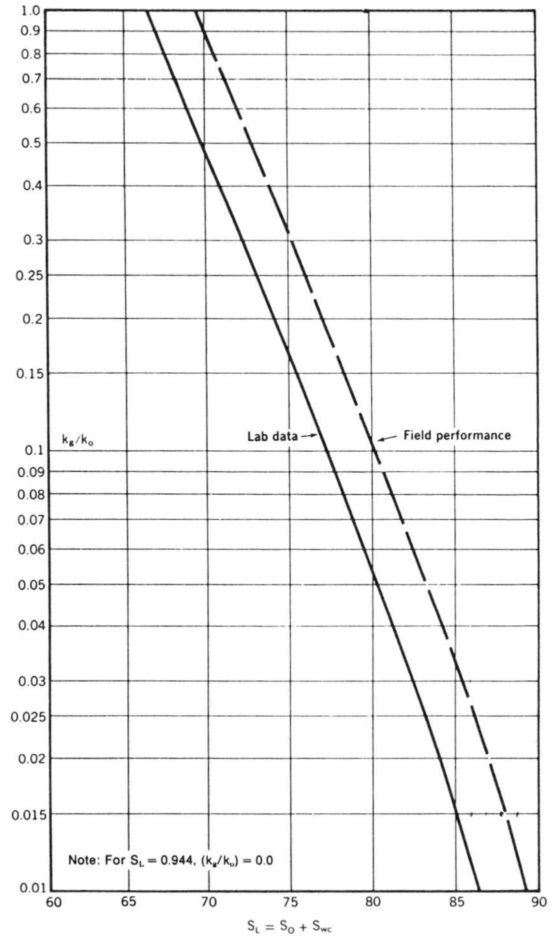


**Figure 4–29.** Comparison of laboratory-measured and field-calculated relative permeability (from Slider, courtesy of PennWell Books).[20]

| $p$ | $B_o$ | $R_s$ | $B_g$ | $\mu_o/\mu_g$ | $R$ |
|---|---|---|---|---|---|
| 3,013 | 1.315 | 650 | 0.000726 | 53.9 | 650 |
| 2,496 | 1.325 | 650 | 0.000796 | 56.6 | 650 |
| 1,302 | 1.233 | 450 | 0.001616 | 102.61 | 2,080 |
| 1,200 | 1.224 | 431 | 0.001807 | 108.96 | ? |

Solution: Using Equation (4.118), $S_o$ is

(a)
$$S_o = \left(1 - \frac{1{,}179{,}000}{10{,}025{,}000}\right)\left(\frac{1.233}{1.315}\right)(1 - 0.22)$$
$$= 0.646$$
$$S_L = S_o + S_w$$
$$= 0.646 + 0.22 = 0.866$$

Using Equation (4.104), calculate $k_g/k_o$.

$$k_g/k_o = \frac{(R - R_s)}{[(\mu_o/\mu_g)(B_o/B_g)]}$$
$$= \frac{(2{,}080 - 450)}{(102.61)\left(\dfrac{1.233}{1.616 \times 10^{-3}}\right)} = 0.0207$$

(b) Using Equation (4.121), calculate $N_{p1{,}200}$ at $p = 1{,}200$ psi.

$$N_{p_n} = \frac{N[B_o - B_{oi} + (R_{si} - R_s)B_g] + G(B_g - B_{gi})}{B_o - B_gR_s + (R_n - R_{n-1})B_g/2} - \frac{B_g[G_{p_{n-1}} - (R_n + R_{n-1})N_{p_{n-1}}/2]}{B_o - B_gR_s + (R_n + R_{n-1})B_g/2} = \frac{A}{B}$$
$$A = 10{,}025{,}000[1.224 - 1.315 + (650 - 431)0.001807] - 0.001807$$
$$\times [1.123 \times 10^9(2{,}080 + R_{1{,}200})1{,}179{,}000/2] = 3{,}241{,}299 + 1{,}065.23R_{1{,}200}$$
$$B = 1.224 - 0.001807 \times 431 + (2{,}080 + R_{1{,}200})0.001807/2 = 2.324 + 0.0009035R_{1{,}200}$$
$$N_{p1{,}200} = \frac{3{,}241{,}299 + 1{,}065.23R_{1{,}200}}{2.324 + 0.0009035R_{1{,}200}}$$

From Figure 4–28, assume an estimated $R_{1{,}200} = 2{,}520$; thus $N_{p1{,}200}$ is

$$N_{p1{,}200} = 1{,}288{,}000 \quad \text{STB}$$

Now calculate $S_{o1{,}200}$ at $p = 1{,}200$ psi.

Using Equation (4.118),

$$S_{o1{,}200} = (1 - 0.22)\left(\frac{1.224}{1.315}\right)\left(1 - \frac{N_{p1{,}200}}{10{,}025{,}000}\right)$$
$$= 0.726\left(1 - \frac{1{,}288{,}000}{10{,}025{,}000}\right)$$
$$= 0.635$$
$$S_L = 0.635 + 0.22 = 0.855$$

From Figure 4–29, $k_g/k_o = 0.0274$ at $S_L = 0.855$,

$$
\begin{aligned}
R_{1,200} &= 431 + \left(\frac{k_g}{k_o}\right)(108.96)\left(\frac{1.224}{0.001807}\right) \\
&= 431 + 73{,}806\left(\frac{k_g}{k_o}\right) \\
&= 431 + (73{,}806)(0.0274) \\
&= 2{,}453
\end{aligned}
$$



The $R_{cal}$ is less than the $R_{est}$; therefore, repeat calculations using the new $R_{est} = R_{cal}$ (i.e., $R_{est} = 2{,}453$ in this case), until $R_{cal} = R_{est}$.

**Second Trial**—$R_{est} = 2{,}453$:

$$
\begin{aligned}
N_{p1,200} &= \frac{3{,}241{,}299 + 1{,}065.23(2{,}453)}{2.324 + 0.0009035(2{,}453)} \\
&= 1{,}289{,}495.196 \\
S_o &= 0.726\left(1 - \frac{1{,}289{,}495.196}{10{,}025{,}000}\right) \\
&= 0.726(1 - 0.1285) = 0.633 \\
S_L &= 0.853 \quad \text{and} \quad k_g/k_o = 0.03 \\
R_{1,200} &= 431 + 73{,}806(0.03) = 2{,}645 \quad \text{psi} \quad > 2{,}453 \quad \text{psi}
\end{aligned}
$$

**Third Trial**—$R_{est} = 2{,}645$:

$$
\begin{aligned}
N_p &= 1{,}285284.758 \\
S_o &= 0.633 \\
S_L &= 0.653 \quad \text{Therefore, } k_g/k_o = 0.03 \\
R_{1,200} &= 2{,}645 = R_{est}
\end{aligned}
$$

Thus $N_{p1,200} \approx 1{,}285{,}285$.

Using rearranged form of Equation (4.105), $G_{p1,200}$ is

$$
\begin{aligned}
\Delta G_p &= \left(\frac{R_{1,302} + R_{1,200}}{2}\right)(\Delta N_p) \\
&= \left(\frac{2{,}080 + 2{,}645}{2}\right)(1{,}285{,}285 - 1{,}179{,}000) \\
&= 251 \quad \text{MMSCF} \\
G_p &= 251 + 1{,}123 \\
&= 1{,}374 \quad \text{MMSCF}
\end{aligned}
$$

## THE TRACY METHOD

Tarner and Tracy solved the same material balance equation for a series of pressure decrements; however, Tracy's method is more convenient and straightforward to implement.[28] The procedure is iterative and usually converges within a few iterations. Both methods should give the same results.

Tracy simplified the following material balance equation below the bubblepoint pressure by introducing pressure functions $\Phi_g$ and $\Phi_o$ as given below.

$$N = \frac{N_p B_o + (G_p - N_p R_s) B_g}{(B_o - B_{oi}) + (R_{si} - R_s) B_g + m B_{oi} \left( \frac{B_g - B_{gi}}{B_{gi}} \right)}$$

The pressure functions are given by

$$\Phi_g = \frac{1}{\left( \frac{B_o - B_{oi}}{B_g} \right) - (R_s - R_{si}) + m B_{oi} \left( \frac{1}{B_{gi}} - \frac{1}{B_g} \right)} \quad \textbf{(4.122)}$$

$$\Phi_o = \left( \frac{B_o}{B_g - R_s} \right) \Phi_g \quad \textbf{(4.123)}$$

With these $\Phi$ functions, the material balance equation becomes

$$N = N_p \Phi_o + G_p \Phi_g \quad \textbf{(4.124)}$$

When applied to pressure decrement from $p_{n-1}$ to $p_n$,

$$N = [(N_p)_{n-1} + \Delta N_p](\Phi_o)_n + [(G_p)_{n-1} + \overline{R} \Delta N_p](\Phi_g)_n \quad \textbf{(4.125)}$$

Here the average producing GOR is given by

$$\overline{R} = \frac{\Delta G_p}{\Delta N_p} = \frac{R_{n-1} + R_n}{2}$$

Solving Equation (4.125) for $\Delta N_p$,

$$\Delta N_p = \frac{N - (N_p)_{n-1}(\Phi_o)_n - (G_p)_{n-1}(\Phi_g)_n}{(\Phi_o)_n + R(\Phi_g)_n} \quad \textbf{(4.126)}$$

The $\Phi$ values are functions only of gas cap size and of pressure, and they have to be calculated only once for each pressure level. The following stepwise procedure has to be used:

1. Estimate the average GOR, $\overline{R}$ for the pressure level from $p_{n-1}$ to $p_n$.
2. Using Equation (4.126), calculate $\Delta N_p$.
3. Calculate $(N_p)_n = (N_p)_{n-1} + \Delta N_p$.
4. Compute oil saturation $(S_o)_n$ using Equation (4.118) and compute $S_L = (S_o)_n + S_{iw}$.
5. Determine $(k_g/k_o)$ corresponding to the liquid saturation $(S_L)$.
6. Compute $R_n$ using Equation (4.104).
7. Compute new estimate of $\overline{R} = (R_{n-1} + R_n)/2$.
8. Compute new estimate of $\Delta N_p$ as in Step 2 and of $(N_p)_n$ as in Step 3.
9. Compute $(G_p)_n$ from $(G_p)_{n-1} + \overline{R}\Delta N_p$.
10. Compute the estimated OOIP (N) using Equation (4.124) or (4.125).
11. If the following criteria is satisfied, go to Step 12. Otherwise go to Step 4 and iterate with the new $(N_p)_n$ calculated in Step 8.

$$0.999 \leq \frac{R_{step-7}}{\overline{R}_{old}} \leq 1.001$$

12. To test material balance, the following criteria should be satisfied:

$$0.999 \leq \frac{N_c}{N_s} \leq 1.001$$

where $N_c$ is the computed OOIP and $N_s$ is the specified OOIP. If this is satisfied, an adequate material balance is obtained; otherwise go to Step 4 and reiterate. In most cases the material balance test will be satisfied once the GOR test is satisfied in Step 11.

The Tracy method was originally derived for use in below the bubblepoint pressure reservoirs; however, to use the method for the full pressure range of an initially undersaturated oil, the following considerations are pertinent:

1. The computed recovery will be a fraction of the OOIP at abandonment pressure rather than at the bubblepoint pressure.
2. Bubblepoint pressure should be used as one of the pressure level $(p_n)$ in order to properly account for the gas evaluation starting at the bubblepoint pressure.
3. The effect of rock and fluid compressibilities must be added to the material balance equation. They can be indirectly added by use of pseudovalues of $B_o$ at pressure below the initial pressure. The following equation is used to calculate the pseudovalue $(B_o^*)$:

$$B_o^* = B_o B_{oi}\left(\frac{S_{iw} + c_f}{1 - S_{iw}}\right)(p_{iR} - p_R) \tag{4.127}$$

4. The method does not require iterations because above the bubblepoint pressure, $R = R_s$ can be estimated more accurately.

## THE MUSKAT AND TAYLOR METHOD

Muskat and Taylor suggested the following differential type of material balance equation that can be used to predict the solution gas drive reservoir behavior.[29] The method is applicable to the tank-type material balance with no initial gas cap. Unlike the Schilthuis and the Tarner techniques, this method of prediction is not self-correcting, and therefore the results should be periodically checked using the regular material balance equations.

$$\frac{\Delta S_o}{\Delta p} = \frac{\dfrac{\Delta B_g S_o}{B_o}\dfrac{\Delta R_s}{\Delta p} + \dfrac{S_o k_g \mu_o}{B_o k_o \mu_g}\dfrac{\Delta B_o}{\Delta p} + (1 - S_o - S_{iw})B_g\dfrac{\Delta(1/B_g)}{\Delta p}}{1 + \dfrac{k_g \mu_o}{k_o \mu_g}} \tag{4.128}$$

Note that $\Delta R_s/\Delta p$, $\Delta B_o/\Delta p$, and $\Delta B_g/\Delta p$ in the above equation represent the slope of a curve of $R_s$ vs. $p$, $B_o$ vs. $p$, and $B_g$ vs. $p$, respectively. The equation will provide direct calculation of the slope of a curve ($\Delta S_o/\Delta p$) for a plot of $S_o$ vs. $p$.

Equation (4.128) is used to calculate a finite change in oil saturation corresponding a finite change in pressure. The slopes $\Delta R_s/\Delta p$, $\Delta B_o/\Delta p$, and $\Delta B_g/\Delta_p$ are evaluated at average pressure. The method is theoretically a trial and error solution in which $S_o$ and $k_g/k_o$ at average pressure are assumed and $\Delta S_o$ calculated until it is found that they are satisfied by the calculated change in $\Delta S_o$.

The method can be used without trial-and-error when it is incorporated into a computer program. Using a computer, such small pressure increments can be used which will not introduce a significant error into calculations because of very small corresponding change in $S_o$. The accuracy is further improved by basing $S_o$ and $k_g/k_o$ on a saturation equal to $S_{On-1} + 0.5\Delta S_{On-1}$.

## RATE-TIME FORECAST FROM MATERIAL BALANCE

In the previous sections, we discussed techniques which can be used to predict performance of solution gas drive reservoirs. This information and the original oil and gas in place calculated using material balance equations are only a part of the information required by the reservoir engineer. Usually, a rate-time forecast also will be required.

In this section we will discuss how to predict the time and rates corresponding to the information provided from an oil material balance previously discussed. The material presented here requires calculating the well productivity index ($J$ or $PI$) and predicting its decline. The following equations are utilized:

Above the bubblepoint pressure:

$$q_o = q_{oi} \frac{p - p_{wf}}{p_i - p_{wf}} \frac{k_{ro} B_{oi} \mu_{oi}}{k_{roi} B_o \mu_o} \tag{4.129}$$

Below the bubblepoint pressure:

$$q_o = q_{oi} \frac{p - p_{wf}}{p_i - p_{wf}} \frac{k_{ro}}{k_{roi}} \frac{B_{oi} \mu_{oi}}{B_{ob} \mu_{ob}}$$

$$k_{ro} = \left( \frac{\%S_o}{100 - \%S_{iw}} \right)^{a+b} \tag{4.129a}$$

where $a$ and $b$ are the coefficients of $k_{ro}$ correlation. The $S_o$ can be calculated using Equation (4.118); $k_{roi}$ is the relative oil permeability at $S_{oi}$ at $p_i$. If $q_i$ is unknown, it is calculated from the initial productivity index. The corresponding time is calculated at the $i$th pressure step as follows:

$$t_i = t_{i-1} + \frac{\ln (q_{oi}/q_{oi-1})(N_{pi} - N_{pi-1})}{(q_{oi} - q_{oi-1})365} \tag{4.130}$$

Gas producing rates are predicted based on the instantaneous values of $R_p$ ($q_g = R_p q_o$).

## GAS-CONDENSATE RESERVOIRS

The properties of gas-condensate reservoirs are materially different from those of dry gas reservoirs discussed in this chapter. The composition of the reservoir fluid is still predominantly methane, although the amounts of the heavier hydrocarbons are relatively increased. These reservoir fluids are commonly referred to as *gas condensate* and the reservoirs are referred to as gas-condensate reservoirs. Gas-condensate production is predominantly gas from which more or less liquid is *condensed* in the surface separators. The term *wet gas* is sometimes used as more or less equivalent to gas condensate.

Although reservoir fluid classification properly depends upon (1) the composition of the hydrocarbon accumulation and (2) the temperature and pressure at which these hydrocarbons exist in the reservoir; however, gas-condensate reservoirs may be approximately identified as those producing light colored or colorless stock-tank liquids with API gravities above 45° API (ranging from 45° to 70° API) at gas/oil ratios in the range of 5,000 to 100,000 SCF/STB. The gas/oil ratios and consequently the API gravity of the produced liquid will vary with the number of separators used and their operating temperature and pressures.

The gas/oil ratio also increases due to (a) pressure decline in oil reservoirs and some gas-condensate reservoirs owing to the reservoir mechanism and (b) decline in separator efficiencies due to decline in wellhead pressures.

When liquid is condensed from the reservoir gas upon reduction of temperature and pressure, *retrograde condensation* takes place. In gas-condensate reservoirs, 85% of the dry gas may be recovered by pressure depletion; however, it is not unusual to lose as much as 50% to 60% of the heavier components due to retrograde condensation.

## RESERVOIR TYPES DEFINED WITH REFERENCE TO PHASE DIAGRAMS

Pressure-temperature phase diagrams are typically utilized for the identification of the various types of reservoirs by locating the initial reservoir temperature and pressure on the phase diagram. A typical pressure-temperature phase diagram of a particular reservoir fluid is shown in Figure 4–30. Each hydrocarbon accumulation will have its own phase diagram derived from the composition of the accumulation.

The area enclosed by the *bubblepoint* and *dew point* lines is the pressure-temperature region in which both gas and liquid phases will coexist. The curves within the two-phase region referred to as *quality lines* show the percentage of the total liquid hydrocarbon volume for any temperature and pressure. These lines intersect the *critical point* and are essentially parallel to the bubblepoint and dew point curves. The bubblepoint curve represents 0% vapor (100% liquid by volume) and the dew point curve 100% vapor (0% liquid by volume). Critical point is that state of the pressure and temperature at which the intensive properties of each phase are identical. The temperature and pressure at this point are referred to as *critical temperature* and *critical pressures*, respectively.

The highest pressure at which the liquid and vapor can coexist in equilibrium is called

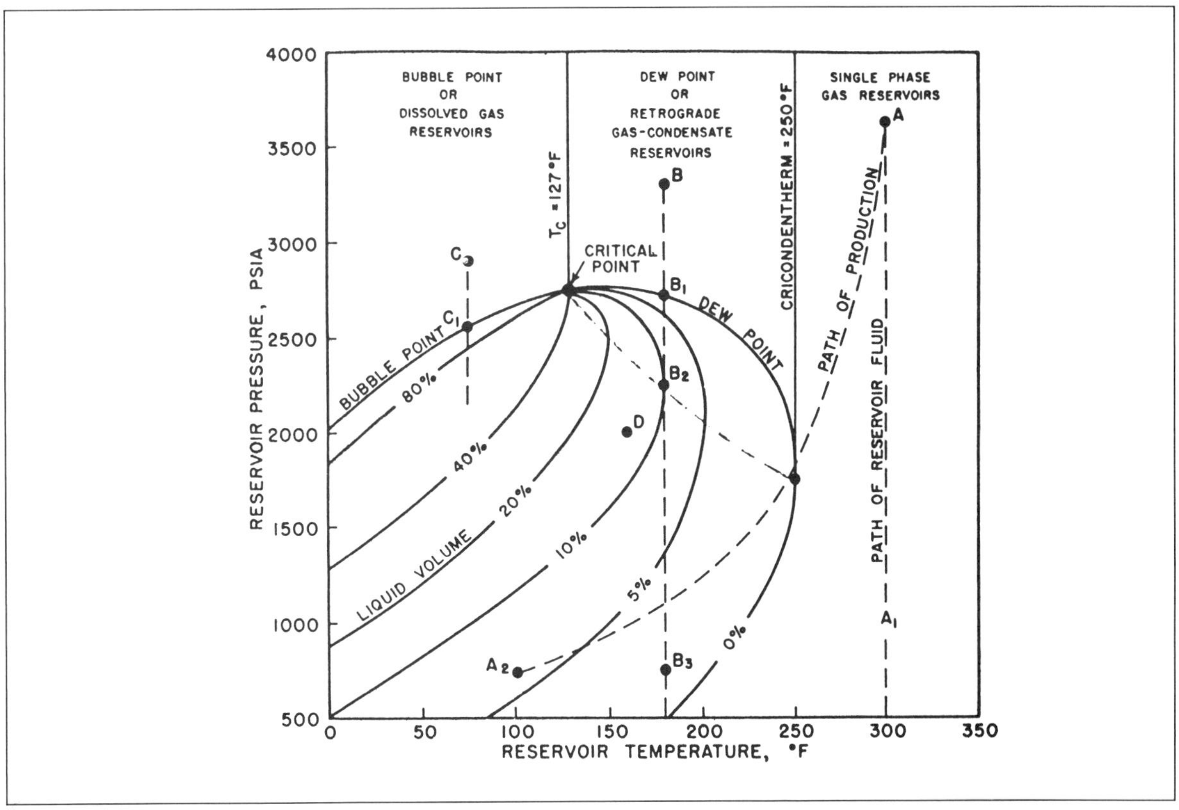


**Figure 4–30.** Pressure-temperature phase diagram for a reservoir fluid (courtesy of Prentice-Hall, Inc.).[6]

*cricondenbar.* This point falls slightly to the right of the critical point. The highest temperature at which liquid and vapor may coexist in equilibrium is called *cricondentherm.*

The critical temperature of the normally occurring condensate terms is lower than the temperature of the reservoir in which they are found. Location of the critical point is most important because it defines the shape of the quality lines which in turn governs the vapor/liquid ratio at a given temperature and pressure within the phase envelope. The shaded area inside the phase envelope is the *retrograde region* where condensation of liquids occurs by lowering pressure or increasing temperature.

In Figure 4–30, four different kinds of reservoirs (Points A, B, C, and D) are shown. Points A, B, C, and D represent initial bottomhole pressure and temperature. The vertical line ($A - A_1$) from these points represents reservoir pressure decline at constant temperature. The curved line ($A - A_2$) represents pressure and temperature changes that occur in the wellbore. Point $A_2$ represents wellhead conditions. Some qualitative interpretation of these reservoirs with respect to the phase diagram follows.

Assuming that the phase diagram represents the fluid of Reservoir A at an initial temperature and pressure of 300°F and 3,700 psia, respectively. Note that Point A is outside the two-phase region; thus it is originally in a single phase as gas and occurs at a temperature above the cricondentherm. Since the temperature does not change with production, the fluid in the reservoir will remain in gaseous state and the composition of the well fluid will not change as the pressure declines along the Path $A - A_1$. However, as shown by $A - A_2$, due to the temperature difference at the surface, production may enter the two-phase region at the surface. This accounts for the condensate production at the surface separators.

Next let us consider Reservoir B containing the same fluid as Figure 4–30 but at an initial temperature and pressure of 180°F and 3,300 psia, respectively. Since the temperature is higher than the critical temperature and the pressure is higher than the dew point pressure, the fluid in the reservoir is initially in the single-phase region as gas. From Point B to $B_1$, the reservoir is in a single phase (gas). Below Point $B_1$, liquid starts to condense out of the reservoir fluid, this type of reservoir is referred to as *dew point reservoir.* This process of retrograde condensation continues until a maximum liquid volume of 10% is reached at Point $B_2$. For qualitative purposes *vaporization* of the retrograde liquid occurs from $B_2$ to the abandonment pressure $B_3$.

If accumulation occurs at Point C with a reservoir pressure and temperature above the bubblepoint pressure and critical temperature, respectively, the reservoir would lie in a single phase (liquid). The reservoir at this point is called *bubblepoint reservoir.* As the pressure declines (with production) below the bubblepoint pressure of $C_1$, bubbles, or free gas, will evolve from the liquid. Eventually the free gas evolved begins to flow to the wellbore in ever increasing quantities, ultimately leaving much unrecovered oil in the reservoir. These types of reservoirs are also referred to as depletion, dissolved, solution gas drive, expansion, and internal gas drive.

When the pressure and temperature lie within the two-phase region as Point D, an oil zone will exist with an overlaying gas cap. In such cases the two phases may be represented by separate phase diagrams. The oil zone will produce as a bubblepoint oil reservoir and the gas cap either as a single-phase gas reservoir or a retrograde gas-condensate reservoir, depending on the gas composition and the temperature and pressure of the gas cap.

## CALCULATION OF VAPOR/LIQUID EQUILIBRIUM

The distribution of a component of a system between vapor and liquid is expressed by the *equilibrium ratio* K, also referred to as *equilibrium constant*. The equilibrium constant is the ratio of the mole fraction of the components in the *vapor phase* to the mole fraction of the component in the *liquid phase*.

$$K_i = \frac{y_i}{x_i} \tag{4.131}$$



where

$K_i$ = equilibrium ratio of component $i$ at pressure $p$ and some temperature $T$
$y_i$ = mole fraction of component $i$ in the vapor phase
$x_i$ = mole fraction of component $i$ in the liquid phase

The K-value of each component in a system is a function of the system temperature, system pressure, and the composition of the vapor and liquid phase. The composition of the system considerably affects the K-values at pressures above 1,000 psi; below 1,000 psi the effect is relatively small. The K-values can be physically measured for each reservoir fluid, it may be estimated from empirical correlations, or it may be selected from a large number of K-value measurements available in the literature.

The effect of composition is correlated by means of convergence pressures in the most extensive set of published equilibrium ratios for hydrocarbons, the GPSA *Engineering Data Book*, a sample of which is as shown in Figure 4–31.[30] Charts like these are tedious to use; reading them incorrectly is easy. Therefore the same data is presented in table form in Table 4–19. The table is based on a log-log plot of K vs. $p$. Over any small pressure interval, a given temperature line may be assumed to be linear for interpolation purposes.

At equilibrium, the mole fractions of the components in both liquid and vapor phases must sum to unity. Since the total number of moles of composite (mixture input to separator) is equal to the sum of the moles in the liquid and vapor phase (output from separator). By material balance the following equations result:

$$\sum_{i=1}^{m} x_i = \sum_{i=1}^{m} \frac{z_i}{L + VK_i} = 1 \tag{4.132}$$

$$\sum_{i=1}^{m} y_i = \sum_{i=1}^{m} \frac{z_i K_i}{L + VK_i} = 1 \tag{4.133}$$

where $z_i$ is the mole fraction of component $i$ in the total mixture, $L$ is the mole fraction of *liquid phase* in the total mixture, and V is the mole fraction of *vapor phase* in the total mixture. These equations require trial-and-error calculations, each time assuming a value for V or $L$ until a value is

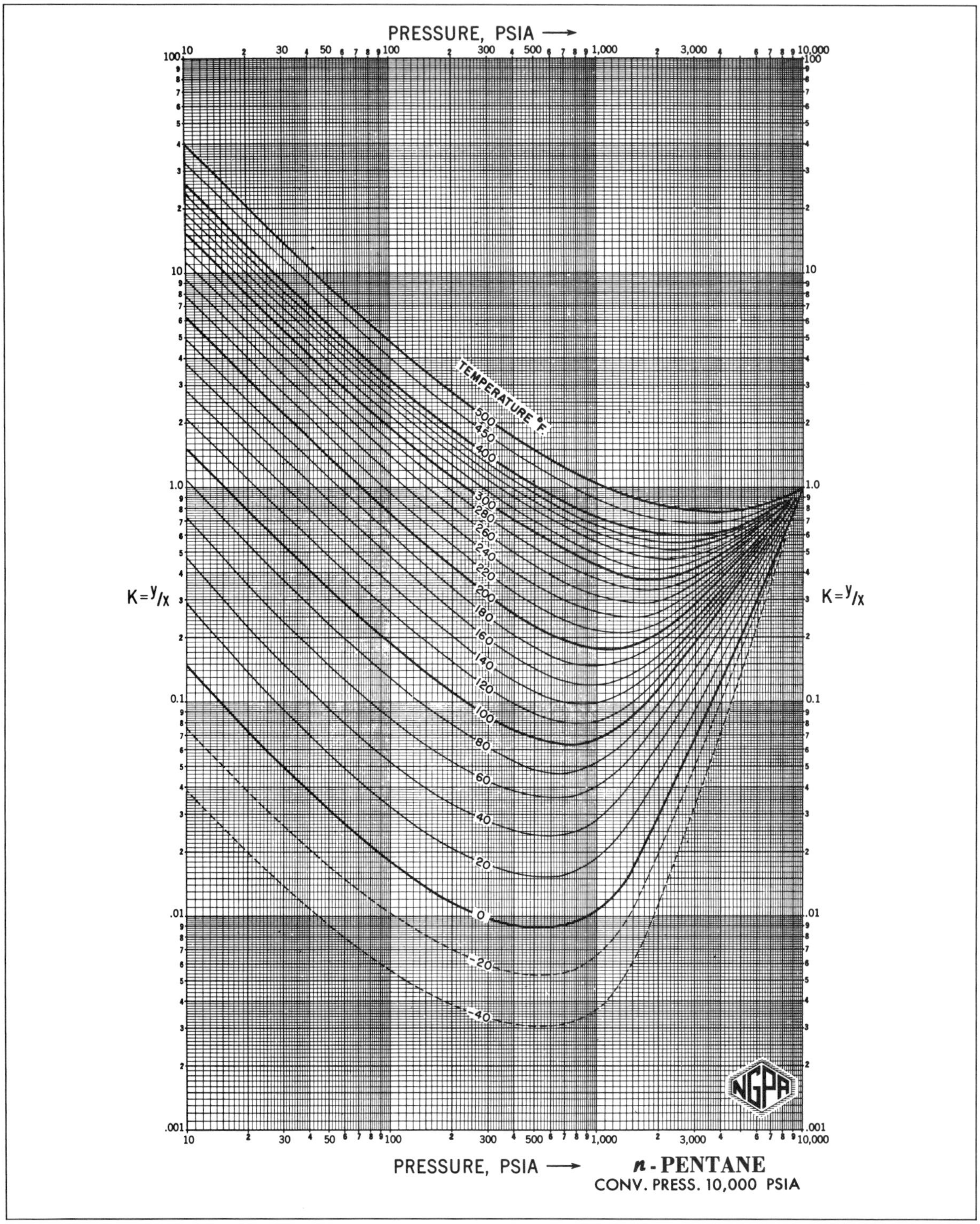


**Figure 4–31.** Equilibrium *K*-value chart (courtesy of GPSA).[30]

## TABLE 4–17. Equilibrium K-values (courtesy of J. M. Campbell & Co.)[31]

Pressure = 50.0 psia

| Comp | −20.0 | 0.0 | 20.0 | 40.0 | 60.0 | 80.0 | 100.0 | 120.0 | 140.0 | 160.0 | 180.0 | 200.0 | 250.0 | 300.0 | 400.0 |
|---|---|---|---|---|---|---|---|---|---|---|---|---|---|---|---|
| | | | | | | | Temperature, Deg F | | | | | | | | |
| N2 | 152.710 | 171.562 | 190.880 | 210.567 | 230.536 | 250.709 | 271.018 | 291.401 | 311.806 | 332.185 | 352.497 | 372.708 | 422.602 | 471.315 | 564.334 |
| CO2 | 9.8472 | 11.7917 | 13.8920 | 16.1287 | 18.4793 | 20.9189 | 23.4210 | 25.9574 | 28.4998 | 31.0194 | 33.4879 | 35.8778 | 41.3422 | 45.8035 | 50.7948 |
| H2S | 1.7350 | 2.4712 | 3.3535 | 4.3706 | 5.5061 | 6.7401 | 8.0514 | 9.4188 | 10.8219 | 12.2423 | 13.6635 | 15.0717 | 18.4630 | 21.5784 | 26.7840 |
| C1 | 32.5598 | 35.4868 | 38.4816 | 41.5235 | 44.5898 | 47.6554 | 50.6944 | 53.6790 | 56.5811 | 59.3721 | 62.0233 | 64.5069 | 69.8099 | 73.5337 | 75.3228 |
| C2 | 2.9782 | 3.9182 | 5.0150 | 6.2647 | 7.6583 | 9.1826 | 10.8206 | 12.5522 | 14.3553 | 16.2063 | 18.0809 | 19.9547 | 24.4833 | 28.5305 | 34.2541 |
| C3 | 0.5224 | 0.7788 | 1.1136 | 1.5353 | 2.0499 | 2.6601 | 3.3662 | 4.1651 | 5.0512 | 6.0166 | 7.0518 | 8.1455 | 11.0573 | 14.0715 | 19.7641 |
| IC4 | 0.1499 | 0.2411 | 0.3694 | 0.5423 | 0.7668 | 1.0488 | 1.3924 | 1.8003 | 2.2730 | 2.8089 | 3.4045 | 4.0548 | 5.8727 | 7.8562 | 11.7456 |
| NC4 | 0.0931 | 0.1554 | 0.2459 | 0.3717 | 0.5392 | 0.7546 | 1.0225 | 1.3464 | 1.7279 | 2.1669 | 2.6613 | 3.2074 | 4.7626 | 6.4973 | 9.9977 |
| IC5 | 0.0271 | 0.0491 | 0.0839 | 0.1356 | 0.2090 | 0.3088 | 0.4398 | 0.6058 | 0.8103 | 1.0556 | 1.3431 | 1.6729 | 2.6751 | 3.8981 | 6.7649 |
| NC5 | 0.0180 | 0.0339 | 0.0597 | 0.0993 | 0.1568 | 0.2365 | 0.3429 | 0.4796 | 0.6500 | 0.8565 | 1.1003 | 1.3818 | 2.2448 | 3.3062 | 5.8134 |
| C6 | 0.0040 | 0.0080 | 0.0151 | 0.0269 | 0.0455 | 0.0734 | 0.1137 | 0.1694 | 0.2443 | 0.3416 | 0.4648 | 0.6168 | 1.1365 | 1.8652 | 3.8432 |
| C7 | 0.0010 | 0.0021 | 0.0042 | 0.0080 | 0.0145 | 0.0249 | 0.0409 | 0.0646 | 0.0984 | 0.1451 | 0.2075 | 0.2887 | 0.5927 | 1.0677 | 2.5335 |
| C8 | 0.0003 | 0.0006 | 0.0013 | 0.0025 | 0.0046 | 0.0082 | 0.0138 | 0.0227 | 0.0359 | 0.0552 | 0.0825 | 0.1201 | 0.2760 | 0.5542 | 1.5788 |
| C9 | 0.0001 | 0.0002 | 0.0005 | 0.0009 | 0.0017 | 0.0030 | 0.0053 | 0.0089 | 0.0146 | 0.0232 | 0.0359 | 0.0542 | 0.1363 | 0.2993 | 1.0058 |
| C10 | 0.0000 | 0.0001 | 0.0001 | 0.0003 | 0.0006 | 0.0011 | 0.0020 | 0.0035 | 0.0060 | 0.0099 | 0.0159 | 0.0248 | 0.0680 | 0.1610 | 0.6242 |

Pressure = 100.0 psia

| Comp | −20.0 | 0.0 | 20.0 | 40.0 | 60.0 | 80.0 | 100.0 | 120.0 | 140.0 | 160.0 | 180.0 | 200.0 | 250.0 | 300.0 | 400.0 |
|---|---|---|---|---|---|---|---|---|---|---|---|---|---|---|---|
| N2 | 79.495 | 89.238 | 99.214 | 109.374 | 119.672 | 130.070 | 140.532 | 151.027 | 161.527 | 172.010 | 182.455 | 192.842 | 218.471 | 243.473 | 291.167 |
| CO2 | 5.3295 | 6.3417 | 7.4274 | 8.5762 | 9.7765 | 11.0158 | 12.2810 | 13.5586 | 14.8353 | 16.0974 | 17.3321 | 18.5267 | 21.2628 | 23.5193 | 26.1842 |
| H2S | 0.9956 | 1.4130 | 1.9117 | 2.4851 | 3.1239 | 3.8168 | 4.5521 | 5.3179 | 6.1029 | 6.8969 | 7.6910 | 8.4772 | 10.3697 | 12.1076 | 15.0127 |
| C1 | 17.7568 | 19.2705 | 20.8078 | 22.3583 | 23.9109 | 25.4538 | 26.9747 | 28.4606 | 29.8989 | 31.2764 | 32.5804 | 33.7985 | 36.3923 | 38.2180 | 39.1711 |
| C2 | 1.5996 | 2.0870 | 2.6512 | 3.2897 | 3.9973 | 4.7673 | 5.5913 | 6.4593 | 7.3610 | 8.2851 | 9.2203 | 10.1554 | 12.4231 | 14.4737 | 17.5031 |
| C3 | 0.2944 | 0.4332 | 0.6127 | 0.8369 | 1.1084 | 1.4286 | 1.7972 | 2.2126 | 2.6717 | 3.1705 | 3.7038 | 4.2659 | 5.7559 | 7.2875 | 10.1298 |
| IC4 | 0.0832 | 0.1318 | 0.1996 | 0.2905 | 0.4081 | 0.5556 | 0.7354 | 0.9493 | 1.1979 | 1.4808 | 1.7967 | 2.1431 | 3.1200 | 4.1975 | 6.3226 |
| NC4 | 0.0530 | 0.0863 | 0.1343 | 0.2003 | 0.2880 | 0.4008 | 0.5416 | 0.7125 | 0.9150 | 1.1497 | 1.4159 | 1.7120 | 2.5637 | 3.5196 | 5.3934 |
| IC5 | 0.0164 | 0.0290 | 0.0485 | 0.0771 | 0.1172 | 0.1714 | 0.2420 | 0.3313 | 0.4410 | 0.5727 | 0.7273 | 0.9050 | 1.4482 | 2.1171 | 3.6998 |
| NC5 | 0.0105 | 0.0193 | 0.0336 | 0.0551 | 0.0863 | 0.1293 | 0.1867 | 0.2604 | 0.3525 | 0.4643 | 0.5968 | 0.7504 | 1.2247 | 1.8126 | 3.1993 |
| C6 | 0.0025 | 0.0049 | 0.0090 | 0.0157 | 0.0262 | 0.0417 | 0.0638 | 0.0942 | 0.1348 | 0.1875 | 0.2539 | 0.3358 | 0.6164 | 1.0131 | 2.1154 |
| C7 | 0.0006 | 0.0013 | 0.0026 | 0.0048 | 0.0085 | 0.0143 | 0.0231 | 0.0360 | 0.0541 | 0.0789 | 0.1120 | 0.1549 | 0.3152 | 0.5677 | 1.3762 |
| C8 | 0.0002 | 0.0004 | 0.0008 | 0.0015 | 0.0028 | 0.0048 | 0.0080 | 0.0129 | 0.0201 | 0.0306 | 0.0453 | 0.0653 | 0.1481 | 0.2959 | 0.8549 |
| C9 | 0.0001 | 0.0002 | 0.0003 | 0.0006 | 0.0011 | 0.0018 | 0.0031 | 0.0052 | 0.0083 | 0.0130 | 0.0198 | 0.0296 | 0.0733 | 0.1599 | 0.5393 |
| C10 | 0.0000 | 0.0000 | 0.0001 | 0.0002 | 0.0004 | 0.0007 | 0.0012 | 0.0020 | 0.0034 | 0.0055 | 0.0087 | 0.0135 | 0.0361 | 0.0850 | 0.3315 |

Pressure = 200.0 psia

| Comp | −20.0 | 0.0 | 20.0 | 40.0 | 60.0 | 80.0 | 100.0 | 120.0 | 140.0 | 160.0 | 180.0 | 200.0 | 250.0 | 300.0 | 400.0 |
|---|---|---|---|---|---|---|---|---|---|---|---|---|---|---|---|
| N2 | 40.640 | 45.548 | 50.565 | 55.667 | 60.832 | 66.040 | 71.275 | 76.521 | 81.765 | 86.995 | 92.202 | 97.376 | 110.125 | 122.543 | 146.183 |
| CO2 | 2.9058 | 3.4274 | 3.9836 | 4.5693 | 5.1787 | 5.8059 | 6.4447 | 7.0886 | 7.7312 | 8.3662 | 8.9872 | 9.5884 | 10.9681 | 12.1125 | 13.4911 |
| H2S | 0.5837 | 0.8214 | 1.1032 | 1.4251 | 1.7818 | 2.1668 | 2.5737 | 2.9960 | 3.4275 | 3.8628 | 4.2971 | 4.7264 | 5.7567 | 6.7001 | 8.2735 |
| C1 | 9.6268 | 10.3640 | 11.1092 | 11.8576 | 12.6040 | 13.3433 | 14.0697 | 14.7775 | 15.4610 | 16.1145 | 16.7321 | 17.3085 | 18.5353 | 19.4013 | 19.8732 |
| C2 | 0.8771 | 1.1335 | 1.4285 | 1.7607 | 2.1278 | 2.5263 | 2.9520 | 3.4003 | 3.8660 | 4.3435 | 4.8272 | 5.3117 | 6.4903 | 7.5620 | 9.1586 |
| C3 | 0.1666 | 0.2434 | 0.3421 | 0.4647 | 0.6125 | 0.7862 | 0.9854 | 1.2091 | 1.4558 | 1.7230 | 2.0081 | 2.3078 | 3.0994 | 3.9084 | 5.3920 |
| IC4 | 0.0524 | 0.0806 | 0.1192 | 0.1700 | 0.2351 | 0.3160 | 0.4141 | 0.5304 | 0.6654 | 0.8192 | 0.9913 | 1.1806 | 1.7192 | 2.3222 | 3.5402 |
| NC4 | 0.0345 | 0.0542 | 0.0818 | 0.1191 | 0.1680 | 0.2301 | 0.3071 | 0.4003 | 0.5106 | 0.6386 | 0.7842 | 0.9470 | 1.4214 | 1.9659 | 3.0783 |
| IC5 | 0.0112 | 0.0191 | 0.0309 | 0.0478 | 0.0711 | 0.1021 | 0.1424 | 0.1929 | 0.2550 | 0.3295 | 0.4171 | 0.5181 | 0.8293 | 1.2171 | 2.1426 |
| NC5 | 0.0072 | 0.0128 | 0.0215 | 0.0344 | 0.0527 | 0.0776 | 0.1107 | 0.1531 | 0.2059 | 0.2703 | 0.3468 | 0.4360 | 0.7141 | 1.0638 | 1.8934 |
| C6 | 0.0018 | 0.0035 | 0.0062 | 0.0105 | 0.0172 | 0.0267 | 0.0401 | 0.0583 | 0.0824 | 0.1133 | 0.1521 | 0.1997 | 0.3623 | 0.5928 | 1.2452 |
| C7 | 0.0005 | 0.0010 | 0.0019 | 0.0034 | 0.0057 | 0.0094 | 0.0148 | 0.0225 | 0.0331 | 0.0475 | 0.0663 | 0.0905 | 0.1801 | 0.3210 | 0.7883 |
| C8 | 0.0002 | 0.0003 | 0.0006 | 0.0011 | 0.0019 | 0.0033 | 0.0053 | 0.0083 | 0.0127 | 0.0189 | 0.0275 | 0.0391 | 0.0861 | 0.1696 | 0.4922 |
| C9 | 0.0001 | 0.0001 | 0.0003 | 0.0005 | 0.0008 | 0.0013 | 0.0022 | 0.0034 | 0.0054 | 0.0082 | 0.0123 | 0.0180 | 0.0432 | 0.0925 | 0.3095 |
| C10 | 0.0000 | 0.0000 | 0.0001 | 0.0002 | 0.0003 | 0.0005 | 0.0009 | 0.0014 | 0.0023 | 0.0036 | 0.0055 | 0.0083 | 0.0214 | 0.0490 | 0.1874 |

Pressure = 300.0 psia

| Comp | −20.0 | 0.0 | 20.0 | 40.0 | 60.0 | 80.0 | 100.0 | 120.0 | 140.0 | 160.0 | 180.0 | 200.0 | 250.0 | 300.0 | 400.0 |
|---|---|---|---|---|---|---|---|---|---|---|---|---|---|---|---|
| N2 | 27.1019 | 30.3246 | 33.6141 | 36.9545 | 40.3317 | 43.7331 | 47.1477 | 50.5660 | 53.9794 | 57.3808 | 60.7639 | 64.1234 | 72.3902 | 80.4286 | 95.7005 |
| CO2 | 2.0622 | 2.4153 | 2.7905 | 3.1844 | 3.5935 | 4.0139 | 4.4416 | 4.8725 | 5.3025 | 5.7273 | 6.1429 | 6.5452 | 7.4692 | 8.2358 | 9.1542 |
| H2S | 0.4371 | 0.6094 | 0.8122 | 1.0421 | 1.2952 | 1.5671 | 1.8531 | 2.1486 | 2.4496 | 2.7522 | 3.0534 | 3.3503 | 4.0606 | 4.7085 | 5.7851 |
| C1 | 6.7105 | 7.1833 | 7.6595 | 8.1363 | 8.6104 | 9.0787 | 9.5376 | 9.9838 | 10.4136 | 10.8237 | 11.2106 | 11.5709 | 12.3356 | 12.8730 | 13.1590 |

## TABLE 4–17. Continued

| Comp | −20.0 | 0.0 | 20.0 | 40.0 | 60.0 | 80.0 | 100.0 | 120.0 | 140.0 | 160.0 | 180.0 | 200.0 | 250.0 | 300.0 | 400.0 |
|---|---|---|---|---|---|---|---|---|---|---|---|---|---|---|---|
| | | | | | | | Temperature, Deg F | | | | | | | | |
| C2 | 0.6337 | 0.8121 | 1.0166 | 1.2463 | 1.4997 | 1.7746 | 2.0684 | 2.3780 | 2.7000 | 3.0306 | 3.3660 | 3.7024 | 4.5226 | 5.2690 | 6.3683 |
| C3 | 0.1252 | 0.1810 | 0.2522 | 0.3401 | 0.4456 | 0.5689 | 0.7100 | 0.8681 | 1.0421 | 1.2303 | 1.4310 | 1.6420 | 2.1990 | 2.7686 | 3.8128 |
| IC4 | 0.0440 | 0.0659 | 0.0952 | 0.1333 | 0.1814 | 0.2405 | 0.3118 | 0.3958 | 0.4930 | 0.6036 | 0.7272 | 0.8635 | 1.2525 | 1.6919 | 2.5955 |
| NC4 | 0.0293 | 0.0449 | 0.0663 | 0.0947 | 0.1314 | 0.1775 | 0.2341 | 0.3021 | 0.3820 | 0.4743 | 0.5791 | 0.6961 | 1.0377 | 1.4338 | 2.2706 |
| IC5 | 0.0099 | 0.0163 | 0.0257 | 0.0390 | 0.0571 | 0.0810 | 0.1117 | 0.1502 | 0.1972 | 0.2536 | 0.3198 | 0.3962 | 0.6318 | 0.9260 | 1.6250 |
| NC5 | 0.0066 | 0.0112 | 0.0183 | 0.0284 | 0.0427 | 0.0619 | 0.0871 | 0.1193 | 0.1593 | 0.2079 | 0.2658 | 0.3333 | 0.5450 | 0.8129 | 1.4483 |
| C6 | 0.0017 | 0.0032 | 0.0055 | 0.0092 | 0.0147 | 0.0224 | 0.0332 | 0.0476 | 0.0664 | 0.0903 | 0.1201 | 0.1564 | 0.2790 | 0.4513 | 0.9385 |
| C7 | 0.0005 | 0.0010 | 0.0018 | 0.0032 | 0.0052 | 0.0084 | 0.0129 | 0.0192 | 0.0279 | 0.0393 | 0.0541 | 0.0729 | 0.1412 | 0.2473 | 0.6006 |
| C8 | 0.0002 | 0.0003 | 0.0006 | 0.0011 | 0.0018 | 0.0030 | 0.0047 | 0.0073 | 0.0109 | 0.0159 | 0.0227 | 0.0318 | 0.0678 | 0.1305 | 0.3721 |
| C9 | 0.0001 | 0.0002 | 0.0003 | 0.0005 | 0.0008 | 0.0013 | 0.0020 | 0.0031 | 0.0048 | 0.0071 | 0.0105 | 0.0150 | 0.0347 | 0.0724 | 0.2362 |
| C10 | 0.0000 | 0.0001 | 0.0001 | 0.0002 | 0.0003 | 0.0005 | 0.0008 | 0.0013 | 0.0021 | 0.0032 | 0.0049 | 0.0072 | 0.0177 | 0.0391 | 0.1440 |
| Pressure = 400.0 psia | | | | | | | | | | | | | | | |
| N2 | 20.1743 | 22.5352 | 24.9413 | 27.3809 | 29.8440 | 32.3217 | 34.8061 | 37.2905 | 39.7688 | 42.2361 | 44.6879 | 47.1205 | 53.0988 | 58.9021 | 69.9041 |
| CO2 | 1.6425 | 1.9112 | 2.1957 | 2.4937 | 2.8024 | 3.1193 | 3.4413 | 3.7656 | 4.0890 | 4.4084 | 4.7210 | 5.0236 | 5.7190 | 6.2964 | 6.9866 |
| H2S | 0.3629 | 0.5012 | 0.6625 | 0.8441 | 1.0428 | 1.2550 | 1.4770 | 1.7056 | 1.9375 | 2.1699 | 2.4004 | 2.6272 | 3.1676 | 3.6585 | 4.4708 |
| C1 | 5.2179 | 5.5650 | 5.9132 | 6.2605 | 6.6045 | 6.9430 | 7.2736 | 7.5940 | 7.9016 | 8.1942 | 8.4695 | 8.7251 | 9.2648 | 9.6408 | 9.8320 |
| C2 | 0.5170 | 0.6564 | 0.8153 | 0.9933 | 1.1891 | 1.4014 | 1.6282 | 1.8672 | 2.1160 | 2.3717 | 2.6315 | 2.8925 | 3.5303 | 4.1121 | 4.9647 |
| C3 | 0.1077 | 0.1532 | 0.2107 | 0.2809 | 0.3646 | 0.4620 | 0.5731 | 0.6972 | 0.8336 | 0.9812 | 1.1387 | 1.3044 | 1.7436 | 2.1956 | 3.0325 |
| IC4 | 0.0413 | 0.0604 | 0.0855 | 0.1175 | 0.1574 | 0.2061 | 0.2643 | 0.3325 | 0.4109 | 0.4999 | 0.5992 | 0.7084 | 1.0203 | 1.3738 | 2.1081 |
| NC4 | 0.0274 | 0.0413 | 0.0600 | 0.0845 | 0.1156 | 0.1543 | 0.2013 | 0.2572 | 0.3224 | 0.3974 | 0.4821 | 0.5763 | 0.8505 | 1.1691 | 1.8569 |
| IC5 | 0.0093 | 0.0151 | 0.0234 | 0.0350 | 0.0507 | 0.0711 | 0.0972 | 0.1296 | 0.1691 | 0.2163 | 0.2716 | 0.3352 | 0.5310 | 0.7748 | 1.3514 |
| NC5 | 0.0065 | 0.0107 | 0.0170 | 0.0259 | 0.0382 | 0.0547 | 0.0760 | 0.1031 | 0.1367 | 0.1774 | 0.2257 | 0.2821 | 0.4590 | 0.6833 | 1.2142 |
| C6 | 0.0018 | 0.0032 | 0.0055 | 0.0089 | 0.0140 | 0.0212 | 0.0309 | 0.0439 | 0.0606 | 0.0816 | 0.1074 | 0.1387 | 0.2431 | 0.3880 | 0.7957 |
| C7 | 0.0005 | 0.0010 | 0.0019 | 0.0032 | 0.0052 | 0.0081 | 0.0123 | 0.0181 | 0.0258 | 0.0359 | 0.0488 | 0.0650 | 0.1227 | 0.2109 | 0.5032 |
| C8 | 0.0002 | 0.0004 | 0.0007 | 0.0012 | 0.0019 | 0.0030 | 0.0047 | 0.0071 | 0.0105 | 0.0150 | 0.0212 | 0.0292 | 0.0603 | 0.1132 | 0.3155 |
| C9 | 0.0001 | 0.0002 | 0.0003 | 0.0005 | 0.0009 | 0.0014 | 0.0021 | 0.0032 | 0.0047 | 0.0069 | 0.0100 | 0.0141 | 0.0313 | 0.0634 | 0.2009 |
| C10 | 0.0000 | 0.0001 | 0.0001 | 0.0002 | 0.0004 | 0.0006 | 0.0009 | 0.0014 | 0.0022 | 0.0032 | 0.0048 | 0.0069 | 0.0162 | 0.0347 | 0.1228 |
| Pressure = 500.0 psia | | | | | | | | | | | | | | | |
| N2 | 15.9584 | 17.7953 | 19.6641 | 21.5563 | 23.4639 | 25.3804 | 27.2998 | 29.2170 | 31.1276 | 33.0278 | 34.9144 | 36.7846 | 41.3745 | 45.8224 | 54.2365 |
| CO2 | 1.4002 | 1.6188 | 1.8494 | 2.0903 | 2.3393 | 2.5943 | 2.8532 | 3.1136 | 3.3731 | 3.6294 | 3.8801 | 4.1230 | 4.6814 | 5.1459 | 5.7039 |
| H2S | 0.3193 | 0.4368 | 0.5726 | 0.7244 | 0.8893 | 1.0644 | 1.2467 | 1.4335 | 1.6223 | 1.8109 | 1.9975 | 2.1804 | 2.6148 | 3.0076 | 3.6546 |
| C1 | 4.3187 | 4.5955 | 4.8719 | 5.1461 | 5.4166 | 5.6815 | 5.9392 | 6.1879 | 6.4257 | 6.6511 | 6.8623 | 7.0577 | 7.4678 | 7.7504 | 7.8862 |
| C2 | 0.4540 | 0.5702 | 0.7021 | 0.8490 | 1.0102 | 1.1846 | 1.3707 | 1.5666 | 1.7706 | 1.9805 | 2.1939 | 2.4085 | 2.9346 | 3.4166 | 4.1255 |
| C3 | 0.1015 | 0.1414 | 0.1910 | 0.2509 | 0.3217 | 0.4034 | 0.4961 | 0.5995 | 0.7128 | 0.8354 | 0.9662 | 1.1041 | 1.4708 | 1.8510 | 2.5629 |
| IC4 | 0.0405 | 0.0580 | 0.0807 | 0.1092 | 0.1444 | 0.1869 | 0.2373 | 0.2959 | 0.3631 | 0.4389 | 0.5233 | 0.6159 | 0.8801 | 1.1798 | 1.8074 |
| NC4 | 0.0269 | 0.0399 | 0.0571 | 0.0794 | 0.1074 | 0.1417 | 0.1830 | 0.2318 | 0.2883 | 0.3527 | 0.4251 | 0.5053 | 0.7376 | 1.0073 | 1.5984 |
| IC5 | 0.0093 | 0.0148 | 0.0226 | 0.0333 | 0.0476 | 0.0662 | 0.0897 | 0.1188 | 0.1540 | 0.1957 | 0.2444 | 0.3002 | 0.4711 | 0.6829 | 1.1825 |
| NC5 | 0.0068 | 0.0110 | 0.0170 | 0.0254 | 0.0369 | 0.0520 | 0.0715 | 0.0961 | 0.1263 | 0.1628 | 0.2060 | 0.2562 | 0.4132 | 0.6116 | 1.0792 |
| C6 | 0.0018 | 0.0033 | 0.0056 | 0.0090 | 0.0140 | 0.0209 | 0.0302 | 0.0424 | 0.0580 | 0.0773 | 0.1010 | 0.1293 | 0.2224 | 0.3499 | 0.7069 |
| C7 | 0.0006 | 0.0011 | 0.0020 | 0.0033 | 0.0053 | 0.0082 | 0.0123 | 0.0178 | 0.0252 | 0.0346 | 0.0465 | 0.0613 | 0.1131 | 0.1909 | 0.4468 |
| C8 | 0.0002 | 0.0004 | 0.0008 | 0.0013 | 0.0021 | 0.0033 | 0.0051 | 0.0075 | 0.0109 | 0.0154 | 0.0214 | 0.0291 | 0.0585 | 0.1076 | 0.2939 |
| C9 | 0.0001 | 0.0002 | 0.0004 | 0.0006 | 0.0010 | 0.0015 | 0.0022 | 0.0033 | 0.0049 | 0.0070 | 0.0099 | 0.0138 | 0.0297 | 0.0584 | 0.1797 |
| C10 | 0.0000 | 0.0001 | 0.0001 | 0.0002 | 0.0004 | 0.0006 | 0.0010 | 0.0015 | 0.0023 | 0.0033 | 0.0048 | 0.0069 | 0.0155 | 0.0320 | 0.1092 |
| Pressure = 600.0 psia | | | | | | | | | | | | | | | |
| N2 | 13.1218 | 14.6064 | 16.1143 | 17.6386 | 19.1732 | 20.7129 | 22.2530 | 23.7896 | 25.3193 | 26.8391 | 28.3467 | 29.8399 | 33.4994 | 37.0394 | 43.7212 |
| CO2 | 1.2460 | 1.4314 | 1.6262 | 1.8289 | 2.0380 | 2.2517 | 2.4682 | 2.6857 | 2.9023 | 3.1160 | 3.3251 | 3.5277 | 3.9937 | 4.3825 | 4.8543 |
| H2S | 0.2918 | 0.3953 | 0.5139 | 0.6453 | 0.7872 | 0.9370 | 1.0921 | 1.2504 | 1.4097 | 1.5682 | 1.7246 | 1.8775 | 2.2391 | 2.5645 | 3.0981 |
| C1 | 3.7210 | 3.9537 | 4.1847 | 4.4127 | 4.6364 | 4.8544 | 5.0655 | 5.2682 | 5.4612 | 5.6433 | 5.8132 | 5.9697 | 6.2957 | 6.5176 | 6.6172 |
| C2 | 0.4172 | 0.5182 | 0.6320 | 0.7580 | 0.8958 | 1.0444 | 1.2026 | 1.3691 | 1.5424 | 1.7206 | 1.9020 | 2.0846 | 2.5334 | 2.9468 | 3.5610 |
| C3 | 0.1017 | 0.1384 | 0.1832 | 0.2366 | 0.2990 | 0.3706 | 0.4513 | 0.5409 | 0.6391 | 0.7451 | 0.8582 | 0.9776 | 1.2965 | 1.6295 | 2.2603 |
| IC4 | 0.0408 | 0.0574 | 0.0785 | 0.1048 | 0.1369 | 0.1753 | 0.2204 | 0.2726 | 0.3321 | 0.3989 | 0.4730 | 0.5540 | 0.7845 | 1.0455 | 1.5949 |
| NC4 | 0.0274 | 0.0400 | 0.0565 | 0.0776 | 0.1038 | 0.1355 | 0.1734 | 0.2176 | 0.2685 | 0.3261 | 0.3905 | 0.4614 | 0.6654 | 0.9013 | 1.4225 |
| IC5 | 0.0096 | 0.0150 | 0.0227 | 0.0332 | 0.0470 | 0.0648 | 0.0871 | 0.1144 | 0.1472 | 0.1859 | 0.2308 | 0.2820 | 0.4377 | 0.6296 | 1.0829 |
| NC5 | 0.0071 | 0.0112 | 0.0170 | 0.0251 | 0.0359 | 0.0501 | 0.0681 | 0.0907 | 0.1183 | 0.1515 | 0.1906 | 0.2360 | 0.3770 | 0.5547 | 0.9723 |

# TABLE 4–17. Continued

| | Temperature, Deg F | | | | | | | | | | | | | | |
|---|---|---|---|---|---|---|---|---|---|---|---|---|---|---|---|
| Comp | −20.0 | 0.0 | 20.0 | 40.0 | 60.0 | 80.0 | 100.0 | 120.0 | 140.0 | 160.0 | 180.0 | 200.0 | 250.0 | 300.0 | 400.0 |
| C6 | 0.0019 | 0.0033 | 0.0056 | 0.0091 | 0.0140 | 0.0207 | 0.0296 | 0.0412 | 0.0559 | 0.0739 | 0.0957 | 0.1215 | 0.2053 | 0.3183 | 0.6333 |
| C7 | 0.0006 | 0.0012 | 0.0021 | 0.0035 | 0.0056 | 0.0086 | 0.0126 | 0.0181 | 0.0253 | 0.0343 | 0.0457 | 0.0596 | 0.1078 | 0.1788 | 0.4096 |
| C8 | 0.0003 | 0.0005 | 0.0008 | 0.0014 | 0.0022 | 0.0035 | 0.0052 | 0.0077 | 0.0109 | 0.0153 | 0.0209 | 0.0281 | 0.0548 | 0.0983 | 0.2604 |
| C9 | 0.0002 | 0.0003 | 0.0004 | 0.0007 | 0.0011 | 0.0016 | 0.0024 | 0.0036 | 0.0052 | 0.0073 | 0.0101 | 0.0139 | 0.0288 | 0.0551 | 0.1646 |
| C10 | 0.0001 | 0.0001 | 0.0002 | 0.0003 | 0.0005 | 0.0007 | 0.0011 | 0.0017 | 0.0025 | 0.0035 | 0.0050 | 0.0070 | 0.0153 | 0.0307 | 0.1011 |
| Pressure = 700.0 psia | | | | | | | | | | | | | | | |
| N2 | 11.0833 | 12.3149 | 13.5638 | 14.8242 | 16.0912 | 17.3607 | 18.6290 | 19.8930 | 21.1499 | 22.3974 | 23.6336 | 24.8570 | 27.8510 | 30.7419 | 36.1862 |
| CO2 | 1.1426 | 1.3043 | 1.4735 | 1.6490 | 1.8294 | 2.0134 | 2.1995 | 2.3861 | 2.5718 | 2.7549 | 2.9339 | 3.1073 | 3.5068 | 3.8412 | 4.2527 |
| H2S | 0.2740 | 0.3675 | 0.4737 | 0.5904 | 0.7156 | 0.8469 | 0.9823 | 1.1197 | 1.2574 | 1.3941 | 1.5283 | 1.6593 | 1.9677 | 2.2440 | 2.6947 |
| C1 | 3.2939 | 3.4966 | 3.6966 | 3.8930 | 4.0845 | 4.2703 | 4.4492 | 4.6202 | 4.7822 | 4.9344 | 5.0757 | 5.2053 | 5.4733 | 5.6533 | 5.7288 |
| C2 | 0.3963 | 0.4865 | 0.5874 | 0.6985 | 0.8194 | 0.9493 | 1.0874 | 1.2324 | 1.3831 | 1.5381 | 1.6959 | 1.8549 | 2.2469 | 2.6099 | 3.1569 |
| C3 | 0.1055 | 0.1399 | 0.1812 | 0.2299 | 0.2862 | 0.3502 | 0.4219 | 0.5013 | 0.5879 | 0.6814 | 0.7811 | 0.8864 | 1.1683 | 1.4644 | 2.0298 |
| IC4 | 0.0417 | 0.0578 | 0.0780 | 0.1029 | 0.1329 | 0.1685 | 0.2100 | 0.2577 | 0.3117 | 0.3721 | 0.4388 | 0.5115 | 0.7175 | 0.9504 | 1.4425 |
| NC4 | 0.0284 | 0.0410 | 0.0572 | 0.0776 | 0.1027 | 0.1328 | 0.1683 | 0.2095 | 0.2564 | 0.3091 | 0.3676 | 0.4317 | 0.6148 | 0.8252 | 1.2922 |
| IC5 | 0.0102 | 0.0159 | 0.0237 | 0.0343 | 0.0481 | 0.0657 | 0.0875 | 0.1141 | 0.1457 | 0.1827 | 0.2253 | 0.2736 | 0.4192 | 0.5973 | 1.0188 |
| NC5 | 0.0077 | 0.0118 | 0.0177 | 0.0257 | 0.0363 | 0.0500 | 0.0673 | 0.0888 | 0.1149 | 0.1460 | 0.1826 | 0.2247 | 0.3550 | 0.5179 | 0.8993 |
| C6 | 0.0020 | 0.0036 | 0.0060 | 0.0095 | 0.0145 | 0.0214 | 0.0304 | 0.0419 | 0.0563 | 0.0738 | 0.0947 | 0.1192 | 0.1977 | 0.3019 | 0.5911 |
| C7 | 0.0007 | 0.0013 | 0.0023 | 0.0037 | 0.0059 | 0.0089 | 0.0131 | 0.0186 | 0.0256 | 0.0345 | 0.0455 | 0.0588 | 0.1041 | 0.1699 | 0.3814 |
| C8 | 0.0003 | 0.0006 | 0.0010 | 0.0016 | 0.0025 | 0.0039 | 0.0057 | 0.0083 | 0.0117 | 0.0162 | 0.0219 | 0.0291 | 0.0554 | 0.0974 | 0.2524 |
| C9 | 0.0002 | 0.0003 | 0.0005 | 0.0080 | 0.0012 | 0.0018 | 0.0027 | 0.0039 | 0.0055 | 0.0077 | 0.0106 | 0.0144 | 0.0288 | 0.0536 | 0.1550 |
| C10 | 0.0001 | 0.0001 | 0.0002 | 0.0003 | 0.0005 | 0.0008 | 0.0013 | 0.0019 | 0.0027 | 0.0038 | 0.0053 | 0.0073 | 0.0154 | 0.0300 | 0.0955 |
| Pressure = 800.0 psia | | | | | | | | | | | | | | | |
| N2 | 9.5481 | 10.5895 | 11.6435 | 12.7056 | 13.7716 | 14.8383 | 15.9025 | 16.9618 | 18.0141 | 19.0574 | 20.0903 | 21.1114 | 23.6069 | 26.0120 | 30.5308 |
| CO2 | 1.0705 | 1.2143 | 1.3641 | 1.5189 | 1.6776 | 1.8390 | 2.0018 | 2.1648 | 2.3268 | 2.4864 | 2.6424 | 2.7935 | 3.1417 | 3.4343 | 3.8000 |
| H2S | 0.2626 | 0.3487 | 0.4455 | 0.5512 | 0.6636 | 0.7809 | 0.9011 | 1.0226 | 1.1439 | 1.2638 | 1.3812 | 1.4954 | 1.7632 | 2.0018 | 2.3893 |
| C1 | 2.9736 | 3.1543 | 3.3315 | 3.5045 | 3.6724 | 3.8343 | 3.9894 | 4.1369 | 4.2760 | 4.4060 | 4.5262 | 4.6360 | 4.8611 | 5.0104 | 5.0689 |
| C2 | 0.3853 | 0.4674 | 0.5585 | 0.6583 | 0.7663 | 0.8820 | 1.0045 | 1.1329 | 1.2662 | 1.4032 | 1.5426 | 1.6832 | 2.0305 | 2.3540 | 2.8488 |
| C3 | 0.1111 | 0.1437 | 0.1822 | 0.2270 | 0.2782 | 0.3361 | 0.4006 | 0.4716 | 0.5490 | 0.6324 | 0.7213 | 0.8152 | 1.0673 | 1.3335 | 1.8456 |
| IC4 | 0.0441 | 0.0601 | 0.0798 | 0.1041 | 0.1329 | 0.1667 | 0.2058 | 0.2504 | 0.3005 | 0.3561 | 0.4172 | 0.4836 | 0.6700 | 0.8794 | 1.3210 |
| NC4 | 0.0300 | 0.0426 | 0.0588 | 0.0790 | 0.1034 | 0.1342 | 0.1663 | 0.2051 | 0.2491 | 0.2981 | 0.3522 | 0.4110 | 0.5778 | 0.7680 | 1.1911 |
| IC5 | 0.0109 | 0.0167 | 0.0248 | 0.0356 | 0.0494 | 0.0669 | 0.0883 | 0.1141 | 0.1446 | 0.1800 | 0.2205 | 0.2660 | 0.4020 | 0.5667 | 0.9577 |
| NC5 | 0.0085 | 0.0129 | 0.0190 | 0.0271 | 0.0378 | 0.0515 | 0.0687 | 0.0898 | 0.1152 | 0.1454 | 0.1805 | 0.2208 | 0.3444 | 0.4979 | 0.8555 |
| C6 | 0.0022 | 0.0038 | 0.0064 | 0.0101 | 0.0153 | 0.0223 | 0.0314 | 0.0429 | 0.0571 | 0.0741 | 0.0943 | 0.1177 | 0.1912 | 0.2871 | 0.5513 |
| C7 | 0.0008 | 0.0014 | 0.0025 | 0.0040 | 0.0063 | 0.0095 | 0.0137 | 0.0193 | 0.0264 | 0.0352 | 0.0460 | 0.0589 | 0.1023 | 0.1643 | 0.3614 |
| C8 | 0.0003 | 0.0006 | 0.0011 | 0.0017 | 0.0027 | 0.0041 | 0.0061 | 0.0087 | 0.0121 | 0.0165 | 0.0220 | 0.0289 | 0.0535 | 0.0919 | 0.2310 |
| C9 | 0.0002 | 0.0003 | 0.0006 | 0.0009 | 0.0014 | 0.0021 | 0.0030 | 0.0043 | 0.0060 | 0.0083 | 0.0112 | 0.0150 | 0.0291 | 0.0529 | 0.1483 |
| C10 | 0.0001 | 0.0002 | 0.0003 | 0.0004 | 0.0006 | 0.0010 | 0.0014 | 0.0021 | 0.0029 | 0.0041 | 0.0056 | 0.0077 | 0.0156 | 0.0296 | 0.0911 |
| Pressure = 900.0 psia | | | | | | | | | | | | | | | |
| N2 | 8.3509 | 9.2440 | 10.1464 | 11.0540 | 11.9638 | 12.8727 | 13.7783 | 14.6787 | 15.5720 | 16.4567 | 17.3318 | 18.1961 | 20.3051 | 22.3338 | 26.1364 |
| CO2 | 1.0197 | 1.1495 | 1.2841 | 1.4226 | 1.5641 | 1.7076 | 1.8521 | 1.9965 | 2.1397 | 2.2807 | 2.4184 | 2.5517 | 2.8593 | 3.1188 | 3.4489 |
| H2S | 0.2558 | 0.3362 | 0.4258 | 0.5227 | 0.6252 | 0.7314 | 0.8397 | 0.9486 | 1.0569 | 1.1635 | 1.2675 | 1.3685 | 1.6041 | 1.8131 | 2.1506 |
| C1 | 2.7233 | 2.8871 | 3.0468 | 3.2019 | 3.3515 | 3.4951 | 3.6319 | 3.7613 | 3.8828 | 3.9958 | 4.0998 | 4.1944 | 4.3867 | 4.5127 | 4.5587 |
| C2 | 0.3818 | 0.4577 | 0.5412 | 0.6320 | 0.7299 | 0.8343 | 0.9445 | 1.0597 | 1.1791 | 1.3017 | 1.4265 | 1.5524 | 1.8638 | 2.1555 | 2.6093 |
| C3 | 0.1195 | 0.1508 | 0.1872 | 0.2289 | 0.2763 | 0.3295 | 0.3884 | 0.4530 | 0.5232 | 0.5988 | 0.6794 | 0.7646 | 0.9936 | 1.2365 | 1.7077 |
| IC4 | 0.0468 | 0.0629 | 0.0827 | 0.1065 | 0.1346 | 0.1673 | 0.2048 | 0.2472 | 0.2946 | 0.3470 | 0.4041 | 0.4659 | 0.6385 | 0.8312 | 1.2379 |
| NC4 | 0.0321 | 0.0451 | 0.0615 | 0.0817 | 0.1058 | 0.1342 | 0.1670 | 0.2042 | 0.2460 | 0.2923 | 0.3429 | 0.3978 | 0.5517 | 0.7259 | 1.1128 |
| IC5 | 0.0116 | 0.0177 | 0.0259 | 0.0368 | 0.0506 | 0.0677 | 0.0885 | 0.1132 | 0.1420 | 0.1751 | 0.2125 | 0.2543 | 0.3772 | 0.5241 | 0.8706 |
| NC5 | 0.0094 | 0.0141 | 0.0205 | 0.0290 | 0.0400 | 0.0540 | 0.0713 | 0.0924 | 0.1177 | 0.1474 | 0.1819 | 0.2213 | 0.3413 | 0.4893 | 0.8342 |
| C6 | 0.0024 | 0.0042 | 0.0069 | 0.0108 | 0.0163 | 0.0236 | 0.0329 | 0.0446 | 0.0587 | 0.0756 | 0.0953 | 0.1179 | 0.1877 | 0.2773 | 0.5226 |
| C7 | 0.0009 | 0.0016 | 0.0027 | 0.0044 | 0.0069 | 0.0102 | 0.0146 | 0.0204 | 0.0276 | 0.0365 | 0.0473 | 0.0601 | 0.1024 | 0.1617 | 0.3485 |
| C8 | 0.0004 | 0.0007 | 0.0012 | 0.0020 | 0.0030 | 0.0045 | 0.0066 | 0.0093 | 0.0128 | 0.0172 | 0.0228 | 0.0296 | 0.0534 | 0.0897 | 0.2195 |
| C9 | 0.0002 | 0.0004 | 0.0007 | 0.0010 | 0.0016 | 0.0023 | 0.0034 | 0.0048 | 0.0066 | 0.0090 | 0.0120 | 0.0158 | 0.0299 | 0.0530 | 0.1439 |
| C10 | 0.0001 | 0.0002 | 0.0003 | 0.0005 | 0.0008 | 0.0011 | 0.0016 | 0.0023 | 0.0033 | 0.0045 | 0.0061 | 0.0082 | 0.0162 | 0.0301 | 0.0896 |

## TABLE 4–17. Continued

| Comp | −20.0 | 0.0 | 20.0 | 40.0 | 60.0 | 80.0 | 100.0 | 120.0 | 140.0 | 160.0 | 180.0 | 200.0 | 250.0 | 300.0 | 400.0 |
|---|---|---|---|---|---|---|---|---|---|---|---|---|---|---|---|
| | | | | | | | Temperature, Deg F | | | | | | | | |
| Pressure = 1000.0 psia | | | | | | | | | | | | | | | |
| N2 | 7.3913 | 8.1658 | 8.9468 | 9.7310 | 10.5158 | 11.2987 | 12.0777 | 12.8512 | 13.6177 | 14.3761 | 15.1254 | 15.8647 | 17.6660 | 19.3953 | 22.6287 |
| CO2 | 0.9836 | 1.1019 | 1.2240 | 1.3491 | 1.4765 | 1.6053 | 1.7347 | 1.9637 | 1.9915 | 2.1171 | 2.2397 | 2.3584 | 2.6323 | 2.8643 | 3.1646 |
| H2S | 0.2524 | 0.3284 | 0.4122 | 0.5021 | 0.5965 | 0.6938 | 0.7924 | 0.8910 | 0.9887 | 1.0845 | 1.1778 | 1.2679 | 1.4774 | 1.6623 | 1.9594 |
| C1 | 2.5221 | 2.6722 | 2.8178 | 2.9583 | 3.0932 | 3.2219 | 3.3440 | 3.4589 | 3.5662 | 3.6655 | 3.7565 | 3.8388 | 4.0050 | 4.1125 | 4.1490 |
| C2 | 0.3836 | 0.4544 | 0.5317 | 0.6152 | 0.7048 | 0.7999 | 0.8999 | 1.0042 | 1.1121 | 1.2228 | 1.3354 | 1.4489 | 1.7301 | 1.9950 | 2.4137 |
| C3 | 0.1305 | 0.1607 | 0.1954 | 0.2347 | 0.2790 | 0.3283 | 0.3826 | 0.4421 | 0.5065 | 0.5757 | 0.6494 | 0.7273 | 0.9369 | 1.1598 | 1.5938 |
| IC4 | 0.0507 | 0.0673 | 0.0874 | 0.1113 | 0.1393 | 0.1714 | 0.2080 | 0.2490 | 0.2945 | 0.3443 | 0.3985 | 0.4567 | 0.6181 | 0.7971 | 1.1737 |
| NC4 | 0.0349 | 0.0485 | 0.0654 | 0.0858 | 0.1100 | 0.1382 | 0.1704 | 0.2067 | 0.2471 | 0.2914 | 0.3396 | 0.3915 | 0.5358 | 0.6979 | 1.0576 |
| IC5 | 0.0123 | 0.0185 | 0.0269 | 0.0377 | 0.0513 | 0.0680 | 0.0880 | 0.1114 | 0.1385 | 0.1692 | 0.2035 | 0.2415 | 0.3515 | 0.4810 | 0.7850 |
| NC5 | 0.0105 | 0.0155 | 0.0223 | 0.0311 | 0.0424 | 0.0565 | 0.0738 | 0.0947 | 0.1195 | 0.1485 | 0.1818 | 0.2196 | 0.3334 | 0.4723 | 0.7934 |
| C6 | 0.0026 | 0.0045 | 0.0075 | 0.0117 | 0.0175 | 0.0250 | 0.0347 | 0.0465 | 0.0608 | 0.0775 | 0.0969 | 0.1189 | 0.1858 | 0.2702 | 0.5010 |
| C7 | 0.0010 | 0.0018 | 0.0030 | 0.0048 | 0.0074 | 0.0109 | 0.0155 | 0.0214 | 0.0287 | 0.0376 | 0.0483 | 0.0609 | 0.1018 | 0.1583 | 0.3341 |
| C8 | 0.0004 | 0.0008 | 0.0013 | 0.0021 | 0.0032 | 0.0048 | 0.0069 | 0.0096 | 0.0130 | 0.0173 | 0.0226 | 0.0290 | 0.0509 | 0.0835 | 0.1975 |
| C9 | 0.0003 | 0.0005 | 0.0008 | 0.0012 | 0.0018 | 0.0027 | 0.0038 | 0.0054 | 0.0073 | 0.0099 | 0.0130 | 0.0170 | 0.0312 | 0.0539 | 0.1416 |
| C10 | 0.0001 | 0.0002 | 0.0004 | 0.0006 | 0.0009 | 0.0013 | 0.0019 | 0.0027 | 0.0037 | 0.0050 | 0.0067 | 0.0089 | 0.0171 | 0.0310 | 0.0891 |
| Pressure = 1200.0 psia | | | | | | | | | | | | | | | |
| N2 | 5.9495 | 6.5459 | 7.1450 | 7.7444 | 8.3422 | 8.9368 | 9.5267 | 10.1109 | 10.6884 | 11.2584 | 11.8204 | 12.3738 | 13.7177 | 15.0026 | 17.3926 |
| CO2 | 0.9381 | 1.0379 | 1.1400 | 1.2439 | 1.3488 | 1.4542 | 1.5595 | 1.6640 | 1.7671 | 1.8682 | 1.9667 | 2.0619 | 2.2817 | 2.4691 | 2.7197 |
| H2S | 0.2538 | 0.3233 | 0.3985 | 0.4778 | 0.5599 | 0.6434 | 0.7271 | 0.8100 | 0.8914 | 0.9706 | 1.0472 | 1.1207 | 1.2902 | 1.4382 | 1.6733 |
| C1 | 2.2091 | 2.3370 | 2.4598 | 2.5770 | 2.6884 | 2.7937 | 2.8924 | 2.9845 | 3.0697 | 3.1479 | 3.2188 | 3.2823 | 3.4087 | 3.4885 | 3.5120 |
| C2 | 0.3983 | 0.4610 | 0.5284 | 0.6004 | 0.6767 | 0.7570 | 0.8408 | 0.9277 | 1.0173 | 1.1088 | 1.2016 | 1.2952 | 1.5274 | 1.7476 | 2.1062 |
| C3 | 0.1621 | 0.1907 | 0.2228 | 0.2587 | 0.2985 | 0.3424 | 0.3904 | 0.4425 | 0.4987 | 0.5590 | 0.6230 | 0.6906 | 0.8726 | 1.0666 | 1.4460 |
| IC4 | 0.0601 | 0.0778 | 0.0988 | 0.1231 | 0.1510 | 0.1824 | 0.2176 | 0.2563 | 0.2987 | 0.3445 | 0.3936 | 0.4458 | 0.5884 | 0.7438 | 1.0687 |
| NC4 | 0.0426 | 0.0577 | 0.0759 | 0.0974 | 0.1223 | 0.1507 | 0.1824 | 0.2176 | 0.2559 | 0.2974 | 0.3419 | 0.3891 | 0.5182 | 0.6603 | 0.9735 |
| IC5 | 0.0161 | 0.0239 | 0.0340 | 0.0467 | 0.0623 | 0.0809 | 0.1026 | 0.1276 | 0.1557 | 0.1870 | 0.2214 | 0.2588 | 0.3646 | 0.4865 | 0.7746 |
| NC5 | 0.0135 | 0.0194 | 0.0270 | 0.0368 | 0.0489 | 0.0637 | 0.0815 | 0.1026 | 0.1271 | 0.1552 | 0.1870 | 0.2225 | 0.3273 | 0.4523 | 0.7371 |
| C6 | 0.0033 | 0.0058 | 0.0094 | 0.0145 | 0.0213 | 0.0300 | 0.0407 | 0.0536 | 0.0687 | 0.0860 | 0.1054 | 0.1271 | 0.1904 | 0.2676 | 0.4763 |
| C7 | 0.0014 | 0.0024 | 0.0040 | 0.0062 | 0.0093 | 0.0135 | 0.0188 | 0.0254 | 0.0334 | 0.0429 | 0.0542 | 0.0671 | 0.1078 | 0.1622 | 0.3274 |
| C8 | 0.0007 | 0.0012 | 0.0020 | 0.0031 | 0.0046 | 0.0066 | 0.0093 | 0.0126 | 0.0168 | 0.0218 | 0.0278 | 0.0350 | 0.0585 | 0.0921 | 0.2059 |
| C9 | 0.0004 | 0.0007 | 0.0011 | 0.0017 | 0.0025 | 0.0036 | 0.0051 | 0.0069 | 0.0093 | 0.0122 | 0.0158 | 0.0201 | 0.0351 | 0.0578 | 0.1421 |
| C10 | 0.0002 | 0.0004 | 0.0006 | 0.0009 | 0.0013 | 0.0018 | 0.0025 | 0.0035 | 0.0047 | 0.0062 | 0.0081 | 0.0104 | 0.0188 | 0.0324 | 0.0867 |
| Pressure = 1400.0 psia | | | | | | | | | | | | | | | |
| N2 | 4.9175 | 5.3867 | 5.8560 | 6.3238 | 6.7887 | 7.2495 | 7.7054 | 8.1555 | 8.5993 | 9.0363 | 9.4661 | 9.8885 | 10.9106 | 11.8834 | 13.6832 |
| CO2 | 0.9186 | 1.0041 | 1.0908 | 1.1782 | 1.2658 | 1.3532 | 1.4401 | 1.5259 | 1.6102 | 1.6926 | 1.7726 | 1.8499 | 2.0284 | 2.1814 | 2.3925 |
| H2S | 0.2639 | 0.3290 | 0.3981 | 0.4697 | 0.5426 | 0.6159 | 0.6884 | 0.7596 | 0.8289 | 0.8957 | 0.9599 | 1.0211 | 1.1609 | 1.2817 | 1.4715 |
| C1 | 1.9827 | 2.0929 | 2.1976 | 2.2966 | 2.3898 | 2.4769 | 2.5579 | 2.6327 | 2.7013 | 2.7637 | 2.8198 | 2.8696 | 2.9672 | 3.0274 | 3.0425 |
| C2 | 0.4256 | 0.4818 | 0.5414 | 0.6044 | 0.6704 | 0.7393 | 0.8107 | 0.8844 | 0.9598 | 1.0366 | 1.1143 | 1.1925 | 1.3866 | 1.5718 | 1.8813 |
| C3 | 0.2027 | 0.2291 | 0.2585 | 0.2910 | 0.3269 | 0.3662 | 0.4089 | 0.4552 | 0.5049 | 0.5580 | 0.6144 | 0.6737 | 0.8334 | 1.0034 | 1.3358 |
| IC4 | 0.0736 | 0.0931 | 0.1156 | 0.1410 | 0.1696 | 0.2012 | 0.2358 | 0.2733 | 0.3137 | 0.3567 | 0.4022 | 0.4501 | 0.5783 | 0.7155 | 0.9990 |
| NC4 | 0.0536 | 0.0707 | 0.0907 | 0.1137 | 0.1397 | 0.1687 | 0.2005 | 0.2350 | 0.2720 | 0.3114 | 0.3530 | 0.3966 | 0.5134 | 0.6395 | 0.9140 |
| IC5 | 0.0199 | 0.0288 | 0.0401 | 0.0539 | 0.0703 | 0.0894 | 0.1111 | 0.1353 | 0.1619 | 0.1907 | 0.2218 | 0.2548 | 0.3456 | 0.4472 | 0.6864 |
| NC5 | 0.0180 | 0.0251 | 0.0340 | 0.0451 | 0.0585 | 0.0745 | 0.0932 | 0.1149 | 0.1396 | 0.1674 | 0.1984 | 0.2324 | 0.3305 | 0.4445 | 0.7003 |
| C6 | 0.0045 | 0.0077 | 0.0124 | 0.0187 | 0.0270 | 0.0373 | 0.0496 | 0.0640 | 0.0804 | 0.0986 | 0.1186 | 0.1403 | 0.2015 | 0.2734 | 0.4661 |
| C7 | 0.0019 | 0.0034 | 0.0054 | 0.0083 | 0.0122 | 0.0173 | 0.0235 | 0.0311 | 0.0401 | 0.0505 | 0.0625 | 0.0760 | 0.1172 | 0.1702 | 0.3273 |
| C8 | 0.0011 | 0.0019 | 0.0030 | 0.0045 | 0.0065 | 0.0092 | 0.0125 | 0.0165 | 0.0214 | 0.0272 | 0.0340 | 0.0418 | 0.0667 | 0.1009 | 0.2130 |
| C9 | 0.0006 | 0.0010 | 0.0017 | 0.0025 | 0.0037 | 0.0052 | 0.0071 | 0.0095 | 0.0124 | 0.0159 | 0.0201 | 0.0251 | 0.0415 | 0.0653 | 0.1497 |
| C10 | 0.0004 | 0.0006 | 0.0010 | 0.0014 | 0.0020 | 0.0027 | 0.0037 | 0.0049 | 0.0064 | 0.0083 | 0.0106 | 0.0133 | 0.0228 | 0.0375 | 0.0935 |
| Pressure = 1600.0 psia | | | | | | | | | | | | | | | |
| N2 | 4.1409 | 4.5145 | 4.8866 | 5.2559 | 5.6215 | 5.9826 | 6.3386 | 6.6892 | 7.0337 | 7.3721 | 7.7042 | 8.0296 | 8.8144 | 9.5578 | 10.9249 |
| CO2 | 0.9131 | 0.9865 | 1.0601 | 1.1337 | 1.2069 | 1.2795 | 1.3512 | 1.4216 | 1.4905 | 1.5577 | 1.6227 | 1.6855 | 1.8303 | 1.9550 | 2.1321 |
| H2S | 0.2821 | 0.3439 | 0.4082 | 0.4737 | 0.5394 | 0.6044 | 0.6681 | 0.7299 | 0.7895 | 0.8466 | 0.9009 | 0.9525 | 1.0691 | 1.1687 | 1.3234 |
| C1 | 1.7992 | 1.8934 | 1.9820 | 2.0651 | 2.1425 | 2.2142 | 2.2802 | 2.3407 | 2.3957 | 2.4452 | 2.4894 | 2.5283 | 2.6035 | 2.6487 | 2.6583 |

**TABLE 4–17. Continued**

| Comp | | | | | | | Temperature, Deg F | | | | | | | | |
|---|---|---|---|---|---|---|---|---|---|---|---|---|---|---|---|
| | −20.0 | 0.0 | 20.0 | 40.0 | 60.0 | 80.0 | 100.0 | 120.0 | 140.0 | 160.0 | 180.0 | 200.0 | 250.0 | 300.0 | 400.0 |
| C2 | 0.4634 | 0.5140 | 0.5670 | 0.6224 | 0.6799 | 0.7394 | 0.8006 | 0.8634 | 0.9274 | 0.9923 | 1.0578 | 1.1236 | 1.2867 | 1.4430 | 1.7101 |
| C3 | 0.2525 | 0.2759 | 0.3020 | 0.3310 | 0.3630 | 0.3980 | 0.4360 | 0.4771 | 0.5212 | 0.5681 | 0.6178 | 0.6700 | 0.8100 | 0.9584 | 1.2470 |
| IC4 | 0.0945 | 0.1166 | 0.1415 | 0.1690 | 0.1992 | 0.2318 | 0.2669 | 0.3043 | 0.3438 | 0.3853 | 0.4286 | 0.4735 | 0.5915 | 0.7152 | 0.9676 |
| NC4 | 0.0699 | 0.0896 | 0.1120 | 0.1372 | 0.1649 | 0.1951 | 0.2275 | 0.2619 | 0.2982 | 0.3362 | 0.3757 | 0.4166 | 0.5238 | 0.6371 | 0.8802 |
| IC5 | 0.0279 | 0.0393 | 0.0533 | 0.0699 | 0.0890 | 0.1105 | 0.1342 | 0.1599 | 0.1875 | 0.2166 | 0.2472 | 0.2791 | 0.3639 | 0.4561 | 0.6718 |
| NC5 | 0.0252 | 0.0340 | 0.0448 | 0.0576 | 0.0728 | 0.0904 | 0.1105 | 0.1332 | 0.1586 | 0.1865 | 0.2170 | 0.2500 | 0.3427 | 0.4475 | 0.6782 |
| C6 | 0.0064 | 0.0108 | 0.0170 | 0.0251 | 0.0354 | 0.0478 | 0.0622 | 0.0785 | 0.0965 | 0.1159 | 0.1367 | 0.1585 | 0.2179 | 0.2850 | 0.4636 |
| C7 | 0.0030 | 0.0050 | 0.0078 | 0.0117 | 0.0167 | 0.0229 | 0.0305 | 0.0394 | 0.0497 | 0.0613 | 0.0743 | 0.0886 | 0.1307 | 0.1831 | 0.3351 |
| C8 | 0.0018 | 0.0029 | 0.0044 | 0.0064 | 0.0090 | 0.0123 | 0.0162 | 0.0210 | 0.0264 | 0.0328 | 0.0400 | 0.0481 | 0.0731 | 0.1060 | 0.2100 |
| C9 | 0.0010 | 0.0017 | 0.0026 | 0.0038 | 0.0055 | 0.0075 | 0.0101 | 0.0131 | 0.0168 | 0.0211 | 0.0261 | 0.0318 | 0.0499 | 0.0751 | 0.1601 |
| C10 | 0.0007 | 0.0011 | 0.0016 | 0.0023 | 0.0032 | 0.0043 | 0.0056 | 0.0072 | 0.0092 | 0.0116 | 0.0144 | 0.0178 | 0.0289 | 0.0452 | 0.1045 |
| Pressure = 2000.0 psia | | | | | | | | | | | | | | | |
| N2 | 3.0401 | 3.2786 | 3.5136 | 3.7446 | 3.9713 | 4.1934 | 4.4108 | 4.6232 | 4.8307 | 5.0333 | 5.2309 | 5.4235 | 5.8841 | 6.3156 | 7.0985 |
| CO2 | 0.9265 | 0.9783 | 1.0293 | 1.0794 | 1.1284 | 1.1764 | 1.2231 | 1.2685 | 1.3125 | 1.3550 | 1.3959 | 1.4351 | 1.5253 | 1.6033 | 1.7165 |
| H2S | 0.3449 | 0.4014 | 0.4576 | 0.5127 | 0.5663 | 0.6178 | 0.6670 | 0.7138 | 0.7580 | 0.7995 | 0.8385 | 0.8750 | 0.9559 | 1.0234 | 1.1256 |
| C1 | 1.5151 | 1.5804 | 1.6407 | 1.6962 | 1.7469 | 1.7931 | 1.8350 | 1.8727 | 1.9064 | 1.9363 | 1.9625 | 1.9853 | 2.0283 | 2.0531 | 2.0570 |
| C2 | 0.5665 | 0.6056 | 0.6457 | 0.6869 | 0.7289 | 0.7717 | 0.8152 | 0.8592 | 0.9036 | 0.9482 | 0.9928 | 1.0374 | 1.1471 | 1.2520 | 1.4356 |
| C3 | 0.3898 | 0.4048 | 0.4230 | 0.4443 | 0.4683 | 0.4950 | 0.5243 | 0.5559 | 0.5898 | 0.6258 | 0.6637 | 0.7033 | 0.8081 | 0.9173 | 1.1244 |
| IC4 | 0.1565 | 0.1839 | 0.2131 | 0.2439 | 0.2761 | 0.3096 | 0.3441 | 0.3796 | 0.4158 | 0.4526 | 0.4899 | 0.5276 | 0.6228 | 0.7185 | 0.9077 |
| NC4 | 0.1276 | 0.1541 | 0.1826 | 0.2127 | 0.2442 | 0.2769 | 0.3104 | 0.3446 | 0.3794 | 0.4145 | 0.4499 | 0.4855 | 0.5750 | 0.6654 | 0.8525 |
| IC5 | 0.0567 | 0.0753 | 0.0964 | 0.1197 | 0.1447 | 0.1710 | 0.1982 | 0.2259 | 0.2538 | 0.2818 | 0.3096 | 0.3373 | 0.4063 | 0.4766 | 0.6385 |
| NC5 | 0.0569 | 0.0715 | 0.0882 | 0.1069 | 0.1276 | 0.1503 | 0.1750 | 0.2015 | 0.2299 | 0.2599 | 0.2916 | 0.3247 | 0.4133 | 0.5088 | 0.7127 |
| C6 | 0.0155 | 0.0247 | 0.0367 | 0.0515 | 0.0689 | 0.0883 | 0.1093 | 0.1313 | 0.1539 | 0.1767 | 0.1995 | 0.2221 | 0.2781 | 0.3366 | 0.4930 |
| C7 | 0.0082 | 0.0127 | 0.0187 | 0.0262 | 0.0352 | 0.0455 | 0.0572 | 0.0701 | 0.0839 | 0.0987 | 0.1143 | 0.1306 | 0.1752 | 0.2266 | 0.3709 |
| C8 | 0.0068 | 0.0098 | 0.0136 | 0.0182 | 0.0235 | 0.0297 | 0.0366 | 0.0443 | 0.0528 | 0.0621 | 0.0721 | 0.0829 | 0.1140 | 0.1523 | 0.2646 |
| C9 | 0.0035 | 0.0053 | 0.0076 | 0.0106 | 0.0142 | 0.0184 | 0.0233 | 0.0289 | 0.0351 | 0.0419 | 0.0495 | 0.0577 | 0.0816 | 0.1117 | 0.2033 |
| C10 | 0.0031 | 0.0043 | 0.0058 | 0.0076 | 0.0097 | 0.0121 | 0.0150 | 0.0183 | 0.0220 | 0.0262 | 0.0310 | 0.0363 | 0.0528 | 0.0750 | 0.1466 |

obtained which satisfies $\Sigma_{i=1}^{m} x_i = 1$ and similarly $\Sigma_{i=1}^{m} y_i = 1$. The composition of the composite fluid or feed must be known together with the pressure and temperature at which the equilibrium exists.

If $\Sigma K_i z_i$ and $\Sigma z_i/K_i$ are both greater than 1.0, the system is in the two-phase region. If $\Sigma K_i z_i$ is less than 1.0 the system is all liquid, compressed above its bubblepoint. If $\Sigma z_i/K_i$ is less than 1.0, then the sample is all vapor and may either be above its upper dew point or below its lower dew point. The $\Sigma z_i K_i$ and $\Sigma z_i/K_i$ cannot both be less than 1.0.

The method is presented here for completeness; however, in practice computer programs are available to carry out the flash calculations using more efficient iterative methods.

## DETERMINATION OF CONVERGENCE PRESSURE AND EQUILIBRIUM RATIOS

Convergence pressure is the pressure for a system at a given temperature when vapor-liquid separation is no longer possible. This pressure can usually be used as the parameter that represents the composition of the vapor and liquid phase in equilibrium. The convergence pressure is, in general, the critical pressure of a system at a given temperature. When the system pressure reaches the convergence pressure at a given temperature, the K-values of all components of the system converge to unity.

The following iterative procedure is suggested for the calculation of convergence pressure:[30]

1. Assume the liquid phase composition or make an approximation. If there is no guide, use the total feed composition.
2. Identify the lightest hydrocarbon component which is present at least 0.1% mole in the liquid phase.
3. Calculate the weight average critical temperature and critical pressure for the remaining heavier components to form a pseudobinary system.
4. Trace the critical locus (on Figure 4–32) of the binary system consisting of the light component and pseudoheavy component. When the averaged pseudoheavy component is between two real hydrocarbons, an interpolation of the two critical loci must be made.
5. Read the convergence pressure ($p_K$) at the temperature corresponding to that of the desired flash conditions.
6. Using $p_K$ from Step 5, together with system temperature and system pressure, obtain K-values for the components from the appropriate convergence pressure in Table 4–19.
7. Make flash calculations with the feed composition and the K-values from Step 6.
8. Repeat Steps 2 to 7 until the assumed and the calculated $p_K$ check within acceptable tolerance.

## BUBBLEPOINT AND DEW POINT PRESSURES

The bubblepoint pressure of a system is the state at which an infinitesimal quantity of gas is in equilibrium with a large quantity of fluid. Thus to calculate the bubblepoint pressure of a system, it is necessary to determine, by trial-and-error, the pressure at which the following relationship is satisfied:

$$\sum_{i=1}^{m} y_i = \sum_{i=1}^{m} K_i z_i = 1 \tag{4.134}$$

On the other hand, the evaluation of a dew point pressure requires a trial-and-error procedure until the following relationship is satisfied:

$$\sum_{i=1}^{m} x_i = \sum_{i=1}^{m} \frac{z_i}{K_i} = 1 \tag{4.135}$$

The primary use of dew point or bubblepoint pressure calculations in petroleum production engineering work is in connection with adjusting equilibrium ratios to observed bubblepoint or dew point pressures.

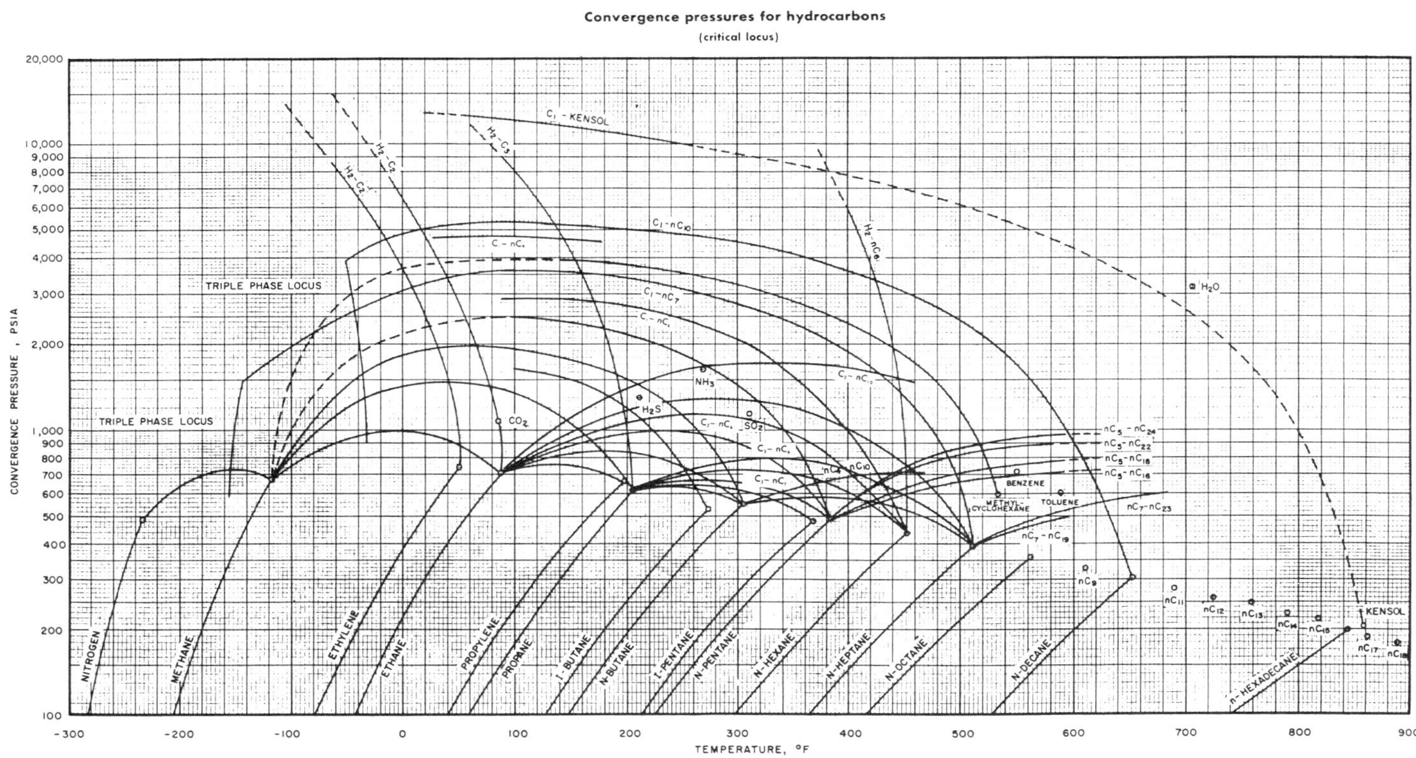


Note: These convergence pressures are the basis for the convergence charts. Later data do not necessarily agree.

Figure 4–32. Convergence pressures for hydrocarbons (courtesy of GPSA).[30]

## TWO-PHASE GAS DEVIATION FACTOR

The two-phase gas deviation factor for the fluid remaining in the reservoir during gas-condensate production can be calculated from the gas law as

$$z_{two\text{-}phase} = \frac{379.4 p_R V_m}{(G - G_p) R T_R} \tag{4.136}$$

where $R = 10.73$, $V_m$ is the molal volume of the reservoir gas in cubic feet of reservoir gas/pound-mole of reservoir gas, $p_R$ is the reservoir pressure in pounds per square inch, and $T_R$ is the reservoir temperature in °R.

If production data are not available to enable the calculation of $z_{two\text{-}phase}$ during the depletion of the gas-condensate reservoir, the gas deviation factor of the fluids remaining in the reservoir may be assumed constant at its initial value. This usually gives a closer approximation to the two-phase gas deviation factor.[15]

## ORIGINAL OIL AND GAS IN PLACE—GAS-CONDENSATE RESERVOIRS

The original oil (condensate) and gas in place in a gas-condensate reservoir may be calculated by recombining the produced gas and oil in the correct ratio to find the average specific gravity (air = 1.0) of the total well fluid. The following equations are used. The original wet gas and condensate liquids in place at standard conditions are given by

$$G_{wg} = \frac{V_R p_R (T_{sc} + 460) z_{sc}}{p_{sc}(T_R + 460) z} \tag{4.137}$$

$$G_L = \frac{G_{wg}}{10^6}(C_L)_{re} \tag{4.138}$$

where

$V_R = 43{,}560 A\phi h(1 - S_{iw})$, ft$^3$
$G_{wg}$ = wet gas in place, SCF
$G_L$ = condensate in place, STB
$(C_L)_{re}$ = condensate content of the wet gas at reservoir conditions, STB/MMSCF

The gas and condensate recoveries can be converted to moles as follows:

$$(n_p)_{wg} = \frac{(G_p)_{wg}}{379.4} \tag{4.139}$$

$$(n_p)_L = \frac{42 (G_p)_L \bar{\rho}_L}{\overline{M}_L} \tag{4.140}$$

where $(n_p)_{wg}$ is the cumulative gas produced in lb-mole, $(G_p)_{wg}$ is the cumulative wet gas produced in SCF, $(n_p)_L$ is the cumulative condensate liquids produced in lb-mole, $(G_p)_L$ is the cumulative volume of condensate liquids produced in barrels, $\bar{\rho}_L$ is the average density (weight-weighted) of all produced condensate liquids in pounds per million/gallons, and $\overline{M}_L$ is the average (mole-weighted) molecular weight of all produced condensate liquids in lbm/lb-mole.

Residue gas recovered (MSCF) to abandonment pressure is given by

$$G_p = 379.4 \left[ \frac{(n_p)_{wg} - (n_p)_L}{10^3} \right] \tag{4.141}$$



The pounds of total well fluid $m_w$ resulting from 1 bbl of tank oil and $R$ SCF of separator or residue gas are

$$\begin{aligned} m_w &= \frac{28.97 R_s \gamma_g}{379.4} 350\gamma_o \\ &= 0.07636 R_s \gamma_g + 350\gamma_o \end{aligned} \tag{4.142}$$

Thus the total moles of fluid in 1 bbl of oil and $R$ ft$^3$ of gas is

$$\begin{aligned} n_t &= \frac{R_s}{379.4} + \frac{350\gamma_o}{M_o} \\ &= 0.002636 R_s + 350 \frac{\gamma_o}{M_o} \end{aligned} \tag{4.143}$$

The molecular weight of the total well fluid ($M_w = m_w/n_t$), specific gravity of well fluid, and the molecular weight of tank oil are given by Equations (2.27), (2.28), and (2.29), respectively.

Since the volume fraction equals the mole fraction in the gas state, the fraction of the total fluid which is produced as gas on the surface is given by

$$f_g = \frac{n_g}{n_g + n_o} = \frac{R_s/379.4}{R_s/379.4 + 350\gamma_o/M_o} \tag{4.144}$$

Recently, Gold et al. presented the following equations which provides an improved estimate of the reservoir gas specific gravity for retrograde-gas reservoirs from field production information. The correlations were developed with the flash liberation results from 234 laboratory fluid analyses.

$$\gamma_w = \frac{R_1\gamma_1 + 4{,}602\gamma_o + G_{pa}}{R_1 + V_{eq}} \tag{4.145}$$

where $G_{pa}$ is the additional gas production (SCF/STB), $V_{eq}$ is the vapor equivalent of primary-separator liquid (SCF/STB), $R_1$ is the primary-separator gas/liquid ratio (SCF/STB), and $\gamma_1$ is the specific gravity of primary-separator gas (air = 1). The $G_{pa}$ and $V_{eq}$ are given by the following equations:[33]

For a two-stage separation system:

$$G_{pa} = R_3\gamma_3 \quad \textbf{(4.146)}$$

$$V_{eq} = \left(133{,}316\frac{\gamma_o}{M_o}\right) + R_3 \quad \textbf{(4.147)}$$

For a three-stage separation system:

$$G_{pa} = R_2\gamma_2 + R_3\gamma_3 \quad \textbf{(4.148)}$$

$$V_{eq} = \left(133{,}316\frac{\gamma_o}{M_o}\right) + R_2 + R_3 \quad \textbf{(4.149)}$$

where $\gamma_o$ = specific gravity of the stock-tank liquid, $R_2$ and $R_3$ are the gas/liquid ratios (SCF/STB) from the secondary separator and the stock tank, respectively, and $\gamma_2$ and $\gamma_3$ are the specific gravities of secondary separator and stock-tank gas, respectively.

Often only the primary-separator gas/liquid ratio, $R_1$, the specific gravity of the primary-separator gas, $\gamma_1$, and the specific gravity of the stock-tank liquid, $\gamma_o$, are known.

The following empirical correlations were developed for the calculation of additional gas production, $G_{pa}$, and vapor equivalent, $V_{eq}$. Two models are given. Model 1 includes stock-tank temperature $T_{s3}$ as a parameter, and Model 2 refers to the model that does not.

Model 1 ($T_{s3}$ included)—three-stage separation:

$$G_{pa} = 1.0347(p_{s1} - 14.65)^{0.97703}\gamma_1^{6.5230}\,\mathrm{API}^{1.3045}T_{s1}^{-1.1860}T_{s2}^{0.42499}T_{s3}^{0.11524} \quad \textbf{(4.150)}$$

$$V_{eq} = 536.56 + 1.4309p_{s1}^{0.79225}\gamma_1^{4.6352}\,\mathrm{API}^{1.2639}T_{s1}^{-0.84710}T_{s2}^{0.17103}T_{s3}^{0.18980} \quad \textbf{(4.151)}$$

Two-stage separation:

$$G_{pa} = 0.15763(p_{s1} - 14.65)^{1.3365}\gamma_1^{7.0393}\,\mathrm{API}^{1.3891}T_{s1}^{-1.4370}T_{s3}^{0.77907} \quad \textbf{(4.152)}$$

$$V_{eq} = 651.55 + 0.091752p_{s1}^{1.0928}\gamma_1^{5.2244}\,\mathrm{API}^{1.7350}T_{s1}^{-1.0843}T_{s3}^{0.41384} \quad \textbf{(4.153)}$$

Model 2 ($T_{s3}$ not included)—three-stage separation:

$$G_{pa} = 2.9922(p_{s1} - 14.65)^{0.97050}\gamma_1^{6.8049}\,\mathrm{API}^{1.0792}T_{s1}^{-1.1960}T_{s2}^{0.55367} \quad \textbf{(4.154)}$$

$$V_{eq} = 535.92 + 2.6231p_{s1}^{0.79318}\gamma_1^{4.6612}\,\mathrm{API}^{1.2094}T_{s1}^{-0.84911}T_{s2}^{0.26987} \quad \textbf{(4.155)}$$

Two-stage separation:

$$G_{pa} = 1.4599(p_{s1} - 14.65)^{1.3394}\gamma_1^{7.0943}\,\mathrm{API}^{1.1436}T_{s1}^{-0.93446} \quad \textbf{(4.156)}$$

$$V_{eq} = 635.53 + 0.36182p_{s1}^{1.0544}\gamma_1^{5.0831}\,\mathrm{API}^{1.5812}T_{s1}^{-0.79130} \quad \textbf{(4.157)}$$

Since $V_{eq}$ accounts for the low-pressure separator gas in addition to the stock-tank liquid production, the following equation can be used to estimate the reservoir fluid withdrawal:

$$q_{sc} = q_{s1}\left(1 + \frac{V_{eq}}{R_1}\right) \tag{4.158}$$

The following equation can be used to convert $q_{sc}$ to reservoir conditions:

$$q_R = q_{sc}\left[\frac{z p_{sc}(T_R + 460)}{\bar{p}_R(T_{sc} + 460)}\right] \tag{4.159}$$

where $z$ is evaluated at $\gamma_w$ calculated by using Equation (4.145), °API is the stock-tank oil gravity.

**Example 4–28.**[6] The initial daily production from a gas-condensate reservoir is 186 bbl of condensate, 3,750 MSCF of high-pressure gas, and 95 MSCF of stock-tank gas. The tank oil gravity is 51.2°API at 60°F. The specific gravity of separator gas is 0.712 and of the stock-tank gas, 1.30. The initial reservoir pressure is 3,480 psi, and reservoir temperature is 220°F. Average hydrocarbon porosity is 17.2%. Assume standard conditions of 14.7 psi and 60°F.

(a) What is the average gravity of the produced gases?
(b) What is the initial gas/oil ratio?
(c) Estimate the molecular weight of the condensate.
(d) Calculate the specific gravity (air = 1.0) for the total production.
(e) Calculate the gas deviation factor of the initial reservoir fluid at initial reservoir pressure.
(f) Calculate the initial moles in place per acre-foot.
(g) Calculate the mole fraction which is gas in the initial reservoir fluid.
(h) Calculate the initial (sales) gas and condensate in place per acre-foot.
(i) The total daily gas-condensate production in SCF and the reservoir voidage.

Solution:

(a) Using Equation (2.30),

$$(\gamma_g)_{avg} = \frac{(3{,}750)(0.712) + (95)(1.3)}{3{,}750 + 95} = 0.727$$

(b)

$$\begin{aligned} R_s &= \frac{q_{SG} + q_{TG}}{q_o} \\ &= \frac{3{,}750 + 95}{186} \times 1{,}000 \\ &= 20{,}672 \quad \approx 20{,}700 \quad \text{SCF/bbl} \end{aligned}$$

(c) Using Equation (2.29),

$$M_o = \frac{6{,}084}{51.2 - 5.9} = 134.31$$

(d) Using Equation (2.28), specific gravity of well fluid is

$$\gamma_{ff} = \frac{R_s(\gamma_g)_{avg} + 4{,}584\gamma_o}{R_s + 132{,}800(\gamma_o/M_o)}$$

$$\gamma_o = \frac{141.5}{51.2 + 131.5} = 0.774 \quad (\text{water} = 1.0)$$

$$\gamma_{ff} = \frac{(20{,}700)(0.727) + (4{,}584)(0.774)}{(20{,}700) + (132{,}800)(0.774/134.3)}$$

$$= \frac{18{,}596.916}{21{,}465.355} = 0.866 \quad (\text{air} = 1.0)$$

(e) From Figure 2–1 using the condensate curves,

$$T_c = 420°\text{R} \qquad p_{pc} = 653 \quad \text{psi}$$

$$T_{pr} = \frac{220 + 460}{420} = 1.619$$

$$p_{pr} = \frac{3{,}480}{653} = 5.329$$

Using Figure 2–3, $z = 0.870$

(f) Initial moles per acre-foot at 17.2% hydrocarbon porosity is

$$n_t = \frac{p_r V_R}{zR(T_R + 460)}$$

$$= \frac{(3{,}480)(43{,}560)(0.172)}{(0.87)(10.73)(220 + 460)}$$

$$= 4{,}107 \quad \text{moles/ac-ft}$$

(g) Using Equation (4.144),

$$f_g = \frac{20{,}700/379.4}{20{,}700/379.4 + 350(0.774/134.3)}$$

$$= \frac{54.560}{56.577} = 0.964$$

(h)

$$\text{initial gas in place} = (4{,}107 \times 379.4 \times 0.964)/1{,}000$$
$$= 1{,}502 \quad \text{MSCF/ac-ft}$$
$$\text{initial oil in place} = \frac{(1{,}502)(10^3)}{20{,}700}$$
$$= 72.56 \quad \text{bbls/ac-ft}$$

(i) Since the gas production is 96.4% of the total moles produced, then

$$\Delta G_p = \frac{\text{Daily Gas}}{0.964} = \frac{3{,}750 + 95}{0.964}$$
$$= 3{,}988.59 \quad \text{MSCF/day}$$

The total daily reservoir voidage by gas law is

$$\Delta V = 3{,}988.59 \times 10^3 \times \frac{680}{520} \times \frac{14.7}{3480} \times 0.87$$
$$\approx 19{,}168 \quad \text{ft}^3\text{/day}$$

For better accuracy the gas volume should be that from all separators including the stock-tank vapors than just the gas from the main high-pressure separator. The gas gravity should be the average of all gases produced.

If the gas and liquid composition from the high pressure separator is available, it should be used to calculate the original oil and gas in place. The following example will clarify the calculations of original oil and gas in place when gas and liquid compositions are available. Table 4–18 gives physical properties of hydrocarbon compounds commonly found in natural gases.

**TABLE 4–18. Physical properties of some hydrocarbon compounds**

| Compound | Liquid Density at 60°F & 14.7 psi lb/gal | Est. Part. Vol. at 60°F & 14.7 psi gal/MSCF | Est. Part. Vol. at 60°F & 14.7 psi gal/lb-mole |
|---|---|---|---|
| Methane | 2.900 | 14.60 | 5.530 |
| Ethane | 4.040 | 19.60 | 7.440 |
| Propane | 4.233 | 27.46 | 10.417 |
| Isobutane | 4.695 | 32.64 | 12.380 |
| n-Butane | 4.872 | 31.44 | 11.929 |
| Isopentane | 5.209 | 36.50 | 13.851 |
| n-Pentane | 5.262 | 36.14 | 13.710 |
| n-Hexane | 5.536 | 41.03 | 15.565 |
| n-Heptane | 5.738 | 46.03 | 17.463 |
| n-Octane | 5.892 | 51.09 | 19.385 |
| n-Nonane | 6.017 | 56.19 | 21.314 |
| n-Decane | 6.121 | 61.27 | 23.245 |

**TABLE 4–19. Data and calculations for Example 4–29.**

| (1) | (2) | (3) | (4) | (5) | (6) | (7) |
|---|---|---|---|---|---|---|
| | Mole Composition | | $p_{ci}$ | $T_{ci}$ | Molecular | lb/mole |
| Component | Gas | Liquid | (psi) | (°R) | Weight | (3 × 6) |
| $CO_2$ | 0.0120 | 0.0000 | 1,070 | 547 | — | 0.0000 |
| $C_1$ | 0.9404 | 0.2024 | 673 | 343 | 16.04 | 3.247 |
| $C_2$ | 0.0305 | 0.0484 | 708 | 550 | 30.07 | 1.455 |
| $C_3$ | 0.0095 | 0.0312 | 617 | 666 | 44.09 | 1.376 |
| $iC_4$ | 0.0024 | 0.0113 | 529 | 735 | 58.12 | 0.657 |
| $nC_4$ | 0.0023 | 0.0196 | 550 | 766 | 58.12 | 1.139 |
| $iC_5$ | 0.0006 | 0.0159 | 484 | 830 | 72.15 | 1.147 |
| $nC_5$ | 0.0003 | 0.0170 | 490 | 846 | 72.15 | 1.227 |
| $C_6$ | 0.0013 | 0.0384 | 440 | 914 | 86.17 | 3.309 |
| $C_7^+$ | 0.0007 | 0.6158 | **300 | **1,227 | 185.0 | 113.923 |
| | 1.0000 | 1.0000 | | | | 127.480 |

** From Figure 4–33 for Mol. wt. of $C_7^+$ = 185 and $\gamma_{C_7^+}$ = 0.8342

**Example 4–29.**[6] Calculate the initial gas and oil in place from the composition of the gas and liquid from a high-pressure separator. Mole composition of separator fluids is given in Table 4–19. Other pertinent data are as follows:

$$\begin{aligned}
\text{Mol. wt. of } C_7^+ \text{ in separator liquid} &= 185.0 \\
\text{Sp. gravity of } C_7^+ \text{ in separator liquid} &= 0.8343 \\
\text{Sp. gravity of separator liquid at 880 psig \& 60°F} &= 0.7675 \\
\text{Separator liquid vol. factor at 880 psig \& 60°F} &= 1.235 \quad \text{RB/STB} \\
\text{Molar Vol. at 15.025 psi and 60°F} &= 371.2 \quad \text{ft}^3\text{/mole}
\end{aligned}$$

$$p_i = 4350 \text{ psi} \qquad p_{sc} = 15.025 \text{ psi}$$
$$T_R = 217°\text{F} \qquad T_{sc} = 60°\text{F}$$
$$(1 - S_{iw})\phi = 0.174 \qquad q_{SG} = 842{,}600 \text{ SCF/day}$$
$$q_o = 31.1 \text{ STB/day}$$

Solution: Calculations are tabulated in Table 4–19.

**Step 1.** Calculate molecular weight of separator liquid as shown in Column 7 (127.480 lb/mole).

**Step 2.** Calculate the moles/bbl for the separator liquid.

$$n_o = \frac{350\gamma_o}{M_o}$$

$$= \frac{350 \times 0.7675}{127.48} = 2.107 \quad \text{moles/bbl}$$

**TABLE 4–19. Continued**

| (8)<br>(2) × 59.11 | (9)<br>(3) × 2.107 | (10)<br>(8) + (9) | (11)<br>(10)/61.217 | (12)<br>$p_{pc}$, psi<br>(11) × (4) | (13)<br>$T_{pc}$, (°R)<br>(11) × (5) |
|---|---|---|---|---|---|
| 0.709 | 0.0000 | 0.7090 | 0.0116 | 12.41 | 6.36 |
| 55.587 | 0.4265 | 56.0135 | 0.9150 | 615.80 | 313.85 |
| 1.803 | 0.1020 | 1.9050 | 0.0311 | 22.02 | 17.11 |
| 0.562 | 0.0657 | 0.6277 | 0.0102 | 6.29 | 6.70 |
| 0.142 | 0.0238 | 0.1658 | 0.0027 | 1.43 | 1.98 |
| 0.136 | 0.0413 | 0.1773 | 0.0029 | 1.60 | 2.22 |
| 0.035 | 0.0335 | 0.0685 | 0.0011 | 0.53 | 0.91 |
| 0.018 | 0.0358 | 0.0538 | 0.0009 | 0.44 | 0.76 |
| 0.077 | 0.0809 | 0.1579 | 0.0026 | 1.14 | 2.38 |
| 0.041 | 1.2975 | 1.3385 | 0.0219 | 6.57 | 26.87 |
| 59.110 | 2.1070 | 61.2170 | 1.0000 | 668.23 | 379.23 |

If the specific gravity of the separator liquid is not available, the mole per barrel figure may be calculated by multiplying the mole fraction of each liquid component by its barrel/mole figure from Table 4–18. The sum of the product of liquid component by its barrel/mole gives the number of barrels of separator liquid per mole of separator liquid, and its reciprocal gives moles/barrels.

**Step 3.** Calculate separator gas/oil ratio.

$$R_s = \frac{q_{SG}}{q_o B_o}$$

$$= \frac{842{,}600}{31.1 \times 1.235} = 21{,}940 \quad \text{SCF/bbl}$$

**Step 4.** Calculate the moles/barrels for separator gas.

$$n_g = \frac{R_s}{\text{Molar Volume}} = \frac{21{,}940}{371.2} = 59.11 \quad \text{moles}$$

Therefore, recombine the separator gas and liquid in the ratio of 59.11 moles of gas to 2.107 moles of liquid (Columns 8 and 9 in Table 4–19).

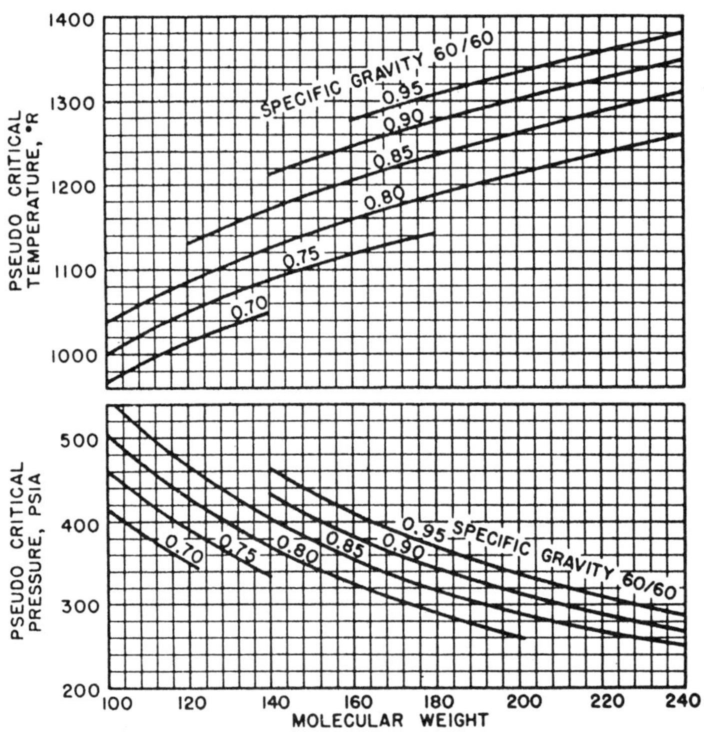


**Figure 4–33.** Correlation charts for estimating $T_{pc}$ and $p_{pc}$ of heptanes plus fractions from molecular weight and specific gravity.[32]

Column 8 = Moles of each component in 59.11 moles of gas.
Column 9 = Moles of each component in 2.107 moles of liquid.
Column 10 = Moles of each component in 59.11 + 2.107 = 61.217 moles of gas and liquid.
Column 11 = Recombined mole composition of total well fluid.

$$p_{pr} = \frac{4350}{668.23} = 6.5097$$

$$T_{pr} = \frac{217 + 460}{379.23} = 1.7852$$

Using Figure 2–3, $z = 0.963$.

$$n_t = \frac{(4{,}350)(43{,}560)(0.174)}{(0.963)(10.73)(677)}$$
$$= 4{,}713 \quad \text{moles/ac-ft}$$

$$f_g = \frac{59.11}{59.11 + 2.107} = 0.966$$

$$G_{wg} = \frac{(0.966)(4{,}713)(371.2)}{1{,}000} = 1{,}690 \quad \text{MSCF/ac-ft}$$

$$G_L = \frac{(1 - 0.966)(4713)}{(2.107)(1.235)} = 61.61 \quad \text{STB/ac-ft}$$

## GAS-CONDENSATE MATERIAL BALANCE



The material balance equation for a gas-condensate reservoir is the same as that for dry gas reservoir, provided that oil zone is completely absent or its contribution is negligible. The material balance equation can thus be written as

$$\frac{p_{sc}G_p}{(T_{sc} + 460)} = \frac{p_i V_i}{z_i(T_R + 460)} - \frac{p_R(V_i - W_e + B_w W_p)}{z(T_R + 460)} \tag{4.160}$$

and

$$G(B_g - B_{gi}) + W_e = G_p B_g + B_w W_p \tag{4.161}$$

Note that $G_p$ should be the cumulative reservoir gas production (total dry gas and gas equivalent of liquid hydrocarbons), and the two-phase gas deviation factors be applied, if available. Equations (4.160) and (4.161) contain $z$-factor at the lower pressure; $z$ is included in the $B_g$ in Equation (4.161). Because this deviation factor applies to the gas-condensate fluid remaining in the reservoir, when the pressure is below the dew point pressure, in retrograde reservoirs it is a *two-phase* gas deviation factor.[6]

## RESERVOIR PERFORMANCE—RETROGRADE GAS-CONDENSATE RESERVOIRS

Reservoir performance of a volumetric retrograde gas-condensate reservoir can be assessed by duplicating the reservoir depletion data in laboratory studies using reservoir fluid samples.

When properly collected bottomhole samples are not available, reservoir performance prediction for gas-condensate reservoir may be made from equilibrium ratios. The following example will clarify the calculations and the data required.

**Example 4–30.**[6] Calculate the volumetric depletion performance of a retrograde gas-condensate reservoir based on the laboratory tests in Table 4–20. Other pertinent data are as follows:

| | |
|---|---|
| $p_i = 2{,}960$ psi (dew point) | $T_{sc} = 60°F$ |
| $p_a = 500$ psi | $p_{sc} = 14.7$ psi |
| $T_R = 195°F$ | $V_{cell} = 947.5$ cm$^3$ |
| $S_{iw} = 0.30$ | $M_{C_7^+} = 114$ lb/lb-mole |
| $\phi = 0.25$ | $\gamma_{C_7^+} = 0.755$ @ 60°F |

**TABLE 4–20. Retrograde condensate fluid composition, volume, and z-Factor for Example 4–30**

| (1) | (2) | (3) | (4) | (5) | (6) | (7) | (8) |
|---|---|---|---|---|---|---|---|
| | Composition of Produced Gas Increments | | | | | | |
| Pressure | Mole Fraction | | | | | | |
| (psi) | $C_1$ | $C_2$ | $C_3$ | $C_4$ | $C_5$ | $C_6$ | $C_7^+$ |
| 2,960 | 0.752 | 0.077 | 0.044 | 0.031 | 0.022 | 0.022 | 0.052 |
| 2,500 | 0.783 | 0.077 | 0.043 | 0.028 | 0.019 | 0.016 | 0.034 |
| 2,000 | 0.795 | 0.078 | 0.042 | 0.027 | 0.017 | 0.014 | 0.027 |
| 1,500 | 0.798 | 0.079 | 0.042 | 0.027 | 0.016 | 0.013 | 0.025 |
| 1,000 | 0.793 | 0.080 | 0.043 | 0.028 | 0.017 | 0.013 | 0.026 |
| 500 | 0.768 | 0.082 | 0.048 | 0.033 | 0.021 | 0.015 | 0.033 |

**TABLE 4–20. Continued**

| (1) Pressure (psi) | (9) Produced Gas cm³ @ 195°F & Cell Pressure | (10) Retrograde Liquid Volume (cm³) | (11) Retrograde Vol. Percent of HC (Volume) | (12) z-Factor at 195°F & Cell Pressure |
|---|---|---|---|---|
| 2,960 | 0.0 | 0.0 | 0.0 | 0.771 |
| 2,500 | 175.3 | 62.5 | 6.6 | 0.794 |
| 2,000 | 227.0 | 77.7 | 8.2 | 0.805 |
| 1,500 | 340.4 | 75.0 | 7.9 | 0.835 |
| 1,000 | 544.7 | 67.2 | 7.1 | 0.875 |
| 500 | 1,080.7 | 56.9 | 6.0 | 0.945 |

Solution: Assume the liquid recovery from gas is 25% of the butanes, 50% of the pentane, 75% of the hexane, and 100% of the heptanes and heavier.

**Step 1.** Calculate increments of gross gas production $\Delta G_p$ in MSCF/ac-ft of net bulk reservoir rock (Table 4–21, Column 2).

$$V_{HC} = 43{,}560 \times 0.25 \times (1 - 0.30)$$
$$= 7{,}623 \quad \text{ft}^3/\text{ac-ft}$$

$$\Delta V_{2{,}960-2{,}500} = 7{,}623 \frac{175.3}{947.5} = 1{,}410 \quad \text{ft}^3/\text{ac-ft}$$

$$\Delta G_{p_{2{,}960-2{,}500}} = \frac{(379.4)(2{,}500)(1{,}410)}{(1{,}000)(0.794)(10.73)(655)} = 240.1 \quad \text{MSCF}$$

**TABLE 4–21. Gas and liquid recoveries in percent and per acre-foot for Example 4–30**

| (1) Pressure (psi) | (2) $\Delta G_p$ (MSCF) | (3) Res. Gas (MSCF) | (4) $\Delta n_L$ (bbl) | (5) = (3)/(4) $R_{si}$ (SCF/bbl) | (6) $\Delta G_p$ % | (7) % | (8) % |
|---|---|---|---|---|---|---|---|
| 2960 | 0 | 0 | 0 | 10,600 | 0 | 0 | 0 |
| 2500 | 240.1 | 225.1 | 15.3 | 14,700 | 15.2 | 15.6 | 10.7 |
| 2000 | 245.2 | 232.3 | 13.1 | 17,730 | 30.7 | 31.7 | 19.8 |
| 1500 | 266.0 | 252.8 | 13.3 | 19,010 | 47.6 | 49.3 | 29.1 |
| 1000 | 270.8 | 256.9 | 14.0 | 18,350 | 64.7 | 67.1 | 38.9 |
| 500 | 248.7 | 233.0 | 15.9 | 14,650 | 80.4 | 83.3 | 50.0 |
| | 1270.8 | 1200.1 | 71.6 | | | | |

Column 2 = Increment of gross gas production, MSCF.
Column 3 = Residue gas in each increment, MSCF.
Column 4 = Liquid in each increment, bbls.
Column 5 = Average gas/oil ratio of each increment, SCF residue gas per barrel.
Column 6 = Cumulative gross gas recovery, percent.
Column 7 = Cumulative residue gas recovery, percent, (7) = (3) × 100/1,441.
Column 8 = Cumulative liquid recovery, percent, (8) = (4) × 100/143.2.

**Step 2.** Calculate the MSCF of residue gas and the barrels of liquid obtained from each increment of gross gas production.

$$\Delta n_L = (0.25)(0.028) + (0.5)(0.019) + (0.75)(0.016) + 0.034 = 0.0625 \quad \text{mole fraction}$$

The MSCF recovered as liquid from 240.1 MSCF:

$$\begin{aligned} \Delta G_L &= \Delta G_{p_{2,960-2,500}} \times \Delta n_L \\ &= 240.1 \times 0.0625 = 15.006 \quad \text{MSCF} \\ \text{Residue Gas} &= 240.1(1 - 0.0625) \\ &= 225.1 \quad \text{MSCF} \qquad \text{(Column 3)} \end{aligned}$$

The gas volume of 15.006 MSCF can be converted to gallons of liquid. The average of the iso and normal compounds is used for $C_4$ and $C_5$.

For $C_7^+$,

$$\frac{114 \text{ lb/lb-mole}}{0.3794 \quad \text{MSCF/lb-mole} \times 8.337 \quad \text{lb/gal} \times 0.755} = 47.71 \quad \text{gal/MSCF}$$

The total liquid recovered from 240.1 MSCF is

$$= [(1.681)(32.04) + (2.281)(36.32) + (2.881)(41.03) + (8.163)(47.71)]/42$$
$$= 15.3 \quad \text{bbls} \qquad \text{(Column 4)}$$

**Step 3.** Calculate the gas/oil ratio for each increment (Column 5).

$$R_{s2960-2500} = \frac{225.1 \times 1{,}000}{15.3} = 14{,}700 \quad \text{SCF/bbl}$$

**Step 4.** Calculate the cumulative percent recoveries of gross gas, residue gas, and liquid. The initial gross gas in place is

$$G_{wg} = \frac{(379.4)(2{,}960)(7{,}623)}{(1{,}000)(0.771)(10.73)(655)}$$
$$= 1580 \quad \text{MSCF/ac-ft}$$

Of this the liquid mole fraction is 0.088 and the total liquid recovery is 3.808 gallons/MSCF of gross gas, calculated in the same way as shown in Step 2 above. Then

$$G = (1 - 0.088)(1{,}580) = 1{,}441 \quad \text{MSCF/ac-ft}$$
$$N = \frac{(3.808)(1{,}580)}{42} = 143.2 \quad \text{bbl/ac-ft}$$

Recoveries at 2,500 psi are then

$$\text{Gross gas recovery} = \frac{(100)(240.1)}{1{,}580} = 15.2\% \qquad \text{(Column 6)}$$
$$\text{Residue gas recovery} = \frac{(100)(225.1)}{1{,}441} = 15.6\% \qquad \text{(Column 7)}$$
$$\text{Liquid recovery} = \frac{(100)(15.3)}{143.2} = 10.7\% \qquad \text{(Column 8)}$$

## GOR BEHAVIOR OF GAS-CONDENSATE RESERVOIRS

When the more complete and sophisticated laboratory compositional studies are unavailable, the empirical correlations presented in this section can be used to forecast GOR behavior of volumetric gas-condensate reservoirs. The data required for the calculations are normally acquired during well testing.

The correlation can be used to predict gas/oil ratio as a function of pressure. The correlation requires the dew point pressure of the gas which may be estimated by using gas composition and the correlation presented in Chapter 2. If there is not enough data to allow the calculation of dew point pressure, the initial reservoir pressure may be used.

$$R_{50\%} = \exp[31.49 - 1.085(10^{-4})p_d - 92.03(\%C_{4+})/T_R + 110.8(\%C_{5+})/T_R + 0.0215T_R + 6.833\gamma_w - 26.98/R_d - 6.632\ln(T_R)]$$

where

$$\%C_{4+} = 6.547 + 25.52\gamma_w + 30.38/R_d + 0.02633R_d - 30.3\gamma_o - 0.00417T_R$$
$$\%C_{5+} = -8.53 + 7.83\gamma_w + 56.25/R_d + 0.0109R_d + 0.7286(°API) - 0.00424T_R$$

For $0.3p_d \leq p_R \leq p_d$:

$$\log R = 2(1 - p_R/p_d)\log(R_{50\%}/R_d) + \log R_d$$

For $p_R < p_d$, R is evaluated at $p_R = 0.3p_d$.

For $p_R \geq p_d$, $R = R_d$.

where $R$ = total surface GOR at $p_R$ (MCF/STB), $R_d$ = total surface GOR at $p_d$ (MCF/STB), $R_{50\%}$ = total surface GOR at 50% of $p_d$ (MCF/STB), and $\gamma_w$ = total well-stream gas gravity at $p_d$.

## JACOBY-KOELLER–BERRY METHOD[34]

Where laboratory studies of properly collected reservoir fluid samples are not available, or for small reservoirs where the cost of analyzing reservoir fluid samples may not be justified, or for preliminary evaluation done in advance of full-scale sampling and laboratory analysis, it is sometimes possible to use empirical correlations developed by Jacoby-Koeller and Berry to predict the future performance of gas-condensate reservoirs.

These authors studied several rich natural gas-condensate systems in the GOR range from about 3,600 to 60,000 SCF/STB, and one natural volatile oil system (2,363 SCF/STB). The following empirical correlations were presented:

Ultimate oil recovery by depletion from the saturation pressure to 500 psi:

$$N_p = -0.061743 + \frac{143.55}{R_i} + 0.00012184T_R + 0.00100114(°API) \qquad \textbf{(4.162)}$$

For the separator gas in place:

$$G = -2{,}229.4 + 148.43\left(\frac{R_i}{100}\right)^{0.20} + \frac{124{,}130}{T_R} + 21.831(°\text{API}) + 0.26356p_{d,b} \quad \textbf{(4.163)}$$

where

$G$ = total primary separator gas initially in place, SCF
$N_p$ = cumulative stock-tank oil production from $p_{d,b}$ to 500 psi, STB/bbl
$p_{d,b}$ = saturation pressure (dew point or bubblepoint), psi
$R_i$ = initial separator gas/oil ratio, SCF/STB
$T_R$ = reservoir temperature in °F

## ULTIMATE RECOVERY OF GAS AND CONDENSATE LIQUIDS BY CYCLING

To maximize ultimate condensate recovery from gas-condensate reservoirs, dry gas cycling has been used in many of these reservoirs for pressure maintenance. In dry gas cycling, as the name implies, condensate liquid is removed from the produced wet gas and the residue or dry gas is returned back to the reservoir through injection wells. The injected gas helps in maintaining the reservoir pressure close to or above the dew point pressure, retarding retrograde condensation and maximizing ultimate recovery.

Cycling recoveries may be broken down into three separate recovery factors or efficiencies. The total reservoir cycling efficiency is the product of (1) displacement efficiency, (2) sweep efficiency, and (3) permeability stratification factor.

$$E_R = E_p \times E_I \times E_D \quad \textbf{(4.164)}$$

where

1. $E_p$ = pattern sweep efficiency, developed from areal efficiency by proper weighting for variations in net pay thickness, porosity, and hydrocarbon saturation.
2. $E_I$ = vertical invasion efficiency, hydrocarbon pore space invaded (affected, contacted) by the injection fluid divided by the hydrocarbon pore space enclosed in all layers behind the injected fluid.
3. $E_D$ = displacement efficiency, volume of hydrocarbons (oil or gas) displaced from individual pores or small groups of pores divided by the volume of hydrocarbon in the same pores just before displacement.

The calculations are made first for the cycling period and then for a blow-down period in which the reservoir would be depleted to some abandonment pressure. The following steps have to be used:

1. Calculate the overall reservoir cycling efficiency using Equation (4.164). $E_p$ is determined from potentiometric model study, $E_I$ is estimated from the knowledge of how extreme the permeability variation may be, and $E_D$ is considered to be unity when cycling is being carried out at or above the dew point.
2. Calculate cumulative wet gas produced during cycling period using Equation (4.165).

$$(\Delta G_p)_{wgm} = G_{wg}(E_R)_m \tag{4.165}$$

where

$(\Delta G_p)_{wg}$ = cumulative reservoir wet gas produced during cycling, SCF
$G_{wg}$ = wet gas in place, SCF
$(E_R)_m$ = overall cycling efficiency, fraction

The subscript $m$ applies only to the cycling (pressure maintenance) period. After breakthrough has occurred in one or more wells, $(\Delta G_p)_{wgm}$ may consist of wet gas diluted by increasing amounts of dry injected gas.

3. Amount of condensate liquids recovered during the cycling period in barrels at standard conditions is given by

$$(\Delta G_p)_{Lm} = \frac{(\Delta G_p)_{wgm}(C_L)_{rem}}{10^6} \tag{4.166}$$

where

$(C_L)_{rem}$ = condensate content of wet gas at the reservoir cycling pressure, measured at standard conditions, STB/MMSCF

4. During the blow-down period, reservoir wet gas to be recovered is given by

$$(\Delta G_p)_{wgd} = (G_p)_{wg}[1 - (E_R)_m] \tag{4.167}$$

where

$(\Delta G_p)_{wgd}$ = cumulative wet gas to the depletion period
$(\Delta G_p)_{wg}$ = cumulative wet gas which would have been produced to abandonment pressure without cycling.

## GAS WELL DELIVERABILITY

In Chapter 7 of Vol. 2 several versions of gas well deliverability tests are discussed. The methods involve preparing a sandface backpressure plot which allows determination of AOF potential, performance coefficient C and exponent $n$.

Once the stabilized backpressure plot is established, a wellhead gas deliverability can be made by plotting wellhead pressure versus gas flow rate. Two points on this plot are usually known; these are $q_{max}$ at zero sand-face pressure and the $\overline{p}_R$ at zero flow rate. Several $q_g$ vs. $p_{tf}$ points are obtained using the stabilized backpressure plot and the wellhead deliverability plot constructed.

In the following sections, we will discuss the effects of well spacing, and equipment capacity (tubing or casing capacity, flowline capacity, pipeline capacity, and compressor capacity) limitations on gas well deliverability.

### RESERVOIR DELIVERABILITY

The exponent $n$ of the backpressure equation is usually constant; however, the flow coefficient C is dependent on well spacing and well completion effectiveness, among other factors. In some reservoirs, both C and $n$ may also vary from well to well. In such cases, an average total reservoir deliverability has to be established using the following relationships:

$$q_{reservoir} = C_{reservoir}(\overline{p}_R^2 - p_{wf}^2)^n \tag{4.168}$$

where

$$C_{reservoir} = \sum_{i=1}^{N} C_i$$

$N$ = number of wells

$C_i$ = performance coefficient of $i$th well

$n$ = the weighted average rate of decline

The average performance coefficient is then given by

$$q_{avg} = C_{avg}(\overline{p}_R^2 - p_{wf}^2)^n \tag{4.169}$$

The average $n$ is calculated by selecting two values of $(\overline{p}_R^2 - p_{wf}^2)$ wide apart (one log cycle for convenience) on each deliverability curve and reading the corresponding flow rates. The $n$ is then calculated using Equation (4.170) as given below.

$$n = \frac{\log q_{avg1} - \log q_{avg2}}{\log(\overline{p}_R^2 - p_{wf}^2)_1 - \log(\overline{p}_R^2 - p_{wf}^2)_2} \tag{4.170}$$

## WELL SPACING EFFECTS

When well spacing is changed, the flow coefficient C will also change; however, the exponent $n$ will stay the same (i.e., the deliverability curve will shift parallel to its existing curve). The vertical shift is proportional to the change in log of the constant C. The new C is calculated using the following equation:

$$C_2 = C_1 \left[ \frac{\ln (r_{d1}/r_w)}{\ln (r_{d2}/r_w)} \right]^n \tag{4.171}$$

where for $t$ in hours,

$$r_d = 0.0704 \left( \frac{kt\bar{p}_R}{\phi\mu} \right)^{0.5} \tag{4.172}$$

Also

$$C_2 = C_1 \left[ \frac{\ln (0.606 r_{e1}/r_w)}{\ln (0.606 r_{e2}/r_w)} \right] \tag{4.172a}$$

where $t$ is in hours and $\phi$ is a fraction.

Equation (4.171) is based on the assumption that during testing there will be no significant change in the produced gas viscosity and $z$-factor. However, to account for depletion accompanying the change in spacing, change in gas viscosity and $z$-factor should be added to Equation (4.171) giving

$$C_2 = C_1 \left[ \frac{\mu_{g1} z_{avg1} \ln (0.606 r_{e1}/r_w)}{\mu_{g2} z_{avg2} \ln (0.606 r_{e2}/r_w)} \right] \tag{4.173}$$

**Example 4–31.** A deliverability curve was obtained for a well with a drainage area of 2,250 acres and an external boundary pressure of 2,800 psi. The $p/z$-plot on the same well showed that the static reservoir pressure after producing cumulative gas of 45,950 MMSCF would be 1,568 psi. When $p_{ws}$ = 1,568 psi, five additional equally spaced wells will be drilled to drain this 4,500-ac reservoir. What will be the bottomhole pressure for the well to produce 4 MMSCF/D? Well radius is 0.25 ft and $n = 0.90$.

Solution: One well radius would be

$$r_e = \left[ \frac{4{,}500/5(43{,}560)}{3.14} \right]^{0.5} = 3{,}533 \text{ ft}$$

From the deliverability curve (for $r_e = 3{,}533$ ft),

$$@ \quad q_g = 1800 \quad \text{MSCF/D}, \qquad \log(\Delta p^2) = 5.851$$

Therefore, $C_2 = 0.00975$, and

$$C_5 = 0.00975 \left[ \frac{\ln\left(0.606 \dfrac{5{,}580}{0.25}\right)}{\ln\left(0.606 \dfrac{3{,}533}{0.25}\right)} \right]$$
$$= 0.01024$$

For $q_g = 4$ MMSCF/D,

$$\begin{aligned}
\log q_g &= \log C + 0.9 \log(\Delta p^2) \\
\log(\Delta p^2) &= \frac{\log(4{,}000) - \log(0.01024)}{0.90} \\
&= 6.213 \\
(\Delta p^2) &= 1.6333 \times 10^6 \\
(\bar{p}_R^2 - p_{wf}^2) &= 1.6333 \times 10^6 \\
p_{wf} &= [(1{,}568)^2 - 1.6333 \times 10^6]^{0.50} \\
&= [(2.4586 - 1.6333) \times 10^6]^{0.50} \\
&= 908 \quad \text{psi}
\end{aligned}$$

The new curve for five well passes through $q_g = 4{,}000$ MSCF/D and $(\Delta p^2) = 1.6333 \times 10^6$ and is parallel to the existing curve with $n = 0.90$.

## EQUIPMENT CAPACITY LIMITATIONS ON DELIVERABILITY

In addition to the reservoir capacity to produce gas, gas deliverability is also affected by the capacity of the surface and subsurface equipment installed on the well. Note that the produced gas has to pass through the tubing, separators, dehydrators, meter run, and flowline to the pipeline. Since pressure drop is a function of flow rate, because of friction some pressure drop is associated with each piece of the equipment.

For a particular set of equipment, pipeline pressure, and state of reservoir depletion, there is some maximum rate that can be produced. This rate is determined by (1) plotting bottomhole flowing pressure versus the production rate based on sand face production and (2) on the same graph plot wellhead flowing pressure versus production rate at the surface through all the facilities. The line drawn through the points in Case 1 represents reservoir flow capacity and Case 2 the tubing and flowline capacity. *The intersection of these two capacity curves gives optimum production rate for the given setup.* At any other rate the capacity of the well to produce is limited by either the reservoir or the equipment capacity.

The following equations are widely used by the industry to calculate pressure drop through various pieces of gas production and transmission equipment.

### Tubing or Casing Capacity:

$$q_g = 20{,}000 \left[ \frac{d^5(p_{wf}^2 - e^s p_{tf}^2)s}{\gamma_g(\bar{T} + 460)\bar{f}H(e^s - 1)} \right]^{0.50} \tag{4.174}$$

where

$s = 0.0375\gamma_g H/(\bar{T} + 460)\bar{z}$
$d$ = inside tubing diameter, in.
$H$ = difference in elevation between $p_{tf}$ and $p_{wf}$, ft
$\bar{f}$ = Moody fraction factor at $\bar{T}$ and $\bar{p}$

A trial-and-error procedure is required to obtain $p_{wf}$.

### Flowline Capacity:

The Weymouth equation for horizontal flow is given below. This equation will provide adequate result for a flowline 1 to 2 miles in length.

$$q_g = \frac{433.49(T_{sc} + 460)}{p_{sc}} \left[ \frac{(p_1^2 - p_2^2)d^{16/3}}{\gamma_g(T_R + 460)L_{miles}\bar{z}} \right]^{0.50} \tag{4.175}$$

Note that the flowline length is given in miles.

### Compressor Capacity:

The adiabatic horsepower required to compress 1 MMSCF/D of natural gas for any given set of conditions is given by

$$\text{hp/MMSCF/D} = \frac{3.027p_{sc}}{(T_{sc} + 460)} \frac{k}{k - 1} (T_{suc} + 460) \left[ \left( \frac{p_{dis}}{p_{suc}} \right)^{z_{suc}\frac{(k-1)}{k}} - 1 \right] \tag{4.176}$$

The total brake horsepower (Bhp) is given by

$$\text{Bhp} = \frac{(\text{hp/MMSCF/D})(q_g)}{E} \tag{4.177}$$

where

$k = C_p/C_v$ = ratio of specific heats for the gas at suction conditions
$p_{dis}$ = discharge pressure, psi
$p_{suc}$ = suction pressure, psi
$E$ = overall compressor efficiency

**Pipeline Capacity:**

The pressure drop through a pipeline is given by the well-known Panhandle equation as follows:

$$q_g = 435.87E\left(\frac{T_{sc} + 460}{p_{sc}}\right)^{1.07881}\left(\frac{p_1^2 - p_2^2}{(T + 460)L\bar{z}}\right)^{0.5394}\left(\frac{1}{\gamma_g}\right)^{0.4606} d^{2.6182} \tag{4.178}$$

where

$E$ = pipeline efficiency factor

## RATE-TIME PREDICTION

Future reservoir performance could be established based only on reservoir performance or on combined reservoir and equipment performance. To properly predict the reservoir performance, it is necessary to consider the capacity of the reservoir to produce, capacity of the equipment, and the state of depletion of the reservoir.

The following step-by-step procedure adopted from Ikoku can be used for the rate-time prediction based on (1) only reservoir performance and (2) combined reservoir and equipment performance:[15]

**Only Reservoir Performance:**

1. Prepare a $p/z$ plot for the reservoir.
2. Conduct a deliverability test and prepare a stabilized deliverability plot.
3. Select several reservoir pressures; small pressure increments are recommended for better accuracy. At each of these pressures, find $G_p$ from the $p/z$ plot.
4. At the selected reservoir pressures, determine the AOF potential. If the contract producing rate does not exceed the allowable producing rate (usually 25% of AOF), then use the contract rate

for this time interval. On the other hand, if the contract rate exceeds the allowable producing rate, then the allowable rate must be used for the interval.

5. Calculate the time required to produce the gas during each interval by

$$\text{Time} = \frac{\text{gas produced during interval}}{q_{avg}}$$

6. Repeat Steps 3 to 5 for consecutively lower values of $\bar{p}_R$ until abandonment pressure is reached. 

**Combined Reservoir and Equipment Performance:**

1. Determine the stage of cumulative gas production at which the gas well will no longer be able to make maximum contract rate.
   (a) Prepare a log-log plot of $(p_{wf}^2 - e^s p_{tf}^2)$ vs. $q_g$; obtain $(p_{wf}^2 - e^s p_{tf}^2)$ for maximum value of $q_g$.
   (b) Prepare a plot of $p_{wf}^2$ vs. $(p_{wf}^2 - e^s p_{tf}^2)$; read $p_{wf}^2$ for the $(p_{wf}^2 - e^s p_{tf}^2)$ obtained in (a).
   (c) Prepare gas well deliverability plot from the stabilized deliverability curve, read $(\bar{p}_R^2 - p_{wf}^2)$ for $q_g$ in Step (a).
   (d) Calculate $\bar{p}_R$ from data obtained in Steps (b) and (c).
   (e) Prepare $p/z$ plot. From the plot obtain cumulative gas production for the $\bar{p}_R$ from Step (d).
2. Determine the cumulative production at a time when the well is no longer able to produce at normal contract rate. Repeat Steps (a) through (e) for the new rate.
3. Divide the rate interval between the normal contract rate and the rate at the economic limit into 10 equal parts, for each part calculate $G_p$.
4. Calculate time required to produce gas by

$$\Delta t_i = \frac{G_{pi} - G_{pi-1}}{q_{gi} - q_{gi-1}} \times 2 \tag{4.179}$$

5. Prepare a plot of $G_p$ and $q_g$ as a function of time.

**Example 4–32.** Average gas well deliverability curve and average $p/z$-plot for a 10-well gas field are given in Figures 4–34 and 4–35. Initial gas production rate from the field will be 3,000 MSCF/D. At the time when deliverability drops to 1,250 MSCF/D per well, compression will be added to be able to again produce 3,000 MSCF/D of reservoir gas per well. Other pertinent data are as follows. Assume the ratio of bottomhole pressure to surface pressure (flowing or shut in) is 1.2.

$H = 5{,}150$ ft
$p_i = 2{,}400$ psi
$T_R = 166$ °F
$A = 1{,}600$ acres
$h = 42$ ft

$GE_{cond} = 752$ SCF/STB
$q_{ST} = 8.0$ STB/MMSCF
$p_{suc} = 1{,}000$ psi
$p_{sc} = 15.025$ psi
$\gamma_g = 0.625$ (air = 1.0)
$z_i = 0.827$

$S_{iw} = 0.24$
$\phi_{avg} = 0.265$
$T_{sc} = 60°F$
$T_{suc} = 80°F$
$k = C_p/C_v = 1.30$
$E_{comp} = 0.85$

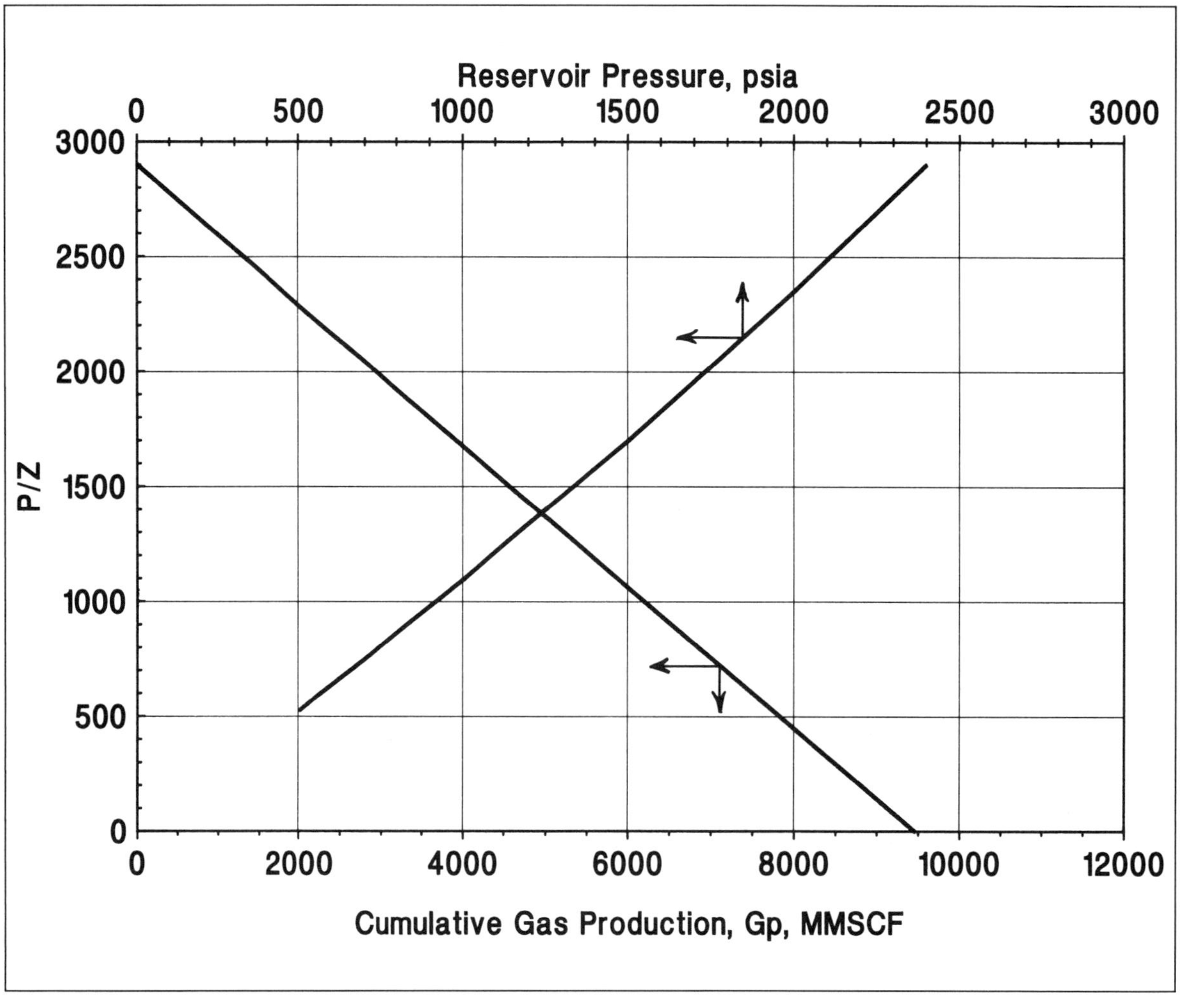


**Figure 4–34.** $p/z$-plot and $P_R$ vs. $p/z$ plot for Example 4–32.

Determine:

(a) The original reservoir gas in place

(b) Cumulative field reservoir gas produced at a rate of 3,000 MSCF/D per well into the 1,000 psi line

(c) Field compressor horsepower requirement

Solution:

(a)

$$G = 43{,}560Ah\phi(1 - S_{iw})\frac{p_R(T_{sc} + 460)}{p_{sc}(T_R + 460)z}$$

$$= 43{,}560(1{,}600)(42)(0.265)(1 - 0.24)\frac{(2{,}400)(60 + 460)}{(15.025)(166 + 460)(0.827)}$$

$$\approx 94{,}600 \quad \text{MMSCF}$$

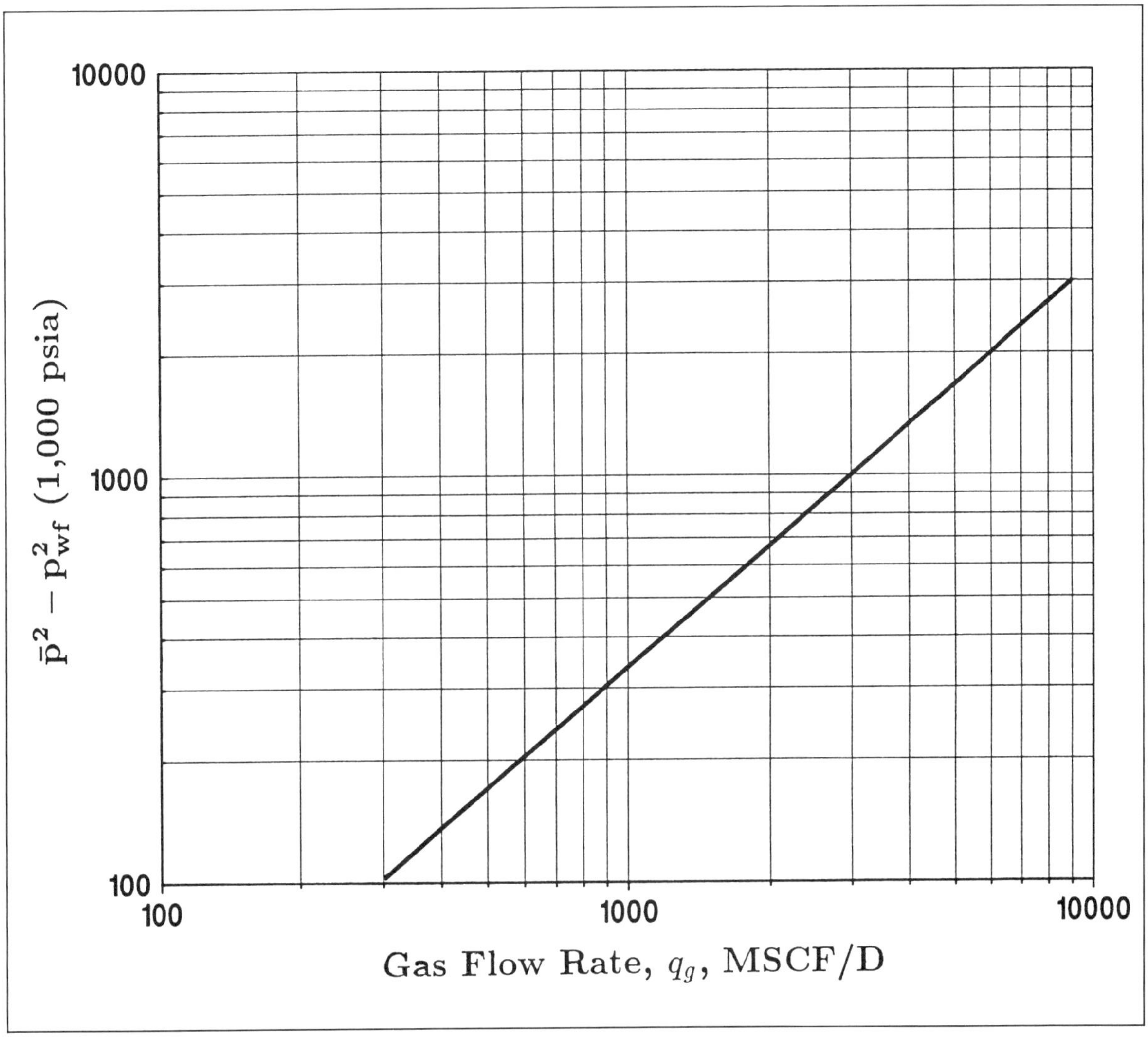


**Figure 4–35.** Stabilized deliverability curve for Example 4–32.

Also from $p/z$-plot $G_{p/z=0} = 9{,}460$ MMSCF/well
(b) From Figure 4–35, at $q_g = 3{,}000$ MSCF/D,

$$(\bar{p}_R^2 - p_{wf}^2) = 1{,}000{,}000 \quad \text{psi}^2$$
$$\bar{p}_R^2 = 1{,}000{,}000 + 1{,}200^2 = 2{,}440{,}000 \quad \text{psi}^2$$
$$\bar{p} = 1{,}562 \quad \text{psi}$$

From Figure 4–34, $p/z = 1{,}780$ and $G_p = 3{,}630$ MMSCF/well
Total field production $(G_p)_{field} = 36{,}300$ MMSCF

(c) Calculate sales gas volume.
Reservoir gas = sales gas + gas equivalent of condensate

$$3.0 \quad \text{MMSCF} = (q_g)_{\text{sales gas}} + (q_g)_{\text{sales gas}}(752 \times 10^{-6})(8.0)$$
$$(q_g)_{sales\ gas} = 2.982 \quad \text{MMSCF/D/well}$$

Determine compressor suction pressure from Figure 4–35, at $q_g = 1{,}250$ MSCF/D.

$$\bar{p}_R^2 - p_{wf}^2 = 420{,}000 \quad \text{psi}^2$$
$$\bar{p}_R^2 = 420{,}000 + 1{,}200^2 = 1{,}860{,}000$$
$$\bar{p}_R = 1{,}364 \quad \text{psi}$$

From Figure 4–35, at $q_g = 3{,}000$ MSCF/D,

$$\bar{p}_R^2 - p_{wf}^2 = 1{,}000{,}000 \quad \text{psi}^2$$
$$p_{wf}^2 = 1{,}364^2 - 1{,}000{,}000 = 860{,}000 \quad \text{psi}^2$$
$$p_{wf} = 928 \quad \text{psi}$$

Compressor suction pressure, $p_{suc} = (928/1.2) = 773$ psi.
Compressor horsepower required $T_{pc}$ and $p_{pc}$ at $\gamma_g = 0.625$ are

$$T_{pc} = 365°\text{R} \qquad p_{pc} = 669 \quad \text{psi}$$
$$T_{pr} = \frac{80 + 460}{365} = 1.479$$
$$p_{pr} = \frac{928}{669} = 1.387$$

Therefore, $z = 0.681$

Using Equation (4.176), hp/MMSCF/D is

$$\text{hp/MMSCF/D} = \frac{(3.027)(15.025)}{(60 + 460)} \frac{1.30}{1.30 - 1}(80 + 460)\left[\left(\frac{1{,}000}{773}\right)^{0.861 \times \left(\frac{1.3-1}{1.3}\right)} - 1\right]$$
$$= 204.663(0.052)$$
$$= 10.642$$

Using Equation (4.177),

$$\text{Field Bhp} = \frac{(10.642)(2.982)(10)}{0.85} = 373 \approx 400 \quad \text{hp}$$

## GAS STORAGE CAPACITY OF SIMPLE PIPELINES

Pipeline storage capacity is often utilized for daily temporary demand fluctuations. The amount of gas which is not required by the utilities at a certain moment is stored in the pipeline by appreciably raising pressure in the main pipeline system. This could easily be achieved by the intermediate compression stations.

The storage capacity is given by the difference between the content of the pipeline under packed and unpacked conditions. Packed pipeline condition refers to minimum withdrawals for a constant supply of gas at maximum discharge pressure. The unpacked condition refers to maximum withdrawal at a minimum discharge pressure at constant supply of gas.

The quantity of gas measured at base conditions stored in the pipeline for any given flow condition is given by

$$V = 19.20\frac{d^2(T_{sc}+460)L_{miles}}{p_{sc}(T+460)}\left[p_1+p_2-\frac{p_1p_2}{p_1+p_2}\right] \quad \textbf{(4.180)}$$

The pressures at each end of the pipeline under both packed and unpacked conditions are calculated by using Equation (4.175) where $q_g$ is in hours.

**Example 4–33.** The average flow condition through a 50-mi long and 10-in. diameter pipeline is 36 MMSCF/D. The gas delivery pressure is 50 psi and minimum flow is 10 MMSCF/D. Calculate the storage capacity of the pipeline for gas gravity of 0.6 and flowing temperature of 60°F. The standard temperature and pressure are 60°F and 14.7 psi, respectively.

Solution: Solving Equation (4.175) for minimum pressure,

$$\frac{36{,}000{,}000}{24} = \frac{(18.062)(520)}{14.7}\left[\frac{(p_1^2-2{,}500)10^{16/3}}{(0.6)(520)(50)(0.942)}\right]^{0.50}$$

$$p_1 = 615.2 \quad \text{psi} \qquad (\text{assume max.} = 650 \text{ psi})$$

Flow rate for packed conditions is 23,000,000 CF/D (36,000,000 + 10,000,000)/2.

Solving for $p_2'$ under packed flow conditions,

$$\frac{23{,}000{,}000}{24} = \frac{(18.062)(520)}{14.7}\left[\frac{(650^2-p_2^2)10^{16/3}}{(0.60)(520)(50)(0.90)}\right]^{0.50}$$

$$p_2' = 525 \quad \text{psi}$$

Thus pressures for the two conditions are

| | | |
|---|---|---|
| Packed flow | $p_1' = 650$ psi | $p_2' = 525$ psi |
| Unpacked flow | $p_1 = 615$ psi | $p_2 = 50$ psi |

Using Equation (4.180),

For unpacked flow (average flow) conditions:

$$V = \frac{(19.20)(100)(520)(50)}{(14.7)(520)}\left[(615 + 50) - \frac{615 \times 50}{615 + 50}\right]$$
$$= 6{,}530.612(618.759)$$
$$= 4{,}040{,}875 \quad ft^3$$

For unpacked flow conditions:

$$V' = 6{,}530.612\left[(650 + 525) - \frac{650 \times 525}{650 + 525}\right]$$
$$= 6{,}530.612(884.574)$$
$$= 5{,}776{,}810 \quad ft^3$$

Thus the storage capacity of pipeline is

$$V' - V = 1{,}736{,}000 \quad ft^3$$

## DECLINE CURVE ANALYSIS

The purpose of decline curve analysis is to determine future production and therefore ultimate recovery for wells with some production history. Since decline curve analysis depends on a curve-fit of past performance, the accuracy is expected to be greater for a well with several months or years of uninterrupted production history than for a well with only a limited amount of history. Decline curve analysis is based on the following assumptions:

1. Sufficient past production performance is available in order to make a reasonable match of this performance to obtain an extrapolation of future performance.
2. The past production history is based on capacity (unrestricted) production with no changes in operational policy such as artificial lift, stimulation, etc. It is assumed that the property will continue to be operated in the same manner in the future.

In its most simple form, decline curve analysis can be performed by finding a curve which approximates the past production history and extrapolating this curve into the future. A more rigorous procedure for analyzing past production history is to fit the history with a mathematical curve. Once the characteristics of this curve are known, they can be used to predict future performance.

Three rate-time decline curves (exponential decline, hyperbolic decline, and harmonic decline, a special case of hyperbolic decline) are discussed in this section. In addition to these rate-time curves, several other decline curves used in the industry (such as rate-cumulative production plot, cumulative oil-cumulative gas curves, water cut versus cumulative production curve, and GOR versus cumulative oil production curve) are also discussed. Note that the rate-time curves are commonly plotted on semilogarithmic graph paper as log $q$ vs. $t$.

To work with rate-time curves, monthly production is first plotted on semilogarithmic graph paper. After production is plotted, a declining pattern is observed and the future trend is forecasted. This future trend is usually extrapolated to the *economic limit* of the property.

The economic limit (EL) is the production rate below which the net revenue from the production will not meet the expenses, including taxes. The economic limit is computed as follows:

$$EL_{oil} = \frac{WI \times LOE}{NRI[P_o + (P_g \times GOR)/1{,}000] \times (1 - T)} \quad \textbf{(4.181)}$$

$$EL_{gas} = \frac{WI \times LOE}{NRI[(P_o \times Y) + P_g] \times (1 - T)} \quad \textbf{(4.182)}$$

where

$EL_{oil}$ = economic limit for oil well, bbls/month
$EL_{gas}$ = economic limit for gas well, MSCF/month
$P_o, P_g$ = oil and gas prices, \$/bbl and \$/MSCF
LOE = lease operating expenses, \$/well/month
WI = working interest, fraction
NRI = net revenue interest, fraction
GOR = gas/oil ratio, SCF/bbl
$Y$ = condensate yield, bbls/MMSCF
$T$ = production and severance taxes, fraction

## EXPONENTIAL DECLINE

Exponential decline is also referred to as *constant percentage decline*. The mathematical expression that defines this type of decline is an exponential equation of the form $y = ae^{bx}$. The exponentially declining production plots as a straight line on semilogarithmic graph paper (production on log scale and time on linear scale). The following equations are used to interpret exponential decline curve:

Oil produced during interval $t = 0$ and $t_o$:

$$\Delta N_p = \frac{(q_i - q_o)}{a} \quad \textbf{(4.183)}$$

Time required to produce $\Delta N_p$:

$$t = \frac{\ln (q_i/q_o)}{a} \tag{4.184}$$

Rate at time $t_o$:

$$\begin{aligned} q_{ot} &= q_i e^{-at_o} \\ d &= \frac{q_i - q_o}{q_i} = 1 - \frac{q_i}{q_o} = 1 - e^{-at} \\ a &= -\ln(1 - d) = \frac{\ln q_i - \ln q_o}{t} \end{aligned} \tag{4.185}$$

where

$q_i$ = rate at the beginning of time period.
$q_o$ = rate at end of time period or EL
$t$ = time period between $q_i$ and $q_o$
$\Delta N_p$ = cumulative production during time period
$a$ = nominal decline rate
$d$ = effective decline rate

Changing effective monthly decline rate to effective yearly decline rate, and vice versa,

$$d_y = 1 - (1 - d_m)^{12}$$

OR

$$d_m = 1 - (1 - d_y)^{1/12} \tag{4.186}$$

**Example 4–34.** Figure 4–36 shows a rate-time plot of actual oil production for years 1985 through 1989 and the forecasted oil production thereafter. Calculate:

(a) The effective and nominal decline rate
(b) Cumulative oil production from 1/1/1990 to economic limit of 200 bbls of oil per month.
(c) The time required to produce the oil calculated in (b).
(d) The production rate at the end of 1993.

Solution: Calculate effective decline rate/year.

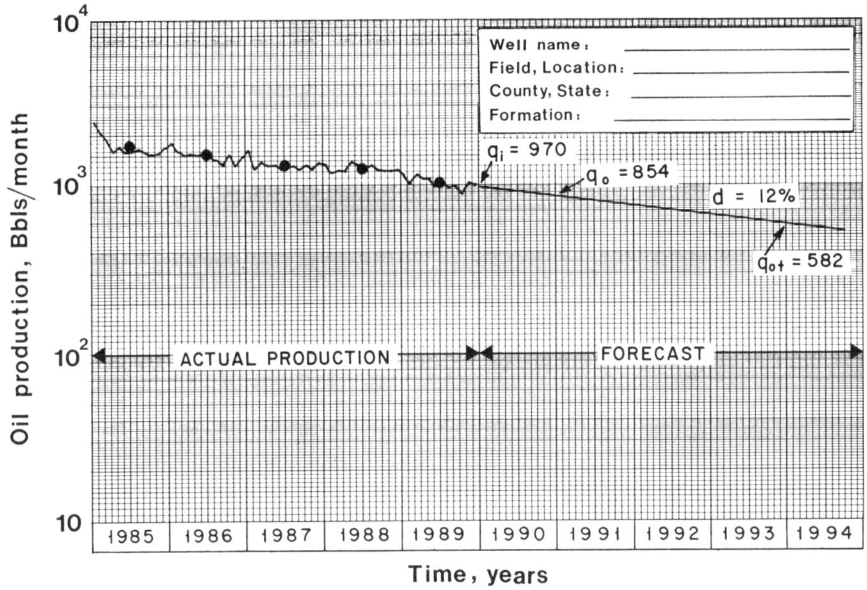


**Figure 4–36.** Rate-time plot showing exponential decline for Example 4–34.

(a)
$$d = 970 - 854/970 \times 100 = 12\%$$
$$a = -\ln(1 - 0.12)$$
$$= 0.1278$$

(b)
$$\Delta N_p = \frac{(q_i - q_{EL})}{a}$$
$$= \frac{(970 - 200)12}{0.1278} = 72{,}300 \quad \text{STB}$$

(c)
$$t = \frac{\ln(970/200)}{0.1278} = 12.36 \quad \text{years}$$

Time to economic limit could also be directly read from the graph.

(d)
$$q_{ot} = q_i e^{-at_o}$$
$$= 970e^{(-0.1278 \times 4)} = 582 \quad \text{STB/month}$$

Note that the raw production data plot may normally exhibit wild fluctuations around the trend of the data. To make the data easier to extrapolate into future, it can generally be smoothed by calculating averages for periods of time and plotting the averages at the middle of the time period used for averaging (yearly averages shown in Figure 4–36). For steeper decline rate, averages over a shorter time period, such as three or six months, should be used. Note that the average rate for the period does not represent the true logarithmic average rate over the period; this is particularly true for steeply declining production.

## HYPERBOLIC DECLINE

In contrast to the exponential decline, hyperbolic decline is a concave upward curve when plotted on semilogarithmic graph paper. As a consequence, the decline characteristic $a$ is not a constant value but rather is the slope of the tangent to the rate-time curve at any point. The slope of this tangent to the curve may be evaluated graphically in the same manner as the exponential decline. The decline characteristic $a$, however, changes with producing time. The curvature of the curve is defined by *hyperbolic exponent b* (also denoted by $n$ in some books), which is constant with time.

In normal petroleum operations, the value of $b$ ranges between 0 and 1.0 with $b = 0$ being an exponential decline and $b = 1$ being harmonic decline (special case of hyperbolic decline). However, it is found that in fractured tight (low-permeability) formations, exponents in excess of 1.0 may be calculated. Care should be exercised in these cases as a large value of $b$ will result in a very unrealistically low decline rate late in the well life. The following equations are used to interpret hyperbolic decline curves:

$$\Delta N_p = \frac{q_i^b}{(1 - b)a_i\,[q_i^{(1-b)} - q_o^{(1-b)}]} \qquad \textbf{(4.187)}$$

$$t = \frac{(q_i/q_o)^b - 1}{ba_i} \qquad \textbf{(4.188)}$$

$$q_{ot} = q_i(1 + ba_it_o)^{-1/b} \qquad \textbf{(4.189)}$$

where

$$d = \frac{q_i - q_o}{q_i}$$

$$a_t = a_i(q_t/q_i)^b = \frac{1}{b}\,[(1 - d_t)^{-b} - 1]$$

$$d_t = 1 - (1 + bat)$$

Several methods are available in literature to evaluate the hyperbolic exponent $b$[35–39]. Three of the most commonly used methods are (1) the french curve method, (2) shifting the curve on log-log graph paper, and (3) type-curve fitting method. Only the first two methods are presented in the following paragraphs.

**French Curve Method.** A french curve is usually a plastic template made in the shape of a hyperbolic curve. This method is preferred by this author because it is easy to use, will give comparable results, and is not very time consuming. The following steps should be used:

1. Plot the monthly production data on a semilogarithmic graph paper.
2. With the help of a french curve, draw a best-fit curve through the plotted data.
3. Select a point $q_i$ on this curve and draw tangent to the curve at this point.
4. Compute the effective decline rate $d$ per year and compute $a_i = -\ln(1 - d)$.
5. Select another point $q_t$ at some point $t$ from the initial rate $q_i$ and try different values of $b$ in Equation (4.189) until you find the $b$ value that will give a $q_t$ value close to the one selected. This trial-and-error calculation can be easily done with the help of a programmable calculator.

These values of $b$, $a_i$, and $q_i$ are then used to forecast production performance. If the $q_i$ is different from the $q_i$ at effective date of evaluation, a new value of $a_i$ must be computed by using the following equation:

$$a_{i_{new}} = \frac{a_{i_{old}}}{(1 + n a_{i_{old}} \Delta t)} \tag{4.190}$$

**Curve Shift on Log-Log Paper.** This method involves a log-log trial-and-error plot of the rate versus time data until a straight line is obtained. The procedure involves plotting the log of the rate versus the log of the time plus some constant, $(t + A)$. The constant A is varied until a straight line is obtained. Over shifting the curve to the right will show upward curvature. The hyperbolic exponent $b$ is then given by the reciprocal of the slope ($b = -1$/slope) of the straight line. The method is illustrated in Figure 4–37. A computer program presented in Chapter 1 can be used also to achieve this.

## HARMONIC DECLINE

Harmonic decline is a special case of hyperbolic decline with hyperbolic exponent $b = 1.0$. In this case the rate-time relationship can also be straightened out on log-log graph paper after shifting, and assumes a slope of 45° (unit slope line). In this specific case a plot of the inverse of the production rate versus time on a linear scale should also yield a straight line. The following equations are used in the same way as with hyperbolic decline:

$$\Delta N_p = \frac{q_i}{a_i} \ln \frac{q_i}{q_o} \tag{4.191}$$

$$t = \frac{(q_i/q_o) - 1}{a} \tag{4.192}$$

$$q_{ot} = \frac{q_i}{1 + a_i t} \tag{4.193}$$

## PRODUCTION FORECAST—LOW-PERMEABILITY (TIGHT) GAS RESERVOIRS

Low-permeability gas reservoirs (tight gas sands) are classified by the Federal Energy Regulatory Commission (FERC) as those reservoirs that have an average in-situ permeability to gas of 0.1 md

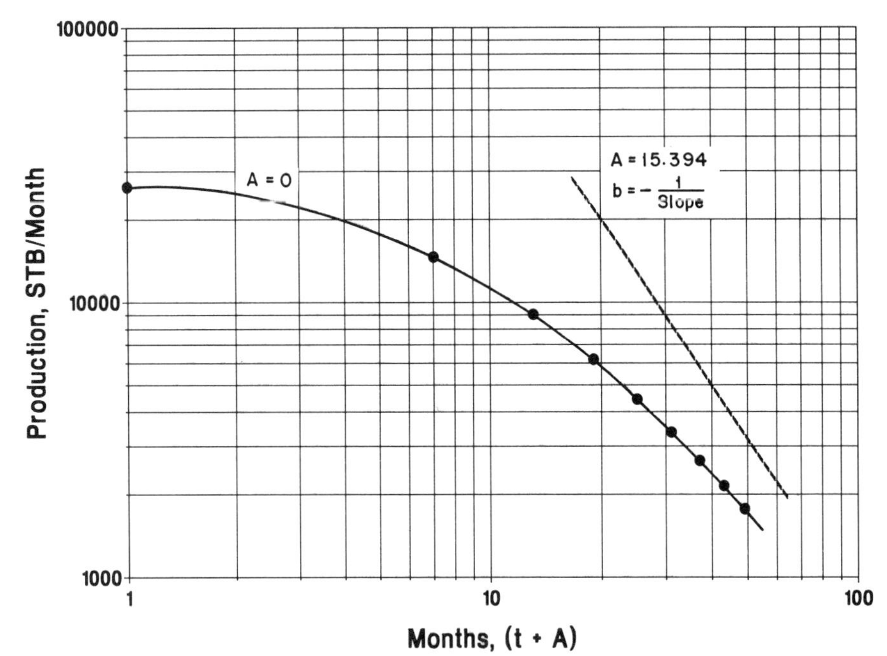


**Figure 4–37.** Log-log extrapolation of hyperbolically declining production.

or less. These tight reservoirs must be hydraulically fractured to achieve economical production rates. Massive hydraulic fracturing (MHF) treatments have made it possible to economically produce these reservoirs. MHF, as the name implies, is a fracturing treatment applied on a massive scale, which may involve the use of 50,000 to 500,000 gallons of treating fluid and 100,000 to 1 million pounds of proppant (sand). The MHF treatment improves production by exposing larger surface areas of these low-permeability formations to flow into the wellbore through the creation of vertical fractures.

The flow regimes resulting from the complex geometry of a hydraulically fractured low-permeability reservoir are:

1. *Fractured linear flow,* where all flow occurs within the fracture in the direction of the wellbore
2. *Bilinear flow,* which is characterized by fracture linear flow as well as formation linear flow
3. *Formation linear flow,* where the significant flow is from the formation into the fracture from a direction perpendicular to the fracture face
4. *Elliptical flow,* which is transitional flow between linear and pseudoradial flow regimes when flow begins to be influenced by portions of the reservoir beyond the fracture tips
5. *Pseudoradial flow,* where fluid flow approaches the wellbore fracture system essentially from all directions

Production from these low-permeability fractured reservoirs when plotted on semilogarithmic graph paper will exhibit hyperbolic decline characteristics with hyperbolic exponent greater than 1.0.

The same production data when plotted on log-log paper (rate vs. time) will show a decline slope of −0.50. Formation damage reduces the early flush production rates, but has a diminishing effect so that the slope will be less than −0.50 initially and then will approach −0.50 after three to six months. Sometimes, early time slope is greater than −0.50, which will indicate the depletion of thin lenses of high permeability near the wellbore. This typical behavior is observed in many low-permeability reservoirs.[40–42] The following equations can be used to forecast the past production performance of these reservoirs:

$$\log q_g = -\frac{1}{2}\log t + \log C \tag{4.194}$$

which gives

$$q_g = C/\sqrt{t} \tag{4.194a}$$

and

$$1/q_g = m\sqrt{t} + C_1 \tag{4.195}$$

Equation (4.194) is used with the production plotted on log-log paper (rate $q_g$ vs. elapsed time $t$) and a half-unit slope line is drawn through the data points to give intercept C at $t = 1$ month. This intercept is then used in Equation (4.194a) to forecast future performance. This method can be used on wells with at least 12 months of production or when sufficient data fall on a half-unit slope line.

The second equation, Equation (4.195), can be used with as little as four months of initial production. The reciprocal production rate ($1/q_g$) is plotted versus the square root of time ($\sqrt{t}$) on coordinate graph paper. The method of linear regression will usually be used to determine the values of $C_1$ and $m$. Feeding these values into Equation (4.195) will generate production rates at any given time $t$.

Note that in most low-permeability gas reservoirs the linear flow characteristics will last for up to three to four years. When using these methods, the production should be scheduled for up to this time period and then switching to an average exponential decline to economic limit. Failing to do this will result in unrealistically long productive life and very high gas recovery.

## RATE-CUMULATIVE PRODUCTION PLOT

In this section we will discuss the use of rate-cumulative plots for estimating the ultimate recovery of oil from a well, or a producing lease. These types of plots have an advantage over *rate-time* plots, because any interruptions due to mechanical or other failures or downtime do not affect the plotting coordinates. The evaluation of workovers and remedial treatments tends to be directly evident on these plots, and any change in recovery can be read easily by extrapolating the plotted data. The rate-cumulative plots have the following characteristics:

1. The rate-cumulative data will form a straight line on Cartesian coordinate graph paper if the rate-time plot of the same data shows a straight line on semilogarithmic graph paper (i.e., exponential decline with $b = 0$). Theoretically the slope of the rate-cumulative curve should be exactly the same as the nominal decline rate calculated from the rate-time plot. However, because of the difference in "eyeballing" the curves and data smoothing, some difference will always exist. A typical rate-cumulative plot corresponding to the rate-time plot in Figure 4–36 is shown in Figure 4–38. Note that the slope of the rate-cumulative curve is 10.25% versus 12% decline of the rate-time curve. Also the ultimate recovery from the rate-cumulative plot at EL = 200 STB/mon is 158,000 bbls versus 155,000 bbls from the rate-time curve.
2. The rate-cumulative data will form a straight line when plotted on semilogarithmic graph paper (with rate plotted on log scale) if the rate-time plot of the same data shows a harmonic decline with $b = 1.0$.
3. If the rate-time plot of the data exhibit hyperbolic decline with $0 < b < 1.0$ then the rate-cumulative data will form a straight line after being shifted (in the $x$ direction) on a log-log graph paper, i.e., to plot log $q$ versus log $(A - N_p)$. The value of A can be found by trial-and-error, using different values until one is found which gives a straight line plot on log-log graph paper with a slope $= [1/(1 - b)]$. Also, for a hyperbolically declining production, the rate-cumulative plot will be a straight line on a special graph paper with the rate scale marked off so that $q^{-b}$ and the cumulative scale are linear.

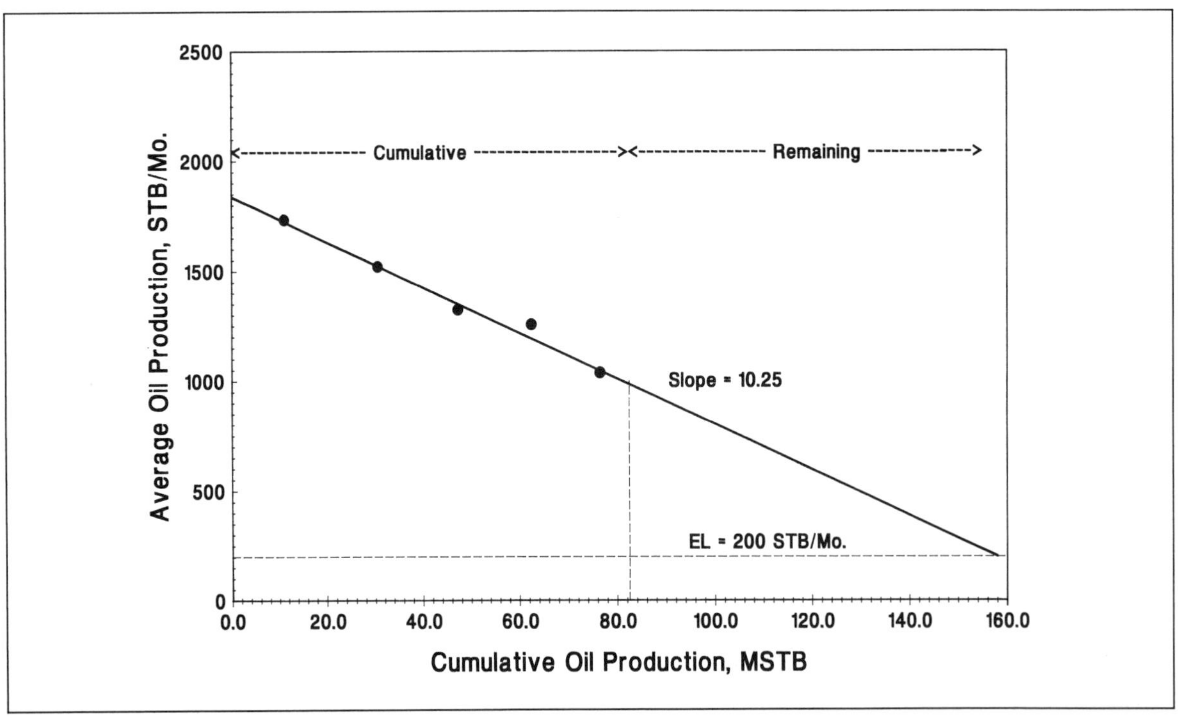


**Figure 4–38.** Rate-cumulative plot for the data in Figure 4–37.

## CUMULATIVE GAS ($G_p$) VERSUS CUMULATIVE OIL ($N_p$) PLOT

The projection of associated gas in oil production or condensate in gas production is not as straight forward as the projection of the primary product. There are several methods of projecting associated gas production from solution gas drive reservoirs. The material balance techniques (Tarner, Tracy, etc.) as presented in the previous sections can be used to predict gas/oil ratio versus cumulative oil production. However, these methods require fluid property and relative permeability data which are often not available.



Another method of projecting associated gas which is in common use is to assume constant gas/oil ratio throughout the productive economic life of the property. This assumption is correct as long as the reservoir is producing at above the bubblepoint pressure; however, the gas/oil ratio increases dramatically as the reservoir pressure drops below the bubblepoint, in which case a very pessimistic gas rate will be predicted. On the other hand, near abandonment, the GOR trends downward so a constant GOR assumption at this stage may be too optimistic.

A log-log plot of cumulative gas production ($G_p$) versus the corresponding cumulative oil production ($N_p$) is most often utilized to project the associated gas production. On this plot $G_p$ and $N_p$ will plot as a straight line as shown in Figure 4–39, with unit slope until the reservoir pressure is above the bubblepoint pressure. Below the bubblepoint pressure, the GOR will increase and the slope on log-log plot will also increase as shown in Figure 4–39.

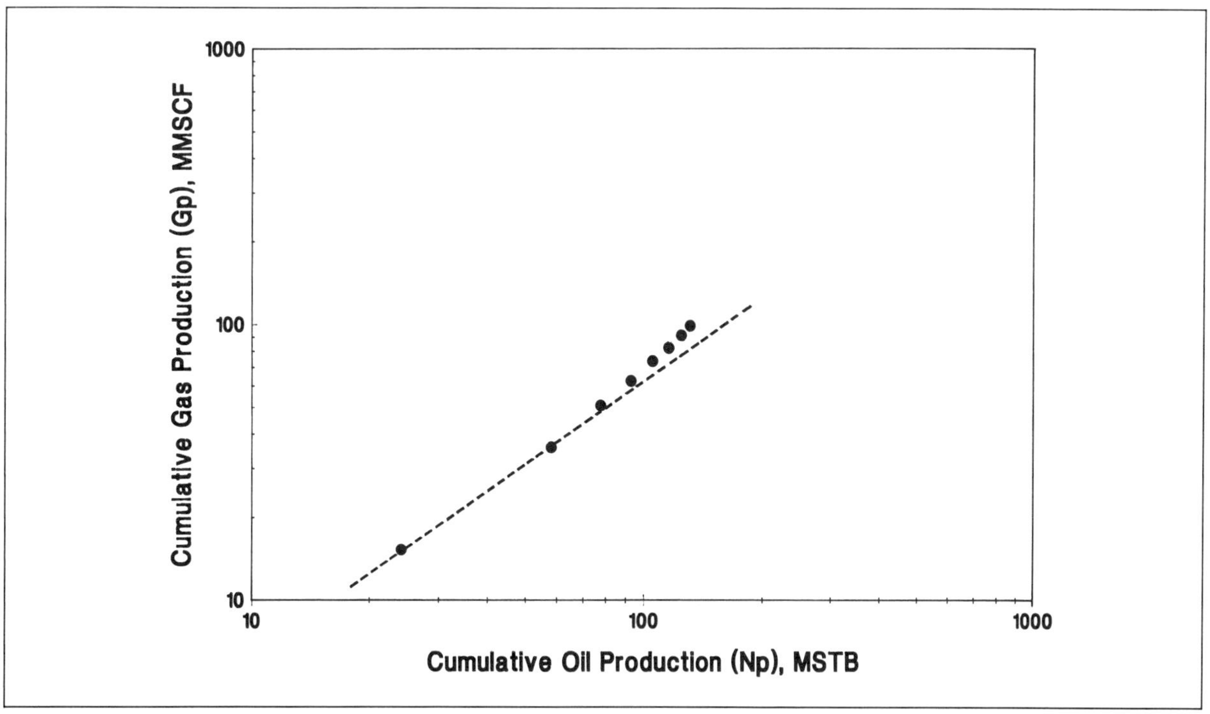


**Figure 4–39.** Cumulative gas ($G_p$) versus cumulative oil ($N_p$) plot.

Another plot of $G_p$ versus $N_p$ which according to this author's experience provides better forecasting tool, is to plot log of $G_p$ versus $N_p$ (on semilogarithmic graph paper) as shown in Figure 4–40. On this plot the data will plot as a straight line after shifting it by a constant A ($G_p$ + A) in the direction of y-axis. When the cumulative oil production at a particular point in time is known, its associated cumulative gas production can be projected using this plot.

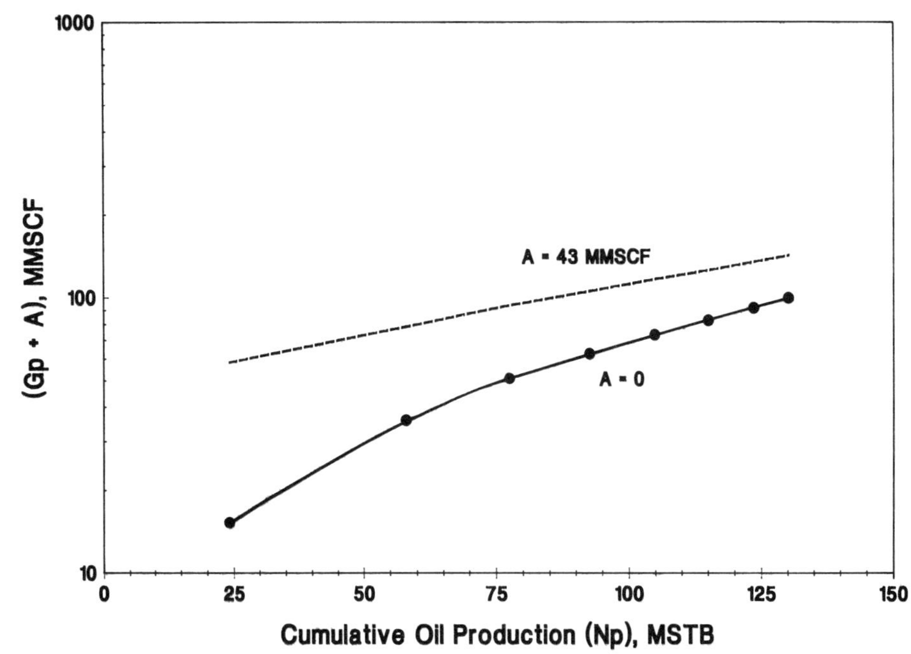


**Figure 4–40.** Semilogarithmic plot of cumulative gas production versus cumulative oil production.

## WATER CUT VERSUS CUMULATIVE OIL PLOT

This plot is most often used in waterdrive reservoirs and reservoirs producing under waterflood. However, its use is not limited only to these two types; the method can be applied to any reservoir showing increasing water production. In the case of waterdrive or waterflood, the oil production is very consistent and maintains a shallow decline while the water production increases relatively faster.

Since it costs money to dispose of the produced water, the variable monthly operating expenses will increase, hence raising the economic limit, which in turn will limit the oil production. The wells showing increasing water production, if forecasted with the regular rate-time curves or rate-cumulative curves without accounting for water production and the associated cost, may not provide realistic results.

To determine the ultimate oil recovery, the water cut which is the percentage of water produced per total fluid produced $[q_w/(q_o + q_w)]$, is plotted versus $N_p$ on semilogarithmic graph paper. The plot will usually provide a straight line, or it will show some relationship between the two variables. After the trend is defined, the cumulative oil production limit is obtained from the trend at about 98% water cut.

This author points out that the arbitrarily chosen 98% water cut may not always provide a true economic limit of cumulative oil production, because at this percentage one does not know what the economic limit would be. Therefore it is suggested that a companion plot of total fluid versus water cut should be plotted on semilogarithmic graph paper so that an idea of the specific ratio of water and oil at a certain water cut can be obtained. Then determine if the percentage of oil in the total fluid is above or below the economic limit. These curves may also be used to forecast the water production and hence the associated variable cost.



A water cut-cumulative oil production plot and a water cut-total fluid production plot for a well producing from Mission Canyon formation in North Dakota are shown in Figures 4–41 and 4–42, respectively. The cumulative oil production at 98% water cut is read directly from Figure 4–41 as 47,250 bbls. At 98% water cut, the total monthly fluid production is read directly from Figure 4–42 as 3,311 bbls/mon, giving oil production of approximately 66 bbls [3,311 × (1 − 0.98)], which meets the requirement of the estimated economic limit.

If the economic limit was more than 66 bbls/mon, then one would have to determine the cumulative at a percentage lower than the 98%, and so on. In fact, one would determine first what

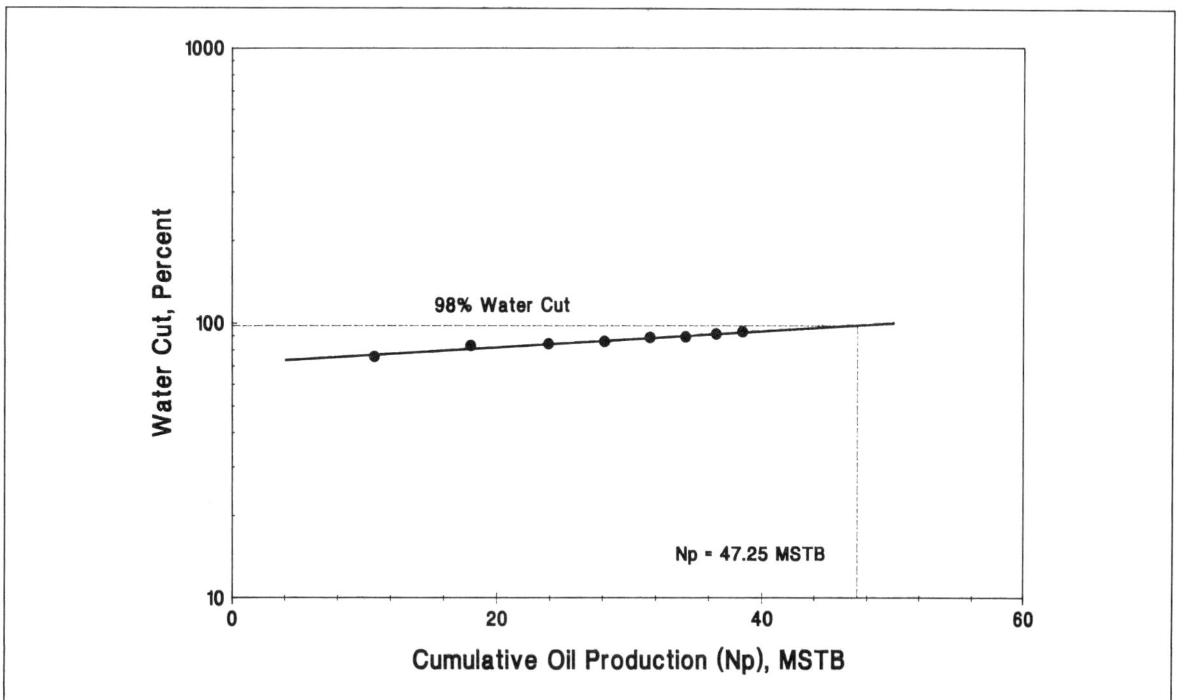


**Figure 4–41.** Water cut versus cumulative oil production plot.

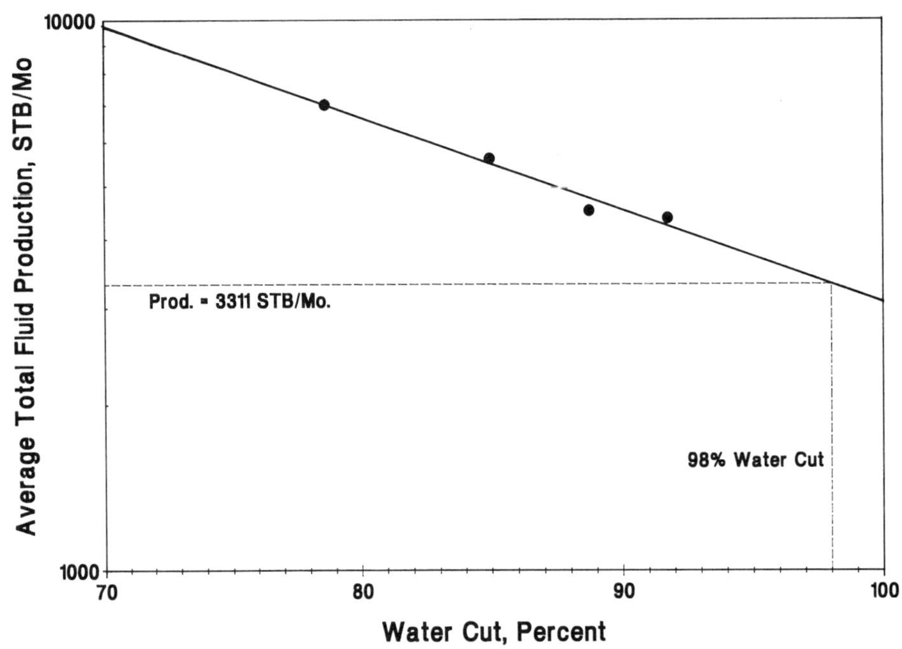


**Figure 4–42.** Water cut versus total fluid production plot for the data in Figure 4–42.

water cut percentage will give the forecasted economic limit, and then determine the ultimate oil recovery at this percentage.

## RESERVOIR ECONOMICS

Economic evaluations of oil and gas properties are made for the sole purpose of providing guidance to management or investors in allocating capital resources. Whenever investment opportunities exceed available capital, projects must be ranked in order of desirability.

Performance of the evaluation engineer is certainly affected by broad areas of government policy, worldwide supply and demand, capital formation, and exploration strategy. However, his concern centers on everyday considerations such as cash flow, present worth, project timing, rate of return, cost of money, future prices, risk factors associated with different categories of reserve estimates, and government regulations.

All economic evaluations involve a look into the future. To do this, the evaluation engineer needs to know the annual production rates, future operating costs and product prices, drilling and completion costs, taxes, inflation rate, cost of money, participation factors, reversionary interests, and future investments required to keep the project going.

Some of the yardsticks used in economic decisions are presented in Chapter 1; these same yardsticks are used in decisions regarding oil and gas investments. In this section we will discuss the ownership interest computation, effect of severance and ad valorem taxes on economic evaluation, future cash flow computation dedicated to specific working interest, and determination of fair market value.

## OWNERSHIP INTEREST COMPUTATION

Although appropriate ownership interest rates for each property for economic evaluation will be provided by the land department, the mechanism of how the interest of each partner in an oil deal is derived needs to be understood. 

The mineral interest in a property is divided into *working (operating) interests* and *nonoperating interests*. A working interest (WI) can be classified as a basic, joint, pooled, or unitized interest. A nonoperating interest can be classified as a royalty, overriding royalty, production payment, or net profits interest. Associated with each working interest there is a *net revenue interest (NRI)* reduced proportionately by *nonoperating working interest*. For an investor in an oil and gas deal, NRI is the key economic figure which measures the percent of profits earned by investor. This NRI is earned by a specific working interest.

When a lease is obtained, a royalty, commonly one-eighth (12.5%) is retained by the land owner/mineral interest owner (*lessor*). The working-interest owner at this stage will earn a maximum of only 87.5% of any oil and gas revenues from the lease while paying 100% of the costs. On the other hand, the lessor receives 12.5% of the oil and gas free of any costs except for the severance, ad valorem, and any other surcharge taxes levied by the state. The various types of interest arrangements and oil deals are defined as[44]

1. *Leasehold interest.* A leasehold interest is conveyed by the interest owner (lessor) to the *lessee.* The leasehold interest owner has the right to explore for and produce minerals.
2. *Primary term.* Leases are usually in effect for a period of one to ten years and as long thereafter as hydrocarbons are produced. When hydrocarbons are produced on a lease, the lease is referred to as *held by production.*
3. *Lease bonus.* Upon acquiring mineral rights, the lessee pays a *sign-up* bonus to the lessor. The amount ranges from $1 per acre to $3,000 per acre, depending on the exploration activity within or nearby the leased property.
4. *Delay rental.* Unless operations are commenced on or before one year from the effective date of the lease, the lease terminates regardless of the length of the primary term of the lease, unless the lessee pays a previously agreed delay rental, which enables the lessee to defer operations for one more year.
5. *Overriding royalty interests (ORRI).* This type of nonoperating interest is similar to a landowner's royalty in that the ORRI owner receives a portion of the production free and clear of all costs. An override is often said to be *carved out* of the WI since it is created from the leasehold's share of the revenue.
6. *Joint working interest.* When more than one working interest owners are involved, each own an individual fraction of the WI in a single lease. A joint WI may result from leases, purchases

or exchanges, or sharing arrangements. One party known as the operator (usually the party with the largest interest) of the property has the responsibility for developing and operating the property.

7. *Pooled working interest.* When two or more properties are combined which may have the same or different WI owner. The NRIs are combined separately at the same time. Pooling may be voluntary or mandatory. It will usually result in more efficient operations and a greater recovery of hydrocarbons.
8. *Unitized working interest.* When two or more WIs in different properties are combined, the resulting WI is referred to as unitized WI. Unitization usually occurs after production starts because the WIs and the associated NRIs are prorated based on the ultimate recovery of hydrocarbons dedicated to each lease. On the other hand, poolings usually occur before development is started.
9. *Production payment (PPI).* Nonoperating interest normally created out of the WI and usually expressed as a certain amount of money, a certain period of time, or a certain quantity of oil or gas. The PPI reverts back to the WI from which it was created after the specified amount of money, period of time, or oil or gas are recovered. It is often used to pay down debt since the amount of money going for debt services is proportional to the producing rate.
10. *Carried interest.* It is a form of deal in which an interest owner does not receive any revenue and does not pay any costs until a certain condition is met. Carried interests are more often carried to the *casing point* or through the *tanks* of the first well. A farmout could be expressed as a carried interest. If a company has a X% WI to the casing point, then the carrying party pays all of the dry-hole costs. At the casing point, the carrying party may decide to put up X% of the completion costs to gain the X% of WI with its associated NRI. A one-third for one-fourth is a typical farmout agreement in which the farmee pays one-third of the costs to some predetermined point in order to receive one-fourth WI.

The computation of ownership interest in oil and gas deals ranges from a simple two-owner arrangement (landowner and investor) to a complex multiple owner arrangement. The best way to determine these interests is to always flow-chart the investors and their associated interests in a systematic manner. The following example will clarify the computations.

**Example 4–35.** Assume a lease is obtained by X Company for a 12.5% landowner's royalty. To reduce the risk exposure in development, the X Company further sells 30% of the deal to Company Y. Compute the NRI of each company.

Solution: Ownership interests of each party are given in Table 4–22.

**Example 4–36.** A lease broker, Z, obtains a lease for a 12.5% landowner's royalty. He sells 100% of the deal to A, retaining a 2.5% ORRI. A sells 100% of the deal to B, retaining 2.5% ORRI. B sells 50% of the deal to C and 25% to D. Compute the ownership interests of each party in the deal.

**TABLE 4–22. Working and net revenue interest computation of Example 4–35**

| | Working Interest Amount Spent (%) | Revenue Interest Amount Earned (%) |
|---|---|---|
| Lessor | 0 | 12.5 |
| Company X | 70 | 61.25 |
| Company Y | 30 | 26.25 |
| Total | 100 | 100.0 |

Net Lease = 1.0 − 0.125 = 0.875
Company X = 0.70 × 0.875 = 0.6125
Company Y = 0.30 × 0.875 = 0.2625

Solution: Ownership interest computations are shown in Table 4–23.

**Example 4–37.** A lease is obtained for a 12.5% landowner royalty. A geologist is retained to study the lease and retaining 2% ORRI. A promoter who sells the deal to investors keeps 5.5% ORRI with 25% back-in working interest after payout (after the WI participants have recovered all their costs from a successful producing well). Compute the before and after payout ownership interests for each party.

Solution: The various interests are shown in Table 4–24.

**TABLE 4–23. WI and NRI computation for Example 4–36**

| | WI (fraction) | NRI (fraction) |
|---|---|---|
| Landowner | 0 | 0.125 |
| Broker Z | 0 | 0.025 |
| Broker A | 0 | 0.025 |
| Company B | 0.25 | 0.206 |
| Company C | 0.50 | 0.413 |
| Company D | 0.25 | 0.206 |
| Total | 1.00 | 1.00 |

## SEVERANCE AND AD VALOREM TAXES

In the oil and gas industry, besides the federal and state income taxes, each state has its own surcharge on oil and gas production. This surcharge is referred to as ad valorem and severance taxes. These taxes are usually applied on a percentage basis or on a dollar per unit basis of the gross value at the point of production.

**TABLE 4–24. WI and NRI comparison for the deal in Example 4–37**

| | Before Payout (%) | | After Payout (%) | |
|---|---|---|---|---|
| | WI | NRI | WI | NRI |
| Lessor | 0 | 12.5 | 0 | 12.5 |
| Geologist | 0 | 2.0 | 0 | 2.0 |
| Promoter | 0 | 5.5 | 25.0 | 21.375 |
| Investor | 100 | 80.0 | 75.0 | 64.125 |
| Total | 100.0 | 100.0 | 100.0 | 100.0 |

These taxes range anywhere from zero in some states to as high as 15% of the gross value in the other states. Therefore it is extremely important to consider the proper amount of these taxes in oil and gas property evaluations. The properties that are in the states with higher taxes will have a significant impact on the value of the property.

Many states have complicated exemptions, limits, or sliding scales. In order to provide a quick reference, a summary of the severance and ad valorem taxes is presented in Table 4–25 and Table 4–26, respectively. American Petroleum Institute's publication "TC-10 State and Local Oil and Gas Severance and Production Taxes" provides a 26-page summary of these taxes. For further and more accurate details, individual state's departments of revenue should be contacted.

The following description relates to the notes in Table 4–25:[43]

1. For wells producing less than 25 BOPD or 200 MSCF/D in unitized fields and all wells that commence production after September 1, 1979. Baldwin county has an additional 1% severance tax.
2. For wells producing more than 25 BOPD or 200 MSCF/D which commenced production on or before September 1, 1979. Baldwin County has an additional 1% severance tax.
3. For oil wells which commenced production after June 30, 1981. For first five years only and subject to oil gravity and economic limit.
4. For oil wells commencing production on or before June 30, 1981. Rate is affected by oil gravity and economic limit.
5. Oil production is 10 BOPD or less during any calendar month.
6. Production is greater than 10 BOPD during any calendar month.
7. Varies, depending on needs of Oil and Gas Division. Usually small when compared to the estimated ad valorem tax rate. 1983 rate = $0.0144/Bbl.
8. For gross income under $25,000. Wells averaging less than 10 BOPD in a taxable year are exempt.
9. For gross income greater than $25,000 and less than $100,000. Wells averaging less than 10 BOPD for taxable year are exempt.

TABLE 4–25. **Oil and gas severance and taxes by state**[43]

| Note | State | Oil | Gas | Note | State | Oil | Gas |
|---|---|---|---|---|---|---|---|
| | Rates are stated as fraction of gross value unless otherwise noted | | | | | | |
| 1 | Alabama | .06 | .06 | 17 | Louisiana | .03125 | $.07 |
| 2 | | .10 | | 18 | | .0625 | — |
| 3 | Alaska | .1225 | .10 | 19 | | .125 | — |
| 4 | | .15 | – | 20 | | Note | — |
| | Arizona | .025 | .025 | 21 | Michigan | .066 | .05 |
| 5 | Arkansas | .04 | $.003 | 22 | | .04 | — |
| 6 | | .05 | — | | Mississippi | .06 | .06 |
| | | 5 mill/STB | — | 23 | Montana | .05 | .0265 |
| 7 | California | varies | varies | 24 | | .06 | — |
| 8 | Colorado | .02 | .02 | 25 | | .0002 | — |
| 9 | | .03 | .03 | 26 | Nebraska | .02 | .02 |
| 10 | | .04 | .04 | 27 | | .03 | .03 |
| 11 | | .05 | .05 | 28 | | .0004 | .0005 |
| 12 | | .0006 | — | 29 | Nevada | $.005 | $.0001 |
| 13 | | .0009 | — | 30 | New Mexico | .0375 | $.152 |
| 14 | Florida | .05 | .05 | 31 | | .0019 | — |
| 15 | | .08 | .05 | 32 | | .0255 | — |
| | Georgia | $.005 | $.0005 | | New York | None | None |
| | Idaho | .02 | .02 | 33 | North Carolina | $.005 | $.0005 |
| | Indiana | .01 | .01 | 34 | North Dakota | .05 | .05 |
| 16 | Kansas | .08 | .08 | 35 | | .115 | — |
| | Kentucky | .045 | .045 | | Ohio | $.1 | $.03 |
| | Oklahoma | .07085 | .07085 | 37 | Utah | .04 | .04 |
| | Oregon | .06 | .06 | | Virginia | None | None |
| | South Dakota | .045 | .045 | 38 | West Virginia | .0434 | .0863 |
| | Tennessee | .03 | .03 | 39 | Wyoming | .06 | .06 |
| | Texas | .046 | .075 | 40 | | .04 | — |
| 36 | | .001875/b | — | 41 | | 0.0001 | — |

$ taxes are based on $/STB and $/MSCF.

10. For gross income greater than $100,000 but less than $300,000. Wells averaging less than 10 BOPD for taxable year are exempt.
11. For gross income $300,000 and over. Stripper production is exempt.
12. A conservation tax, for production on or after January 1, 1982.
13. A conservation tax, for production after January 1, 1979 and before July 1, 1981.

**Table 4–26. Approximate ad valorem tax rates (fraction of gross value)[43]**

| State | Oil | Gas | State | Oil | Gas |
|---|---|---|---|---|---|
| Alabama | None | None | Nebraska | 0.06 | 0.06 |
| Alaska | 0.002 | 0.002 | Nevada | 0.06 | 0.06 |
| Arizona | 0.05 | 0.05 | New Mexico | 0.05 | 0.05 |
| Arkansas | 0.04 | 0.04 | New York | 0.10 | 0.10 |
| California | 0.09 | 0.09 | North Carolina | 0.05 | 0.05 |
| Colorado | 0.03 | 0.03 | North Dakota | None | None |
| Florida | 0.02 | 0.02 | Ohio | 0.07 | 0.07 |
| Georgia | 0.05 | 0.05 | Oklahoma | None | None |
| Idaho | 0.05 | 0.05 | Oregon | None | None |
| Illinois | — | 0.08 | Pennsylvania | 0.08 | 0.08 |
| Indiana | 0.07 | 0.07 | South Carolina | 0.05 | 0.05 |
| Kansas | 0.08 | 0.08 | South Dakota | 0.02 | 0.02 |
| Kentucky | 0.06 | 0.06 | Tennessee | None | None |
| Louisiana | None | None | Texas | 0.04 | 0.04 |
| Maryland | 0.07 | 0.07 | Utah | 0.05 | 0.05 |
| Michigan | 0.02 | 0.02 | Virginia | None | None |
| Minnesota | 0.05 | 0.05 | West Virginia | None | None |
| Mississippi | None | None | Wyoming | 0.07 | 0.07 |
| Montana | 0.09 | 0.09 | | | |

Ad valorem taxes are often based on assessed value of equipment and property and are administered on county basis.

14. For wells producing less than 100 BOPD that is all produced by tertiary methods.
15. For wells producing more than 100 BOPD and not utilizing tertiary methods.
16. Any ad valorem taxes paid may be used as credit against the severance tax up to 3.67% of the gross value. Thus the effective severance tax rate may only be 4.33% if the ad valorem taxes paid are at least 3.67% of the gross value.
17. For 10 BOPD or less.
18. For 25 BOPD or less and at least 50% salt water.
19. For wells producing more than 25 BOPD and any condensate production.
20. A working interest in newly discovered fields are exempt from 50% of the severance tax for 24 months on the first 100 BOPD. The 24-month period starts when regular production begins.
21. All production except stripper production.
22. Stripper production as per definition of the State of Michigan.
23. After April 1, 1985.
24. Until March 31, 1985.
25. Conservation tax. Montana also has a Resource Indemnity Trust tax of $25 + 0.5% of gross value over $5,000.
26. Wells producing 10 BOPD or less.

27. Wells producing more than 10 BOPD.
28. Conservation tax.
29. Conservation tax.
30. Severance tax, gas rate changes yearly.
31. Conservation tax.
32. Emergency school tax.
33. Conservation tax.
34. Production tax on all crude oil and gas.
35. Extraction tax of 6.5% on all crude oil except stripper production and royalty interests on first 100 BOPD in addition to 5% production tax.
36. Oil conservation tax.
37. Occupation tax effective 1/1/84. Wells producing less than 20 BOPD are exempt. Production for the first six months is also exempt. The state has a conservation tax of 0.002.
38. The 0.0434 tax is the business and occupation tax. The gas tax is on the value over $5,000 per year.
39. For production where average daily rate is in excess of 10 BOPD for preceding calendar year.
40. For production where average daily rate is less than or equal to 10 BOPD for preceding calendar year.
41. Oil and gas conservation tax. Rate varies; maximum rate is 0.0004.

## APPRAISAL OF OIL AND GAS PROPERTIES

Engineering appraisal is an estimate of future recoverable oil and the cash profit to be derived from the production and sale of that oil and gas. Oil and gas reserve evaluation reports are most often utilized by:

1. Financial institutions to determine the collateral value of the properties for financing
2. Management of the corporations making decisions regarding any further development of the fields and future corporate planning
3. Sellers and buyers for determining the value of the properties at which they could be sold or purchased
4. Investors to determine the prosperity of a certain company which is a potential investment
5. Attorneys and courts as evidence in any litigation or disputes over a property
6. Publicly traded companies reporting their oil and gas reserves position in their annual reports and 10-K reports to the Securities and Exchange Commission (SEC) and their shareholders
7. Internal Revenue Service (IRS) for estate taxes, gift taxes, and other estate settlements
8. General partners to report to their limited partners the estimated value of their respective interests

Preparation of the reserve evaluation report involves (1) estimating the remaining recoverable oil and gas reserves and the manner in which they will be produced, (2) forecasting the associated lease operating and overhead expenses and any other future development costs, (3) forecasting the future product prices at which the oil and gas will be sold, (4) estimating the future tax consequences (for after-tax reports), and (5) using appropriate severance/production taxes and ownership interests at which the properties have to be appraised.

After all the information is gathered and thoroughly analyzed, cash flow analysis computer programs are then utilized to generate cash flows for individual properties. A sample of a typical before–federal income tax cash flow for a hypothetical oil property is shown in Table 4–27. The cash flow was generated using OGRE (Oil and Gas Reserve Evaluation) computer program by David P. Cook & Associates, Inc. The profitability criteria and decision analysis tools given in Chapter 1 are then utilized for various decision making.

## DATA REQUIRED FOR EVALUATION OF OIL AND GAS PROPERTIES

Oil and gas property evaluation reports, depending on their use, most often are prepared by a third-party oil and gas consultant. The quality of the report and the cost of preparing the report are both dependent on the quality of the data that are provided to the consultant. This author has experienced that even most of the very established oil and gas companies fail to provide appropriate data.[45]

To make a sound evaluation of a given oil and gas property, the evaluation engineer requires certain basic data. The following check list may serve as a reminder when collecting such data:[3]

1. Ownership maps, geological structure maps, isopach maps, and geological cross sections, etc.
2. Lease location data
3. Complete set of well logs
4. Core analysis data for the zones cored and analyzed
5. Reservoir fluid sample analysis data
6. Chronological history of all well operations including original drilling and completion
7. Tabulation of monthly oil, water, and gas production by lease, by wells, and by pay zones since original completion along with a tabulation of any bottomhole pressure surveys, etc.
8. Summary of allowable formula and current daily allowable rates for each well
9. Gross oil and gas prices accompanied by a summary of the oil and gas sales contract agreement with the respective purchasers and the term of agreement
10. Severance and local taxes actually paid to the local authorities accompanied by randomly selected payment vouchers for confirmation
11. Tabulation of actual historical gross operating expenses per well per month for each property. Identify the routine operating expenditure versus the nonroutine expenditure, such as equipment replacement, workover, stimulation, etc.
12. In case of undeveloped or nonproducing reserves, estimate of completed well costs or recompletion costs for reserves behind the casing.

# TABLE 4–27. Typical cash flow of oil property

PROVED PRODUCING RESERVES
SUMMARY FORECAST OF FUTURE ANNUAL
PRODUCTION AND NET CASH FLOW

RESERVES AND ECONOMICS

AS OF JANUARY 1, 1985

| -END- MO-YR | ---GROSS PRODUCTION--- OIL, MBBL | GAS, MMCF | ----NET PRODUCTION---- OIL, MBBL | GAS, MMCF | --PRICES-- OIL $/B | GAS $/M | NET OPER REVENUES | --------OPERATIONS, M$-------- SEV + ADV + WF TAXES | NET OPER EXPENSES | CAPITAL COSTS, M$ | CASH FLOW BTAX, M$ | 10.00 PCT CUM, DISC BTAX, M$ |
|---|---|---|---|---|---|---|---|---|---|---|---|---|
| 12-85 | 33.137 | 60.685 | 17.303 | 40.665 | 27.16 | 4.41 | 649.351 | 120.652 | 48.126 | 0.000 | 480.573 | 458.381 |
| 12-86 | 21.983 | 53.607 | 12.082 | 35.467 | 27.20 | 4.50 | 488.183 | 83.652 | 48.126 | 0.000 | 356.405 | 767.424 |
| 12-87 | 17.647 | 50.024 | 9.889 | 33.247 | 27.22 | 4.40 | 415.479 | 67.200 | 48.126 | 0.000 | 300.153 | 1004.030 |
| 12-88 | 15.287 | 43.884 | 8.596 | 28.716 | 27.22 | 4.51 | 363.613 | 55.093 | 48.126 | 0.000 | 260.394 | 1190.634 |
| 12-89 | 13.455 | 39.641 | 7.566 | 25.779 | 27.22 | 4.53 | 322.712 | 45.225 | 48.126 | 0.000 | 229.361 | 1340.056 |
| 12-90 | 11.839 | 36.260 | 6.657 | 23.533 | 27.22 | 4.50 | 287.141 | 33.778 | 48.126 | 0.000 | 205.237 | 1461.607 |
| 12-91 | 10.419 | 32.202 | 5.859 | 20.638 | 27.22 | 4.58 | 254.007 | 22.674 | 48.126 | 0.000 | 183.207 | 1560.247 |
| 12-92 | 9.168 | 29.438 | 5.155 | 18.823 | 27.22 | 4.56 | 226.071 | 20.085 | 48.126 | 0.000 | 157.860 | 1637.513 |
| 12-93 | 8.068 | 26.936 | 4.537 | 17.188 | 27.22 | 4.53 | 201.337 | 17.803 | 48.126 | 0.000 | 135.408 | 1697.764 |
| 12-94 | 7.100 | 24.675 | 3.993 | 15.720 | 27.22 | 4.50 | 179.416 | 15.787 | 48.126 | 0.000 | 115.503 | 1744.487 |
| 12-95 | 6.248 | 22.127 | 3.513 | 13.960 | 27.22 | 4.55 | 159.102 | 13.919 | 48.126 | 0.000 | 97.057 | 1780.178 |
| 12-96 | 5.498 | 20.273 | 3.092 | 12.769 | 27.22 | 4.52 | 141.856 | 12.346 | 48.126 | 0.000 | 81.384 | 1807.386 |
| 12-97 | 4.839 | 18.093 | 2.721 | 11.261 | 27.22 | 4.59 | 125.715 | 10.873 | 48.126 | 0.000 | 66.716 | 1827.661 |
| 12-98 | 4.258 | 16.574 | 2.394 | 10.296 | 27.22 | 4.56 | 112.112 | 9.643 | 48.126 | 0.000 | 54.343 | 1842.675 |
| 12-99 | 3.747 | 15.196 | 2.107 | 9.428 | 27.22 | 4.53 | 100.065 | 8.561 | 48.126 | 0.000 | 43.378 | 1853.570 |
| 12- 0 | 3.297 | 13.947 | 1.854 | 8.644 | 27.22 | 4.50 | 89.347 | 7.601 | 48.126 | 0.000 | 33.620 | 1861.247 |
| 12- 1 | 2.074 | 11.316 | 0.907 | 6.626 | 27.00 | 4.97 | 57.421 | 4.461 | 30.126 | 0.000 | 22.834 | 1865.987 |
| 12- 2 | 1.825 | 10.289 | 0.798 | 5.990 | 27.00 | 4.98 | 51.362 | 3.957 | 30.126 | 0.000 | 17.279 | 1869.248 |
| 12- 3 | 1.606 | 9.358 | 0.703 | 5.416 | 27.00 | 4.99 | 45.980 | 3.513 | 30.126 | 0.000 | 12.341 | 1871.366 |
| 12- 4 | 1.413 | 6.961 | 0.618 | 3.540 | 27.00 | 5.16 | 34.961 | 2.678 | 24.195 | 0.000 | 8.088 | 1872.628 |
| S TOT | 182.908 | 541.486 | 100.344 | 347.706 | 27.20 | 4.53 | 4305.231 | 559.501 | 884.589 | 0.000 | 2861.141 | 1872.628 |
| AFTER | 1.519 | 24.019 | 0.664 | 10.209 | 27.00 | 5.45 | 73.567 | 3.730 | 57.162 | 0.000 | 12.675 | 1874.247 |
| TOTAL | 184.427 | 565.505 | 101.008 | 357.915 | 27.20 | 4.56 | 4378.798 | 563.231 | 941.751 | 0.000 | 2873.816 | 1874.247 |
| CUM. | 155.559 | 2822.307 | | | | | | | | | | |
| ULT. | 339,986 | 3387.812 | | | | | | | | | | |

| | |
|---|---|
| NET OIL REVENUES (M$) | 2747.542 |
| NET GAS REVNUES (M$) | 1631.256 |
| TOTAL REVENUES (M$) | 4378.798 |

| | | | |
|---|---|---|---|
| BTAX RATE OF RETURN (PCT) | 100.00 | PROJECT LIFE (YEARS) | 25.513 |
| BTAX PAYOUT YEARS | 0.00 | DISCOUNT RATE (PCT) | 10.000 |
| BTAX PAYOUT YEARS (DISC) | 0.00 | GROSS OIL WELLS | 2.000 |
| BTAX NET INCOME/INVEST | 0.00 | GROSS GAS WELLS | 2.000 |
| BTAX NET INCOME/INVEST (DISC) | 0.00 | GROSS WELLS | 4.000 |

---------PRESENT WORTH PROFILE---------

| DISC RATE | PW OF NET BTAX, M$ | DISC RATE | PW OF NET BTAX, M$ |
|---|---|---|---|
| 0.0 | 2873.816 | 30.0 | 1139.759 |
| 2.0 | 2597.226 | 35.0 | 1046.784 |
| 5.0 | 2267.991 | 40.0 | 970.719 |
| 8.0 | 2013.407 | 45.0 | 907.321 |
| 10.0 | 1874.247 | 50.0 | 853.658 |
| 12.0 | 1754.119 | 60.0 | 767.685 |
| 15.0 | 1602.242 | 70.0 | 701.717 |
| 18.0 | 1476.865 | 80.0 | 649.407 |
| 20.0 | 1404.883 | 90.0 | 606.810 |
| 25.0 | 1255.897 | 100.0 | 571.396 |

13. Tabulation of the ownership interests including reversionary interests and the most latest payout amounts or status
14. Copy of all previously prepared geological or evaluation reports that contain data pertinent to the current evaluation of the property
15. Summary of lease and assignment provisions, lease facilities, operating agreements, and unitization agreements, etc.



## DETERMINATION OF FAIR MARKET VALUE

One of the most important objectives of the oil and gas property evaluation report is to determine a fair market value of a certain property or properties. A fair market value is the value of a property at which the property will change hands and it is usually determined through mutual transactions between two parties. The value at which a willing seller will sell his/her property and a willing buyer will buy the property is referred to as the fair market value of that property.

Although there have been significant technological advances in the oil and gas industry in the past several years, unfortunately no unique method has yet been developed which will consistently determine the fair market value. Some methods currently used by the industry are presented here which may be used as a starting point for negotiating the value of a property.

1. *A specified fraction of the discounted net future cash flow.* In this method the value is determined as two-thirds of the discounted cash flow discounted at 15%. In some cases the industry has also used a 100% of the discounted cash flow discounted at 25%.
2. *Specified years of future net revenue.* The fair market value determined by using this technique is based on the concept of payback period. A payback criteria is first established and then the cumulative net revenue (undiscounted) for this period is considered as a fair market value. The payback period currently used in the industry is based on two-and-one-half to four years with an average of approximately three years.
3. *Price per barrel of equivalent barrel of net oil reserves in the ground.* This method is more frequently utilized by the industry. The net remaining equivalent oil reserves are multiplied by a certain dollar value to determine the fair market value. The net gas reserves are converted into equivalent barrels of oil on either a heating value basis or price ratio basis, typically 6 MSCF = 1 bbl is used. The range of dollar per equivalent barrel of oil currently is between $5.00 to $8.00 (usually one-third of the posted wellhead price). Some transactions in the past have taken place using dollar per barrel of oil equivalent in the range of $6.00 to $12.51 with a mean of $8.44 and a median of $7.87.
4. *A specified price per equivalent net barrel per day of the production rate for the first year of production from the effective date of transaction.* In this technique the first year's equivalent net oil production is divided by 365 to determine the average equivalent net barrels of oil per day. This rate is then multiplied by a dollar value. In the past, transactions have taken place at $14,000 to $20,000 per equivalent barrel per day.

5. *Profit-to-investment ratio.* This is the ratio of the net future cash flow to the purchase price of the property. A frequently used ratio of 3 to 1 is used (i.e., the net future cash flow is divided by 3 to determine the value of the property).

All of the above methods are dependent on the time pattern of the generated production, associated cash flows, and the economic conditions of the industry at the time of purchase. An article by Gustavson compares the values determined by the different methods and shows the effect of different patterns of production (decline type) on the calculated values.[46]



The industry will also apply risk factors to the generated cash flows or the calculated values. The risk factors are based on the confidence level one has developed in the basic data used in evaluation and the generated numbers. It is always recommended that at the time of a transaction, a buyer/seller should do a quick statistical survey of the current market in order to determine the trend at which properties are traded. These statistical parameters can then be utilized in further transactions. The most commonly used risk factors are within the following ranges:

| Reserve Category | Risk Factor |
|---|---|
| Proved producing reserves | 0.90–1.00 |
| Proved undeveloped reserves | 0.75–0.90 |
| Probable reserves | 0.35–0.60 |
| Possible reserves | 0.00–0.10 |

## REFERENCES


1 "A Statistical Study of Recovery Efficiency," *API Bull. D-14*, Oct. 1967.

2 R. K. Guthrie and M. H. Greenberger, "The Use of Multiple Correlation Analysis for Interpreting Petroleum Engineering Data, *API 901-31-G*, API, Dallas, March 1955.

3 H. B. Bradley, *Petroleum Engineering Handbook*, Society of Petroleum Engineers of AIME, Richardson, Texas, 1987.

4 R. J. Schilthuis, "Active Oil and Reservoir Energy," *Trans.* AIME, 148, 1936, pp. 33–52.

5 D. Havlena and A. S. Odeh, "The Material Balance Equation as an Equation of a Straight Line," *Journal of Petroleum Tech.*, Trans. AIME, 228, Aug. 1963, pp. 896–900.

6 B. C. Craft and M. F. Hawkins, *Applied Petroleum Reservoir Engineering*, Prentice-Hall Inc., Englewood Cliffs, NJ, 1959.

7 L. P. Dake, *Fundamentals of Reservoir Engineering*, Elsevier Scientific Publishing Co., Amsterdam, 1978.

9 M. J. Rzasa and D. L. Katz, "Calculation of Static Pressure Gradient in Gas Wells," *Trans.*, AIME, 160, 1945, pp. 100–106.

10 M. H. Cullender and R. V. Smith, "Practical Solution of Gas-Flow Equations for Wells and Pipelines with Large Temperature Gradients," *Trans.*, AIME, 207, 1956, pp. 281–287.

11 D. N. Meehan and E. L. Vogel, *HP-41 Reservoir Engineering Manual*, PennWell Publishing Co., Tulsa, Oklahoma, 1982.

[12] *Theory and Practice of the Testing of Gas Wells*, Third Edition, Energy Resources and Conservation Board, Calgary, Alberta, Canada, 1978.

[13] R. V. Smith, "Determining Friction Factor for Measuring Productivity of Gas Wells," *Trans.*, AIME, 189, 1950, p. 73.

[14] M. H. Cullender and C. W. Binckley, *Phillips Petroleum Company Report*, presented to the Railroad Commission of Texas Hearing, Amarillo, Texas, Nov. 9, 1950.

[15] Chi U. Ikoku, *Natural Gas Reservoir Engineering*, John Wiley and Sons, Inc., New York, 1984.

[16] R. J. Bruns, M. J. Fetkovitch, and V. C. Meitzen, "The Effect of Water Influx on P/Z-Cumulative Gas Production Curves," *Journal of Petroleum Tech.*, March 1965, pp. 287–291.

[17] D. Havlena and A. S. Odeh, "The Material Balance as an Equation of a Straight Line—Part II, Field Cases," *Journal of Petroleum Tech.*, July, 1964, pp. 815–822.

[18] D. J. Hammerlindl, "Predicting Gas Reserves in Abnormally Pressured Reservoirs," Paper presented at the 64th Annual Fall Meeting of SPE of AIME, New Orleans, Oct. 1971.

[19] R. H. Roach, "Analyzing Geopressured Reservoirs—A Material Balance Technique," *SPE Paper* 9968, Society of Petroleum Engineers of AIME, Dallas, Texas, Dec. 1981.

[20] H. C. Slider, *Practical Petroleum Reservoir Engineering Methods*, PennWell Publishing Co., Tulsa, Oklahoma, 1976.

[21] A. F. Van Everdingen and W. Hurst, "The Application of the Laplace Transformation to Flow Problems in Reservoirs," *Trans.*, AIME, 186, 1949, pp. 305–324.

[22] R. D. Carter and G. W. Tracy, "An Improved Method for Calculating Water Influx," *Journal of Petroleum Tech.*, Trans., AIME, 219, Dec. 1960, pp. 68–60.

[23] K. H. Coats, "A Mathematical Model for Water Movement About Bottomwater-Drive Reservoirs," *Trans.*, AIME, 225, March 1962, pp. 44–52.

[24] D. R. Allard and S. M. Chen, "Calculation of Water Influx for Bottomwater Drive Reservoirs," *SPE Reservoir Engineering*, May 1988, pp. 369–379.

[25] A. F. van Everdingen, E. H. Timmerman, and J. J. McMahon, "Application of the Material Balance Equation to a Partial Water-Drive Reservoirs," *Journal of Petroleum Tech.*, Trans., AIME, 198, Feb. 1953, pp. 51–60.

[26] M. J. Fetkovitch, "A Simplified Approach to Water Influx Calculations—Finite Aquifer Systems," *Journal of Petroleum Tech.*, July 1971, pp. 814–828.

[27] J. Tarner, "How Different Size Gas Caps and Pressure Maintenance Programs Affect Amount of Recoverable Oil," *Oil Weekly*, June 12, 1944, pp. 32–44.

[28] G. W. Tracy, "Simplified Form of The Material Balance Equation," *Trans.*, AIME, 204, 1955, pp. 243–246.

[29] M. Muskat and M. O. Taylor, "Effect of Reservoir Fluid and Rock Characteristics on Production Histories of Gas-Drive Reservoirs," *Trans.*, AIME, 165, 1946, pp. 78–93.

[30] *Engineering Data Book*, Gas Processors Suppliers Association, Ninth Edition, 1972, Tulsa, Oklahoma.

[31] J. M. Campbell, *Gas Conditioning and Processing*, Campbell Petroleum Series, J. M. Campbell and Co., 1976.

[32] T. A. Mathews, C. H. Roland, and D. L. Katz, "High Pressure Gas Measurements," *Proc.* NGAA, 1942, p. 41.

[33] D. K. Gold, W. D. McCain, Jr., and J. W. Jennings, "An Improved Method for the Determination of the Reservoir-Gas Specific Gravity for Retrograde Gases," *Journal of Pet. Techn.*, July 1989, pp. 747–752.

[34] R. H. Jacoby, R. C. Koeller, and V. J. Berry, Jr., "Effect of Composition and Temperature on Phase Behavior and Depletion Performance of Rich Gas-Condensate Systems," *Journal of Petroleum Tech.*, Trans, AIME, 216, July 1958, pp. 58–63.

[35] J. J. Arps, "Analysis of Decline Curves," *Trans.*, AIME, 160, 1944, pp. 228–247.

[36] M. J. Fetkovitch, "Decline Curve Analysis Using Type Curves," *Journal of Petroleum Tech.*, June 1980, pp. 1065–1077.

[37] D. A. Rowland and Cheng Lin, "New Linear Method Gives Constants of Hyperbolic Decline," *Oil and Gas Journal*, Jan. 14, 1985, pp. 86–89.

[38] L. C. Luther, "Linearization and Regression Analysis Technique Predicts Hyperbolic Decline in Reserves," *Oil and Gas Journal*, Aug. 26, 1985, pp. 78–79.

[39] D. R. Long and M. J. Davis, "A New Approach to the Hyperbolic Curve, *Journal of Petroleum Tech.*, July 1988, pp. 909–912.

[40] D. B. Neal and M. A. Mian, "Early-Time Tight Gas Production Forecasting Technique Improves Reserves and Reservoir Description," *SPE Formation Evaluation*, March 1989, pp. 25–32.

[41] M. A. Mian, "J-Sand (Tight) Performance Predicted by Type Curves," *Oil and Gas Journal*, May 30, 1983, pp. 109–112.

[42] M. A. Mian, "Predicting the Performance of Tight Gas Reservoirs," *World Oil*, August 1, 1984, pp. 47–50.

[43] R. S. Thompson and J. D. Wright, *Oil Property Evaluation*, Thompson-Wright Associates, Golden, Colorado, 1984.

[44] R. A. Gallun and J. W. Stevenson, *Fundamentals of Oil and Gas Accounting*, PennWell Publishing Company, Tulsa, Oklahoma, 1983.

[45] M. A. Mian, "Creating Quality, Cost-Effective Property Reports," *World Oil*, Sept. 1985, pp. 71–74.

[46] J. B. Gustavson, "Comparison of Methods for Determining Fair Market Values," *SPE Paper 13769*, presented at the 1985 SPE Hydrocarbon Economics and Evaluation Symposium, Texas, March 14–15.

# CHAPTER FIVE

# SECONDARY OIL RECOVERY



Using primary recovery techniques with the existing natural energy of the producing reservoir, a substantial amount of oil remains unrecoverable. Various methods of improving the ultimate oil recovery have been developed, all of which involve maintaining reservoir pressure by injecting some fluid into the reservoir, thus augmenting the natural reservoir energy. The increased ultimate recovery is due primarily to one or both of the following factors:[1]

1. Decreasing the depletion drive index by maintaining reservoir pressure as much as possible
2. Replacing the natural displacing energy with a more efficient displacing force (i.e., by replacing a gas cap drive with an artificial waterdrive)

Reservoir pressure can be maintained by (1) gas injection, (2) water injection, (3) miscible fluid injection, or (4) combinations of the above. Waterflooding is generally the dominating pressure maintenance method (preferably in water wet rocks) than gas flooding. Its dominance is due to (1) the general availability of water, (2) the relative ease with which water is injected, (3) the ability with which water spreads through an oil-bearing formation, and (4) water's efficiency in displacing oil.[2]

The water viscosity is of the order of 50 times the gas viscosity, and it occupies the less conductive portions of the pore spaces, whereas the gas occupies the more conductive portions.[3] In *oil-wet* reservoirs, the displacing water will invade the more conductive portions first, resulting in lower displacement efficiency. The distribution of oil, gas, and water in the pore spaces at any set of fluid saturations is determined by (1) wettability characteristics of the rock and (2) the surface or interfacial tension between the *immiscible* phases at the gas/oil and water/oil interfaces.

In this chapter we will discuss the mechanism of fluid displacement by describing fractional flow and the frontal advance equations, followed by waterflood oil recovery performance, prediction of water injection rates, oil producing rates, producing water/oil ratios, and cumulative oil recovery calculations at various stages of the waterflood.

## THE DISPLACEMENT OF OIL AND GAS

This section describes how to calculate the oil and gas recovery resulting from displacement by water injection. As mentioned above, all of the expansive forces prevalent in reservoir systems that cause oil displacement can essentially be described by the fractional flow and frontal advance equations; these two concepts are presented in this section. The gravity segregation effect on fluid displacement is also discussed. Other topics presented in this section include determination of the frontal saturation by material balance, displacement in stratified reservoirs, and factors influencing oil production performance.



### THE FRACTIONAL FLOW EQUATION

In 1941 Leverette in his pioneering paper presented the concept of *fractional flow*.[4] Starting with the well-known Darcy's law, he obtained the following equation for water/oil system:

$$f_w = \frac{1 + 0.001127 \dfrac{k_o A}{q_t \mu_o} \left[ \dfrac{\partial P_c}{\partial L} - 0.433(\rho_w - \rho_o) \sin\theta \right]}{1 + \dfrac{\mu_w}{\mu_o} \dfrac{k_o}{k_w}} \quad \textbf{(5.1)}$$

where $f_w$ is the fractional flow of the displacing fluid; $q_t = (q_w + q_o)$ is the total flow rate in bbls; $\mu_o$ and $\mu_w$ are the viscosities of oil and water in centipoise, respectively; $k_o$ and $k_w$ are the effective permeabilities (in millidarcies) to oil and water, respectively; A is the cross-sectional area in square feet; $\rho_o$ and $\rho_w$ are the densities in grams per cubic centimeters for oil and water, respectively; $\theta$ is the dip angle (positive updip); $L$ is the distance along the direction of movement in feet; and $P_c$ is the capillary pressure in psi.

After neglecting the effects of the capillary pressure gradient and the dip angle of the reservoir (horizontal system), the terms $\partial P_c/\partial L$ and $0.433\Delta\rho \sin\theta$ become insignificant and Equation (5.1) reduces to

$$f_w = \frac{1}{1 + \left(\dfrac{k_o}{k_w}\right)\left(\dfrac{\mu_w}{\mu_o}\right)} \quad \textbf{(5.2)}$$

or in terms of relative permeabilities,

$$f_w = \frac{1}{1 + \left(\frac{k_{ro}}{k_{rw}}\right)\left(\frac{\mu_w}{\mu_o}\right)} \tag{5.2a}$$

The $f_w$ term is a function of $S_w$ inasmuch as both $k_{ro}$ and $k_{rw}$ are themselves functions of $S_w$. As $S_w$ increases, the value of $k_{ro}$ declines, whereas that of $k_{rw}$ rises, resulting in the increase of the value of $f_w$.[2] A typical fractional flow curve as a function of $S_w$ is shown in Figure 5–1.

The relative permeability ratio is usually plotted as a straight line versus $S_w$ on semilogarithmic graph paper as shown in Figure 5–2 (i.e., log $(k_o/k_w)$ versus $S_w$). Like many relative permeability ratio curves, the central or main portion of the curve is quite linear. It is represented by the following equation:

$$\frac{k_o}{k_w} = ae^{-bS_w} \tag{5.3}$$



where $a$ is the intercept and $b$ is the slope of the straight line portion. Substituting this relationship in Equation (5.2) yields

$$f_w = \frac{1}{1 + \frac{\mu_w}{\mu_o} ae^{-bS_w}} \tag{5.2b}$$

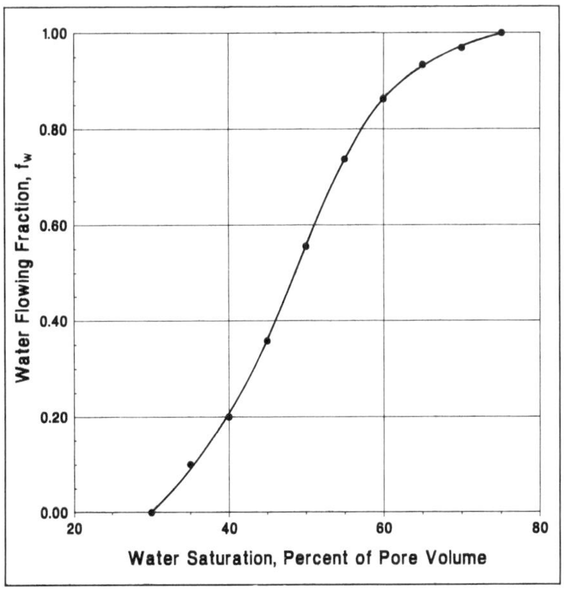


**Figure 5–1.** Plot of $f_w$ versus $S_w$.

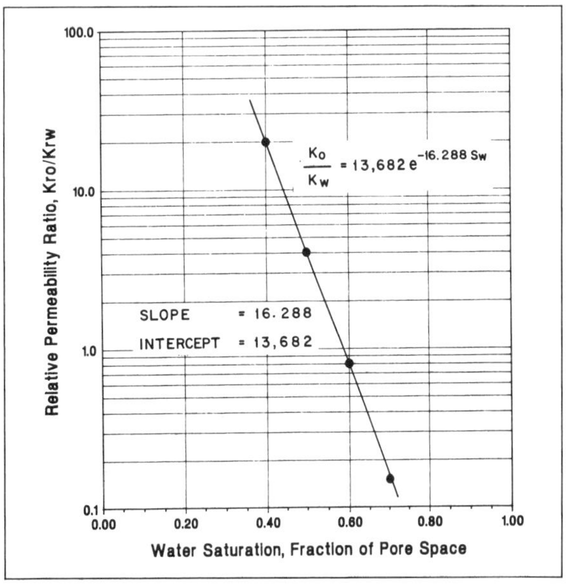


**Figure 5–2.** Semilogarithmic plot of relative permeability ratio versus $S_w$.

while the producing watercut at the surface is given by

$$f_{ws} = \frac{1}{1 + \dfrac{B_w}{B_o}\left(\dfrac{1}{f_w} - 1\right)} = \frac{\dfrac{q_w}{B_w}}{\dfrac{q_w}{B_w} + \dfrac{q_o}{B_o}} \tag{5.4}$$

## THE FRONTAL ADVANCE EQUATION

The fractional flow equation discussed in the previous section is used to calculate the fraction of the total flow which is water at any point in the reservoir, provided the water saturation at that point is known. The frontal advance equation typically known as the Buckley-Leverette equation is used to determine when a given $S_w$ plane reaches a particular point in the linear system.[5] The following equation was derived:



$$X_{S_w} = \frac{5.615 q_t t}{A\phi}\left(\frac{\partial f_w}{\partial S_w}\right)_{S_w} \tag{5.5}$$

where $X_{S_w}$ is the total distance that the plane of given $S_w$ moves. Equation (5.5) assumes that (1) there is no mass transfer between phases and (2) the phases are incompressible. The equation expresses the linear advance of a surface of constant saturation during an increment of time, $t$, under the application of a constant displacing fluid injection rate, $q_t$. Displacement of fluids by frontal advance creates *stabilized* saturation distribution at the front and *unstabilized* saturation distribution conditions behind the front.[6] The term *stabilized* zone refers to the region at the leading edge of the displacement process where saturations moved at *equal velocities*.[7] The *nonstabilized zone* refers to the saturation interval where all points of saturation continue to get farther apart.

The derivative $\partial f_w/\partial S_w$ in Equation (5.5) could be determined accurately for any value of $S_w$ from a plot of $f_w$ versus $S_w$ by drawing a tangent to the curve and graphically taking slopes at values of $S_w$ as shown in Figure 5–3. The tangent to the fractional flow curve has the following properties:[2]

1. Tangent to the fractional flow curve at connate water saturation, $S_{iw}$, and $f_w = 0$ yields at the point of tangency the water saturation at the front, $S_{wf}$. Thus the stabilized zone includes all water saturations from $S_w$ to $S_{wf}$.
2. Extrapolating the tangent to $f_w = 1.0$ yields the value of the average water saturation at breakthrough, $\bar{S}_{wbt} = \bar{S}_{wf}$.
3. The value of fractional flow, $f_w$, corresponding to $S_{wf}$ at the point of tangency is the producing watercut at breakthrough.
4. The inverse of the slope of the tangent gives the cumulative injected fluid in pore volumes.
5. The value of the breakthrough saturation, $S_{wf}$, can be checked by applying an overall material balance. If the tangent construction is correct, the value of $Q_{ibt}$ from the following two equations

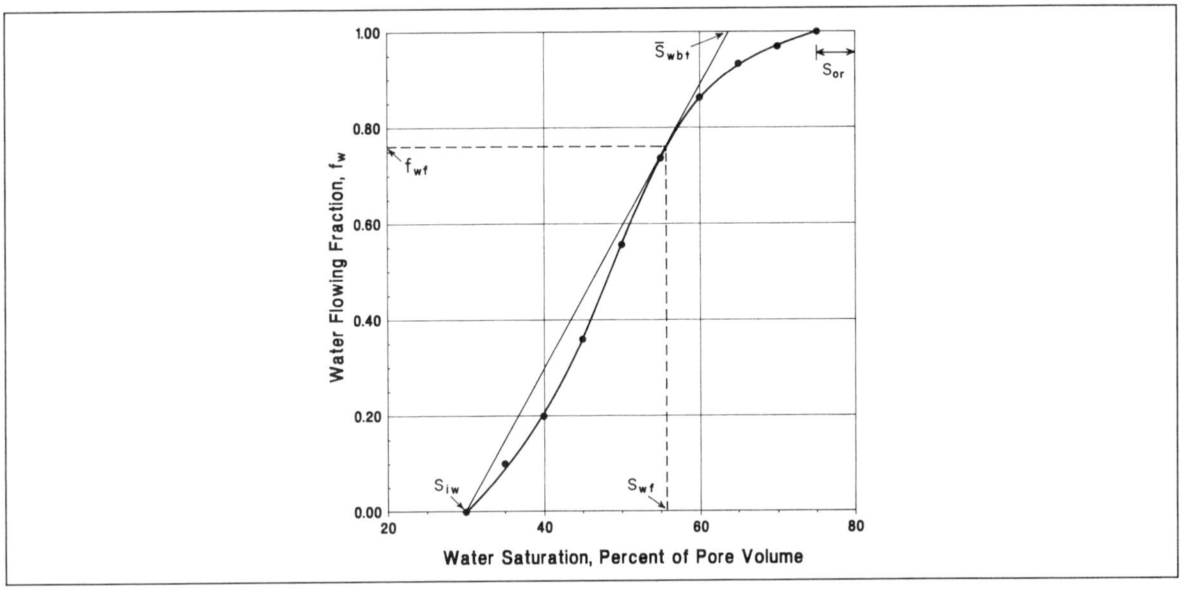


**Figure 5–3.** Plot of $f_w$ versus $S_w$ showing tangent to the fractional flow curve at $S_{iw}$.



will be same. Otherwise a trial-and-error procedure can be used to find $S_{wf}$ that satisfies both equations.

$$Q_{ibt} = \frac{W_i}{A\phi L} = (\bar{S}_{wf} - S_{iw}) \tag{5.6}$$

and

$$Q_{ibt} = \frac{W_i}{A\phi L} = \frac{\bar{S}_{wf} - S_{wf}}{1 - f_{S_{wf}}} \tag{5.7}$$

The derivative $\partial f_w / \partial S_w$ may also be obtained mathematically by using the following relationship between relative permeability ratio and $S_w$:

$$\frac{\partial f_w}{\partial S_w} = \frac{(\mu_w/\mu_o)bae^{-bS_w}}{[1 + (\mu_w/\mu_o)ae^{-bS_w}]^2} = \frac{(\mu_w/\mu_o)b(k_o/k_w)}{[1 + (\mu_w/\mu_o)(k_o/k_w)]^2} = b(f_w^2 - f_w) \tag{5.8}$$

In 1952 Welge extended the work of Buckley and Leverette to derive a simplified method for calculating fractional flow and recovery performance after water breakthrough.[8] This method consists of integrating the saturation distribution over the distance from the injection point to the front, thus obtaining the $\bar{S}_w$ behind the front. The equations developed by Welge are

$\overline{S}_w$ for the interval $x_1 \leq x \leq x_2$:

$$\overline{S}_w = \frac{x_2 S_{w2} - x_1 S_{w1}}{x_2 - x_1} - \frac{q_t t}{A\phi L} \frac{(f_{w2} - f_{w1})}{(x_2 - x_1)} \tag{5.9}$$

where $\overline{S}_w$ is the average water saturation (fraction of PV), $S_{w2}$ is the water saturation at the producing end of the system, $q_t t$ represents the total volume of water injected ($q_i$), and $f_{o2}$ is the fraction of oil flowing at the producing end of the system.

When $x_1 = 0$ and sufficient time has passed for water to have arrived at the end of the core ($x_2 = L$), then the average water saturation in the system is



$$\overline{S}_w = S_{w2} - \frac{q_t t}{A\phi L} (f_{w2} - f_{w1}) \tag{5.10}$$

Usually, $f_{w1} = 1.0$ at $x = 0$; therefore, Equation (5.10) becomes

$$\overline{S}_w = S_{w2} + \frac{q_t t}{A\phi L} (1 - f_{w2}) = S_{w2} + \frac{1 - f_{w2}}{\left(\frac{\partial f_w}{\partial S_w}\right)_{S_{w2}}} \tag{5.11}$$

Also,

$$\left(\frac{\partial f_w}{\partial S_w}\right)_{S_{w2}} = \frac{1 - f_{w2}}{\overline{S}_w - S_{w2}} \tag{5.12}$$

$$Q_i = \frac{1}{\left(\frac{\partial f_w}{\partial S_w}\right)_{S_{w2}}} \tag{5.13}$$

**Example 5–1.**[3] The following relative permeability data were measured on a sandstone core as a function of its water saturation:

| $S_w$ | 0 | 10 | 20 | 30 | 40 | 50 | 60 | 70 | 75 | 80 | 90 | 100 |
|---|---|---|---|---|---|---|---|---|---|---|---|---|
| $k_{ro}$ | 1.0 | 1.0 | 1.0 | 0.94 | 0.80 | 0.44 | 0.16 | 0.045 | 0 | 0 | 0 | 0 |
| $k_{rw}$ | 0 | 0 | 0 | 0 | 0.04 | 0.11 | 0.20 | 0.30 | 0.36 | 0.44 | 0.68 | 1.0 |

(a) Plot the $k_{ro}$ and $k_{rw}$ versus $S_w$ on Cartesian coordinate graph paper.

(b) Plot the relative permeability ratio versus $S_w$ on semilogarithmic graph paper and calculate the constants $a$ and $b$.

(c) If $\mu_o = 3.4$ cp, $\mu_w = 0.68$ cp, $B_o = 1.5$ RB/STB, and $B_w = 1.05$ RB/STB, what is the surface water cut of a well completed in the transition zone where the water saturation is 50%?
(d) What is the reservoir water cut in Part (c) above?
(e) Calculate $f_w$ and plot versus $S_w$.
(f) Calculate $\partial f_w/\partial S_w$ at several values of $S_w$ using Equation (5.8).
(g) Calculate the distances of advance of the constant saturation fronts at 100, 200, and 400 days. Plot on Cartesian coordinate graph paper versus $S_w$. Assume $\phi = 25\%$, $A = 26{,}400$ ft$^2$, $B_o = 1.5$ RB/STB, and $q = 900$ STB/d.

Solution:

(a) Plot of $k_{ro}$ and $k_{rw}$ versus $S_w$ is shown in Figure 5–4.
(b) Plot of relative permeability ratio versus $S_w$ is shown in Figure 5–2. The constants $a$ and $b$ are calculated from two points on the curve as follows:



$$(k_o/k_w)_1 = 20 \qquad (k_o/k_w)_2 = 0.80$$

$$S_{w1} = 40\% \qquad S_{w2} = 60\%$$

$$b = \frac{\ln\left[\dfrac{(k_o/k_w)_2}{(k_o/k_w)_1}\right]}{S_{w2} - S_{w1}} \tag{5.14}$$

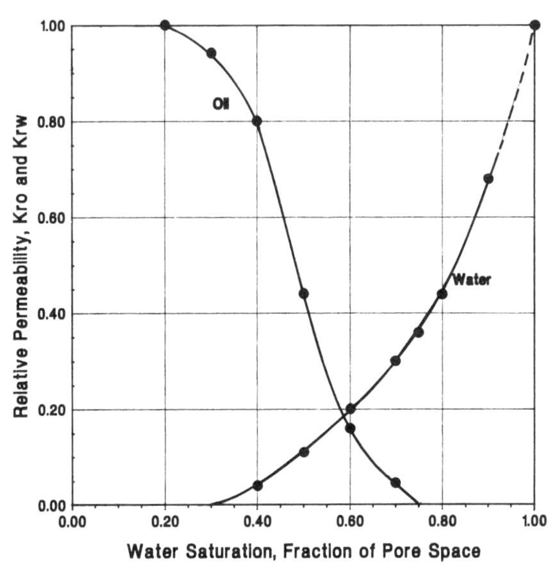


**Figure 5–4.** Water/oil relative permeability curves for Example 5–1.

$$= \frac{\ln\left(\frac{0.80}{20}\right)}{0.6 - 0.4} = -16.094$$

$$a = \frac{\left(\frac{k_o}{k_w}\right)_2}{e^{bS_{w2}}} \quad \textbf{(5.15)}$$

$$= \frac{0.80}{e^{-16.094 \times 0.6}} = 12{,}497 \approx 12{,}500$$

(c) From Figure 5–4 at $S_w = 50\%$, $k_{ro} = 0.44$ and $k_{rw} = 0.11$. Using Equation (5.2),

$$f_w \frac{1}{1 + \left(\frac{0.44}{0.11}\right)\left(\frac{0.68}{3.4}\right)} = 0.556$$

Using Equation (5.4),

$$f_{ws} = \frac{1}{1 + \frac{1.05}{1.5}\left(\frac{1}{0.556} - 1\right)} = 0.641$$

(d) Reservoir water cut $f_w$ is 0.556 as calculated in Part (1).
(e) The $f_w$s are calculated using Equation (5.2) at each $S_w$ as shown in Table 5–1 and Figure 5–3.

**TABLE 5–1. Calculation of $f_w$ and $\partial f_w/\partial S_w$ for the data in Example 5–1**

| $S_w$ (%) | $k_o/k_w$ Eq. (5.3) | $f_w$ | $\partial f_w/\partial S_w$ Eq. (5.8) | $X_{sw}$ (ft) 100 days | $X_{sw}$ (ft) 200 days | $X_{sw}$ (ft) 400 days |
|---|---|---|---|---|---|---|
| 30 | ∞ | 0.000 | 0 | 0 | 0 | 0 |
| 35 | 44.633 | 0.101 | 1.462 | 112 | 224 | 448 |
| 40 | 19.955 | 0.200 | 2.576 | 197 | 394 | 788 |
| 45 | 8.922 | 0.359 | 3.705 | 284 | 567 | 1,136 |
| 50 | 3.989 | 0.556 | 3.975 | 304 | 609 | 1,216 |
| 55 | 1.783 | 0.737 | 3.121 | 239 | 478 | 959 |
| 60 | 0.797 | 0.863 | 1.904 | 146 | 292 | 584 |
| 65 | 0.356 | 0.934 | 0.992 | 76 | 152 | 304 |
| 70 | 0.159 | 0.969 | 0.484 | 37 | 74 | 148 |
| 75 | 0.000 | 1.000 | 0 | 0 | 0 | 0 |

The tangent drawn through $S_{iw}$ = 30% touches the fractional flow curve at $S_{wf} = S_{wbt}$ = .5625 pore volumes. Thus the $S_w$ at the front prior to breakthrough is 56.25% of pore volume.

(f) Using Equation (5.8), $\partial f_w/\partial S_w$ is calculated as shown in Table 5–1.

$$\left(\frac{\partial f_w}{\partial S_w}\right)_{S_w} = b(f_w^2 - f_w)$$

$$\left(\frac{\partial f_w}{\partial S_w}\right)_{S_w=40} = -16.1(0.2^2 - 0.2) = 2.576$$

The plot of $(\partial f_w/\partial S_w)_{S_w}$ versus $S_w$ is shown in Figure 5–5.

(g) Advance of the saturation front is calculated using Equation (5.5) as



$$X_{S_w} = \frac{(5.615)(1.5)(600)t}{(0.25)(26{,}400)}\left(\frac{\partial f_w}{\partial S_w}\right)_{S_w}$$

$$= 0.766t\left(\frac{\partial f_w}{\partial S_w}\right)_{S_w}$$

For 100 days, $t$ = 100; therefore,

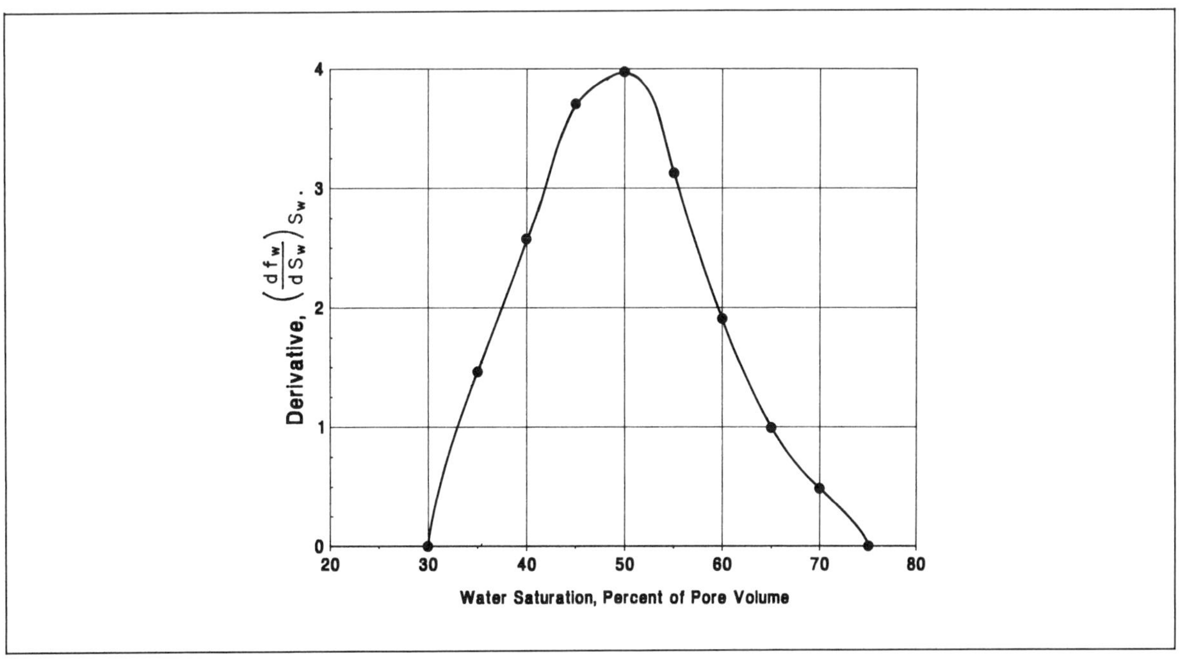


**Figure 5–5.** Plot of $(\partial f_w/\partial S_w)_{sw}$ versus $S_w$ for the data in Table 5–1.

$$X_{S_w} = (0.766)(100)\left(\frac{\partial f_w}{\partial S_w}\right)_{S_w}$$

$$= 76.568\left(\frac{\partial f_w}{\partial S_w}\right)_{S_w}$$

Therefore, $X_{S_w}$ at $S_w = 45\%$ and $(\partial f_w/\partial S_w)_{S_w=45\%} = 3.705$

$$X_{S_w=45\%} = 76.568(3.705) = 283.684 \quad \text{ft}$$

$X_{S_w}$ at the other water saturations is calculated in the same manner. For 200 and 400 days, the $X_{S_w}$ for 100 days is simply multiplied by 2 and 4, respectively. Fluid distribution at 100, 200, and 400 days is shown in Figure 5–6.



## DETERMINATION OF THE FRONTAL SATURATION BY MATERIAL BALANCE

It can be seen in Figure 5–6 that the fluid distribution curves are characteristically double-valued or triple-valued while the saturation can be only one value at any place and time. This shows

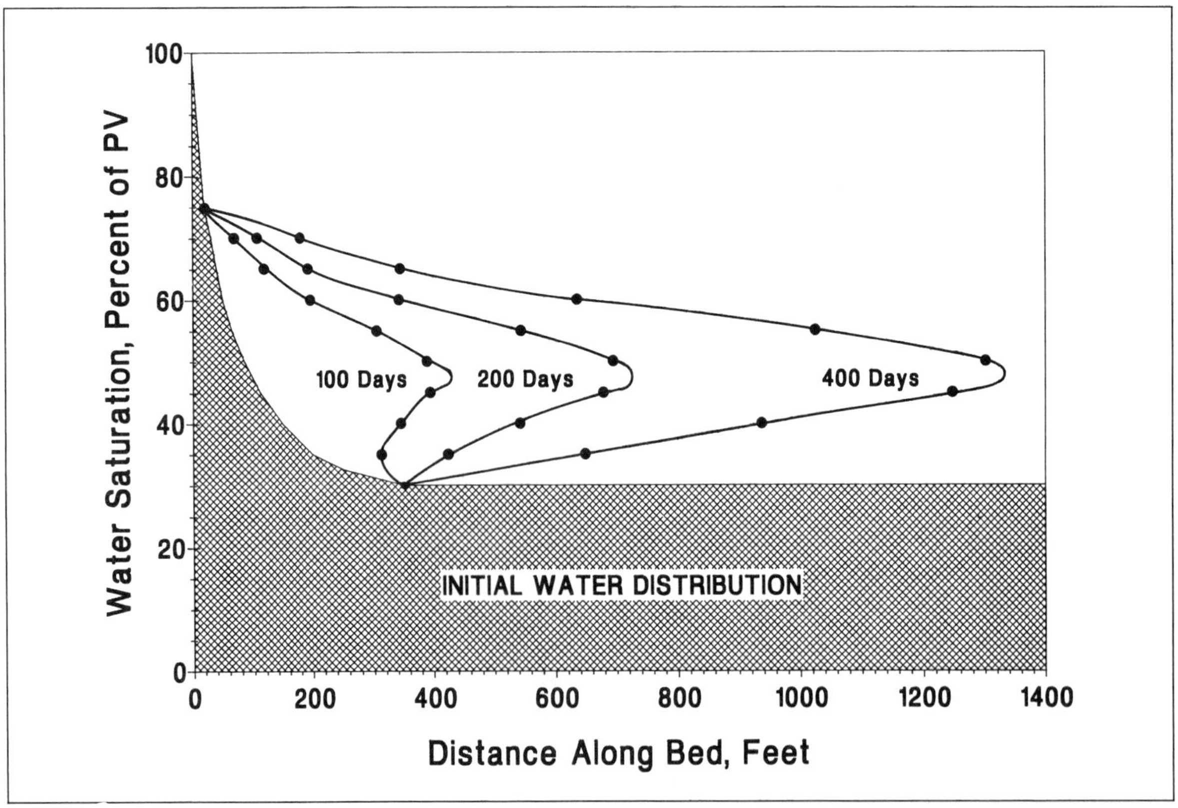


**Figure 5–6.** Fluid distribution at initial conditions and at 100, 200, and 400 days.

that it would be meaningless to evaluate the position of every saturation represented on the fractional flow curve. Consequently, it has to be determined as to how much of the calculated saturation profile has to be used. This can be accomplished by material balance satisfying the following equation:

$$5.615q_t t = \phi A \sum_{0}^{x_f} (S_w - S_{wi})\Delta X \tag{5.16}$$

where $A\phi\Delta x(S_w - S_{wi})$ is the increase in the number of cubic feet of the displacing phase in $\Delta X$. The equation is generally solved graphically. As the $(S_w - S_{wi})\Delta X$ terms are summed, a sum will be found which will satisfy Equation (5.16). Eventually, adding a new term to the summation will cause the summation to exceed the value necessary to satisfy the equation. At this point it will be required to adjust $\Delta X$ for the last term in order to satisfy the equation. This then fixes the position of the front at this time.



**Example 5–2.** Using the material balance technique discussed, determine the frontal advance position for the saturation profile for the 400 days in Example 5–1.

Solution: First calculate the left side of Equation (5.16) as

$$\frac{5.615q_t t}{\phi A} = \frac{5.615(1.5)(600)(400)}{(0.25)(26{,}400)} = 306.273 \quad \text{ft}$$

Figure 5–7 shows the initial $S_{iw}$ distribution and the apparent $S_w$ distribution as calculated in Example 5–1 and plotted in Figure 5–5. The dotted lines on Figure 5–7 are solution construction lines approximating the two curves by step functions. The area under the step functions is considered as area under the curve.[9] Now the summation term in Equation (5.16) can be numerically evaluated beginning with the area closest to the influx face. The calculations are shown in Table 5–2.

$$\begin{aligned} 306.27 &= 275.58 + (0.555 - 0.30)\Delta X \\ \Delta X &= 120.35 \\ X_f &= 920 + 120.35 \approx 1{,}040 \quad \text{ft} \end{aligned}$$

Welge has shown that the position of the front may be obtained by drawing a secant from the connate water saturation as shown in Figure 5–3.[8] The front position is then found by dropping a perpendicular from $S_{wf}$, which in this case comes to 940 ft at $S_{wf} = 56\%$.

## MOBILITY RATIO

Displacement efficiency of a linear process is dependent upon fluid and rock properties. As these parameters change, it becomes difficult to compare performance. The concept of *mobility ratio* is used to correlate displacement data. Where one fluid displaces another, the standard notation for

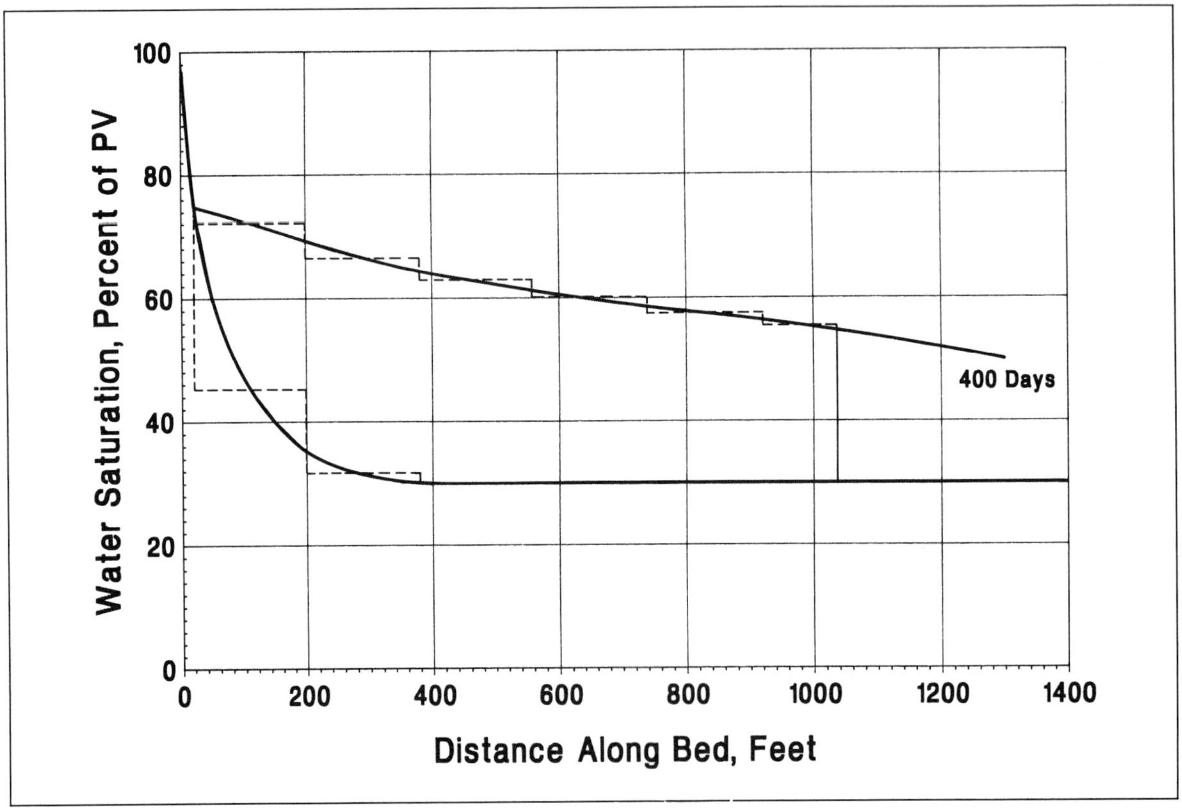


**Figure 5–7.** Graphical integration for Example 5–2.

the mobility ratio is the mobility of the *displacing* fluid behind the front, $\lambda_D$, and the mobility of the *displaced* phase ahead of the front, $\lambda_d$.[3,7] For waterflood under piston-like displacement:

$$M = \frac{\lambda_D}{\lambda_d} \tag{5.17}$$

**TABLE 5–2. Graphical integration of Figure 5–7**

| $\Delta X$ | $\Sigma\Delta X$ | $S_w$ | $S_{wi}$ | $\Delta X(S_w - S_{wi})$ | $\Sigma\Delta X(S_w - S_{wi})$ |
|---|---|---|---|---|---|
| 20 | 20 | 0.750 | 0.750 | 0 | 0 |
| 180 | 200 | 0.720 | 0.432 | 51.84 | 51.84 |
| 180 | 380 | 0.660 | 0.319 | 61.38 | 113.22 |
| 180 | 560 | 0.628 | 0.300 | 59.04 | 172.26 |
| 180 | 740 | 0.600 | 0.300 | 54.00 | 226.26 |
| 180 | 920 | 0.574 | 0.300 | 49.32 | 275.58 |
| 180 | 1,100 | 0.555 | 0.300 | 45.90 | 321.48 |

where

$$\lambda_D = \lambda_w = \left(\frac{k_w}{\mu_o}\right)_{S_{or}} \quad \textbf{(5.18)}$$

$$\lambda_d = \lambda_o = \left(\frac{k_o}{\mu_o}\right)_{S_{iw}} \quad \textbf{(5.19)}$$

and the mobility ratio is given by

$$M = \frac{(k_w/\mu_w)_{S_{or}}}{(k_o/\mu_o)_{S_{iw}}} = \frac{(k_{rw}/\mu_w)_{S_{or}}}{(k_{ro}/\mu_o)_{S_{iw}}} \quad \textbf{(5.20)}$$

The oil and water permeabilities are the endpoints of permeability/water-saturation curves for a waterflood. Note that the water permeability is the one in the water-contacted portion of the reservoir, and the oil permeability is the one in the oil bank.[2] 

Mobility ratio based on average saturation behind the flood front (not endpoints), also referred to as the *shock front mobility ratio,* is computed from the relative permeability at $S_{wf}$.

$$M_{\bar{S}} = \frac{(k_{ro}/\mu_o)_{S_{wf}} + (k_{rw}/\mu_w)_{\bar{S}_{wf}}}{(k_{ro}/\mu_o)_{S_{iw}}} \quad \textbf{(5.20a)}$$

Also for immiscible displacement process:

$$M_{\bar{S}i} = \frac{(\lambda_i)_{\bar{S}_i}}{(\lambda_d)_{\bar{S}_d}} \quad \textbf{(5.21)}$$

For miscible displacement, the total mobility ratio is given by

$$M_t = \frac{\mu_d}{\mu_D} \quad \textbf{(5.22)}$$

where $\lambda_i$ is the mobility of the injected fluid evaluated at the average saturation ($\bar{S}_i$) of the injected fluid at breakthrough and $\lambda_d$ is evaluated at the saturation ($\bar{S}_d$) of the displaced fluid ahead of the front.

Hagoort has shown that the Buckley-Leverette displacement can be regarded as stable for $M_{\bar{S}} \leq 1.0$.[11] If this condition is not met, severe viscous channelling of water through the oil will take place, resulting in earlier breakthrough than predicted by Welge method. Some of the characteristics of mobility ratio are:

1. $M_{\bar{S}} = 1.0$ indicates that both oil and water are moving at the same relative rate.
2. $M_{\bar{S}} < 1.0$ indicates that water moves slower than the oil, resulting in higher displacement efficiency and higher water saturations at breakthrough.

3. $M_{\bar{S}} > 1.0$ indicates that water behind the front is moving faster than the oil ahead of the front, resulting in reduced displacement efficiency.
4. When the mobility ratio increases, large volumes of fluid must be injected resulting in higher produced water/oil ratio while retaining the same displacement efficiency.

## MINIMUM CONDITIONS REQUIRED TO ESTABLISH A FLOOD FRONT

A flood front is characterized by a saturation discontinuity between $S_{iw}$ and $S_{wf}$ at breakthrough. When the velocity of the flood front saturation exceeds that of the initial saturation, an oil bank forms.[7] Velocities of flood front saturation and the initial saturation can be computed using the following equations:



$$v_{S_{wf}} = \frac{q_t}{A\phi}\left(\frac{\partial f_w}{\partial S_w}\right)_{S_{wf}} \tag{5.23}$$

and

$$v_{S_w} = \frac{q_t}{A\phi} f_{iw} \tag{5.24}$$

The flood front contain all velocities in the range $v_{S_{iw}} < v_{S_w} < v_{S_{wf}}$. A flood front can form when

$$f_{iw} < \left(\frac{\partial f_w}{\partial S_w}\right)_{S_{wf}} \tag{5.25}$$

## DISPLACEMENT UNDER SEGREGATED FLOW CONDITIONS

When gravity segregation occurs in a waterflood, the water runs under the oil. Dake has presented a linear displacement model based on the concept of vertical equilibrium with gravity forces dominating the displacement process.[10]

**Horizontal Reservoirs.** The displacement model for gravity-segregated flow in a homogeneous, linear system is based on the following assumptions:[7,10]

1. An interface forms between oil and water zones.
2. Piston-like displacement occurs (i.e., capillary forces are neglected).
3. The change in saturation across the interface is a step function from $S_w = 1 - S_{or}$ in the water zone to $S_{iw}$ in the oil zone.
4. Only water flows in the water zone and oil in the oil zone.
5. Vertical equilibrium is assumed (i.e., fluids are distributed instantaneously in the vertical cross section by gravity forces).

At any vertical cross section, thickness averaged properties are given by

$$S_{wz} = h_b(1 - S_{or}) + (1 - h_b)S_{iw} \tag{5.26}$$
$$k_{roz} = (1 - h_b)k_{ro} \tag{5.27}$$

and

$$k_{rwz} = h_b k_{rw} \tag{5.28}$$

where $k_{ro}$ is the endpoint oil-relative permeability at $S_{iw}$, $k_{rw}$ is the endpoint water relative permeability at $S_{or}$, and $h_b$ is fraction of the reservoir thickness flooded to $S_{or}$ at a particular vertical cross-section. The parameter $h_b$ is also the thickness-averaged dimensionless water saturation, $S_{wDz}$, given by

$$S_{wDz} = \frac{S_w - S_{iw}}{1 - S_{or} - S_{iw}} \tag{5.29}$$



The corresponding fractional flow equations and the displacement performance equations are given below.

$$f_{wz} = \frac{\dfrac{k_{rwz}}{\mu_w}}{\dfrac{k_{rwz}}{\mu_w} + \dfrac{k_{roz}}{\mu_o}} = \frac{Mh_b}{1 + (M - 1)h_b} \tag{5.30}$$

$$Q_{i2} = \frac{1}{\left(\dfrac{\partial f_{wz}}{\partial S_{wz}}\right)_{S_{wz2}}} \tag{5.31}$$

and

$$\bar{S}_{wz2} = S_{wz2} + (1 - f_{w2})Q_{i2} \tag{5.32}$$

Also,

$$Q_{i2} = \frac{(1 - S_{or} - S_{iw})[1 - h_b(M - 1)]^2}{M} \tag{5.33}$$

In this displacement model, the $S_{wbt} = S_{iw}$, which is equivalent to $h_b = 0$; therefore,

$$Q_{ibt} = \frac{1 - S_{or} - S_{iw}}{M} \tag{5.34}$$

Note that this model is valid only for $M > 1.0$. Waterflood performance can be predicted by variation of $h_b$ from 0 to 1.0 at the end of the system and using appropriate equations of fractional flow and saturations.

**Gravity Segregation in Dipping Reservoirs.** The gravity segregation model presented by Dake can be adapted to displacement in a dipping reservoir as shown in Figure 5–8.[10] The following equations are applicable:

$$f_{wy} = \frac{Mh_b - h_b(h_b - 1)G}{(M - 1)h_b + 1} \quad \textbf{(5.35)}$$

where the subscript $y$ refers to thickness-averaged properties in the dip-normal direction, $y$, and G is the dimensionless gravity number given by Equation (5.36). All movable oil is displaced when $h_b = 1.0$.

$$G = \frac{4.9 \times 10^{-4} k k_{rw} A(\gamma_w - \gamma_o) \sin\theta}{\mu_w q_t} \quad \textbf{(5.36)}$$

$$Q_{i2} = \frac{(1 - S_{or} - S_{iw})[1 + h_b(M - 1)]^2}{M - G[1 - 2h_b + h_b^2(M - 1)]} \quad \textbf{(5.37)}$$

At breakthrough, $S_w = S_{wi}$ or $h_b = 0$, then

$$Q_{ibt} = \frac{1 - S_{or} - S_{iw}}{M - G} \quad \textbf{(5.38)}$$

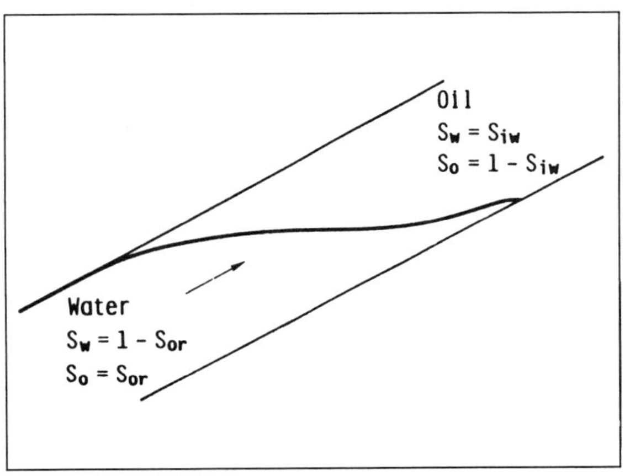


**Figure 5–8.** Unstable displacement with gravity segregation in a dipping reservoir (after Dake).[10]

**Gravity-Stabilized Displacement in Dipping Reservoir.** When displacement velocities are less than a critical value, Dietz presented the following equation which can be used to predict the critical velocity required to propagate a stable interface through a linear system under gravity forces.[12] The model assumes piston-like displacement and negligible capillary effects.

$$q_c = \frac{4.9 \times 10^{-4} k k_{rw} A(\gamma_w - \gamma_o) \sin\theta}{\mu_w(M - 1)} \quad \textbf{(5.39)}$$

where $k$ is in millidarcies and $k_{rw}$ is the endpoint water-relative permeability. When the interface is stable, the velocities of oil and water are equal at every point on the interface. Thus the interface is linear, as shown in Figure 5–9. For as long as $q < q_{crit}$, the interface is displaced at a constant velocity.

The angle that represents the slope (slope is negative for stable displacement) of the interface $\beta$ can be computed using the following equation:

$$\frac{dy}{dx} = -\tan\beta = \frac{(M - 1 - G)}{G}\tan\theta \quad \textbf{(5.40)}$$

The effect of gravity segregation in waterdrive oil reservoirs is usually of much less concern because of the higher displacement efficiency and lower oil/water density differences, whereas the converse is generally true for gas/oil systems.[3] Gravity segregation is applicable to the recovery of updip or *attic* oil in active waterdrive reservoirs possessing good gravity segregation characteristics. High-pressure gas is injected at the interface which migrates updip and displaces oil downdip, which may be produced from the same wells in which gas was injected.

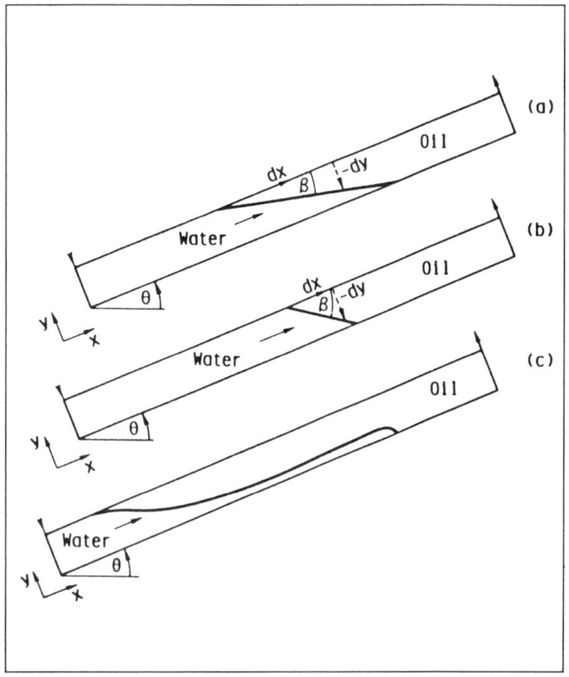


**Figure 5–9.** Stable and unstable displacement in gravity segregated displacement: (a) Stable: $G > m - 1$, $M > 1$, and $\beta < \theta$; (b) Stable: $G > M - 1$, $M < 1$, $\beta > \theta$; and (c) Unstable: $G < M - 1$ (courtesy of Elsevier).[10]



## OIL RECOVERY CALCULATIONS

The following steps are to be followed for calculating oil recovery in practice:[10]

1. Construct the fractional flow curve allowing for gravity effects, if necessary, and neglecting capillary gradient effect.
2. From the fractional flow curve determine $f_{wf}$, $S_{wf}$, $\bar{S}_w$, and $S_{or}$ as shown in Figure 5–3. The following equations are then used to calculate the oil recovery and time at which breakthrough occurs:

$$N_{pd_{bt}} = Q_{id_{bt}} = q_{id}t_{bt} = (\bar{S}_{wbt} - S_{iw}) = \frac{1}{\left(\dfrac{\partial f_w}{\partial S_w}\right)_{S_{wbt}}} \quad \textbf{(5.41)}$$

$$t_{bt} = \frac{Q_{id_{bt}}}{q_{id}} \quad \textbf{(5.42)}$$

where $Q_{idbt} = q_i/LA\phi$ is the dimensionless pore volumes injected; for convenience all volumes are expressed as dimensionless pore volumes.

3. Using Welge's technique, choose $S_{w2}$ at 5%, increments above the saturation at breakthrough ($S_{w2} > S_{wbt} = S_{wf} < 1 - S_{or}$). At each $S_{w2}$ determine the corresponding $f_{w2}$ and $\bar{S}_{w2}$. The oil recovery is then calculated using the following equation:

$$N_{pd} = \bar{S}_{w2} - S_{iw} \tag{5.43}$$

where

$$\bar{S}_{w2} = S_{w2} + (1 - f_{w2})\frac{1}{\left(\dfrac{\partial f_w}{\partial S_w}\right)_{S_{w2}}}$$



4. The reciprocal of the slope of the fractional flow curve, for each $S_{w2}$ determined in Step 3, gives $Q_{id}$ (the number of pore volumes of water injected). This allows calculation of the time scale attached to the recovery.
5. As an alternative to Step 3, the following equation can be used directly to calculate the oil recovery:

$$N_{pd} = (S_{w2} - S_{iw}) + (1 - f_{w2})Q_{id} \tag{5.44}$$

6. Generally, the displacement takes place at a fixed pressure (i.e., above the bubblepoint pressure), then

$$N_{pd} = \frac{N_p}{N}\frac{B_o}{B_{oi}}(1 - S_{iw}) \tag{5.45}$$

OR

$$N_p = N\left[\frac{B_{oi}}{B_o}\frac{N_{pd}}{(1 - S_{iw})}\right] \tag{5.45a}$$

For horizontal displacement under segregated flow conditions for unstable flow ($M > 1.0$) the following recovery equation is used:

$$N_{pD} = \frac{1}{M_1}\left(2\sqrt{Q_{iD}M} - Q_{iD} - 1\right) \tag{5.46}$$

in which all volumes are expressed in *movable oil volumes*. At the time of water breakthrough, Equation (5.46) reduces to

$$N_{pD_{bt}} = \frac{1}{M} \tag{5.47}$$

This shows that for stable piston-like displacement at $M = 1$, $N_{pD_{bt}} = 1.0$. When the total amount of oil has been recovered, $N_{pD} = 1.0$ (MOV), $Q_{iD_{max}} = 1.0$.

Generally, for unstable displacement in a dipping reservoir, when $G < M - 1$, the following recovery equation develops:

$$N_{pD} = \frac{1}{M-1}\left[\sqrt{Q_{iD}M\left(1 - \frac{G}{M-1}\right)\left(1 - \frac{Q_{iD}G}{M-1}\right)} - Q_{iD}\left(1 - \frac{M+1}{M-1}G\right) - 1\right] \tag{5.48}$$

For a horizontal reservoir ($G = 0$), Equation (5.48) reduces to the form of Equation (5.46). At the breakthrough conditions ($N_{pD} = Q_{iD}$), the following equation results: 

$$N_{pD_{bt}} = \frac{1}{M - G} \tag{5.49}$$

while for the maximum recovery ($N_{pD} = 1.0$),

$$Q_{iD_{max}} = \frac{M}{G + 1} \tag{5.50}$$

The production rates at each $S_{w2}$ and the water/oil ratio (WOR) are calculated using the following equations. The WOR is a measure of the efficiency of the displacement at a point in the process.

$$q_{w2} = \frac{f_{w2}q_t}{b_w} \tag{5.51}$$

$$q_{o2} = \frac{(1 - f_{w2})q_t}{B_o} \tag{5.52}$$

$$F_{wo} = \left(\frac{f_{w2}}{f_{o2}}\right)\left(\frac{B_o}{B_w}\right) \tag{5.53}$$

**Example 5–3.** Water is being injected at a constant rate of 1,000 B/D/well in a direct line drive in a reservoir which has the following rock and fluid properties. Distance between injection wells is 625 ft and the distance between injectors and producers is 2,000 ft.

| | |
|---|---|
| $\phi = 0.18$ | $\mu_o = 5$ cp |
| $h = 40$ ft | $\mu_w = 0.5$ cp |

| $S_w$ | 20 | 25 | 30 | 35 | 40 | 45 | 50 | 55 | 60 | 65 | 70 | 75 | 80 |
|---|---|---|---|---|---|---|---|---|---|---|---|---|---|
| $k_{ro}$ | .8 | .610 | .470 | .37 | .285 | .220 | .163 | .12 | .081 | .05 | .027 | .01 | 0 |
| $k_{rw}$ | 0 | .002 | .009 | .02 | .033 | .051 | .075 | .10 | .132 | .17 | .208 | .251 | .30 |

Assuming diffused flow conditions and that the injection starts simultaneously with oil production from the reservoir, determine

(a) $S_{iw}$, $S_{or}$, $S_{wf} = S_{wbt}$, $\bar{S}_{wbt}$, and $f_{wbt}$
(b) Is the displacement stable or unstable?
(c) The time when breakthrough occurs
(d) The cumulative oil production as a function of both the cumulative water injected and the time



Solution:

(a) The fractional flow curve is shown in Figure 5–10 with

$$S_{iw} = 20\% \qquad S_{wbt} = 45\%$$
$$S_{or} = 20\% \qquad \bar{S}_{wbt} = 55\%$$
$$f_{wbt} = 0.699$$

(b) To find if the displacement is stable, calculate the shock front mobility ratio using Equation (5.20a).

$$M_{\bar{S}} = \frac{(0.22/5) + (0.05/0.5)}{(0.8/5)}$$
$$= 0.913$$

Since $M_{\bar{S}} < 1.0$, then displacement is stable.

(c) Using Equation (5.41), cumulative pore volume injected is

$$N_{pDbt} = Q_{iDbt} = (0.55 - 0.20) = 0.35$$

Time when breakthrough occurs is calculated using Equation (5.42).

$$t_{bt} = \frac{Q_{iDbt} \times 625 \times 40 \times 2000 \times 0.18}{100 \times 5.615 \times 365 \quad \text{days/year}}$$
$$= 4.391 \times Q_{ibt}$$
$$= 4.391 \times 0.35$$
$$= 1.537 \quad \text{years}$$

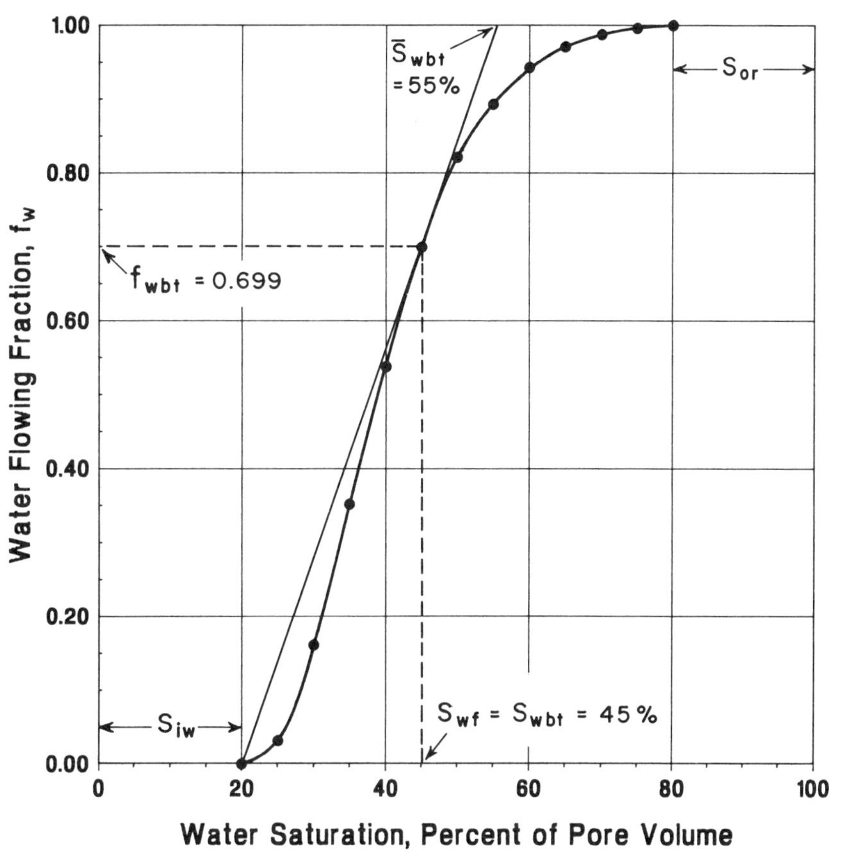


**Figure 5–10.** Fractional flow curve for the data in Example 5–3.

(d) The oil recovery after breakthrough is calculated using Equation (5.44). Using Welge's method, the derivative $(\partial f_w/\partial S_w)_{S_{w2}}$ at 5% increments of $S_{w2}$ are evaluated from Figure 5–11. The corresponding values of $Q_{iD} = 1/(\partial f_w/\partial S_w)_{S_{w2}}$, $N_{pD}$, and $t_D$ are shown in Table 5–3.

**Example 5–4.**[10] Rework Example 5–3 assuming (1) the displacement takes place under segregated flow conditions and (2) the reservoir has a dip angle of 25°. In Part (b), calculate the critical rate for water displacing oil updip. In each case tabulate $Q_{iD}$, $N_{pD}$, and $t_D$. Assume $\gamma_w = 1.04$, $\gamma_o = 0.81$, and $k = 2{,}000$ md.

Solution:

(a) Using Equation (5.20), calculate the endpoint mobility ratio.

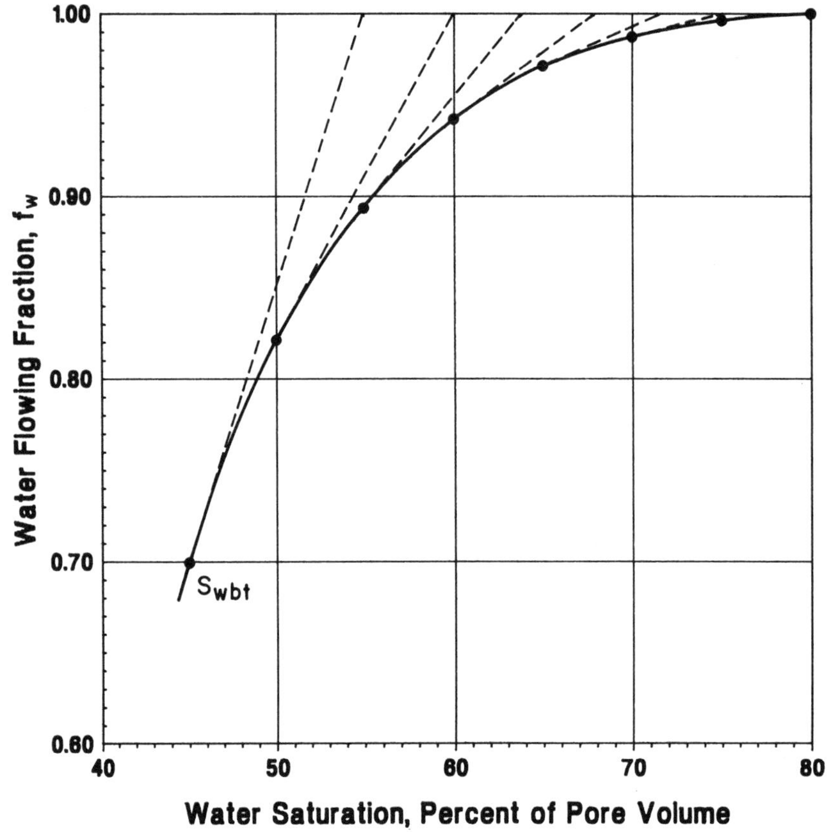


**Figure 5–11.** Welge graphical technique to determine the oil recovery after breakthrough (expanded section of Fig. 5–10).

$$M = \frac{\frac{0.3}{0.5}}{\frac{0.8}{5}} = 3.75$$

Note that the displacement is unstable at M > 1.0.

Using Equation (5.34), breakthrough occurs when

$$Q_{ibt} = \frac{(1 - 0.20 - 0.20)}{3.75} = 0.16(\text{PV})$$

Since 1 MOV = (1 − 0.20 − 0.20) = 0.6 PV and the maximum oil recovery when

TABLE 5–3. **Oil recovery calculations for Example 5–3**

| $S_{w2}$ | $f_{w2}$ Fig. 5–11 | $(\partial f_w/\partial S_w)_{S_{w2}}$ | $Q_{iD}$ (PV) Eq. (5.13) | $\bar{S}_{w2}$ Fig. 5–11 | $N_{pD}$ (PV) Eq. (5.43) | $t$ (years) Eq. (5.42) |
|---|---|---|---|---|---|---|
| .45 | .699 | — | — | — | — | — |
| .50 | .821 | 1.790 | 0.559 | .600 | .400 | 2.455 |
| .55 | .893 | 1.216 | 0.822 | .638 | .438 | 3.609 |
| .60 | .942 | 0.725 | 1.379 | .680 | .480 | 6.055 |
| .65 | .971 | 0.439 | 2.278 | .716 | .516 | 10.003 |
| .70 | .987 | 0.260 | 3.846 | .750 | .550 | 16.888 |
| .75 | .996 | 0.182 | 5.495 | .772 | .572 | 24.129 |



$$Q_{iD_{max}} = M = 3.75(\text{MOV}) = (3.75)(0.6) = 2.25(\text{PV})$$

The oil recovery as a function of $Q_{iD}$ between breakthrough and $Q_{iD_{max}}$ is calculated using Equation (5.46) as given in Table 5–4.

$$N_{pD} = \frac{1}{3.75 - 1}\left(2\sqrt{(0.30 \times 3.75)} - 0.3 - 1.0\right) = 0.299$$

(b) The critical rate for bypassing is calculated using Equation (5.39).

TABLE 5–4. **Oil recovery as a function of $Q_{iD}$ for Example 5–4**[10]

| $Q_{iD}$ (MOV) | $N_{pD}$ Eq. (5.46) | $Q_{id}$ (PV) | $N_{pd}$ (PV) | $t = 4.391(Q_{id})$ (Years) |
|---|---|---|---|---|
| 0.267 | 0.267 | 0.16 | 0.160 | 0.702 |
| 0.300 | 0.299 | 0.18 | 0.179 | 0.790 |
| 0.500 | 0.450 | 0.30 | 0.270 | 1.371 |
| 1.000 | 0.681 | 0.60 | 0.409 | 2.634 |
| 1.500 | 0.816 | 0.90 | 0.489 | 3.951 |
| 2.000 | 0.901 | 1.20 | 0.540 | 5.268 |
| 3.000 | 0.985 | 1.80 | 0.591 | 7.902 |
| 3.750 | 1.000 | 2.20 | 0.600 | 9.878 |

$$q_c = \frac{(4.9 \times 10^{-4})(2{,}000)(0.30)(625)(40)(1.04 - 0.81)\,\text{Sin}\,25}{0.5(3.75 - 1)}$$
$$\approx 520 \quad \text{RB/D}$$

Using Equation (5.36), the dimensionless gravity term is

$$G = 1.43 \qquad \text{at } q_t = 1000 \text{ RB/D}$$

Using Equation (5.38), recovery at breakthrough is

$$Q_{iD_{bt}} = N_{pD_{bt}} = \frac{1 - 0.2 - 0.2}{3.75 - 1.43}$$
$$= 0.259(\text{PV})$$



Using Equation (5.50), the maximum $Q_{iD_{max}}$ required to recover one MOV is

$$Q_{iD_{max}} = \frac{3.75}{1.43 + 1} = 1.543(\text{MOV})$$
$$= 0.926(\text{PV})$$

Using Equation (5.48), $N_{pD}$ is

$$N_{pD} = \frac{1}{3.75 - 1}\left[2\sqrt{0.5 \times 3.75\left(1 - \frac{1.43}{2.75}\right)\left(1 - \frac{0.5 \times 1.43}{2.75}\right)} - 0.5\left(1 - \frac{3.75 + 1}{3.75 - 1} \times 1.43\right) - 1\right]$$
$$= 0.364\,(2\sqrt{1.875 \times 0.48 \times 0.74} + 0.735 - 1)$$
$$= 0.498$$

Similarly, oil recovery at other $Q_{iD}$s are calculated as shown in Table 5–5.

**TABLE 5–5. Oil recovery calculations for Example 5–4 (part 2)[10]**

| $Q_{id}$ (MOV) | $N_{pd}$ (MOV) | $Q_{iD}$ (PV) | $N_{pD}$ (PV) | $t = 4.391(Q_{id})$ (Years) |
|---|---|---|---|---|
| 0.431 | 0.431 | 0.259 | 0.259 | 1.137 |
| 0.500 | 0.498 | 0.300 | 0.298 | 1.317 |
| 0.750 | 0.697 | 0.450 | 0.418 | 1.976 |
| 1.000 | 0.847 | 0.600 | 0.508 | 2.634 |
| 1.250 | 0.950 | 0.750 | 0.570 | 3.293 |
| 1.543 | 1.000 | 0.926 | 0.600 | 4.064 |

## WATERFLOOD CALCULATIONS

A waterflood displacement process may be described by the fractional flow and frontal advance concepts as discussed in the previous section. While some systems upon which waterflooding is initiated do exhibit linear flow characteristics (i.e., peripheral floods, line-drive floods, and end-to-end floods) the majority of systems deviate significantly from such a simple representation.[13]

Two-dimensional displacement is observed in linear systems if (1) permeability variation is observed in the reservoir, (2) if gravity segregation of injected and displaced fluids occurs, and/or (3) when capillary forces are large relative to viscous forces.[7]

In this section we will discuss areal sweep efficiency, water injectivity calculations in regular patterns, various widely used and generally accepted flood performance evaluation methods, and methods of forecasting mature waterfloods.

### WATERFLOOD PATTERN SELECTION



The first step in waterflood design involves selection of an appropriate waterflood pattern. In making the choice of the pattern, all available information about the reservoir must be considered. The proposed waterflood pattern should follow the following guidelines:[2]

1. Provide desired production capacity.
2. Provide sufficient water injection rate to yield desired oil productivity.
3. Maximize oil recovery and minimize water production.
4. Take advantage of reservoir heterogeneities, such as directional permeability, regional permeability differences, formation fractures, and formation dip, etc.
5. The existing wells should be utilized as much as possible with minimum requirement for drilling new wells.
6. In case of existing directional permeability or fractures, the pattern should be arranged such that the line of adjacent injectors follow the direction of directional permeability and/or fractures. Fractures or directional permeability in the injector-producer direction will result in early water breakthrough and in subsequent large volumes of produced water.
7. If the existing wells are on a square spacing (as in most cases), five-spots and nine-spots are the most common flooding patterns.
8. Other factors that must be considered include the desired flood life, well spacing, response time, availability of water, well productivity, and the injectivity into the reservoir.

Figure 5–12 illustrates several flooding patterns used commonly in the industry. Models of displacement performances refer in most cases to an *ideal regular five-spot pattern*. A peripheral flood (as compared to the pattern flood) can yield maximum oil recovery with a minimum of produced water. The water production in peripheral floods can be delayed by shutting in wells that suffer water breakthrough until only the last row of producers remain. However, in most cases the peripheral flood involves a small number of injection wells as compared to the number of producing wells. As a result it takes longer time to fill up, resulting in the delay of flush oil production.

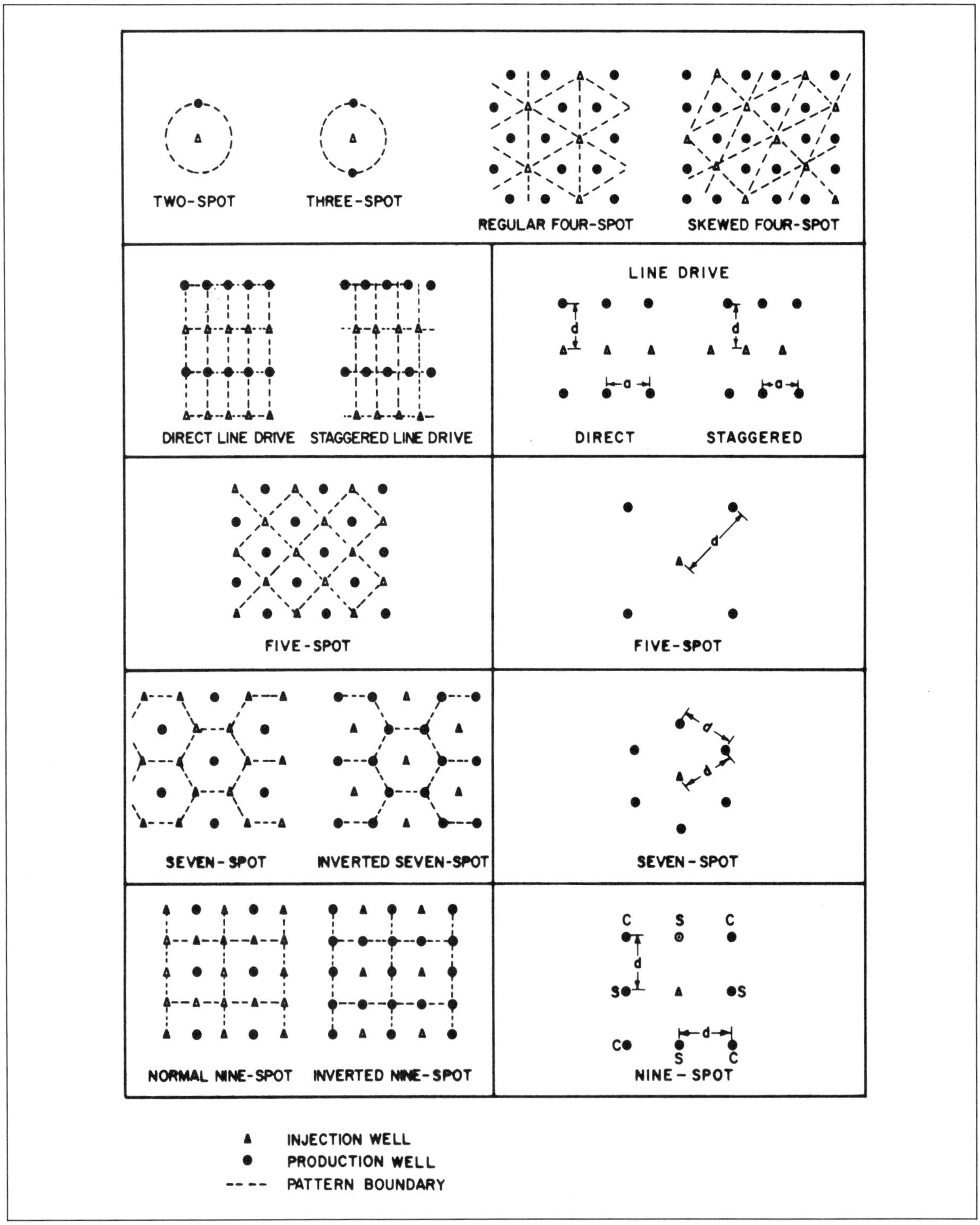


**Figure 5–12.** Flooding Patterns (© SPE, SPE Monograph Series, Vol. 3, 1971).[2]

## EVALUATING AREAL SWEEP EFFICIENCY

In waterflooding operations, the volume of the reservoir that is ultimately to be swept by the injected fluid is of major importance. Volumetric sweep efficiency is made up of the areal sweep efficiency resulting from the injection pattern and the vertical sweep efficiency resulting from formation and fluid characteristics.

Areal sweep efficiency for usual injection patterns has been studied for two phases of waterflooding: the period from the start of water injection until breakthrough occurs and the period after breakthrough until abandonment.

Dyes, Caudle, and Erickson presented a study of the influence of fluid mobilities on the sweep efficiency resulting from injection of gas or water.[14] Their study was extended to cover the production period following breakthrough of the injected fluid. They studied several waterflood patterns including the five-spot, staggered, and direct line drive. The following equations were developed:



For direct line drive:

$$E_{Abt} = 0.5472 + 0.3959 \log (M + 0.3) \quad \textbf{(5.54)}$$

For staggered line drive when $M \geq 1.0$:

$$E_{Abt} = 0.794 + 0.1179 \log (M - 0.5) \quad \textbf{(5.55)}$$

For staggered line drive when $M < 1.0$:

$$E_{Abt} = 0.515 + 0.7807 \log (M + 1) \quad \textbf{(5.56)}$$

For five-spot:

$$E_{Abt} = 0.626 + 0.3836 \log (M + 0.4) \quad \textbf{(5.57)}$$

Craig et al. obtained experimental data in horizontal laboratory models representing a quadrant of a five-spot water and gas pattern floods.[15] Areal sweep efficiencies were determined from X-ray shadowgraphs taken during various stages of the displacement as illustrated in Figure 5–13. The empirical correlations developed are presented below.

$$E_A = E_{Abt} + 0.633 \log \frac{W_i}{W_{ibt}} \quad \textbf{(5.58)}$$

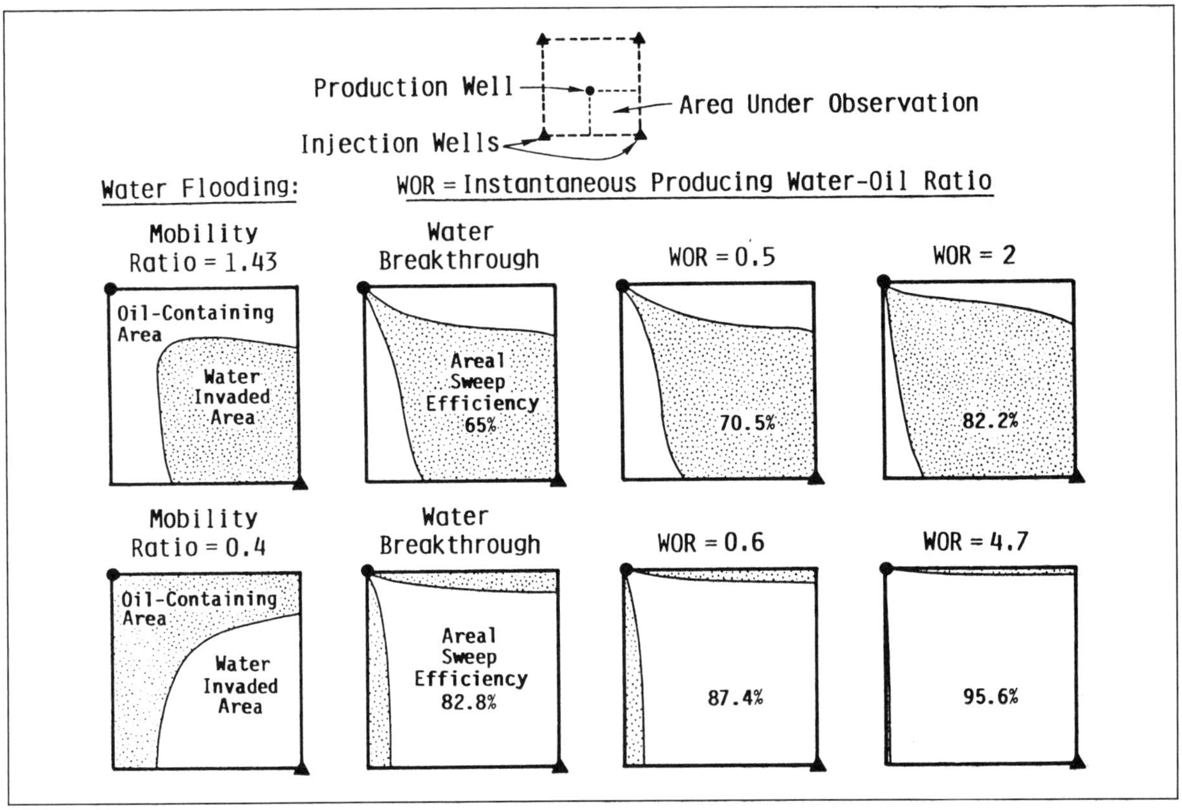


**Figure 5–13.** X-ray shadowgraphs of flood progess in scaled five-spot pattern (© SPE, Trans. AIME, 1955).[15]

OR

$$E_A = E_{Abt} + 0.274 \ln \frac{W_i}{W_{ibt}} \tag{5.58a}$$

where $E_{Abt}$ (fraction) is the areal sweep efficiency at breakthrough, $E_A$ (fraction) is the fraction of the area that has been swept to $\bar{S}_{wf}$, and $W_i/W_{ibt}$ is the ratio of cumulative volume of water injected to the cumulative volume of water injected at breakthrough. The $W_i/W_{ibt}$ values are given in Table 5–6.

If $\%E_A < 100$, $Q_i/Q_{ibt}$ is read or interpolated from Table 5–6 at $\%E_{Abt}$ and $W_i/W_{ibt}$. If $\%E_A \geq 100$, then

$$Q_i/Q_{ibt} = \frac{Q_i}{Q_{ibt(i-1)}} + \frac{W_{i(inc)}}{W_{ibt}} \frac{\%E_{Abt}}{100} \tag{5.59}$$

Areal sweep efficiencies at breaththrough were also correlated with shock front mobility ratio $M_{\bar{S}}$ as given by the following equation:

$$E_{Abt} = 0.54602036 + \frac{0.3170817}{M_{\bar{S}}} + \frac{0.30222997}{e^{M_{\bar{S}}}} - 0.00509693 M_{\bar{S}} \tag{5.60}$$

**TABLE 5–6.[15] $Q_i/Q_{ibt}$ values for various values of $E_{Abt}$ (%) (after Craig et al.)**

| | $E_{Abt}$ percent | | | | | | | | | |
|---|---|---|---|---|---|---|---|---|---|---|
| $W_i/W_{ibt}$ | 50. | 51. | 52. | 53. | 54. | 55. | 56. | 57. | 58. | 59. |
| 1.0 | 1.000 | 1.000 | 1.000 | 1.000 | 1.000 | 1.000 | 1.000 | 1.000 | 1.000 | 1.000 |
| 1.2 | 1.190 | 1.191 | 1.191 | 1.191 | 1.191 | 1.191 | 1.191 | 1.191 | 1.192 | 1.192 |
| 1.4 | 1.365 | 1.366 | 1.366 | 1.367 | 1.368 | 1.368 | 1.369 | 1.369 | 1.370 | 1.370 |
| 1.6 | 1.529 | 1.530 | 1.531 | 1.532 | 1.533 | 1.535 | 1.536 | 1.536 | 1.537 | 1.538 |
| 1.8 | 1.684 | 1.686 | 1.688 | 1.689 | 1.691 | 1.693 | 1.694 | 1.696 | 1.697 | 1.699 |
| 2.0 | 1.832 | 1.834 | 1.837 | 1.839 | 1.842 | 1.844 | 1.846 | 1.849 | 1.851 | 1.853 |
| 2.2 | 1.974 | 1.977 | 1.981 | 1.984 | 1.987 | 1.990 | 1.993 | 1.996 | 1.999 | 2.001 |
| 2.4 | 2.111 | 2.115 | 2.119 | 2.124 | 2.127 | 2.131 | 2.135 | 2.139 | 2.142 | 2.146 |
| 2.6 | 2.244 | 2.249 | 2.254 | 2.259 | 2.264 | 2.268 | 2.273 | 2.277 | 2.282 | 2.286 |
| 2.8 | 2.373 | 2.379 | 2.385 | 2.391 | 2.397 | 2.402 | 2.407 | 2.413 | 2.418 | 2.422 |
| 3.0 | 2.500 | 2.507 | 2.513 | 2.520 | 2.526 | 2.533 | 2.539 | 2.545 | 2.551 | 2.556 |
| 3.2 | 2.623 | 2.631 | 2.639 | 2.646 | 2.653 | 2.660 | 2.667 | 2.674 | 2.681 | 2.687 |
| 3.4 | 2.744 | 2.752 | 2.761 | 2.770 | 2.778 | 2.786 | 2.793 | 2.801 | 2.808 | 2.816 |
| 3.6 | 2.862 | 2.872 | 2.881 | 2.891 | 2.900 | 2.909 | 2.917 | 2.926 | 2.934 | 2.942 |
| 3.8 | 2.978 | 2.989 | 3.000 | 3.010 | 3.020 | 3.030 | 3.039 | 3.048 | 3.057 | 3.066 |
| 4.0 | 3.093 | 3.105 | 3.116 | 3.127 | 3.138 | 3.149 | 3.159 | 3.169 | 3.179 | 3.189 |
| 4.2 | 3.205 | 3.218 | 3.231 | 3.243 | 3.254 | 3.266 | 3.277 | 3.288 | 3.299 | 3.309 |
| 4.4 | 3.316 | 3.330 | 3.343 | 3.357 | 3.369 | 3.382 | 3.394 | 3.406 | 3.417 | 3.428 |
| 4.6 | 3.426 | 3.441 | 3.455 | 3.469 | 3.483 | 3.496 | 3.509 | 3.521 | 3.534 | 3.546 |
| 4.8 | 3.534 | 3.550 | 3.565 | 3.580 | 3.594 | 3.609 | 3.622 | 3.636 | 3.649 | |
| 5.0 | 3.641 | 3.657 | 3.674 | 3.689 | 3.705 | 3.720 | 3.735 | | | |
| 5.2 | 3.746 | 3.764 | 3.781 | 3.798 | 3.814 | 3.830 | | | | |
| 5.4 | 3.851 | 3.869 | 3.887 | 3.905 | 3.922 | | | | | |
| 5.6 | 3.954 | 3.973 | 3.993 | 4.011 | | | | | | |
| 5.8 | 4.056 | 4.077 | 4.097 | | | | | | | |
| 6.0 | 4.157 | 4.179 | | | | | | | | |
| 6.2 | 4.257 | | | | | | | | | |
| Values of $W_i/W_{ibt}$ at which $E_A$ = 100 percent | | | | | | | | | | |
| | 6.164 | 5.944 | 5.732 | 5.527 | 5.330 | 5.139 | 4.956 | 4.779 | 4.608 | 4.443 |

| | $E_{Abt}$ percent | | | | | | | | | |
|---|---|---|---|---|---|---|---|---|---|---|
| $W_i/W_{ibt}$ | 60. | 61. | 62. | 63. | 64. | 65. | 66. | 67. | 68. | 69. |
| 1.0 | 1.000 | 1.000 | 1.000 | 1.000 | 1.000 | 1.000 | 1.000 | 1.000 | 1.000 | 1.000 |
| 1.2 | 1.192 | 1.192 | 1.192 | 1.192 | 1.192 | 1.192 | 1.193 | 1.193 | 1.193 | 1.193 |
| 1.4 | 1.371 | 1.371 | 1.371 | 1.372 | 1.372 | 1.373 | 1.373 | 1.373 | 1.374 | 1.374 |

**TABLE 5–6.[15] Continued**

| | $E_{Abt}$ percent | | | | | | | | | |
|---|---|---|---|---|---|---|---|---|---|---|
| $W_i/W_{ibt}$ | 60. | 61. | 62. | 63. | 64. | 65. | 66. | 67. | 68. | 69. |
| 1.6 | 1.539 | 1.540 | 1.541 | 1.542 | 1.543 | 1.543 | 1.544 | 1.545 | 1.546 | 1.546 |
| 1.8 | 1.700 | 1.702 | 1.703 | 1.704 | 1.706 | 1.707 | 1.708 | 1.709 | 1.710 | 1.711 |
| 2.0 | 1.855 | 1.857 | 1.859 | 1.861 | 1.862 | 1.864 | 1.866 | 1.868 | 1.869 | 1.871 |
| 2.2 | 2.004 | 2.007 | 2.009 | 2.012 | 2.014 | 2.016 | 2.019 | 2.021 | 2.023 | 2.025 |
| 2.4 | 2.149 | 2.152 | 2.155 | 2.158 | 2.161 | 2.164 | 2.167 | 2.170 | 2.173 | 2.175 |
| 2.6 | 2.290 | 2.294 | 2.298 | 2.301 | 2.305 | 2.308 | 2.312 | 2.315 | 2.319 | 2.322 |
| 2.8 | 2.427 | 2.432 | 2.436 | 2.441 | 2.445 | 2.449 | 2.453 | 2.457 | 2.461 | 2.465 |
| 3.0 | 2.562 | 2.567 | 2.572 | 2.577 | 2.582 | 2.587 | 2.592 | 2.597 | 2.601 | 2.606 |
| 3.2 | 2.693 | 2.700 | 2.705 | 2.711 | 2.717 | 2.723 | 2.728 | 2.733 | 2.738 | 2.744 |
| 3.4 | 2.823 | 2.830 | 2.836 | 2.843 | 2.849 | 2.855 | 2.862 | 2.867 | 2.873 | |
| 3.6 | 2.950 | 2.957 | 2.965 | 2.972 | 2.979 | 2.986 | 2.993 | | | |
| 3.8 | 3.075 | 3.083 | 3.091 | 3.099 | 3.107 | | | | | |
| 4.0 | 3.198 | 3.207 | 3.216 | 3.225 | | | | | | |
| 4.2 | 3.319 | 3.329 | | | | | | | | |
| 4.4 | 3.439 | | | | | | | | | |
| Values of $W_i/W_{ibt}$ at which $E_A$ = 100 percent | | | | | | | | | | |
| | 4.235 | 4.132 | 3.984 | 3.842 | 3.704 | 3.572 | 3.444 | 3.321 | 3.203 | 3.088 |

| | $E_{Abt}$ percent | | | | | | | | | |
|---|---|---|---|---|---|---|---|---|---|---|
| $W_i/W_{ibt}$ | 70. | 71. | 72. | 73. | 74. | 75. | 76. | 77. | 78. | 79. |
| 1.0 | 1.000 | 1.000 | 1.000 | 1.000 | 1.000 | 1.000 | 1.000 | 1.000 | 1.000 | 1.000 |
| 1.2 | 1.193 | 1.193 | 1.193 | 1.193 | 1.193 | 1.193 | 1.193 | 1.194 | 1.194 | 1.194 |
| 1.4 | 1.374 | 1.375 | 1.375 | 1.375 | 1.376 | 1.376 | 1.376 | 1.377 | 1.377 | 1.377 |
| 1.6 | 1.547 | 1.548 | 1.548 | 1.549 | 1.550 | 1.550 | 1.551 | 1.551 | 1.552 | 1.552 |
| 1.8 | 1.713 | 1.714 | 1.715 | 1.716 | 1.717 | 1.718 | 1.719 | 1.720 | 1.720 | 1.721 |
| 2.0 | 1.872 | 1.874 | 1.875 | 1.877 | 1.878 | 1.880 | 1.881 | 1.882 | 1.884 | 1.885 |
| 2.2 | 2.027 | 2.029 | 2.031 | 2.033 | 2.035 | 2.037 | 2.039 | 2.040 | 2.042 | 2.044 |
| 2.4 | 2.178 | 2.180 | 2.183 | 2.185 | 2.188 | 2.190 | 2.192 | 2.195 | 1.197 | |
| 2.6 | 2.325 | 2.328 | 2.331 | 2.334 | 2.337 | 2.340 | | | | |
| 2.8 | 2.469 | 2.473 | 2.476 | 2.480 | | | | | | |
| 3.0 | 2.610 | 2.614 | | | | | | | | |
| Values of $W_i/W_{ibt}$ at which $E_A$ = 100 percent | | | | | | | | | | |
| | 2.978 | 2.872 | 2.769 | 2.670 | 2.575 | 2.483 | 2.394 | 2.309 | 2.226 | 2.147 |

**TABLE 5–6. Continued**

| $W_i/W_{ibt}$ | $E_{Abt}$ percent 80. | 81. | 82. | 83. | 84. | 85. | 86. | 87. | 88. | 89. |
|---|---|---|---|---|---|---|---|---|---|---|
| 1.0 | 1.000 | 1.000 | 1.000 | 1.000 | 1.000 | 1.000 | 1.000 | 1.000 | 1.000 | 1.000 |
| 1.2 | 1.194 | 1.194 | 1.194 | 1.194 | 1.194 | 1.194 | 1.194 | 1.194 | 1.194 | 1.194 |
| 1.4 | 1.377 | 1.378 | 1.378 | 1.378 | 1.378 | 1.379 | 1.379 | 1.379 | 1.379 | 1.379 |
| 1.6 | 1.553 | 1.553 | 1.554 | 1.555 | 1.555 | 1.555 | 1.556 | 1.556 | 1.557 | 1.557 |
| 1.8 | 1.722 | 1.723 | 1.724 | 1.725 | 1.725 | 1.726 | 1.727 | 1.728 | | |
| 2.0 | 1.886 | 1.887 | 1.888 | 1.890 | | | | | | |
| 2.2 | 2.045 | | | | | | | | | |
| Values of $W_i/W_{ibt}$ at which $E_A$ = 100 percent | | | | | | | | | | |
| | 2.070 | 1.996 | 1.925 | 1.856 | 1.790 | 1.726 | 1.664 | 1.605 | 1.547 | 1.492 |

| $W_i/W_{ibt}$ | $E_{Abt}$ percent 90. | 91. | 92. | 93. | 94. | 95. | 96. | 97. | 98. | 99. |
|---|---|---|---|---|---|---|---|---|---|---|
| 1.0 | 1.000 | 1.000 | 1.000 | 1.000 | 1.000 | 1.000 | 1.000 | 1.000 | 1.000 | 1.000 |
| 1.2 | 1.194 | 1.195 | 1.195 | 1.195 | 1.195 | 1.195 | 1.195 | 1.195 | 1.195 | 1.195 |
| 1.4 | 1.380 | 1.380 | 1.380 | 1.380 | 1.381 | | | | | |
| 1.6 | 1.558 | | | | | | | | | |
| Values of $W_i/W_{ibt}$ at which $E_A$ = 100 percent | | | | | | | | | | |
| | 1.439 | 1.387 | 1.338 | 1.290 | 1.244 | 1.199 | 1.157 | 1.115 | 1.075 | 1.037 |

Some additional published design charts and correlations developed by Dyes et al. and Kimbler et al. are given in Figures 5–14 and 5–15, respectively.[14,16]

## DETERMINING WATER INJECTIVITY FOR REGULAR PATTERNS

Advance information on injection well injectivity and producing well producibility provides valuable information in order to design an efficient secondary recovery project. An accurate, or at least order-of-magnitude injection rates or injection pressures over the life of the projected flood should be developed. This is also important for the proper design/sizing of injection equipment and pumps.

While injectivity can best be determined from small scale pilot operations, empirical methods for estimating water injectivity before an actual test for regular pattern floods have been presented by Muskat and Deppe for a mobility ratio of unity and by Prats et al.[17–19] While the published correlations

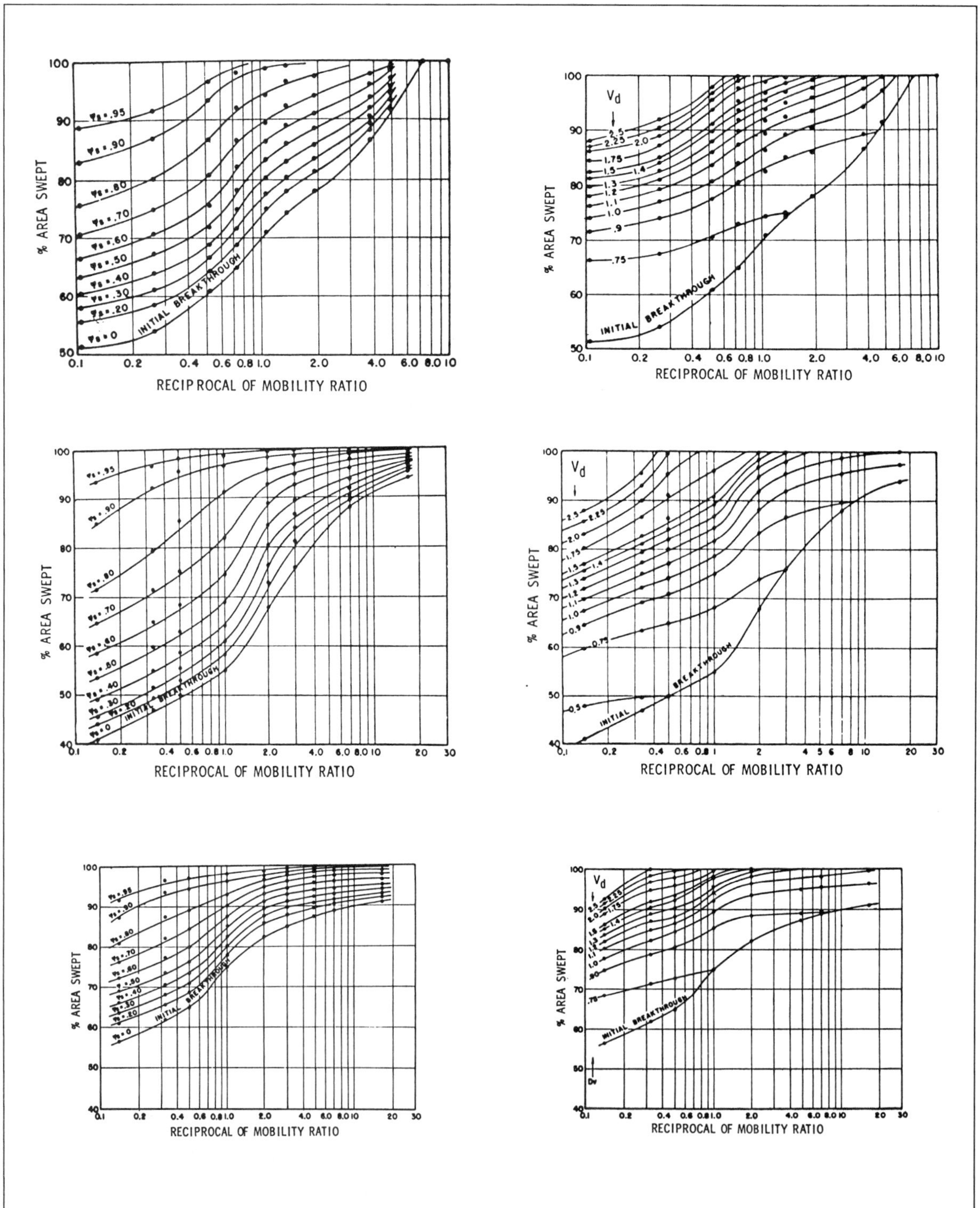


**Figure 5–14.** Effect of mobility ratio on oil production and displacable volumes injected (© SPE, SPE Monograph Series, Vol. 3, 1971).[2]

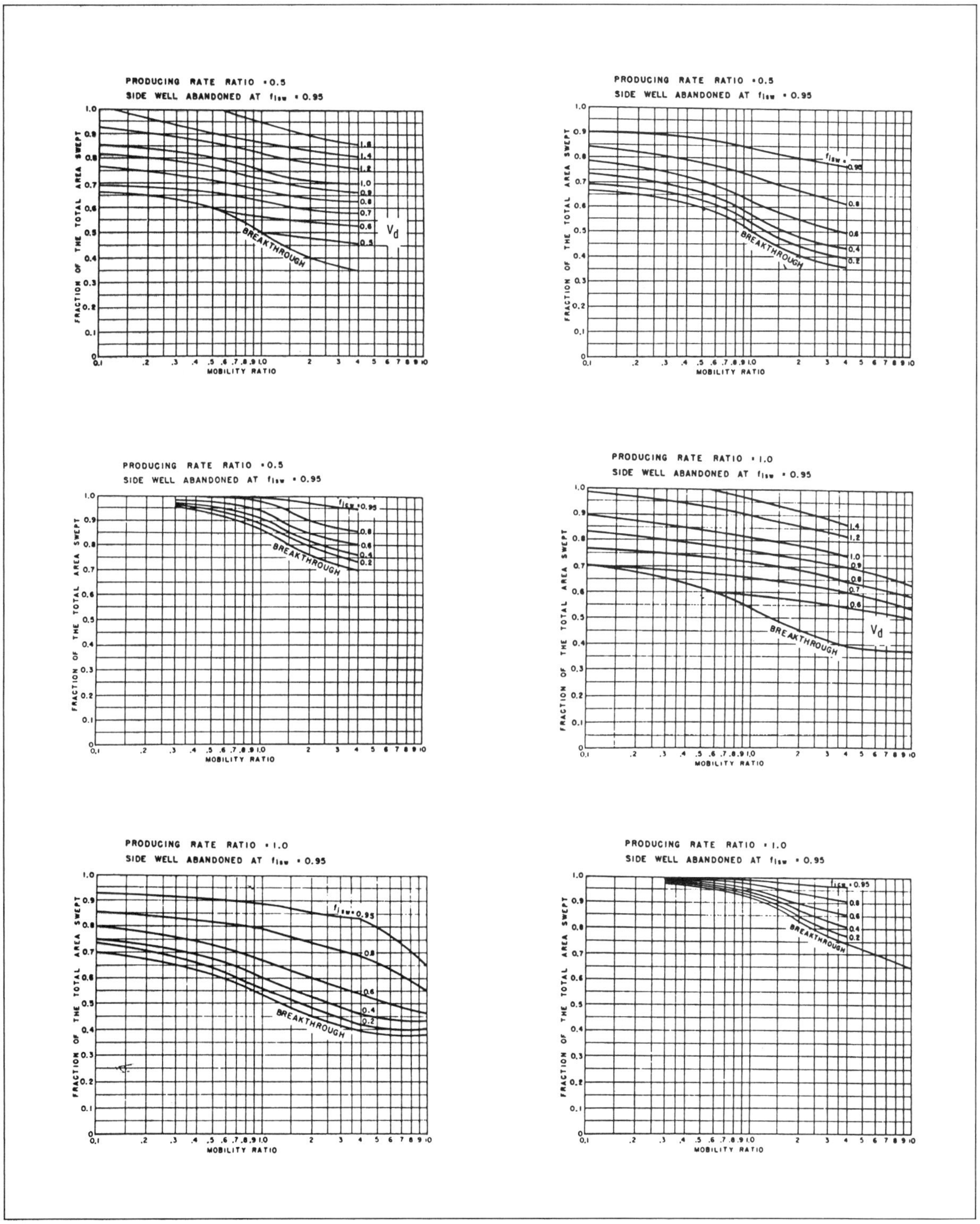


**Figure 5–15.** Sweepout pattern efficiency as a function of mobility ratio (© SPE, SPE Monograph Series, Vol. 3, 1971).[2]

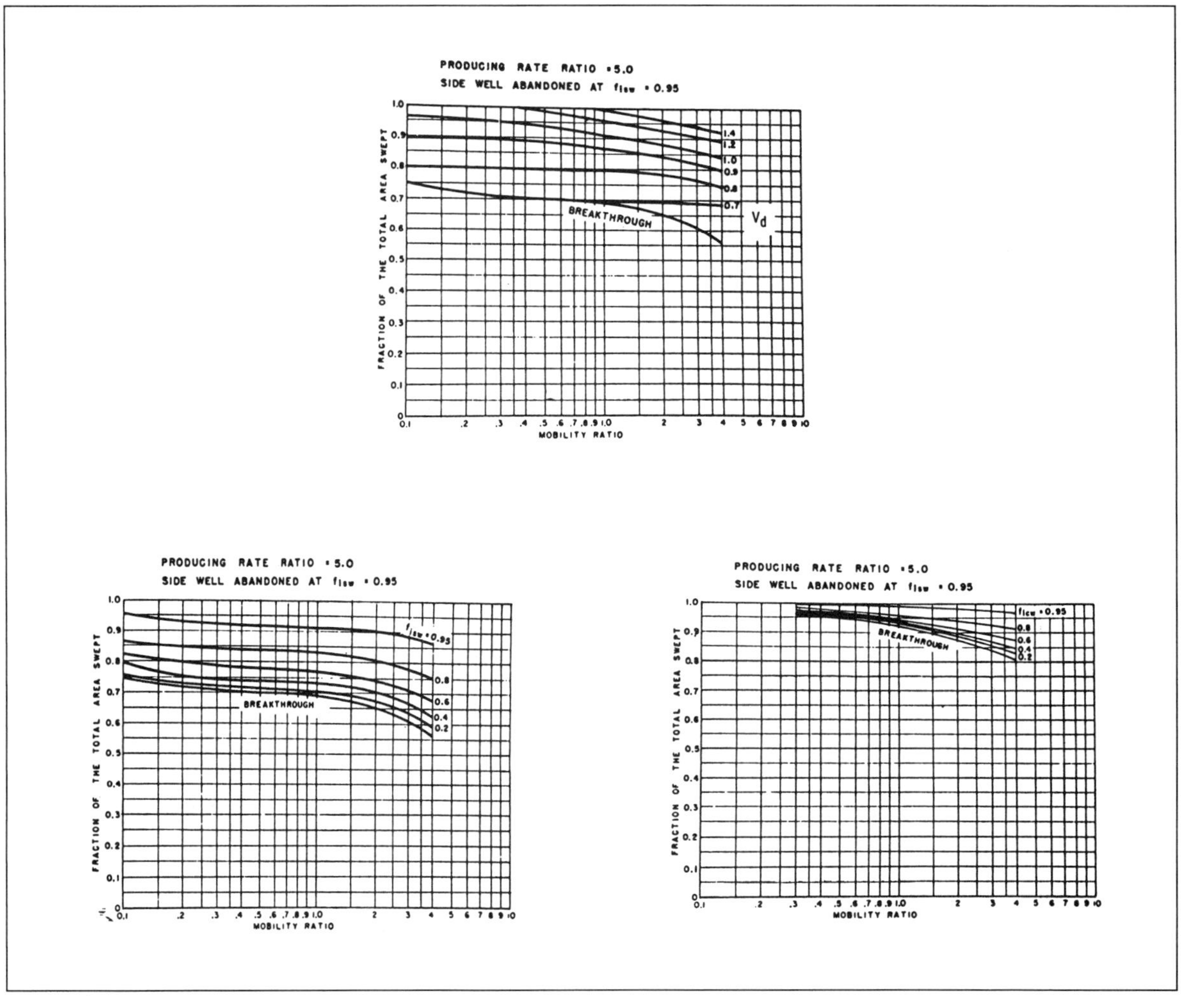


**Figure 5–15. Continued**

are for an ideal homogeneous reservoir, they still can provide reasonable first estimates of injectivity for the design of pilot tests or for screening of waterflood projects. The following equations can be used:

For direct line drive when $\left(\frac{d}{a} \geq 1\right)^{17}$:

$$i = \frac{0.001538kk_{ro}h\Delta p}{\mu_o\left(\log\frac{a}{r_w} + 0.682\frac{d}{a} - 0.902\right)} \tag{5.61}$$

For staggered line drive when $\left(\dfrac{d}{a} \geq 1\right)$:[17]

$$i = \frac{0.001538kk_{ro}h\Delta p}{\mu_o\left(\log\dfrac{a}{r_w} + 0.682\dfrac{d}{a} - 0.798\right)} \tag{5.62}$$

For five-spot pattern:[17]

$$i = \frac{0.001538kk_{ro}h\Delta p}{\mu_o\left(\log\dfrac{d}{r_w} - 0.2688\right)} \tag{5.63}$$

For seven-spot pattern:[18]

$$i = \frac{0.002051kk_{ro}h\Delta p}{\mu_o\left(\log\dfrac{d}{r_w} - 0.2472\right)} \tag{5.64}$$

For inverted nine-spot pattern, corner well:[18]

$$i = \frac{0.001538kk_{ro}h\Delta p_{i,c}}{\mu_o\left(\dfrac{1+R}{2+R}\right)\left(\log\dfrac{d}{r_w} - 0.1183\right)} \tag{5.65}$$

For inverted nine-spot pattern, side well:[19]

$$i = \frac{0.003076kk_{ro}h\Delta p_{i,s}}{\mu_o\left[\left(\dfrac{3+R}{2+R}\right)\left(\log\dfrac{d}{r_w} - 0.1183\right) - \dfrac{0.301}{2+R}\right]} \tag{5.66}$$

where $i$ is the injection rate in B/D, $k$ is the absolute permeability in md, $k_{ro}$ is the relative permeability of reservoir fluid (fraction), $\Delta p$ is the pressure difference between injection and producing well in psi, $\mu_o$ is the oil viscosity in cp, $r_w$ is the wellbore radius in feet, $a$ and $d$ are pattern dimensions in feet (see Figure 5–12), and $R$ is the ratio of corner to side well producing rate (fraction).

Caudle and Witte used these concepts to develop a correlation between injection rate, mobility ratio, and areal sweep efficiency from data obtained for miscible displacement in a scaled five-spot pattern.[20] This correlation is given by the conductance ratio (ratio of the injection rate of the displacement to the injection rate at unit mobility ratio) as

$$\gamma = \frac{i}{i_b} \tag{5.67}$$

Figure 5–16 shows the correlation between the conductance ratio, area swept ($E_A$), and mobility ratio M. Figure 5–16 may be represented by the following equation. If $\%E_A = 0$, $\gamma = 1$, and if $\%E_A = 1$, $\gamma = M$.

If $M \geq 1$,

$$\gamma = M^{E_A} \tag{5.68}$$

If $M < 1$,

$$\gamma = D + EM + F \log M \tag{5.69}$$



where

$$D = 1.412546 - 0.9881E_A$$
$$E = -0.40586 + 0.985715E_A$$
$$F = 1.094353 - 0.812527E_A$$
$$E_A = \text{swept area, fraction}$$

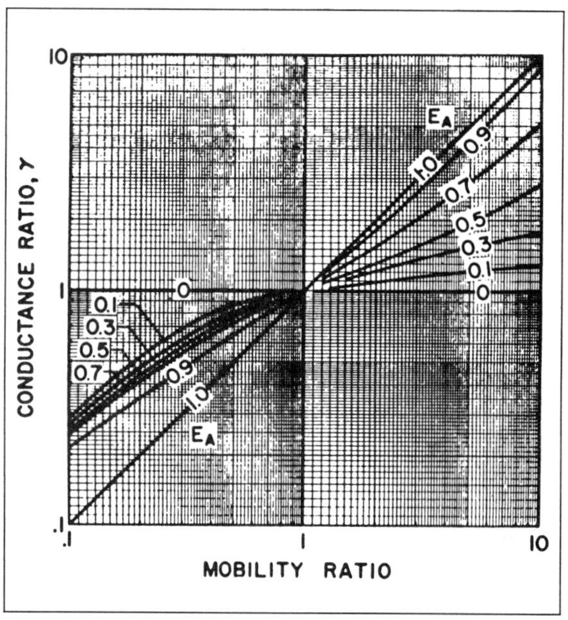


**Figure 5–16.** Correlation of conductance ratio for five-spot patterns—miscible displacement (© SPE, JPT, Trans. AIME, **216,** Dec. 1959).[20]

Craig used Figure 5–16 to estimate water injection rates by evaluating the mobility ratio at $M_{\bar{S}}$ before breakthrough.[2] After breakthrough, the mobility ratio of the five-spot was computed from the following equation:

$$M_{\bar{S}_{w5}} = \frac{\left(\dfrac{k_{rw}}{\mu_w}\right)_{\bar{S}_{w5}}}{\left(\dfrac{k_{ro}}{\mu_o}\right)_{S_{iw}}} \tag{5.70}$$

where $\bar{S}_{w5}$ is the average water saturation in the swept region given by

$$\bar{S}_{w5} = S_{w2} + f_{o2}Q_i^* \tag{5.71}$$

where $Q_i^*$ equals the volume of water injected divided by the pore volume of the five-spot swept by the injected water.

Injection rates can be estimated in the presence of altered permeability (presence of skin) in the immediate area of the wellbore by introducing the skin factor into Equations (5.61) through (5.66). The skin factor is introduced as illustrated in the following equation for a five-spot pattern:

$$i = \frac{0.001538kk_{ro}h\Delta p}{\mu_o\left(\frac{d}{r_w} - \frac{s_i + s_p}{2} - 0.2688\right)} \tag{5.72}$$

where $s_i$ is the skin factor for the injection well and $s_p$ is the skin factor of the production well.

Prats et al. have developed an analytical technique which can be used to calculate injection rates for an enclosed five-spot well pattern where oil, gas, and water saturations are present (i.e., water is displacing oil and a rather high gas saturation exists initially).[19]

$$i = \frac{0.001127I_Dk_wh(\Delta p)}{\mu_w} \tag{5.73}$$



where

1. From the start of the flood until oil bank interference:

$$I_D = \frac{4\pi}{\ln\left(\frac{r_1}{r_w}\right)^2 + M_{wo}M_{og}\left[2\ln\left(\frac{L}{r_w}\right)^2 - 3.856 - \ln\left(\frac{r_2}{r_w}\right)^2\right] + M_{wo}\ln\left(\frac{r_2}{r_1}\right)^2} \tag{5.74}$$

where

$r_1 = \sqrt{W_{iD}A/\pi + r_w^2}$ (outer radius of the water bank)

$r_2 = \sqrt{W_{iD}AF/\pi + r_w^2}$ (outer radius of the oil bank)

$W_{iD} = W_i \Big/ L^2h\phi(1 - S_{or} - S_{gr} - S_{iw})$

$F = 1 + (S_o - S_{or}) \Big/ (S_g - S_{gr})$

$L$ = length of a side of the five-spot pattern

$M_{wo}$ = water/oil mobility ratio

$M_{og}$ = oil/gas mobility ratio

2. From interference of the oil bank to oil breakthrough:

$$I_D = 4\pi \Big/ \ln\left(\frac{r_1}{r_w}\right)^2 + M_{wo}M_{og}\ln\left(\frac{r_3}{r_w}\right)^2 + M_{wo}\left[2\ln\left(\frac{L}{r_w}\right)^2 - 3.856 - \ln\left(\frac{r_3}{r_w}\right)^2 - \ln\left(\frac{r_1}{r_w}\right)^2\right] \tag{5.75}$$

where $r_3$ is the radius of the gas region around the producing well.

$$r_3 = \sqrt{\frac{L^2}{\pi}(1 - W_{iD}F) + r_w^2}$$

3. After oil breakthrough to the producing well,

$$I_D = \frac{1}{\dfrac{1}{I_{D'}} + \dfrac{1}{2\pi}[1 + M_{wo} - f_w(M_{wo} - 1)] \ln \dfrac{3.788 \times 10^{-4} L}{r_w}} \qquad \textbf{(5.76)}$$

where $I_{D'}$ is read from Figure 5–17.



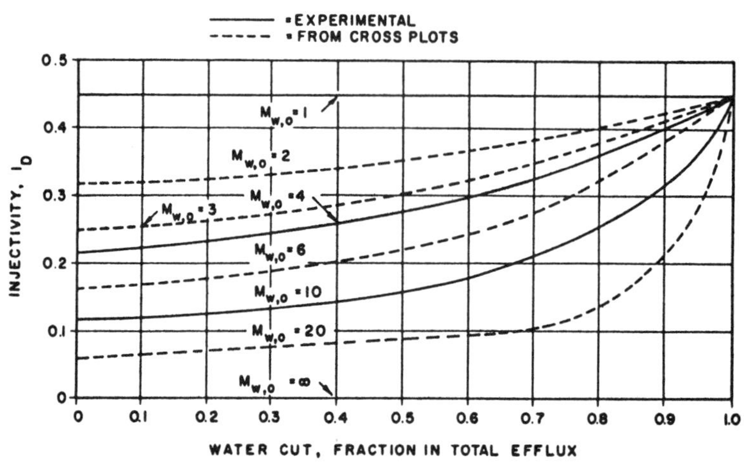


**Figure 5–17.** Dimensionless injectivity as a function of the fractional water cut and the water oil mobility ratio (© SPE, Trans. AIME, **216,** 1959).[17]

## EVALUATING FLOOD PERFORMANCE—CRAIG-GEFFEN-MORSE METHOD

The method also commonly referred to as the CGM method is used for the prediction of waterflood performance. The method lends itself to hand calculations with a pocket calculator, or it can be easily adapted to computer solution. The development in this section considers the displacement of oil by water in a five-spot pattern with no initial gas saturation.

At breakthrough, oil recovery is given by

$$N_{pbt} = E_{Abt}(\bar{S}_{wf} - S_{iw})V_p \quad \textbf{(5.77)}$$

Production after breakthrough is given by

$$N_p = E_A(\bar{S}_{w5} - S_{iw})V_p \quad \textbf{(5.78)}$$

where $\bar{S}_{w5}$ is the average water saturation in a region swept by the injected fluid. A new variable, $Q_i^*$, is introduced to represent the number of water-contacted PVs in the five-spot pattern.

At breakthrough,

$$Q_{ibt}^* = \frac{W_{ibt}}{E_{Abt}V_p} = \bar{S}_{wf} - S_{iw}$$
$$\left(\frac{\partial f_w}{\partial S_w}\right)_{S_{w5}} = \frac{1}{Q_i^*} \quad \textbf{(5.79)}$$

Thus when $Q_{i2} = Q_i^*$,

$$\left(\frac{\partial f_w}{\partial S_w}\right)_{S_{w2}} = \left(\frac{\partial f_w}{\partial S_w}\right)_{S_{w5}}$$

and

$$\bar{S}_{w5} = S_{w2} + f_{o2}Q_i^* \quad \textbf{(5.71)}$$

where $f_{o2}$ and $S_{w2}$ are obtained from the frontal advance solution at

$$\left(\frac{\partial f_w}{\partial S_w}\right)_{S_{w2}} = \frac{1}{Q_i^*} \quad \textbf{(5.80)}$$

The value of $W_i$ required to sweep the pattern completely is given by

$$W_{i100} = W_{ibt}e^{(1-E_{Abt})/0.274} \quad \textbf{(5.81)}$$

If $Q_{i100}^*$ represents the value of $Q_i^*$ at $E_A = 1.0$. Then $Q_i^*$ for $W_i > W_{i100}$ is given by

$$Q_i^* = Q_{i100}^* + \frac{W_i - W_{i100}}{V_p} \quad \textbf{(5.82)}$$

Numerical value of $Q_i^*/Q_{ibt}^*$ may be obtained from Table 5–6 for any $E_{Abt}$ and $W_i$, or from the following equation:

$$\frac{Q_i^*}{Q^*_{ibt}} = 1 + a_1 e^{-al}\left[E(a_2) - E(a_1)\right] \quad \textbf{(5.83)}$$

where

$$a_1 = 3.65E_{Abt}$$

$$a_2 = a_1 + \ln\left(\frac{W_i}{W_{ibt}}\right)$$

$$W_{ibt} \leq W_i \leq W_{i100}$$

$$E(a) = 0.57721557 + \ln(a) + \sum_{n=1}^{\infty}\frac{x^n}{(nn!)}$$

Until $E_A = 1.0$, the WOR is computed using the following equation; thereafter, $F_{wo} = f_{w2}/f_{o2}$.

$$F_{wo} = \frac{(1 - f_{o2})(1 - dN_{pu}/dW_i)}{f_{o2}(1 - dN_{pu}/dW_i) + dN_{pu}/dW_i} \quad \textbf{(5.84)}$$

where

$$\frac{dN_{pu}}{dW_i} = \frac{0.274W_{ibt}}{W_i}\left(\frac{1}{E_{ibt}}\right)\left(\frac{S_{wf} - S_{iw}}{\bar{S}_{wf} - S_{iw}}\right) \quad \textbf{(5.85)}$$

The volume of oil that is displaced by the expanding flood front when the water saturation is changed from $S_{iw}$ to $S_{wf}$ is given by

$$N_{pu} = N_{pbt} + 0.274\frac{W_{ibt}}{E_{Abt}}\left(\frac{S_{wf} - S_{iw}}{\bar{S}_{wf} - S_{iw}}\right)\ln\left(\frac{W_i}{W_{ibt}}\right) \quad \textbf{(5.86)}$$

Note that when $E_A = 1.0$, $W_i = W_{i100}$ and no further oil can be displaced by the expanding flood front.

**Example 5–5.**[7] A five-spot waterflood is under consideration for a reservoir developed on a 10-ac pattern with thickness of 20 ft. The reservoir is horizontal and has a porosity of 0.15 and an initial water saturation of 0.363, which is considered immobile. Water will be injected at a rate of 338 B/D. Viscosity of the oil and water are 2.0 and 1.0 cp, respectively. Relative permeability data corresponding to the displacement of oil by water are given by the following correlations. The residual oil saturation is 0.205. Oil and water FVFs are 1.0.

$$k_{ro} = (1 - S_{wD})^{2.56} \quad \textbf{(5.87)}$$

$$k_{rw} = 0.78S_{wD}^{3.72} \quad \textbf{(5.88)}$$

where

$$S_{wD} = \frac{(S_w - S_{iw})}{(1 - S_{or} - S_{iw})} \tag{5.89}$$

which represents normalized water saturation. Perform the oil recovery calculations using the CGM model where the initial gas saturation is zero.

Solution: The first step, of course, is to calculate $k_{ro}$, $k_{rw}$, and $f_w$ and construct the fractional flow curve as shown in Figure 5–18.



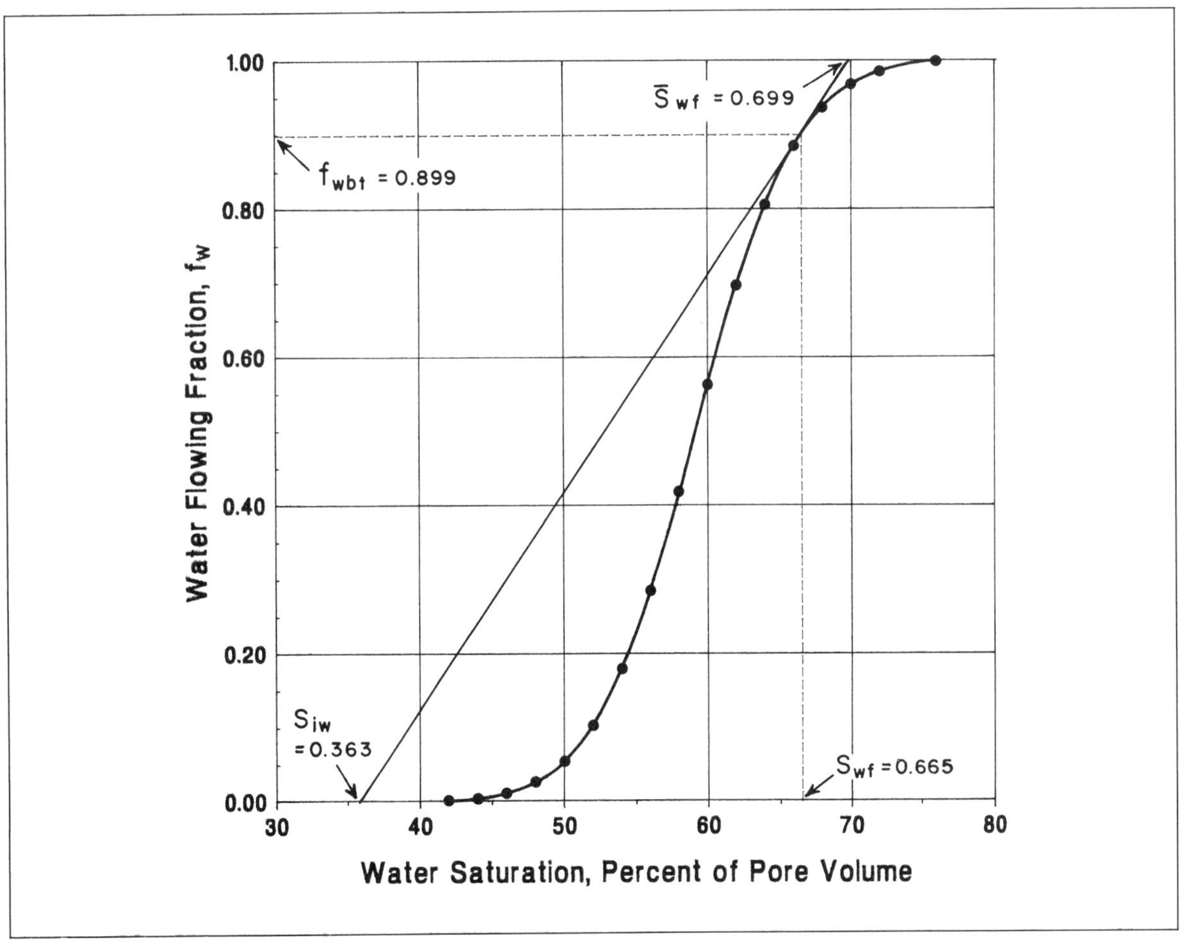


**Figure 5–18.** Fractional flow curve for Example 5–5.

From Figure 5–18,

$$S_{wf} = 0.665$$
$$\overline{S}_{wf} = 0.699$$
$$f_{wbt} = 0.899$$

Thus using Equation (5.29),

$$\overline{S}_D = \frac{S_{wf} - S_{iw}}{1 - S_{or} - S_{iw}} = \frac{0.699 - 0.363}{1 - 0.205 - 0.363} = 0.78$$

Using Equation (5.88),

$$k_{rw} = 0.78(0.78)^{3.72} = 0.31$$

Using Equation (5.20a),

$$M_{\bar{S}} = \frac{0.31/1.0}{1.0/2} = 0.62$$

Using Equation (5.6), areal sweep efficiency at breakthrough is

$$E_{Abt} = 0.54602036 + \frac{0.03170817}{0.62} + \frac{0.30222997}{e^{0.62}} - 0.00509693(0.62)$$
$$\approx 0.763$$

Compute $V_p$, $W_{ibt}$, $W_{i100}$, and $Q^*_{ibt}$.

$$V_p = \frac{43{,}560Ah\phi}{5.615} = \frac{43{,}560(10)(20)(0.15)}{5.615} = 232{,}734 \quad \text{bbls}$$

Using Equation (5.79), water injected at breakthrough is

$$W_{ibt} = E_{Abt} + V_p(\overline{S}_{wf} - S_{iw}) = 0.763(232{,}734)(0.699 - 0.363) = 59{,}666 \quad \text{bbls}$$

**TABLE 5–7. Relative permeabilities and fractional flows for Example 5–5**

| Index | $S_w$ (fraction) | $S_{wD}$ Eq. (5.29) | $k_{ro}$ Eq. (5.87) | $k_{rw}$ Eq. (5.88) | $f_w$ Eq. (5.2a) |
|---|---|---|---|---|---|
| 1 | .363 | 0 | 1.00 | 0 | 0 |
| 2 | .400 | .086 | .795 | 0 | 0 |
| 3 | .440 | .178 | .605 | .001 | .004 |
| 4 | 1.480 | .271 | .445 | .006 | .026 |
| 5 | .520 | .363 | .315 | .018 | .103 |
| 6 | .560 | .456 | .210 | .042 | .285 |
| 7 | .600 | .549 | .131 | .084 | .562 |
| 8 | .640 | .641 | .073 | .149 | .805 |
| 9 | .680 | .734 | .034 | .247 | .936 |
| 10 | .720 | .826 | .011 | .384 | .985 |
| 11 | .760 | .919 | .002 | .570 | .999 |
| 12 | .795 | ∞ | 0 | .780 | 1.000 |



Using Equation (5.81), $W_i$ required to completely sweep the pattern is

$$W_{i100} = 59{,}666e^{(1-0.763)/0.274}$$
$$= 141{,}702 \quad \text{bbls}$$

Using Equation (5.79), the number of water-contacted PVs are

$$Q^*_{ibt} = \overline{S}_{wf} - S_{iw} = 0.699 - 0.363 = 0.336$$

The maximum value of $W_i$, $(W_i)_{max}$, is estimated from the frontal advance solution at, for instance, $f_{w2} = 0.995$, at which $S_{w2} = 0.74$ and $(\partial f_w/\partial S_w)_{S_{w2}} = 0.306$. Thus

$$(Q_i)_{max} = \frac{1}{0.306} = 3.27$$

and

$$(W_i)_{max} = (Q_i)_{max}V_p$$
$$= 3.27 \times 232{,}734$$
$$= 761{,}040$$

The remaining sequence of calculations are as follows. The tabulated result as shown in Table 5–8 was generated.

**TABLE 5–8. Displacement performance of a five-spot waterflood for Example 5–5**

| (1) $W_i$ (Bbl) | (2) $W_i / W_{ibt}$ | (3) $E_A$ (frac) | (4) $Q_i^*$ | (5) $S_{w2}$ (frac) | (6) $f_{o2}$ | (7) $\bar{S}_{w5}$ (frac) | (8) $N_p$ (bbl) | (9) $F_{wo}$ (bbl/bbl) |
|---|---|---|---|---|---|---|---|---|
| For $W_{ibt} < W_i < W_{i100}$ | | | | | | | | |
| 59,666 | 1.000 | 0.763 | 0.336 | .664 | .104 | .699 | 59,666 | 1.5 |
| 68,781 | 1.153 | 0.802 | 0.386 | .670 | .087 | .704 | 63,649 | 1.9 |
| 77,896 | 1.306 | 0.836 | 0.434 | .675 | .075 | .708 | 67,125 | 2.3 |
| 87,011 | 1.458 | 0.867 | 0.480 | .679 | .065 | .710 | 70,018 | 2.7 |
| 96,126 | 1.611 | 0.894 | 0.524 | .683 | .058 | .713 | 72,822 | 3.1 |
| 105,241 | 1.764 | 0.919 | 0.568 | .686 | .053 | .716 | 75,501 | 3.4 |
| 114,356 | 1.917 | 0.942 | 0.610 | .688 | .048 | .717 | 77,609 | 3.8 |
| 123,471 | 2.069 | 0.963 | 0.651 | .691 | .044 | .720 | 80,012 | 4.2 |
| 132,586 | 2.222 | 0.982 | 0.691 | .693 | .041 | .721 | 81,819 | 4.6 |
| 141,701 | 2.375 | 1.000 | 0.371 | .695 | .038 | .723 | 83,784 | 25.1 |
| For $W_{i100} < W_i < (W_i)_{max}$ | | | | | | | | |
| 210,516 | 3.528 | 1.000 | 1.027 | .707 | .025 | .733 | 86,112 | 39.0 |
| 279,332 | 4.682 | 1.000 | 1.322 | .715 | .019 | .740 | 87,741 | 51.6 |
| 348,147 | 5.835 | 1.000 | 1.618 | .721 | .014 | .744 | 88,672 | 70.4 |
| 416,962 | 6.988 | 1.000 | 1.914 | .726 | .012 | .749 | 89,835 | 82.3 |
| 485,778 | 8.142 | 1.000 | 2.209 | .730 | .010 | .752 | 90,534 | 99.0 |
| 554,593 | 9.295 | 1.000 | 2.505 | .733 | .008 | .753 | 90,766 | 124.0 |
| 623,408 | 10.448 | 1.000 | 2.801 | .736 | .007 | .756 | 91,464 | 141.9 |
| 692,224 | 11.602 | 1.000 | 3.096 | .738 | .006 | .757 | 91,697 | 165.7 |
| 761,039 | 12.755 | 1.000 | 3.392 | .741 | .005 | .758 | 91,930 | 199.0 |

1. Column 1. The interval from $W_{ibt}$ to $W_{i100}$ was subdivided into nine increments so that 10 values of $W_i$ would be available from $W_{ibt}$ to $W_{i100}$.
2. Column 2. The interval from $W_{i100}$ to $(W_i)_{max}$ is also subdivided into nine increments so that 10 values of $W_i$ would be available from $W_{i100}$ to $(W_i)_{max}$.
3. Column 2. Compute the ratio of $W_i/W_{ibt}$.
4. Column 3. The areal sweep efficiency was evaluated for $W_{ibt} < W_i < W_{i100}$, using Equation (5.58).
5. Column 4. Values of $Q_i^*$ were computed for $W_{ibt} < W_i < W_{i100}$ using Equation (5.83), and values of $Q_i^*$for $W_{i100} < W_i < (W_i)_{max}$ were computed using Equation (5.82).
6. Column 5. The $S_{w2}$ was read from Figure 5–19 for each $1/Q_i^* = (\partial f_w/\partial S_w)$.
7. Column 6. The value of $f_{w2}$ is read from Figure 5–18 and $f_{o2} = 1 - f_{w2}$.

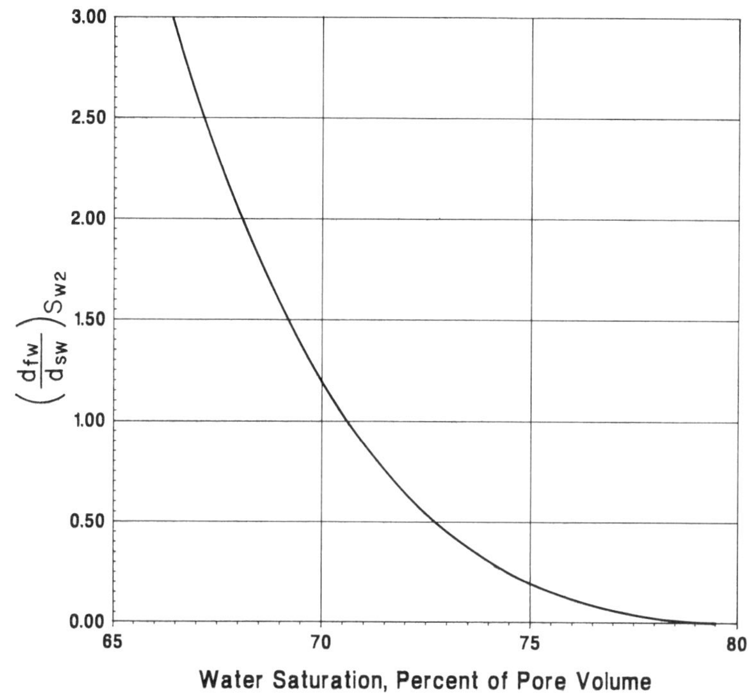


**Figure 5–19.** Plot of $(\partial f_w/\partial s_w)$ vs. $S_w$ for Example 5–5.



8. Column 7. The average water saturation in the swept region, $(\bar{S}_{w5})$, was computed by using Equation (5.71).
9. Column 8. Using Equation (5.78), compute $N_p$.
10. Column 9. The water/oil ratio was computed with Equation (5.84) until $E_A = 1.0$; thereafter, $F_{wo} = (1 - f_{o2})/f_{o2}$.

**Example 5–6.**[7] Estimate the injection rate for the waterflood project in Example 5–5 at 40,000 and 114,900 bbls of water injected. The pressure drop is 500 psi between the injection well and production well. Permeability of the oil at interstitial water saturation is 100 md. An effective wellbore radius is 0.5 ft.

Solution:

In a 10-ac five-spot pattern, $d$ is given by

$$d = \frac{660}{\sqrt{2}} = 466.69 \quad \text{ft}$$

Using Equation (5.63), the base flow rate is

$$i_b = \frac{0.001538(100)(20)(500)}{2\left(\log\dfrac{466.69}{0.5} - 0.2688\right)}$$

$$= \frac{1{,}538}{5.4025} = 284.68 \quad \text{STB/D}$$

Since $W_i = 40{,}000 < W_{ibt} = 59{,}666$ (Table 5–8), thus $E_A < E_{Abt}$ and $M_{\bar{S}} = 0.62$ as calculated in Example 5–5.



$$E_A = \frac{W_i}{V_p(\bar{S}_{wf} - S_{iw})}$$

$$= \frac{40{,}000}{232{,}734(0.699 - 0.363)}$$

$$= 0.51$$

Using Equation (5.69), $i_b$ is corrected for $M_{\bar{S}} < 1.0$ as

$$D = 1.412546 - 0.9881(0.51) = 0.9086$$
$$E = -0.40586 + 0.985715(0.51) = 0.0969$$
$$F = 1.094353 - 0.812527(0.51) = 0.68$$
$$\gamma = 0.9086 + 0.0969(0.62) + 0.68 \log(0.62)$$
$$= 0.8275 \approx 0.83$$

Using Equation (5.67),

$$i = 0.83(284.68)$$
$$= 236.28 \quad \text{STB/D}$$

At $W_i = 114{,}900$, breakthrough has occurred, and $E_A = 0.943$ from Equation (5.58), and $\bar{S}_{w5} = 0$

$$S_{wD} = \frac{0.718 - 0.363}{1 - 0.205 - 0.363} = 0.822$$

Using Equation (5.88),

$$(k_{rw})_{\bar{S}_{w5}} = 0.78(0.822)^{3.72}$$
$$= 0.376$$

Using Equation (5.70),

$$M_{\bar{S}_{w5}} = \frac{0.376/1.0}{1.0/2.0} = 0.752$$

Using Equation (5.69), $\gamma = 0.80$; therefore, the injection rate is 227.7 STB/D.

## EVALUATING FLOOD PERFORMANCE—STILES METHOD

One of the waterflood performance calculation techniques that has been used rather widely is credited to Stiles.[21] So far, displacement has only been considered in homogeneous reservoirs. In *stratified* reservoirs in which there is a defined variation in reservoir parameters with thickness, in the dip-normal direction, the description of displacement is necessarily more complex. 

Stiles has presented a method for calculating oil recovery in these stratified reservoirs when the layers are isolated from one another by impermeable layers so that there is no pressure communication between the layers. Stiles' method is based on the following assumptions:

1. Linear geometry and unit mobility ratio
2. Distance of advance of the flood front in any layer is proportional to the absolute permeability of that layer
3. No vertical communication or cross-flow between layers
4. Production in any layer at the outflow end changes abruptly from oil to water; that is, piston-like displacement
5. Water cut at the producing well depends upon the millidarcy-feet of capacity flowing oil at any time and the millidarcy-feet that have broken through to water production
6. All layers have the same $\phi$, same $k_{ro}$ ahead of the flood front, and the same $k_{ro}$ behind the flood front
7. All layers undergo the same change in oil saturation due to the water displacement
8. There is steady-state flow (i.e., each barrel of water input displaces a barrel of fluid)

In using this technique, it is convenient to divide the total formation into layers of constant thickness, $h_1$, $h_2$, etc., with their corresponding absolute permeabilities, $k_1$, $k_2$, etc., and arrange them in descending order of permeability. In such an arrangement, it is evident that the topmost layer would flood out first, the second layer second, and so on. With these assumptions, the coverage (fraction of the vertical cross section swept to $S_{or}$ by water) is given by

$$C = \sum_{i=1}^{j} \frac{k_i}{k} + \sum_{i=j+1}^{n} \frac{\frac{k_i}{k_j} h_i}{h} \tag{5.89}$$

where the subscript $j$ represents the last layer flooded out and the index $n$ is the total number of layers. The WOR corresponding to the floodout of layer $j$ is given by

$$F_{wo} = \frac{\sum_{i=1}^{j} k_i h_i}{\sum_{i=j+1}^{n} k_i h_i} \tag{5.90}$$

The following equations closely follows those given by Craft and Hawkins.[3] When the $j$th layer has just broken through to water production, the fractional recovery from all the beds is

$$R = \frac{k_j h_j + (C_t - C_j)}{h_t k_j} \tag{5.91}$$

where



$$(C_t - C_j) = \Delta h_k k_k + \Delta h_1 k_1 + \cdot \Delta h_n k_n \tag{5.92}$$

where $C_t$ is the total capacity of the formation in millidarcy-feet and $C_j$ is the millidarcy-feet which have been completely flooded, the difference, $(C_t - C_j)$, being the flow capacity still producing oil at the outflow end.

The *surface water cut* may be expressed by the following equation:

$$f_w = \frac{AC_j}{AC_j + (C_t - C_j)} \tag{5.93}$$

where

$$A = \frac{\mu_o}{\mu_w} \times \frac{k_{rw}}{k_{ro}} \times B_o \tag{5.94}$$

and $C_j = \Sigma k \Delta h_w$ is the millidarcy-feet producing water at any stage of displacement, and $(C_t - C_j) = \Sigma k \Delta h_o$.

Similarly, the *reservoir* water cut is given by

$$f'_w = \frac{A'C_j}{A'C_j + (C_t - C_j)} \tag{5.95}$$

where $A'$ is the endpoint mobility ratio given by

$$A' = \frac{\mu_o}{\mu_w} \times \frac{k_w}{k_o}$$

## EVALUATING FLOOD PERFORMANCE—MUSKAT METHOD

Note that when the mobility ratio is not unity, the displacement calculations in stratified reservoirs are more complex than those developed by Stiles.[21] Muskat developed solutions for displacement calculations in stratified reservoirs where mobility ratios are other than unity, where the permeability distribution is linear, and where it is exponential.[22]

The permeability distribution is said to be linear when the plot of absolute permeability versus cumulative thickness is close to a straight line on Cartesian coordinate graph paper. The permeability distribution is said to be exponential when the plot of $k$ versus $\Sigma h$ approximates a straight line on semilogarithmic graph paper as shown in Figure 5–20. In both cases the permeabilities have to be arranged in ascending order. The exponential distribution (with the exception of the highest and lowest permeability) may be represented by the following equation:

$$k = k_{min} e^{bX/h_t} \tag{5.96}$$



where $h_t$ is the total formation thickness and $k$ is the permeability of the $x$th foot.

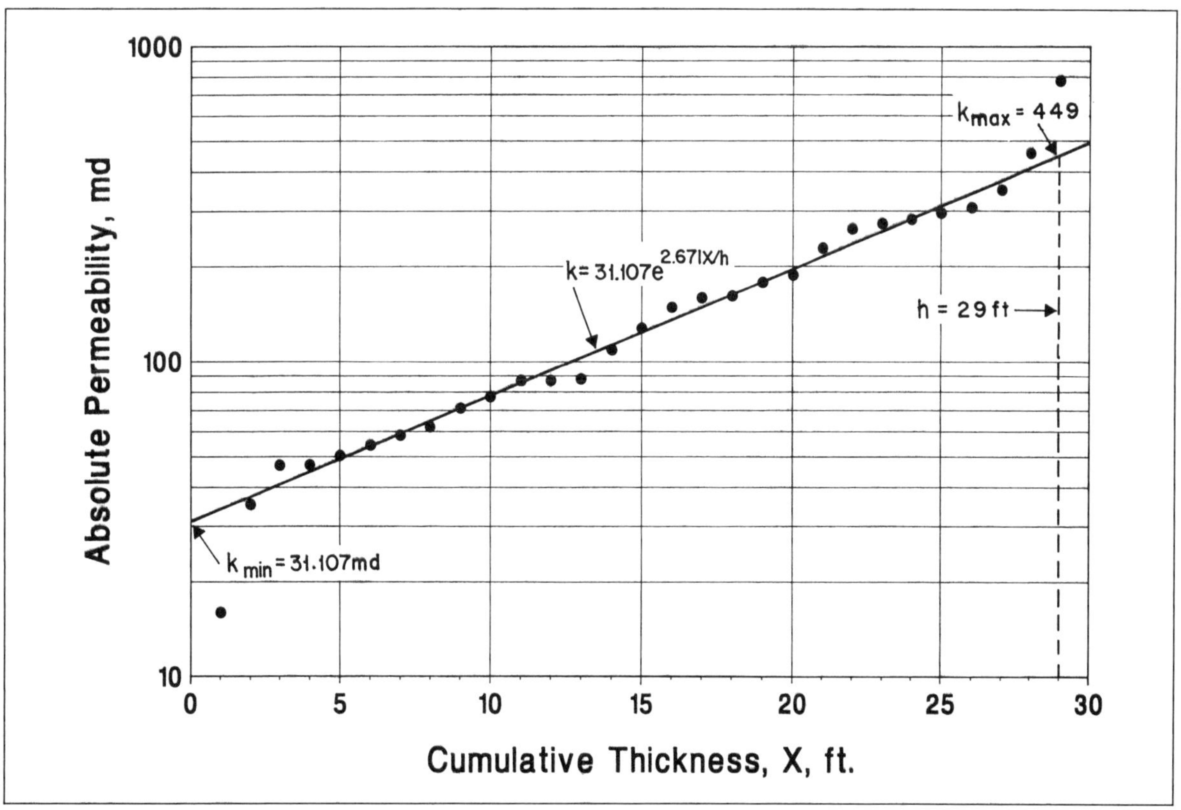


**Figure 5–20.** Plot of absolute permeability versus cumulative thickness.

Muskat's equations for the fractional recovery $R$ and the producing *reservoir* water/oil ratio at any time $t$ for the exponential distribution case are

$$R = 1 + \frac{1}{(M-1)b}\ln\left(\frac{r}{\bar{t}}\right) - \frac{2}{(M-1)b} + \frac{M}{(M-1)b} \times \left\{2A - \ln\left[\frac{(1-M)(A-1)}{(1+M)(A-1)}\right]\right\} \tag{5.97}$$

$$\left(\frac{q_w}{q_o}\right)_{res} = \frac{(M^2-1)(\bar{t}-1)}{2(AM-1)} \tag{5.98}$$

where

$$A = \sqrt{1 - \left[1 - \left(\frac{1-1}{M^2}\right)\frac{\bar{t}}{r}\right]}$$

$$M = \frac{k_w}{\mu_w} \times \frac{\mu_o}{k_o} = \frac{\lambda_w}{\lambda_o} \quad \text{(endpoint mobility ratio)}$$

$b$ = slope of the straight line (Equation [5.96]).

$$r = \frac{k_{max}}{k_{min}} = e^b$$

$\bar{t} = t/t_B$, where $t$ is the real time and $t_B$ is the breakthrough time of the most permeable layer.

**Example 5–7.**[3] Table 5–9 illustrates the application of Stiles' method for predicting waterflood performance for a five-spot pattern flood where stratification exists. The data are due to Stiles,[21] while the tabulation is similar to that presented by Craft and Hawkins.[3] Column (1) and (2) present data. Column (2) gives the absolute permeabilities measured on each foot of a formation of 29 ft net thickness, arranged in *descending order* of permeability. Other reservoir and fluid properties are as follows:

| | |
|---|---|
| $\mu_o = 4.34$ cp | $S_{iw} = 0.24$ |
| $\mu_w = 0.82$ cp | $S_{or} = 0.225$ |
| $B_{oi} = 1.215$ RB/STB | $k_{ro} = 0.80$ |
| $B_o = 1.073$ RB/STB | $k_{rw} = 0.20$ |
| $A = 100$ ac-ft | $\phi = 0.19$ |

Calculate the oil saturation after primary depletion when the primary recovery is 11,015 STB from the 100 ac-ft unit and the displacement performance using the Stiles method.

Solution:

Using Equation (4.1), the original oil in place is

$$N = \frac{7{,}758(0.19)(1 - 0.24)(100)}{1.215}$$
$$\approx 92{,}200 \quad \text{STB}$$

The oil saturation after primary depletion by Equation (4.21) is

$$S_{op} = \frac{(92{,}200 - 11{,}015)(1.073)(1 - 0.24)}{(92{,}200)(1.215)}$$
$$\approx 0.59$$

The oil recovery by complete flooding ($E_A = 100\%$) of the 100 ac-ft unit to $S_{or} = 0.225$ is

$$N_{pf} = \frac{7758(0.19)(0.59 - 0.225)(100)}{1.073}$$
$$\approx 50{,}100 \quad \text{STB}$$

The various columns of Table 5–9 are derived as

1. Column (1). Footage interval, 29-ft thick formation
2. Column (2). The absolute permeability measured on each foot of a formation
3. Column (3). Since all $\Delta h_j$ values are 1 ft, the cumulative flow capacity is simply the summation of Column (2).
4. Column (4). Cumulative flow capacity of each layer
5. Column (5). Recoveries as each foot is flooded out, using Equation (5.91) as

For the first high-permeability layer:

$$R_1 = \frac{(1 \times 776) + (5075 - 776)}{(29)(776)}$$
$$= 0.226$$

Similarly, for the second permeability layer:

$$R_2 = \frac{(2 \times 454) + (5075 - 1230)}{(29)(454)}$$
$$= 0.361 \quad \text{and so on}$$

6. Column (6). Surface flowing fraction, $(f_w)_s$, is calculated using Equation (5.93) as

**TABLE 5–9. Waterflood performance prediction in a stratified reservoir**[3]

| (1) $\Delta h_i$ ft | (2) $k_i$ md | (3) $\Sigma k_i \Delta h_i$ md-ft | (4) $k_i h_i$ md-ft | (5) R (frac.) | (6) $(f_w)_s$ (frac.) | (7) $N_p$ STB |
|---|---|---|---|---|---|---|
| 1 | 776 | 776 | 776 | .226 | .204 | 11,323 |
| 2 | 454 | 1,230 | 908 | .361 | .312 | 18,086 |
| 3 | 349 | 1,579 | 1,047 | .449 | .391 | 22,495 |
| 4 | 308 | 1,887 | 1,232 | .495 | .457 | 24,800 |
| 5 | 295 | 2,182 | 1,475 | .511 | .517 | 25,601 |
| 6 | 282 | 2,464 | 1,692 | .526 | .573 | 26,353 |
| 7 | 273 | 2,737 | 1,911 | .537 | .624 | 26,904 |
| 8 | 262 | 2,999 | 2,096 | .549 | .672 | 27,505 |
| 9 | 228 | 3,227 | 2,052 | .590 | .713 | 29,559 |
| 10 | 187 | 3,414 | 1,870 | .651 | .745 | 32,615 |
| 11 | 178 | 3,592 | 1,958 | .667 | .775 | 33,417 |
| 12 | 161 | 3,753 | 1,932 | .697 | .801 | 34,920 |
| 13 | 159 | 3,912 | 2,067 | .700 | .827 | 35,070 |
| 14 | 148 | 4,060 | 2,072 | .719 | .850 | 36,022 |
| 15 | 127 | 4,187 | 1,905 | .758 | .870 | 37,976 |
| 16 | 109 | 4,296 | 1,744 | .798 | .887 | 39,980 |
| 17 | 88 | 4,384 | 1,496 | .857 | .900 | 42,936 |
| 18 | 87 | 4,471 | 1,566 | .860 | .913 | 43,086 |
| 19 | 87 | 4,558 | 1,653 | .860 | .926 | 43,086 |
| 20 | 77 | 4,635 | 1,540 | .887 | .937 | 44,439 |
| 21 | 71 | 4,706 | 1,491 | .903 | .948 | 45,240 |
| 22 | 62 | 4,768 | 1,364 | .929 | .957 | 46,543 |
| 23 | 58 | 4,826 | 1,334 | .941 | .965 | 47,144 |
| 24 | 54 | 4,880 | 1,296 | .952 | .973 | 47,695 |
| 25 | 50 | 4,930 | 1,250 | .962 | .980 | 48,196 |
| 26 | 47 | 4,977 | 1,222 | .968 | .986 | 48,497 |
| 27 | 47 | 5,024 | 1,269 | .968 | .993 | 48,497 |
| 28 | 35 | 5,059 | 980 | .981 | .998 | 49,148 |
| 29 | 16 | 5,075 | 464 | 1.000 | 1.000 | 50,100 |

$$A = \frac{4.34}{0.82} \times \frac{0.20}{0.80} \times 1.073$$
$$= 1.420$$

$$(f_{ws})_1 = \frac{(1.42)(776)}{(1.42)(776) + (5075 - 776)}$$
$$= 0.204$$

TABLE 5–9. Continued

| (1) $\Delta h_i$ ft | (8) $\Delta N_p$ STB | (9) $f'_w$ (frac.) | (10) $q_{sc}$ STB/D | (11) $\Delta t$ days | (12) $t$ days | (13) $W_i$ bbls |
|---|---|---|---|---|---|---|
| 1 | 11,323 | 0 | 93.2 | 121.5 | 372.1 | 37,210 |
| 2 | 6,763 | .192 | 75.3 | 89.8 | 461.9 | 46,190 |
| 3 | 4,409 | .297 | 65.5 | 67.3 | 529.2 | 52,920 |
| 4 | 2,305 | .373 | 58.4 | 39.5 | 568.7 | 56,870 |
| 5 | 801 | .439 | 52.3 | 15.3 | 584.0 | 58,400 |
| 6 | 752 | .499 | 46.7 | 16.1 | 600.1 | 60,001 |
| 7 | 551 | .555 | 41.5 | 13.3 | 613.4 | 61,340 |
| 8 | 601 | .607 | 36.3 | 16.4 | 629.8 | 62,980 |
| 9 | 2,054 | .656 | 32.1 | 64.0 | 693.8 | 69,380 |
| 10 | 3,056 | .697 | 28.2 | 108.4 | 802.2 | 80,220 |
| 11 | 802 | .731 | 25.1 | 32.0 | 834.2 | 83,420 |
| 12 | 1,503 | .762 | 22.2 | 67.7 | 901.9 | 90,190 |
| 13 | 150 | .789 | 19.7 | 7.6 | 909.5 | 90,950 |
| 14 | 952 | .816 | 17.1 | 55.7 | 965.2 | 96,520 |
| 15 | 1,954 | .841 | 14.8 | 132.0 | 1,097.2 | 109,720 |
| 16 | 2,004 | .862 | 12.9 | 155.3 | 1,252.5 | 125,250 |
| 17 | 2,956 | .879 | 11.3 | 261.6 | 1,514.1 | 151,410 |
| 18 | 150 | .893 | 10.0 | 15.0 | 1,529.1 | 152,910 |
| 19 | 0 | .907 | 8.7 | 0 | 1,529.1 | 152,910 |
| 20 | 1,353 | .921 | 7.4 | 182.8 | 1,711.9 | 171,170 |
| 21 | 801 | .933 | 6.2 | 129.2 | 1,841.1 | 184,110 |
| 22 | 1,303 | .944 | 5.2 | 250.6 | 2,091.7 | 209,170 |
| 23 | 601 | .953 | 4.4 | 136.6 | 2,228.3 | 222,830 |
| 24 | 551 | .962 | 3.5 | 157.4 | 2,385.7 | 238,570 |
| 25 | 501 | .971 | 2.7 | 185.6 | 2,571.3 | 257,130 |
| 26 | 301 | .978 | 2.1 | 143.3 | 2,714.6 | 271,460 |
| 27 | 0 | .985 | 1.4 | 0 | 2,714.6 | 271,460 |
| 28 | 651 | .992 | 0.7 | 930.0 | 3,644.6 | 364,460 |
| 29 | 952 | .998 | 0.2 | 3,760.0 | 7,404.6 | 740,460 |



Similarly, $(f_{ws})_2$ is

$$(f_{ws})_2 = \frac{(1.42)(1{,}230)}{(1.42)(1{,}230) + (5{,}075 - 1{,}230)}$$
$$= 0.312 \quad \text{and so on.}$$

7. Column (7). Cumulative oil recovered assuming 100% sweep efficiency. Column (7) = Column (5) × $N_{pf}$.
8. Column (8). Contains the differences of Column (7) (i.e., the barrels recovered between the successive depletion of the layers).
9. Column (9). The *reservoir* flowing fractions are calculated using Equation (5.95).
10. Column (10). Gives the surface rates of production which prevail while each foot is going to *water*, using Equation (5.52),

$$(q_{sc}) = \frac{100(1 - 0.192)}{1.073} = 75.3 \quad \text{STB/D}$$

11. Column (11). Column (8) ÷ Column (10), which gives the time in days between water breakthrough in successive layers.
12. Column (12). The summation of Column (11), $\sum_{i=1}^{29} (\Delta t + 250)$. The 250 days is the *fillup* time calculated as follows, assuming reservoir gas saturation $S_g = 0.17$:

$$t_{fillup} = \frac{7{,}758(0.19)(0.17)(100)}{100} \approx 250.6 \text{ days}$$

13. Column (13). The cumulative water injected at 100 BWPD (i.e., Column (12) × 100).

**Example 5–8.**[3] A flood is planned for an 80-ac lease which has an average sand thickness of 10 ft and the following permeability profile as determined from core analysis on five wells.

| Subsea Depth, ft | | Absolute Permeability, md |
|---|---|---|
| From | To | |
| 2,050 | 2,051 | 35 |
| 2,051 | 2,052 | 51 |
| 2,052 | 2,053 | 27 |
| 2,053 | 2,054 | 116 |
| 2,054 | 2,055 | 60 |
| 2,055 | 2,056 | 237 |
| 2,056 | 2,057 | 519 |
| 2,057 | 2,058 | 98 |
| 2,058 | 2,059 | 281 |
| 2,059 | 2,060 | 164 |

Other reservoir fluid and rock data are

$$\phi = 0.25 \qquad N_p = 140.5 \text{ STB/ac-ft}$$
$$S_{iw} = 0.23 \qquad B_{oi} = 1.251 \text{ RB/STB (no initial gas cap)}$$
$$\mu_o = 4.5 \text{ cp} \qquad B_o = 1.085 \text{ RB/STB (at depletion)}$$
$$\mu_w = 0.79 \text{ cp} \qquad k_{ro} = 0.85 \text{ (in oil bank)}$$
$$S_{or} = 0.156 \qquad k_{rw} = 0.25 \text{ (behind oil bank)}$$

Using the Stiles' method, perform the following calculations:

(a) Arrange the beds in order of decreasing permeability.
(b) Calculate the fractional recovery and surface water cut after breakthrough in each successive layer.
(c) Plot the surface cut versus fractional recovery.
(d) Calculate the initial oil in place in the 80-ac unit.
(e) Calculate the oil saturation after primary depletion.
(f) Calculate the recoverable oil by flooding using sweep efficiency of 82%.
(g) Calculate the cumulative oil recovery after each successive breakthrough.
h) Calculate the increments of oil recovery between breakthrough in successive layers.
(i) Calcute the reservoir water cut before each breakthrough.
(j) At a total injection rate of 1,000 BWPD for the whole 80-ac unit, calculate the fillup volume and fillup time, $B_w = 1.0$.
(k) Calculate the daily oil producing rate in surface units between successive breakthrough in layers.
(l) Calculate the days between successive breakthrough in layers.
(m) Calculate the cumulative time in days including fillup time.
(n) Calculate the cumulative water injected up to each breakthrough.

Solution:

(a)The beds are arranged in order of decreasing permeability in Column (2), Table 5–10.
(b) Column (5). Calculate $R$ using Equation (5.91).
Column (6). Calculate $(f_w)_s$ using Equation (5.93).

Using Equation (5.91),

$$R_1 = \frac{(1 \times 519) + (1{,}588 - 519)}{(10)(519)}$$

Using Equation (5.93),

$$\begin{aligned} A &= \frac{4.5}{0.79} \times \frac{0.25}{0.85} \times 0.185 \\ &= 1.818 \\ (f_{ws})_1 &= \frac{(1.818)(519)}{(1.818)(519) + (1588 - 519)} \\ &= 0.469 \end{aligned}$$

(c) Plot of the surface water cut $f_{ws}$ versus the fractional recovery $R$ as shown in Figure 5–21.

**TABLE 5–10. Waterflood performance prediction for Example 5–8**

| (1) $\Delta h_j$ ft | (2) $k_i$ md | (3) $\Sigma k_i \Delta h_i$ md-ft | (4) $k_j h_j$ md-ft | (5) R (frac.) | (6) $(f_w)_s$ (frac.) | (7) $N_p$ STB |
|---|---|---|---|---|---|---|
| 1 | 519 | 519 | 519 | .306 | .469 | 155,723 |
| 2 | 281 | 800 | 562 | .480 | .649 | 244,272 |
| 3 | 237 | 1,037 | 711 | .532 | .774 | 270,735 |
| 4 | 164 | 1,201 | 656 | .636 | .849 | 323,660 |
| 5 | 116 | 1,317 | 580 | .734 | .898 | 373,533 |
| 6 | 98 | 1,415 | 588 | .777 | .937 | 395,415 |
| 7 | 60 | 1,475 | 420 | .888 | .960 | 451,903 |
| 8 | 51 | 1,526 | 408 | .922 | .978 | 469,206 |
| 9 | 35 | 1,561 | 315 | .977 | .991 | 497,195 |
| 10 | 27 | 1,588 | 270 | 1.000 | 1.000 | 508,900 |

| (1) $\Delta h_j$ ft | (8) $\Delta N_p$ STB | (9) $f_w$ (frac.) | (10) $q_{sc}$ STB/D | (11) $\Delta t$ (days) | (12) t days | (13) $W_i$ bbls |
|---|---|---|---|---|---|---|
| 1 | 155,723 | .448 | 508.76 | 306 | 586 | 586,000 |
| 2 | 88,549 | .630 | 341.01 | 260 | 846 | 846,000 |
| 3 | 26,463 | .759 | 222.12 | 119 | 965 | 965,000 |
| 4 | 52,925 | .839 | 148.39 | 357 | 1,322 | 1,322,000 |
| 5 | 49,873 | .891 | 100.46 | 496 | 1,818 | 1,818,000 |
| 6 | 21,882 | .932 | 62.67 | 349 | 2,167 | 2,167,000 |
| 7 | 56,488 | .956 | 132.72 | 426 | 2,593 | 2,593,000 |
| 8 | 17,303 | .976 | 22.12 | 782 | 3,375 | 3,375,000 |
| 9 | 27,989 | .990 | 9.22 | 3,036 | 6,411 | 6,411,000 |
| 10 | 11,705 | 1.000 | 0 | 0 | 6,411 | 6,411,000 |

(d)

$$N = \frac{7{,}758(0.25)(1 - 0.23)(10)(80)}{1.251}$$

$$= 955{,}000 \quad \text{STB}$$

(e) Using Equation (4.21),

$$S_{op} = \frac{[955{,}000 - (140.5 \times 10 \times 80)](1.085)(1 - 0.23)}{(955{,}000)(1.251)}$$

$$\approx 0.59$$

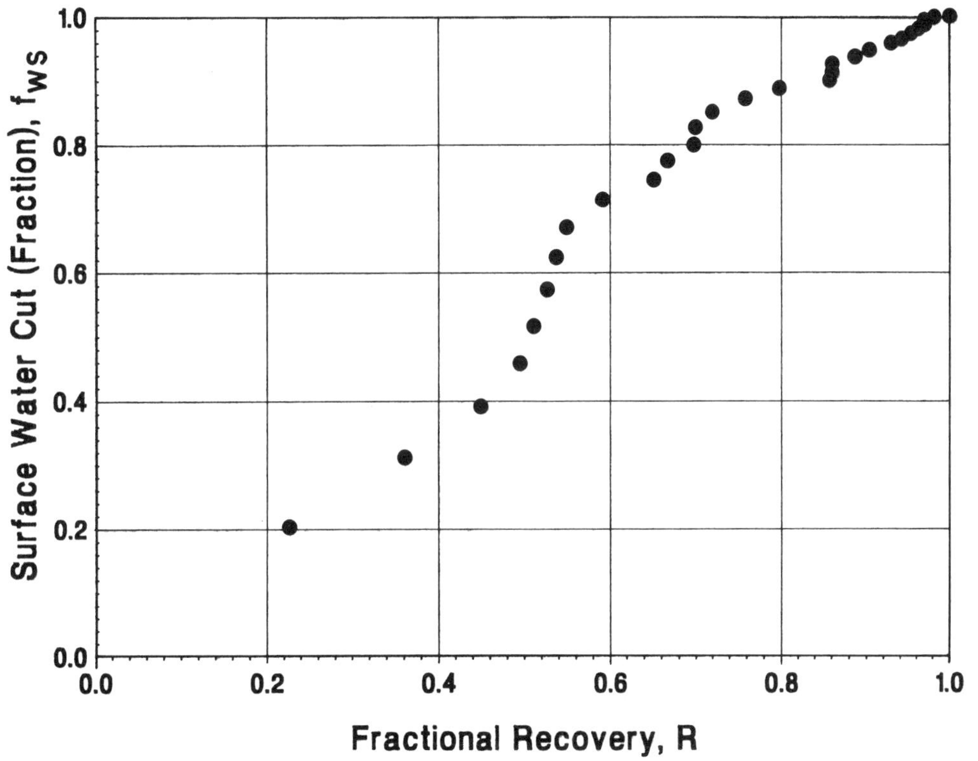


**Figure 5–21.** Plot of $f_{ws}$ vs. $R$ for Example 5–8.



(f) Recoverable oil by flooding at $E_A = 0.82$.

$$N_{pf} = \frac{7{,}758(0.25)(0.59 - 0.156)(80)(10)}{1.085} \times 0.82$$
$$\approx 508{,}900 \quad \text{STB}$$

(g) Column (7). Cumulative oil recovery after each successive breakthrough is calculated by multiplying 508,900 by $R$ (Column [5]).

(h) Column (8). $\Delta N_p = (N_p)_{i+1} - N_{pi}$.

(i) Column (9). Reservoir flowing fractions are calculated using Equation (5.95), where $A' = 1.675$.

(j) Fillup volume and fillup time at 1,000 BWPD are calculated as follows:

$$(W_i)_{fillup} = \frac{7{,}758(\phi)(V_p)(1 - S_{iw} - S_{op})}{B_w}$$

$$= \frac{7{,}758(0.25)(80 \times 10)(1 - 0.23 - 0.59)}{1}$$

$$\approx 280{,}000$$

$$t_{fillup} = \frac{280{,}000}{1{,}000} = 280 \quad \text{days}$$

(k) Column (10). $q_o = \dfrac{1{,}000(1 - f_{ws})}{B_o}$

(l) Column (11). Column (8) ÷ Column (10).

(m) Column (12). Summation of Column (11) + $t_{fillup}$.

(n) Column (13). Column (12) × 1,000.



## EVALUATING FLOOD PRFORMANCE—DYKSTRA-PARSONS METHOD

The Stiles method used to predict waterflood preformance in stratified reservoirs was presented in the previous section. While the Stiles method treats permeability stratification, its limitation is its inability to consider mobility ratios other than unity. Dykstra and Parsons presented a model for estimation of the displacement performance of a layered reservoir by subdivision of the reservoir into $n$ noncommunicating layers based on the log-normal distribution (i.e., with no crossflow between the layers).[23] A constant pressure drop is maintained across the layers. Piston-like displacement was assumed for the waterflood.

Dykstra and Parsons used the commonly found log-normal permeability distribution of reservoir rock. Their calculation method permits any mobility ratio. The Stiles method allows the various permeable layers to vary in thickness, while the Dykstra-Parsons method requires all layers to be constant thickness. They indicate that for certain reservoir conditions, recovery may be maximized by allowing some gas saturation to evolve as the result of primary production prior to waterflooding.

Based on the Dykstra and Parsons method, equations for coverage and producing water/oil ratio are given below.

$$C = \left( X + \frac{(n - X)M}{M - 1} - \frac{1}{M - 1} \sum_{i=X+1}^{n} \sqrt{M^2 + \frac{k_i}{k_x}(1 - M^2)} \right) \Bigg/ n \tag{5.99}$$

$$F_{wo} = \frac{\sum_{i=1}^{n} k_i}{\sum_{i=x+1}^{n} \dfrac{k_i}{\sqrt{M^2 + \dfrac{k_i}{k_x}(1 - M^2)}}} \tag{5.100}$$

where

$M$ = endpoint mobility ratio

$n$ = number of layers of equal thickness

$i$ = $i$th layer
$X$ = $x$th layer just flooded
$k_i$ = heighest permeability when $i = 1$
$k_n$ = lowest permeability when $i = n$
$k_x$ = peremability of layer just flooded

Permeability data are log-normally distributed if the graph of permeability versus cumulative frequency distribution on log-normal probability paper can be approximated by a straight line. The following steps are to be used for the construction of the log-normal permeability plot:

1. Arrange the permeability data in the descending order of magnitude.
2. Calculate the fraction of the samples with permeability greater than the particular sample to obtain the *cumulative percent greater than* (cumulative frequency distribution).
3. Plot the data on log-probability graph paper.
4. Draw a best-fit straight line through the plotted data with more weight placed on the central data points.



The midpoint of the permeability distribution (50th percentile) is the log mean or geometric mean permeability. Dykstra and Parsons recognized that a straight line drawn through the data is also a measure of the dispersion or the heterogeneity of the reservoir rock. They introduced a term called *permeability variation* calculated as follows:

$$V = \frac{k_{50} - k_{84.1}}{k_{50}} \tag{5.101}$$

The term V is also called the Dykstra-Parsons *coefficient of permeability variation*. A homogeneous reservoir will have a permeability variation approaching close to zero while an extremely heterogeneous reservoir would have a permeability variation approaching 1.0. A log-normal permeability plot for the data in Table 5–9 is illustrated in Figure 5–22.

Johnson presented a simplified graphical treatment of the Dykstra and Parsons method.[24] The technique is semiempirical and is based upon the correlation of V, $S_{iw}$, M, and the fractional recovery ($E_R$) of oil in place at a specified $F_{wo}$. The correlations are illustrated in Figures 5–23 through 5–26.

Also volumetric sweep efficiency and volume percent ultimate oil recovery are given by

$$\%E_v = 100\left(\frac{1 - V^2}{M}\right) \tag{5.102}$$

$$\%RF = \frac{\dfrac{\%E_v(100 - \%S_{w_{avg}})}{B_o} + \dfrac{(100 - \%E_v)\%S_{io}}{B_o}}{\dfrac{\%S_{io}}{B_{oi}}} \tag{5.103}$$

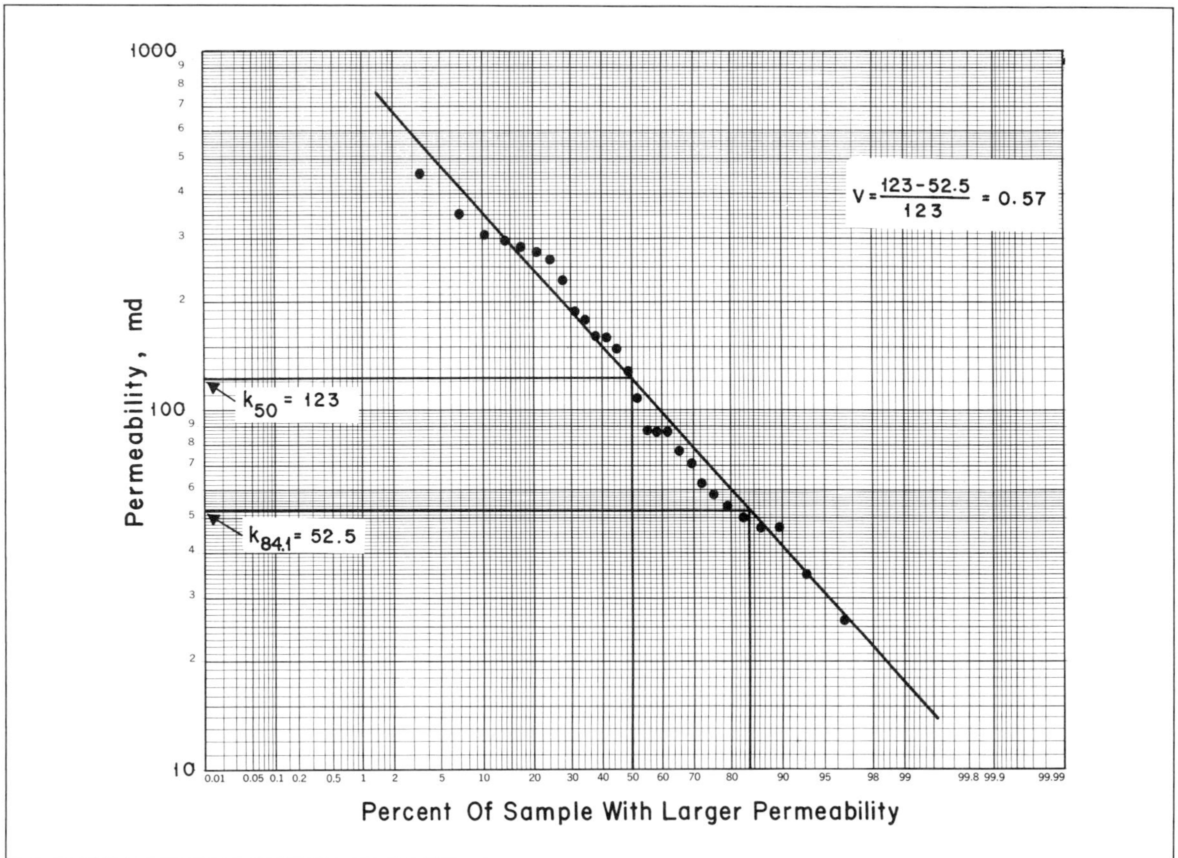


**Figure 5–22.** Log-normal permeability plot of the data in Table 5–9.

## EFFECT OF LAYER SELECTION ON VOLUMETRIC SWEEP CALCULATIONS

In preparing to use a waterflood prediction method the vertical variation in permeabilities has to be first defined. The result is a description of the properties of a number of layers that constitute the reservoir.

Selecting a large number of layers rather than a much smaller will increase performance prediction time without significantly affecting the magnitude of the predicted performance. On the other hand, selecting a smaller number of layers can sometimes result in predicted performance approaching that of a uniform system.

Craig studied the effect of the number of layers on oil performance prediction for a five-spot waterflood at varying permeability variation and mobility ratio.[25] The performance calculations were based on a reservoir composed of one to 100 layers, with permeability variations ranging 0.4–0.8. He found that the calculated recovery at water breakthrough was dependent upon the number of layers selected; however, the water and oil production performance at WORs above about three varied

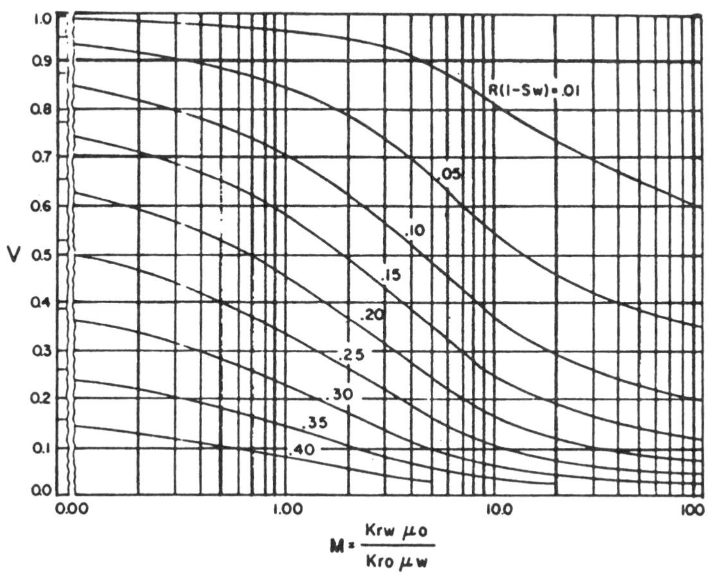


**Figure 5–23.** Permeability variation versus mobility ratio, showing lines of constant $E_R(1 - S_{iw})$ for a $F_{wo} = 1.0$ (© SPE, Trans. AIME, 1956).[24]

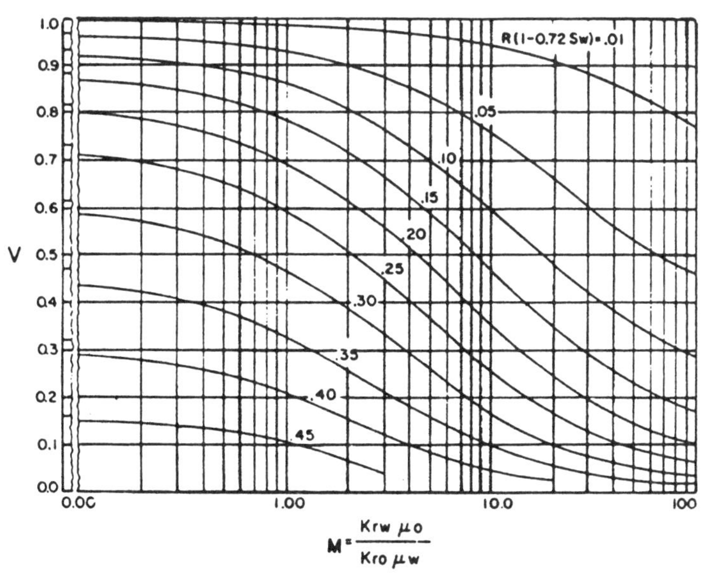


**Figure 5–24.** Permeability variation versus mobility ratio, showing lines of constant $E_R(1 - 0.72S_{iw})$ for a $F_{wo} = 5.0$ (© SPE, Trans. AIME, 1956).[24]

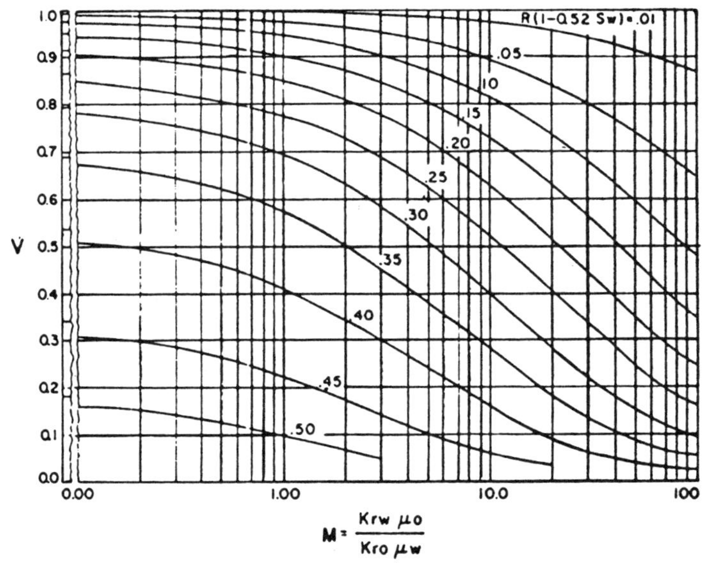


**Figure 5–25.** Permeability variation versus mobility ratio, showing lines of constant $E_R(1 - 0.52S_{iw})$ for a $F_{wo} = 25.0$ (© SPE, Trans. AIME, 1956).[24]

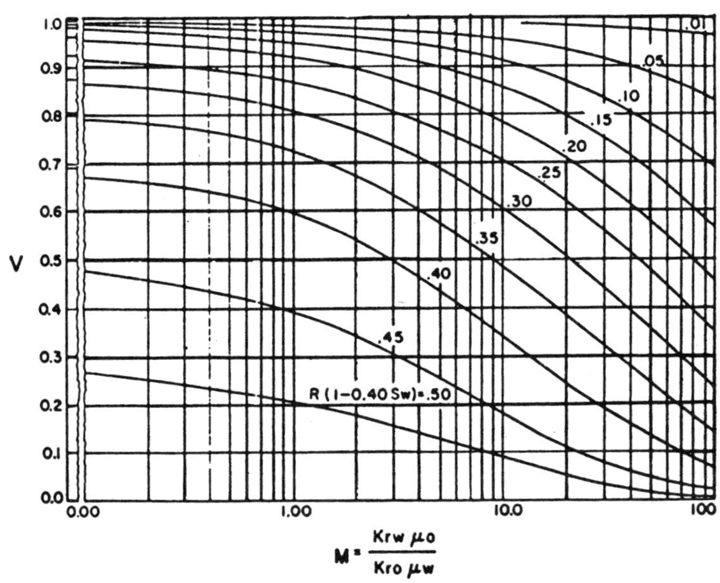


**Figure 5–26.** Permeability variation versus mobility ratio, showing lines of constant $E_R(1 - 0.4S_{iw})$ for a $F_{wo} = 100$ (© SPE, Trans. AIME, 1956).[24]

less widely. It was also shown that in some ranges of mobility ratio, the calculated after-breakthrough

The minimum number of equal-thickness layers that will yield predicted waterflood performance equivalent to that for a 100-layer reservoir at WORs above 2.5, 5.0, and 10.0 are given in Table 5–11.

**TABLE 5–11. Minimum number of equal-thickness layers required to obtain performance of a 100-layer five-spot waterflood[25]**

| Mobility Ratio | Permeability Variation 0.1 | 0.2 | 0.3 | 0.4 | 0.5 | 0.6 | 0.7 | 0.8 |
|---|---|---|---|---|---|---|---|---|
| At Producing WORs Above 2.5. | | | | | | | | |
| 0.05 | 1 | 1 | 2 | 4 | 10 | 20 | 20 | 20 |
| 0.10 | 1 | 1 | 2 | 4 | 10 | 20 | 100 | 100 |
| 0.20 | 1 | 1 | 2 | 4 | 10 | 20 | 100 | 100 |
| 0.50 | 1 | 2 | 2 | 4 | 10 | 20 | 100 | 100 |
| 1.00 | 1 | 3 | 3 | 4 | 10 | 20 | 100 | 100 |
| 2.00 | 2 | 4 | 4 | 10 | 20 | 50 | 100 | 100 |
| 5.00 | 2 | 5 | 10 | 20 | 50 | 100 | 100 | 100 |
| At Producing WORs Above 5.0. | | | | | | | | |
| 0.05 | 1 | 1 | 2 | 4 | 5 | 10 | 10 | 20 |
| 0.10 | 1 | 1 | 2 | 4 | 10 | 10 | 10 | 100 |
| 0.20 | 1 | 1 | 2 | 4 | 10 | 10 | 20 | 100 |
| 0.50 | 1 | 2 | 2 | 4 | 10 | 10 | 20 | 100 |
| 1.00 | 1 | 2 | 3 | 4 | 10 | 10 | 20 | 100 |
| 2.00 | 2 | 3 | 4 | 5 | 10 | 10 | 50 | 100 |
| 5.00 | 2 | 4 | 5 | 10 | 20 | 100 | 100 | 100 |
| At Producing WORs Above 10. | | | | | | | | |
| 0.05 | 1 | 1 | 1 | 2 | 4 | 5 | 10 | 20 |
| 0.10 | 1 | 1 | 1 | 2 | 5 | 5 | 10 | 20 |
| 0.20 | 1 | 1 | 2 | 3 | 5 | 5 | 10 | 20 |
| 0.50 | 1 | 1 | 2 | 3 | 5 | 5 | 10 | 20 |
| 1.00 | 1 | 1 | 2 | 3 | 5 | 10 | 10 | 50 |
| 2.00 | 1 | 2 | 3 | 4 | 10 | 10 | 20 | 100 |
| 5.0 | 1 | 3 | 4 | 5 | 10 | 100 | 100 | 100 |

## FORECASTING MATURE WATERFLOODS

The techniques presented in the previous sections and most techniques related to forecasting waterfloods deals with predicting the waterflood performance early in the life of a flood. In this section a technique is presented which can be used to forecast mature waterfloods with stabilized operations.

The technique involves plotting $RF_{oil}$ versus variable $X = \frac{1}{f_w} - \ln\left(\frac{1}{f_w} - 1\right)$ on a Cartesian coordinate graph paper. Linear regression analysis is performed on the plotted data to determine the correlation variables. These variables are then used in extrapolating any future value of water cut or cumulative production, and in generating a simple rate-time forecast. The technique was first introduced by Ershagi and Omoregie.[26] The following equations are utilized:

$$X = \frac{1}{f_w} - \ln\left(\frac{1}{f_w} - 1\right) \tag{5.104}$$

$$F_w = \frac{f_w}{1 - f_w} \tag{5.105}$$

$$\%RF_{oil} = 100\left(\frac{N_p}{N}\right) = 100(AX + B) \tag{5.106}$$

where A and B are determined by regression as

$$B = \frac{\Sigma X^2 \Sigma RF_{oil} - \Sigma X \Sigma X\,(RF_{oil})}{n\Sigma X^2 - (\Sigma X)^2}$$

$$A = \frac{\Sigma RF_{oil} - nB}{\Sigma X}$$

where $n$ is the number of $f_w$, $\%RF_{oil}$ points used in the regression.

$$q_{oi} = (1 - f_{wi})q_o + q_w \tag{5.107}$$

$$(q_o)_{avg} = \frac{(q_o)_{i-1} + q_{oi}}{2} \tag{5.108}$$

$$\Delta t_i = \frac{\Delta N_{pi}}{365(q_o)_{avgi}} \tag{5.109}$$

$$t_i = t_{i-1} + \Delta t_i \tag{5.110}$$

The $k_o/k_w$ curve can also be obtained using the coefficients A and B as

$$b = \frac{1}{A(S_{iw} - 1)} \tag{5.111}$$

and

$$a = \frac{\mu_o}{\mu_w} e^{-b[B(1 - S_{iw}) + S_{iw}]} \tag{5.112}$$

The $k_o/k_w$ is then calculated from $k_o/k_w = ae^{bS_w}$ as previously discussed.

For any given $N_p$, $f_w$ can be calculated iteratively using Newton Raphson methods as follows:

$$(f_w)_{i+1} = f_{wi} + \frac{(X_i - X)f_{wi}^2(1 - f_{wi})}{1 - 2f_{wi}^2} \tag{5.113}$$

where X is evaluated at the required $N_p$ as the initial guess.

## WATERFLOOD DESIGN EQUATIONS

The design of a waterflood involves both technical and economic evaluations. The economic evaluation is based on the estimated predicted waterflood performance. This section presents equations for designing waterfloods and the associated performance to be used in economic analysis. 

In a five-spot pattern, it is assumed that the flow is steady, incompressible, and radial from the injection well to $r_{ei}$ (the outer radius of the injection segment); then fluid flows radially from the outer radius of the production segment, $r_{ep}$, to the production well.[7]

The location of the flood-front saturation, $r_f$, is given by

$$r_f = \sqrt{\frac{W_i}{\pi\phi h(\bar{S}_{wf} - S_{iw})}} \tag{5.114}$$

where

$$W_i = (\pi r_f^2 \phi h)(\bar{S}_{wf} - S_{iw})$$

For $r_f < r_e$, when the displacement is piston-like, $\bar{S}_{wf} = 1 - S_{or}$, then

$$r_f = \sqrt{\frac{W_i}{\pi\phi h(S_{oi} - S_{or})}} \tag{5.115}$$

when $S_{iw} = (1 - S_{oi})$.

The injection rate when $M \neq 1.0$ is given by

$$i = \frac{2\pi\lambda_o h(p_w - p_p)}{\frac{1}{M}\ln\frac{r_f}{r_w} + \ln\frac{r_e}{r_f} + \ln\frac{r_e}{r_{wp}}} \tag{5.116}$$

where $r_{wp}$ is the radius of the production well (ft), $r_e$ is the effective radius (ft), and $\lambda_o$ is the mobility of the oil phase (darcy/cp).

In expressing the injection rate in terms of a conductance in order to compare the effects of the mobility ratio on the injection ratio, the following equation results:

$$\gamma = \frac{i}{i_b} = \frac{2\left[\ln\left(\frac{d}{r_w}\right) - 0.572\right]}{\frac{1}{M}\ln\left(\frac{d\sqrt{2E_A/\pi}}{r_w}\right) + \ln\left(\frac{d}{r_w\sqrt{2E_A/\pi}}\right)} \tag{5.117}$$

when $r_f >> r_w$, and

$$E_A = \frac{\pi(r_f^2 - r_w^2)}{2d^2}$$

and

$$r_f = d\sqrt{\frac{2E_A}{\pi}}$$

For $E_A \leq 0.5$, the conductance an be expessed in terms of $W_i$ as

$$W_i = (\pi r_f^2 \phi h)(\bar{S}_{wf} - S_{iw}) \tag{5.118}$$

Waterfloods are frequently initiated after some of the oil has been produced by solution gas drive (i.e., a mobile gas saturation is present when water injection begins). For a five-spot model, injection rate during fillup to the interference point when the flow resistance is neglected ahead of the oil bank is given by

$$i = \frac{2\pi\lambda_o(p_{wi} - p_{wp})}{\frac{1}{M}\ln\left(\frac{r_f}{r_w}\right) + \ln\left(\frac{r_{ob}}{r_f}\right)} \tag{5.119}$$

where $r_{ob}$ is the radius of the oil bank (ft), $r_w \leq r_{ob} \leq d\sqrt{2}$. Both $r_f$ and $r_{ob}$ may be defined by material balance on the injected water.

$$W_i = \pi(r_f^2 - r^2{}_w)(\bar{S}_w - S_{iw})\phi h \tag{5.120}$$

Rearranging for $r_f$,

$$r_f = \sqrt{\frac{W_i}{\pi\phi h(\bar{S}_w - S_{iw})} + r_w^2} \tag{5.121}$$

At interference, $r_{ob} = d/\sqrt{2}$; therefore the volume of water required to reach interference is given by

$$r_{ob} = \sqrt{\frac{W_i}{\pi\phi h(S_{gi} - S_{gt})} + r_w^2} \tag{5.122}$$

where $S_{gt}$ is the trapped gas saturation.

If the injection rate is constant, the interference time can be computed from $W_i$. Neglecting the wellbore radius, Equation (5.122) reduces to

$$W_{ii} = \frac{\pi d^2}{2}\phi h(S_{gi} - S_{gt}) \tag{5.123}$$



The volume of water injected at fillup is given by

$$W_{if} = V_p S_{gi} \tag{5.124}$$

When the trapped gas is redissolved in the oil with negligible change in volume, $r_f$ at fillup is obtained from

$$r_f = \sqrt{\frac{2d^2 S_{gi}}{\pi(\bar{S}_w - S_{iw})} + r_w^2} \tag{5.125}$$

The following set of equations can be used in computer programs to solve the fractional flow curve:

$$f(S_{wbt})_i = \frac{b(f_w^2 - f_w)}{(S_{wbt} - S_{iw})} - \frac{100 f_w}{(S_{wbt} - S_w)^2} \tag{5.126}$$

$$f'(S_{wbt})_i = \frac{-b(f_w^2 - f_w)(1 - S_{iw})}{(S_{wbt} - S_{iw})^2} + \frac{200 f_w(S_{wbt} - S_{iw})}{(S_{wbt} - S_{iw})^4} \tag{5.127}$$

where

$$(S_{wbt})_{i+1} = (S_{wbt})_i + \frac{f(S_{wbt})_i}{f'(S_{wbt})_i}$$

The $S_{wbt}$ is solved for iteratively using Newton Raphson Technique, Equation (5.126), and (5.127).

$$\%\bar{S}_{wbt} = \frac{\%S_{wbt} - \%S_{iw}}{f_{wbt}} + \%S_{iw} \tag{5.128}$$

$$f_{w2} = \frac{1}{1 + \dfrac{\mu_w}{\mu_o} a e^{bS_{w2}}} \tag{5.129}$$

OR

$$f_{w2} = \frac{1 + \sqrt{1 + \frac{4}{b}(\partial f_w/\partial S_w)}}{2} \tag{5.129a}$$

where $S_{w2}$ is in fraction

$$\%\bar{S}_{w2} = \%S_{w2} + 100Q_i(1 - f_{w2}) \tag{5.130}$$

where

$$\%S_{w2} = \frac{100}{b}\ln\left[\left(\frac{1}{f_{w2}} - 1\right)\frac{\mu_w}{\mu_o}\frac{1}{a}\right]$$



where

$$Q_i = \frac{1}{\left(\frac{\partial f_w}{\partial S_w}\right)_{S_{w2}}} = \frac{1}{b(f_w^2 - f_w)} \tag{5.131}$$

## WATERFLOOD CALCULATIONS—COMPREHENSIVE EXAMPLE

This comprehensive example is adopted from SPE Monograph 3 using the method proposed by Craig.[2] The sample reservoir presented here has been subject to some pressure depletion by solution gas drive, then to gas injection for a brief period of time. The rock and fluid properties of the sample reservoir are given below. The flooding pattern is five-spot with pattern area as 40 ac; wellbore radius is 1.0 ft.

| $S_w$ (fraction) | .10 | .30 | .40 | .45 | .50 | .55 | .60 | .65 | .70 |
|---|---|---|---|---|---|---|---|---|---|
| $k_{ro}$(fraction) | 1.0 | .373 | .210 | .148 | .100 | .061 | .033 | .012 | 0 |
| $k_{rw}$(fraction) | 0 | .070 | .169 | .226 | .300 | .376 | .476 | .600 | .740 |
| $f_w$(fraction) | 0 | .273 | .617 | .753 | .857 | .925 | .967 | .990 | 1.000 |

$A = 20$ ac
$h = 50$ ft
$k_{avg} = 10$ md
$\phi = 20\%$
$S_{iw} = 10\%$
$S_{gi} = 15\%$
$S_{or} = 25\%$
$V = 0.5$

$\mu_o = 1.0$ cp
$\mu_w = 0.50$ cp
$p_i = 1{,}000$ psi
$RF_p = 10.4\%$
$B_{oi} = 1.29$ RB/STB
$B_o = 1.20$ RB/STB
$(k_{ro})_{end} = 1.0$
$(k_{rw})_{S_{wbt}} = 0.40$

Solution: The waterflood displacement performance for the example reservoir is calculated as shown in Table 5–12. The $S_{w2}$ increment is 2.55% from $S_{wbt}$ to $(1 - S_{or})$. The fractional flow curve and a plot of $(\partial f_w/\partial S_w)$ versus $S_w$ are shown in Figures 5–27 and 5–28, respectively.

**Calculation of Mobility Ratio:** Using Equation (5–20a),

$$M_{\overline{S}} = \frac{0.4}{0.5} \times \frac{1.0}{1.0} = 0.80$$

**Calculation of Ultimate Waterflood Recovery:** To calculate the ultimate waterflood recovery to a producing water cut of, for instance, 98%, we find that this water cut corresponds to an average water saturation of 66.6% PV from Table 5–12.

The initial stock-tank oil in place in a barrel of total pore volume is

$$\frac{S_{oi}}{B_{oi}} = \frac{1 - \overline{S}_w}{B_{oi}} = \frac{1 - 0.010}{1.29} = 0.698 \quad \text{STB}$$

At $f_w = 98\%$, the stock-tank oil remaining in 1 bbl of total pore volume in the swept portion of the reservoir is

$$\frac{S_o}{B_o} = \frac{1 - \overline{S}_w}{B_o} = \frac{1 - 0.666}{1.2} = 0.278 \quad \text{STB}$$

The oil remaining in the swept portion per barrel of total pore volume is

**TABLE 5–12. Waterflood displacement performance-comprehensive example**

| $S_{w2}$ (frac.) | $f_{w2}$ (frac.) | $\left(\frac{\partial f_w}{\partial S_w}\right)$ | $Q_i = \frac{1}{\frac{\partial f_w}{\partial S_w}}$ (PV) | $\overline{S}_{w2} = S_{w2} + Q_i f_{o2}$ (frac.) |
|---|---|---|---|---|
| .469 = $S_{wbt}$ | .798 = $f_{wbt}$ | 2.160 | .463 = $Q_{ibt}$ | .563 = $\overline{S}_{wbt}$ |
| .495 | .848 | 1.751 | .572 | .582 |
| .520 | .888 | 1.410 | .711 | .600 |
| .546 | .920 | 1.130 | .887 | .617 |
| .572 | .946 | .851 | 1.176 | .636 |
| .597 | .965 | .649 | 1.540 | .652 |
| .622 | .980 | .477 | 2.100 | .666 |
| .649 | .990 | .317 | 3.157 | .681 |
| .674 | .996 | .195 | 5.130 | .694 |
| .700 | 1.000 | .102 | 9.800 | .700 |

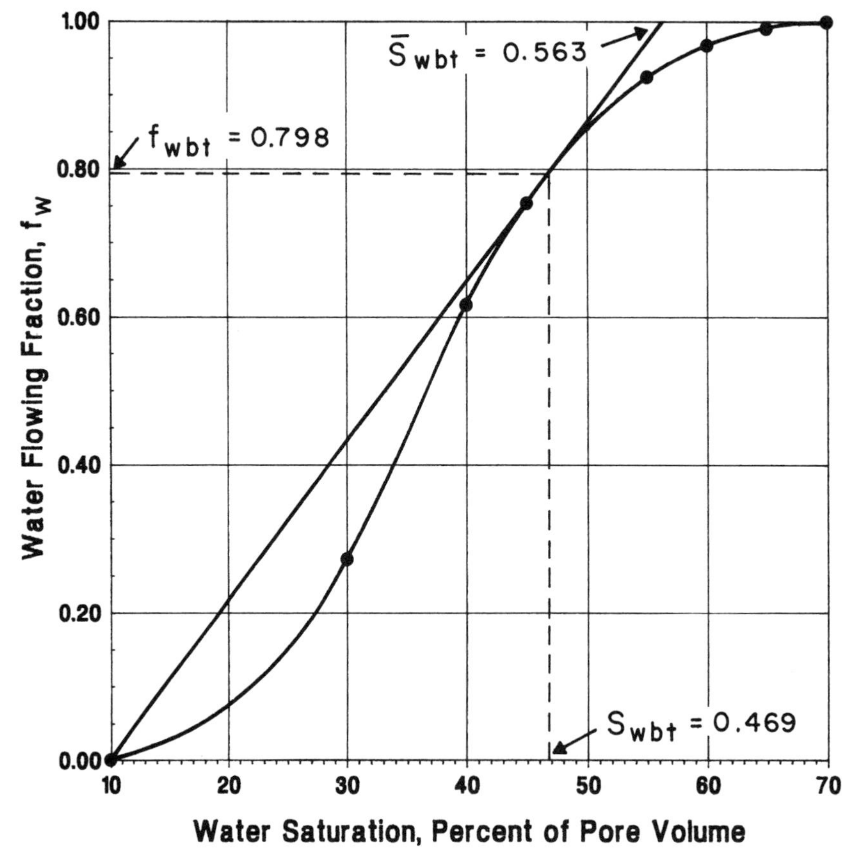


**Figure 5–27.** Fractional flow curve for the comprehensive example.

$$\frac{S_{oi}}{B_o} = \frac{1 - S_{iw}}{B_o} = \frac{1 - 0.10}{1.2} = 0.75 \quad \text{STB}$$

Approximating the volumetric sweep efficiency by the term $(1 - V^2)/M$ (where $V = 0.5$ is the permeability variation for the reservoir), we find this equals 0.9375. The total oil remaining is then given by

$$0.9735 \times 0.278 + (1 - 0.9375) \times 0.75 = 0.3075 \quad \text{STB}$$

Thus the total oil recovery is $(0.698 - 0.3075)/0.698 = 0.559$ of the original oil in place. Since the primary recovery was 10.4%; therefore the recovery due to waterflooding is 45.5%.

The reservoir is divided into 10 equal-thickness layers based on $V = 0.5$ and $M = 0.8$. The permeabilities read from *log-normal permeability* plot for each five-spot layer are as follows:

| Layer | 1 | 2 | 3 | 4 | 5 | 6 | 7 | 8 | 9 | 10 |
|---|---|---|---|---|---|---|---|---|---|---|
| Permeability, md | 31.5 | 20.5 | 16.0 | 13.1 | 10.9 | 8.2 | 7.7 | 6.3 | 4.9 | 3.2 |
| Thickness, ft | 5 | 5 | 5 | 5 | 5 | 5 | 5 | 5 | 5 | 5 |

**Initial Calculations:** Calculations are first performed for the first five-spot interval with permeability of 31.5 md.

1.
$$V_p = 7{,}758\phi Ah$$
$$= 7{,}758(0.20)(40)(5)$$
$$= 310{,}320 \quad \text{bbls}$$

2.
$$N_f = \frac{V_p \times S_{oi}}{B_o}$$
$$= \frac{310{,}320 \times 0.75}{1.2} = 193{,}950 \quad \text{bbl}$$



3. Areal sweep efficiency at breakthrough from Equation (5.60) and $M_{\bar{S}} = 0.8$.

$$E_{Abt} = 0.54602036 + \frac{0.03170817}{0.8} + \frac{0.30222997}{e^{0.8}} - 0.00509693(0.8)$$
$$= 0.717$$

4. Water injected at breakthrough is given by Equation (5.79).

$$W_{ibt} = 0.717(310{,}320)(0.563 - 0.10)$$
$$\approx 103{,}020 \quad \text{bbls}$$

Using Equation (5.79), the number of water-contacted PVs are

$$Q^*_{i\,bt} = \bar{S}_{wbt} - S_{iw}$$
$$= 0.563 - 0.10 = 0.463$$

5. The maximum value of initial gas saturation at which prediction method is accurate is calculated from the following equation:

$$S^*_{gi} = C(\bar{S}_{oi} - \bar{S}_{obt})$$

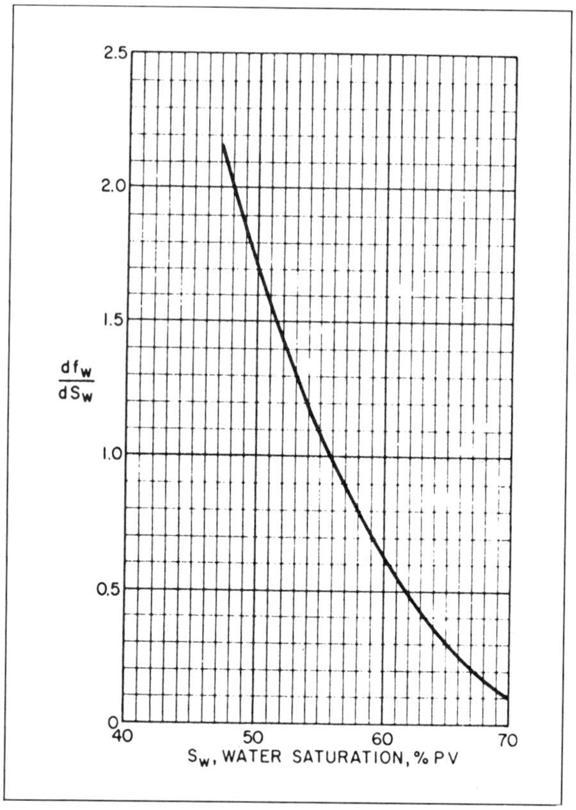


**Figure 5–28.** Plot of $(\partial f_w/\partial S_w)$ versus $S_w$ for the comprehensive example.

where

$$C = 0.62 + \frac{0.46836}{M}$$

$$S^*_{gi} = \left(0.62 + \frac{0.46836}{0.8}\right)[0.75 - (1 - 0.563)]$$
$$= 0.3773 \quad \text{OR} \quad 37.73\%$$

Since $S_{gi} = 15\%$ is less than $S^*_{gi} = 37.73\%$, the prediction method is applicable.

6. Injected water at interference from Equation (5.123) is

$$W_{ii} = [\pi(5)(0.20)(0.15)(660)^2] \Big/ 5.615$$
$$= 36{,}558 \quad \text{bbls}$$



where $r_{ei} = 660$ ft is half the distance between adjacent injectors.

7. Using Equation (5.124), injected water at fillup is

$$W_{if} = 310{,}320 \times 0.15$$
$$= 46{,}548 \quad \text{bbls}$$

8. Areal sweep efficiency prior to water breakthrough is

$$E_A = \frac{W_i}{V_p(\bar{S}_{wbt} - S_{iw})}$$
$$= \frac{W_i}{310{,}320(0.563 - 0.10)} = \frac{W_i}{143{,}678}$$

9. Outer radius of oil bank prior to interference is given by Equation (5.123) as

$$r_e = \sqrt{\frac{5.615W_i}{\pi(5)(0.20)(0.15)}}$$
$$= \sqrt{11.915W_i}$$

10. Outer radius of waterflood front prior to interference

$$r = r_e\left(\frac{S_{gi}}{\bar{S}_{wbt} - S_{iw}}\right)^{0.50}$$
$$= r_e\left(\frac{0.15}{0.563 - 0.10}\right)^{0.50} = 0.5692r_e$$

11. The water injection rate prior to interference for the layer with 31.5 md permeability and $\Delta p = 3{,}000$ psi is

$$i_{wi} = 7.07 \times 10^{-3} kh\Delta p \left(\frac{\mu_w}{k_{rw}} \ln \frac{r}{r_w} + \frac{\mu_o}{k_{ro}} \ln \frac{r_e}{r}\right)^{-1}$$

$$= 7.07 \times 10^{-3} \times 5.0 \times 31.5 \times 3{,}000 \left(\frac{\mu_w}{k_{rw}} \ln \frac{r}{r_w} + \frac{\mu_o}{k_{ro}} \ln \frac{r_e}{r}\right)^{-1}$$

$$= 3{,}340 \left(\frac{\mu_w}{k_{rw}} \ln \frac{r}{r_w} + \frac{\mu_o}{k_{ro}} \ln \frac{r_e}{r_w}\right)^{-1}$$

12. Using Equation (5.63), the base injection rate is

$$i_b = \frac{0.001538(31.5)(1.0)(5)(3{,}000)}{1.0\left(\log \dfrac{932}{1.0} - 0.2688\right)}$$

$$= 269.10 \quad \text{B/D}$$

13. The areal sweep efficiency at fillup is calculated at $W_{if} = 46{,}548$ (Step 7) and using Step 8.

$$E_A = \frac{W_i}{143{,}678} = \frac{46{,}548}{143{,}678} = 0.324 \quad \text{OR} \quad 32.4\%$$

14. Using Equation (5.69), the conductance ratio is

$$\gamma = 1.0924 - 0.0865(0.8) + 0.8311 \log(0.8)$$

$$\approx 0.943$$

15. The water injection rate at fillup is

$$i_{wf} = i_b \times \gamma = 269.1 \times 0.943$$

$$= 253.76 \quad \text{B/D}$$

16. The incremental time occurring from interference to fillup is given by

$$\Delta t = \frac{W_{if} - W_{ii}}{0.5(i_{wi} + i_{wf})}$$

$$= \frac{46{,}548 - 36{,}558}{0.5(418.9 + 253.76)}$$

$$= 29.7 \quad \text{days}$$

Thus the time to fillup is 80.18 + 29.7 = 109.9 days. The $i_{wi}$ is calculated in Step 11 for various $W_i$. Performance *prior to interference* is given in Table 5–13.

17. Performance from fillup to breakthrough is given in Table 5–14.
18. Performance after breakthrough is given in Table 5–15.

**Performance of Remaining Layers:** To compute the performance of other layers that may differ in porosity, thickness, and permeability, the following equations are used:

$$t_n = t_1 \frac{k_1}{k_n} \frac{\phi_n}{\phi_1}$$

where the subscript 1 refers to the first layer, $n$ refers to the $n$th layer, and $t_n$ is the time to inject the same amount of water in PVs in the $n$th layer, at which time the oil producing rate, $q_{on}$, from the $n$th layer is given by



$$q_{on} = q_{o1} \frac{k_n}{k_1} \frac{h_n}{h_1}$$

and the water injection rate

$$i_{wn} = i_{w1} \frac{k_n}{k_1} \frac{h_n}{h_1}$$

The composite performance is calculated by

$$q_{ot} = \sum_{i=1}^{n} q_{oi}$$

$$i_{wt} = \sum_{i=1}^{n} i_{wi}$$

**TABLE 5–13. Performance prior to interference—comprehensive example**

| (1) | (2) | (3) | (4) | (5) | (6) | (7) | (8) | (9) |
|---|---|---|---|---|---|---|---|---|
| $W_i$ (assume) | $r_e$ (Step 9) | $r$ (Step 10) | $\frac{\mu_w}{k_{rw}} \ln \frac{r_e}{r}$ | $\frac{\mu_o}{k_{ro}} \ln \frac{r_e}{r}$ | $i_w$ (Step 11) | $(i_w)_{avg}$ | $\Delta t = \frac{\Delta W_i}{(i_w)_{avg}}$ | $t = \Sigma(\Delta t)$ days |
| 500 | 77.2 | 43.9 | 4.7274 | .5645 | 631.3 | 631.3 | .79 | .79 |
| 5,000 | 244.1 | 138.9 | 6.1672 | .5645 | 496.2 | 563.8 | 7.98 | 8.77 |
| 10,000 | 345.2 | 196.5 | 6.6008 | .5645 | 466.2 | 481.2 | 10.39 | 19.16 |
| 15,000 | 422.8 | 240.7 | 6.8544 | .5645 | 450.3 | 458.3 | 10.91 | 30.07 |
| 20,000 | 488.2 | 277.9 | 7.0341 | .5645 | 439.6 | 445.0 | 11.24 | 41.31 |
| 25,000 | 545.8 | 310.7 | 7.1735 | .5645 | 431.7 | 435.7 | 11.48 | 52.79 |
| 30,000 | 597.9 | 340.3 | 7.2873 | .5645 | 425.5 | 428.6 | 11.67 | 64.61 |
| 35,000 | 645.8 | 367.6 | 7.3837 | .5645 | 420.3 | 422.9 | 11.82 | 76.43 |
| 36,558 | 660.0 | 375.7 | 7.4110 | .5645 | 418.9 | 419.6 | 3.75 | 80.18 |

TABLE 5–14. Performance from fillup to breakthrough—comprehensive example

| (1) | (2) | (3) | (4) | (5) | (6) | (7) | (8) | (9) |
|---|---|---|---|---|---|---|---|---|
| $W_i$ (assume) | $E_A$ (Step 8) | $\gamma$ Eq. (5.69) | $i_w =$ $i_b \times (3)$ | $(i_w)_{avg}$ | $\Delta t = \dfrac{\Delta W_i}{(i_w)_{avg}}$ | $t = \Sigma\Delta t$ days | $W_i - W_{if}$ $(1) - W_{if}$ | $N_p = (8) \div B_o$ bbl |
| 46,548 | .324 | .943 | 253.8 | — | — | 109.90 | 0 | 0 |
| 50,000 | .348 | .940 | 253.0 | 253.4 | 13.62 | 123.52 | 3,452 | 2,877 |
| 60,000 | .418 | .931 | 250.5 | 251.8 | 39.71 | 163.23 | 14,452 | 11,210 |
| 70,000 | .487 | .923 | 248.4 | 249.5 | 40.08 | 203.31 | 23,452 | 19,543 |
| 80,000 | .557 | .915 | 246.2 | 247.3 | 40.44 | 243.75 | 33,452 | 27,877 |
| 90,000 | .626 | .906 | 243.8 | 245.0 | 40.82 | 284.57 | 43,452 | 36,210 |
| 100,000 | .696 | .898 | 241.7 | 242.8 | 41.19 | 325.76 | 53,452 | 44,543 |
| 103,020 | .717 | .895 | 240.8 | 241.3 | 12.52 | 338.28 | 56,472 | 47,060 |

TABLE 5–15. Performance after breakthrough—comprehensive example

| (1) | (2) | (3) | (4) | (5) | (6) | (7) | (8) |
|---|---|---|---|---|---|---|---|
| $W_i$ | $\frac{W_i}{W_{ibt}}$ | $E_A$ Eq. (5.58) | $\frac{Q_i^*}{Q_{ibt}^*}$ Table 5–7 | $Q_i$ (4) × $Q_{ibt}$ | $\frac{\partial f_w}{\partial S_w} = 1/Q_i$ | $S_{w2}$ Fig. 5–28 | $f_{o2} = 1 - f_{w2}$ Fig. 5–27 |
| 103,020 | 1.0 | .717 | 1.000 | .463 | 2.160 | .470 | .200 |
| 123,620 | 1.2 | .767 | 1.193 | .552 | 1.812 | .492 | .157 |
| 144,230 | 1.4 | .809 | 1.375 | .636 | 1.572 | .507 | .130 |
| 164,830 | 1.6 | .846 | 1.548 | .717 | 1.395 | .524 | .107 |
| 185,440 | 1.8 | .879 | 1.715 | .794 | 1.259 | .534 | .095 |
| 206,040 | 2.0 | .906 | 1.875 | .869 | 1.151 | .543 | .080 |
| 257,550 | 2.5 | .969 | 2.256 | 1.046 | 0.956 | .562 | .063 |
| 309,060 | 3.0 | 1.000 | 2.619 | 1.214 | 0.824 | .575 | .051 |
| 412,080 | 4.0 | 1.000 | 3.336 | 1.545 | 0.647 | .597 | .037 |
| 515,100 | 5.0 | 1.000 | 4.053 | 1.877 | 0.533 | .611 | .027 |
| 618,120 | 6.0 | 1.000 | 4.700 | 2.208 | 0.453 | .622 | .202 |

TABLE 5–15. Continued

| (9) | (10) | (11) | (12) | (13) | (14) | (15) | (16) |
|---|---|---|---|---|---|---|---|
| $\bar{S}_{w2}$ Eq. (5.130) | $\lambda = \frac{E_A}{\Delta W_i/W_{ibt}}$ | $\Delta N_{pu}$ Eq. (5.85) | $1 - \Delta N_{pu}$ 1.0 − (11) | $\Delta N_{ps}$ (8) × (12) | $\Delta N_{pu} + \Delta N_{ps}$ (11) + (13) | $WOR_p$ [1 − (14)]/(14) | WOR $B_o$ × (15) |
| .563 | .2749 | .3056 | .6944 | .1388 | .4444 | 1.25 | 1.50 |
| .579 | .2291 | .2545 | .7455 | .1170 | .3715 | 1.69 | 2.03 |
| .590 | .1963 | .2182 | .7818 | .1016 | .3198 | 2.13 | 2.55 |
| .601 | .1718 | .1910 | .8090 | .0866 | .2776 | 2.60 | 3.12 |
| .610 | .1527 | .1697 | .8303 | .0789 | .2486 | 3.02 | 3.63 |
| .613 | .1375 | .1528 | .8472 | .0678 | .2206 | 3.53 | 4.24 |
| .628 | .1100 | .1223 | .8777 | .0552 | .1775 | 4.63 | 5.56 |
| .637 | 0 | 0 | 1.0000 | .0510 | .0510 | 18.61 | 22.30 |
| .653 | 0 | 0 | 1.0000 | .0370 | .0370 | 26.03 | 31.20 |
| .660 | 0 | 0 | 1.0000 | .0270 | .0270 | 36.00 | 43.20 |
| .664 | 0 | 0 | 1.0000 | .0200 | .0200 | 49.00 | 58.80 |

**TABLE 5–15. Continued**

| (17) | (18) | (19) | (20) | (21) | (22) | (23) | (24) | (25) |
|---|---|---|---|---|---|---|---|---|
| $\frac{\bar{S}_{w_2} - S_{iw}}{(9) - S_{iw}}$ | $\frac{[(3) \times (17)] - S_{gi}}{PV}$ | $\frac{(18) \div S_{oi}}{(frac.)}$ | $\frac{(19) \times STOIP}{bbl}$ | $\frac{k_{rw} @}{S_w}$ | $\frac{\lambda}{Eq. (5.6)}$ | $\Delta t = \frac{\Delta W_i}{(i_w)_{avg}}$ | $\frac{t = \Sigma \Delta t}{days}$ | $q_o = (14) \times \frac{i_w}{B_o}$ |
| .463 | .1817 | .2423 | 47,000 | .400 | .91 | — | 334.4 | 90.7 |
| .479 | .2176 | .2901 | 56,260 | .430 | .94 | 82.7 | 417.1 | 78.6 |
| .490 | .2467 | .3289 | 63,790 | .450 | .96 | 80.6 | 497.7 | 69.6 |
| .501 | .2731 | .3641 | 70,620 | .480 | .98 | 79.0 | 576.5 | 61.5 |
| .510 | .2973 | .3964 | 74,940 | .500 | 1.00 | 77.3 | 563.8 | 56.5 |
| .513 | .3143 | .4191 | 81,280 | .510 | 1.02 | 75.9 | 729.7 | 51.5 |
| .528 | .3610 | .4813 | 93,350 | .542 | 1.08 | 183.0 | 912.7 | 43.1 |
| .536 | .3869 | .5159 | 100,050 | .560 | 1.12 | 349.0 | 1,261.7 | 12.8 |
| .553 | .4034 | .5379 | 104,330 | .600 | 1.20 | 330.0 | 1,591.5 | 9.9 |
| .560 | .4101 | .5468 | 106,050 | .625 | 1.25 | 312.0 | 1,903.5 | 7.6 |
| .564 | .4139 | .5519 | 107,040 | .635 | 1.27 | 305.0 | 2,208.5 | 5.7 |

$\Delta N_{ps}$ = incremental oil produced from previously swept region.

## REFERENCES


[1] F. W. Cole, *Reservoir Engineering Manual*, Gulf Publishing Co., Houston, Texas, 1969.

[2] F. F. Craig, Jr., *The Reservoir Engineering Aspects of Waterflooding*, SPE Monograph, Volume 3, Society of Petroleum Engineers of AIME, Richardson, Texas, 1971.

[3] B. C. Craft and M. F. Hawkins, *Applied Petroleum Reservoir Engineering*, Prentice-Hall, Inc., Englewood Cliffs, New Jersy, 1959.

[4] M. C. Leverett, "Capillary Behavior in Porous Solids," *Trans.* AIME, Vol. 142, 1941, pp. 152–169.

[5] S. E. Buckley and M. C. Leverett, "Mechanism of Fluid Displacement in Sands," *Trans.*, AIME, Vol. 146, 1942, p. 107.

[6] S. J. Pirson, *Oil Reservoir Engineering*, 2nd Edition, McGraw-Hill Book Co., New York, 1958.

[7] G. P. Willhite, *Waterflooding*, SPE Textbook Series, Vol. 3, Society of Petroleum Engineers of AIME, Richardson, Texas, 1986.

[8] H. J. Welge, "A Simplified Method for Computing Oil Recoveries by Gas or Water Drive," *Trans.*, AIME, Vol. 195, 1952, p. 91.

[9] H. C. Slider, *Practical Petroleum Reservoir Engineering Methods*, PennWell Publishing Co., Tulsa, Oklahoma, 1976.

[10] L. P. Dake, *Fundamentals of Reservoir Engineering*, Elsevier Scientific Publishing Company, Amsterdam, The Netherlands, 1978.

[11] J. Hagoort, "Displacement Stability of Water Drives in Water Wet Connate Water Bearing Reservoirs," *Soc. of Pet. Engs. Journal, Trans.*, AIME, 1974, pp. 63–74.

[12] D. N. Dietz, "A Technical Approach to the Problem of Encroaching and Bypassing Edge Water," *Proc.*, Akad. Wetenschappen, Amsterdam, V56-B, 1953.

[13] C. R. Smith, *Mechanics of Secondary Oil Recovery*, Robert E. Krieger Publishing Co., Inc., Huntington, New York, 1975.

[14] A. B. Dyes, B. H. Caudle, and R. A. Erickson, "Oil Production after Breakthrough—As Influenced by Mobility Ratio," *Trans.*, AIME, Vol. 201, 1954, pp. 81–86.

[15] F. F. Craig, T. M. Geffen, and R. A. Morse, "Oil Recovery Performance of Pattern Gas or Water Injection Operations from Model Tests," *Trans.*, AIME, 1955, p. 7.

[16] O. K. Kimbler, B. H. Caudle, and H. E. Cooper, Jr., "Areal Sweepout Behavior in a Nine-Spot Injection Pattern," *Trans.*, AIME, Vol. 231, 1964, pp. 199–202.

[17] N. Muskat, *Physical Principles of Oil Production*, McGraw Hill Book Co., New York, 1950.

[18] J. C. Deppe, "Injection Rate—The Effect of Mobility Ratio, Areal Sweep and Pattern," *Society of Petroleum Engineers Journal*, June 1951, pp. 81–91.

[19] M. Prats, C. S. Mathews, R. L. Jewett, and J. D. Baker, "Prediction of Injection Rate and Production History for Multifluid Five-Spot Floods," *Trans.*, AIME, Vol. 216, 1959, pp. 98–105.

[20] B. H. Caudle and M. D. Witte, "Production Potential Changes During Sweep-Out in a Five-Spot System," *Journal of Petroleum Tech., Trans.*, AIME, Vol. 216, Dec. 1959, pp. 63–65.

[21] W. E. Stiles, "Use of Permeability Distribution in Waterflood Calculations," *Trans.*, AIME, Vol. 186, 1949.

[22] M. Muskat, "The Effect of Permeability Stratification in Complete Water-Drive System," *Trans.*, AIME, Vol. 189, 1950.

[23] H. Dykstra and R. L. Parsons, "The Prediction of Oil Recovery by Waterflood," *Secondary Recovery of Oil in the United States*, Second Edition, American Petroleum Institute, New York, 1950.

[24] C. E. Johnson, Jr., "Prediction of Oil Recovery by Waterflood—A Simplified Graphical Treatment of the Dykstra-Parsons Method," *Trans.*, AIME, Vol. 207, 1956, pp. 345–346.

[25] F. F. Craig, Jr., "Effect of Permeability Variation and Mobility Ratio on Five-Spot Oil Recovery Performance Calculations," *Journal of Petroleum Technology*, Oct. 1970, pp. 1239–1245.

[26] I. Ershagi and O. Omoregie, "A Method for Extrapolation of Cut Vs. Recovery Curves," *Journal of Petroleum Technology*, Feb. 1978, pp. 203–204.

APPENDIX A

# DISCRETE COMPOUNDING INTEREST FACTORS, i = 1% TO i = 95%



Single Payment Compound Amount Factor:

$$F/P_{i,n} = (1 + i)^n \tag{A.1}$$

Single Payment Present Worth Factor:

$$P/F_{i,n} = \frac{1}{(1 + i)^n} \tag{A.2}$$

Uniform Series Compound Amount Factor:

$$F/A_{i,n} = \frac{(1 + i)^n - 1}{i} \tag{A.3}$$

Sinking Fund Factor:

$$A/F_{i,n} = \frac{i}{(1 + i)^n - 1} \tag{A.4}$$

Capital Recovery Factor:

$$A/P_{i,n} = \frac{i(1 + i)^n}{(1 + i)^n - 1} \tag{A.5}$$

Uniform Series Present Worth Factor:

$$P/A_{i,n} = \frac{(1 + i)^n - 1}{i(1 + i)^n} \tag{A.6}$$

Arithmetic Gradient Series Factor:

$$A/G_{i,n} = \frac{1}{i} - \frac{n}{i}\,(A/F_{i-n}) \tag{A.7}$$

where

$n$ = Number of compounding/discounting periods
$i$ = Compound/Discount interest factor



i = 1%

| n | F/P | P/F | F/A | A/F | A/P | P/A | A/G |
|---|---|---|---|---|---|---|---|
| 1 | 1.0100 | 0.9901 | 1.0000 | 1.00000 | 1.01000 | 0.9901 | −0.0001 |
| 2 | 1.0201 | 0.9803 | 2.0100 | 0.49751 | 0.50751 | 1.9704 | 0.4975 |
| 3 | 1.0303 | 0.9706 | 3.0301 | 0.33002 | 0.34002 | 2.9410 | 0.9933 |
| 4 | 1.0406 | 0.9610 | 4.0604 | 0.24628 | 0.25628 | 3.9020 | 1.4875 |
| 5 | 1.0510 | 0.9515 | 5.1010 | 0.19604 | 0.20604 | 4.8534 | 1.9800 |
| 6 | 1.0615 | 0.9420 | 6.1520 | 0.16255 | 0.17255 | 5.7955 | 2.4709 |
| 7 | 1.0721 | 0.9327 | 7.2135 | 0.13863 | 0.14863 | 6.7282 | 2.9602 |
| 8 | 1.0829 | 0.9235 | 8.2857 | 0.12069 | 0.13069 | 7.6517 | 3.4477 |
| 9 | 1.0937 | 0.9143 | 9.3685 | 0.10674 | 0.11674 | 8.5660 | 3.9336 |
| 10 | 1.1046 | 0.9053 | 10.4622 | 0.09558 | 0.10558 | 9.4713 | 4.4179 |
| 11 | 1.1157 | 0.8963 | 11.5668 | 0.08645 | 0.09645 | 10.3676 | 4.9005 |
| 12 | 1.1268 | 0.8874 | 12.6825 | 0.07885 | 0.08885 | 11.2551 | 5.3814 |
| 13 | 1.1381 | 0.8787 | 13.8093 | 0.07241 | 0.08241 | 12.1337 | 5.8607 |
| 14 | 1.1495 | 0.8700 | 14.9474 | 0.06690 | 0.07690 | 13.0037 | 6.3383 |
| 15 | 1.1610 | 0.8613 | 16.0969 | 0.06212 | 0.07212 | 13.8650 | 6.8143 |
| 16 | 1.1726 | 0.8528 | 17.2579 | 0.05794 | 0.06794 | 14.7179 | 7.2886 |
| 17 | 1.1843 | 0.8444 | 18.4304 | 0.05426 | 0.06426 | 15.5622 | 7.7613 |
| 18 | 1.1961 | 0.8360 | 19.6147 | 0.05098 | 0.06098 | 16.3983 | 8.2323 |
| 19 | 1.2081 | 0.8277 | 20.8109 | 0.04805 | 0.05805 | 17.2260 | 8.7017 |
| 20 | 1.2202 | 0.8195 | 22.0190 | 0.04542 | 0.05542 | 18.0455 | 9.1693 |
| 21 | 1.2324 | 0.8114 | 23.2392 | 0.43030 | 0.05303 | 18.8570 | 9.6354 |
| 22 | 1.2447 | 0.8034 | 24.4716 | 0.04086 | 0.05086 | 19.6604 | 10.0998 |
| 23 | 1.2572 | 0.7954 | 25.7163 | 0.03889 | 0.04889 | 20.4558 | 10.5625 |
| 24 | 1.2697 | 0.7876 | 26.9735 | 0.03707 | 0.04707 | 21.2434 | 11.0236 |
| 25 | 1.2824 | 0.7798 | 28.2432 | 0.03541 | 0.04541 | 22.0231 | 11.4830 |
| 26 | 1.2953 | 0.7720 | 29.5256 | 0.03387 | 0.04387 | 22.7952 | 11.9409 |
| 27 | 1.3082 | 0.7644 | 30.8209 | 0.03245 | 0.04245 | 23.5596 | 12.3970 |
| 28 | 1.3213 | 0.7568 | 32.1291 | 0.03112 | 0.04112 | 24.3164 | 12.8515 |
| 29 | 1.3345 | 0.7493 | 33.4504 | 0.02990 | 0.03990 | 25.0658 | 13.3044 |
| 30 | 1.3478 | 0.7419 | 34.7849 | 0.02875 | 0.03875 | 25.8077 | 13.7556 |
| 35 | 1.4166 | 0.7059 | 41.6603 | 0.02400 | 0.03400 | 29.4086 | 15.9871 |
| 40 | 1.4889 | 0.6717 | 48.8864 | 0.02046 | 0.03046 | 32.8347 | 18.1776 |
| 45 | 1.5648 | 0.6391 | 56.4811 | 0.01771 | 0.02771 | 36.0945 | 20.3273 |
| 50 | 1.6446 | 0.6080 | 64.4632 | 0.01551 | 0.02551 | 39.1961 | 22.4363 |
| 55 | 1.7285 | 0.5785 | 72.8524 | 0.01373 | 0.02373 | 42.1472 | 24.5049 |
| 60 | 1.8167 | 0.5504 | 81.6696 | 0.01224 | 0.02224 | 44.9550 | 26.5333 |
| 65 | 1.9094 | 0.5237 | 90.9366 | 0.01100 | 0.02100 | 47.6266 | 28.5216 |
| 70 | 2.0068 | 0.4983 | 100.6763 | 0.00993 | 0.01993 | 50.1685 | 30.4702 |
| 75 | 2.1091 | 0.4741 | 110.9128 | 0.00902 | 0.01902 | 52.5870 | 32.3793 |
| 80 | 2.2167 | 0.4511 | 121.6715 | 0.00822 | 0.01822 | 54.8882 | 34.2492 |

**i = 2%**

| n | F/P | P/F | F/A | A/F | A/P | P/A | A/G |
|---|---|---|---|---|---|---|---|
| 1 | 1.0200 | 0.9804 | 1.0000 | 1.00000 | 1.02000 | 0.9804 | −0.0000 |
| 2 | 1.0404 | 0.9612 | 2.0200 | 0.49505 | 0.51505 | 1.9416 | 0.4949 |
| 3 | 1.0612 | 0.9423 | 3.0604 | 0.32676 | 0.34676 | 2.8839 | 0.9867 |
| 4 | 1.0824 | 0.9238 | 4.1216 | 0.24262 | 0.26262 | 3.8077 | 1.4751 |
| 5 | 1.1041 | 0.9057 | 5.2040 | 0.19216 | 0.21216 | 4.7135 | 1.9603 |
| 6 | 1.1262 | 0.8880 | 6.3081 | 0.15853 | 0.17853 | 5.6014 | 2.4421 |
| 7 | 1.1487 | 0.8706 | 7.4343 | 0.13451 | 0.15451 | 6.4720 | 2.9207 |
| 8 | 1.1717 | 0.8535 | 8.5829 | 0.11651 | 0.13651 | 7.3255 | 3.3959 |
| 9 | 1.1951 | 0.8368 | 9.7546 | 0.10252 | 0.12252 | 8.1622 | 3.8679 |
| 10 | 1.2190 | 0.8203 | 10.9497 | 0.09133 | 0.11133 | 8.9826 | 4.3366 |
| 11 | 1.2434 | 0.8043 | 12.1687 | 0.08218 | 0.10218 | 9.7868 | 4.8020 |
| 12 | 1.2682 | 0.7885 | 13.4120 | 0.07456 | 0.09456 | 10.5753 | 5.2641 |
| 13 | 1.2936 | 0.7730 | 14.6803 | 0.06812 | 0.08812 | 11.3484 | 5.7229 |
| 14 | 1.3195 | 0.7579 | 15.9739 | 0.06260 | 0.08260 | 12.1062 | 6.1785 |
| 15 | 1.3459 | 0.7430 | 17.2934 | 0.05783 | 0.07783 | 12.8492 | 6.6307 |
| 16 | 1.3728 | 0.7284 | 18.6392 | 0.05365 | 0.07365 | 13.5777 | 7.0798 |
| 17 | 1.4002 | 0.7142 | 20.0120 | 0.04997 | 0.06997 | 14.2918 | 7.5255 |
| 18 | 1.4282 | 0.7002 | 21.4122 | 0.04670 | 0.06670 | 14.9920 | 7.9680 |
| 19 | 1.4568 | 0.6864 | 22.8405 | 0.04378 | 0.06378 | 15.6784 | 8.4072 |
| 20 | 1.4859 | 0.6730 | 24.2973 | 0.04116 | 0.06116 | 16.3514 | 8.8431 |
| 21 | 1.5157 | 0.6598 | 25.7832 | 0.03878 | 0.05878 | 17.0112 | 9.2759 |
| 22 | 1.5460 | 0.6468 | 27.2989 | 0.03663 | 0.05663 | 17.6580 | 9.7053 |
| 23 | 1.5769 | 0.6342 | 28.8449 | 0.03467 | 0.05467 | 18.2922 | 10.1316 |
| 24 | 1.6084 | 0.6217 | 30.4218 | 0.03287 | 0.05287 | 18.9139 | 10.5546 |
| 25 | 1.6406 | 0.6095 | 32.0302 | 0.03122 | 0.05122 | 19.5234 | 10.9743 |
| 26 | 1.6734 | 0.5976 | 33.6708 | 0.02970 | 0.04970 | 20.1210 | 11.3909 |
| 27 | 1.7069 | 0.5859 | 35.3442 | 0.02829 | 0.04829 | 20.7069 | 11.8042 |
| 28 | 1.7410 | 0.5744 | 37.0511 | 0.02699 | 0.04699 | 21.2812 | 12.2143 |
| 29 | 1.7758 | 0.5631 | 38.7921 | 0.02578 | 0.04578 | 21.8443 | 12.6213 |
| 30 | 1.8114 | 0.5521 | 40.5679 | 0.02465 | 0.04465 | 22.3964 | 13.0250 |
| 35 | 1.9999 | 0.5000 | 49.9943 | 0.02000 | 0.04000 | 24.9986 | 14.9960 |
| 40 | 2.2080 | 0.4529 | 60.4017 | 0.01656 | 0.03656 | 27.3554 | 16.8884 |
| 45 | 2.4378 | 0.4102 | 71.8924 | 0.01391 | 0.03391 | 29.4901 | 18.7032 |
| 50 | 2.6916 | 0.3715 | 84.5790 | 0.01182 | 0.03182 | 31.4236 | 20.4418 |
| 55 | 2.9717 | 0.3365 | 98.5861 | 0.01014 | 0.03014 | 33.1747 | 22.1056 |
| 60 | 3.2810 | 0.3048 | 114.0510 | 0.00877 | 0.02877 | 34.7608 | 23.6960 |
| 65 | 3.6225 | 0.2761 | 131.1255 | 0.00763 | 0.02763 | 36.1974 | 25.2146 |
| 70 | 3.9995 | 0.2500 | 149.9771 | 0.00667 | 0.02667 | 37.4986 | 26.6631 |
| 75 | 4.4158 | 0.2265 | 170.7909 | 0.00586 | 0.02586 | 38.6771 | 28.0433 |
| 80 | 4.8754 | 0.2051 | 193.7709 | 0.00516 | 0.02516 | 39.7445 | 29.3571 |

**i = 3%**

| n | F/P | P/F | F/A | A/F | A/P | P/A | A/G |
|---|---|---|---|---|---|---|---|
| 1 | 1.0300 | 0.9709 | 1.0000 | 1.00000 | 1.03000 | 0.9709 | −0.0000 |
| 2 | 1.0609 | 0.9426 | 2.0300 | 0.49261 | 0.52261 | 1.9135 | 0.4926 |
| 3 | 1.0927 | 0.9151 | 3.0909 | 0.32353 | 0.35353 | 2.8286 | 0.9803 |
| 4 | 1.1255 | 0.8885 | 4.1836 | 0.23903 | 0.26903 | 3.7171 | 1.4631 |
| 5 | 1.1593 | 0.8626 | 5.3091 | 0.18835 | 0.21835 | 4.5797 | 1.9409 |
| 6 | 1.1941 | 0.8375 | 6.4684 | 0.15460 | 0.18460 | 5.4172 | 2.4138 |
| 7 | 1.2299 | 0.8131 | 7.6625 | 0.13051 | 0.16051 | 6.2303 | 2.8818 |
| 8 | 1.2668 | 0.7894 | 8.8923 | 0.11246 | 0.14246 | 7.0197 | 3.3450 |
| 9 | 1.3048 | 0.7664 | 10.1591 | 0.09843 | 0.12843 | 7.7861 | 3.8032 |
| 10 | 1.3439 | 0.7441 | 11.4639 | 0.08723 | 0.11723 | 8.5302 | 4.2565 |
| 11 | 1.3842 | 0.7224 | 12.8078 | 0.07808 | 0.10808 | 9.2526 | 4.7049 |
| 12 | 1.4258 | 0.7014 | 14.1920 | 0.07046 | 0.10046 | 9.9540 | 5.1485 |
| 13 | 1.4685 | 0.6810 | 15.6178 | 0.06403 | 0.09403 | 10.6350 | 5.5872 |
| 14 | 1.5126 | 0.6611 | 17.0863 | 0.05853 | 0.08853 | 11.2961 | 6.0210 |
| 15 | 1.5580 | 0.6419 | 18.5989 | 0.05377 | 0.08377 | 11.9379 | 6.4500 |
| 16 | 1.6047 | 0.6232 | 20.1569 | 0.04961 | 0.07961 | 12.5611 | 6.8742 |
| 17 | 1.6528 | 0.6050 | 21.7616 | 0.04595 | 0.07595 | 13.1661 | 7.2936 |
| 18 | 1.7024 | 0.5874 | 23.4144 | 0.04271 | 0.07271 | 13.7535 | 7.7081 |
| 19 | 1.7535 | 0.5703 | 25.1169 | 0.03981 | 0.06981 | 14.3238 | 8.1179 |
| 20 | 1.8061 | 0.5537 | 26.8704 | 0.03722 | 0.06722 | 14.8775 | 8.5229 |
| 21 | 1.8603 | 0.5375 | 28.6765 | 0.03487 | 0.06487 | 15.4150 | 8.9231 |
| 22 | 1.9161 | 0.5219 | 30.5368 | 0.03275 | 0.06275 | 15.9369 | 9.3186 |
| 23 | 1.9736 | 0.5067 | 32.4529 | 0.03081 | 0.06081 | 16.4436 | 9.7093 |
| 24 | 2.0328 | 0.4919 | 34.4265 | 0.02905 | 0.05905 | 16.9355 | 10.0954 |
| 25 | 2.0938 | 0.4776 | 36.4593 | 0.02743 | 0.05743 | 17.4131 | 10.4768 |
| 26 | 2.1566 | 0.4637 | 38.5530 | 0.02594 | 0.05594 | 17.8768 | 10.8535 |
| 27 | 2.2213 | 0.4502 | 40.7096 | 0.02456 | 0.05456 | 18.3270 | 11.2255 |
| 28 | 2.2879 | 0.4371 | 42.9309 | 0.02329 | 0.05329 | 18.7641 | 11.5930 |
| 29 | 2.3566 | 0.4243 | 45.2188 | 0.02211 | 0.05211 | 19.1885 | 11.9558 |
| 30 | 2.4273 | 0.4120 | 47.5754 | 0.02102 | 0.05102 | 19.6004 | 12.3141 |
| 35 | 2.8139 | 0.3554 | 60.4621 | 0.01654 | 0.04654 | 21.4872 | 14.0375 |
| 40 | 3.2620 | 0.3066 | 75.4012 | 0.01326 | 0.04326 | 23.1148 | 15.6502 |
| 45 | 3.7816 | 0.2644 | 92.7198 | 0.01079 | 0.04079 | 24.5187 | 17.1556 |
| 50 | 4.3839 | 0.2281 | 112.7968 | 0.00887 | 0.03887 | 25.7298 | 18.5575 |
| 55 | 5.0821 | 0.1968 | 136.0716 | 0.00735 | 0.03735 | 26.7744 | 19.8600 |
| 60 | 5.8916 | 0.1697 | 163.0534 | 0.00613 | 0.03613 | 27.6756 | 21.0674 |
| 65 | 6.8300 | 0.1464 | 194.3327 | 0.00515 | 0.03515 | 28.4529 | 22.1841 |
| 70 | 7.9178 | 0.1263 | 230.5940 | 0.00434 | 0.03434 | 29.1234 | 23.2145 |
| 75 | 9.1789 | 0.1089 | 272.6307 | 0.00367 | 0.03367 | 29.7018 | 24.1634 |
| 80 | 10.6409 | 0.0940 | 321.3629 | 0.00311 | 0.03311 | 30.2008 | 25.0353 |

**i = 4%**

| n | F/P | P/F | F/A | A/F | A/P | P/A | A/G |
|---|---|---|---|---|---|---|---|
| 1 | 1.0400 | 0.9615 | 1.0000 | 1.00000 | 1.04000 | 0.9615 | −0.0000 |
| 2 | 1.0816 | 0.9246 | 2.0400 | 0.49020 | 0.53020 | 1.8861 | 0.4902 |
| 3 | 1.1249 | 0.8890 | 3.1216 | 0.32035 | 0.36035 | 2.7751 | 0.9738 |
| 4 | 1.1699 | 0.8548 | 4.2465 | 0.23549 | 0.27549 | 3.6299 | 1.4510 |
| 5 | 1.2167 | 0.8219 | 5.4163 | 0.18463 | 0.22463 | 4.4518 | 1.9216 |
| 6 | 1.2653 | 0.7903 | 6.6330 | 0.15076 | 0.19076 | 5.2421 | 2.3857 |
| 7 | 1.3159 | 0.7599 | 7.8983 | 0.12661 | 0.16661 | 6.0021 | 2.8433 |
| 8 | 1.3686 | 0.7307 | 9.2142 | 0.10853 | 0.14853 | 6.7327 | 3.2944 |
| 9 | 1.4233 | 0.7026 | 10.5828 | 0.09449 | 0.13449 | 7.4353 | 3.7391 |
| 10 | 1.4802 | 0.6756 | 12.0061 | 0.08329 | 0.12329 | 8.1109 | 4.1772 |
| 11 | 1.5395 | 0.6496 | 13.4863 | 0.07415 | 0.11415 | 8.7605 | 4.6090 |
| 12 | 1.6010 | 0.6246 | 15.0258 | 0.06655 | 0.10655 | 9.3851 | 5.0343 |
| 13 | 1.6651 | 0.6006 | 16.6268 | 0.06014 | 0.10014 | 9.9856 | 5.4533 |
| 14 | 1.7317 | 0.5775 | 18.2919 | 0.05467 | 0.09467 | 10.5631 | 5.8658 |
| 15 | 1.8009 | 0.5553 | 20.0236 | 0.04994 | 0.08994 | 11.1184 | 6.2721 |
| 16 | 1.8730 | 0.5339 | 21.8245 | 0.04582 | 0.08582 | 11.6523 | 6.6720 |
| 17 | 1.9479 | 0.5134 | 23.6975 | 0.04220 | 0.08220 | 12.1657 | 7.0656 |
| 18 | 2.0258 | 0.4936 | 25.6454 | 0.03899 | 0.07899 | 12.6593 | 7.4530 |
| 19 | 2.1068 | 0.4746 | 27.6712 | 0.03614 | 0.07614 | 13.1339 | 7.8341 |
| 20 | 2.1911 | 0.4564 | 29.7781 | 0.03358 | 0.07358 | 13.5903 | 8.2091 |
| 21 | 2.2788 | 0.4388 | 31.9692 | 0.03128 | 0.07128 | 14.0292 | 8.5779 |
| 22 | 2.3699 | 0.4220 | 34.2479 | 0.02920 | 0.06920 | 14.4511 | 8.9406 |
| 23 | 2.4647 | 0.4057 | 36.6179 | 0.02731 | 0.06731 | 14.8568 | 9.2973 |
| 24 | 2.5633 | 0.3901 | 39.0826 | 0.02559 | 0.06559 | 15.2470 | 9.6479 |
| 25 | 2.6658 | 0.3751 | 41.6459 | 0.02401 | 0.06401 | 15.6221 | 9.9925 |
| 26 | 2.7725 | 0.3607 | 44.3117 | 0.02257 | 0.06257 | 15.9828 | 10.3312 |
| 27 | 2.8834 | 0.3468 | 47.0842 | 0.02124 | 0.06124 | 16.3296 | 10.6640 |
| 28 | 2.9987 | 0.3335 | 49.9675 | 0.02001 | 0.06001 | 16.6631 | 10.9909 |
| 29 | 3.1186 | 0.3207 | 52.9662 | 0.01888 | 0.05888 | 16.9837 | 11.3120 |
| 30 | 3.2434 | 0.3083 | 56.0849 | 0.01783 | 0.05783 | 17.2920 | 11.6274 |
| 35 | 3.9461 | 0.2534 | 73.6521 | 0.01358 | 0.05358 | 18.6646 | 13.1198 |
| 40 | 4.8010 | 0.2083 | 95.0254 | 0.01052 | 0.05052 | 19.7928 | 14.4765 |
| 45 | 5.8412 | 0.1712 | 121.0292 | 0.00826 | 0.04826 | 20.7200 | 15.7047 |
| 50 | 7.1067 | 0.1407 | 152.6669 | 0.00655 | 0.04655 | 21.4822 | 16.8122 |
| 55 | 8.6464 | 0.1157 | 191.1589 | 0.00523 | 0.04523 | 22.1086 | 17.8070 |
| 60 | 10.5196 | 0.0951 | 237.9903 | 0.00420 | 0.04420 | 22.6235 | 18.6972 |
| 65 | 12.7987 | 0.0781 | 294.9679 | 0.00339 | 0.04339 | 23.0467 | 19.4909 |
| 70 | 15.5716 | 0.0642 | 364.2898 | 0.00275 | 0.04275 | 23.3945 | 20.1961 |
| 75 | 18.9452 | 0.0528 | 448.6305 | 0.00223 | 0.04223 | 23.6804 | 20.8206 |
| 80 | 23.0498 | 0.0434 | 551.2438 | 0.00181 | 0.04181 | 23.9154 | 21.3718 |

**i = 5%**

| n | F/P | P/F | F/A | A/F | A/P | P/A | A/G |
|---|---|---|---|---|---|---|---|
| 1 | 1.0500 | 0.9524 | 1.0000 | 1.00000 | 1.05000 | 0.9524 | −0.0000 |
| 2 | 1.1025 | 0.9070 | 2.0500 | 0.48781 | 0.53781 | 1.8594 | 0.4878 |
| 3 | 1.1576 | 0.8638 | 3.1525 | 0.31721 | 0.36721 | 2.7232 | 0.9675 |
| 4 | 1.2155 | 0.8227 | 4.3101 | 0.23201 | 0.28201 | 3.5459 | 1.4390 |
| 5 | 1.2763 | 0.7835 | 5.5256 | 0.18098 | 0.23098 | 4.3295 | 1.9025 |
| 6 | 1.3401 | 0.7462 | 6.8019 | 0.14702 | 0.19702 | 5.0757 | 2.3579 |
| 7 | 1.4071 | 0.7107 | 8.1420 | 0.12282 | 0.17282 | 5.7864 | 2.8052 |
| 8 | 1.4775 | 0.6768 | 9.5491 | 0.10472 | 0.15472 | 6.4632 | 3.2445 |
| 9 | 1.5513 | 0.6446 | 11.0265 | 0.09069 | 0.14069 | 7.1078 | 3.6758 |
| 10 | 1.6289 | 0.6139 | 12.5779 | 0.07950 | 0.12950 | 7.7217 | 4.0991 |
| 11 | 1.7103 | 0.5847 | 14.2068 | 0.07039 | 0.12039 | 8.3064 | 4.5144 |
| 12 | 1.7959 | 0.5568 | 15.9171 | 0.06283 | 0.11283 | 8.8632 | 4.9219 |
| 13 | 1.8856 | 0.5303 | 17.7129 | 0.05646 | 0.10646 | 9.3936 | 5.3215 |
| 14 | 1.9799 | 0.5051 | 19.5986 | 0.05102 | 0.10102 | 9.8986 | 5.7133 |
| 15 | 2.0789 | 0.4810 | 21.5785 | 0.04634 | 0.09634 | 10.3796 | 6.0973 |
| 16 | 2.1829 | 0.4581 | 23.6574 | 0.04227 | 0.09227 | 10.8378 | 6.4736 |
| 17 | 2.2920 | 0.4363 | 25.8403 | 0.03870 | 0.08870 | 11.2741 | 6.8423 |
| 18 | 2.4066 | 0.4155 | 28.1323 | 0.03555 | 0.08555 | 11.6896 | 7.2033 |
| 19 | 2.5269 | 0.3957 | 30.5389 | 0.03275 | 0.08275 | 12.0853 | 7.5569 |
| 20 | 2.6533 | 0.3769 | 33.0659 | 0.03024 | 0.08024 | 12.4622 | 7.9029 |
| 21 | 2.7860 | 0.3589 | 35.7192 | 0.02800 | 0.07800 | 12.8211 | 8.2416 |
| 22 | 2.9253 | 0.3419 | 38.5051 | 0.02597 | 0.07597 | 13.1630 | 8.5729 |
| 23 | 3.0715 | 0.3256 | 41.4304 | 0.02414 | 0.07414 | 13.4886 | 8.8970 |
| 24 | 3.2251 | 0.3101 | 44.5019 | 0.02247 | 0.07247 | 13.7986 | 9.2139 |
| 25 | 3.3863 | 0.2953 | 47.7270 | 0.02095 | 0.07095 | 14.0939 | 9.5237 |
| 26 | 3.5557 | 0.2812 | 41.1133 | 0.01956 | 0.06956 | 14.3752 | 9.8265 |
| 27 | 3.7334 | 0.2678 | 54.6690 | 0.01829 | 0.06829 | 14.6430 | 10.1224 |
| 28 | 3.9201 | 0.2551 | 58.4024 | 0.01712 | 0.06712 | 14.8981 | 10.4114 |
| 29 | 4.1161 | 0.2429 | 62.3225 | 0.01605 | 0.06605 | 15.1411 | 10.6936 |
| 30 | 4.3219 | 0.2314 | 66.4386 | 0.01505 | 0.06505 | 15.3724 | 10.9691 |
| 35 | 5.5160 | 0.1813 | 90.3200 | 0.01107 | 0.06107 | 16.3742 | 12.2498 |
| 40 | 7.0400 | 0.1420 | 120.7993 | 0.00828 | 0.05828 | 17.1591 | 13.3774 |
| 45 | 8.9850 | 0.1113 | 159.6995 | 0.00626 | 0.05626 | 17.7741 | 14.3644 |
| 50 | 11.4674 | 0.0872 | 209.3470 | 0.00478 | 0.05478 | 18.2559 | 15.2232 |
| 55 | 14.6356 | 0.0683 | 272.7113 | 0.00367 | 0.05367 | 18.6335 | 15.9664 |
| 60 | 18.6791 | 0.0535 | 353.5818 | 0.00283 | 0.05283 | 18.9293 | 16.6062 |
| 65 | 23.8398 | 0.0419 | 456.7954 | 0.00219 | 0.05219 | 19.1611 | 17.1541 |
| 70 | 30.4262 | 0.0329 | 588.5249 | 0.00170 | 0.05170 | 19.3427 | 17.6212 |
| 75 | 38.8324 | 0.0258 | 756.6488 | 0.00132 | 0.05132 | 19.4850 | 18.0176 |
| 80 | 49.5611 | 0.0202 | 971.2221 | 0.00103 | 0.05103 | 19.5965 | 18.3526 |

**i = 6%**

| n | F/P | P/F | F/A | A/F | A/P | P/A | A/G |
|---|---|---|---|---|---|---|---|
| 1 | 1.0600 | 0.9434 | 1.0000 | 1.00000 | 1.06000 | 0.9434 | −0.0000 |
| 2 | 1.1236 | 0.8900 | 2.0600 | 0.48544 | 0.54544 | 1.8334 | 0.4854 |
| 3 | 1.1910 | 0.8396 | 3.1836 | 0.31411 | 0.37411 | 2.6730 | 0.9612 |
| 4 | 1.2625 | 0.7921 | 4.3746 | 0.22859 | 0.28859 | 3.4651 | 1.4272 |
| 5 | 1.3382 | 0.7473 | 5.6371 | 0.17740 | 0.23740 | 4.2124 | 1.8836 |
| 6 | 1.4185 | 0.7050 | 6.9753 | 0.14336 | 0.20336 | 4.9173 | 2.3304 |
| 7 | 1.5036 | 0.6651 | 8.3938 | 0.11914 | 0.17914 | 5.5824 | 2.7676 |
| 8 | 1.5938 | 0.6274 | 9.8975 | 0.10104 | 0.16104 | 6.2098 | 3.1952 |
| 9 | 1.6895 | 0.5919 | 11.4913 | 0.08702 | 0.14702 | 6.8017 | 3.6133 |
| 10 | 1.7908 | 0.5584 | 13.1808 | 0.07587 | 0.13587 | 7.3601 | 4.0220 |
| 11 | 1.8983 | 0.5268 | 14.9716 | 0.06679 | 0.12679 | 7.8869 | 4.4213 |
| 12 | 2.0122 | 0.4970 | 16.8699 | 0.05928 | 0.11928 | 8.3838 | 4.8112 |
| 13 | 2.1329 | 0.4688 | 18.8821 | 0.05296 | 0.11296 | 8.8527 | 5.1920 |
| 14 | 2.2609 | 0.4423 | 21.0150 | 0.04758 | 0.10758 | 9.2950 | 5.5635 |
| 15 | 2.3966 | 0.4173 | 23.2759 | 0.04296 | 0.10296 | 9.7122 | 5.9260 |
| 16 | 2.5403 | 0.3936 | 25.6725 | 0.03895 | 0.09895 | 10.1059 | 6.2794 |
| 17 | 2.6928 | 0.3714 | 28.2128 | 0.03544 | 0.09544 | 10.4773 | 6.6240 |
| 18 | 2.8543 | 0.3503 | 30.9056 | 0.03236 | 0.09236 | 10.8276 | 6.9597 |
| 19 | 3.0256 | 0.3305 | 33.7599 | 0.02962 | 0.08962 | 11.1581 | 7.2867 |
| 20 | 3.2071 | 0.3118 | 36.7855 | 0.02718 | 0.08718 | 11.4699 | 7.6051 |
| 21 | 3.3996 | 0.2942 | 39.9927 | 0.02500 | 0.08500 | 11.7641 | 7.9151 |
| 22 | 3.6035 | 0.2775 | 43.3922 | 0.02305 | 0.08305 | 12.0416 | 8.2166 |
| 23 | 3.8197 | 0.2618 | 46.9958 | 0.02128 | 0.08128 | 12.3034 | 8.5099 |
| 24 | 4.0489 | 0.2470 | 50.8155 | 0.01968 | 0.07968 | 12.5504 | 8.7951 |
| 25 | 4.2919 | 0.2330 | 54.8644 | 0.01823 | 0.07823 | 12.7834 | 9.0722 |
| 26 | 4.5494 | 0.2198 | 59.1563 | 0.01690 | 0.07690 | 13.0032 | 9.3414 |
| 27 | 4.8223 | 0.2074 | 63.7057 | 0.01570 | 0.07570 | 13.2105 | 9.6029 |
| 28 | 5.1117 | 0.1956 | 68.5280 | 0.01459 | 0.07459 | 13.4062 | 9.8568 |
| 29 | 5.4184 | 0.1846 | 73.6397 | 0.01358 | 0.07358 | 13.5907 | 10.1032 |
| 30 | 5.7435 | 0.1741 | 79.0580 | 0.01265 | 0.07265 | 13.7648 | 10.3422 |
| 35 | 7.6861 | 0.1301 | 111.4346 | 0.00897 | 0.06897 | 14.4982 | 11.4319 |
| 40 | 10.2857 | 0.0972 | 154.7616 | 0.00646 | 0.06646 | 15.0463 | 12.3590 |
| 45 | 13.7646 | 0.0727 | 212.7430 | 0.00470 | 0.06470 | 15.4558 | 13.1413 |
| 50 | 18.4201 | 0.0543 | 290.3351 | 0.00344 | 0.06344 | 15.7619 | 13.7964 |
| 55 | 24.6502 | 0.0406 | 394.1708 | 0.00254 | 0.06254 | 15.9905 | 14.3411 |
| 60 | 32.9876 | 0.0303 | 533.1263 | 0.00188 | 0.06188 | 16.1614 | 14.7909 |
| 65 | 44.1448 | 0.0227 | 719.0803 | 0.00139 | 0.06139 | 16.2891 | 15.1601 |
| 70 | 59.0757 | 0.0169 | 967.9284 | 0.00103 | 0.06103 | 16.3845 | 15.4613 |
| 75 | 79.0566 | 0.0126 | 1300.9440 | 0.00077 | 0.06077 | 16.4558 | 15.7058 |
| 80 | 105.7955 | 0.0095 | 1746.5920 | 0.00057 | 0.06057 | 16.5091 | 15.9033 |

**i = 7%**

| n | F/P | P/F | F/A | A/F | A/P | P/A | A/G |
|---|---|---|---|---|---|---|---|
| 1 | 1.0700 | 0.9346 | 1.0000 | 1.00000 | 1.07000 | 0.9346 | −0.0000 |
| 2 | 1.1449 | 0.8734 | 2.0700 | 0.48309 | 0.55309 | 1.8080 | 0.4831 |
| 3 | 1.2250 | 0.8163 | 3.2149 | 0.31105 | 0.38105 | 2.6243 | 0.9549 |
| 4 | 1.3108 | 0.7629 | 4.4399 | 0.22523 | 0.29523 | 3.3872 | 1.4155 |
| 5 | 1.4926 | 0.7130 | 5.7507 | 0.17389 | 0.24389 | 4.1002 | 1.8649 |
| 6 | 1.5007 | 0.6663 | 7.1533 | 0.13980 | 0.20980 | 4.7665 | 2.3032 |
| 7 | 1.6058 | 0.6228 | 6.6540 | 0.11555 | 0.18555 | 5.3893 | 2.7304 |
| 8 | 1.7182 | 0.5820 | 10.2598 | 0.09747 | 0.16747 | 5.9713 | 3.1465 |
| 9 | 1.8385 | 0.5439 | 11.9780 | 0.08349 | 0.15349 | 6.5152 | 3.5517 |
| 10 | 1.9672 | 0.5083 | 13.8164 | 0.07238 | 0.14238 | 7.0236 | 3.9461 |
| 11 | 2.1049 | 0.4751 | 15.7836 | 0.06336 | 0.13336 | 7.4987 | 4.3296 |
| 12 | 2.2522 | 0.4440 | 17.8884 | 0.05590 | 0.12590 | 7.9427 | 4.7025 |
| 13 | 2.4098 | 0.4150 | 20.1406 | 0.04965 | 0.11965 | 8.3576 | 5.0648 |
| 14 | 2.5785 | 0.3878 | 22.5505 | 0.04435 | 0.11434 | 8.7455 | 5.4167 |
| 15 | 2.7590 | 0.3624 | 25.1290 | 0.03979 | 0.10979 | 9.1079 | 5.7583 |
| 16 | 2.9522 | 0.3387 | 27.8880 | 0.03586 | 0.10586 | 9.4466 | 6.0897 |
| 17 | 3.1588 | 0.3166 | 30.8402 | 0.03243 | 0.10243 | 9.7632 | 6.4110 |
| 18 | 3.3799 | 0.2959 | 33.9990 | 0.02941 | 0.09941 | 10.0591 | 6.7225 |
| 19 | 3.6165 | 0.2765 | 37.3789 | 0.02675 | 0.09675 | 10.3356 | 7.0242 |
| 20 | 3.8697 | 0.2584 | 40.9954 | 0.02439 | 0.09439 | 10.5940 | 7.3163 |
| 21 | 4.1406 | 0.2415 | 44.8651 | 0.02229 | 0.09229 | 10.8355 | 7.5990 |
| 22 | 4.4304 | 0.2257 | 49.0057 | 0.02041 | 0.09041 | 11.0612 | 7.8725 |
| 23 | 4.7405 | 0.2109 | 53.4360 | 0.01871 | 0.08871 | 11.2722 | 8.1368 |
| 24 | 5.0724 | 0.1971 | 58.1766 | 0.01719 | 0.08719 | 11.4693 | 8.3923 |
| 25 | 5.4274 | 0.1842 | 63.2489 | 0.01581 | 0.08581 | 11.6536 | 8.6391 |
| 26 | 5.8073 | 0.1722 | 68.6763 | 0.01456 | 0.08456 | 11.8258 | 8.8773 |
| 27 | 6.2139 | 0.1609 | 74.4837 | 0.01343 | 0.08343 | 11.9867 | 9.1072 |
| 28 | 6.6488 | 0.1504 | 80.6975 | 0.01239 | 0.08239 | 12.1371 | 9.3289 |
| 29 | 7.1142 | 0.1406 | 87.3464 | 0.01145 | 0.08145 | 12.2777 | 9.5427 |
| 30 | 7.6122 | 0.1314 | 94.4606 | 0.01059 | 0.08059 | 12.4090 | 9.7487 |
| 35 | 10.6766 | 0.0937 | 138.2366 | 0.00723 | 0.07723 | 12.9477 | 10.6687 |
| 40 | 14.9744 | 0.0668 | 199.6346 | 0.00501 | 0.07501 | 13.3317 | 11.4233 |
| 45 | 21.0024 | 0.0476 | 285.7485 | 0.00350 | 0.07350 | 13.6055 | 12.0360 |
| 50 | 29.4569 | 0.0339 | 406.5277 | 0.00246 | 0.07246 | 13.8007 | 12.5287 |
| 55 | 41.3149 | 0.0242 | 575.9266 | 0.00174 | 0.07174 | 13.9399 | 12.9215 |
| 60 | 57.9462 | 0.0173 | 813.5174 | 0.00123 | 0.07123 | 14.0392 | 13.2321 |
| 65 | 81.2725 | 0.0123 | 1146.7510 | 0.00087 | 0.07087 | 14.1099 | 13.4760 |
| 70 | 113.9889 | 0.0088 | 1614.1280 | 0.00062 | 0.07062 | 14.1604 | 13.6662 |
| 75 | 159.8753 | 0.0063 | 2269.6470 | 0.00044 | 0.07044 | 14.1964 | 13.8136 |
| 80 | 224.2333 | 0.0045 | 3189.0470 | 0.00031 | 0.07031 | 14.2220 | 13.9273 |

i = 8%

| n | F/P | P/F | F/A | A/F | A/P | P/A | A/G |
|---|---|---|---|---|---|---|---|
| 1 | 1.0800 | 0.9259 | 1.0000 | 1.00000 | 1.08000 | 0.9259 | 0.0000 |
| 2 | 1.1664 | 0.8573 | 2.0800 | 0.48077 | 0.56077 | 1.7833 | 0.4808 |
| 3 | 1.2597 | 0.7938 | 3.2464 | 0.30803 | 0.38803 | 2.5771 | 0.9488 |
| 4 | 1.3605 | 0.7350 | 4.5061 | 0.22192 | 0.30192 | 3.3121 | 1.4040 |
| 5 | 1.4693 | 0.6806 | 5.8666 | 0.17046 | 0.25046 | 3.9927 | 1.8465 |
| 6 | 1.5869 | 0.6302 | 7.3359 | 0.13632 | 0.21632 | 4.6229 | 2.2764 |
| 7 | 1.7138 | 0.5835 | 8.9228 | 0.11207 | 0.19207 | 5.2064 | 2.6937 |
| 8 | 1.8509 | 0.5403 | 10.6366 | 0.09401 | 0.17401 | 5.7466 | 3.0985 |
| 9 | 1.9990 | 0.5002 | 12.4876 | 0.08008 | 0.16008 | 6.2469 | 3.4910 |
| 10 | 2.1589 | 0.4632 | 14.4866 | 0.06903 | 0.14903 | 6.7101 | 3.8713 |
| 11 | 2.3316 | 0.4289 | 16.6455 | 0.06008 | 0.14008 | 7.1390 | 4.2395 |
| 12 | 2.5182 | 0.3971 | 18.9771 | 0.05269 | 0.13269 | 7.5361 | 4.5958 |
| 13 | 2.7196 | 0.3677 | 21.4953 | 0.04652 | 0.12652 | 7.9038 | 4.9402 |
| 14 | 2.9372 | 0.3405 | 24.2149 | 0.04130 | 0.12130 | 8.2442 | 5.2731 |
| 15 | 3.1722 | 0.3152 | 27.1521 | 0.03683 | 0.11683 | 8.5595 | 5.5945 |
| 16 | 3.4259 | 0.2919 | 30.3243 | 0.03298 | 0.11298 | 8.8514 | 5.9046 |
| 17 | 3.7000 | 0.2703 | 33.7503 | 0.02963 | 0.10963 | 9.1216 | 6.2038 |
| 18 | 3.9960 | 0.2502 | 37.4503 | 0.02670 | 0.10670 | 9.3719 | 6.4920 |
| 19 | 4.3157 | 0.2317 | 41.4463 | 0.02413 | 0.10413 | 9.6036 | 6.7697 |
| 20 | 4.6610 | 0.2145 | 45.7620 | 0.02185 | 0.10185 | 9.8182 | 7.0370 |
| 21 | 5.0338 | 0.1987 | 50.4230 | 0.01983 | 0.09983 | 10.0168 | 7.2940 |
| 22 | 5.4365 | 0.1839 | 55.4568 | 0.01803 | 0.09803 | 10.2007 | 7.5412 |
| 23 | 5.8715 | 0.1703 | 60.8934 | 0.01642 | 0.09642 | 10.3711 | 7.7786 |
| 24 | 6.3412 | 0.1577 | 66.7648 | 0.01498 | 0.09498 | 10.5288 | 8.0066 |
| 25 | 6.8485 | 0.1460 | 73.1060 | 0.01368 | 0.09368 | 10.6748 | 8.2254 |
| 26 | 7.3964 | 0.1352 | 79.9545 | 0.01251 | 0.09251 | 10.8100 | 8.4352 |
| 27 | 7.9881 | 0.1252 | 87.3509 | 0.01145 | 0.09145 | 10.9352 | 8.6363 |
| 28 | 8.6271 | 0.1159 | 95.3389 | 0.01049 | 0.09049 | 11.0511 | 8.8289 |
| 29 | 9.3173 | 0.1073 | 103.9661 | 0.00962 | 0.08962 | 11.1584 | 9.0133 |
| 30 | 10.0627 | 0.0994 | 113.2833 | 0.00883 | 0.08883 | 11.2578 | 9.1897 |
| 35 | 14.7854 | 0.0676 | 172.3170 | 0.00580 | 0.08580 | 11.6546 | 9.9611 |
| 40 | 21.7245 | 0.0460 | 259.0569 | 0.00386 | 0.08386 | 11.9246 | 10.5699 |
| 45 | 31.9205 | 0.0313 | 386.5062 | 0.00259 | 0.08259 | 12.1084 | 11.0447 |
| 50 | 46.9017 | 0.0213 | 573.7712 | 0.00174 | 0.08174 | 12.2335 | 11.4107 |
| 55 | 68.9140 | 0.0145 | 848.9248 | 0.00118 | 0.08118 | 12.3186 | 11.6902 |
| 60 | 101.2573 | 0.0099 | 1253.2160 | 0.00080 | 0.08080 | 12.3766 | 11.9015 |
| 65 | 148.7802 | 0.0067 | 1847.2520 | 0.00054 | 0.08054 | 12.4160 | 12.0602 |
| 70 | 218.6069 | 0.0046 | 2720.0870 | 0.00037 | 0.08037 | 12.4428 | 12.1783 |
| 75 | 321.2053 | 0.0031 | 4002.5670 | 0.00025 | 0.08025 | 12.4611 | 12.2658 |
| 80 | 471.9560 | 0.0021 | 5886.9510 | 0.00017 | 0.08017 | 12.4735 | 12.3301 |

i = 9%

| n | F/P | P/F | F/A | A/F | A/P | P/A | A/G |
|---|---|---|---|---|---|---|---|
| 1 | 1.0900 | 0.9174 | 1.0000 | 1.00000 | 1.09000 | 0.9174 | 0.0000 |
| 2 | 1.1881 | 0.8417 | 2.0900 | 0.47847 | 0.56847 | 1.7591 | 0.4785 |
| 3 | 1.2950 | 0.7722 | 3.2781 | 0.30505 | 0.39505 | 2.5313 | 0.9426 |
| 4 | 1.4116 | 0.7084 | 4.5731 | 0.21867 | 0.30867 | 3.2397 | 1.3925 |
| 5 | 1.5386 | 0.6499 | 5.9847 | 0.16709 | 0.25709 | 3.8897 | 1.8282 |
| 6 | 1.6771 | 0.5963 | 7.5233 | 0.13292 | 0.22292 | 4.4859 | 2.2498 |
| 7 | 1.8280 | 0.5470 | 9.2004 | 0.10869 | 0.19869 | 5.0330 | 2.6574 |
| 8 | 1.9926 | 0.5019 | 11.0285 | 0.09067 | 0.18067 | 5.5348 | 3.0512 |
| 9 | 2.1719 | 0.4604 | 13.0210 | 0.07680 | 0.16680 | 5.9952 | 3.4312 |
| 10 | 2.3674 | 0.4224 | 15.1929 | 0.06582 | 0.15582 | 6.4177 | 3.7978 |
| 11 | 2.5804 | 0.3875 | 17.5603 | 0.05695 | 0.14695 | 6.8052 | 4.1510 |
| 12 | 2.8127 | 0.3555 | 20.1407 | 0.04965 | 0.13965 | 7.1607 | 4.4910 |
| 13 | 3.0658 | 0.3262 | 22.9534 | 0.04357 | 0.13357 | 7.4869 | 4.8182 |
| 14 | 3.3417 | 0.2992 | 26.0192 | 0.03843 | 0.12843 | 7.7862 | 5.1326 |
| 15 | 3.6425 | 0.2745 | 29.3609 | 0.03406 | 0.12406 | 8.0607 | 5.4346 |
| 16 | 3.9703 | 0.2519 | 33.0034 | 0.03030 | 0.12030 | 8.3126 | 5.7245 |
| 17 | 4.3276 | 0.2311 | 36.9737 | 0.02705 | 0.11705 | 8.5436 | 6.0024 |
| 18 | 4.7171 | 0.2120 | 41.3014 | 0.02421 | 0.11421 | 8.7556 | 6.2687 |
| 19 | 5.1417 | 0.1945 | 46.0185 | 0.02173 | 0.11173 | 8.9501 | 6.5236 |
| 20 | 5.6044 | 0.1784 | 51.1602 | 0.01955 | 0.10955 | 9.1285 | 6.7675 |
| 21 | 6.1088 | 0.1637 | 56.7646 | 0.01762 | 0.10762 | 9.2922 | 7.0006 |
| 22 | 6.6586 | 0.1502 | 62.8734 | 0.01590 | 0.10590 | 9.4424 | 7.2232 |
| 23 | 7.2579 | 0.1378 | 69.5320 | 0.01438 | 0.10438 | 9.5802 | 7.4357 |
| 24 | 7.9111 | 0.1264 | 76.7899 | 0.01302 | 0.10302 | 9.7066 | 7.6384 |
| 25 | 8.6231 | 0.1160 | 84.7010 | 0.01181 | 0.10181 | 9.8226 | 7.8316 |
| 26 | 9.3992 | 0.1064 | 93.3241 | 0.01072 | 0.10072 | 9.9290 | 8.0156 |
| 27 | 10.2451 | 0.0976 | 102.7233 | 0.00973 | 0.09973 | 10.0266 | 8.1906 |
| 28 | 11.1672 | 0.0895 | 112.9684 | 0.00885 | 0.09885 | 10.1161 | 8.3571 |
| 29 | 12.1722 | 0.0822 | 124.1356 | 0.00806 | 0.09806 | 10.1983 | 8.5154 |
| 30 | 13.2677 | 0.0754 | 136.3078 | 0.00734 | 0.09734 | 10.2737 | 8.6657 |
| 35 | 20.4140 | 0.0490 | 215.7111 | 0.00464 | 0.09464 | 10.5668 | 9.3083 |
| 40 | 31.4095 | 0.0318 | 337.8831 | 0.00296 | 0.09296 | 10.7574 | 9.7957 |
| 45 | 48.3274 | 0.0207 | 525.8600 | 0.00190 | 0.09190 | 10.8812 | 10.1603 |
| 50 | 74.3577 | 0.0134 | 815.0855 | 0.00123 | 0.09123 | 10.9617 | 10.4295 |
| 55 | 114.4085 | 0.0087 | 1260.0950 | 0.00079 | 0.09079 | 11.0140 | 10.6261 |
| 60 | 176.0318 | 0.0057 | 1944.7980 | 0.00051 | 0.09051 | 11.0480 | 10.7683 |
| 65 | 270.8468 | 0.0037 | 2998.2980 | 0.00033 | 0.09033 | 11.0701 | 10.8702 |
| 70 | 416.7314 | 0.0024 | 4619.2390 | 0.00022 | 0.09022 | 11.0845 | 10.9427 |
| 75 | 641.1931 | 0.0016 | 7113.2570 | 0.00014 | 0.09014 | 11.0938 | 10.9940 |
| 80 | 986.5552 | 0.0010 | 10950.6100 | 0.00009 | 0.09009 | 11.0999 | 11.0299 |

i = 10%

| n | F/P | P/F | F/A | A/F | A/P | P/A | A/G |
|---|---|---|---|---|---|---|---|
| 1 | 1.1000 | 0.9091 | 1.0000 | 1.00000 | 1.10000 | 0.9091 | 0.0000 |
| 2 | 1.2100 | 0.8264 | 2.1000 | 0.47619 | 0.57619 | 1.7355 | 0.4762 |
| 3 | 1.3310 | 0.7513 | 3.3100 | 0.30211 | 0.40211 | 2.4869 | 0.9366 |
| 4 | 1.4641 | 0.6830 | 4.6410 | 0.21547 | 0.31547 | 3.1699 | 1.3812 |
| 5 | 1.6105 | 0.6209 | 6.1051 | 0.16380 | 0.26380 | 3.7908 | 1.8101 |
| 6 | 1.7716 | 0.5645 | 7.7156 | 0.12961 | 0.22961 | 4.3553 | 2.2236 |
| 7 | 1.9487 | 0.5132 | 9.4872 | 0.10541 | 0.20541 | 4.8684 | 2.6216 |
| 8 | 2.1436 | 0.4665 | 11.4359 | 0.08744 | 0.18744 | 5.3349 | 3.0045 |
| 9 | 2.3579 | 0.4241 | 13.5795 | 0.07364 | 0.17364 | 5.7590 | 3.3724 |
| 10 | 2.5937 | 0.3855 | 15.9374 | 0.06275 | 0.16275 | 6.1446 | 3.7255 |
| 11 | 2.8531 | 0.3505 | 18.5312 | 0.05396 | 0.15396 | 6.4951 | 4.0641 |
| 12 | 3.1384 | 0.3186 | 21.3843 | 0.04676 | 0.14676 | 6.8137 | 4.3884 |
| 13 | 3.4523 | 0.2897 | 24.5227 | 0.04078 | 0.14078 | 7.1034 | 4.6988 |
| 14 | 3.7975 | 0.2633 | 27.9750 | 0.03575 | 0.13575 | 7.3667 | 4.9955 |
| 15 | 4.1772 | 0.2394 | 31.7725 | 0.03147 | 0.13147 | 7.6061 | 5.2789 |
| 16 | 4.5950 | 0.2176 | 35.9497 | 0.02782 | 0.12782 | 7.8237 | 5.5493 |
| 17 | 5.0545 | 0.1978 | 40.5447 | 0.02466 | 0.12466 | 8.0216 | 5.8071 |
| 18 | 5.5599 | 0.1799 | 45.5992 | 0.02193 | 0.12193 | 8.2014 | 6.0526 |
| 19 | 6.1159 | 0.1635 | 51.1591 | 0.01955 | 0.11955 | 8.3649 | 6.2861 |
| 20 | 6.7275 | 0.1486 | 57.2750 | 0.01746 | 0.11746 | 8.5136 | 6.5081 |
| 21 | 7.4003 | 0.1351 | 64.0025 | 0.01562 | 0.11562 | 8.6487 | 6.7189 |
| 22 | 8.1403 | 0.1228 | 71.4028 | 0.01401 | 0.11401 | 8.7715 | 6.9189 |
| 23 | 8.9543 | 0.1117 | 79.5431 | 0.01257 | 0.11257 | 8.8832 | 7.1085 |
| 24 | 9.8497 | 0.1015 | 88.4974 | 0.01130 | 0.11130 | 8.9847 | 7.2881 |
| 25 | 10.8347 | 0.0923 | 98.3471 | 0.01017 | 0.11017 | 9.0770 | 7.4580 |
| 26 | 11.9182 | 0.0839 | 109.1818 | 0.00916 | 0.10916 | 9.1609 | 7.6187 |
| 27 | 13.1100 | 0.0763 | 121.1000 | 0.00826 | 0.10826 | 9.2372 | 7.7704 |
| 28 | 14.4210 | 0.0693 | 134.2100 | 0.00745 | 0.10745 | 9.3066 | 7.9137 |
| 29 | 15.8631 | 0.0630 | 148.6310 | 0.00673 | 0.10673 | 9.3696 | 8.0489 |
| 30 | 17.4494 | 0.0573 | 164.4941 | 0.00608 | 0.10608 | 9.4269 | 8.1762 |
| 35 | 28.1025 | 0.0356 | 271.0245 | 0.00369 | 0.10369 | 9.6442 | 8.7086 |
| 40 | 45.2593 | 0.0221 | 442.5929 | 0.00226 | 0.10226 | 9.7791 | 9.0962 |
| 45 | 72.8905 | 0.0137 | 718.9053 | 0.00139 | 0.10139 | 9.8628 | 9.3740 |
| 50 | 117.3909 | 0.0085 | 1163.9100 | 0.00086 | 0.10086 | 9.9148 | 9.5704 |
| 55 | 189.0593 | 0.0053 | 1880.5930 | 0.00053 | 0.10053 | 9.9471 | 9.7075 |
| 60 | 304.4819 | 0.0033 | 3034.8190 | 0.00033 | 0.10033 | 9.9672 | 9.8023 |
| 65 | 490.3712 | 0.0020 | 4893.7130 | 0.00020 | 0.10020 | 9.9796 | 9.8672 |
| 70 | 789.7478 | 0.0013 | 7887.4790 | 0.00013 | 0.10013 | 9.9873 | 9.9113 |
| 75 | 1271.8970 | 0.0008 | 12708.9700 | 0.00008 | 0.10008 | 9.9921 | 9.9410 |
| 80 | 2048.4030 | 0.0005 | 20474.0300 | 0.00005 | 0.10005 | 9.9951 | 9.9609 |

i = 11%

| n | F/P | P/F | F/A | A/F | A/P | P/A | A/G |
|---|---|---|---|---|---|---|---|
| 1 | 1.1100 | 0.9009 | 1.0000 | 1.00000 | 1.11000 | 0.9009 | 0.0000 |
| 2 | 1.2321 | 0.8116 | 2.1100 | 0.47393 | 0.58393 | 1.7125 | 0.4739 |
| 3 | 1.3676 | 0.7312 | 3.3421 | 0.29921 | 0.40921 | 2.4437 | 0.9306 |
| 4 | 1.5181 | 0.6587 | 4.7097 | 0.21233 | 0.32233 | 3.1024 | 1.3700 |
| 5 | 1.6851 | 0.5935 | 6.2278 | 0.16057 | 0.27057 | 3.6959 | 1.7923 |
| 6 | 1.8704 | 0.5346 | 7.9129 | 0.12638 | 0.23638 | 4.2305 | 2.1976 |
| 7 | 2.0762 | 0.4817 | 9.7833 | 0.10222 | 0.21222 | 4.7122 | 2.5863 |
| 8 | 2.3045 | 0.4339 | 11.8594 | 0.08432 | 0.19432 | 5.1461 | 2.9585 |
| 9 | 2.5580 | 0.3909 | 14.1640 | 0.07060 | 0.18060 | 5.5370 | 3.3144 |
| 10 | 2.8394 | 0.3522 | 16.7220 | 0.05980 | 0.16980 | 5.8892 | 3.6544 |
| 11 | 3.1518 | 0.3173 | 19.5614 | 0.05112 | 0.16112 | 6.2065 | 3.9788 |
| 12 | 3.4985 | 0.2858 | 22.7132 | 0.04403 | 0.15403 | 6.4924 | 4.2879 |
| 13 | 3.8833 | 0.2575 | 26.2116 | 0.03815 | 0.14815 | 6.7499 | 4.5822 |
| 14 | 4.3104 | 0.2320 | 30.0949 | 0.03323 | 0.14323 | 6.9819 | 4.8619 |
| 15 | 4.7846 | 0.2090 | 34.4054 | 0.02907 | 0.13907 | 7.1909 | 5.1275 |
| 16 | 5.3109 | 0.1883 | 39.1900 | 0.02552 | 0.13552 | 7.3792 | 5.3794 |
| 17 | 5.8951 | 0.1696 | 44.5009 | 0.02247 | 0.13247 | 7.5488 | 5.6180 |
| 18 | 6.5436 | 0.1528 | 50.3959 | 0.01984 | 0.12984 | 7.7016 | 5.8439 |
| 19 | 7.2633 | 0.1377 | 56.9395 | 0.01756 | 0.12756 | 7.8393 | 6.0574 |
| 20 | 8.0623 | 0.1240 | 64.2029 | 0.01558 | 0.12558 | 7.9633 | 6.2590 |
| 21 | 8.9492 | 0.1117 | 72.2652 | 0.01384 | 0.12384 | 8.0751 | 6.4491 |
| 22 | 9.9336 | 0.1007 | 81.2143 | 0.01231 | 0.12231 | 8.1757 | 6.6283 |
| 23 | 11.0263 | 0.0907 | 91.1479 | 0.01097 | 0.12097 | 8.2664 | 6.7969 |
| 24 | 12.2392 | 0.0817 | 102.1742 | 0.00979 | 0.11979 | 8.3481 | 6.9555 |
| 25 | 13.5855 | 0.0736 | 114.4133 | 0.00874 | 0.11874 | 8.4217 | 7.1045 |
| 26 | 15.0799 | 0.0663 | 127.9988 | 0.00781 | 0.11781 | 8.4881 | 7.2443 |
| 27 | 16.7386 | 0.0597 | 143.0787 | 0.00699 | 0.11699 | 8.5478 | 7.3754 |
| 28 | 18.5799 | 0.0538 | 159.8173 | 0.00626 | 0.11626 | 8.6016 | 7.4982 |
| 29 | 20.6237 | 0.0485 | 178.3972 | 0.00561 | 0.11561 | 8.6501 | 7.6131 |
| 30 | 22.8923 | 0.0437 | 199.0209 | 0.00502 | 0.11502 | 8.6938 | 7.7206 |
| 35 | 38.5749 | 0.0259 | 341.5896 | 0.00293 | 0.11293 | 8.8552 | 8.1594 |
| 40 | 65.0009 | 0.0154 | 581.8262 | 0.00172 | 0.11172 | 8.9511 | 8.4659 |
| 45 | 109.5303 | 0.0091 | 986.6388 | 0.00101 | 0.11101 | 9.0079 | 8.6763 |
| 50 | 184.5649 | 0.0054 | 1668.7720 | 0.00060 | 0.11060 | 9.0417 | 8.8185 |
| 55 | 311.0025 | 0.0032 | 2818.2050 | 0.00035 | 0.11035 | 9.0617 | 8.9135 |
| 60 | 524.0573 | 0.0019 | 4755.0680 | 0.00021 | 0.11021 | 9.0736 | 8.9762 |
| 65 | 883.0671 | 0.0011 | 8018.7930 | 0.00012 | 0.11012 | 9.0806 | 9.0172 |
| 70 | 1488.0190 | 0.0007 | 13518.3600 | 0.00007 | 0.11007 | 9.0848 | 9.0438 |
| 75 | 2507.3990 | 0.0004 | 22785.4500 | 0.00004 | 0.11004 | 9.0873 | 9.0610 |
| 80 | 4225.1140 | 0.0002 | 38401.0400 | 0.00003 | 0.11003 | 9.0888 | 9.0720 |

**i = 12%**

| n | F/P | P/F | F/A | A/F | A/P | P/A | A/G |
|---|---|---|---|---|---|---|---|
| 1 | 1.1200 | 0.8929 | 1.0000 | 1.00000 | 1.12000 | 0.8929 | 0.0000 |
| 2 | 1.2544 | 0.7972 | 2.1200 | 0.47170 | 0.59170 | 1.6901 | 0.4717 |
| 3 | 1.4049 | 0.7118 | 3.3744 | 0.29635 | 0.41635 | 2.4018 | 0.9246 |
| 4 | 1.5735 | 0.6355 | 4.7793 | 0.20923 | 0.32923 | 3.0373 | 1.3589 |
| 5 | 1.7623 | 0.5674 | 6.3528 | 0.15741 | 0.27741 | 3.6048 | 1.7746 |
| 6 | 1.9738 | 0.5066 | 8.1152 | 0.12323 | 0.24323 | 4.1114 | 2.1720 |
| 7 | 2.2107 | 0.4523 | 10.0890 | 0.09912 | 0.21912 | 4.5638 | 2.5515 |
| 8 | 2.4760 | 0.4039 | 12.2997 | 0.08130 | 0.20130 | 4.9676 | 2.9131 |
| 9 | 2.7731 | 0.3606 | 14.7757 | 0.06768 | 0.18768 | 5.3283 | 3.2574 |
| 10 | 3.1058 | 0.3220 | 17.5487 | 0.05698 | 0.17698 | 5.6502 | 3.5847 |
| 11 | 3.4785 | 0.2875 | 20.6546 | 0.04842 | 0.16842 | 5.9377 | 3.8953 |
| 12 | 3.8960 | 0.2567 | 24.1331 | 0.04144 | 0.16144 | 6.1944 | 4.1897 |
| 13 | 4.3635 | 0.2292 | 28.0291 | 0.03568 | 0.15568 | 6.4235 | 4.4683 |
| 14 | 4.8871 | 0.2046 | 32.3926 | 0.03087 | 0.15087 | 6.6282 | 4.7317 |
| 15 | 5.4736 | 0.1827 | 37.2797 | 0.02682 | 0.14682 | 6.8109 | 4.9803 |
| 16 | 6.1304 | 0.1631 | 42.7533 | 0.02339 | 0.14339 | 6.9740 | 5.2147 |
| 17 | 6.8660 | 0.1456 | 48.8837 | 0.02046 | 0.14046 | 7.1196 | 5.4353 |
| 18 | 7.6900 | 0.1300 | 55.7497 | 0.01794 | 0.13794 | 7.2497 | 5.6427 |
| 19 | 8.6128 | 0.1161 | 63.4397 | 0.01576 | 0.13576 | 7.3658 | 5.8375 |
| 20 | 9.6463 | 0.1037 | 72.0525 | 0.01388 | 0.13388 | 7.4694 | 6.0202 |
| 21 | 10.8038 | 0.0926 | 81.6987 | 0.01224 | 0.13224 | 7.5620 | 6.1913 |
| 22 | 12.1003 | 0.0826 | 92.5026 | 0.01081 | 0.13081 | 7.6446 | 6.3514 |
| 23 | 13.5523 | 0.0738 | 104.6029 | 0.00956 | 0.12956 | 7.7184 | 6.5010 |
| 24 | 15.1786 | 0.0659 | 118.1553 | 0.00846 | 0.12846 | 7.7843 | 6.6406 |
| 25 | 17.0001 | 0.0588 | 133.3339 | 0.00750 | 0.12750 | 7.8431 | 6.7708 |
| 26 | 19.0401 | 0.0525 | 150.3340 | 0.00665 | 0.12665 | 7.8957 | 6.8921 |
| 27 | 21.3249 | 0.0469 | 169.3740 | 0.00590 | 0.12590 | 7.9426 | 7.0049 |
| 28 | 23.8839 | 0.0419 | 190.6989 | 0.00524 | 0.12524 | 7.9844 | 7.1098 |
| 29 | 26.7499 | 0.0374 | 214.5828 | 0.00466 | 0.12466 | 8.0218 | 7.2071 |
| 30 | 29.9599 | 0.0334 | 241.3327 | 0.00414 | 0.12414 | 8.0552 | 7.2974 |
| 35 | 52.7996 | 0.0189 | 431.6636 | 0.00232 | 0.12232 | 8.1755 | 7.6577 |
| 40 | 93.0510 | 0.0107 | 767.0915 | 0.00130 | 0.12130 | 8.2438 | 7.8988 |
| 45 | 163.9876 | 0.0061 | 1358.2300 | 0.00074 | 0.12074 | 8.2825 | 8.0572 |
| 50 | 289.0022 | 0.0035 | 2400.0190 | 0.00042 | 0.12042 | 8.3045 | 8.1597 |
| 55 | 509.3206 | 0.0020 | 4236.0060 | 0.00024 | 0.12024 | 8.3170 | 8.2251 |
| 60 | 897.5969 | 0.0011 | 7471.6420 | 0.00013 | 0.12013 | 8.3241 | 8.2664 |
| 65 | 1581.8720 | 0.0006 | 13173.9400 | 0.00008 | 0.12008 | 8.3281 | 8.2922 |
| 70 | 2787.8000 | 0.0004 | 23223.3400 | 0.00004 | 0.12004 | 8.3303 | 8.3082 |
| 75 | 4913.0550 | 0.0002 | 40933.8000 | 0.00002 | 0.12002 | 8.3316 | 8.3181 |
| 80 | 8658.4820 | 0.0001 | 72145.7000 | 0.00001 | 0.12001 | 8.3324 | 8.3241 |

**i = 13%**

| n | F/P | P/F | F/A | A/F | A/P | P/A | A/G |
|---|---|---|---|---|---|---|---|
| 1 | 1.1300 | 0.8850 | 1.0000 | 1.00000 | 1.13000 | 0.8850 | 0.0000 |
| 2 | 1.2769 | 0.7831 | 2.1300 | 0.46948 | 0.59948 | 1.6681 | 0.4695 |
| 3 | 1.4429 | 0.6931 | 3.4069 | 0.29352 | 0.42352 | 2.3612 | 0.9187 |
| 4 | 1.6305 | 0.6133 | 4.8498 | 0.20619 | 0.33619 | 2.9745 | 1.3479 |
| 5 | 1.8424 | 0.5428 | 6.4803 | 0.15431 | 0.28431 | 3.5172 | 1.7571 |
| 6 | 2.0820 | 0.4803 | 8.3227 | 0.12015 | 0.25015 | 3.9975 | 2.1468 |
| 7 | 2.3526 | 0.4251 | 10.4047 | 0.09611 | 0.22611 | 4.4226 | 2.5171 |
| 8 | 2.6584 | 0.3762 | 12.7573 | 0.07839 | 0.20839 | 4.7988 | 2.8685 |
| 9 | 3.0040 | 0.3329 | 15.4157 | 0.06487 | 0.19487 | 5.1317 | 3.2014 |
| 10 | 3.3946 | 0.2946 | 18.4197 | 0.05429 | 0.18429 | 5.4262 | 3.5162 |
| 11 | 3.8359 | 0.2607 | 21.8143 | 0.04584 | 0.17584 | 5.6869 | 3.8134 |
| 12 | 4.3345 | 0.2307 | 25.6502 | 0.03899 | 0.16899 | 5.9176 | 4.0936 |
| 13 | 4.8980 | 0.2042 | 29.9847 | 0.03335 | 0.16335 | 6.1218 | 4.3573 |
| 14 | 5.5348 | 0.1807 | 34.8827 | 0.02867 | 0.15867 | 6.3025 | 4.6050 |
| 15 | 6.2543 | 0.1599 | 40.4175 | 0.02474 | 0.15474 | 6.4624 | 4.8375 |
| 16 | 7.0673 | 0.1415 | 46.6717 | 0.02143 | 0.15143 | 6.6039 | 5.0552 |
| 17 | 7.9861 | 0.1252 | 53.7390 | 0.01861 | 0.14861 | 6.7291 | 5.2589 |
| 18 | 9.0243 | 0.1108 | 61.7251 | 0.01620 | 0.14620 | 6.8399 | 5.4491 |
| 19 | 10.1974 | 0.0981 | 70.7494 | 0.01413 | 0.14413 | 6.9380 | 5.6265 |
| 20 | 11.5231 | 0.0868 | 80.9468 | 0.01235 | 0.14235 | 7.0248 | 5.7917 |
| 21 | 13.0211 | 0.0768 | 92.4699 | 0.01081 | 0.14081 | 7.1016 | 5.9454 |
| 22 | 14.7138 | 0.0680 | 105.4910 | 0.00948 | 0.13948 | 7.1695 | 6.0881 |
| 23 | 16.6266 | 0.0601 | 120.2048 | 0.00832 | 0.13832 | 7.2297 | 6.2205 |
| 24 | 18.7881 | 0.0532 | 136.8314 | 0.00731 | 0.13731 | 7.2829 | 6.3431 |
| 25 | 21.2305 | 0.0471 | 155.6195 | 0.00643 | 0.13643 | 7.3300 | 6.4566 |
| 26 | 23.9905 | 0.0417 | 176.8500 | 0.00565 | 0.13565 | 7.3717 | 6.5614 |
| 27 | 27.1093 | 0.0369 | 200.8405 | 0.00498 | 0.13498 | 7.4086 | 6.6582 |
| 28 | 30.6335 | 0.0326 | 227.9497 | 0.00439 | 0.13439 | 7.4412 | 6.7474 |
| 29 | 34.6158 | 0.0289 | 258.5831 | 0.00387 | 0.13387 | 7.4701 | 6.8296 |
| 30 | 39.1159 | 0.0256 | 293.1990 | 0.00341 | 0.13341 | 7.4957 | 6.9052 |
| 35 | 72.0684 | 0.0139 | 546.6803 | 0.00183 | 0.13183 | 7.5856 | 7.1998 |
| 40 | 132.7814 | 0.0075 | 1013.7030 | 0.00099 | 0.13099 | 7.6344 | 7.3888 |
| 45 | 244.6410 | 0.0041 | 1874.1620 | 0.00053 | 0.13053 | 7.6609 | 7.5076 |
| 50 | 450.7352 | 0.0022 | 3459.5020 | 0.00029 | 0.13029 | 7.6752 | 7.5811 |
| 55 | 830.4503 | 0.0012 | 6380.3880 | 0.00016 | 0.13016 | 7.6830 | 7.6260 |
| 60 | 1530.0500 | 0.0007 | 11761.9300 | 0.00009 | 0.13009 | 7.6873 | 7.6531 |
| 65 | 2819.0180 | 0.0004 | 21677.0700 | 0.00005 | 0.13005 | 7.6896 | 7.6692 |
| 70 | 5193.8580 | 0.0002 | 39945.0700 | 0.00003 | 0.13003 | 7.6908 | 7.6788 |
| 75 | 9569.3450 | 0.0001 | 73602.6600 | 0.00001 | 0.13001 | 7.6915 | 7.6845 |
| 80 | 17630.9000 | 0.0001 | 135614.6000 | 0.00001 | 0.13001 | 7.6919 | 7.6878 |

**i = 14%**

| n | F/P | P/F | F/A | A/F | A/P | P/A | A/G |
|---|---|---|---|---|---|---|---|
| 1 | 1.1400 | 0.8772 | 1.0000 | 1.00000 | 1.14000 | 0.8772 | 0.0000 |
| 2 | 1.2996 | 0.7695 | 2.1400 | 0.46729 | 0.60729 | 1.6467 | 0.4673 |
| 3 | 1.4815 | 0.6750 | 3.4396 | 0.29073 | 0.43073 | 2.3216 | 0.9129 |
| 4 | 1.6890 | 0.5921 | 4.9211 | 0.20320 | 0.34320 | 2.9137 | 1.3370 |
| 5 | 1.9254 | 0.5194 | 6.6101 | 0.15128 | 0.29128 | 3.4331 | 1.7399 |
| 6 | 2.1950 | 0.4556 | 8.5355 | 0.11716 | 0.25716 | 3.8887 | 2.1218 |
| 7 | 2.5023 | 0.3996 | 10.7305 | 0.09319 | 0.23319 | 4.2883 | 2.4832 |
| 8 | 2.8526 | 0.3506 | 13.2328 | 0.07557 | 0.21557 | 4.6389 | 2.8246 |
| 9 | 3.2519 | 0.3075 | 16.0853 | 0.06217 | 0.20217 | 4.9464 | 3.1463 |
| 10 | 3.7072 | 0.2697 | 19.3373 | 0.05171 | 0.19171 | 5.2161 | 3.4490 |
| 11 | 4.2262 | 0.2366 | 23.0445 | 0.04339 | 0.18339 | 5.4527 | 3.7333 |
| 12 | 4.8179 | 0.2076 | 27.2708 | 0.03667 | 0.17667 | 5.6603 | 3.9998 |
| 13 | 5.4924 | 0.1821 | 32.0887 | 0.03116 | 0.17116 | 5.8424 | 4.2491 |
| 14 | 6.2613 | 0.1597 | 37.5811 | 0.02661 | 0.16661 | 6.0021 | 4.4819 |
| 15 | 7.1379 | 0.1401 | 43.8424 | 0.02281 | 0.16281 | 6.1422 | 4.6990 |
| 16 | 8.1373 | 0.1229 | 50.9804 | 0.01962 | 0.15962 | 6.2651 | 4.9011 |
| 17 | 9.2765 | 0.1078 | 59.1176 | 0.01692 | 0.15692 | 6.3729 | 5.0888 |
| 18 | 10.5752 | 0.0946 | 68.3941 | 0.01462 | 0.15462 | 6.4674 | 5.2630 |
| 19 | 12.0557 | 0.0829 | 78.9692 | 0.01266 | 0.15266 | 6.5504 | 5.4243 |
| 20 | 13.7435 | 0.0728 | 91.0250 | 0.01099 | 0.15099 | 6.6231 | 5.5734 |
| 21 | 15.6676 | 0.0638 | 104.7684 | 0.00954 | 0.14954 | 6.6870 | 5.7111 |
| 22 | 17.8610 | 0.0560 | 120.4360 | 0.00830 | 0.14830 | 6.7429 | 5.8381 |
| 23 | 20.3616 | 0.0491 | 138.2971 | 0.00723 | 0.14723 | 6.7921 | 5.9549 |
| 24 | 23.2122 | 0.0431 | 158.6587 | 0.00630 | 0.14630 | 6.8351 | 6.0624 |
| 25 | 26.4619 | 0.0378 | 181.8709 | 0.00550 | 0.14550 | 6.8729 | 6.1610 |
| 26 | 30.1666 | 0.0331 | 208.3328 | 0.00480 | 0.14480 | 6.9061 | 6.2514 |
| 27 | 34.3899 | 0.0291 | 238.4994 | 0.00419 | 0.14419 | 6.9352 | 6.3342 |
| 28 | 39.2045 | 0.0255 | 272.8893 | 0.00366 | 0.14366 | 6.9607 | 6.4100 |
| 29 | 44.6931 | 0.0224 | 312.0938 | 0.00320 | 0.14320 | 6.9830 | 6.4791 |
| 30 | 50.9502 | 0.0196 | 356.7870 | 0.00280 | 0.14280 | 7.0027 | 6.5423 |
| 35 | 98.1002 | 0.0102 | 693.5729 | 0.00144 | 0.14144 | 7.0700 | 6.7824 |
| 40 | 188.8836 | 0.0053 | 1342.0260 | 0.00075 | 0.14075 | 7.1050 | 6.9300 |
| 45 | 363.6792 | 0.0027 | 2590.5660 | 0.00039 | 0.14039 | 7.1232 | 7.0188 |
| 50 | 700.2331 | 0.0014 | 4994.5230 | 0.00020 | 0.14020 | 7.1327 | 7.0714 |
| 55 | 1348.2390 | 0.0007 | 9623.1380 | 0.00010 | 0.14010 | 7.1376 | 7.1020 |
| 60 | 2595.9200 | 0.0004 | 18535.1400 | 0.00005 | 0.14005 | 7.1401 | 7.1197 |
| 65 | 4998.2210 | 0.0002 | 35694.4400 | 0.00003 | 0.14003 | 7.1414 | 7.1299 |
| 70 | 9623.6490 | 0.0001 | 68733.2100 | 0.00001 | 0.14001 | 7.1421 | 7.1356 |
| 75 | 18529.5100 | 0.0001 | 132346.5000 | 0.00001 | 0.14001 | 7.1425 | 7.1388 |
| 80 | 35676.9900 | 0.0000 | 254828.6000 | 0.00000 | 0.14000 | 7.1427 | 7.1406 |

**i = 15%**

| n | F/P | P/F | F/A | A/F | A/P | P/A | A/G |
|---|---|---|---|---|---|---|---|
| 1 | 1.1500 | 0.8696 | 1.0000 | 1.00000 | 1.15000 | 0.8696 | −0.0000 |
| 2 | 1.3225 | 0.7561 | 2.1500 | 0.46512 | 0.61512 | 1.6257 | 0.4651 |
| 3 | 1.5209 | 0.6575 | 3.4725 | 0.28798 | 0.43798 | 2.2832 | 0.9071 |
| 4 | 1.7490 | 0.5718 | 4.9934 | 0.20027 | 0.35027 | 2.8550 | 1.3263 |
| 5 | 2.0114 | 0.4972 | 6.7424 | 0.14832 | 0.29832 | 3.3522 | 1.7228 |
| 6 | 2.3131 | 0.4323 | 8.7537 | 0.11424 | 0.26424 | 3.7845 | 2.0972 |
| 7 | 2.6600 | 0.3759 | 11.0668 | 0.09036 | 0.24036 | 4.1604 | 2.4499 |
| 8 | 3.0590 | 0.3269 | 13.7268 | 0.07285 | 0.22285 | 4.4873 | 2.7813 |
| 9 | 3.5179 | 0.2843 | 16.7858 | 0.05957 | 0.20957 | 4.7716 | 3.0922 |
| 10 | 4.0456 | 0.2472 | 20.3037 | 0.04925 | 0.19925 | 5.0188 | 3.3832 |
| 11 | 4.6524 | 0.2149 | 24.3493 | 0.04107 | 0.19107 | 5.2337 | 3.6549 |
| 12 | 5.3503 | 0.1869 | 29.0017 | 0.03448 | 0.18448 | 5.4206 | 3.9082 |
| 13 | 6.1528 | 0.1625 | 34.3519 | 0.02911 | 0.17911 | 5.5831 | 4.1438 |
| 14 | 7.0757 | 0.1413 | 40.5047 | 0.02469 | 0.17469 | 5.7245 | 4.3624 |
| 15 | 8.1371 | 0.1229 | 47.5804 | 0.02102 | 0.17102 | 5.8474 | 4.5650 |
| 16 | 9.3576 | 0.1069 | 55.7175 | 0.01795 | 0.16795 | 5.9542 | 4.7522 |
| 17 | 10.7613 | 0.0929 | 65.0751 | 0.01537 | 0.16537 | 6.0472 | 4.9251 |
| 18 | 12.3755 | 0.0808 | 75.8364 | 0.01319 | 0.16319 | 6.1280 | 5.0843 |
| 19 | 14.2318 | 0.0703 | 88.2118 | 0.01134 | 0.16134 | 6.1982 | 5.2307 |
| 20 | 16.3665 | 0.0611 | 102.4436 | 0.00976 | 0.15976 | 6.2593 | 5.3651 |
| 21 | 18.8215 | 0.0531 | 118.8101 | 0.00842 | 0.15842 | 6.3125 | 5.4883 |
| 22 | 21.6447 | 0.0462 | 137.6317 | 0.00727 | 0.15727 | 6.3587 | 5.6010 |
| 23 | 24.8915 | 0.0402 | 159.2764 | 0.00628 | 0.15628 | 6.3988 | 5.7040 |
| 24 | 28.6252 | 0.0349 | 184.1679 | 0.00543 | 0.15543 | 6.4338 | 5.7979 |
| 25 | 32.9190 | 0.0304 | 212.7930 | 0.00470 | 0.15470 | 6.4642 | 5.8834 |
| 26 | 37.8568 | 0.0264 | 245.7120 | 0.00407 | 0.15407 | 6.4906 | 5.9612 |
| 27 | 43.5353 | 0.0230 | 283.5688 | 0.00353 | 0.15353 | 6.5135 | 6.0319 |
| 28 | 50.0656 | 0.0200 | 327.1041 | 0.00306 | 0.15306 | 6.5335 | 6.0960 |
| 29 | 57.5755 | 0.0174 | 377.1697 | 0.00265 | 0.15265 | 6.5509 | 6.1541 |
| 30 | 66.2118 | 0.0151 | 434.7452 | 0.00230 | 0.15230 | 6.5660 | 6.2066 |
| 35 | 133.1755 | 0.0075 | 881.1702 | 0.00113 | 0.15113 | 6.6166 | 6.4019 |
| 40 | 267.8636 | 0.0037 | 1779.0910 | 0.00056 | 0.15056 | 6.6418 | 6.5168 |
| 45 | 538.7693 | 0.0019 | 3585.1290 | 0.00028 | 0.15028 | 6.6543 | 6.5830 |
| 50 | 1083.6580 | 0.0009 | 7217.7170 | 0.00014 | 0.15014 | 6.6605 | 6.6205 |
| 55 | 2179.6220 | 0.0005 | 14524.1500 | 0.00007 | 0.15007 | 6.6636 | 6.6414 |
| 60 | 4383.9990 | 0.0002 | 29220.0000 | 0.00003 | 0.15003 | 6.6651 | 6.6530 |
| 65 | 8817.7870 | 0.0001 | 58778.5900 | 0.00002 | 0.15002 | 6.6659 | 6.6593 |
| 70 | 17735.7200 | 0.0001 | 118231.5000 | 0.00001 | 0.15001 | 6.6663 | 6.6627 |
| 75 | 35672.8700 | 0.0000 | 237812.5000 | 0.00000 | 0.15000 | 6.6665 | 6.6646 |
| 80 | 71750.8800 | 0.0000 | 478332.6000 | 0.00000 | 0.15000 | 6.6666 | 6.6656 |

**i = 20%**

| n | F/P | P/F | F/A | A/F | A/P | P/A | A/G |
|---|---|---|---|---|---|---|---|
| 1 | 1.2000 | 0.8333 | 1.0000 | 1.00000 | 1.20000 | 0.8333 | 0.0000 |
| 2 | 1.4400 | 0.6944 | 2.2000 | 0.45455 | 0.65455 | 1.5278 | 0.4545 |
| 3 | 1.7280 | 0.5787 | 3.6400 | 0.27473 | 0.47473 | 2.1065 | 0.8791 |
| 4 | 2.0736 | 0.4823 | 5.3680 | 0.18629 | 0.38629 | 2.5887 | 1.2742 |
| 5 | 2.4883 | 0.4019 | 7.4416 | 0.13438 | 0.33438 | 2.9906 | 1.6405 |
| 6 | 2.9860 | 0.3349 | 9.9299 | 0.10071 | 0.30071 | 3.3255 | 1.9788 |
| 7 | 3.5832 | 0.2791 | 12.9159 | 0.07742 | 0.27742 | 3.6046 | 2.2902 |
| 8 | 4.2998 | 0.2326 | 16.4991 | 0.06061 | 0.26061 | 3.8372 | 2.5756 |
| 9 | 5.1598 | 0.1938 | 20.7989 | 0.04808 | 0.24808 | 4.0310 | 2.8364 |
| 10 | 6.1917 | 0.1615 | 25.9587 | 0.03852 | 0.23852 | 4.1925 | 3.0739 |
| 11 | 7.4301 | 0.1346 | 32.1504 | 0.03110 | 0.23110 | 4.3271 | 3.2893 |
| 12 | 8.9161 | 0.1122 | 39.5805 | 0.02526 | 0.22526 | 4.4392 | 3.4841 |
| 13 | 10.6993 | 0.0935 | 48.4966 | 0.02062 | 0.22062 | 4.5327 | 3.6597 |
| 14 | 12.8392 | 0.0779 | 59.1959 | 0.01689 | 0.21689 | 4.6106 | 3.8175 |
| 15 | 15.4070 | 0.0649 | 72.0351 | 0.01388 | 0.21388 | 4.6755 | 3.9588 |
| 16 | 18.4884 | 0.0541 | 87.4421 | 0.01144 | 0.21144 | 4.7296 | 4.0851 |
| 17 | 22.1861 | 0.0451 | 105.9306 | 0.00944 | 0.20944 | 4.7746 | 4.1976 |
| 18 | 26.6233 | 0.0376 | 128.1167 | 0.00781 | 0.20781 | 4.8122 | 4.2975 |
| 19 | 31.9480 | 0.0313 | 154.7400 | 0.00646 | 0.20646 | 4.8435 | 4.3861 |
| 20 | 38.3376 | 0.0261 | 186.6880 | 0.00536 | 0.20536 | 4.8696 | 4.4643 |
| 21 | 46.0051 | 0.0217 | 225.0256 | 0.00444 | 0.20444 | 4.8913 | 4.5334 |
| 22 | 55.2061 | 0.0181 | 271.0307 | 0.00369 | 0.20369 | 4.9094 | 4.5941 |
| 23 | 66.2474 | 0.0151 | 326.2369 | 0.00307 | 0.20307 | 4.9245 | 4.6475 |
| 24 | 79.4969 | 0.0126 | 392.4843 | 0.00255 | 0.20255 | 4.9371 | 4.6943 |
| 25 | 95.3962 | 0.0105 | 471.9811 | 0.00212 | 0.20212 | 4.9476 | 4.7352 |
| 26 | 114.4755 | 0.0087 | 567.3773 | 0.00176 | 0.20176 | 4.9563 | 4.7709 |
| 27 | 137.3706 | 0.0073 | 681.8529 | 0.00147 | 0.20147 | 4.9636 | 4.8020 |
| 28 | 164.8447 | 0.0061 | 819.2233 | 0.00122 | 0.20122 | 4.9697 | 4.8291 |
| 29 | 197.8136 | 0.0051 | 984.0681 | 0.00102 | 0.20102 | 4.9747 | 4.8527 |
| 30 | 237.3764 | 0.0042 | 1181.8820 | 0.00085 | 0.20085 | 4.9789 | 4.8731 |

**i = 25%**

| n | F/P | P/F | F/A | A/F | A/P | P/A | A/G |
|---|---|---|---|---|---|---|---|
| 1 | 1.2500 | 0.8000 | 1.0000 | 1.00000 | 1.25000 | 0.8000 | 0.0000 |
| 2 | 1.5625 | 0.6400 | 2.2500 | 0.44444 | 0.69444 | 1.4400 | 0.4444 |
| 3 | 1.9531 | 0.5120 | 3.8125 | 0.26230 | 0.51230 | 1.9520 | 0.8525 |
| 4 | 2.4414 | 0.4096 | 5.7656 | 0.17344 | 0.42344 | 2.3616 | 1.2249 |
| 5 | 3.0518 | 0.3277 | 8.2070 | 0.12185 | 0.37185 | 2.6893 | 1.5631 |
| 6 | 3.8147 | 0.2621 | 11.2588 | 0.08882 | 0.33882 | 2.9514 | 1.8683 |
| 7 | 4.7684 | 0.2097 | 15.0735 | 0.06634 | 0.31634 | 3.1611 | 2.1424 |
| 8 | 5.9605 | 0.1678 | 19.8419 | 0.05040 | 0.30040 | 3.3289 | 2.3872 |
| 9 | 7.4506 | 0.1342 | 25.8023 | 0.03876 | 0.28876 | 3.4631 | 2.6048 |
| 10 | 9.3132 | 0.1074 | 33.2529 | 0.03007 | 0.28007 | 3.5705 | 2.7971 |
| 11 | 11.6415 | 0.0859 | 42.5661 | 0.02349 | 0.27349 | 3.6564 | 2.9663 |
| 12 | 14.5519 | 0.0687 | 54.2077 | 0.01845 | 0.26845 | 3.7251 | 3.1145 |
| 13 | 18.1899 | 0.0550 | 68.7596 | 0.01454 | 0.26454 | 3.7801 | 3.2437 |
| 14 | 22.7374 | 0.0440 | 86.9495 | 0.01150 | 0.26150 | 3.8241 | 3.3559 |
| 15 | 28.4217 | 0.0352 | 109.6868 | 0.00912 | 0.25912 | 3.8593 | 3.4530 |
| 16 | 35.5271 | 0.0281 | 138.1086 | 0.00724 | 0.25724 | 3.8874 | 3.5366 |
| 17 | 44.4089 | 0.0225 | 173.6357 | 0.00576 | 0.25576 | 3.9099 | 3.6084 |
| 18 | 55.5112 | 0.0180 | 218.0446 | 0.00459 | 0.25459 | 3.9279 | 3.6698 |
| 19 | 69.3889 | 0.0144 | 273.5558 | 0.00366 | 0.25366 | 3.9424 | 3.7222 |
| 20 | 86.7362 | 0.0115 | 342.9447 | 0.00292 | 0.25292 | 3.9539 | 3.7667 |
| 21 | 108.4202 | 0.0092 | 429.6809 | 0.00233 | 0.25233 | 3.9631 | 3.8045 |
| 22 | 135.5253 | 0.0074 | 538.1011 | 0.00186 | 0.25186 | 3.9705 | 3.8365 |
| 23 | 169.4066 | 0.0059 | 673.6263 | 0.00148 | 0.25148 | 3.9764 | 3.8634 |
| 24 | 211.7583 | 0.0047 | 843.0329 | 0.00119 | 0.25119 | 3.9811 | 3.8861 |
| 25 | 264.6978 | 0.0038 | 1054.7910 | 0.00095 | 0.25095 | 3.9849 | 3.9052 |
| 26 | 330.8723 | 0.0030 | 1319.4890 | 0.00076 | 0.25076 | 3.9879 | 3.9212 |
| 27 | 413.5903 | 0.0024 | 1650.3610 | 0.00061 | 0.25061 | 3.9903 | 3.9346 |
| 28 | 516.9879 | 0.0019 | 2063.9520 | 0.00048 | 0.25048 | 3.9923 | 3.9457 |
| 29 | 646.2349 | 0.0015 | 2580.9400 | 0.00039 | 0.25039 | 3.9938 | 3.9551 |
| 30 | 807.7936 | 0.0012 | 3227.1740 | 0.00031 | 0.25031 | 3.9950 | 3.9628 |

### i = 30%

| n | F/P | P/F | F/A | A/F | A/P | P/A | A/G |
|---|---|---|---|---|---|---|---|
| 1 | 1.3000 | 0.7692 | 1.0000 | 1.00000 | 1.30000 | 0.7692 | −0.0000 |
| 2 | 1.6900 | 0.5917 | 2.3000 | 0.43478 | 0.73478 | 1.3609 | 0.4348 |
| 3 | 2.1970 | 0.4552 | 3.9900 | 0.25063 | 0.55063 | 1.8161 | 0.8271 |
| 4 | 2.8561 | 0.3501 | 6.1870 | 0.16163 | 0.46163 | 2.1662 | 1.1783 |
| 5 | 3.7129 | 0.2693 | 9.0431 | 0.11058 | 0.41058 | 2.4356 | 1.4903 |
| 6 | 4.8268 | 0.2072 | 12.7560 | 0.07839 | 0.37839 | 2.6427 | 1.7654 |
| 7 | 6.2748 | 0.1594 | 17.5828 | 0.05687 | 0.35687 | 2.8021 | 2.0063 |
| 8 | 8.1573 | 0.1226 | 23.8577 | 0.04192 | 0.34192 | 2.9247 | 2.2156 |
| 9 | 10.6045 | 0.0943 | 32.0150 | 0.03124 | 0.33124 | 3.0190 | 2.3963 |
| 10 | 13.7858 | 0.0725 | 42.6195 | 0.02346 | 0.32346 | 3.0915 | 2.5512 |
| 11 | 17.9216 | 0.0558 | 56.4053 | 0.01773 | 0.31773 | 3.1473 | 2.6833 |
| 12 | 23.2981 | 0.0429 | 74.3269 | 0.01345 | 0.31345 | 3.1903 | 2.7952 |
| 13 | 30.2875 | 0.0330 | 97.6250 | 0.01024 | 0.31024 | 3.2233 | 2.8895 |
| 14 | 39.3737 | 0.0254 | 127.9124 | 0.00782 | 0.30782 | 3.2487 | 2.9685 |
| 15 | 51.1859 | 0.0195 | 167.2862 | 0.00598 | 0.30598 | 3.2682 | 3.0344 |
| 16 | 66.5416 | 0.0150 | 218.4720 | 0.00458 | 0.30458 | 3.2832 | 3.0892 |
| 17 | 86.5041 | 0.0116 | 285.0136 | 0.00351 | 0.30351 | 3.2948 | 3.1345 |
| 18 | 112.4553 | 0.0089 | 371.5177 | 0.00269 | 0.30269 | 3.3037 | 3.1718 |
| 19 | 146.1919 | 0.0068 | 483.9729 | 0.00207 | 0.30207 | 3.3105 | 3.2025 |
| 20 | 190.0494 | 0.0053 | 630.1648 | 0.00159 | 0.30159 | 3.3158 | 3.2275 |
| 21 | 247.0643 | 0.0040 | 820.2141 | 0.00122 | 0.30122 | 3.3198 | 3.2480 |
| 22 | 321.1835 | 0.0031 | 1067.2780 | 0.00094 | 0.30094 | 3.3230 | 3.2646 |
| 23 | 417.5385 | 0.0024 | 1388.4620 | 0.00072 | 0.30072 | 3.3253 | 3.2781 |
| 24 | 542.8001 | 0.0018 | 1806.0000 | 0.00055 | 0.30055 | 3.3272 | 3.2890 |
| 25 | 705.6400 | 0.0014 | 2348.8000 | 0.00043 | 0.30043 | 3.3286 | 3.2979 |
| 26 | 917.3321 | 0.0011 | 3054.4400 | 0.00033 | 0.30033 | 3.3297 | 3.3050 |
| 27 | 1192.5320 | 0.0008 | 3971.7720 | 0.00025 | 0.30025 | 3.3305 | 3.3107 |
| 28 | 1550.2910 | 0.0006 | 5164.3030 | 0.00019 | 0.30019 | 3.3312 | 3.3153 |
| 29 | 2015.3780 | 0.0005 | 6714.5940 | 0.00015 | 0.30015 | 3.3317 | 3.3189 |
| 30 | 2619.9910 | 0.0004 | 8729.9710 | 0.00011 | 0.30011 | 3.3321 | 3.3219 |

### i = 35%

| n | F/P | P/F | F/A | A/F | A/P | P/A | A/G |
|---|---|---|---|---|---|---|---|
| 1 | 1.3500 | 0.7407 | 1.0000 | 1.00000 | 1.35000 | 0.7407 | 0.0000 |
| 2 | 1.8225 | 0.5487 | 2.3500 | 0.42553 | 0.77553 | 1.2894 | 0.4255 |
| 3 | 2.4604 | 0.4064 | 4.1725 | 0.23966 | 0.58966 | 1.6959 | 0.8029 |
| 4 | 3.3215 | 0.3011 | 6.6329 | 0.15076 | 0.50076 | 1.9969 | 1.1341 |
| 5 | 4.4840 | 0.2230 | 9.9544 | 0.10046 | 0.45046 | 2.2200 | 1.4220 |
| 6 | 6.0534 | 0.1652 | 14.4384 | 0.06926 | 0.41926 | 2.3852 | 1.6698 |
| 7 | 8.1722 | 0.1224 | 20.4919 | 0.04880 | 0.39880 | 2.5075 | 1.8811 |
| 8 | 11.0324 | 0.0906 | 28.6640 | 0.03489 | 0.38489 | 2.5982 | 2.0597 |
| 9 | 14.8937 | 0.0671 | 39.6964 | 0.02519 | 0.37519 | 2.6653 | 2.2094 |
| 10 | 20.1066 | 0.0497 | 54.5902 | 0.01832 | 0.36832 | 2.7150 | 2.3338 |
| 11 | 27.1439 | 0.0368 | 74.6967 | 0.01339 | 0.36339 | 2.7519 | 2.4364 |
| 12 | 36.6442 | 0.0273 | 101.8406 | 0.00982 | 0.35982 | 2.7792 | 2.5205 |
| 13 | 49.4697 | 0.0202 | 138.4848 | 0.00722 | 0.35722 | 2.7994 | 2.5889 |
| 14 | 66.7841 | 0.0150 | 187.9544 | 0.00532 | 0.35532 | 2.8144 | 2.6443 |
| 15 | 90.1585 | 0.0111 | 254.7385 | 0.00393 | 0.35393 | 2.8255 | 2.6889 |
| 16 | 121.7139 | 0.0082 | 344.8970 | 0.00290 | 0.35290 | 2.8337 | 2.7246 |
| 17 | 164.3138 | 0.0061 | 466.6109 | 0.00214 | 0.35214 | 2.8398 | 2.7530 |
| 18 | 221.8237 | 0.0045 | 630.9246 | 0.00158 | 0.35158 | 2.8443 | 2.7756 |
| 19 | 299.4620 | 0.0033 | 852.7484 | 0.00117 | 0.35117 | 2.8476 | 2.7935 |
| 20 | 404.2736 | 0.0025 | 1152.2100 | 0.00087 | 0.35087 | 2.8501 | 2.8075 |
| 21 | 545.7694 | 0.0018 | 1556.4840 | 0.00064 | 0.35064 | 2.8519 | 2.8186 |
| 22 | 736.7886 | 0.0014 | 2102.2530 | 0.00048 | 0.35048 | 2.8533 | 2.8272 |
| 23 | 994.6648 | 0.0010 | 2839.0420 | 0.00035 | 0.35035 | 2.8543 | 2.8340 |
| 24 | 1342.7970 | 0.0007 | 3833.7070 | 0.00026 | 0.35026 | 2.8550 | 2.8393 |
| 25 | 1812.7760 | 0.0006 | 5176.5040 | 0.00019 | 0.35019 | 2.8556 | 2.8433 |
| 26 | 2447.2480 | 0.0004 | 6989.2800 | 0.00014 | 0.35014 | 2.8560 | 2.8465 |
| 27 | 3303.7850 | 0.0003 | 9436.5280 | 0.00011 | 0.35011 | 2.8563 | 2.8490 |
| 28 | 4460.1100 | 0.0002 | 12740.3100 | 0.00008 | 0.35008 | 2.8565 | 2.8509 |
| 29 | 6021.1480 | 0.0002 | 17200.4200 | 0.00006 | 0.35006 | 2.8567 | 2.8523 |
| 30 | 8128.5500 | 0.0001 | 23221.5700 | 0.00004 | 0.35004 | 2.8568 | 2.8535 |

**i = 40%**

| n | F/P | P/F | F/A | A/F | A/P | P/A | A/G |
|---|---|---|---|---|---|---|---|
| 1 | 1.4000 | 0.7143 | 1.0000 | 1.00000 | 1.40000 | 0.7143 | 0.0000 |
| 2 | 1.9600 | 0.5102 | 2.4000 | 0.41667 | 0.81667 | 1.2245 | 0.4167 |
| 3 | 2.7440 | 0.3644 | 4.3600 | 0.22936 | 0.62936 | 1.5889 | 0.7798 |
| 4 | 3.8416 | 0.2603 | 7.1040 | 0.14077 | 0.54077 | 1.8492 | 1.0923 |
| 5 | 5.3782 | 0.1859 | 10.9456 | 0.09136 | 0.49136 | 2.0352 | 1.3580 |
| 6 | 7.5295 | 0.1328 | 16.3238 | 0.06126 | 0.46126 | 2.1680 | 1.5811 |
| 7 | 10.5414 | 0.0949 | 23.8534 | 0.04192 | 0.44192 | 2.2628 | 1.7664 |
| 8 | 14.7579 | 0.0678 | 34.3947 | 0.02907 | 0.42907 | 2.3306 | 1.9185 |
| 9 | 20.6611 | 0.0484 | 49.1526 | 0.02034 | 0.42034 | 2.3790 | 2.0422 |
| 10 | 28.9255 | 0.0346 | 69.8137 | 0.01432 | 0.41432 | 2.4136 | 2.1419 |
| 11 | 40.4957 | 0.0247 | 98.7392 | 0.01013 | 0.41013 | 2.4383 | 2.2215 |
| 12 | 56.6940 | 0.0176 | 139.2349 | 0.00718 | 0.40718 | 2.4559 | 2.2845 |
| 13 | 79.3716 | 0.0126 | 195.9289 | 0.00510 | 0.40510 | 2.4685 | 2.3341 |
| 14 | 111.1202 | 0.0090 | 275.3004 | 0.00363 | 0.40363 | 2.4775 | 2.3729 |
| 15 | 155.5683 | 0.0064 | 386.4206 | 0.00259 | 0.40259 | 2.4839 | 2.4030 |
| 16 | 217.7956 | 0.0046 | 541.9890 | 0.00185 | 0.40185 | 2.4885 | 2.4262 |
| 17 | 304.9139 | 0.0033 | 759.7846 | 0.00132 | 0.40132 | 2.4918 | 2.4441 |
| 18 | 426.8795 | 0.0023 | 1064.6990 | 0.00094 | 0.40094 | 2.4941 | 2.4577 |
| 19 | 597.6313 | 0.0017 | 1491.5780 | 0.00067 | 0.40067 | 2.4958 | 2.4682 |
| 20 | 836.6838 | 0.0012 | 2089.2090 | 0.00048 | 0.40048 | 2.4970 | 2.4761 |
| 21 | 11171.3570 | 0.0009 | 2925.8930 | 0.00034 | 0.40034 | 2.4979 | 2.4821 |
| 22 | 1639.9000 | 0.0006 | 4097.2510 | 0.00024 | 0.40024 | 2.4985 | 2.4866 |
| 23 | 2295.8610 | 0.0004 | 5737.1510 | 0.00017 | 0.40017 | 2.4989 | 2.4900 |
| 24 | 3214.2050 | 0.0003 | 8033.0130 | 0.00012 | 0.40012 | 2.4992 | 2.4925 |
| 25 | 4499.8880 | 0.0002 | 11247.2200 | 0.00009 | 0.40009 | 2.4994 | 2.4944 |
| 26 | 6299.8440 | 0.0002 | 15747.1100 | 0.00006 | 0.40006 | 2.4996 | 2.4959 |
| 27 | 8819.7810 | 0.0001 | 22046.9500 | 0.00005 | 0.40005 | 2.4997 | 2.4969 |
| 28 | 12347.7000 | 0.0001 | 30866.7400 | 0.00003 | 0.40003 | 2.4998 | 2.4977 |
| 29 | 17286.7700 | 0.0001 | 43214.4300 | 0.00002 | 0.40002 | 2.4999 | 2.4983 |
| 30 | 24201.4900 | 0.0000 | 60501.2100 | 0.00002 | 0.40002 | 2.4999 | 2.4988 |

**i = 45%**

| n | F/P | P/F | F/A | A/F | A/P | P/A | A/G |
|---|---|---|---|---|---|---|---|
| 1 | 1.4500 | 0.6897 | 1.0000 | 1.00000 | 1.45000 | 0.6897 | 0.0000 |
| 2 | 2.1025 | 0.4756 | 2.4500 | 0.40816 | 0.85816 | 1.1653 | 0.4082 |
| 3 | 3.0486 | 0.3280 | 4.5525 | 0.21966 | 0.66966 | 1.4933 | 0.7578 |
| 4 | 4.4205 | 0.2262 | 7.6011 | 0.13156 | 0.58156 | 1.7195 | 1.0528 |
| 5 | 6.4097 | 0.1560 | 12.0216 | 0.08318 | 0.53318 | 1.8755 | 1.2980 |
| 6 | 9.2941 | 0.1076 | 18.4314 | 0.05426 | 0.50426 | 1.9831 | 1.4988 |
| 7 | 13.4765 | 0.0742 | 27.7255 | 0.03607 | 0.48607 | 2.0573 | 1.6612 |
| 8 | 19.5409 | 0.0512 | 41.2020 | 0.02427 | 0.47427 | 2.1085 | 1.7907 |
| 9 | 28.3343 | 0.0353 | 60.7428 | 0.01646 | 0.46646 | 2.1438 | 1.8930 |
| 10 | 41.0847 | 0.0243 | 89.0771 | 0.01123 | 0.46123 | 2.1681 | 1.9728 |
| 11 | 59.5728 | 0.0168 | 130.1618 | 0.00768 | 0.45768 | 2.1849 | 2.0344 |
| 12 | 86.3806 | 0.0116 | 189.7347 | 0.00527 | 0.45527 | 2.1965 | 2.0817 |
| 13 | 125.2519 | 0.0080 | 276.1153 | 0.00362 | 0.45362 | 2.2045 | 2.1176 |
| 14 | 181.6153 | 0.0055 | 401.3672 | 0.00249 | 0.45249 | 2.2100 | 2.1447 |
| 15 | 263.3421 | 0.0038 | 582.9824 | 0.00172 | 0.45172 | 2.2138 | 2.1650 |
| 16 | 381.8461 | 0.0026 | 846.3245 | 0.00118 | 0.45118 | 2.2164 | 2.1802 |
| 17 | 553.6768 | 0.0018 | 1228.1710 | 0.00081 | 0.45081 | 2.2182 | 2.1915 |
| 18 | 802.8315 | 0.0012 | 1781.8470 | 0.00056 | 0.45056 | 2.2195 | 2.1998 |
| 19 | 1164.1060 | 0.0009 | 2584.6790 | 0.00039 | 0.45039 | 2.2203 | 2.2059 |
| 20 | 1687.9530 | 0.0006 | 3748.7850 | 0.00027 | 0.45027 | 2.2209 | 2.2104 |
| 21 | 2447.5320 | 0.0004 | 5436.7380 | 0.00018 | 0.45018 | 2.2213 | 2.2136 |
| 22 | 3548.9220 | 0.0003 | 7884.2710 | 0.00013 | 0.45013 | 2.2216 | 2.2160 |
| 23 | 5145.9370 | 0.0002 | 11433.1900 | 0.00009 | 0.45009 | 2.2218 | 2.2178 |
| 24 | 7461.6090 | 0.0001 | 16579.1300 | 0.00006 | 0.45006 | 2.2219 | 2.2190 |
| 25 | 10819.3300 | 0.0001 | 24040.7400 | 0.00004 | 0.45004 | 2.2220 | 2.2199 |
| 26 | 15688.0400 | 0.0001 | 34860.0800 | 0.00003 | 0.45003 | 2.2221 | 2.2206 |
| 27 | 22747.6500 | 0.0000 | 50548.1100 | 0.00002 | 0.45002 | 2.2221 | 2.2210 |
| 28 | 32984.1000 | 0.0000 | 73295.7600 | 0.00001 | 0.45001 | 2.2222 | 2.2214 |
| 29 | 47826.9400 | 0.0000 | 106279.9000 | 0.00001 | 0.45001 | 2.2222 | 2.2216 |
| 30 | 69349.0700 | 0.0000 | 154106.8000 | 0.00001 | 0.45001 | 2.2222 | 2.2218 |

i = 50%

| n | F/P | P/F | F/A | A/F | A/P | P/A | A/G |
|---|---|---|---|---|---|---|---|
| 1 | 1.5000 | 0.6667 | 1.0000 | 1.00000 | 1.50000 | 0.6667 | −0.0000 |
| 2 | 2.2500 | 0.4444 | 2.5000 | 0.40000 | 0.90000 | 1.1111 | 0.4000 |
| 3 | 3.3750 | 0.2963 | 4.7500 | 0.21053 | 0.71053 | 1.4074 | 0.7368 |
| 4 | 5.0625 | 0.1975 | 8.1250 | 0.12308 | 0.62308 | 1.6049 | 1.0154 |
| 5 | 7.5938 | 0.1317 | 13.1875 | 0.07583 | 0.57583 | 1.7366 | 1.2417 |
| 6 | 11.3906 | 0.0878 | 20.7812 | 0.04812 | 0.54812 | 1.8244 | 1.4226 |
| 7 | 17.0859 | 0.0585 | 32.1719 | 0.03108 | 0.53108 | 1.8829 | 1.5648 |
| 8 | 25.6289 | 0.0390 | 49.2578 | 0.02030 | 0.52030 | 1.9220 | 1.6752 |
| 9 | 38.4434 | 0.0260 | 74.8867 | 0.01335 | 0.51335 | 1.9480 | 1.7596 |
| 10 | 57.6650 | 0.0173 | 113.3301 | 0.00882 | 0.50882 | 1.9653 | 1.8235 |
| 11 | 86.4976 | 0.0116 | 170.9951 | 0.00585 | 0.50585 | 1.9769 | 1.8713 |
| 12 | 129.7463 | 0.0077 | 257.4927 | 0.00388 | 0.50388 | 1.9846 | 1.9068 |
| 13 | 194.6195 | 0.0051 | 387.2390 | 0.00258 | 0.50258 | 1.9897 | 1.9329 |
| 14 | 291.9293 | 0.0034 | 581.8585 | 0.00172 | 0.50172 | 1.9931 | 1.9519 |
| 15 | 437.8939 | 0.0023 | 873.7877 | 0.00114 | 0.50114 | 1.9954 | 1.9657 |
| 16 | 656.8408 | 0.0015 | 1311.6820 | 0.00076 | 0.50076 | 1.9970 | 1.9756 |
| 17 | 985.2612 | 0.0010 | 1968.5220 | 0.00051 | 0.50051 | 1.9980 | 1.9827 |
| 18 | 1477.8920 | 0.0007 | 2953.7840 | 0.00034 | 0.50034 | 1.9986 | 1.9878 |
| 19 | 2216.8380 | 0.0005 | 4431.6760 | 0.00023 | 0.50023 | 1.9991 | 1.9914 |
| 20 | 3325.2570 | 0.0003 | 6648.5120 | 0.00015 | 0.50015 | 1.9994 | 1.9940 |
| 21 | 4987.8850 | 0.0002 | 9973.7690 | 0.00010 | 0.50010 | 1.9996 | 1.9958 |
| 22 | 7481.8280 | 0.0001 | 14961.6500 | 0.00007 | 0.50007 | 1.9997 | 1.9971 |
| 23 | 11222.7400 | 0.0001 | 22443.4800 | 0.00004 | 0.50004 | 1.9998 | 1.9980 |
| 24 | 16834.1100 | 0.0001 | 33666.2200 | 0.00003 | 0.50003 | 1.9999 | 1.9986 |
| 25 | 25251.1700 | 0.0000 | 50500.3300 | 0.00002 | 0.50002 | 1.9999 | 1.9990 |
| 26 | 37876.7500 | 0.0000 | 75751.5000 | 0.00001 | 0.50001 | 1.9999 | 1.9993 |
| 27 | 56815.1300 | 0.0000 | 113628.2000 | 0.00001 | 0.50001 | 2.0000 | 1.9995 |
| 28 | 85222.6900 | 0.0000 | 170443.4000 | 0.00001 | 0.50001 | 2.0000 | 1.9997 |
| 29 | 127834.0000 | 0.0000 | 255666.1000 | 0.00000 | 0.50000 | 2.0000 | 1.9998 |
| 30 | 191751.1000 | 0.0000 | 383500.1000 | 0.00000 | 0.50000 | 2.0000 | 1.9998 |

i = 60%

| n | F/P | P/F | F/A | A/F | A/P | P/A | A/G |
|---|---|---|---|---|---|---|---|
| 1 | 1.6000 | 0.6250 | 1.0000 | 1.00000 | 1.60000 | 0.6250 | 0.0000 |
| 2 | 2.5600 | 0.3906 | 2.6000 | 0.38462 | 0.98462 | 1.0156 | 0.3846 |
| 3 | 4.0960 | 0.2441 | 5.1600 | 0.19380 | 0.79380 | 1.2598 | 0.6977 |
| 4 | 6.5536 | 0.1526 | 9.2560 | 0.10804 | 0.70804 | 1.4124 | 0.9464 |
| 5 | 10.4858 | 0.0954 | 15.8096 | 0.06325 | 0.66325 | 1.5077 | 1.1396 |
| 6 | 16.7772 | 0.0596 | 26.2954 | 0.03803 | 0.63803 | 1.5673 | 1.2864 |
| 7 | 26.8435 | 0.0373 | 43.0726 | 0.02322 | 0.62322 | 1.6046 | 1.3958 |
| 8 | 42.9497 | 0.0233 | 69.9161 | 0.01430 | 0.61430 | 1.6279 | 1.4760 |
| 9 | 68.7195 | 0.0146 | 112.8658 | 0.00886 | 0.60886 | 1.6424 | 1.5338 |
| 10 | 109.9512 | 0.0091 | 181.5852 | 0.00551 | 0.60551 | 1.6515 | 1.5749 |
| 11 | 175.9218 | 0.0057 | 291.5364 | 0.00343 | 0.60343 | 1.6572 | 1.6038 |
| 12 | 281.4750 | 0.0036 | 467.4582 | 0.00214 | 0.60214 | 1.6607 | 1.6239 |
| 13 | 450.3599 | 0.0022 | 748.9331 | 0.00134 | 0.60134 | 1.6630 | 1.6377 |
| 14 | 720.5758 | 0.0014 | 1199.2930 | 0.00083 | 0.60083 | 1.6644 | 1.6472 |
| 15 | 1152.9210 | 0.0009 | 1919.8690 | 0.00052 | 0.60052 | 1.6652 | 1.6536 |
| 16 | 1844.6740 | 0.0005 | 3072.7900 | 0.00033 | 0.60033 | 1.6658 | 1.6580 |
| 17 | 2951.4790 | 0.0003 | 4917.4640 | 0.00020 | 0.60020 | 1.6661 | 1.6609 |
| 18 | 4722.3650 | 0.0002 | 7868.9420 | 0.00013 | 0.60013 | 1.6663 | 1.6629 |
| 19 | 7555.7840 | 0.0001 | 12591.3100 | 0.00008 | 0.60008 | 1.6664 | 1.6642 |
| 20 | 12089.2600 | 0.0001 | 20147.0900 | 0.00005 | 0.60005 | 1.6665 | 1.6650 |

i = 65%

| n | F/P | P/F | F/A | A/F | A/P | P/A | A/G |
|---|---|---|---|---|---|---|---|
| 1 | 1.6500 | 0.6061 | 1.0000 | 1.00000 | 1.65000 | 0.6061 | 0.0000 |
| 2 | 2.7225 | 0.3673 | 2.6500 | 0.37736 | 1.02736 | 0.9734 | 0.3774 |
| 3 | 4.4921 | 0.2226 | 5.3725 | 0.18613 | 0.83613 | 1.1960 | 0.6794 |
| 4 | 7.4120 | 0.1349 | 9.8646 | 0.10137 | 0.75137 | 1.3309 | 0.9146 |
| 5 | 12.2298 | 0.0818 | 17.2766 | 0.05788 | 0.70788 | 1.4127 | 1.0932 |
| 6 | 20.1792 | 0.0496 | 29.5065 | 0.03389 | 0.68389 | 1.4622 | 1.2256 |
| 7 | 33.2957 | 0.0300 | 49.6856 | 0.02013 | 0.67013 | 1.4923 | 1.3217 |
| 8 | 54.9379 | 0.0182 | 82.9813 | 0.01205 | 0.66205 | 1.5105 | 1.3901 |
| 9 | 90.6475 | 0.0110 | 137.9192 | 0.00725 | 0.65725 | 1.5215 | 1.4381 |
| 10 | 149.5683 | 0.0067 | 228.5667 | 0.00438 | 0.65438 | 1.5282 | 1.4712 |
| 11 | 246.7878 | 0.0041 | 378.1350 | 0.00264 | 0.65264 | 1.5322 | 1.4937 |
| 12 | 407.1998 | 0.0025 | 624.9228 | 0.00160 | 0.65160 | 1.5347 | 1.5089 |
| 13 | 671.8798 | 0.0015 | 1032.1230 | 0.00097 | 0.65097 | 1.5362 | 1.5191 |
| 14 | 1108.6020 | 0.0009 | 1704.0030 | 0.00059 | 0.65059 | 1.5371 | 1.5258 |
| 15 | 1829.1930 | 0.0005 | 2812.6040 | 0.00036 | 0.65036 | 1.5376 | 1.5303 |
| 16 | 3018.1680 | 0.0003 | 4641.7970 | 0.00022 | 0.65022 | 1.5380 | 1.5332 |
| 17 | 4979.9780 | 0.0002 | 7659.9650 | 0.00013 | 0.65013 | 1.5382 | 1.5350 |
| 18 | 8216.9640 | 0.0001 | 12639.9400 | 0.00008 | 0.65008 | 1.5383 | 1.5363 |
| 19 | 13557.9900 | 0.0001 | 20856.9100 | 0.00005 | 0.65005 | 1.5383 | 1.5371 |
| 20 | 22370.6900 | 0.0000 | 34414.9000 | 0.00003 | 0.65003 | 1.5384 | 1.5376 |

**i = 70%**

| n | F/P | P/F | F/A | A/F | A/P | P/A | A/G |
|---|---|---|---|---|---|---|---|
| 1 | 1.7000 | 0.5882 | 1.0000 | 1.00000 | 1.70000 | 0.5882 | 0.0000 |
| 2 | 2.8900 | 0.3460 | 2.7000 | 0.37037 | 1.07037 | 0.9343 | 0.3704 |
| 3 | 4.9130 | 0.2035 | 5.5900 | 0.17889 | 0.87889 | 1.1378 | 0.6619 |
| 4 | 8.3521 | 0.1197 | 10.5030 | 0.09521 | 0.79521 | 1.2575 | 0.8845 |
| 5 | 14.1986 | 0.0704 | 18.8551 | 0.05304 | 0.75304 | 1.3280 | 1.0497 |
| 6 | 24.1376 | 0.0414 | 33.0537 | 0.03025 | 0.73025 | 1.3694 | 1.1693 |
| 7 | 41.0339 | 0.0244 | 57.1912 | 0.01749 | 0.71749 | 1.3938 | 1.2537 |
| 8 | 69.7576 | 0.0143 | 98.2251 | 0.01018 | 0.71018 | 1.4081 | 1.3122 |
| 9 | 118.5879 | 0.0084 | 167.9827 | 0.00595 | 0.70595 | 1.4165 | 1.3520 |
| 10 | 201.5994 | 0.0050 | 286.5706 | 0.00349 | 0.70349 | 1.4215 | 1.3787 |
| 11 | 342.7190 | 0.0029 | 488.1700 | 0.00205 | 0.70205 | 1.4244 | 1.3964 |
| 12 | 582.6223 | 0.0017 | 830.8890 | 0.00120 | 0.70120 | 1.4261 | 1.4079 |
| 13 | 990.4580 | 0.0010 | 1413.5110 | 0.00071 | 0.70071 | 1.4271 | 1.4154 |
| 14 | 1683.7790 | 0.0006 | 2403.9700 | 0.00042 | 0.70042 | 1.4277 | 1.4203 |
| 15 | 2862.4240 | 0.0003 | 4087.7480 | 0.00024 | 0.70024 | 1.4281 | 1.4233 |
| 16 | 4866.1210 | 0.0002 | 6950.1720 | 0.00014 | 0.70014 | 1.4283 | 1.4253 |
| 17 | 8272.4050 | 0.0001 | 11816.2900 | 0.00008 | 0.70008 | 1.4284 | 1.4265 |
| 18 | 14063.0900 | 0.0001 | 20088.7000 | 0.00005 | 0.70005 | 1.4285 | 1.4273 |
| 19 | 23907.2500 | 0.0000 | 34151.7900 | 0.00003 | 0.70003 | 1.4285 | 1.4278 |
| 20 | 40642.3300 | 0.0000 | 58059.0400 | 0.00002 | 0.70002 | 1.4285 | 1.4281 |

**i = 75%**

| n | F/P | P/F | F/A | A/F | A/P | P/A | A/G |
|---|---|---|---|---|---|---|---|
| 1 | 1.7500 | 0.5714 | 1.0000 | 1.00000 | 1.75000 | 0.5714 | −0.0000 |
| 2 | 3.0625 | 0.3265 | 2.7500 | 0.36364 | 1.11364 | 0.8980 | 0.3636 |
| 3 | 5.3594 | 0.1866 | 5.8125 | 0.17204 | 0.92204 | 1.0845 | 0.6452 |
| 4 | 9.3789 | 0.1066 | 11.1719 | 0.08951 | 0.83951 | 1.1912 | 0.8559 |
| 5 | 16.4131 | 0.0609 | 20.5508 | 0.04866 | 0.79866 | 1.2521 | 1.0089 |
| 6 | 28.7229 | 0.0348 | 36.9639 | 0.02705 | 0.77705 | 1.2869 | 1.1169 |
| 7 | 50.2651 | 0.0199 | 65.6868 | 0.01522 | 0.76522 | 1.3068 | 1.1912 |
| 8 | 87.9639 | 0.0114 | 115.9518 | 0.00862 | 0.75862 | 1.3182 | 1.2413 |
| 9 | 153.9368 | 0.0065 | 203.9157 | 0.00490 | 0.75490 | 1.3247 | 1.2745 |
| 10 | 269.3894 | 0.0037 | 357.8525 | 0.00279 | 0.75279 | 1.3284 | 1.2961 |
| 11 | 471.4315 | 0.0021 | 627.2419 | 0.00159 | 0.75159 | 1.3305 | 1.3100 |
| 12 | 825.0050 | 0.0012 | 1098.6730 | 0.00091 | 0.75091 | 1.3317 | 1.3188 |
| 13 | 1443.7590 | 0.0007 | 1923.6780 | 0.00052 | 0.75052 | 1.3324 | 1.3243 |
| 14 | 2526.5780 | 0.0004 | 3367.4370 | 0.00030 | 0.75030 | 1.3328 | 1.3278 |
| 15 | 4421.5110 | 0.0002 | 5894.0150 | 0.00017 | 0.75017 | 1.3330 | 1.3299 |
| 16 | 7737.6450 | 0.0001 | 10315.5300 | 0.00010 | 0.75010 | 1.3332 | 1.3313 |
| 17 | 13540.8800 | 0.0001 | 18053.1700 | 0.00006 | 0.75006 | 1.3332 | 1.3321 |
| 18 | 23696.5400 | 0.0000 | 31594.0500 | 0.00003 | 0.75003 | 1.3333 | 1.3326 |
| 19 | 41468.9400 | 0.0000 | 55290.5800 | 0.00002 | 0.75002 | 1.3333 | 1.3329 |
| 20 | 72570.6500 | 0.0000 | 96759.5200 | 0.00001 | 0.75001 | 1.3333 | 1.3331 |

**i = 80%**

| n | F/P | P/F | F/A | A/F | A/P | P/A | A/G |
|---|---|---|---|---|---|---|---|
| 1 | 1.8000 | 0.5556 | 1.0000 | 1.00000 | 1.80000 | 0.5556 | 0.0000 |
| 2 | 3.2400 | 0.3086 | 2.8000 | 0.35714 | 1.15714 | 0.8642 | 0.3571 |
| 3 | 5.8320 | 0.1715 | 6.0400 | 0.16556 | 0.96556 | 1.0357 | 0.6291 |
| 4 | 10.4976 | 0.0953 | 11.8720 | 0.08423 | 0.88423 | 1.1309 | 0.8288 |
| 5 | 18.8957 | 0.0529 | 22.3696 | 0.04470 | 0.84470 | 1.1838 | 0.9706 |
| 6 | 34.0122 | 0.0294 | 41.2653 | 0.02423 | 0.82423 | 1.2132 | 1.0682 |
| 7 | 61.2220 | 0.0163 | 75.2775 | 0.01328 | 0.81328 | 1.2296 | 1.1338 |
| 8 | 110.1996 | 0.0091 | 136.4995 | 0.00733 | 0.80733 | 1.2387 | 1.1767 |
| 9 | 198.3594 | 0.0050 | 246.6992 | 0.00405 | 0.80405 | 1.2437 | 1.2044 |
| 10 | 357.0469 | 0.0028 | 445.0585 | 0.00225 | 0.80225 | 1.2465 | 1.2219 |
| 11 | 642.6844 | 0.0016 | 802.1055 | 0.00125 | 0.80125 | 1.2481 | 1.2329 |
| 12 | 1156.8320 | 0.0009 | 1444.7900 | 0.00069 | 0.80069 | 1.2489 | 1.2396 |
| 13 | 2082.2970 | 0.0005 | 2601.6220 | 0.00038 | 0.80038 | 1.2494 | 1.2438 |
| 14 | 3748.1360 | 0.0003 | 4683.9190 | 0.00021 | 0.80021 | 1.2497 | 1.2463 |
| 15 | 6746.6450 | 0.0001 | 8432.0550 | 0.00012 | 0.80012 | 1.2498 | 1.2478 |
| 16 | 12143.9600 | 0.0001 | 15178.7000 | 0.00007 | 0.80007 | 1.2499 | 1.2487 |
| 17 | 21859.1300 | 0.0000 | 27322.6600 | 0.00004 | 0.80004 | 1.2499 | 1.2492 |
| 18 | 39346.4300 | 0.0000 | 49181.7900 | 0.00002 | 0.80002 | 1.2500 | 1.2495 |
| 19 | 70823.5900 | 0.0000 | 88528.2200 | 0.00001 | 0.80001 | 1.2500 | 1.2497 |
| 20 | 127482.5000 | 0.0000 | 159351.8000 | 0.00001 | 0.80001 | 1.2500 | 1.2498 |

### i = 85%

| n | F/P | P/F | F/A | A/F | A/P | P/A | A/G |
|---|---|---|---|---|---|---|---|
| 1 | 1.8500 | 0.5405 | 1.0000 | 1.00000 | 1.85000 | 0.5405 | 0.0000 |
| 2 | 3.4225 | 0.2922 | 2.8500 | 0.35088 | 1.20088 | 0.8327 | 0.3509 |
| 3 | 6.3316 | 0.1579 | 6.2725 | 0.15943 | 1.00943 | 0.9907 | 0.6138 |
| 4 | 11.7135 | 0.0854 | 12.6041 | 0.07934 | 0.92934 | 1.0760 | 0.8031 |
| 5 | 21.6700 | 0.0461 | 24.3176 | 0.04112 | 0.89112 | 1.1222 | 0.9346 |
| 6 | 40.0895 | 0.0249 | 45.9876 | 0.02174 | 0.87175 | 1.1471 | 1.0230 |
| 7 | 74.1656 | 0.0135 | 86.0771 | 0.01162 | 0.86162 | 1.1606 | 1.0808 |
| 8 | 137.2063 | 0.0073 | 160.2427 | 0.00624 | 0.85624 | 1.1679 | 1.1177 |
| 9 | 253.8317 | 0.0039 | 297.4491 | 0.00336 | 0.85336 | 1.1718 | 1.1409 |
| 10 | 469.5888 | 0.0021 | 551.2808 | 0.00181 | 0.85181 | 1.1740 | 1.1551 |
| 11 | 868.7392 | 0.0012 | 1020.8700 | 0.00098 | 0.85098 | 1.1751 | 1.1638 |
| 12 | 1607.1680 | 0.0006 | 1889.6090 | 0.00053 | 0.85053 | 1.1757 | 1.1690 |
| 13 | 2973.2610 | 0.0003 | 3496.7770 | 0.00029 | 0.85029 | 1.1761 | 1.1721 |
| 14 | 5500.5330 | 0.0002 | 6470.0380 | 0.00015 | 0.85015 | 1.1763 | 1.1739 |
| 15 | 10175.9900 | 0.0001 | 11970.5700 | 0.00008 | 0.85008 | 1.1764 | 1.1750 |
| 16 | 18825.5800 | 0.0001 | 22146.5600 | 0.00005 | 0.85005 | 1.1764 | 1.1756 |
| 17 | 34827.3200 | 0.0000 | 40972.1400 | 0.00002 | 0.85002 | 1.1764 | 1.1760 |
| 18 | 64430.5500 | 0.0000 | 75799.4600 | 0.00001 | 0.85001 | 1.1765 | 1.1762 |
| 19 | 119196.5000 | 0.0000 | 140230.0000 | 0.00001 | 0.85001 | 1.1765 | 1.1763 |
| 20 | 220513.6000 | 0.0000 | 259426.5000 | 0.00000 | 0.85000 | 1.1765 | 1.1764 |

### i = 90%

| n | F/P | P/F | F/A | A/F | A/P | P/A | A/G |
|---|---|---|---|---|---|---|---|
| 1 | 1.9000 | 0.5263 | 1.0000 | 1.00000 | 1.90000 | 0.5263 | 0.0000 |
| 2 | 3.6100 | 0.2770 | 2.9000 | 0.34483 | 1.24483 | 0.8033 | 0.3448 |
| 3 | 6.8590 | 0.1458 | 6.5100 | 0.15361 | 1.05361 | 0.9491 | 0.5991 |
| 4 | 13.0321 | 0.0767 | 13.3690 | 0.07480 | 0.97480 | 1.0259 | 0.7787 |
| 5 | 24.7610 | 0.0404 | 26.4011 | 0.03788 | 0.93788 | 1.0662 | 0.9007 |
| 6 | 47.0459 | 0.0213 | 51.1621 | 0.01955 | 0.91955 | 1.0875 | 0.9808 |
| 7 | 89.3872 | 0.0112 | 98.2080 | 0.01018 | 0.91018 | 1.0987 | 1.0319 |
| 8 | 169.8357 | 0.0059 | 187.5952 | 0.00533 | 0.90533 | 1.1046 | 1.0637 |
| 9 | 322.6878 | 0.0031 | 357.4309 | 0.00280 | 0.90280 | 1.1077 | 1.0831 |
| 10 | 613.1070 | 0.0016 | 680.1187 | 0.00147 | 0.90147 | 1.1093 | 1.0948 |
| 11 | 1164.9030 | 0.0009 | 1293.2260 | 0.00077 | 0.90077 | 1.1102 | 1.1017 |
| 12 | 2213.3160 | 0.0005 | 2458.1290 | 0.00041 | 0.90041 | 1.1106 | 1.1057 |
| 13 | 4205.3010 | 0.0002 | 4671.4450 | 0.00021 | 0.90021 | 1.1108 | 1.1080 |
| 14 | 7990.0730 | 0.0001 | 8876.7460 | 0.00011 | 0.90011 | 1.1110 | 1.1094 |
| 15 | 15181.1400 | 0.0001 | 16866.8200 | 0.00006 | 0.90006 | 1.1110 | 1.1101 |
| 16 | 28844.1600 | 0.0000 | 32047.9600 | 0.00003 | 0.90003 | 1.1111 | 1.1106 |
| 17 | 54803.9100 | 0.0000 | 60892.1200 | 0.00002 | 0.90002 | 1.1111 | 1.1108 |
| 18 | 104127.4000 | 0.0000 | 115696.0000 | 0.00001 | 0.90001 | 1.1111 | 1.1109 |
| 19 | 197842.1000 | 0.0000 | 219823.5000 | 0.00000 | 0.90000 | 1.1111 | 1.1110 |
| 20 | 375900.1000 | 0.0000 | 417665.6000 | 0.00000 | 0.90000 | 1.1111 | 1.1111 |

### i = 95%

| n | F/P | P/F | F/A | A/F | A/P | P/A | A/G |
|---|---|---|---|---|---|---|---|
| 1 | 1.9500 | 0.5128 | 1.0000 | 1.00000 | 1.95000 | 0.5128 | −0.0000 |
| 2 | 3.8025 | 0.2630 | 2.9500 | 0.33898 | 1.28898 | 0.7758 | 0.3390 |
| 3 | 7.4149 | 0.1349 | 6.7525 | 0.14809 | 1.09809 | 0.9107 | 0.5850 |
| 4 | 14.4590 | 0.0692 | 14.1674 | 0.07058 | 1.02058 | 0.9798 | 0.7554 |
| 5 | 28.1951 | 0.0355 | 28.6264 | 0.03493 | 0.98493 | 1.0153 | 0.8688 |
| 6 | 54.9804 | 0.0182 | 56.8214 | 0.01760 | 0.96760 | 1.0335 | 0.9415 |
| 7 | 107.2118 | 0.0093 | 111.8018 | 0.00894 | 0.95894 | 1.0428 | 0.9867 |
| 8 | 209.0629 | 0.0048 | 219.0136 | 0.00457 | 0.95457 | 1.0476 | 1.0142 |
| 9 | 407.6727 | 0.0025 | 428.0765 | 0.00234 | 0.95234 | 1.0500 | 1.0305 |
| 10 | 794.9618 | 0.0013 | 835.7491 | 0.00120 | 0.95120 | 1.0513 | 1.0400 |
| 11 | 1550.1750 | 0.0006 | 1630.7110 | 0.00061 | 0.95061 | 1.0520 | 1.0455 |
| 12 | 3022.8420 | 0.0003 | 3180.8860 | 0.00031 | 0.95031 | 1.0523 | 1.0487 |
| 13 | 5894.5430 | 0.0002 | 6203.7280 | 0.00016 | 0.95016 | 1.0525 | 1.0504 |
| 14 | 11494.3600 | 0.0001 | 12098.2700 | 0.00008 | 0.95008 | 1.0525 | 1.0514 |
| 15 | 22414.0000 | 0.0000 | 23592.6300 | 0.00004 | 0.95004 | 1.0526 | 1.0520 |
| 16 | 43707.3000 | 0.0000 | 46006.6300 | 0.00002 | 0.95002 | 1.0526 | 1.0523 |
| 17 | 85229.2400 | 0.0000 | 89713.9200 | 0.00001 | 0.95001 | 1.0526 | 1.0524 |
| 18 | 166197.0000 | 0.0000 | 174943.2000 | 0.00001 | 0.95001 | 1.0526 | 1.0525 |
| 19 | 324084.2000 | 0.0000 | 341140.2000 | 0.00000 | 0.95000 | 1.0526 | 1.0526 |
| 20 | 631964.3000 | 0.0000 | 665224.4000 | 0.00000 | 0.95000 | 1.0526 | 1.0526 |

# BASIC SUBROUTINES FOR PVT PROPERTIES' CORRELATIONS



In this appendix subroutines written in BASIC language and their respective equations are presented which can be used to calculate the PVT properties of oil, gas, and water. The application of these subroutines can be seen in the programs (VOLUME, GASMBE, and PSEUDO) presented in Appendix C. The correlations presented are the ones widely used in the industry. The equations presented represent some of the correlation charts presented in Chapter 2.

## A. PSEUDOCRITICAL TEMPERATURE AND PRESSURE FROM GAS GRAVITY

Condensate Gases:

$$(T_c)_{HC} = 187 + 330(\gamma_g)_{HC} - 71.5(\gamma_g)^2_{HC}$$
$$(p_c)_{HC} = 706 - 51.7(\gamma_g)_{HC} - 11.1(\gamma_g)^2_{HC}$$

Miscellaneous Gases:

$$(T_c)_{HC} = 168 + 325(\gamma_g)_{HC} - 12.5(\gamma_g)^2_{HC}$$
$$(p_c)_{HC} = 677 + 15(\gamma_g)_{HC} - 37.5(\gamma_g)^2_{HC}$$

where

$$(\gamma_g)_{HC} = \frac{\gamma_g - 0.9672y_{N_2} - 1.5195y_{CO_2} - 1.1765y_{H_2S}}{1 - y_{N_2} - y_{CO_2} - y_{H_2S}}$$

Where $y_{N_2}$, $y_{H_2S}$, and $y_{CO_2}$ are mole fractions of $N_2$, $H_2S$, and $CO_2$, respectively.

$$T_c = (1 - y_{N_2} - y_{CO_2} - y_{H_2S})(T_c)_{HC} + 227.3y_{N_2} + 547.6y_{CO_2} + 672.4y_{H_2S}$$
$$p_c = (1 - y_{N_2} - y_{CO_2} - y_{H_2S})(p_c)_{HC} + 493y_{N_2} + 1071y_{CO_2} + 1306y_{H_2S}$$

Wichert Azia Correction:

$$CWA = 120[(y_{CO_2} + y_{H_2S})^{0.9} - (y_{CO_2} + y_{H_2S})^{1.6}] + 15(y_{H_2S}^{0.5} - y_{H_2S}^{4})$$
$$T_c^* = T_c - CWA$$
$$p_c^* = \frac{p_c T_c^*}{T_c + y_{H_2S}(1 - y_{H_2S})CWA}$$

Range of validity of the equations:

$$0 \le N_2 < 1.0$$
$$0 \le CO_2 < 1.0$$
$$0 \le H_2S < 1.0$$
$$0 \le N_2 + CO_2 + H_2S < 1.0$$

Condensate fluids:

$$0.56 < \gamma_g < 1.30$$

Miscellaneous gases:

$$0.56 < \gamma_g < 1.71$$

Wichert Aziz corrections:

$$0 \le CO_2 + H_2S < 0.80$$

## B. THE REAL GAS DEVIATION, OR z-FACTOR[2,3]

$$z = \frac{0.27p_r}{\rho_r T_r}$$

The pseudoreduced density, $\rho_r$, is found iteratively using Newton-Raphson iteration.

$$f(\rho_r) = A\rho_r^6 + B\rho_r^3 + C\rho_r^2 + D\rho_r + E\rho_r^3(1 + F\rho_r^2)e^{(-F\rho_r^2)} - G$$
$$f'(\rho_r) = 6A\rho_r^5 + 3B\rho_r^2 + 2C\rho_r + D + E\rho_r^2\,[3 + F\rho_r^2(3 - 2F\rho_r^2)]e^{(-F\rho_r^2)}$$
$$\rho_{r_{i+1}} = \rho_{ri} - \frac{f(\rho_r)}{f'(\rho_r)}$$

$A = 0.06423$

$B = 0.5353T_r - 0.6123$

$E = 0.6816/T_r^2$

$\rho_{ro} = 0.27p_r/T_r$ initial guess

$C = 0.3151T_r - 1.0467 - [0.5783/T_r^2]$

$D = T_r$

$F = 0.6845$

$G = 0.27(p_r)$

Range of validity:

$$1.05 \le T_r < 3.0$$
$$0 \le p_r < 30$$



## C. GAS ISOTHERMAL COMPRESSIBILITY[3,4]

$$c_g = \frac{c_r}{p_c}$$

where

$$c_r = \frac{1}{p_r[1 + (\rho_r/z)(\partial z/\partial \rho_r)]}$$

and

$$\frac{\partial z}{\partial \rho_r} = \frac{1}{\rho_r T_r}\,[5A\rho_r^5 + 2B\rho_r^2 + C\rho_r + 2E\rho_r^2(1 + F\rho_r^2 - F^2\rho_r^4)e^{(-F\rho_r^2)}]$$

The coefficients A, B, C, D, E, and F are the same as those given in $z$-factor calculation and $\rho_r$ is calculated iteratively in the same way as in $z$-factor calculation.

## D. GAS VISCOSITY[5,6]

$$\mu_g = Ae^{(B\rho^C)}$$

where

$$A = \frac{(9.40 + 0.02M_g)(T_R + 460)^{1.5}}{[209 + 19M_g + (T_R + 460)]10^4}$$

$$B = 3.5 + \frac{986}{(T_R + 460)} + 0.01M_g$$

$$C = 2.4 - 0.2B$$

$$\rho' = \frac{p_R M_g}{zR(T_R + 460)} \quad \text{gm/cm}^3$$

R = Universal gas constant, 669.8

## E. OIL ISOTHERMAL COMPRESSIBILITY[7,8]

Above the bubblepoint pressure:

$$c_o = \frac{-1{,}433 + 5R_{si} + 17.2T_R - 1{,}180\gamma_{gs} + 12.61°\,\text{API}}{10^5 p_R}$$

where gas gravity corrected for separator conditions is

$$\gamma_{gs} = \gamma_g \left[1 + 5.912(10^{-5})°\text{API}(T_{sp}) \log\left(\frac{p_{sp}}{114.7}\right)\right]$$

Below the bubblepoint pressure:

$$c_{ob} = \frac{-1}{B_{ob}}\frac{\partial B_{ob}}{\partial p_R} + \frac{B_g}{B_{ob}}\frac{\partial R_{sb}}{\partial p_R}$$

where $B_g$ is in STB/SCF and $B_{ob}$ is the oil FVF below the bubblepoint.

$$\frac{\partial B_{ob}}{\partial R_{sb}} = B + C(T_R - 60)\left(\frac{°\text{API}}{\gamma_{gs}}\right)$$

$$\frac{\partial R_{sb}}{\partial p_R} = \frac{DR_{sb}}{p_R}$$

| For °API ≤ 30 | For °API > 30 |
|---|---|
| $A = 1.751 \times 10^{-5}$ | $A = 1.10 \times 10^{-5}$ |
| $B = 4.677 \times 10^{-4}$ | $B = 4.670 \times 10^{-4}$ |
| $C = -1.811 \times 10^{-8}$ | $C = 1.337 \times 10^{-9}$ |
| $D = 1.0937$ | $D = 1.1870$ |

## F. OIL FORMATION VOLUME FACTOR[7,8]

Above the bubblepoint pressure:

$$B_o = B_{obp} e^{[c_o(p_{bp} - p_R)]}$$

Below the bubblepoint pressure:

$$B_{ob} = 1 + A(T_R - 60)\left(\frac{°API}{\gamma_{gs}}\right) + \left[B + C(T_R - 60)\left(\frac{°API}{\gamma_{gs}}\right)\right]R_{sb}$$

At the bubblepoint pressure:

$$B_{obp} = B_{obp} \text{ at } p_R = p_{bp}$$



The values of coefficients A, B, and C are the same as those used in the calculation of $c_o$. The bubblepoint pressure, $p_{bp}$, is given by:

$$p_{bp} = \left[\frac{R_s}{A(\gamma_{gs})e^{[C°API/(T_R+460)]}}\right]^{1/B}$$

| For °API ≤ 30 | For °API > 30 |
|---|---|
| A = 0.0362 | A = 0.0178 |
| B = 1.0937 | B = 1.1870 |
| C = 25.7240 | C = 23.9310 |

Range of validity:

$$76 < T_{sp} < 150°F$$
$$30 < p_{sp} < 535 \text{ psi}$$
$$15.3 < °API < 59.5$$

Above the bubblepoint pressure:

$$0.511 < \gamma_g < 1.351$$
$$111 < p_R < 9{,}485 \text{ psi}$$

At or below the bubblepoint pressure:
For 15.3 < °API ≤ 30

$$0.511 < \gamma_g < 1.351$$
$$14.7 < p_R < 4{,}542 \text{ psi}$$

For $30.6 < °API < 59.5$

$$0.530 < \gamma_g < 1.259$$
$$14.7 < p_R < 6{,}025 \text{ psi}$$

## G. OIL VISCOSITY[8,9]



Dead oil:

$$\mu_{od} = 10^A - 1$$

where

$$A = BT_R^{-1.163}$$
$$B = 10^C$$
$$C = 3.0324 - 0.02023\,(°API)$$

Above the bubblepoint pressure (live oil):

$$\mu_o = \mu_{obp}\left[\frac{p_R}{(p_{bp})^A}\right]$$
$$A = 2.6p_R^{1.187}\,e^{[-8.98(10^{-5})pR - 11.513]}$$

Below the bubblepoint pressure (live oil):

$$\mu_{ob} = A(\mu_{od})^B$$
$$A = 10.715(R_{sb} + 100)^{-0.515}$$
$$B = 5.44(R_{sb} + 150)^{-0.338}$$

At the bubblepoint pressure, $\mu_{obp} = \mu_{ob}$ at $R_{sb} = R_{si}$.

Range of validity,
Dead oil:

$$16 < °API < 58$$
$$70 < T_R < 295\ °F$$

Live oil:

$$76 < T_{sp} < 150°F$$
$$30 < p_{sp} < 535 \text{ psi}$$

Above the bubblepoint pressure:

$$15.3 < °API < 59.5° \text{ API}$$
$$0.511 < \gamma_g < 1.351$$
$$111 < p_R < 9{,}485 \text{ psi}$$

At or below the bubblepoint pressure:



$$20 < R_{sb} < 2{,}070 \text{ SCF/Bbl}$$
$$14.7 < p_R < 5{,}265 \text{ psi}$$
$$70 < T_R < 295° \text{ F}$$
$$16 < °API < 58° \text{ API}$$

## H. GAS/OIL RATIO[8]

$$R_{sb} = A(\gamma_{gs}) p_R^B e^{[C°API/(T_R + 460)]}$$

| For °API ≤ 30°API | For °API > 30°API |
|---|---|
| A = 0.0362 | A = 0.0178 |
| B = 1.0937 | B = 1.1870 |
| C = 25.7240 | C = 23.9310 |

Range of validity:
For $15.3 < °API \leq 30$

$$0.511 < \gamma_f < 1.351$$
$$14.7 < p_R < 4{,}542 \text{ psi}$$

For $30.6 < °API < 59.5$

$$0.530 < \gamma_g < 1.259$$
$$14.7 < p_R < 6{,}025 \text{ psi}$$

and

$$76 < T_{sp} < 150°F$$
$$30 < p_{sp} < 535 \text{ psi}$$

## I. TWO-PHASE FORMATION VOLUME FACTOR[7]

Above the bubblepoint pressure:

$$B_t = B_{tbp} e^{[c_o(p_{bp} - p_R)]}$$

Below the bubblepoint pressure:

$$B_{tb} = B_{ob} + (R_{si} - R_{sb})B_g$$

where $B_g$ is in RB/SCF.
At the bubblepoint pressure:

$$B_{tbp} = B_{tb} \text{ at } p_R = p_{bp} \text{ (i.e., } R_{sb} = R_{si})$$

## J. WATER ISOTHERMAL COMPRESSIBILITY[10,11]

$$(c_w)_{H_2O} = (A + BT_R + CT_R^2)/10^6$$

where

$$A = 3.8546 - 0.000134p_R$$
$$B = -0.01052 + 4.77(10^{-7})p_R$$
$$C = 3.9267(10^{-5}) - 8.8(10^{-5})p_R$$

Gas saturation correction:

$$(c_w)_{gas} = (c_w)_{H_2O}[1 + 8.9(10^{-3})R_{sw}]$$

Salinity correction:

$$(c_w)_{brine} = (c_w)_{H_2O/Gas}\{[-0.052 + 2.7(10^{-4})T_R - 1.14(10^{-6})T_R^2 + 1.121(10^{-9})T_R^3]\%NaCl^{0.7} + 1\}$$

Range of validity:

$$80 < T_R < 250°F$$
$$1{,}000 < p_R < 6{,}000 \text{ psi}$$
$$0 < \%NaCl < 25$$

## K. WATER FORMATION VOLUME FACTOR

$$(B_w)_{H_2O} = A + B_{pR} + Cp_R^2$$

Gas-free water:

$$A = 0.9947 + 5.8(10^{-6})T_R + 1.02(10^{-6})T_R^2$$
$$B = -4.228(10^{-6}) + 1.8367(10^{-8})T_R - 6.77(10^{-11})T_R^2$$
$$C = 1.3(10^{-10}) - 1.3855(10^{-12})T_R + 4.285(10^{-15})T_R^2$$



Gas-saturated water:

$$A = 0.911 + 6.35(10^{-5})T_R + 8.5(10^{-5})T_R^2$$
$$B = -1.093(10^{-6}) - 3.497(10^{-9})T_R + 4.57(10^{-12})T_R^2$$
$$C = -5(10^{-11}) + 6.429(10^{-13})T_R - 1.43(10^{-15})T_R^2$$

Salinity correction:

$$(B_w)_{brine} = (B_w)_{H_2O}\{[5.1(10^{-8})p_R] + [5.47(10^{-6}) - 1.95(10^{-10})p_R(T_R - 60)] + [-3.23(10^{-8}) + 8.5(10^{-13})p_R(T_R - 60)^2]\%NaCl + 1\}$$

Range of validity:

$$100 < T_R < 250°F$$
$$1{,}000 < p_R < 5{,}000 \text{ psi}$$
$$0 \leq \%NaCl < 25$$

## L. WATER VISCOSITY[11–14]

$$(\mu_w)_{H_2O} = 241.4(10^{-4})10^{[247.8/(T_R' - 140)]} \times [1 + (p_R' - p_{sat}')1.0467(10^{-6})(T_R' - 305)]$$

where the $T_R'$ is in °K and $p_R'$ and $p_{sat}'$ are in bars.

Salinity correction:

$$(\mu_w)_{brine} = (\mu_w)_{H_2O}\{1 - 1.87(10^{-3})\%NaCl^{0.5} + 2.18(10^{-4})\%NaCl^{2.5} + (T_R^{0.5} - 0.0135T_R) \times [2.76(10^{-3})\%NaCl - 3.44(10^{-4}\%NaCl^{1.5}]\}$$

Saturation (vapour) pressure:

$$p'_{sat} = 22{,}088 \exp\left[\frac{374.136 - T''_R}{T'_R}\sum_{i=1}^{8} A_i(0.65 - 0.01T''_R)^{i-1}\right]$$

$$p_{sat}\text{in psi} = \frac{p'_{sat}}{6.894757293}$$



where $p'_{sat}$ is the saturation pressure in Kpa, $T'$ is in °K, and $T''$ is in °C.

$$A_1 = -7.419242 \qquad A_5 = 0.001094098$$
$$A_2 = -0.29721 \qquad A_6 = 0.00439993$$
$$A_3 = -0.1155286 \qquad A_7 = 0.002520658$$
$$A_4 = -0.008685635 \qquad A_8 = 0.0005218684$$

Range of validity: For $\mu_w$

$$32 < T_R < 572°F$$
$$p_{sat} < p_R < 11{,}600 \text{ psi}$$
$$0 \leq \%NaCl < 25$$

For $p_{sat}$

$$32 < T_R < 705°F$$
$$0.0887 < p_R < 3{,}203.6 \text{ psi}$$

## M. GAS/WATER RATIO

$$(R_{sw})_{H_2O} = A + Bp_R + Cp_R^2$$

$$\left(\frac{\partial R_{sw}}{\partial p_R}\right)_{H_2O} = B + 2Cp_R$$

where

$$A = 2.12 + 3.45(10^{-3})T_R - 3.59(10^{-5})T_R^2$$
$$B = 0.0107 - 5.26(10^{-5})T_R + 1.48(10^{-7})T_R^2$$
$$C = -8.75(10^{-7}) + 3.9(10^{-9})T_R - 1.02(10^{-11})T_R^2$$

```
1000  CLEAR: CLS
1010  REM ****** SUBROUTINES TO CALCULATE OIL PVT PROPERTIES ******
1020  '
1030  INPUT "PVT Properties for Oil, Gas, or Water (O, G, W) ........................ ",PVT$
1040  IF LEFT$ (PVT$,1) = "g" OR LEFT$ (PVT$,1) = G" THEN 1100
1050  IF LEFT$ (PVT$,1) = "W" OR LEFT$ (PVT$,1) = "w" THEN 2900
1060  INPUT "Separator Temperature - Deg. F ............................................ ",TS
1070  INPUT "Separator Pressure - psi ................................................... ",PS
1080  INPUT "Oil Gravity - Deg. API ..................................................... ",OG
1090  INPUT"Initial Gas-Oil-Ratio - SCF/STB .............................................. ",RSI
1100  INPUT "Mole fractions of H2S, CO2, and N2 ......................................... ",Y1,Y2,Y3
1110  INPUT "Standard Temperature - Deg. F ............................................. ",ST
1120  INPUT "Standard Pressure - psi .................................................... ",SP
1130  INPUT "Reservoir Temperature - Deg. F ............................................ ",T
1140  INPUT "Reservoir Pressure - psi ................................................... ",P
1150  INPUT "Gas Gravity (air = 1) ...................................................... ",GG
1160  IF LEFT$(PVT$,1) = "o" OR LEFT$(PVT$,1) = "O" THEN 1190
1170  INPUT "Type of Gas (Wet or Dry) ................................................... ",GAS$
1180  GOTO 2000
1190  '
1200  REM ****** BUBBLEPOINT PRESSURE CALCULATION ******
1210  '
1220  IF PS = 0 THEN PS = 114.7
1230  IF OG>30 THEN 1260
1240  A = .0362: B = 1.0937: C = 25.724
1250  GOTO 1270
1260  A = .0178: B = 1.187: C = 23.931
1270  GGS = GG*(1 + 5.912E-05*OG*TS*((LOG(PS/114.7))/2.302585))
1280  PBP = ((RSI)/(A*GGS*EXP(C*OG/(T + 459.67))))^(1/B)
1290  If P>PBP THEN 1320
1300  RSB = A*GGS*(P^B)*EXP(C*(OG/(T + 459.67))): M2 = (B*RSB/P)
1310  GOTO 1330
1320  RSB = RSI
1330  '
1340  REM ****** OIL VISCOSITY CORRELATIONS ******
1350  '
1360  C1 = 3.0324-.02023*OG
1370  B1 = 10^C1: A1 = B1*T^(-1.163): UOD = (10^A1)-1
1380  A2 = 10.715*(RSB + 100)^(-.515)
1390  B2 = 5.44*(RSB + 150)^(-.338)
1400  UOB = A2*(UOD)^(B2)
1410  IF P>PBP THEN 1420: GOTO 1460
1420  A2 = 10.715*(RSI + 100)^(-.515): B2 = 5.44*(RSI + 150)^(-.338)
1430  UOBP = A2*(UOD)^(B2)
1440  A3 = 2.6*P^1.187*EXP(-.0000898*P-11.513)
1450  UO = UOBP*(P/PBP)^A3
1460  IF P>PBP THEN Z1 = UO: GOTO 1490
1470  IF P = PBP THEN Z1 = UOBP: GOTO 1490
1480  IF P<PBP THEN Z1 = UOB
1490  '
1500  REM ****** OIL FORMATION VOLUME FACTOR CALCULATION ******
1510  '
1520  CO = (-1433 + 5*RSI + 17.2*T-1180*GGS + 12.61*OG)/((10^5)*P)
1530  IF OG>30 THEN 1560
1540  A4 = 1.751*10^-5: B4 = 4.677*10^-4: C4: = -1.811*10^-8
1550  GOTO 1570
1560  A4 = 1.1*10^-5: B4 = 4.67*10^-4: C4 = 1.337*10^-9
1570  BOBP = 1 + A4*(T-60)*(OG/GGS) + ((B4 + C4*(T-60)*(OG/GGS))*RSI)
1580  IF P>PBP THEN 1610
1590  BOB = 1 + A4*(T-60)*(OG/GGS) + ((B4 + C4*(T-60)*(OG/GGS))*RSB)
1600  GOTO 1620
1610  BO = BOBP*EXP(CO*(PBP-P))
1620  IF P>PBP THEN Z3 = BO: GOTO 1640
1630  IF P<PBP THEN Z3 = BOB
1640  '
1650  REM ****** OIL ISOTHERMAL COMPRESSIBILITY CALCULATION ******
1660  '
1670  CO = (-1433 + 5*RSI + 17.2*T-1180*GGS + 12.61*OG)/((10^5)*P)
1780  GOSUB 2220
1690  GOSUB 2400
1700  GOSUB 2460
1710  BG = (SP*Z*(T + 459.67))/(5.615*P*(ST + 459.67))
1720  M1 = B4 + C4*(T-60)*(OG/GGS)
1730  M3 = M1*M2
1740  IF P> = PBP THEN 1760
1750  COB = ((-1/BOB)*M3) + ((BG/BOB)*M2)
1760  IF P>PBP THEN Z4 = CO: GOTO 1780
1770  IF P<PBP THEN Z4 = COB
1780  '
1790  REM ****** TWO-PHASE FORMATION VOLUME FACTOR ******
1800  '
1810  BTBP = BOBP
1820  IF P>PBP THEN 1850
1830  BTB = BOB + (RSI-RSB)*BG
1840  GOTO 1860
1850  BT = BTBP*EXP(CO*(PBP-P))
1960  IF P>PBP THEN Z2 = BT: GOTO 1880
1870  IF P<PBP THEN Z2 = BTB
1880  '
1890  PRINT: PRINT
1900  PRINT USING "Bubblepoint Pressure - psi                        #,###.###";PBP
1910  PRINT USING "Corrected Gas Gravity (air = 1)                     #.####";GGS
1920  PRINT USING "Gas-Oil-Ratio - SCF/STB                          #,###.###";RSB
1930  PRINT USING "Oil FVF at Bubblepoint Pressure - RB/STB           ##.####";BOBP
1940  PRINT USING "Oil FVF at Reservoir Pressure - RB/STB             ##.####";Z3
1950  PRINT USING "Oil Viscosity at Bubblepoint Pressure - cp        #.#####";UOBP
1960  PRINT USING "Oil Viscosity at Reservoir Pressure - cp          #.#####";Z1
1970  PRINT USING "Oil Isothermal Compressibility - 1/psi        ##.####^^^^";Z4
1980  PRINT USING "Two-Phase FVF at Reservoir Pressure - RB/STB       #.####";Z2
1990  END
2000  '
2010  REM ****** SUBROUTINES FOR CALCULATING GAS PVT PROPERTIES ******
2020  '
2030  GOSUB 2220
2040  GOSUB 2400
2050  GOSUB 2460
2060  GOSUB 2650
2070  GOSUB 2700
2080  GOSUB 2790
2090  PRINT: PRINT
2100  PRINT USING "Pseudocritical Temperature - Deg. R         #,###.###";TC
2110  PRINT USING "Pseudocritical Pressure - psi              ##,###.###";PC
2120  PRINT USING "Corrected Pseudocritical Temp. - Deg. R     #,###.###";TC2
2130  PRINT USING "Corrected Pseudocritical Pressure - psi     #,###.###";PC2
2140  PRINT USING "Gas Deviation Factor - Z-Factor               #.#####";Z
2150  PRINT USING "Gas Formation Volume Factor - SCF/CF     #,###.####";GFV
2160  PRINT USING "Gas Isothermal Compressibility - 1/psi   ##.####^^^^";CG
2170  PRINT USING "Gas Viscosity - cp                            #.#####";UG
2180  PRINT USING "Wichert-Aziz Correction                       ###.###";CWA
2190  END
2200  '
2210  REM ****** CALCULATING GAS PSEUDOCRITICAL PROPERTIES ******
2220  '
2230  GHC1 = (GG-.9672*Y3-1.5195*Y2-1.1765*Y1)
2240  GHC2 = (1-Y3-Y2-Y1)
2250  GHC = GHC1/GHC2
2260  CWA = 120*(((Y2 + Y1)^.9)-((Y2 + Y1)^1.6)) + 15*((Y1^.5-Y1^4))
2270  IF LEFT$ (GAS$,1) = "D" or LEFT$(GAS$,1) = "d" THEN 2310
2280  TC1 = 187 + 330*GHC-71.5*GHC^2
2290  PC1 = 706-51.7*GHC-11.1*GHC^2
2300  GOTO 2330
2310  TC1 = 168 + 325*GHC-12.5*GHC^2
2320  PC1 = 677 + 15*GHC-37.5*GHC^2
2330  TC = (GHC2*TC1) + (227.3*Y3 + 547.6*Y2 + 672.4*Y1)
2340  PC = (GHC2*PC1) + (493*Y3 + 1071*Y2 + 1306*Y1)
2350  TC2 = TC - CWA
2360  PC2 = (PC*TC2)/(TC + ((Y1*(1-Y1))*CWA))
2370  RETURN
2380  '
2390  REM ****** CALCULATING GAS PSEUDOREDUCED PROPERTIES ******
2400  '
2410  TR = (T + 459.67)/TC2
2420  PR = P/PC2
2430  RETURN
2440  '
2450  REM ****** GAS Z-FACTOR CALCULATION ******
2460  '
2470  DR = (.27*PR)/TR
2480  A5 = .06423
2490  B5 = (.5353*TR)-.6123
2500  C5 = (.3151*TR)-1.0467-(.5783/TR^2)
2510  D5 = TR
2520  E = .6816/TR^2
2530  F = .6845
2540  G = .27*PR
2550  DR1 = A5*DR^6 + B5*DR^3 + C5*DR^2 + D5*DR
2560  DR1 = DR1 + E*DR^3*(1 + F*DR^2)*EXP(-F*DR^2)-G
2570  DR2 = (6*A5*DR^5) + (3*B5*DR^2) + (2*C5*DR) + D5
2580  DR2 = DR2 + E*DR^2*(3 + F*DR^2*(3-2*F*DR^2))*EXP(-F*DR^2)
2590  DR = DR-(DR1/DR2)
2600  IF ABS (DR1)>.00001 THEN 2550
2610  Z = G/(DR*TR)
2620  RETURN
2630  '
2640  REM ****** GAS FORMATION VOLUME FACTOR CALCULATION ******
2650  '
2660  GFVF = ((ST + 459.67)*P)/(SP*Z*(T + 459.67))
2670  RETURN
2680  '
2690  REM ****** GAS ISOTHERMAL COMPRESSIBILITY CALCULATION ******
2700  '
2710  DZ = 5*A*DR^5 + 2*B*DR^2 + C*DR
2720  DZ = DZ + 2*E*DR^2*(1 + F*DR^2-F^2*DR^4)*EXP(-F*DR^2)
2730  DZ = DZ*(1/(DR*TR))
2740  CR = (1/PR)*(1/(1 + (DR/Z)*(DR)))
2750  CG = CR/PC2
2760  RETURN
2770  '
2780  REM ****** GAS VISCOSITY CALCULATION ******
2790  '
2800  MW = 28.964*GG
2810  A1 = (9.399999 + .02*MW)*(T + 459.67)^1.5
2820  A1 = A1/((209 + 19*MW + (T + 459.67))*10^4)
2830  B1 = 3.5 + (986/(T + 459.67)) + (.01*MW)
2840  C1 = 2.4-.2*B1
2850  DR3 = (P*MW)/(Z*669.8*(T + 459.67))
2860  UG = A1*EXP(B1*(DR3^C1))
2870  RETURN
2880  '
2890  REM ****** SUBROUTINES FOR WATER PVT PROPERTIES ******
```

```
2900 '
2910 INPUT "Gas Saturated Water - RSW >0 (Y/N) .................. ",GS$
2920 INPUT "Water Salinity, ppm ............................................ ",PPM
2930 NA=PPM/10000
2940 IF PPM>0 THEN 2960
2950 INPUT "Percent Sodium Chloride - %NaCl ....................... ",NA
2960 INPUT "Reservoir Temperature - Deg. F ............................ ",T
2970 INPUT "Reservoir Pressure - psi ........................................ ",P
2980 GOSUB 3130
2990 GOSUB 3230
3000 GOSUB 3400
3010 GOSUB 3520
3020 GOSUB 3640
3030 GOSUB 3710
3040 PRINT: PRINT
3050 PRINT USING "Gas-to-Water Ratio - SCF/STB                 #,###.####";RSW
3060 PRINT USING "Water FVF - RB/STB                            ###.####";BW
3070 PRINT USING "Water Viscosity - cp                          ##.#####";UW
3080 PRINT USING "Water Isothermal Compressibility - 1 psi     ##.####^^^^";CW
3090 PRINT USING "Saturation (Vapor) Pressure @T - psi          #,###.###";SATP
3100 PRINT USING "Water Density - Lbs/gal                       #,###.###";DEN
3110 END
3120 '
3130 'GAS-TO-WATER RATIO
3140 '
3150 IF LEFT$(GS$,1)="N" OR LEFT$(GS$,1)="n" THEN 3250
3160 A=2.12+3.45*(10^-3)*T-3.59*(10^-5)*T^2
3170 B=.0107-5.26*(10^-5)*T+1.48*(10^-7)*T^2
3180 C=-8.75*(10^-7)+3.9*(10^-9)*T-1.02*(10^-11)*T^2
3190 RSW=A+B*P+C*P^2
3200 RSW=RSW*(1-(7.53*(10^-2)-1.73*(10^-4)*T)*NA)
3210 RETURN
3220 '
3230 ' WATER FORMATION VOLUME FACTOR
3240 '
3250 IF LEFT$(GS$,1)="N" OR LEFT$(GS$,1)="n" THEN 3300
3260 A1=.9911+6.35*(10^-5)*T+8.5*(10^-7)*T^2
3270 B1=-1.093*(10^-6)-3.497*(10^-9)*T+4.57*(10^-12)*T^2
3280 C1=-5*(10^-11)+6.429*(10^-13)*T-1.43*(10^-15)*T^2
3290 GOTO 3330
3300 A1=.9947+5.8*(10^-6)*T+1.02*(10^-6)*T^2
3310 B1=-4.228*(10^-6)+1.8376*(10^-8)*T-6.77*(10^-11)*T^2
3320 C1=1.3*(10^-10)-1.3855*(10^-12)*T+4.285*(10^-15)*T^2
3330 BW=A1+B1*P+C1*P^2
3340 Z=5.1*(10^-8)*P
3350 Z1=(5.47*(10^-6)-1.95*(10^-10)*P)*(T-60)
3360 Z2=(-3.23*(10^-8)+8.5*(10^-13)*P)*(T-60)^2
3370 BW=BW*((Z+Z1+Z2)*NA+1)
3380 RETURN
3390 '
3400 ' WATER ISOTHERMAL COMPRESSIBILITY
3410 '
3420 A2=3.8546-.000134*P
3430 B2=-.01052+4.77*(10^-7)*P
3440 C2=3.9267*(10^-5)-8.8*(10^-10)*P
3450 CW=(A2+B2*T+C2*T^2)/10^6
3460 IF LEFT$(GS$,1)="N" OR LEFT$(GS$,1)="n" THEN 3480
3470 CW=CW*(1+.0089*RSW)
3480 Z3=(-.052+2.7*(10^-4)*T-1.14*(10^-6)*T^2)
3490 Z4=(((Z3+1.121*(10^-9)*T^3)*(NA^.7))+1): CW=CW*Z4
3500 RETURN
3510 '
3520 ' SATURATION (VAPOUR) PRESSURE CALCULATION
3530 '
3540 FOR I=1 to 8: READ A(I): NEXT I
3550 T1=(5/9)*(T-32): T2=(T1+273.15)
3560 FOR I=1 to 8
3570 D=A(I)*((.65-.01*T1)^(I-1)): D7=D+D7: NEXT I
3580 PSAT=22088*EXP(((374.136-T1)/T2)*D7)
3590 SATP=(PSAT/6.894757293#)
3600 D1=10^(247.8/(T2-140)): D2=241.4*(10^-4)
3610 PSAT=((PSAT*10^3)/10^5): P2=(P*6.894757293#*10^3)/10^5
3620 RETURN
3630 '
3640 ' WATER VISCOSITY CALCULATION
3650 '
3660 UW=(D1*D2)*(1+(P2-PSAT)*1.0467*(10^-6)*(T2-305))
3670 D3=(1-1.87*(10^-3)*NA^.5): D4=(2.18*(10^-4)*NA^2.5)
3680 D5=(T^.5-.0135*T): D6=2.76*(10^-3)*NA
3690 UW=UW*(D3+D4+D5*(D6-(3.44*(10^-4)*NA^1.5)))
3700 '
3710 ' ****** WATER DENSITY CALCULATION ******
3720 '
3730 IF T>=68 AND T<=212 THEN 3750
3740 IF T>=212 AND T<=392 THEN 3770
3750 IF P>=1436 AND P<=15939 THEN 3780
3760 IF P>=1436 and P<=26092 THEN 3820
3770 IF P>=4336 AND P<=26092 THEN 3800
3780 DEN=8.3555*EXP(2.7384*10^-6*P-1.5353*10^-4*(T-60)-7.469*10^-7*(T-59)^2)
3790 GOTO 3830
3800 DEN=8.4172*EXP(3.1403*10^-6*P-2.8348*10^-4*(T-60)-3.7549*10^-7*(T-60)^2)
3810 GOTO 3830
3820 DEN=8.361901*EXP(3.0997*10^-6*P-2.2139*10^-4*(T-60)-5.0123*10^-7*(T-60)^2)
3830 RETURN
3840 '
3850 DATA -7.419242,-0.29721,-0.1155286,-0.008685635,0.001094098
3860 DATA 0.00439993,0.002520658,0.0005218684
```

Salinity correction:

$$SC = 1 - [7.53(10^{-2}) - 1.73(10^{-4})T_R]\ \%\text{NaCl}$$

$$(R_{sw})_{brine} = (R_{sw})_{H_2O} \times SC$$

$$\left(\frac{\partial R_{sw}}{\partial p_R}\right)_{brine} = \left(\frac{\partial R_{sw}}{\partial p_R}\right)_{H_2O} \times SC$$

Range of validity:

$$90 < T_R < 250°\text{F}$$

$$500 < p_R < 5{,}000 \text{ psi}$$

$$0 \leq \%\text{NaCl} < 3$$

## REFERENCES


1. N. L. Carr, R. Kobayashi, and D. B. Burrows, "Viscosity of Hydrocarbon Gases Under Pressure," *Trans.*, AIME, 1954, pp. 264–272.
2. J. R. Dempsey, "Computer Routine Treats Gas Viscosity as a Variable," *Oil and Gas Journal*, Aug. 16, 1965, p. 141.

[3] P. M. Dranchuk, R. A. Purvis, and D. B. Robinson, "Computer Calculations of Natural Gas Compressibility Factors Using the Standing and Katz Correlation," *Institute of Petroleum Technical Series,* No. IP 74–008 1974.

[4] A. L. Lee, M. H. Gonzalez, and B. E. Eakin, "The Viscosity of Natural Gases," *Journal of Petroleum Technology,* Aug. 1966, pp. 997–1000.

[5] M. B. Standing, *Volumetric and Phase Behavior of Oil Field Hydrocarbon Systems,* SPE Reprint, Eighth edition, 1977, pp. 26, 122.

[6] M. Vasquez, and H. D. Beggs, "Correlations for Fluid Physical Property Prediction," *Journal of Petroleum Technology,* June 1980, pp. 968–970.

[7] E. Wichert, and K. Aziz, "Calculate Zs for Sour Gases," *Hydrocarbon Processing,* May 1972, pp. 51, 119.

[8] L. Yarborough, and K. R. Hall, "How to Solve Equations of State for Z-Factors," *Oil and Gas Journal,* Feb. 18, 1974, p. 86.

[9] G. Takacs, "Comparison Made for Computer Z-Factor Calculations," *Oil and Gas Journal,* Dec. 20, 1976, pp. 64–66.

[10] A. S. Trube, "Compressibility of Natural Gases," *Trans.,* AIME, Vol. **210,** 1957, p. 335.

[11] M. H. Gonzalez and A. L. Lee, "Graphical Viscosity Correlation for Hydrocarbons," *AICHE Journal,* March 1968, pp. 242–244.

[12] C. Beal, "The Viscosity of Air, Water, Natural Gas, Crude Oil and its Associated Gases at Oil-Field Temperatures and Pressures," *Trans.,* AIME, 1946, pp. 94–115.

[13] H. D. Beggs and J. R. Robinson, "Evaluating the Viscosity of Crude Oil Systems," *Journal of Petroleum Technology—Forum,* Sept. 1975, pp. 1140–1141.

[14] J. N. Chew and C. A. Connally, "A Viscosity Correlation for Gas-Saturated Crude Oils," *Trans.,* AIME, 1959, pp. 23–25.

[15] D. N. Meehan, "A Correlation for Water Compressibility," *Petroleum Engineer,* Nov. 1980, pp. 125–126.

[16] D. N. Meehan, "Estimating Water Viscosity at Reservoir Conditions," *Petroleum Engineer,* July 1980, pp. 117–118.

[17] D. Numbere, W. E. Brigham, and M. B. Standing, *Correlations for Physical Properties of Petroleum Reservoir Brines,* Petroleum Research Institute, Stanford University, Nov. 1977, pp. 8–16.

[18] D. N. Meehan and W. K. Lyons, "Calculations Programmable for Gas Compressibility," *Oil and Gas Journal,* Oct. 8, 1979, pp. 74–78.

[19] H. J. Ramey, "Rapid Methods for Estimating Reservoir Compressibilities," *Journal of Petroleum Technology,* April 1964, pp. 447–454.

# FREQUENTLY USED BASIC COMPUTER PROGRAMS



Four frequently used computer programs written in BASIC language are presented in this Appendix. The programs include (1) VOLUME, (2) GASMBE, (3) PSEUDO, and (4) KG/KO. The programs can be used on IBM or any IBM-compatible microcomputers. The comprehensive programs are based on the information presented in Chapters 2 and 4.

The properties common to all four programs include their flexibility and practical utility, user friendliness, and the professionally organized layout of their output. The printed output from each one of these programs can be directly used in well files and professional presentations and reports; no retyping of the information will be required. The printouts can be obtained on both Epson printers and the HP Laser printers.

I make no claims that the programs presented here are paragons of the programming art. Many step-saving and time-saving tricks could have been added to shorten program length and running time. The functions of these four programs are summarized.

## VOLUME

VOLUME (Table C–1) performs oil and gas volumetric calculations. Table C–2 illustrates the typical output for a hypothetical oil well volumetric calculations. Once the program is executed, data input screens will appear and the user will be prompted for data input. Some of the program's capabilities and working features of VOLUME are as follows:

1. For both oil and gas wells, the PVT properties can be either directly input or the program will calculate these properties if the basic data required to calculate these are input. The empirical correlations used to calculate PVT properties are presented in Appendix B.

2. For gas well volumetric calculations, the program also provides an option to calculate gas PVT properties using gas composition if it is available. To use this option a data file called MIAN2 as shown in Table C–3 has to be first created using the EDLIN text line editor feature of DOS. MIAN2 lists gas properties ($\gamma_{gi}$, $T_{ci}$, and $p_{ci}$) for 19 different gas constituents (also see Table 2–2).
3. The recovery factors used to calculate recoverable oil or gas can either be specified or the program will use the API correlations presented in Chapter 4 (Eqs. 4.2, 4.3, 4.9, 4.10, and 4.11) to calculate the recovery factors based on various reservoir drive mechanisms. The gas reservoir recovery factor may also be calculated by specifying abandonment pressure.
4. The program includes an option to correct gas PVT properties for sour gases ($H_2S$, $N_2$, and $CO_2$).



## GASMBE

GASMBE (Table C–4) performs gas well material balance calculations, typically referred to as $p/z$ calculations. Typical output from GASMBE is shown in Table C–5. The program allows for creating a data file to which data on several wells may be added. The data of any well in the data file may be called for editing/updating. The various functions and properties of GASMBE are as follows:

1. The program requires gas PVT data to be either input by the user or they may be internally calculated using the correlations presented in Appendix B.
2. For each well in the data file, up to 15 pressure test points and the corresponding cumulative gas, condensate, and water production may be input.
3. Shut-in bottomhole pressures for each test may be specified or they may be calculated from wellhead shut-in pressures using the Cullender and Smith method (see Chapter 4).
4. Recoverable gas and recovery factors for up to five abandonment pressures may be calculated.
5. The program calculates original gas in place (OGIP) at each test point and at the same time performs linear regression on user-selected data points ($p/z$ vs. $G_p$) to calculate OGIP.
6. The program allows for the following various options for processing and printing the data file.
   a. Process all data file, i.e., all wells in the data file starting from Well #1 to Well #$n$.
   b. Batch processing of the data file, i.e., processing Well #$n$ to Well #$n+1$ within the data file.
   c. Processing any 10 user-specified wells within the data file.

## PSEUDO

PSEUDO (Table C–6) is used to generate gas pseudopressures given gas PVT properties ($\gamma_g$, $T_c$, $p_c$, and $T_R$). Gas viscosity and $z$-factor are calculated using the correlations presented in Appendix B. The program generates pseudopressures given the minimum (base), maximum, and increment pressure ranges at equal increments of up to 200 increments of pressures. PSEUDO has the following program capabilities:

1. Generating up to 200 pseudopressures at equal increments between the minimum (base) and the maximum pressures. By selecting this option, the program prompts the user to input gas PVT data and pressure ranges. It then generates pseudopressures and stores them in a user specified data file with an extension *.PSD. The pseudopressures generated and stored in *.PSD can be called for printing.
2. The program allows for creating and editing a user-specified data file with an extension *.DAT. Up to 48 pressures can be specified in this option which can then be converted to their respective pseudopressures by using the *.PSD data file. Linear interpolation is used to convert pressures into their respective pseudopressures. This feature will be very useful in performing gas well transient test analysis using pseudopressures (refer to Chapter 7, Volume II).

## KG/KO

KG/KO (Table C–7) is a very comprehensive and very useful program for calculating relative permeability ratios in gas/oil and water/oil systems. The program will calculate $K_g/K_o$ ratios from field performance given (1) reservoir pressure, (2) produced GOR, (3) cumulative oil production, (4) oil formation volume factor, (5) gas formation volume factor, (6) solution GOR, and (7) oil and gas viscosities. Up to 16 performance data points may be specified. A data file is created to store the performance data for later editing and updating. Calculations are based on the procedure outlined in Example 4–27.

Another option provided by this program is to calculate gas/oil and water/oil relative permeability ratios using empirical correlations. In this option, the program will generate relative permeability ratios for several combinations of empirical correlations as presented in Chapter 2. This option is useful in running various correlations and selecting the one which closely represents the relative permeability ratios obtained in the laboratory.

# TABLE C–1. VOLUME—oil and gas volumetrics

```
1000 REM VOLUME
1010 CLEAR: KEY OFF
1020 DIM GG(30),TC(30),PC(30),C(30),A(30)
1030 CLS: COLOR 0,10
1040 PRINT "                    OIL AND GAS RESERVES - VOLUMETRICS";
1050 PRINT "                    ": PRINT
1060 COLOR 14,0
1070 PRINT "Well Name: .................................... "
1080 PRINT "Field, Location: .............................. "
1090 PRINT "County, State: ................................ "
1100 PRINT "Operator: ..................................... "
1110 PRINT "Formation, Depth: ............................. "
1120 PRINT "Reserve Category: ............................. "
1130 PRINT "Oil OR Gas Well ............................... "
1140 PRINT
1150 PRINT "  Reservoir      Standard      Separator"
1160 PRINT " Temp.  Pres.  Temp.  Pres.  Temp.  Pres.  Gas Type";
1170 PRINT "  Initial GOR  Cond. Yield"
1180 PRINT "  (F)   (psi)   (F)   (psi)   (F)   (psi)   (Wet/Dry)";
1190 PRINT "  (SCF/STB)    (STB/MMSCF)"
1200 PRINT " ------------------------------------------------------------------------------";
1210 PRINT " -----------------------------------"
1220 COLOR 15,0
1230 LOCATE 3,30: LINE INPUT "",WN$
1240 LOCATE 4,30: LINE INPUT "",FL$
1250 LOCATE 5,30: LINE INPUT "",CS$
1260 LOCATE 6,30: LINE INPUT "",O$
1270 LOCATE 7,30: LINE INPUT "",FD$
1280 LOCATE 8,30: LINE INPUT "",RC$
1290 LOCATE 9,30: LINE INPUT "",OG$
1300 LOCATE 15,3: INPUT "",T
1310 LOCATE 15,10: INPUT "",P
1320 LOCATE 15,18: INPUT "",ST
1330 LOCATE 15,24: INPUT "",SP
1340 IF LEFT$(OG$,1)="o" OR LEFT$(OG$,1)="O" THEN 1370
1350 LOCATE 15,50: INPUT "",WD$
1360 LOCATE 15,72: INPUT "",CY: GOTO 1400
1370 LOCATE 15,32: INPUT "",TS
1380 LOCATE 15,40: INPUT "",PS: IF PS=0 THEN PS=114.7
1390 LOCATE 15,59: INPUT "",RSI
1400 PRINT: COLOR 14,0
1410 IF LEFT$(OG$,1)="G" OR LEFT$(OG$,1)="g" THEN 1450
1420 PRINT "PHI (%)  SW (%)  H (Ft.)  DA (Acres)  RF (%)  GG  OG (API)";
1430 PRINT "  BOI (RB/STB)": COLOR 15,0
1440 GOTO 1470
1450 PRINT "PHI (%)  SW (%)  H (Ft.)  DA (Acres)  RF (%)  RF TYPE  GG";
1460 PRINT "  Pab (psia)  SGR (%)": GOTO 3160
1470 LOCATE 18,2: INPUT " ",PHI: PHI=PHI/100
1480 LOCATE 18,10: INPUT " ",SW: SW=SW/100: SO=1-SW
1490 LOCATE 18,19: INPUT " ",H: LOCATE 18,29: INPUT " ",DA
1500 LOCATE 18,38: INPUT " ",RF: LOCATE 18,46: INPUT " ",GG
1510 LOCATE 18,54: INPUT " ",OG: LOCATE 18,65: INPUT " ",BOI
1520 IF BOI>0 THEN Z2=BOI
1530 IF RF>0 THEN 1780
1540 PRINT: COLOR 14,0
1550 PRINT "Type of Recovery Factor - Water or Gas Drive ...."
1560 COLOR 15,0: LOCATE 20,15: INPUT "",RF$
1570 IF LEFT$(RF$,1)="W" or LEFT$(RF$,1)="w" THEN 1680
1580 COLOR 14,0
1590 PRINT: PRINT "BOBP (RB/STB)    UOBP (CP)    PBP (PSIA)    Pab (PSIA)";
1600 PRINT "    K (MD)"
1610 COLOR 15,0
1620 LOCATE 23,3: INPUT " ",BOBP
1630 LOCATE 23,18: INPUT " ",UOBP
1640 LOCATE 23,31: INPUT " " ,PBP
1650 LOCATE 23,47: INPUT " ",PAB
1660 LOCATE 23,59: INPUT " ",K
1670 GOTO 1770
1680 COLOR 14,0
1690 PRINT: PRINT "% NaCl       UOI (CP)       Pab (PSIA)        K (MD)"
1700 LOCATE 23,2: INPUT " ",NA
1710 COLOR 15,0
1720 LOCATE 23,11: INPUT " ",UOI
1730 LOCATE 23,24: INPUT " ",UW
1740 LOCATE 23,39: INPUT " ",PAB
1750 LOCATE 23,53: INPUT " ",K
1760 IF UOI>0 THEN Z1=UOI
1770 IF RF=0 THEN 1840
1780 COLOR 14,0: LOCATE 21,1: PRINT "NOTES:"
1790 COLOR 15,0
1800 LOCATE 21,8: LINE INPUT "",F1$
1810 LOCATE 22,8: LINE INPUT "",F2$
1820 LOCATE 23,8: LINE INPUT "",F3$
1830 LOCATE 24,8: LINE INPUT "",F4$: GOTO 1900
1840 CLS: COLOR 0,10
1850 PRINT "                    OIL AND GAS RESERVES - VOLUMETRICS";
1860 PRINT "                    ": PRINT
1870 COLOR 14,0: LOCATE 4,1: PRINT "NOTES:": COLOR 15,0
1880 LOCATE 4,8: LINE INPUT "",F1$: LOCATE 5,8: LINE INPUT "",F2$
1890 LOCATE 6,8: LINE INPUT "",F3$: LOCATE 7,8: LINE INPUT "",F4$
1900 CLS: COLOR 14,0: LOCATE 12,29: PRINT "PRINTING - PLEASE WAIT"
1910 IF GG=0 THEN GG=1: IF OG=0 THEN OG=1
1920 GGS=GG*(1+5.912*(10^-5)*OG*TS*(LOG(PS/114.7)/(2.302585))
1930 IF PBP>0 THEN 1990
1940 IF OG>30 THEN 1970
1950 A=.0362: B=1.0937: C=25.724
1960 GOTO 1980
1970 A=.0178: B=1.187: C=23.931
1980 PBP=((RSI)/(A*GGS*EXP(C*OG/(T+459.67))))^(1/B)
1990 IF P>=PBP THEN 2020
2000 RSB=A*GGS*(P^B)*EXP(C*(OG/(T+459.67)))
2010 GOTO 2030
2020 RSB=RSI
2030 C1=3.0324-.02023*OG: B1=10^C1
2040 A1=B1*T^(-1.163): UOD=(10^A1)-1
2050 A2=10.715*(RSB+100)^(-.515): B2=5.44*(RSB+150)^(-.338)
2060 UOB=A2*(UOD)^(B2)
2070 A2=10.715*(RSI+100)^(-.515): B2=5.44*(RSI+150)^(-.338)
2080 IF UOBP>0 THEN 2160
2090 UOBP=A2*(UOD)^(B2)
2100 A3=2.6*P^1.187*EXP(-.0000898*P-11.513)
2110 IF UOI>0 THEN 2160
2120 UO=UOBP*(P/PBP)^A3
2130 IF P>PBP THEN Z1=UO: GOTO 2160
2140 IF P=PBP THEN Z1=UOBP: GOTO 2160
2150 IF P<PBP THEN Z1=UOB
2160 CO=(-1433+5*RSI+17.2*T-1180*GGS+12.61*OG)/((10^5)*P)
2170 IF OG>30 THEN 2200
2180 A4=1.751*10^-5: B4=4.677*10^-4: C4=1.811*10^-8
2190 GOTO 2210
2200 A4=1.1*10^-5: B4=4.67*10^-4: C4=1.337*10^-9
2210 IF LEFT$ (RF$,1)="W" THEN 2250
2220 IF RF>0 THEN 2250
2230 IF BOBP>0 THEN 2260
2240 GOTO 2250
2250 BOBP=1+A4*(T-60)*(OG/GGS)+((B4+C4*(T-60)*(OG/GGS))*RSI)
2260 IF Z2>0 THEN 2350
2270 IF P>PBP THEN 2300
2280 BOB= 1+A4*(T-60)*(OG/GGS)+((B4+C4*(T-60)*(OG/GGS))*RSB)
2290 GOTO 2320
2300 BO=BOBP*EXP(CO*(PBP-P))
2310 IF Z2>0 THEN 2350
2320 IF P=PBP THEN Z2=BOBP: GOTO 2350
2330 IF P>PBP THEN Z2=BO: GOTO 2350
2340 IF P<PBP THEN Z2=BOB
2350 IF RF>0 THEN 2600
2360 IF LEFT$(RF$,1)="G" OR LEFT$(RF$,1)="g" THEN 2510
2370 IF UW>0 THEN 2470
2380 FOR I=1 TO 8: READ A(I): NEXT: D7=0: D=0
2390 T1=(5/9)*(T-32): T2=(T1+273.15)
2400 FOR I=1 to 8
2410 D=A(I)*((.65-.01*T1)^(I-1)): D7=D+D7: NEXT I
2420 PSAT=22088*EXP(((374.136-T1)/T2)*D7)
2430 SATP=(PSAT/6.894757293#)
2440 D1=10^(247.8/(T2-140)): D2=241.4*(10^-4)
2450 PSAT=((PSAT*10^3)/10^5): P2=(P*6.894757293#*10^3)/10^5
2460 UW=(D1*D2)*(1+(P2-PSAT)*1.0467*(10^-6)*(T2-305))
2470 IF NA=0 THEN 2510
2480 D3=(1-1.87*(10^-3)*NA^.5): D4=(2.18*(10^-4)*NA^2.5)
2490 D5=(T^.5-.0135*T): D6=2.76*(10^-3)*NA
2500 UW=UW*(D3+D4+D5*(D6-(3.44*(10^-4)*NA^1.5)))
2510 IF LEFT$(RF$,1)="W" OR LEFT$(RF$,1)="w" THEN 2560
2520 RFG=41.815*(PHI*SO/BOBP)^.1611
2530 RFG=RFG*((K/1000)/UOBP)^9.790001E-02
2540 RFG=RFG*(SW^.3722)*(PBP/PAB)^.1741
2550 GOTO 2580
2560 RFW=54.898*(((PHI*SO/Z2)^.0422)*(((K/1000)/Z1)*UW)^.077)
2570 RFW=RFW*(SW^-.1903)*((P/PAB)^-.2159)
2580 IF LEFT$(RF$,1)="W" OR LEFT$(RF$,1)="w" THEN RF=RFW: GOTO 2600
2590 IF LEFT$(RF$,1)="G" OR LEFT$(RF$,1)="g" THEN RF=RFG
2600 VO=(7758*SO*PHI)/Z2: AF=DA*H: OOIP=VO*AF: RF=RF/100
2610 N=OOIP*RF
2620 NG=N*RSI
2630 DATA -7.419242,-0.29721,-0.1155286,-0.008685635,0.001094098
2640 DATA 0.00439993,0.002520658,0.0005218684
2650 LPRINT TAB(31) "OIL WELL VOLUMETRICS": LPRINT: LPRINT
2660 GOSUB 5150
2670 IF GG=1 THEN GG=0: IF OG=1 THEN OG=0
2680 LPRINT TAB(11) "Gas Gravity Corrected for Separator Conditions";
2690 LPRINT USING "      #.####";GGS: LPRINT
2700 LPRINT TAB(11) "Oil Gravity - Deg. API";
2710 LPRINT USING "                              ###.##";OG: LPRINT
2720 LPRINT TAB(11) "Reservoir Temperature - Deg. F";
2730 LPRINT USING "                       ###";T: LPRINT
2740 LPRINT TAB(11) "Reservoir Pressure - PSIA";
2750 LPRINT USING "                            ##,###";P: LPRINT
2760 LPRINT TAB(11) "Separator Temperature - Deg. F";
2770 LPRINT USING "                       ###";TS: LPRINT
2780 IF GG>1 THEN 2800
2790 IF GG=1 THEN PBP=0
2800 IF OG=1 THEN PBP=0
2810 LPRINT TAB(11) "Separator Pressure - PSIA";
```

## TABLE C–1. Continued

```
2820  IF PS=114.7 THEN PS=0
2830  LPRINT USING "                              #,###";PS: LPRINT
2840  IF PBP=0 THEN 2880
2850  LPRINT TAB(11) "Bubble Point Pressure - PSIA";
2860  LPRINT USING "                       ##,###.###";PBP: :LPRINT
2870  GOTO 2890
2880  LPRINT TAB(11) "Bubble Point Pressure is not Calculated"
2890  LPRINT TAB(11) "Oil Formation Volume Factor - RB/STB";
2900  LPRINT USING "                   #.####";Z2: LPRINT
2910  IF LEFT$(RF$,1)="W" OR LEFT$(RF$,1)="w"THEN 2960
2920  IF LEFT$(RF$,1)="G" OR LEFT$(RF$,1)="g" THEN 2990
2930  LPRINT TAB(11) "Recovery Factor - Percent";
2940  LPRINT USING "                           ##.###";RF*100: LPRINT
2950  GOTO 3010
2960  LPRINT TAB(11) "Recovery Factor (Water Drive) - Percent";
2970  LPRINT USING "               ##.###";RF*100: LPRINT
2980  GOTO 3010
2990  LPRINT TAB(11) "Recovery Factor (Gas Drive) - Percent";
3000  LPRINT USING "               ##.###";RF*100: LPRINT
3010  LPRINT TAB(11) "Original Oil in Place - STB/Acre-Foot";
3020  LPRINT USING "  ##,###,###.###";VO: LPRINT
3030  LPRINT TAB(11) "Drainage Area - Acre-Foot";
3040  LPRINT USING "                         ###,###";AF: LPRINT
3050  LPRINT TAB(11) "Original Oil in Place - STB            ";
3060  LPRINT USING "   #,###,###,###,###";VO*AF: LPRINT
3070  LPRINT TAB(11) "Ultimate Oil Recovery - STB";
3080  LPRINT USING "                  ###,###,###";N: LPRINT
3090  LPRINT TAB(11) "Initial Gas-Oil Ratio - SCF/STB";
3100  LPRINT USING "                    ##,###.#";RSI: LPRINT
3110  LPRINT TAB(11) "Ultimate Associated Gas Recovery - MSCF";
3120  LPRINT USING "           ##,###,###";NG/1000
3130  GOSUB 5310
3140  FOR R=1 TO 9: LPRINT: NEXT R
3150  GOTO 4910
3160  LOCATE 20,1: PRINT "  Gas Comp. (Y/N)     Tc (R)     Pc (psi)     H2S (%)
  CO2 (%)     N2 (%)"
3170  COLOR 15,0: LOCATE 18,1: INPUT " ",PHI: PHI=PHI/100
3180  LOCATE 18,10: INPUT " ",SW: SW=SW/100: SO=1-SW
3190  LOCATE 18,19: INPUT " ",H: LOCATE 18,29: INPUT " ",DA
3200  LOCATE 18,39: INPUT " ",RF: RF=RF/100
3210  IF RF>0 THEN 3230
3220  LOCATE 18,49: INPUT " ",RF$
3230  LOCATE 18,55: INPUT " ",GG: LOCATE 18,62: INPUT " ",PAB
3240  IF LEFT$(RF$,1)="G" OR LEFT$(RF$,1)="g" THEN 3260
3250  LOCATE 18,74: INPUT " ",SGR: SGR=SGR/100
3260  LOCATE 21,8: INPUT " ",GC$
3270  IF LEFT$(GC$,1)="Y" OR LEFT$(GC$,1)="y" THEN 3320
3280  LOCATE 21,20: INPUT " ",TC: LOCATE 21,34: INPUT " ",PC
3290  LOCATE 21,46: INPUT " ",Y1: Y1=Y1/100
3300  LOCATE 21,58: INPUT " ",Y2: Y2=Y2/100
3310  LOCATE 21,70: INPUT " ",Y3: Y3=Y3/100: GOTO 3750
3320  CLS: COLOR 0,10
3330  PRINT "                        OIL AND GAS RESERVES - VOLUMETRICS";
3340  PRINT "                              ": PRINT
3350  COLOR 14,0
3360  PRINT "ENTER GAS COMPOSITION (IN MOLE PERCENT)": PRINT
3370  PRINT "Nitrogen, N2 ........................................ "
3380  PRINT "Carbon Dioxide, CO2 ................................. "
3390  PRINT "Hydrogen Sulphide, H2S .............................. "
3400  PRINT "Methane, C1 ......................................... "
3410  PRINT "Ethane, C2 .......................................... "
3420  PRINT "Propane, C3 ......................................... "
3430  PRINT "Isobutane, I-C4 ..................................... "
3440  PRINT "N-Butane, N-C4 ...................................... "
3450  PRINT "Isopentane, I-C5 .................................... "
3460  PRINT "N-Pentane, N-C5 ..................................... "
3470  PRINT "N-Hexane, N-C6 ...................................... "
3480  PRINT "N-Heptane, N-C7 ..................................... "
3490  PRINT "N-Octane, N-C8 ...................................... "
3500  PRINT "N-Nonane, N-C9 ...................................... "
3510  PRINT "N-Decane, N-C10 ..................................... "
3520  PRINT "Oxygen, O2 .......................................... "
3530  PRINT "Hydrogen, H2 ........................................ "
3540  PRINT "Helium, He .......................................... "
3550  PRINT "Water, H2O .......................................... ": COLOR 15,0
3560  LOCATE 5,31: INPUT "",C(1): C(1)=C(1)/100
3570  LOCATE 6,31: INPUT "",C(2): C(2)=C(2)/100
3580  LOCATE 7,31: INPUT "",C(3): C(3)=C(3)/100
3590  LOCATE 8,31: INPUT "",C(4): C(4)=C(4)/100
3600  LOCATE 9,31: INPUT "",C(5): C(5)=C(5)/100
3610  LOCATE 10,31: INPUT "",C(6): C(6)=C(6)/100
3620  LOCATE 11,31: INPUT "",C(7): C(7)=C(7)/100
3630  LOCATE 12,31: INPUT "",C(8): C(8)=C(8)/100
3640  LOCATE 13,31: INPUT "",C(9): C(9)=C(9)/100
3650  LOCATE 14,31: INPUT "",C(10): C(10)=C(10)/100
3660  LOCATE 15,31: INPUT "",C(11): C(11)=C(11)/100
3670  LOCATE 16,31: INPUT "",C(12): C(12)=C(12)/100
3680  LOCATE 17,31: INPUT "",C(13): C(13)=C(13)/100
3690  LOCATE 18,31: INPUT "",C(14): C(14)=C(14)/100
3700  LOCATE 19,31: INPUT "",C(15): C(15)=C(15)/100
3710  LOCATE 20,31: INPUT "",C(16): C(16)=C(16)/100
3720  LOCATE 21,31: INPUT "",C(17): C(17)=C(17)/100
3730  LOCATE 22,31: INPUT "",C(18): C(18)=C(18)/100
3740  LOCATE 23,31: INPUT "",C(19): C(19)=C(19)/100
3750  CLS: COLOR 0,10
3760  PRINT "                        OIL AND GAS RESERVES - VOLUMETRICS";
3770  PRINT "                              ": PRINT
3780  COLOR 14,0: LOCATE 4,1: PRINT "NOTES:": COLOR 15,0
3790  LOCATE 4,8: LINE INPUT "",F1$: LOCATE 5,8: LINE INPUT "",F2$
3800  LOCATE 6,8: LINE INPUT "",F3$: LOCATE 7,8: LINE INPUT "",F4$
3810  CLS: COLOR 14,0: LOCATE 12,29: PRINT "PRINTING - PLEASE WAIT"
3820  IF LEFT$(GC$,1)="N" OR LEFT$(GC$,1)="n" THEN 3980
3830  OPEN "MIAN2" FOR INPUT AS #1
3840  FOR I=1 to 19
3850  INPUT #1,GG(I),TC(I),PC(I)
3860  NEXT I: GG=0: CLOSE #1
3870  FOR I=1 to 19
3880  TC=TC+TC(I)*C(I)
3890  PC=PC+PC(I)*C(I)
3900  GG=GG+GG(I)*C(I)
3910  SUM=SUM+C(I)
3920  NEXT I
3930  TC=TC*1/SUM: PC=PC*1/SUM: GG=GG*1/SUM
3940  CWA=120*(((C(2)+C(3))^.9)-((C(2)+C(3))^1.6))
3950  CWA=CWA+15*((C(3)^.5-C(3)^4))
3960  TC2=TC-CWA: PC2=(PC*TC2)/(TC+((C(3)*(1-C(3)))*CWA))
3970  GOTO 4140
3980  GHC1=(GG-.9672*Y3-1.5195*Y2-1.1765*Y1)
3990  GHC2=(1-Y3-Y2-Y1): GHC=GHC1/GHC2
4000  CWA=120*(((Y2+Y1)^.9)-((Y2+Y1)^1.6))+15*((Y1^.5-Y1^4))
4010  IF TC>0 THEN 4080
4020  IF LEFT$(WD$,1)="D" OR LEFT$(WD$,1)="d" THEN 4040
4030  TC1=187+330*GHC-71.5*GHC^2: GOTO 4050
4040  TC1=168+325*GHC-12.5*GHC^2
4050  IF PC>0 THEN 4090
4060  IF LEFT$(WD$,1)="D" OR LEFT$(WD$,1)="d" THEN 4080
4070  PC1=706-51.7*GHC-11.1*GHC^2: GOTO 4090
4080  PC1=677+15*GHC-37.5*GHC^2
4090  IF TC>0 THEN 4110
4100  TC=(GHC2*TC1)+(227.3*Y3+547.6*Y2+672.4*Y1)
4110  IF PC>0 THEN 4130
4120  PC=(GHC2*PC1)+(493*Y3+1071*Y2+1306*Y1)
4130  TC2=TC-CWA: PC2=(PC*TC2)/(TC+((Y1*(1-Y1))*CWA))
4140  TR=(T+459.67)/TC2
4150  PR=P/PC2
4160  DR=(.27*PR)/TR: A=.06423: B=(.5353*TR)-.6123
4170  C=(.3151*TR)-1.0467-(.5783/TR^2): D=TR: E=.6816/TR^2
4180  F=.6845: G=.27*PR
4190  DR1=A*DR^6+B*DR^3+C*DR^2+D*DR
4200  DR1=DR1+E*DR^3*(1+F*DR^2)*EXP(-F*DR^2)-G
4210  DR2=(6*A*DR^5)+(3*B*DR^2)+(2*C*DR)+D
4220  DR2=DR2+E*DR^2*(3+F*DR^2*(3-2*F*DR^2))*EXP(-F*DR^2)
4230  DR=DR-(DR1/DR2)
4240  IF ABS(DR1)>.00001 THEN 4190
4250  ZI=G/(DR*TR)
4260  IF P=PAB THEN 4290
4270  BGI=((ST+459.67)*P)/(SP*ZI*(T+459.67))
4280  Z1=ZI
4290  IF RF>0 THEN 4400
4300  IF P=PAB THEN ZAB=ZI: GOTO 4320
4310  P1=P: P=PAB: GOTO 4150
4320  IF LEFT$(RF$,1)="W" OR LEFT$(RF$,1)="w" THEN 4340
4330  RFG=((P1/Z1)-(PAB/ZAB))/(P1/Z1): GOTO 4370
4340  IF SGR>0 THEN 4360
4350  SGR=62.5-1.3125*(PHI*100): SGR=SGR/100
4360  RFW=((SO*P1/Z1)-(SGR*PAB/ZAB))/(SO*P1/Z1)
4370  IF LEFT$(RF$,1)="G" OR LEFT$(RF$,1)="g" THEN RF=RFG: GOTO 4410
4380  IF LEFT$(RF$,1)="W" OR LEFT$(RF$,1)="w" THEN RF=RFW
4390  GOTO 4410
4400  P1=P
4410  VG=(43560!*PHI*(1-SW)*BGI)/1000000!: AF=DA*H: OGIP=VG*AF
4420  RG=OGIP*RF: RO=RG*CY
4430  LPRINT TAB(31) "GAS WELL VOLUMETRICS": LPRINT: LPRINT
4440  GOSUB 5150
4450  LPRINT TAB(11) "Initial Reservoir Pressure - PSIA";
4460  LPRINT USING "                  ##,###";P1: LPRINT
4470  LPRINT TAB(11) "Reservoir Temperature - Deg. F";
4480  LPRINT USING "                         ###";T: LPRINT
4490  LPRINT TAB(11) "Standard Temperature - Deg. F";
4500  LPRINT USING "                               ##";ST: LPRINT
4510  LPRINT TAB(11) "Standard Pressure - PSIA";
4520  LPRINT USING "                             ##.###";SP: LPRINT
4530  LPRINT TAB(11) "Gas Specific Gravity (Air = 1)";
4540  LPRINT USING "                         #.####";GG: LPRINT
4550  IF PAB=0 THEN 4590
4560  LPRINT TAB(11) "Abandonment Pressure - PSIA";
4570  LPRINT USING "                            #,###";PAB: LPRINT
4580  GOTO 4600
4590  LPRINT TAB(11) "Abandonment Pressure is not Specified": LPRINT
4600  LPRINT TAB(11) "Pseudo-Critical Temperature - Deg. R";
4610  LPRINT USING "                 ###.###";TC2: LPRINT
4620  LPRINT TAB(11) "Pseudo-Critical Pressure - PSIA";
```

## TABLE C–1. Continued

```
4630 LPRINT USING "                    #,###.###";PC2: LPRINT
4640 LPRINT TAB(11); "Initial Gas Deviation Factor - Z";
4650 LPRINT USING "                       #.####";Z1: LPRINT
4660 LPRINT TAB(11) "Gas Formation Volume Factor - SCF/CF";
4670 LPRINT USING "                 ###.###";BGI: LPRINT
4680 LPRINT TAB(11) "Original Gas in Place - MMSCF/Acre-Foot";
4690 LPRINT USING "  #,###,###.####";VG: LPRINT
4700 IF LEFT$(RF$,1)="W" OR LEFT$(RF$,1)="w" THEN 4770
4710 IF LEFT$(RF$,1)="G" OR LEFT$(RF$,1)="g" THEN 4750
4720 LPRINT TAB(11) "Recovery Factor - Percent";
4730 LPRINT USING "                       ##.###";RF*100: LPRINT
4740 GOTO 4790
4750 LPRINT TAB(11) "Recovery Factor (Gas Drive) - Percent";
4760 LPRINT USING "          ##.###";RF*100: LPRINT GOTO 4790
4770 LPRINT TAB(11) "Recovery Factor (Water Drive) - Percent";
4780 LPRINT USING "                ##.###";RF*100: LPRINT
4790 LPRINT TAB(11) "Drainage Area - Acre-Foot";
4800 LPRINT USING "                      ###,###";AF: LPRINT
4810 LPRINT TAB(11) "Original Gas in Place - MMSCF                    ";
4820 LPRINT USING "  ###,###,###,###";VG*AF: LPRINT
4830 LPRINT TAB(11) "Ultimate Gas Recovery - MMSCF";
4840 LPRINT USING "          ###,###,###";RG: LPRINT
4850 LPRINT TAB(11) "Oil/Condensate Yield - STB/MMSCF";
4860 LPRINT USING "           #,###.###";CY: LPRINT
4870 LPRINT TAB(11) "Ultimate Oil/Condensate Recovery - STB";
4880 LPRINT USING "      ###,###,###";RO
4890 GOSUB 5310
4900 FOR Q=1 to 5: LPRINT: NEXT Q
4910 CLS: COLOR 0,10: PRINT
4920 PRINT "                    OIL AND GAS RESERVES - VOLUMETRICS";
4930 PRINT "                        ": PRINT
4940 COLOR 14,0
4950 LOCATE 9,25: PRINT "1 - RUN THE SAME PROGRAM AGAIN"
4960 LOCATE 11,25: PRINT "2 - GO TO MAIN MENU"
4970 LOCATE 13,25: PRINT "3 - END THE SESSION"
4980 LOCATE 15,25: PRINT "ENTER YOUR SELECTION: "
4990 LOCATE 25,47: PRINT " ";: LOCATE 15,47,1: SEL=0
5000 I$=INKEY$: IF I$="" THEN 5000 ELSE PRINT I$;
5010 IF MID$(I$,1,1)="1" THEN SEL=1
5020 IF MID$(I$,1,1)="2" THEN SEL=2
5030 IF MID$(I$,1,1)="3" THEN SEL=3
5040 IF SEL<1 OR SEL>3 THEN BEEP: GOTO 5060
5050 ON SEL GOTO 1010,5130,5140
5060 FOR I=1 TO 60
5070 COLOR 4,0
5080 LOCATE 15,25: PRINT "INVALID SELECTION - PLEASE TRY AGAIN!"
5090 NEXT I
5100 LOCATE 15,25: PRINT "                                     "
5110 COLOR 14,0
5120 GOTO 4980
5130 CHAIN "PEPAC"
5140 CLS: COLOR 7,0: END
5150 LPRINT TAB(11) "Well Name:          ";WN$
5160 LPRINT TAB(11) "Field, Location:    ";FL$
5170 LPRINT TAB(11) "County, State:      ";CS$
5180 LPRINT TAB(11) "Operator:           ";O$: LPRINT
5190 LPRINT TAB(11) "Formation, Depth: ";FD$
5200 LPRINT TAB(11) "Reserve Category:  ";RC$
5210 LPRINT: LPRINT: LPRINT
5220 LPRINT TAB(11) "Average Net Pay Over Drainage Area - Feet";
5230 LPRINT USING "                    ###";H: LPRINT
5240 LPRINT TAB(11) "Estimated Drainage Area - Acres";
5250 LPRINT USING "                          #,###";DA: LPRINT
5260 LPRINT TAB(11) "Porosity - Percent";
5270 LPRINT USING "                             ##.#";PHI*100: LPRINT
5280 LPRINT TAB(11) "Connate Water Saturation - Percent";
5290 LPRINT USING "               ##.#";SW*100: LPRINT
5300 RETURN
5310 LPRINT: LPRINT: LPRINT
5320 LPRINT TAB(11) F1$
5330 LPRINT TAB(11) F2$
5340 LPRINT TAB(11) F3$
5350 LPRINT TAB(11) F4$
5360 RETURN
```

**TABLE C–2. Oil well volumetrics output of VOLUME**

## OIL WELL VOLUMETRICS

| | |
|---|---|
| Well Name: | Mian #2–25 |
| Field, Location: | Wattenberg Field, NWNW-25-T5N-R65W |
| County, State: | Weld County, Colorado |
| Operator: | Mian Oil Company |
| Formation, Depth: | Codell, 7,010–7,020 ft |
| Reserve Category: | Proved Developed Producing |

| | |
|---|---|
| Average Net Pay Over Drainage Area—ft | 10 |
| Estimated Drainage Area–ac | 40 |
| Porosity—Percent | 18.0 |
| Connate Water Saturation—Percent | 35.0 |
| Gas Gravity Corrected for Separator Conditions | 0.7000 |
| Oil Gravity—°API | 32.00 |
| Reservoir Temperature—°F | 160 |
| Reservoir Pressure—psia | 2,500 |
| Separator Temperature—°F | 0 |
| Separator Pressure—psia | 0 |
| Bubblepoint Pressure—psia | 1,734.498 |
| Oil Formation Volume Factor—RB/STB | 1.1835 |
| Recovery Factor—Percent | 15.000 |
| Original Oil in Place—STB/ac-foot | 766.949 |
| Drainage Area—ac-foot | 400 |
| Original Oil in Place—STB | 306,780 |
| Ultimate Oil Recovery—STB | 46,017 |
| Initial Gas/Oil Ratio—SCF/STB | 300.0 |
| Ultimate Associated Gas Recovery—MSCF | 13,805 |

This well is drilled and completed into prolific Codell formation in the Wattenberg field of Weld County, Colorado. The data and well location are hypothetic.

**TABLE C–3. Contents of MIAN2 data file for VOLUME**

```
0.9672,227.3,493.0
1.5195,547.6,1071
1.1765,672.4,1306
0.5539,343.04,667.8
1.0382,549.76,707.8
1.5225,665.68,616.3
2.0068,734.65,529.1
2.0068,765.32,550.7
2.4911,828.77,490.4
2.4911,845.4,488.6
2.9753,913.4,436.9
3.4596,972.5,396.8
3.9439,1023.89,360.6
4.4282,1070.35,332
4.9125,1111.8,304
1.1048,278.6,736.9
0.0696,59.9,188.1
0.138,9.5,33.2
0.622,1165.3,3208
```

## TABLE C–4. GASMBE-gas reservoir material balance calculations

```
1000 REM GASMBE
1010 CLOSE #1
1020 CLEAR: KEY OFF
1030 DIM WN$(9),WHP(15),BHP(15),WHT(15),GP(15),NP(15),WP(15)
1040 DIM WHP$(15),WHT$(15),BHP$(15),NP$(15),GP$(15),GE(15),BHPZ(15)
1050 DIM TWN$(8),TWHP$(15),TBHP$(15),TWHT$(15),TNP$(15),TGP$(15)
1060 DIM TDAT$(15),TUSE$(15),DAT$(15),USE$(15),Z(30),PAB$(5)
1070 DIM TPAB$(5),PZA(15),RF(5),RR(5),TWP$(15),WP$(15),OGP(15)
1080 DIM GVCOND(15),HGAS(15),GEE(15),BG(15),PAB(15)
1090 W1$ = "MAIN MENU"
1100 TB2 = LEN(W1$)
1110 TB2 = (80-TB2)/2
1120 GOSUB 7580: COLOR 15,0
1130 PRINT: PRINT: PRINT:PRINT TAB(TB2);W1$
1140 LOCATE 9,26: PRINT "1 ... Create a New File"
1150 LOCATE 11,26: PRINT "2 ... Add Wells to Existing File"
1160 LOCATE 13,26: PRINT "3 ... Edit Well File"
1170 LOCATE 15,26: PRINT "4 ... List Contents of Data File"
1180 LOCATE 17,26: PRINT "5 ... Analyze Well File/Files"
1190 LOCATE 19,26: PRINT "6 ... Go to Main Menu"
1200 LOCATE 21,26: PRINT "7 ... End the Session"
1210 COLOR 0,7:LOCATE 23,25: PRINT "ENTER YOUR SELECTION = = = >"
1220 COLOR 0,7: LOCATE 23,51: PRINT "     ":LOCATE 23,52
1230 I$ = INKEY$: IF I$ = "" THEN 1230 ELSE PRINT I$;
1240 COLOR 15,0
1250 IF MID$(I$,1,1) = "1" THEN SEL = 1
1260 IF MID$(I$,1,1) = "2" THEN SEL = 2
1270 IF MID$(I$,1,1) = "3" THEN SEL = 3
1280 IF MID$(I$,1,1) = "4" THEN SEL = 4
1290 IF MID$(I$,1,1) = "5" THEN SEL = 5
1300 IF MID$(I$,1,1) = "6" THEN SEL = 6
1310 IF MID$(I$,1,1) = "7" THEN SEL = 7
1320 IF SEL <1 OR SEL>7 THEN BEEP: GOTO 1350
1330 ON SEL GOSUB 1450,1640,2960,5430,5140,7490,1400
1340 GOTO 1020
1350 FOR I = 1 TO 50
1360 COLOR 4,0: LOCATE 23,25: PRINT "INVALID SELECTION, PLEASE TRY AGAIN! "
1370 NEXT I
1380 LOCATE 23,25: PRINT "                                        "
1390 GOTO 1210
1400 CLS
1410 FOR I = 1 TO 20
1420 LOCATE 12,29: PRINT "***END OF SESSION***"
1430 NEXT I
1440 CLOSE: CLS: COLOR 7,0: END
1450 '
1460 GOSUB 2850
1470 '
1480 ON ERROR GOTO 1570
1490 OPEN FILE$ FOR INPUT AS #1
1500 CLOSE #1
1510 ON ERROR GOTO 0
1520 COLOR 4,0
1530 LOCATE 22,18:PRINT "FILE ALREADY EXISTS WITH THIS NAME: ";FILE$
1540 LOCATE 23,15: PRINT "CREATING NEW FILE WILL DESTROY CONTENTS OF OLD FILE"
1550 LOCATE 24,30:INPUT "CONTINUE (Y/N) = = = = > ",AN$
1560 IF LEFT$(AN$,1) = "N" OR LEFT$(AN$,1) = "n" THEN RETURN ELSE 1620
1570 IF ERR = 53 THEN RESUME 1610
1580 RESUME 1590
1590 ON ERROR GOTO 0
1600 RETURN
1610 ON ERROR GOTO 0
1620 OPEN FILE$ FOR OUTPUT AS #1
1630 GOTO 1700
1640 '
1650 'APPEND TO FILE
1660 IF SEL < >2 THEN 1690
1670 GOSUB 2850
1680 OPEN FILE$ FOR APPEND AS #1
1690 '
1700 GOSUB 2640: COLOR 15,0
1710 LOCATE 3,75: INPUT "",WELNO$
1720 LOCATE 3,34: LINE INPUT "",WN$(1)
1730 LOCATE 4,34: LINE INPUT "",WN$(2)
1740 LOCATE 5,34: LINE INPUT "",WN$(3)
1750 LOCATE 6,34: LINE INPUT "",WN$(4)
1760 LOCATE 7,34: LINE INPUT "",WN$(5)
1770 LOCATE 9,45: INPUT "",GG$
1780 LOCATE 10,45: INPUT "", OG$
1790 LOCATE 11,45: INPUT "", PPM$: PPM = VAL(PPM$)
1800 IF PPM>0 THEN 1830
1810 LOCATE 12,45: INPUT "",NA$
1820 GOTO 1840
1830 NA = PPM/10000: NA$ = STR$(NA)
1840 LOCATE 13,45: INPUT "",T$
1850 LOCATE 14,45: INPUT "",ST$: ST = VAL(ST$)
1860 LOCATE 15,45: INPUT "",SP$: SP = VAL(SP$)
1870 IF ST = 0 THEN ST$ = STR$(60): IF SP = 0 THEN SP$ = STR$(14.7)
1880 LOCATE 18,3: INPUT "",TC$: LOCATE 18,13: INPUT "",PC$
1890 LOCATE 18,23: INPUT "",Y1$: LOCATE 18,31: INPUT "",Y2$
1900 LOCATE 18,39: INPUT "",Y3$: LOCATE 18,52: INPUT "",GT$
1910 LOCATE 18,64: INPUT "",TVD$: LOCATE 18,74: INPUT "",MD$
1920 LOCATE 20,8: LINE INPUT "",WN$(6)
1930 LOCATE 21,8: LINE INPUT "",WN$(7)
1940 LOCATE 22,8: LINE INPUT "",WN$(8)
1950 GOSUB 7620
1960 FOR I = 1 to 15
1970 LOCATE I + 5,2: INPUT "",DAT$(I)
1980 LOCATE I + 5,14: INPUT "",WHP$(I)
1990 LOCATE I + 5,23: INPUT "",WHT$(I)
2000 LOCATE I + 5,30: INPUT "",BHP$(I)
2010 LOCATE I + 5,39: INPUT "",GP$(I)
2020 LOCATE I + 5,50: INPUT "",NP$(I)
2030 LOCATE I + 5,59: INPUT "",WP$(I)
2040 LOCATE I + 5,71: INPUT "",USE$(I)
2050 NEXT I
2060 PRINT
2070 COLOR 14,0
2080 PRINT " Pab (1)     Pab (2)     Pab (3)     Pab (4)     Pab (5)     Pi (Psi)"
2090 COLOR 15,0
2100 LOCATE 23,3: INPUT "",PAB$(1)
2110 LOCATE 23,14: INPUT "",PAB$(2)
2120 LOCATE 23,25: INPUT "",PAB$(3)
2130 LOCATE 23,36: INPUT "",PAB$(4)
2140 LOCATE 23,47: INPUT "",PAB$(5)
2150 LOCATE 23,59: INPUT "",PP$
2160 PRINT #1,WELNO$
2170 FOR I = 1 to 8
2180 PRINT #1,WN$(I)
2190 NEXT I
2200 WRITE #1,GG$,OG$,PPM$,NA$
2210 WRITE #1,T$,ST$,SP$
2220 WRITE #1,TVD$,MD$,TC$,PC$
2230 WRITE #1,Y1$,Y2$,Y3$,GT$
2240 FOR I = 1 to 15
2250 WRITE #1,WHP$(I),WHT$(I),BHP$(I),NP$(I)
2260 WRITE #1,GP$(I),DAT$(I),USE$(I),WP$(I)
2270 NEXT I
2280 WRITE #1,PAB$(1),PAB$(2),PAB$(3),PAB$(4),PAB$(5),PP$
2290 CLS:LOCATE 12,19: INPUT "DO YOU WANT TO ADD ANOTHER WELL (Y/N) = = = > ",ADD$
2300 IF LEFT$(ADD$,1) = "Y" OR LEFT$(ADD$,1) = "y" THEN 2330
2310 CLOSE #1
2320 RETURN
2330 GOTO 1700
2340 '
2350 Y1 = Y1/100: Y2 = Y2/100: Y3/Y3 = 100
2360 GHC1 = (GG-.9672*Y3-1.5195*Y2-1.1765*Y1)
2370 GHC2 = (1-Y3-Y2-Y1): GHC = GHC1/GHC2
2380 CWA = 120*(((Y2 + Y1)^.9)-((Y2 + Y1)^1.6)) + 15*((Y1^.5-Y1^4))
2390 IF TC>0 OR PC>0 THEN 2480
2400 IF LEFT$(GT$,1) = "D" OR LEFT$(GT$,1) = "d" THEN 2440
2410 TC1 = 187 + 330*GHC-71.5*GHC^2
2420 PC1 = 706-51.7*GHC-11.1*GHC^2
2430 GOTO 2460
2440 TC1 = 168 + 325*GHC-12.5*GHC^2
2450 PC1 = 677 + 15*GHC-37.5*GHC^2
2460 TC = (GHC2*TC1) + (227.3*Y3 + 547.6*Y2 + 672.4*Y1)
2470 PC = (GHC2*PC1) + (493*Y3 + 1071*Y2 + 1306*Y1)
2480 TC2 = TC-CWA
2490 PC2 = (PC*TC2)/(TC + ((Y1*(1-Y1))*CWA))
2500 RETURN
2510 '
2520 DR = (.27*PR)/TR: A = .06423: B = (.5353*TR)-.6123
2530 C = (.3151*TR)-1.0467-(.5783/TR^2): D = TR
2540 E = .6816/TR^2: F = .6845: G = .27*PR
2550 DR1 = A*DR^6 + B*DR^3 + C*DR^2 + D*DR
2560 DR1 = DR1 + E*DR^3*(1 + F*DR^2)*EXP(-F*DR^2)-G
2570 DR2 = (6*A*DR^5) + (3*B*DR^2) + (2*C*DR) + D
2580 DR2 = DR2 + E*DR^2*(3 + F*DR^2*(3-2*F*DR^2))*EXP(-F*DR^2)
2590 DR = DR-(DR1/DR2)
2600 IF ABS(DR1)>.00001 THEN 2550
2610 Z = G/(DR*TR)
2620 RETURN
2630 '
2640 GOSUB 7580: COLOR 7,0: COLOR 14,0
2650 PRINT: PRINT "Well Name ................................................ ";
2660 PRINT "                                   Well Number: "
2670 PRINT "Field/Location ............................. "
2680 PRINT "County, State .............................. "
2690 PRINT "Operator ................................... "
2700 PRINT "Formation, Depth ........................... "
2710 PRINT
2720 PRINT "Gas Specific Gravity (Air = 1) ..................................... "
2730 PRINT "Oil/Condensate Gravity - Degs. API ................................. "
2740 PRINT "Water Salinity - ppm NaCl .......................................... "
2750 PRINT "Water Salinity - Percent NaCl ...................................... "
2760 PRINT "Reservoir Temperature - Degs. F .................................... "
2770 PRINT "Standard Temperature - Degs. F ..................................... "
2780 PRINT "Standard Pressure - Psi ............................................ "
2790 PRINT
2800 PRINT " Tc (R)    Pc (Psi)    %H2S    %CO2    %N2    Gas (Wet/Dry)";
2810 PRINT "     TVD (ft)     MD (ft)"
2820 PRINT:PRINT
2830 PRINT "Notes:"
```

## TABLE C–4. Continued

```
2840  RETURN
2850  '
2860  CLS
2870  LOCATE 3,21: PRINT "THE FOLLOWING INFORMATION IS REQUIRED FOR"
2880  LOCATE 4,20: PRINT "CREATING, EDITING, APPENDING, AND ANALYZING"
2890  LOCATE 5,35: PRINT "WELL FILES"
2900  LOCATE 11,20: INPUT "What Disk is the Data File on: ",DISC$
2910  DIS$=DISC$
2920  CLS: FILES DISC$+":*.DAT"
2930  LOCATE 19,20: INPUT "What is the Data File Name: ",FILE$
2940  FILE$=DISC$+":"+FILE$+".DAT"
2950  RETURN
2960  GOSUB 2860
2970  GOSUB 4570
2980  PRINT:PRINT:PRINT "                          ";
2990  INPUT "ENTER WELL NO. TO BE EDITED ===> ",WELNO$
3000  OPEN DISC$+":SEQTEMP.DAT" FOR OUTPUT AS #2
3010  LOCATE 1,75: INPUT #1, TWELNO$
3020  FOR I=1 to 8
3030  LINE INPUT #1,TWN$(I)
3040  NEXT I
3050  INPUT #1,TGG$,TOG$,TPPM$,TNA$
3060  INPUT #1,TT$,TST$,TSP$
3070  INPUT #1,TTVD$,TMD$,TTC$,TPC$
3080  INPUT #1,TY1$,TY2$,TY3$,TGT$
3090  FOR I=1 to 15
3100  INPUT #1,TWHP$(I),TWHT$(I),TBHP$(I),TNP$(I)
3110  INPUT #1,TGP$(I),TDAT$(I),TUSE$(I),TWP$(I)
3120  NEXT I
3130  INPUT #1,TPAB(1),TPAB$(2),TPAB$(3),TPAB$(4),TPAB$(5),TPP$
3140  IF TWELNO$=WELNO$ THEN 3360
3150  IF EOF(1) THEN 3310
3160  IF TWELNO$="" THEN 3300
3170  PRINT #2,TWELNO$
3180  FOR I=1 TO 8
3190  PRINT #2,TWN$(I)
3200  NEXT I
3210  WRITE #2,TGG$,TOG$,TPPM$,TNA$
3220  WRITE #2,TT$,TST$,TSP$
3230  WRITE #2,TTVD$,TMD$,TTC$,TPC$
3240  WRITE #2,TY1$,TY2$,TY3$,TGT$
3250  FOR I=1 to 15
3260  WRITE #2,TWHP$(I),TWHT$(I),TBHP$(I),TNP$(I)
3270  WRITE #2,TGP$(I),TDAT$(I),TUSE$(I),TWP$(I)
3280  NEXT I
3290  WRITE #2,TPAB$(1),TPAB$(2),TPAB$(3),TPAB$(4),TPAB$(5),TPP$
3300  GOTO 3010
3310  PRINT "END OF FILE, NO MATCH FOUND FOR";WELNO$
3320  CLOSE #2
3330  KILL DISC$+":SEQTEMP.DAT"
3340  CLOSE #1
3350  RETURN
3360  '
3370  GOSUB 2640: COLOR 15,0
3380  LOCATE 3,75: INPUT "",WELNO$
3390  LOCATE 3,34: PRINT TWN$(1)
3400  LOCATE 4,34: PRINT TWN$(2)
3410  LOCATE 5,34: PRINT TWN$(3)
3420  LOCATE 6,34: PRINT TWN$(4)
3430  LOCATE 7,34: PRINT TWN$(5)
3440  LOCATE 9,45: PRINT TGG$
3450  LOCATE 10,45: PRINT TOG$
3460  LOCATE 11,45: PRINT TPPM$
3470  LOCATE 12,44: PRINT TNA$
3480  LOCATE 13,45: PRINT TT$
3490  LOCATE 14,45: PRINT TST$
3500  LOCATE 15,45: PRINT TSP$
3510  LOCATE 18,3: PRINT TTC$
3520  LOCATE 18,13: PRINT TPC$
3530  LOCATE 18,23: PRINT TY1$
3540  LOCATE 18,31: PRINT TY2$
3550  LOCATE 18,39: PRINT TY3$
3560  LOCATE 18,52: PRINT TGT$
3570  LOCATE 18,64: PRINT TTVD$
3580  LOCATE 18,74: PRINT TMD$
3590  LOCATE 20,8: PRINT TWN$(6)
3600  LOCATE 21,8: PRINT TWN$(7)
3610  LOCATE 22,8: PRINT TWN$(8)
3620  COLOR 4,0
3630  LOCATE 3,34: LINE INPUT "",WN$(1)
3640  IF WN$(1) = "" THEN WN$(1)=TWN$(1)
3650  LOCATE 4,34: LINE INPUT "",WN$(2)
3660  IF WN$(2) = "" THEN WN$(2)=TWN$(2)
3670  LOCATE 5,34: LINE INPUT "",WN$(3)
3680  IF WN$(3) = "" THEN WN$(3)=TWN$(3)
3690  LOCATE 6,34: LINE INPUT "",WN$(4)
3700  IF WN$(4) = "" THEN WN$(4)=TWN$(4)
3710  LOCATE 7,34: LINE INPUT "",WN$(5)
3720  IF WN$(5) = "" THEN WN$(5)=TWN$(5)
3730  LOCATE 9,45: INPUT "",GG$
3740  IF GG$ = "" THEN GG$=TGG$
3750  LOCATE 10,45: INPUT "",OG$
3760  IF OG$ = "" THEN OG$=TOG$
3770  LOCATE 11,45: INPUT "",PPM$
3780  IF PPM$ = "" THEN PPM$=TPPM$
3790  LOCATE 12,45: INPUT "",NA$
3800  IF NA$ = "" THEN NA$=TNA$
3810  LOCATE 13,45: INPUT "",T$
3820  IF T$ = "" THEN T$=TT$
3830  LOCATE 14,45: INPUT "",ST$
3840  IF ST$ = "" THEN ST$=TST$
3850  LOCATE 15,45: INPUT "",SP$
3860  IF SP$ = "" THEN SP$=TSP$
3870  LOCATE 18,3: INPUT "",TC$
3880  IF TC$ = "" THEN TC$=TTC$
3890  LOCATE 18,13: INPUT "",PC$
3900  IF PC$ = "" THEN PC$=TPC$
3910  LOCATE 18,23: INPUT "",Y1$
3920  IF Y1$ = "" THEN Y1$=TY1$
3930  LOCATE 18,31: INPUT "",Y2$
3940  IF Y2$ = "" THEN Y2$=TY2$
3950  LOCATE 18,39: INPUT "",Y3$
3960  IF Y3$ = "" THEN Y3$=TY3$
3970  LOCATE 18,52: INPUT "",GT$
3980  IF GT$ = "" THEN GT$=TGT$
3990  LOCATE 18,64: INPUT "",TVD$
4000  IF TVD$="" THEN TVD$=TTVD$
4010  LOCATE 18,74: INPUT "",MD$
4020  IF MD$="" THEN MD$=TMD$
4030  LOCATE 20,8: LINE INPUT "",WN$(6)
4040  IF WN$(6) = "" THEN WN$(6)=TWN$(6)
4050  LOCATE 21,8: LINE INPUT "",WN$(7)
4060  IF WN$(7) = "" THEN WN$(7)=TWN$(7)
4070  LOCATE 22,8: LINE INPUT "",WN$(8)
4080  IF WN$(8) = "" THEN WN$(8)=TWN$(8)
4090  GOSUB 7620
4100  FOR I=1 TO 15
4110  LOCATE I+5,2: PRINT TDAT$(I)
4120  LOCATE I+5,14: PRINT TWHP$(I)
4130  LOCATE I+5,23: PRINT TWHT$(I)
4140  LOCATE I+5,30: PRINT TBHP$(I)
4150  LOCATE I+5,39: PRINT TGP$(I)
4160  LOCATE I+5,50: PRINT TNP$(I)
4170  LOCATE I+5,59: PRINT TWP$(I)
4180  LOCATE I+5,71: PRINT TUSE$(I)
4190  NEXT I
4200  COLOR 14,0
4210  PRINT " Pab (1)     Pab (2)     Pab (3)     Pab (4)     Pab (5)     Pi (Psi)"
4220  COLOR 15,0
4230  LOCATE 23,3: PRINT TPAB$(1)
4240  LOCATE 23,14: PRINT TPAB$(2)
4250  LOCATE 23,25: PRINT TPAB$(3)
4260  LOCATE 23,36: PRINT TPAB$(4)
4270  LOCATE 23,47: PRINT TPAB$(5)
4280  LOCATE 23,59: PRINT TPP$
4290  COLOR 4,0
4300  FOR I=1 TO 15
4310  LOCATE I+5,2: INPUT "",DAT$(I)
4320  IF DAT$(I) = "" THEN DAT$(I)=TDAT$(I)
4330  LOCATE I+5,14: INPUT "",WHP$(I)
4340  IF WHP$(I) = "" THEN WHP$(I)=TWHP$(I)
4350  LOCATE I+5,23: INPUT "",WHT$(I)
4360  IF WHT$(I) = "" THEN WHT$(I)=TWHT$(I)
4370  LOCATE I+5,30: INPUT "",BHP$(I)
4380  IF BHP$(I) = "" THEN BHP$(I)=TBHP$(I)
4390  LOCATE I+5,39: INPUT "",GP$(I)
4400  IF GP$(I) = "" THEN GP$(I)=TGP$(I)
4410  LOCATE I+5,50: INPUT "",NP$(I)
4420  IF NP$(I) = "" THEN NP$(I)=TNP$(I)
4430  LOCATE I+5,59: INPUT "",WP$(I)
4440  IF WP$(I)="" THEN WP$(I)=TWP$(I)
4450  LOCATE I+5,71: INPUT "",USE$(I)
4560  IF USE$(I) = "" THEN USE$(I)=TUSE$(I)
4470  NEXT I
4480  LOCATE 23,3: INPUT "",PAB$(1): IF PAB$(1)="" THEN PAB$(1)=TPAB$(1)
4490  LOCATE 23,14: INPUT "",PAB$(2): IF PAB$(2)="" THEN PAB$(2)=TPAB$(2)
4500  LOCATE 23,25: INPUT "",PAB$(3): IF PAB$(3)="" THEN PAB$(3)=TPAB$(3)
4510  LOCATE 23,36: INPUT "",PAB$(4): IF PAB$(4)="" THEN PAB$(4)=TPAB$(4)
4520  LOCATE 23,47: INPUT "",PAB$(5): IF PAB$(5)="" THEN PAB$(5)=TPAB$(5)
4530  LOCATE 23,59: INPUT "",PP$: IF PP$="" THEN PP$=TPP$
4540  COLOR 7,0
4550  IF WELNO$="" THEN 4770
4560  GOTO 4640
4570  OPEN FILES$ FOR INPUT AS #1
4580  '
4590  IF NOT EOF(1) THEN 4630
4600  PRINT: PRINT "INPUT FILE IS EMPTY"
4610  CLOSE #1
4620  GOTO 1020
4630  RETURN
4640  PRINT #2, WELNO$
4650  FOR I=1 TO 8
4660  PRINT #2,WN$(I)
4670  NEXT I
```

# TABLE C–4. Continued

```
4680 WRITE #2,GG$,OG$,PPM$,NA$
4690 WRITE #2,T$,ST$,SP$
4700 WRITE #2,TVD$,MD$,TC$,PC$
4710 WRITE #2,Y1$,Y2$,Y3$,GT$
4720 FOR I=1 TO 15
4730 WRITE #2,WHP$(I),WHT$(I),BHP$(I),NP$(I)
4740 WRITE #2,GP$(I),DAT$(I),USE$(I),WP$(I)
4750 NEXT I
4760 WRITE #2, PAB$(1),PAB$(2),PAB$(3),PAB$(4),PAB$(5),PP$
4770 IF EOF(1) THEN 5050
4780 INPUT #1,WELNO$
4790 FOR K=1 TO 8
4800 LINE INPUT #1,WN$(K)
4810 NEXT K
4820 INPUT #1,GG$,OG$,PPM$,NA$
4830 INPUT #1,T$,ST$,SP$
4840 INPUT #1,TVD$,MD$,TC$,PC$
4850 INPUT #1,Y1$,Y2$,Y3$,GT$
4860 FOR J=1 TO 15
4870 INPUT #1,WHP$(J),WHT$(J),BHP$(J),NP$(J)
4880 INPUT #1,GP$(J),DAT$(J),USE$(J),WP$(J)
4890 NEXT J
4900 INPUT #1, PAB$(1),PAB$(2),PAB$(3),PAB$(4),PAB$(5),PP$
4910 PRINT #2,WELNO$
4920 FOR M=1 to 8
4930 PRINT #2,WN$(M)
4940 NEXT M
4950 WRITE #2,GG$,OG$,PPM$,NA$
4960 WRITE #2,T$,ST$,SP$
4970 WRITE #2,TVD$,MD$,TC$,PC$
4980 WRITE #2,Y1$,Y2$,Y3$,GT$
4990 FOR I=1 to 15
5000 WRITE #2,WHP$(I),WHT$(I),BHP$(I),NP$(I)
5010 WRITE #2,GP$(I),DAT$(I),USE$(I),WP$(I)
5020 NEXT I
5030 Write #2, PAB$(1),PAB$(2),PAB$(3),PAB$(4),PAB$(5),PP$
5040 GOTO 4770
5050 CLOSE #2: CLOSE #1
5060 '
5070 KILL FILE$
5080 '
5090 NAME DISC$+":SEQTEMP.DAT" AS FILE$
5100 CLOSE #1: CLS
5110 LOCATE 12,20: INPUT "DO YOU WANT TO EDIT ANOTHER WELL ===> ",ED$
5120 IF LEFT(ED$,1)="Y" OR LEFT$(ED$,1)="y" THEN 2970
5130 RETURN
5140 GOSUB 7580: COLOR 15,0
5150 LOCATE 3,23: PRINT "The Following Well File Processing"
5160 LOCATE 4,29: PRINT "Options are Available"
5170 LOCATE 8,23: PRINT "1 - Process All Data Base"
5180 LOCATE 9,23: PRINT "2 - Process Well Files in Batch"
5190 LOCATE 10,23: PRINT "3 - Process Particular Well Files (Max=10)
5200 LOCATE 11,23: PRINT "4 - Go to Main Menu"
5210 COLOR 0,7: LOCATE 15,23: PRINT "ENTER YOUR SELECTION ===>"
5220 COLOR 0,7: LOCATE 15,49: PRINT "      ": LOCATE 15,51
5230 I$=INKEY$: IF I$="" THEN 5230 ELSE PRINT I$;
5240 COLOR 15,0
5250 IF MID$(I$,1,1)="1" THEN PEL=1
5260 IF MID$(I$,1,1)="2" THEN PEL=2
5270 IF MID$(I$,1,1)="3" THEN PEL=3
5280 IF MID$(I$,1,1)="4" THEN PEL=4
5290 IF PEL<1 or PEL>4 THEN BEEP: GOTO 5310
5300 ON PEL GOTO 5430,5350,5390,1020
5310 COLOR 4,0: FOR I=1 TO 60
5320 LOCATE 15,23: PRINT "INVALID SELECTION, PLEASE TRY AGAIN!"
5330 NEXT I
5340 LOCATE 15,23: PRINT "                                    ": GOTO 5210
5350 LOCATE 18,23: PRINT "Process Well No.        to ";"Well No."
5360 LOCATE 18,40: INPUT "",W1
5370 LOCATE 18,56: INPUT "",W2: W2=W2+1
5380 GOTO 5430
5390 LOCATE 18,20: PRINT "Enter Well Numbers on the Following Line"
5400 LOCATE 19,33: PRINT "(Maximum = 10)
5410 LOCATE 22,15: LINE INPUT "Well Nos.: ",LIN$
5420 CLS: GOSUB 6650
5430 OX=0: GOSUB 2860
5440 OPEN FILE$ FOR INPUT AS #1
5450 IF EOF(1) THEN 7040
5460 FOR Q=1 TO 15
5470 GE(Q)=0: GVCOND(Q)=0: GP(Q)=0: HGAS(Q)=0: BHPZ(Q)=0: GEE(Q)=0
5480 BG(Q)=0: Z(J)=0: BHP(Q)=0: NP(Q)=0: BG=0: PAB(Q)=0: PZA(Q)=0
5490 WHT(Q)=0: WHP(Q)=0: USE$(Q)="": WP(Q)=0: PP=0
5500 NEXT Q
5510 INPUT #1,WELNO$
5520 FOR I=1 to 8
5530 LINE INPUT #1, WN$(I)
5540 NEXT I
5550 INPUT #1,GG$,OG$,PPM$,NA$
5560 INPUT #1,T$,ST$,SP$
5570 INPUT #1,TVD$,MD$,TC$,PC$
5580 INPUT #1,Y1$,Y2$,Y3$,GT$
5590 FOR I=1 to 15
5600 INPUT #1,WHP$(I),WHT$(I),BHP$(I),NP$(I)
5610 INPUT #1,GP$(I),DAT$(I),USE$(I),WP$(I)
5620 WHP(I)=VAL(WHP$(I)): WHT(I)=VAL(WHT$(I)): BHP(I)=VAL(BHP$(I))
5630 NP(I)=VAL(NP$(I)): GP(I)=VAL(GP$(I)): WP(I)=VAL(WP$(I))
5640 NEXT I
5650 INPUT #1,PAB$(1),PAB$(2),PAB$(3),PAB$(4),PAB$(5),PP$
5660 IF SEL <>4 THEN 5690
5670 GOSUB 7330
5680 GOTO 5450
5690 IF PEL>2 THEN 5740
5700 IF PEL=1 THEN 5780
5710 IF VAL(WELNO$)<W1 THEN 5450
5720 IF VAL(WELNO$)>=W2 THEN CLOSE #1: GOTO 1020
5730 GOTO 5780
5740 FOR N%=1 to I%
5750 IF VAL(WELNO$)=VALUE(N%) THEN BIBI$=WELNO$
5760 NEXT N%
5770 IF BIBI$ <> WELNO$ THEN 5450
5780 GG=VAL(GG$): OG=VAL(OG$): PPM=VAL(PPM$): NA=VAL(NA$): PP=VAL(PP$)
5790 T=VAL(T$): ST=VAL(ST$): SP=VAL(SP$): ID=VAL(ID$): TVD=VAL(TVD$)
5800 MD=VAL(MD$): TC=VAL(TC$): PC=VAL(PC$): Y1=VAL(Y1$): Y2=VAL(Y2$)
5810 Y3=VAL(Y3$): PAB(1)=VAL(PAB$(1)): PAB(2)=VAL(PAB$(2)): WELNO=VAL(WELNO$)
5820 PAB(3)=VAL(PAB$(3)): PAB(4)=VAL(PAB$(4)): PAB(5)=VAL(PAB$(5))
5830 GOSUB 2350
5840 TC=TC2: PC=PC2
5850 BHPZ=0: SGE=0: PZGP=0: GPS=0: N=0
5860 W2$="GAS MATERIAL BALANCE"
5870 XYZ=0: TB3=LEN(W2$): TB3=(80-TB3)/2
5880 LPRINT TAB(TB3);W2$
5890 LPRINT: LPRINT: LPRINT
5900 LPRINT TAB(11) "Well Name:                  ";WN$(1)
5910 LPRINT TAB(11) "Field, Location:            ";WN$(2)
5920 LPRINT TAB(11) "County, State:              ";WN$(3)
5930 LPRINT TAB(11) "Operator:                   ";WN$(4)
5940 LPRINT TAB(11) "Formation, Depth:           ";WN$(5)
5950 LPRINT: LPRINT
5960 FOR J=1 TO 15
5970 XYZ=XYZ+1
5980 IF BHP(J)=0 THEN GOSUB 6720
5990 IF BHP(J)=0 THEN 6330
6000 IF NP(J)>0 THEN GOSUB 6930
6010 GE(J)=GVCOND(J)+GP(J)+HGAS(J)
6020 P=BHP(J): PR=P/PC: TR=(T+459.67)/TC
6030 GOSUB 2520
6040 Z(J)=Z: BHPZ(J)=BHP(J)/Z(J)
6050 IF LEFT$(USE$(J),1)="N" OR LEFT$(USE$(J),1)="n" THEN 6080
6060 BHPZ=BHPZ+BHPZ(J): SGE=SGE+GE(J): GEE(J)=GE(J)^2
6070 PZGP=PZGP+(BHPZ(J)*GE(J)): GPS=GPS+GEE(J): N=N+1
6080 BG(J)=(Z(J)*(T+459.67)*SP)/(BHP(J)*(ST+459.67)
6090 IF XYZ>1 THEN 6320
6100 LPRINT TAB(11) "Gas Specific Gravity (Air = 1)              ";
6110 LPRINT USING "              #.####";GG
6120 LPRINT TAB(11) "Oil/Condensate Gravity - Degs. API          ";
6130 LPRINT USING "                ##.##";OG
6140 LPRINT TAB(11) "Water Salinity - Percent NaCl               ";
6150 LPRINT USING "                ##.##";NA
6160 LPRINT TAB(11) "Reservoir Temperature - Degs. F             ";
6170 LPRINT USING "              ###.##";T
6180 LPRINT TAB(11) "Standard Temperature - Degs. F              ";
6190 LPRINT USING "              ###.##";ST
6200 LPRINT TAB(11) "Standard Pressure - PSIA                    ";
6210 LPRINT USING "              ##.###";SP
6220 LPRINT TAB(11) "True Vertical Depth - Feet                  ";
6230 LPRINT USING "             ##,###";TVD
6240 LPRINT TAB(11) "Measured Depth - Feet                       ";
6250 LPRINT USING "             ##,###";MD
6260 LPRINT TAB(11) "Pseudo-Critical Temperature - Degs. R       ";
6270 LPRINT USING "         #,###.###";TC
6280 LPRINT TAB(11) "Pseudo-Critical Pressure - PSIA             ";
6290 LPRINT USING "         #,###.###";PC
6300 LPRINT: LPRINT
6310 GOSUB 7080
6320 '
6330 NEXT J
6340 PZB=BHPZ/N: GEB=SGE/N
6350 XY=(N*(PZGP)-(BHPZ*SGE))/(N*(GPS)-(SHGE^2)
6360 YY=PZB-XY*(GEB): OGIP=YY/-XY
6370 IF PP=0 THEN 6420
6380 P=PP: PR=P/PC: TR=T/TC
6390 GOSUB 2510
6400 PP1=PP/Z
6410 YY=PP1
6420 OGIP=YY/-XY
6430 BG=(T+459.67)*SP/(YY*(ST+459.67))
6440 GOSUB 7150
6450 FOR I=1 TO 5
6460 XYZ=XYZ+1
6470 P=PAB(I)
6480 IF P=0 THEN 6540
6490 PR=P/PC
6500 GOSUB 2520
6510 Z(I)=Z: PZA(I)=P/Z(I)
```

## TABLE C–4. Continued

```
6520 RR(I) = (YY-PZA(I))/-XY
6530 RF(I) = (RR(I)/OGIP)*100
6540 GOSUB 7220
6550 NEXT I
6560 LPRINT
6570 LPRINT TAB(18)"*** ORIGINAL GAS IN PLACE, MMCF                    ";
6580 LPRINT USING "###,###.###";OGIP;: LPRINT " ***"
6590 LPRINT: LPRINT
6600 LPRINT TAB(11) WN$(6)
6610 LRPINT TAB(11) WN$(7)
6620 LPRINT TAB(11) WN$(8)
6630 FOR D = 1 TO 7: LPRINT: NEXT D
6640 GOTO 5450
6650 NOLD% = 1
6660 FOR I% = 1 TO 10
6670 N% = INSTR(NOLD%,LIN$," ")
6680 IF N% = 0 THEN VALUE(I%) = VAL(MID$(LIN$,NOLD%)): RETURN
6690 IF N% = NOLD% THEN NOLD% = NOLD% + 1: GOTO 6670
6700 VALUE(I%) = VAL(MID$(LIN$,NOLD%,N%-NOLD%))
6710 NOLD% = N% + 1: NEXT
6720 IF WHP(J) = 0 THEN RETURN
6730 TA(J) = (T + WHT(J))/2
6740 TA1(J) = ((T + 459.67) + (WHT(J) + 459.67))/2
6750 T(1) = WHT(J) + 459.67: T(2) = T + 459.67
6760 BHP(J) = WHP(J) + (WHP(J)*MD/40000!)
6770 P = WHP(J): PR = P/PC: TR = T(1)/TC
6780 GOSUB 2520
6790 Z(16) = Z
6800 P = BHP(J): PR = P/PC: TR = T(2)/TC
6810 GOSUB 2520
6820 Z(17) = Z
6830 ZA = (Z(16) + Z(17))/2
6840 AAA = (GG*TVD)/(26.67*TA1(J)*ZA)
6850 L1 = WHP(J)^2*EXP(AAA)
6860 L2 = ((GG*TA(J))*MD*(EXP(AAA)-1))/(40000!*AAA)
6870 L3 = L1 + L2
6880 L4% = (L3 + BHP(J)^2)/(2*BHP(J))
6890 P1% = BHP(J)
6900 IF L4% = P1% THEN RETURN
6910 BHP(J) = (L3 + BHP(J)^2)/(2*BHP(J))
6920 GOTO 6800
6930 OGG = 141.5/(131.5 + OG)
6940 GCON = 2.99965*(1.03-OGG)
6950 GVCOND(J) = GCON*NP(J)/1000
6960 IF WP(J) = 0 THEN RETURN
6970 A5 = 3.4 + 60630!/BHP(J)
6980 B5 = 3.2147 + (2.657*BHP(J)/10000)-(2.27*BHP(J)^2/100000000#)
6990 SC = 1-(4.893/1000)*NA-(1.757/10000)*NA^2
7000 WGR = A5*(T/100)^B5*SC
7010 HGAS(J) = (WGR/1000)*(GP(J)*1000)/(42*8.337)
7020 HGAS(J) = HGAS(J)*(208.9369*(ST + 459.67)/SP)/1000000!
7030 RETURN
7040 CLS: FOR M = 1 TO 25
7050 LOCATE 12,30: PRINT "***   END OF FILE   ***"
7060 NEXT M
7070 CLOSE #1: GOTO 1020
7080 '
7090 LPRINT "     Test Date          BHP                BHP/Z           Cumulative";
7100 LPRINT "         OGIP          Pressure"
7110 LPRINT "     (MM/DD/YY)        (Psi)  Z-Factor    (Psi)           GE (MMCF)";
7120 LPRINT "         (MMCF)          Used"
7130 LPRINT
7140 RETURN
7150 FOR J = 1 TO 15
7160 IF BHP(J) = 0 THEN LPRINT: GOTO 7200
7170 LPRINT USING "      \              \ ##.###      #.###      ##,###      ###,###,###";DAT$(J);BHP(J);Z(J);BHPZ(J);GE(J);
7180 OGP(J) = (GE(J)*BG(J))/(BG(J)-BG)
7190 LPRINT USING " ###,###.###        \ \";OGP(J);USE$(J)
7200 NEXT J
7210 RETURN
7220 IF XYZ>16 THEN 7300
7230 LPRINT
7240 W3$ = "RECOVERABLE GAS AND RECOVERY FACTOR"
7250 TB4 = LEN(W3$): TB4 = (80-TB4)/2
7260 LPRINT TAB(TB4);W3$: LPRINT
7270 LPRINT "     Pab                   Pab/Z    Recovery      Recovery Factor"
7280 LPRINT "     (Psi)      Z-Factor   (Psi)     (MMCF)          Percent     "
7290 LPRINT
7300 IF P = 0 THEN LPRINT: GOTO 7320
7310 LPRINT USING "        ##,###      #.###      ##,###      ###,###.###      ##.##";PAB(I);Z(I);PZA(I);RR(I);RF(I)
7320 RETURN
7330 OX = OX + 1: IF OX>1 THEN 7440
7340 WW$ = "GAS MATERIAL BALANCE"
7350 WW1$ = "CONTENTS OF DATA FILE " + FILE$ + " "
7360 TB5 = LEN(WW$): TB5 = (80-TB5)/2
7370 TB6 = LEN(WW1$): TB6 = (80-TB6)/2
7380 IF OX1 = 1 THEN 7420
7390 LPRINT TAB(TB5);WW$
7400 LPRINT: LPRINT TAB(TB6);WW1$
7410 LPRINT: LPRINT: LPRINT
7420 LPRINT "        FILE No.          WELL/LEASE NAME"
7430 LPRINT
7440 LPRINT USING "            \     \            \                        \";WELNO$,WN$(1)
7450 IF OX = 50 THEN GOTO 7500
7460 IF EOF(1) THEN 7470 ELSE 7480
7470 FOR X = 1 TO 58-OX: LPRINT: NEXT X
7480 RETURN
7490 CHAIN "PEPAC"
7500 GOSUB 7520
7510 GOTO 7460
7520 OX = 0: OX1 = 1
7530 FOR X = 1 TO 8: LPRINT: NEXT X
7540 LPRINT TAB(TB5);WW$
7550 LPRINT: LPRINT TAB(TB6);WW1$
7560 LPRINT: LPRINT: LPRINT
7570 RETURN
7580 CLS: COLOR 0,10
7590 PRINT "                         GAS MATERIAL BALANCE CALCULATIONS";
7600 PRINT "                         "
7610 RETURN
7620 GOSUB 7580: COLOR 7,0: COLOR 14,0
7630 PRINT TAB(44)"Cumulative to Test"
7640 PRINT TAB(39)"..........................................."
7650 PRINT "Test Date     WHSIP     WHT    BHP     Gas  ";
7660 PRINT "      Cond.     Water"
7670 PRINT "(MM/DD/YY)   (Psi)      (F)     (Psi)   (MMcf)";
7680 PRINT "      (Bbls)     (Bbl)      Use"
7690 COLOR 15,0
7700 RETURN
```

**TABLE C–5. Output from GASMBE program**

**GAS MATERIAL BALANCE**

| | |
|---|---|
| Well Name: | Smart C #1 |
| Field, Location: | Camerick Gas Area, Sec. 11-T1N-R18E |
| County, State: | Texas County, Oklahoma |
| Operator: | Graham Michaelis Drilling Co. |
| Formation, Depth: | Morrow, 6,428 – 6,448 ft |

| | |
|---|---|
| Gas Specific Gravity (Air = 1) | 0.7000 |
| Oil/Condensate Gravity — °API | 0.00 |
| Water Salinity — Percent NaCl | 0.00 |
| Reservoir Temperature — °F | 150.00 |
| Standard Temperature — °F | 60.00 |
| Standard Pressure — psia | 14.650 |
| True Vertical Depth — ft | 6,438 |
| Measured Depth — ft | 6,438 |
| Pseudocritical Temperature — °R | 389.375 |
| Pseudocritical Pressure — psia | 669.125 |

| Test Date (MM/DD/YY) | BHP (Psi) | Z-Factor | BHP/Z (Psi) | Cumulative GE (MMCF) | OGIP (MMCF) | Pressure Used |
|---|---|---|---|---|---|---|
| 07/15/66 | 1,204 | 0.8602 | 1,400 | 64.321 | 19,626.310 | Y |
| 07/15/67 | 1,134 | 0.8668 | 1,308 | 279.084 | 4,086.034 | Y |
| 07/15/68 | 1,067 | 0.8733 | 1,222 | 457.803 | 3,524.262 | Y |
| 07/15/69 | 1,003 | 0.8798 | 1,140 | 633.851 | 3,369.721 | Y |
| 07/22/70 | 797 | 0.9019 | 884 | 895.982 | 2,416.914 | N |
| 07/21/71 | 857 | 0.8953 | 957 | 1,207.012 | 3,791.944 | Y |
| 06/22/72 | 801 | 0.9015 | 889 | 1,336.780 | 3,640.067 | Y |
| 06/22/73 | 740 | 0.9084 | 815 | 1,472.608 | 3,507.291 | Y |
| 06/21/74 | 671 | 0.9164 | 732 | 1,703.173 | 3,559.099 | Y |
| 06/13/75 | 595 | 0.9253 | 643 | 1,830.610 | 3,376.996 | Y |
| 06/24/76 | 531 | 0.9330 | 569 | 2,093.820 | 3,520.788 | N |
| 06/21/77 | 465 | 0.9410 | 494 | 2,371.344 | 3,658.865 | Y |
| 06/09/78 | 408 | 0.9481 | 430 | 2,604.033 | 3,754.742 | Y |

## RECOVERABLE GAS AND RECOVERY FACTOR

| Pab (Psi) | Z-Factor | Pab/Z (Psi) | Recovery (MMCF) | Recovery Factor Percent |
|---|---|---|---|---|
| 408 | 0.9481 | 430 | 2,510.103 | 69.35 |
| 300 | 0.9616 | 312 | 2,815.163 | 77.78 |
| 200 | 0.9742 | 205 | 3,090.188 | 85.38 |
| 150 | 0.9806 | 153 | 3,225.056 | 89.11 |
| 100 | 0.9871 | 101 | 3,358.184 | 92.79 |

*** ORIGINAL GAS IN PLACE, MMCF 3,619.305 ***

This well was drilled to a total depth of 6,875 ft. The well is completed into Morrow formation at a depth of 6,428 ft. It is shut-in after producing 3,238.957 MMCF through October 1983.

## TABLE C–6. PSEUDO-calculating gas pseudo pressures

```
1000 REM PSEUDO
1010 CLEAR: KEY OFF
1020 DIM P$(50),TP$(50),P(200),PP(200),Z(200),UG(200), PPB(200),MP(200),PR(200)
1030 DIM PZ(200),MZ(200)
1040 GOSUB 1860
1050 COLOR 14,0
1060 LOCATE 7,35: PRINT "MAIN MENU"
1070 LOCATE 9,25: PRINT "1 ... Create a New File"
1080 LOCATE 11,25: PRINT "2 ... Edit an Existing File"
1090 LOCATE 13,25: PRINT "3 ... Generate Pseudo Pressures"
1100 LOCATE 15,25: PRINT "4 ... Convert Pressures to Pseudo Pressures"
1110 LOCATE 17,25: PRINT "5 ... Print Pseudo Pressure Data File"
1120 LOCATE 19,25: PRINT "6 ... Go to Main Menu"
1130 LOCATE 21,25: PRINT "7 ... End the Session"
1140 COLOR 0,7: LOCATE 23,25: PRINT "ENTER YOUR SELECTION = = = >"
1150 LOCATE 23,51: PRINT "   ": LOCATE 23,52
1160 I$ = INKEY$: IF I$ = "" THEN 1160 ELSE PRINT I$
1170 IF MID$((I$,1,1) = "1" THEN SEL = 1
1180 IF MID$(I$,1,1) = "2" THEN SEL = 2
1190 IF MID$(I$,1,1) = "3" THEN SEL = 3
1200 IF MID$(I$,1,1) = "4" THEN SEL = 4
1210 IF MID$(I$,1,1) = "5" THEN SEL = 5
1220 IF MID$(I$,1,1) = "6" THEN SEL = 6
1230 IF MID$(I$,1,1) = "7" THEN SEL = 7
1240 IF SEL<1 OR SEL>7 THEN BEEP: GOTO 1260
1250 ON SEL GOSUB 1380,2050,2650,3990,3380,1310,1320
1260 COLOR 4,0: FOR I = 1 TO 60
1270 LOCATE 23,25: PRINT "INVALID SELECTION, PLEASE TRY AGAIN!"
1280 NEXT I
1290 LOCATE 23,25: PRINT "                                    "
1300 GOTO 1140
1310 CHAIN "PEPAC"
1320 CLOSE #1: CLOSE #2
1330 COLOR 7,0
1340 CLS: FOR I = 1 TO 60
1350 LOCATE 12,29: PRINT "*** END OF SESSION ***"
1360 NEXT I
1370 CLS: END
1380 COLOR 7,0
1390 GOSUB 1860
1400 GOSUB 1770
1410 ON ERROR GOTO 1520
1420 OPEN FILE$ FOR INPUT AS #1
1430 CLOSE #1
1440 ON ERROR GOTO 0
1450 BEEP: COLOR 4,0
1460 LOCATE 22,18: PRINT "FILE ALREADY EXISTS WITH THIS NAME: "FILE$
1470 LOCATE 23,15: PRINT "CREATING NEW FILE WILL DESTROY CONTENTS OF OLD FILE"
1480 LOCATE 24,30: INPUT "CONTINUE (Y/N) = = = > ",AN$: CLS
1490 IF SEL>3 OR SEL<3 THEN 1510
1500 IF LEFT$(AN$,1) = "N" OR LEFT$(AN$,1) = "n" THEN 1010 ELSE 2800
1510 IF LEFT$(AN$,1) = "N" OR LEFT$(AN$,1) = "n" THEN 1010 ELSE 1570
1520 IF ERR = 53 THEN RESUME 1560
1530 RESUME 1540
1540 ON ERROR GOTO 0
1550 RETURN
1560 ON ERROR GOTO 0
1570 OPEN FILE$ FOR OUTPUT AS #1
1580 GOSUB 1860
1590 GOSUB 1900
1600 COLOR 15,0
1610 FOR I = 1 TO 5
1620 LOCATE I + 2,30: LINE INPUT "",WL$(I)
1630 NEXT I
1640 FOR Z = 1 TO 8: LOCATE Z + 11,3: INPUT "",P$(Z): NEXT Z
1650 FOR Z = 9 TO 16: LOCATE Z + 3,17: INPUT "",P$(Z): NEXT Z
1660 FOR Z = 17 TO 24: LOCATE Z - 5,31: INPUT "",P$(Z): NEXT Z
1670 FOR Z = 25 TO 32: LOCATE Z - 13,45: INPUT "",P$(Z): NEXT Z
1680 FOR Z = 33 TO 40: LOCATE Z - 21,59: INPUT "",P$(Z): NEXT Z
1690 FOR Z = 41 TO 48: LOCATE Z - 29,73: INPUT "",P$(Z): NEXT Z
1700 FOR I = 6 TO 8
1710 LOCATE I + 15,8: LINE INPUT "",WL$(I)
1720 NEXT I
1730 FOR I = 1 TO 8: PRINT #1, WL$(I): NEXT I
1740 FOR Z = 1 TO 48: WRITE #1, P$(Z): NEXT Z
1750 CLOSE #1: GOTO 1010
1760 GOSUB 1860
1770 COLOR 14,0
1780 LOCATE 5,20: PRINT "THE FOLLOWING INFORMATION IS REQUIRED FOR"
1790 LOCATE 6,20: PRINT "CREATING, ANALYZING, AND EDITTING DATA FILE"
1800 LOCATE 14,20: INPUT "What Disk is the Data File on: ",DISC$
1810 DIS$ = DISC$
1820 CLS: FILES DISC$ + ":*.DAT"
1830 LOCATE 19,20: INPUT "What is the Data File Name: ",FILE$
1840 FILE$ = DISC$ + ":" + FILE$ + ".DAT"
1850 RETURN
1860 CLS: COLOR 0,11
1870 PRINT "                    GAS PSEUDO PRESSURE CALCULATIONS";
1880 PRINT "                    "
1890 RETURN
1900 COLOR 14,0
1910 PRINT
1920 PRINT "Well Name ....................................  "
1930 PRINT "Field, Location ...............................  "
1940 PRINT "County, State .................................  "
1950 PRINT "Operator ......................................  "
1960 PRINT "Formation, Depth ..........................  "
1970 LOCATE 9,14: PRINT "INPUT PRESSURES (IN";
1980 COLOR 10,0: PRINT " INCREASING";
1990 COLOR 14,0: PRINT "ORDER) TO BE CONVERTED"
2000 LOCATE 10,18: PRINT "TO THEIR CORRESPONDING PSEUDO PRESSURES - PSI"
2010 LOCATE 11,1: PRINT "----------------------------------------------------------------"
2020 LOCATE 11,41: PRINT "----------------------------------------------------------------"
2030 LOCATE 21,1: PRINT "NOTES"
2040 RETURN
2050 COLOR 7,0
2060 GOSUB 1860
2070 GOSUB 1770
2080 GOSUB 2570
2090 OPEN DISC$ + ":SEQTEMP.DAT" FOR OUTPUT AS #2
2100 FOR I = 1 TO 8
2110 LINE INPUT #1, TWL$(I): NEXT I
2120 FOR Z = 1 TO 48: INPUT #1, TP$(Z): NEXT Z
2130 GOSUB 1860
2140 GOSUB 1900
2150 COLOR 15,0
2160 FOR I = 1 TO 5: LOCATE I + 2,30: PRINT TWL$(I): NEXT I
2170 FOR Z = 1 TO 8: LOCATE Z + 11,3: PRINT TP$(Z): NEXT Z
2180 FOR Z = 9 TO 16: LOCATE Z + 3,17: PRINT TP$(Z): NEXT Z
2190 FOR Z = 17 TO 24: LOCATE Z - 5,31: PRINT TP$(Z): NEXT Z
2200 FOR Z = 25 TO 32: LOCATE Z - 13,45: PRINT TP$(Z): NEXT Z
2210 FOR Z = 33 TO 40: LOCATE Z - 21,59: PRINT TP$(Z): NEXT Z
2220 FOR Z = 41 TO 48: LOCATE Z - 29,73: PRINT TP$(Z): NEXT Z
2230 FOR I = 6 TO 8: LOCATE I + 15,8: PRINT TWL$(I): NEXT I
2240 COLOR 4,0
2250 FOR I = 1 TO 5
2260 LOCATE I + 2,30: LINE INPUT "",WL$(I)
2270 IF WL$(I) = "" THEN WL$(I) = TWL$(I)
2280 NEXT I
2290 FOR Z = 1 TO 8: LOCATE Z + 11,3: INPUT "",P$(Z)
2300 IF P$(Z) = "" THEN P$(Z) = TP$(Z)
2310 NEXT Z
2320 FOR Z = 9 TO 16: LOCATE Z + 3,17: INPUT "",P$(Z)
2330 IF P$(Z) = "" THEN P$(Z) = TP$(Z)
2340 NEXT Z
2350 FOR Z = 17 TO 24: LOCATE Z - 5,31: INPUT "",P$(Z)
2360 IF P$(Z) = "" THEN P$(Z) = TP$(Z)
2370 NEXT Z
2380 FOR Z = 25 TO 32: LOCATE Z - 13,45: INPUT "",P$(Z)
2390 IF P$(Z) = "" THEN P$(Z) = TP$(Z)
2400 NEXT Z
2410 FOR Z = 33 TO 40: LOCATE Z - 21,59: INPUT "",P$(Z)
2420 IF P$(Z) = "" THEN P$(Z) = TP$(Z)
2430 NEXT Z
2440 FOR Z = 41 TO 48: LOCATE Z - 29,73: INPUT "",P$(Z)
2450 IF P$(Z) = "" THEN P$(Z) = TP$(Z)
2460 NEXT Z
2470 FOR I = 6 TO 8: LOCATE I + 15,8: LINE INPUT "",WL$(I)
2480 IF WL$(I) = "" THEN WL$(I) = TWL$(I)
2490 NEXT I
2500 FOR I = 1 TO 8
2510 PRINT #2, WL$(I): NEXT I
2520 FOR Z = 1 TO 48: WRITE #2, P$(Z): NEXT Z
2530 CLOSE #1: CLOSE #2
2540 KILL FILE$
2550 NAME DISC$ + ":SEQTEMP.DAT" AS FILE$
2560 CLOSE #1: GOTO 1010
2570 OPEN FILE$ FOR INPUT AS #1
2580 IF NOT EOF(1) THEN 2640
2590 CLS: COLOR 4,0
2600 FOR I = 1 TO 60
2610 LOCATE 12,30: PRINT "INPUT FILE IS EMPTY": NEXT I
2620 CLOSE #1
2630 GOTO 1010
2640 RETURN
2650 COLOR 7,0
2660 GOSUB 1860
2670 COLOR 14,0
2680 LOCATE 5,18: PRINT "THE PSEUDO PRESSURES GENERATED BY THIS OPTION"
2690 LOCATE 6,18: PRINT "ARE STORED IN A DATA FILE WITH EXTENSION *.PSD"
2700 LOCATE 7,19: PRINT "TO BE USED FOR FUTURE PRESSURE CONVERSIONS"
2710 LOCATE 14,20: INPUT "What Disk is the Data File on: ",DISC$
2720 DIS$ = DISC$
2730 CLS: FILES DISC$ + ":*.PSD"
2740 LOCATE 19,20: INPUT "What is the Data File Name: ",FILE$
2750 FILE$ = DISC$ + ":" + FILES$ + ".PSD"
2760 ON ERROR GOTO 2780
2770 GOSUB 1420
2780 IF ERR = 53 THEN RESUME 2790
2790 ON ERROR GOTO 0
2800 OPEN FILE$ FOR OUTPUT AS #1
2810 GOSUB 1860
```

## TABLE C–6. Cont'd.

```
2820  COLOR 14,0
2830  LOCATE 3,1: PRINT "Reference:"
2840  LOCATE 10,20: PRINT "Corrected"
2850  LOCATE 11,3: PRINT "Gas Gravity    Tc    Pc    Pressure Range (psi) Reservoir Temp."
2860  LOCATE 12,5: PRINT "(Air=1)       (R)   (PSI)    Min.    Max.    Inc.     (F)"
2870  LOCATE 13,3: PRINT "-----------------------------------------------------------------"
2880  LOCATE 13,43: PRINT " ---------------------------------------------------"
2890  COLOR 15,0
2900  FOR I=1 TO 5
2910  LOCATE I+2,12: LINE INPUT " ",REF$(I): NEXT I
2920  LOCATE 14,6: INPUT " ",GG: LOCATE 14,17: INPUT " ",TC
2930  LOCATE 14,26: INPUT " ",PC: LOCATE 14,36: INPUT " ",PMIN
2940  LOCATE 14,44: INPUT " ",PMAX
2950  LOCATE 14,53: INPUT " ",PINC
2960  X=(PMAX=PMIN)/PINC: IF X>200 THEN BEEP
2970  IF X<=200 THEN 3070 ELSE 2980
2980  COLOR 4,0
2990  LOCATE 18,24: PRINT "MAXIMUM NUMBER OF POINTS ALLOWED"
3000  LOCATE 19,26: PRINT "ARE 200! PLEASE INCREASE Pinc."
3010  LOCATE 22,28: PRINT "PRESS ANY KEY TO CONTINUE"
3020  AB$=INKEY$: IF AB$="" THEN 3020
3030  LOCATE 18,24: PRINT "                                    "
3040  LOCATE 19,26: PRINT "                                    "
3050  LOCATE 22,28: PRINT "                                    "
3060  COLOR 15,0: GOTO 2950
3070  LOCATE 14,67: INPUT " ",T
3080  FOR I=1 TO 5: PRINT #1,REF$(I): NEXT I
3090  TR=(T+459.67)/TC
3100  WRITE #1,X
3110  WRITE #1,GG,TC,PC,T
3120  A=.06423: B=(.5353*TR)-.6123: D=TR: F=.6845
3130  C=(.3151*TR)-1.0467-(.5783/TR^2): E=.6816/TR^2
3140  MW=28.964*GG
3150  A1=(9.399999+.02*MW)*(T+459.67)^1.5
3160  A1=A1/((209+19*MW+(T+459.67))*10^4)
3170  B1=3.5+(986/(T+459.67))+(.01*MW): C1=2.4-.2*B1
3180  CLS: COLOR 4,0
3190  FOR I=1 TO X
3200  LOCATE 12,30: PRINT "WORKING, PLEASE WAIT"
3210  P(I)=P(I-1)+PINC
3220  PR(I)=P(I)/PC
3230  DR=(.27*PR(I))/TR: G=.27*PR(I)
3240  DR1=A*DR^6+B*DR^3+C*DR^2+D*DR+E*DR^3*(1+F*DR^2)*EXP(-F*DR^2)-G
3250  DR2=(6*A*DR^5)+(3*B*DR^2)+(2*C*DR)+D
3260  DR2=DR2+E*DR^2*(3+F*DR^2*(3-2*F*DR^2))*EXP(-F*DR^2)
3270  DR-DR=(DR1/DR2)
3280  IF ABS(DR1)>.00001 THEN 3240
3290  Z(I)=G/(DR*TR): PZ(I)=P(I)/Z(I)
3300  DR3=(P(I)*MW)/(Z(I)*669.8*(T+459.67))
3310  UG(I)=A1*EXP(B1*(DR3^C1))
3320  PP(I)=2*P(I)/(UG(I)*Z(I))
3330  PPB(I)=((PP(I)+PP(I-1))/2)*(P(I)-P(I-1)): PPB(I)=PPB(I)/1000000!
3340  MP(I)=MP(I-1)+PPB(I)
3350  WRITE #1,P(I),Z(I),UG(I),PP(I),PPB(I),MP(I)
3360  NEXT I
3370  CLOSE #1: GOTO 1010
3380  COLOR 7,0
3390  GOSUB 1860
3400  COLOR 14,0
3410  LOCATE 5,19: PRINT " THIS OPTION RECALLS AND PRINTS CONTENTS OF"
3420  LOCATE 6,21: PRINT " *.PSD DATA FILE GENERATED IN OPTION #3"
3430  LOCATE 14,20: INPUT "What Disk is the Data File on: ",DISC$
3440  DIS$=DISC$
3450  CLS: FILES DISC$+":*.PSD"
3460  LOCATE 19,20: INPUT "What is the Data File Name: ",FILE$
3470  FILE$=DISC$+":"+FILE$+".PSD"
3480  GOSUB 2570
3490  GOSUB 3500: GOTO 3550
3500  FOR I=1 TO 5: LINE INPUT #1, REF$(I): NEXT I
3510  INPUT #1, X
3520  INPUT #1,GG,TC,PC,T
3530  FOR I=1 TO X: INPUT #1,P(I),Z(I),UG(I),PP(I),PPB(I),MP(I)
3540  NEXT I: CLOSE #1: RETURN
3550  CLS: COLOR 14,0: LOCATE 12,30: PRINT "PRINTING, PLEASE WAIT"
3560  GOSUB 3830
3570  FOR I=1 TO 5: LPRINT REF$(I): NEXT I
3580  LPRINT
3090  LPRINT USING "Gas Specific Gravity (Air=1)          #.###";GG
3600  LPRINT USING "Pseudocritical Temperature (R)    #,###.###";TC
3610  LPRINT USING "Pseudocritical Pressure (psi)     #,###.###";PC
3620  LPRINT USING "Reservoir Temperature (F)           ###.###";T
3630  LPRINT: LPRINT
3640  GOSUB 3880
3650  FOR I=1 TO X
3660  IF I=36 THEN GOSUB 3950 ELSE 3680
3670  GOSUB 3830: GOSUB 3880
3680  IF I=81 THEN GOSUB 3950 ELSE 3700
3690  GOSUB 3830: GOSUB 3880
3700  IF I=126 THEN GOSUB 3950 ELSE 3720
3710  GOSUB 3830: GOSUB 3880
3720  IF I=172 THEN GOSUB 3950 ELSE 3740
3730  GOSUB 3830: GOSUB 3880
3740  LPRINT USING "###     ##.###      #.####     #.#####";I;P(I),Z(I),UG(I);
3750  LPRINT USING      ###,###.###     ##,###.###   ###,###.###";PP(I),PPB(I),M
3760  NEXT I
3770  IF I<=35 THEN Z1=(35-I)+11: GOTO 3970
3780  IF I<=80 THEN Z1=(80-I)+11: GOTO 3970
3790  IF I<=125 THEN Z1=(125-I)+11: GOTO 3970
3800  IF I<=171 THEN Z1=(171-I)+11: GOTO 3970
3810  IF I<200 OR I=200 THEN Z1=(200-I)+34
3820  GOTO 3970
3830  LPRINT TAB(27) "PSEUDO PRESSURES' DATA FILE"
3840  LPRINT: LPRINT: LPRINT
3850  LPRINT "File Name: ",FILE$
3860  LPRINT
3870  RETURN
3880  LPRINT "          Pressure          Viscosity";
3890  LPRINT"                    (2P/MU*Z)dP          m(p)x10^6"
3900  LPRINT "No.      (psi)          Z          (cp) ";
3910  LPRINT "        2P/MU*Z        (10^6)         (psi)^2/cp "
3920  LPRINT " ------------------------------------------------------------------------------------------";
3930  LPRINT " ------------------------------------"
3940  RETURN
3950  FOR Z=1 TO 12: LPRINT: NEXT Z
3960  RETURN
3970  FOR ZZ=1 TO Z1: LPRINT: NEXT ZZ
3980  GOTO 1010
3990  COLOR 7,0: GOSUB 1860
4000  COLOR 14,0
4010  LOCATE 5,20: PRINT "THE FOLLOWING INFORMATION IS REQUIRED TO"
4020  LOCATE 6,21: PRINT "CONVERT PRESSURES TO PSEUDO PRESSURES"
4030  LOCATE 9, 16: PRINT "Input Name of the Pseudo Pressures Data File:"
4040  LOCATE 11,16: PRINT "Input Name of the Pressures' Data File:"
4050  COLOR 15,0: LOCATE 9,62: INPUT " ",FILE$
4060  LOCATE 11,62: INPUT " ",FILE1$
4070  FILE$=FILE$+".PSD": FILE1$=FILE1$+".DAT"
4080  GOSUB 2570: GOSUB 3500
4090  OPEN FILE1$ FOR INPUTE AS #2
4100  IF NOT EOF(2) THEN 4150
4110  CLS: COLOR 4,0
4120  LOCATE 12,21:PRINT FILE1$:"IS EMPTY - PLEASE TRY AGAIN"
4130  LOCATE 22,28: PRINT "PRESS ANY KEY TO CONTINUE"
4140  CB$=INKEY$: IF CB$="" THEN 4140 ELSE 1010
4150  FOR I=1 TO 8
4160  LINE INPUT #2, WL$(I): NEXT I
4170  FOR Z=1 TO 48: INPUT #2, P$(Z): NEXT Z
4180  CLOSE #1: CLOSE #2
4190  LPRINT TAB(19) "PRESSURES CONVERTED TO THEIR CORRESPONDING"
4200  LPRINT TAB(32) "PSEUDO PRESSURES": LPRINT: LPRINT: LPRINT: LPRINT
4210  FOR I=1 to 5: LPRINT WL$(I): NEXT I
4220  LPRINT
4230  LPRINT TAB(1) "Data Files Used: ";FILE$
4240  LPRINT TAB(18) FILE1$
4250  LPRINT: LPRINT
4260  LPRINT "  Pressures      Pseudo Pressures      ";
4270  LPRINT "  Pressures      Pseudo Pressures"
4280  LPRINT "    (psi)        (psi)^2/cp × 10^6     ";
4290  LPRINT "    (psi)        (psi)^2/cp × 10^6"
4300  LPRINT " ---------------------------------------------------";
4310  LPRINT " -------------------------------------------"
4320  LPRINT
4330  FOR I=1 TO 48
4340  PZ(I)=VAL(P$(I))
4350  IF PZ(I)=0 THEN 4430
4360  FOR Z=1 TO X
4370  IF PZ(I)=P(Z) THEN MZ(I)=MP(Z)"
4380  IF PZ(I) < P(Z) THEN 4400
4390  GOTO 4420
4400  MZ(I)=MP(Z-1)+PZ(I)-P(Z-1))*((MP(Z)-MP(Z-1))/(P(Z)-P(Z-1)))
4410  GOTO 4430
4420  NEXT Z
4430  NEXT I: K=0
4440  FOR I=1 to 24
4450  LPRINT USING " ##,###.##       #,###,###.###          ";PZ(I+K);MZ(I+K
4460  LPRINT USING " ##,###.##          #,###,###.### ";PZ(I+1+K);MZ(I+1+K)
4470  K=K+1
4480  NEXT I
4490  LPRINT: LPRINT: LPRINT
4500  LPRINT WL$(6)
4510  LPRINT WL$(7)
4520  LPRINT WL$(8)
4530  FOR I=1 to 16: LPRINT: NEXT I
4540  GOTO 1010
```

## TABLE C–7. KG/KO-calculating gas/oil and water/oil relative permeabilities

```
1000 REM KG/KO
1010 CLEAR: KEY OFF
1020 DIM P$(16),NP$(16),R$(16),RS$(16),BO$(16),BG$(16),OVIS$(16),GVIS$(16)
1030 DIM P(16),NP(16),R(16),RS(16),BO(16),BG(16),OVIS(16),GVIS(16),TGVIS$(16)
1040 DIM TP$(16),TNP$(16),TR$(16),TRS$(16),TBO$(16),TBG$(16),TOVIS$(16),KRG(20)
1050 DIM SO(20),KGKO(20),SL(16),MGMO(16),SG(20),SW(20),VALUE(20),KRO(20),KRW(20)
1060 GOSUB 1710
1070 COLOR 14,0
1080 LOCATE 7,35: PRINT "MAIN MENU"
1090 LOCATE 9,25: PRINT "1 ... Create a New File"
1100 LOCATE 11,25: PRINT "2 ... Edit an Existing Data File"
1110 LOCATE 13,25: PRINT "3 ... Generate KG/KO Ratios"
1120 LOCATE 15,25: PRINT "4 ... Go to Main Menu"
1130 LOCATE 17,25: PRINT "5 ... End the Session"
1140 COLOR 0,14: LOCATE 21,25: PRINT "ENTER YOUR SELECTION = = = > "
1150 LOCATE 21,51: PRINT "     ": LOCATE 21,52
1160 I$ = INKEY$: IF I$ = "" THEN 1160 ELSE PRINT I$
1170 IF MID$(I$,1) = "1" THEN SEL = 1
1180 IF MID$(I$,1) = "2" THEN SEL = 2
1190 IF MID$(I$,1) = "3" THEN SEL = 3
1200 IF MID$(I$,1) = "4" THEN SEL = 4
1210 IF MID$(I$,1) = "5" THEN SEL = 5
1220 IF SEL<1 OR SEL>5 THEN BEEP: GOTO 1240
1230 ON SEL GOTO 1350,2010,2570,1290,1300
1240 COLOR 4,0: FOR I = 1 TO 60
1250 LOCATE 21,25: PRINT "INVALID SELECTION, PLEASE TRY AGAIN!"
1260 NEXT I
1270 LOCATE 21,25: PRINT "                                              "
1280 GOTO 1140
1290 CHAIN "PEPAC"
1300 CLOSE #1
1310 COLOR 7,0: CLS
1320 FOR I = 1 TO 60
1330 LOCATE 12,29: PRINT "*** END OF SESSION ***"
1340 NEXT I: CLS: END
1350 COLOR 7,0: GOSUB 1710
1360 GOSUB 1760
1370 ON ERROR GOTO 1460
1380 OPEN FILE$ FOR INPUT AS #1
1390 CLOSE #1
1400 ON ERROR GOTO 0
1410 BEEP: COLOR 4,0
1420 LOCATE 22,18: PRINT "FILE ALREADY EXISTS WITH THIS NAME: ";FILE$
1430 LOCATE 23,15: PRINT "CREATING NEW FILE WILL DESTROY CONTENTS OF OLD FILE"
1440 LOCATE 24,30: INPUT "CONTINUE (Y/N) = = = > ",AN$: CLS
1450 IF LEFT$(AN$,1) = "N" OR LEFT$(AN$, 1) = "n" THEN 1010 ELSE 1510
1460 IF ERR = 53 THEN RESUME 1500
1470 RESUME 1480
1480 ON ERROR GOTO 0
1490 RETURN
1500 ON ERROR GOTO 0
1510 OPEN FILE$ FOR OUTPUT AS #1
1520 GOSUB 1850
1530 COLOR 15,0
1540 FOR I = 1 TO 5: LOCATE I + 3,12: LINE INPUT "",WN$(I): NEXT I
1550 LOCATE 10,35: INPUT "",SW$: LOCATE 11,35: INPUT "",N$
1560 LOCATE 12,35: INPUT "",BOI$
1570 GOSUB 1920
1580 COLOR 15,0
1590 FOR I = 1 TO 16
1600 LOCATE I + 6,2: INPUT "",P$(I): LOCATE I + 6,13: INPUT "",R$(I)
1610 LOCATE I + 6,23: INPUT "",NP$(I): LOCATE I + 6,32: INPUT "",BO$(I)
1620 LOCATE I + 6,41: INPUT "",BG$(I): LOCATE I + 6,53: INPUT "",RS$(I)
1630 LOCATE I + 6,63: INPUT "",OVIS$(I): LOCATE I + 6,72: INPUT "",GVIS$(I)
1640 NEXT I
1650 FOR I = 1 TO 5: PRINT #1,WN$(I): NEXT I
1660 WRITE #1,SW$,N$,BOI$
1670 FOR I = 1 TO 16
1680 WRITE #1,P$(I),R$(I),NP$(I),BO$(I),BG$(I),RS$(I),OVIS$(I),GVIS$(I)
1690 NEXT I: CLOSE #1
1700 GOTO 1010
1710 CLS: COLOR 0,10
1720 PRINT "                         ";
1730 PRINT "KG/KO RATIOS FROM FIELD DATA";
1740 PRINT "                         "
1750 COLOR 7,0: RETURN
1760 COLOR 14,0
1770 LOCATE 5,20: PRINT "THE FOLLOWING INFORMATION IS REQUIRED FOR"
1780 LOCATE 6,20: PRINT "CREATING, EDITING, AND ANALYZING DATA FILE"
1790 LOCATE 14,20: INPUT "What Disk is the Data File on: ",DISC$
1800 DIS$ = DISC$
1810 CLS: FILES DISC$ + ":*.DAT"
1820 LOCATE 19,20: INPUT "What is the Data File Name: ",FILES$
1830 FILE$ = DISC$ + ":" + FILES$ + ".DAT"
1840 RETURN
1850 GOSUB 1710
1860 COLOR 14,0: PRINT: PRINT
1870 PRINT "Reference:"
1880 LOCATE 10,1: PRINT "Water Saturation - fraction ........................................ "
1890 LOCATE 11,1: PRINT "Original Oil in Place - STB ....................................... "
1900 LOCATE 12,1: PRINT "Initial Oil FVF - RB/STB ........................................... "
1910 RETURN
1920 GOSUB 1710
1930 COLOR 14,0
1940 PRINT: PRINT: PRINT "Pressure Prod. GOR      Np      Oil FVF ";
1950 PRINT " Gas FVF    Sol. GOR    Oil Vis.    Gas Vis."
1960 PRINT " (psi)      (SCF/STB)      (MMSTB)      (RB/STB)";
1970 PRINT " (SCF/STB)      (SCF/STB)      (cp)      (cp) "
1980 PRINT " ------------------------------------------------------------";
1990 PRINT " ------------------------------------------------------------"
2000 RETURN
2010 COLOR 7,0: GOSUB 1710
2020 GOSUB 1760
2030 GOSUB 2440
2040 GOSUB 2510
2050 OPEN DISC$ + ":SEQTEMP.DAT" FOR OUTPUT AS #2
2060 GOSUB 1850: COLOR 15,0
2070 FOR I = 1 TO 5: LOCATE I + 3,12: PRINT TWN$(I): NEXT I
2080 LOCATE 10,35: PRINT TSW$: LOCATE 11,35: PRINT TN$
2090 LOCATE 12,35: PRINT TBOI$
2100 COLOR 4,0
2110 FOR I = 1 TO 5
2120 LOCATE I + 3,12: INPUT "",WN$(I)
2130 IF WN$(I) = "" THEN WN$(I) = TWN$(I)
2140 NEXT I
2150 LOCATE 10,35: INPUT "",SW$: IF SW$ = "" THEN SW$ = TSW$
2160 LOCATE 11,35: INPUT "",N$: IF N$ = "" THEN N$ = TN$
2170 LOCATE 12,35: INPUT "",BOI$: IF BOI$ = "" THEN BOI$ = TBOI$
2180 GOSUB 1920: COLOR 15,0
2190 FOR I = 1 TO 16
2200 LOCATE I + 6,2: PRINT TP$(I): LOCATE I + 6,13: PRINT TR$(I)
2210 LOCATE I + 6,23: PRINT TNP$(I): LOCATE I + 6,32: PRINT TBO$(I)
2220 LOCATE I + 6,41: PRINT TBG$(I): LOCATE I + 6,53: PRINT TRS$(I)
2230 LOCATE I + 6,63: PRINT TOVIS$(I): LOCATE I + 6,72: PRINT TGVIS$(I)
2240 NEXT I: COLOR 4,0
2250 FOR I = 1 TO 16
2260 LOCATE I + 6,2: INPUT "",P$(I): IF P$(I) = "" THEN P$(I) = TP$(I)
2270 LOCATE I + 6,13: INPUT "",R$(I): IF R$(I) = "" THEN R$(I) = TR$(I)
2280 LOCATE i + 6,23: INPUT "",NP$(I): IF NP$(I) = "" THEN NP$(I) = TNP$(I)
2290 LOCATE I + 6,32: INPUT "",BO$(I): IF BO$(I) = "" THEN BO$(I) = TBO$(I)
2300 LOCATE I + 6,41: INPUT "",BG$(I): IF BG$(I) = "" THEN BG$(I) = TBG$(I)
2310 LOCATE I + 6,53: INPUT "",RS$(I): IF RS$(I) = "" THEN RS$(I) = TRS$(I)
2320 LOCATE I + 6,63: INPUT "",OVIS$(I): IF OVIS$(I) = "" THEN OVIS$(I) = TOVIS$(I)
2330 LOCATE I + 6,72: INPUT "",GVIS$(I): IF GVIS$(I) = "" THEN GVIS$(I) = TGVIS$(I)
2340 NEXT I
2350 FOR I = 1 TO 5: PRINT #2,WN$(I): NEXT I
2360 WRITE #2,SW$,N$,BOI$
2370 FOR I = 1 TO 16
2380 WRITE #2,P$(I),R$(I),NP$(I),BO$(I),BG$(I),RS$(I),OVIS$(I),GVIS$(I)
2390 NEXT I
2400 CLOSE #1: CLOSE #2
2410 KILL FILE$
2420 NAME DISC$ + ":SEQTEMP.DAT" AS FILE$
2430 CLOSE #1: GOTO 1010
2440 OPEN FILE$ FOR INPUT AS #1
2450 IF NOT EOF(1) THEN 2500
2460 CLS: COLOR 4,0
2470 FOR I = 1 TO 60
2480 LOCATE 12,30: PRINT "INPUT FILE IS EMPTY": NEXT I
2490 CLOSE #1: GOTO 1010
2500 RETURN
2510 FOR I = 1 TO 5: LINE INPUT #1, TWN$(I): NEXT I
2520 INPUT #1, TSW$,TN$,TBOI$
2530 FOR I = 1 TO 16
2540 INPUT #1,TP$(I),TR$(I),TNP$(I),TBO$(I),TBG$(I),TRS$(I),TOVIS$(I),TGVIS$(I)
2550 NEXT I: CLOSE #1
2560 RETURN
2570 COLOR 7,0: GOSUB 1710: COLOR 14,0
2580 LOCATE 6,22: PRINT "THE FOLLOWING OPTIONS ARE AVAILABLE"
2590 LOCATE 7,22: PRINT STRING$(35,205)
2600 LOCATE 9,25: PRINT "1 ... Use Field Performance"
2610 LOCATE 11,25: PRINT "2 ... Use Empirical Correlations"
2620 LOCATE 13,25: PRINT "3 ... Go to Main Menu"
2630 COLOR 0,4: LOCATE 15,25: PRINT "ENTER YOUR SELECTION = = = >"
2640 LOCATE 15,51: PRINT "   ": COLOR 7,0
2650 LOCATE 19,2: PRINT "Note that Option #2 Handles Empirical Correlations for Estimating Two - Phase"
2660 LOCATE 20,6: PRINT "Relative Permeability in Consolidated Rocks = = By Honar pour, et. al.,"
2670 LOCATE 21,30: PRINT "and Two Other Options"
2680 LOCATE 23,6: PRINT "Option #1 Requires a DATA FILE Created in Option #1 of the Main Menu"
2690 LOCATE 15,52
2700 II$ = INKEY$: IF II$ = "" THEN 2700 ELSE PRINT II$
2710 IF MID$(II$,1) = "1" THEN OPT = 1
2720 IF MID$(II$,1) = "2" THEN OPT = 2
2730 IF MID$(II$,1) = "3" THEN OPT = 3
2740 ON OPT GOTO 2750,3080,1010
2750 COLOR 7,0: GOSUB 1710: GOSUB 1760
2760 GOSUB 2440: GOSUB 2510
2770 LPRINT TAB(20) "Kg/Ko RATIOS FROM PRODUCTION PERFORMANCE"
2780 LPRINT: LPRINT: LPRINT
2790 FOR I = 1 TO 5
2800 LPRINT TAB(5) TWN$(I): NEXT I
2810 LPRINT: LPRINT TAB(5) "Data File Used: ";FILE$: LPRINT: LPRINT"
```

# TABLE C–7. Cont'd.

```
2820 LPRINT USING "  Water Saturation - Fraction ....      #.###";VAL(TSW$)
2830 LPRINT USING "  Original Oil in Place - STB ...............................................
                                                ###,###,###";VAL(TN$)
2840 LPRINT USING "  Initial Oil FVF - RB/STB ....          #.###";VAL(TBOI$)
2850 LPRINT: LPRINT
2860 FOR I=1 TO 16
2870 K=K+1
2880 IF VAL(TP$(I))=0 THEN 3060
2890 SO(I)=(1-(VAL(TNP$(I))/(VAL(TN$)/1000000!)))*(VAL(TBO$(I))/VAL(TBOI$))
2900 SO(I)=SO(I)*(1-VAL(TSW$))*100: SL(I)=SO(I)+VAL(TSW$)*100
2910 KGKO(I)=((VAL(TR$(I))-VAL(TRS$(I)))/(VAL(TBO$(I))*VAL(TBG$(I))))
2920 KGKO(I)=KGKO(I)*(VAL(TGVIS$(I))/VAL(TOVIS$(I)))
2930 MGMO(I)=VAL(TGVIS$(I))/VAL(TOVIS$(I))
2940 IF K>1 THEN 3030
2950 LPRINT "     Pres.     R     Rs ";
2960 LPRINT "     Np        Bo    Bg     ";
2970 LPRINT "            SL         "
2980 LPRINT "       psi  SCF/STB SCF/STB";
2990 LPRINT " MMSTB    RB/STB  STB/SCF";
3000 LPRINT " MUg/MUo   %      Kg/Ko "
3010 LPRINT TAB(5) "---------------------------------------------------------------------------";
3020 LPRINT " -----------------------------------"
3030 LPRINT USING "     ##,###   ##,###   #,###";VAL(TP$(I)),VAL(TR$(I)),VAL(TRS$(I));
3040 LPRINT USING " #,###.###   #.###   #.######";VAL(TNP$(I)),VAL(TBO$(I)),1/VAL(TBG$(I));
3050 LPRINT USING " #.### ###.##   ##.####";MGMO(I),SL(I),KGKO(I)
3060 NEXT I
3070 FOR I=1 TO 55-K: LPRINT: NEXT I
3080 GOSUB 4250
3090 COLOR 14,0: PRINT: PRINT
3100 LOCATE 4,22: PRINT "THE FOLLOWING OPTIONS ARE AVAILABLE"
3110 LOCATE 5,22: PRINT STRINGS$(35,205)
3120 LOCATE 6,22: PRINT "1 ... Honarpour et. al. Correlations"
3130 LOCATE 8,22: PRINT "2 ... Other Correlations"
3140 LOCATE 10,22: PRINT "3 ... Go to the Previous Menu"
3150 COLOR 0,4: LOCATE 12,22: PRINT "ENTER YOUR SELECTION ===>"
3160 LOCATE 12,48: PRINT "     ": COLOR 7,0
3170 LOCATE 12,50
3180 BB$=INKEY$: IF BB$="" THEN 3180 ELSE PRINT BB$
3190 IF MID$(BB$,1)="1" THEN OPT=1
3200 IF MID$(BB$,1)="2" THEN OPT=2
3210 IF MID$(BB$,1)="3" THEN OPT=3
3220 ON OPT GOTO 3230,4620,2570
3230 GOSUB 4250
3240 GOSUB 5030
3250 GOSUB 4250
3260 COLOR 14,0: LOCATE 4,25: PRINT "Honarpour et. al. Correlations"
3270 LOCATE 5,25: PRINT STRING$(30,205)
3280 LOCATE 6,1: INPUT "Water-Oil System (Y/N) ===> ",YN$
3290 IF LEFT$(YN$,1)="N" OR LEFT$(YN$,1)="n" THEN X$="Gas/Oil"
3300 IF LEFT$(YN$,1)="Y" OR LEFT$(YN$,1)="y" THEN X$="Water/Oil"
3310 IF LEFT$(YN$,1)="N" OR LEFT$(YN$,1)="n" THEN 3800
3320 LOCATE 7,18: PRINT "1 ... Krw = Equation #1, Sandstone - Water Wet"
3330 LOCATE 8,18: PRINT "2 ... Krw = Equation #2, Sandstone - Oil/Intermediate"
3340 LOCATE 9,18: PRINT "3 ... Krw = Equation #3, Carbonates - Water Wet"
3350 LOCATE 10,18:PRINT "4 ... Krw = Equation #4, Carbonates - Oil/Intermediate"
3360 COLOR 0,4: LOCATE 11,18: PRINT "ENTER YOUR SELECTION ====>   ": COLOR 7,0
3370 COLOR 14,0
3380 LOCATE 13,18: PRINT "1 ... Kro - Equation #1 - Sandstone"
3390 LOCATE 14,18: PRINT "2 ... Kro - Equation #2 - Carbonates"
3400 COLOR 0,4: LOCATE 15,18: PRINT "ENTER YOUR SELECTION ===>   ": COLOR 7,0
3410 COLOR 14,0
3420 LOCATE 17,1: PRINT "Irreducible Water Saturation - fraction ........................ "
3430 LOCATE 18,1: PRINT "Residual Oil Saturation to Water - fraction ................ "
3440 LOCATE 19,1: PRINT "Porosity - fraction ....................................................... "
3450 LOCATE 20,1: PRINT "Absolute Permeability to Air - MD ............................. "
3460 LOCATE 22,1: PRINT "Sw - fraction (MAX 20): "
3470 COLOR 15,0
3480 LOCATE 11,45: INPUT "",KRWE: LOCATE 15,45: INPUT "",KROE
3490 LOCATE 17,50: INPUT "",SWI: LOCATE 18,50: INPUT "",SORW
3500 LOCATE 19,50: INPUT "",PHI
3510 IF KRWE=3 THEN 3520 ELSE 3530
3520 LOCATE 20,50: INPUT "",KA
3530 LOCATE 22,25: LINE INPUT "",LIN$
3540 GOSUB 4290
3550 GOSUB 4360
3560 XX=1-SWI-SORW
3570 FOR N%=1 TO I%
3580 SW(I)=VALUE(N%)
3590 ON KRWE GOTO 3600,3630,3660,3680
3600 KRW(I)=.035388*((SW(I)-SWI)/XX)-.010874*((SW(I)-SORW)/XX)^(2.9)
3610 KRW(I)=KRW(I)+.56556*(SW(I))^(3.6)*(SW(I)-SWI)
3620 GOTO 3700
3630 KRW(I)=1.5814*((SW(I)-SWI)/(1-SWI))^(1.91)
3640 KRW(I)=KRW(I)-1.2484*PHI*(1-SWI)*(SW(I)-SWI)
3650 GOTO 3700
3660 KRW(I)=.0020525*((SW(I)-SWI)/PHI^(2.15))-.051371*(SW(I)-SWI)*(1/KA)^(.43)
3670 GOTO 3700
3680 KRW(I)=.29986*((SW(I)-SWI)/(1-SWI))-.32797*((SW(I)-SORW)/XX)^2*(SW(I)-SWI)
3690 KRW(I)=KRW(I)+.413259*((SW(I)-SWI)/XX)^4
3700 IF KROE>1 THEN 3740
3710 KRO(I)=.76067*((((1-SW(I))/(1-SWI))-SORW)/(1-SORW))^1.8*((1-SW(I)-SORW)/XX)^2
```

# TABLE C–7. Cont'd.

```
3720 KRO(I)=KRO(I)+2.6318*PHI*(1-SORW)*(1-SW(I)-SORW)
3730 GOTO 3750
3740 KRO(I)=1.2624*((1-SW(I)-SORW)/(1-SORW))*((1-SW(I)-SORW)/XX)^2
3750 LPRINT USING "                                  #.###     ##.######";SW(I),KRW(I);
3760 LPRINT USING "      ##.######   ##,###,###.##";KRO(I),KRO(I)/KRW(I)
3770 NEXT N%
3780 FOR I=1 TO 42-N%: LPRINT: NEXT I
3790 GOTO 1010
3800 LOCATE 7,18: PRINT "1 ... Kro - Equation #1 - Sandstone"
3810 LOCATE 8,18: PRINT "2 ... Kro - Equation #2 - Carbonates"
3820 COLOR 0,4: LOCATE 9,18: PRINT "ENTER YOUR SELECTION ===>    "
3830 COLOR 7,0: COLOR 14,0
3840 LOCATE 11,18: PRINT "1 ... Krg - Equation #1 - Sandstone"
3850 LOCATE 12,18: PRINT "2 ... Krg - Equation #2 - Carbonates"
3860 COLOR 0,4: LOCATE 13,18: PRINT "ENTER YOUR SELECTION ===>      "
3870 COLOR 7,0: COLOR 14,0
3880 LOCATE 15,1: PRINT "Irreducible Water Saturation - fraction ......................... "
3890 LOCATE 16,1: PRINT "Residual Oil Saturation to Gas - fraction .................... "
3900 LOCATE 17,1: PRINT "Critical Gas Saturation - fraction ................................. "
3910 LOCATE 18,1: PRINT "Gas Relative Permeability at Sor - fraction ................... "
3920 LOCATE 19,1: PRINT "Porosity - fraction ......................................................... "
3930 LOCATE 20,1: PRINT "Absolute Permeability to Air - MD .............................. "
3940 COLOR 15,0
3950 LOCATE 9,45: INPUT "",KROE
3960 LOCATE 13,45: INPUT "",KRGE
3970 LOCATE 15,50: INPUT "",SWI: LOCATE 16,50: INPUT "",SORG
3980 LOCATE 17,50: INPUT "",SGC: LOCATE 18,50: INPUT "",KRG
3990 LOCATE 19,50: INPUT "",PHI: LOCATE 20,50: INPUT "",KA
4000 GOSUB 4250: COLOR 14,0
4010 LOCATE 2,22: PRINT "So (fraction)              Sg (fraction)"
4020 LOCATE 3,22: PRINT " ----------------------------------------------- "
4030 COLOR 15,0
4040 FOR I=1 TO 20
4050 LOCATE I+3,26: INPUT "",SO(I): LOCATE I+3,49: INPUT "",SG(I)
4060 NEXT I
4070 GOSUB 4360
4080 FOR I=1 TO 20
4090 IF SO(I)=0 THEN 4220
4100 XXX=XXX+1
4110 KRO(I)=.98372*((1-SO(I))/(1-SWI))^4*((1-SO(I)-SORG)/(1-SWI-SORG))^2
4120 GOTO 4140
4130 KRO(I)=.93752*((1-SO(I))/(1-SWI))^4*((1-SO(I)-SORG)/(1-SWI-SORG))^2
4140 IF KRGE=2 THEN 4170
4150 KRG(I)=1.1072*((SG(I)-SGC)/(1-SWI))^2*KRG+2.7794*((SORG*(SG(I)-SGC))/(1-SWI))*KRG
4160 GOTO 4200
4170 KRG(I)=1.8655*(((SG(I)-SGC)*SG(I))/(1-SWI))*KRG
4180 KRG(I)=KRG(I)+8.0053*(((SG(I)-SGC)*SORG^2)/(1-SWI))
4190 KRG(I)=KRG(I)-.02589*(SG(I)-SGC)*((1-SWI-SORG-SGC)/(1-SWI))^2*(1-((1-SWI-SORG-SGC)/(1-
        SWI)))^2*(KA/PHI)^(.5)
4200 LPRINT USING "          #.###        #.###        ##.######";SG(I),SO(I),KRG(I);
4210 LPRINT USING "   ##.######   #,###,###.##";KRO(I),KRG(I)/KRO(I)
4220 NEXT I
4230 FOR I=1 TO 39-XXX: LPRINT: NEXT I
4240 GOTO 1010
4250 CLS: COLOR 0,10
4260 PRINT "            Relative Permeability Ratios from Empirical Correlations ";
4270 PRINT "            "
4280 RETURN
4290 NOLD%=1
4300 FOR I%=1 TO 20
4310 N%=INSTR(NOLD%,LIN$, " ")
4320 IF N%=0 THEN VALUE(I%)=VAL(MID$(LIN$,NOLD%)): RETURN
4330 IF N%=NOLD% THEN NOLD%=NOLD%+1: GOTO 4310
4340 VALUE(I%)=VAL(MID$(LIN$,NOLD%,N%-NOLD%))
4350 NOLD%=N%+1: NEXT
4360 LPRINT TAB(12) "RELATIVE PERMEABILITY RATIOS FROM EMPIRICAL COR-
        RELATIONS"
4370 LPRINT TAB(21) "USING HONARPOUR ET. AL.'S CORRELATIONS"
4380 LPRINT: LPRINT: LPRINT: LPRINT
4390 FOR I=1 TO 5
4400 LPRINT TAB(5) WN$(I): NEXT I
4410 LPRINT: LPRINT: LPRINT
4420 LPRINT USING "     Irreducible Water Saturation - frac-
        tion ....................................................................     #.###";SWI
4430 LPRINT USING "     Porosity - fraction .......................     #.###";PHI
4440 LPRINT USING "     Absolute Permeability to Air -
        MD ..............................................................     ##,###.##";KA
4450 IF LEFT$(YN$,1)="N" OR LEFT$(YN$,1)="n" THEN 4500
4460 LPRINT USING "     Residual Oil Saturation to Water - frac-
        tion ....................................................................     #.###";SORW
4470 LPRINT: LPRINT TAB(5) "Fluid System: ";X$
4480 LPRINT TAB(5) "Equations Used: Kro ...";KROE;" Krw ...";KRWE
4490 GOTO 4550
4500 LPRINT USING "     Residual Oil Saturation to Gas - fraction ... #.###";SORG
4510 LPRINT USING "     Critical Gas Saturation - fraction ............ #.###";SGC
4520 LPRINT USING "     Krg at Sor - fraction ..................... ###.####";KRG
4530 LPRINT: LPRINT TAB(5) "Fluid System: ";X$
4540 LPRINT TAB(5) "Equations Used: Kro ...";KROE;"     Krg ...";KRGE
4550 LPRINT: LPRINT
4560 IF LEFT$(YN$,1)="N" OR LEFT$(YN$,1)="n" THEN 4590
4570 LPRINT TAB(20) " Sw          Krw          Kro          Ko/Kw "
4580 LPRINT TAB(19) " ---------------------------------------------------------- ": GOTO 4610
4590 LPRINT TAB(10) " Sg          So          Krg          Kro          Kg/Ko   "
4600 LPRINT TAB(10) " -----------------------------------------------------------------------"
4610 RETURN
4620 GOSUB 4250
4630 COLOR 14,0: LOCATE 4,31: PRINT "Other Correlations"
4640 LOCATE 5,20: PRINT STRING$(38,205)
4650 LOCATE 7,20: PRINT "1 ... Ko/Kw = ae^(bSw)"
4660 LOCATE 8,20: PRINT "2 ... Ko/Kw = A1(1-Swd)^m/A2Swd^n"
4670 LOCATE 10,20: PRINT "3 ... Kg/Ko = (1-S*)^a(1-S*^b)/S*^(a+b)"
4680 LOCATE 11,20: PRINT "4 ... Kg/Ko = A(0.04335+0.4556A)"
4690 LOCATE 13,20: PRINT "5 ... Go to Main Menu"
4700 COLOR 0,4: LOCATE 15,20: PRINT "ENTER YOUR SELECTION ===>"
4710 LOCATE 15,46: PRINT "        ": COLOR 7,0
4720 LOCATE 20,20: PRINT "Swd = (Sw-Siw)/(1-Sor-Siw)"
4730 LOCATE 21,20: PRINT "S* = So/(1-Siw)"
4740 LOCATE 22,20: PRINT "A = (100-%Sgc-%Siw-%So)/(%So-C)"
4750 LOCATE 15,48
4760 CC$=INKEY$: IF CC$="" THEN 4760 ELSE PRINT CC$
4770 IF MID$(CC$,1)="1" THEN EQU=1
4780 IF MID$(CC$,1)="2" THEN EQU=2
4790 IF MID$(CC$,1)="3" THEN EQU=3
4800 IF MID$(CC$,1)="4" THEN EQU=4
4810 IF MID$(CC$,1)="5" THEN EQU=5
4820 ON EQU GOTO 4830,5110,5470,5740,1010
4830 GOSUB 4250: GOSUB 5030
4840 COLOR 14,0: LOCATE 11,28: PRINT "Equation Used: ae^(bSw)"
4850 LOCATE 12,28: PRINT STRING$(24,205)
4860 LOCATE 14,28: PRINT "Correlation Coefficients: a=                    b="
4870 LOCATE 20,5: PRINT "Sw - fraction (MAX-20):"
4880 COLOR 15,0: LOCATE 14,56: INPUT "",A: LOCATE 14,69: INPUT "",B
4890 LOCATE 20,29: LINE INPUT "",LIN$
4900 GOSUB 4290
4910 GOSUB 5070
4920 LPRINT TAB(5) "Equation Used: Ko/Kw = ae^(bSw)"
4930 LPRINT USING "     Coefficients Used: a=#,###.####   b=#,###.####";A,B
4940 LPRINT:LPRINT
4950 LPRINT TAB(30) " Sw            Ko/Kw "
4960 LPRINT TAB(30) " ----------------------------"
4970 FOR N%=1 TO I%: SW(I)=VALUE(N%)
4980 KOKW(I)=A*EXP(B*SW(I))
4990 LPRINT USING "                              #.###     #,###.######";SW(I),KOKW(I)
5000 NEXT N%
5010 FOR I=1 TO 51-N%: LPRINT: NEXT I
5020 GOTO 1010
5030 COLOR 14,0
5040 LOCATE 5,5: PRINT "Reference:": COLOR 15,0
5050 FOR I=1 TO 5: LOCATE I+4,17: INPUT "",WN$(I): NEXT I
5060 RETURN
5070 LPRINT TAB(12) "RELATIVE PERMEABILITY RATIOS FROM EMPIRICAL CORRELATIONS"
5080 LPRINT: LPRINT: LPRINT
5090 FOR I=1 TO 5: LPRINT TAB(5) WN$(I): NEXT I
5100 LPRINT: RETURN
5110 GOSUB 4250: GOSUB 5030
5120 COLOR 14,0: LOCATE 11,25: PRINT "Equation Used: Ko/Kw = A1(1-Swd)^m/A2Swd^n"
5130 LOCATE 12,20: PRINT STRING$(52,205)
5140 LOCATE 14,20: PRINT "Initial Water Saturation - fraction ..............  "
5150 LOCATE 15,20: PRINT "Residual Oil Saturation - fraction ..............  "
5160 LOCATE 16,20: PRINT "Correlation Coefficients - A1 .....................  "
5170 LOCATE 17,20: PRINT "                              A2 ......................."
5180 LOCATE 18,20: PRINT "                              m .......................  "
5190 LOCATE 19,20: PRINT "                              n .......................  "
5200 LOCATE 21,5: PRINT "Sw - fraction (MAX-20): "
5210 COLOR 15,0: LOCATE 14,64: INPUT "",SIW: LOCATE 15,64: INPUT "",SOR
5220 LOCATE 16,64: INPUT "",A1: LOCATE 17,64: INPUT "",A2
5230 LOCATE 18,64: INPUT "",M: LOCATE 19,64: INPUT "",N
5240 LOCATE 21,29: LINE INPUT "",LIN$
5250 GOSUB 4290: GOSUB 5070
5260 LPRINT TAB(5) "Equation Used: Ko/Kw = A1(1-Swd)^m/A2Swd^n"
5270 LPRINT TAB(5) "                 Where     Swd = (Sw-Siw)/(1-Sor-Siw)"
5280 LPRINT
5290 LPRINT USING "     Initial Water Saturation - fraction                #.###";SIW
5300 LPRINT USING "     Residual Water Saturation - fraction               #.###";SOR
5310 LPRINT USING "     Correlation Coefficients: A1                    ###.###";A1
5320 LPRINT USING "                               A2                    ###.###";A2
5330 LPRINT USING "                               m                     ###.###";M
5340 LPRINT USING "                               n                     ###.###";N
5350 LPRINT: LPRINT: LPRINT
5360 LPRINT TAB(15) " Sw                Kro                Krw                Ko/Kw "
5370 LPRINT TAB(15) "---------------------------------------------------------------------------- "
5380 FOR N%=1 TO I%: SW(I)=VALUE(N%)
5390 SWD=(SW(I)-SIW)/(1-SOR-SIW)
5400 KRO(I)=A1*(1-SWD)^M: KRW(I)=A2*SWD^N
5410 IF KRW(I)=0 THEN 5440
5420 If KRW(I)<.00001 THEN 5440
5430 KOKW(I)=KRO(I)/KRW(I)
5440 LPRINT USING "     #.###       ##.####       ##.####       ##,###.#####
        ";SW(I),KRO(I),KRW(I),KOKW(I)
5450 NEXT N%
5460 FOR I=1 TO 43-N%: LPRINT: NEXT I: GOTO 1010
5470 GOSUB 4250: GOSUB 5030
```

## TABLE C–7. Cont'd.

```
5480 COLOR 14,0: LOCATE 11,20: PRINT "Equation Used: Kg/Ko = (1-S*)^a(1-S*^b)/S*^(a+b)"
5490 LOCATE 12,20: PRINT STRING$(47,205)
5500 LOCATE 14,20: PRINT "Initial Water Saturation - fraction ....... "
5510 LOCATE 15,20: PRINT "Correlation Coefficient: a .................... "
5520 LOCATE 15,20: PRINT "                          b .................. "
5530 LOCATE 21,5: PRINT "So - fraction (MAX=20):"
5540 COLOR 15,0: LOCATE 14,62: INPUT "",SIW
5550 LOCATE 15,62: INPUT "",AA: LOCATE 16,62: INPUT "",BB
5560 LOCATE 21,29: LINE INPUT "",LIN$
5570 GOSUB 4290: GOSUB 5070
5580 LPRINT TAB(5) "Equation Used: Kg/Ko = (1-S*)^a(1-S*^b)/S*^(a+b)"
5590 LPRINT TAB(5) "Where              S* = So/(1-Siw)"
5600 LPRINT
5610 LPRINT USING "     Initial Water Saturation - fraction      #.###";SIW
5620 LPRINT USING "        Correlation Coefficients: a          ##.###";AA
5630 LPRINT USING "                                  b          ##.###";BB
5640 LPRINT: LPRINT: LPRINT
5650 LPRINT TAB(23) " So            SL            Kg/Ko "
5660 LPRINT TAB(23) " ------------------------------------------------ "
5670 FOR N%=1 TO I%: SO(I)=VALUE(N%)
5680 S=SO(I)/(1-SIW)
5690 KGKO(I)=((1-S)^AA*(1-S^BB))/(S^(AA+BB)
5700 SL(I)=SO(I)+SIW
5710 LPRINT USING "    #.###  #.###  ###.######";SO(I),SL(I),KGKO(I)
5720 NEXT N%
5730 FOR I=1 TO 46-N%: LPRINT: NEXT I: GOTO 1010
5740 GOSUB 4250: GOSUB 5030
5750 COLOR 14,0: LOCATE 11,24: PRINT "Equation Used: Kg/Ko = A(0.04335+0.4556A)"
5760 LOCATE 12,20: PRINT STRING$(48,205)
5770 LOCATE 14,20: PRINT "Initial Water Saturation - fraction ---- ";
5780 LOCATE 15,20: PRINT "Critical Gas Saturation - fraction ----- ";
5790 LOCATE 16,20: PRINT "Correlation Coefficient - C ----------- ";
5800 LOCATE 21,5: PRINT "So - fraction (MAX=20):"
5810 COLOR 15,0: LOCATE 14,62: INPUT "",SIW
5820 LOCATE 15,62: INPUT "",SGC: LOCATE 16,62: INPUT "",CC
5830 LOCATE 21,29: LINE INPUT "",LIN$
5840 GOSUB 4290: GOSUB 5070
5850 LPRINT TAB(5) "Equation Used: Kg/Ko = A(0.04335+0.4556A)"
5860 LPRINT TAB(5) "          Where    A = (100-%Sgc-%Siw-%So)/(%So-C)"
5870 LPRINT
5880 LPRINT USING "     Initial Water Saturation - fraction     #.###";SIW
5890 LPRINT USING "     Critical Gas Saturation  - fraction     #.###";SGC
5900 LPRINT USING "    Correlation Coefficient   - C        ###.###";CC
5910 LPRINT: LPRINT
5920 LPRINT TAB(23) " So         SL         Kg/Ko "
5930 LPRINT TAB(23) " ------------------------------------ "
5940 FOR N%=1 TO I%: SO(I)=VALUE(N%)
5950 AAA=(100-SGC*100-SIW*100-SO(I)*100)/(SO(I)*100-CC)
5960 KGKO(I)=AAA*(.04335+.4556*AAA): SL(I)=SO(I)+SIW
5961 IF KGKO(I)<0 THEN KGKO(I)=0
5970 LPRINT USING "    #.###  #.###  ###.######";SO(I),SL(I),KGKO(I)
5980 NEXT N%
5990 FOR I=1 TO 47-N%: LPRINT: NEXT I: GOTO 1010
```

# INDEX



## A



## B

## E



## F

## G

## H



## I



## J



## L



## M

## R

## S



## T

## U





## V



## W